DEVELOPMENTS IN BRIDGE DESIGN AND CONSTRUCTION

Developments in Bridge Design and Construction

Edited by

K. C. ROCKEY, M.Sc., Ph.D., C.Eng., F.I.C.E., M.I.Mech.E.

J. L. BANNISTER, B.Sc.Tech., M.Sc., C.Eng., F.I.Struct.E., A.M.C.T.
and

H. R. EVANS, M.Sc., Ph.D.

all of the Department of Civil and Structural Engineering,
University College, Cardiff

CROSBY LOCKWOOD & SON LTD
26 Old Brompton Road, London SW7

First published 1971 by Crosby Lockwood & Son Ltd
26 Old Brompton Road, London SW7

Printed by photo-lithography and made in Great Britain at
the Pitman Press, Bath

ISBN 0.258.96824.9

CONTENTS

PREFACE

W.G.N. Geddes, B.Sc., C.Eng., F.I.Struct.E., F.I.C.E., F.I.W.E., F.I.E.S.,
President, Institution of Structural Engineers

It gives me great pleasure to write a preface to this Volume of Proceedings of the International Conference on Developments in Bridge Design and Construction which was held at the University College, Cardiff, 29 March to 2 April 1971.

The Conference aroused great interest attracting papers from world-wide sources and delegates attended from a great many countries. Recent problems in design and erection of bridges deepened the interest in discussion and debate.

Ever-increasing demands of transportation in all countries has led to a continuing expansion in road systems and the number of bridges of all types has multiplied accordingly. The papers ranged widely between theoretical and research considerations to the problems of constructional detail,and study of the various subjects presented and of the discussions generated will prove to be of absorbing interest. University College are to be congratulated on their successful achievement in arranging such a stimulating and topical Conference.

October 1971   W. G. N. GEDDES

DESIGN OF CANTILEVER SLABS
FOR SPINE BEAM BRIDGES

PROF. F. SAWKO and J. H. MILLS

University of Liverpool

SYNOPSIS

The geometry and structural action of spine beam bridge decks is such that the over-hanging slabs can be treated as semi-infinite cantilever slabs and can be designed for transverse action only. The design of these cantilever slabs under the action of point loads is the topic of this paper.

Following a review of published solutions, approximate solutions, based on finite element results, are presented for the distribution of moment along the support, the distribution of shear along the support, and the sagging moment under the load.

Load positions producing maximum shear are discussed and approximations for edge beam effects are described. Design curves for 45 units of *HB* load are presented in the Appendix.

Introduction

Structural design of spine beam bridges presents many difficulties because of the complex nature of interaction of individual elements. It is fortuitous that the problem of design of cantilever slabs can almost be isolated and treated as an entity on its own. This is made possible only after a detailed investigation of the load carrying characteristics of all structural elements.

The structural action of spine beam bridges is composed of three main modes, interacting with each other to a certain extent. These are the longitudinal behaviour, torsional behaviour and transverse behaviour. Cantilever slabs contribute to the longitudinal stiffness because of their large flange area. They add little to the torsional behaviour, since their stiffness is small compared to that of the cell. Transversely their function is to transmit the loading applied to the cantilevers to the spine of the bridge. It is thus evident, that in design of cantilever slabs only the longitudinal and transverse effects have to be considered.

It is further possible, and usual, to go a stage further and to design these slabs for transverse action only, and to incorporate the slab cross-sectional area thus obtained as part of longitudinal action. The design of cantilever slabs subjected to the Ministry of Transport[1, 2] loading is the topic of this paper.

Slab geometry and loading

Cantilever slabs are usually continuous over the longitudinal span of the bridge and it is therefore sufficiently accurate to treat them as semi-infinite in the longitudinal

(y) direction and of constant width in the transverse (x) direction (Fig. 1). Their thickness generally varies transversely from a maximum at the cantilever root to a minimum close to the edge, where an edge beam is provided to support the safety railings. Cantilever slabs are usually constructed of reinforced concrete, but pre-stressing is sometimes adopted for long cantilevers. A typical cross-section defining dimensional coefficients is shown in Fig. 1(c).

The design loading for cantilever slabs is defined in B.S.153 Part 3A LOADS[1] and in the Ministry of Transport Memorandum No. 771[2]. In both cases the loading consists of either uniformly distributed load or a series of concentrated point loads.

The design of cantilever slabs under the action of uniformly distributed loading is trivial, as the slab can be idealised to a beam of an arbitrary width under plain strain conditions. For concentrated point loading, however, the full effect of slab action and slab beam interaction has to be considered. In elastic design, the superposition-ing of stresses is permitted, and the analytical problem reduced to that of the analysis of a semi-infinite cantilever slab subjected to an arbitrarily placed point load.

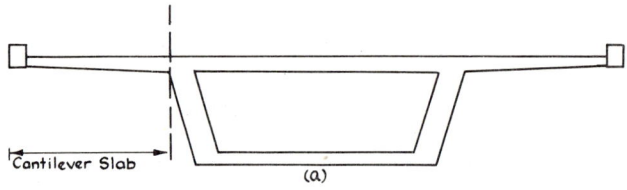

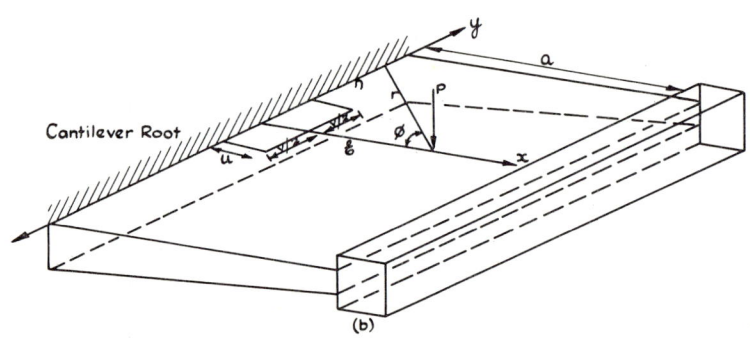

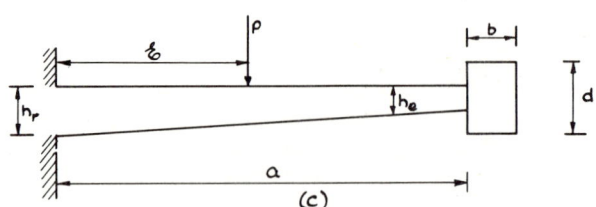

FIG. 1   GENERAL ARRANGEMENT AND DEFINITION
        OF TERMS.

Review of Published Work

Published solutions for the analysis of semi-infinite cantilever slabs under the action of an arbitrarily placed concentrated load have tended to concentrate on the distribution of hogging moment along the cantilever root and there is little published data for the other design criteria. Thus the distribution of transverse bending moment due to a point load and the distribution of the shear force along the cantilever root due to a point load have not been tackled. It is particularly surprising that the analysis for shear has received such little attention as in some cases shear is the design criteria for the depth at the root.

The published solutions for the distribution of moment along the cantilever root can be categorised as shown in Table 1.

Jaramillo[3] *– constant thickness slab with no edge beam*

Jaramillo's solution is an exact solution expressed in terms of improper integrals for the deflections and moments due to a transverse concentrated load acting at an arbitrary point of an infinitely long cantilever plate of constant width and thickness, Fig. 2(*a*). The solution is transformed into a series form by means of contour integration. Although this is the most rigorous solution for this section type the presentation

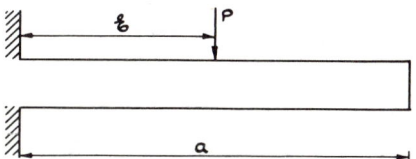

(a) Cross Section and Loading Analysed by Jaramillo [3]

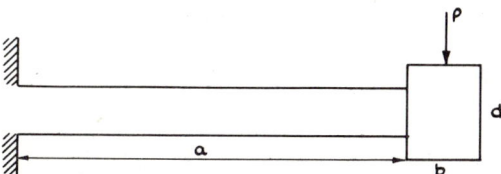

(b) Cross Section and Loading Analysed by Reismann and Cheng. [5]

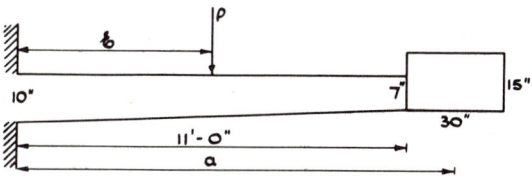

(c) Cross Section and Loading Analysed by Lee [6]

FIG. 2 CROSS SECTIONS AND LOADING TREATED IN PUBLISHED SOLUTIONS

of the formulae and the general computation is difficult to understand and this severely restricts its use. However, Jaramillo has presented the following curves using a Poisson's ratio value of 0·30.

(1) Dependence of the maximum bending moment at the root ($y/a = 0.0$) on load position (ξ/a)
(2) Dependence of the distribution of the bending moment along the clamped edge ($x/a = 0$) on relative load position (ξ/a)

The form of the second curve makes interpolation difficult and the value of Poisson's ratio of 0·30 is more appropriate to steel plates than to concrete slabs for which Poisson's ratio is generally taken between 0·15 and 0·20. These curves were used to verify the finite element approach to the problem and are discussed in more detail later.

Westergaard[4] – infinite cantilever slab with concentrated loading

This is perhaps the best known and most widely used solution to the distribution of the bending moment along the root of a cantilever plate due to a transverse concentrated load acting at an arbitrary point of the cantilever. The solution assumes the cantilever plate to have constant thickness and to be infinite in both the transverse (x) direction and the longitudinal (y) direction. Westergaard's formula, equation (1), which is independent of cantilever length, therefore strictly only applies to large cantilevers loaded close to the root.

$$M_x = -\frac{P}{\pi}\left(\frac{1}{1+\left(\frac{y}{\xi}\right)^2}\right) \tag{1}$$

It is argued by Somerville *et al.*[7] that this is not a serious restriction as the area of the edge beam can be replaced by extending the cantilever slab, and that this would then tend to approximate to the assumed conditions. Point load tests carried out on the cantilevers of the Mancunian Way model by Somerville, Roll and Caldwell[7] showed that Westergaard's formula underestimated the maximum transverse bending moment at the root. In the report on the testing[7] this was put down to cracking along the root as the loads used were approximately three times the design load in order that reasonable experimental accuracy could be obtained.

However, equation (1) indicates that for $y = 0$, i.e. directly opposite the load, the value of $M_x = -\frac{P}{\pi}$ and is independent of the load position (ξ). This is contradictory to Jaramillo's solution which, although tending to $-\frac{P}{\pi}$ for $\left(\frac{\xi}{a}\right)$ tending to zero (tending to Westergaard's assumption of the length a tending to infinity), shows a marked increase in M_x with increase in $\frac{\xi}{a}$ (approximately $1.60 \times \left(-\frac{P}{\pi}\right)$ for $\frac{\xi}{a} = 1.0$).

It was this discrepancy between Westergaard and Jaramillo which led the authors to attempt to provide a formula for the distribution of bending moment along the root for this section type, which would be straightforward to use and have a reasonable accuracy when compared with Jaramillo's exact solution.

Reismann and Cheng[5] – constant thickness slab with an edge beam

The solution published by Reismann and Cheng is for the analysis of a cantilever plate strip of finite width and infinite length rigidly clamped along one of its longitudinal edges (y) with the opposite edge monolithically attached to a beam. The loading considered is a single concentrated load acting on the beam, Fig. 2(*b*). The analysis is similar to Jaramillo's and once again the computation is too difficult for general use. The results presented in reference (5) include the dependence of the distribution of the bending moment along the clamped edge (*x* = 0) with the rela-

tive load position $\left(\frac{\xi}{a}\right)$ and two factors relating the beam and slab bending and

torsional inertias are derived for a Poisson's ratio value of 0·30.

The results show that the edge beam effectively modifies the cantilever length and hence for a given value of load position (ξ) reduces the maximum moment at the root (at *y* = 0, directly opposite the load). The stiffer the edge beam, the greater the reduction of the maximum moment, but in order to maintain statical equilibrium of the system the distribution must become flatter. The results presented in their paper are rather limited and again not suited to concrete slabs due to the Poisson's ratio value used. They have, however, been used to verify the finite element approach to this type of section.

Equivalent Slab Techniques

As previously mentioned the effect of an edge beam is to modify the effective cantilever length. The most common method used to obtain the new slab length is to replace the edge beam by extending the slab so that the additional area is equal to that of the edge beam.

Jaramillo's solution can then be used on this equivalent slab and a reasonable degree of accuracy obtained. Westergaard's solution cannot be used with this method as Westergaard's formula is independent of the cantilever length.

Varying Thickness Slab with an Edge Beam

There is no published solution for this section type which can be described as exact. Three alternative approaches can be used for an approximate solution.

(i) The variation in section properties can be ignored and the problem treated as a constant thickness slab with an edge beam using the methods outlined above.

(ii) The equivalent slab technique can be used to account for both the edge beam and the varying thickness.

(iii) An approximate numerical procedure (e.g. grillage or finite element) can be used to produce a set of design curves suitable for a wide range of sections.

The first two approaches have not been reported in the literature. The third approach has been used by Lee who adopted a grillage idealisation of the section shown in Fig. 2(*c*). Among the results he has published in reference (6) is the distribution of bending moment along the cantilever root due to an arbitrarily placed concentrated load. The results are for the specific example shown in Fig. 2(*c*), and are for a Poisson's ratio value of 0·25. Lee recommended that the results can be applied generally with care.

The authors have analysed the example using the finite element method and discovered large discrepancies between the two approaches. It is clear that there is a need for a simple design rule for spine beam cantilever slabs and this is described in subsequent sections.

Approximate Solution for Uniform Cantilever Slabs

(a) Hogging moment at cantilever root

The following approach has been used to obtain design charts for this case. The finite element solution of the problem for a Poisson's ratio value of 0·30 has been verified against Jaramillo's exact solution. The finite element solution is then employed to obtain the necessary constants for Poisson's ratio of 0·15 and to verify the approximate solution derived by the authors.

The finite element analysis was carried out using the computer program developed by Sawko and Cope[8] which is based on the Zienkiewicz rectangular plate bending element.[9] The accuracy of this element for slab bending has been verified extensively and results reported in literature.[8,9]

The idealisation used in the analysis is shown in Fig. 3. Westergaard's formula indicates that for reasonable accuracy $L/a \geqslant 10$, but the results of a trial run indicated that a value of $L/a \geqslant 5$ was sufficient and this determined the length of the slab to be used in the analysis. Symmetry conditions were used to reduce the size of the problem and the mesh was loaded with point loads at the nodes 2–9 (see Fig. 3). The

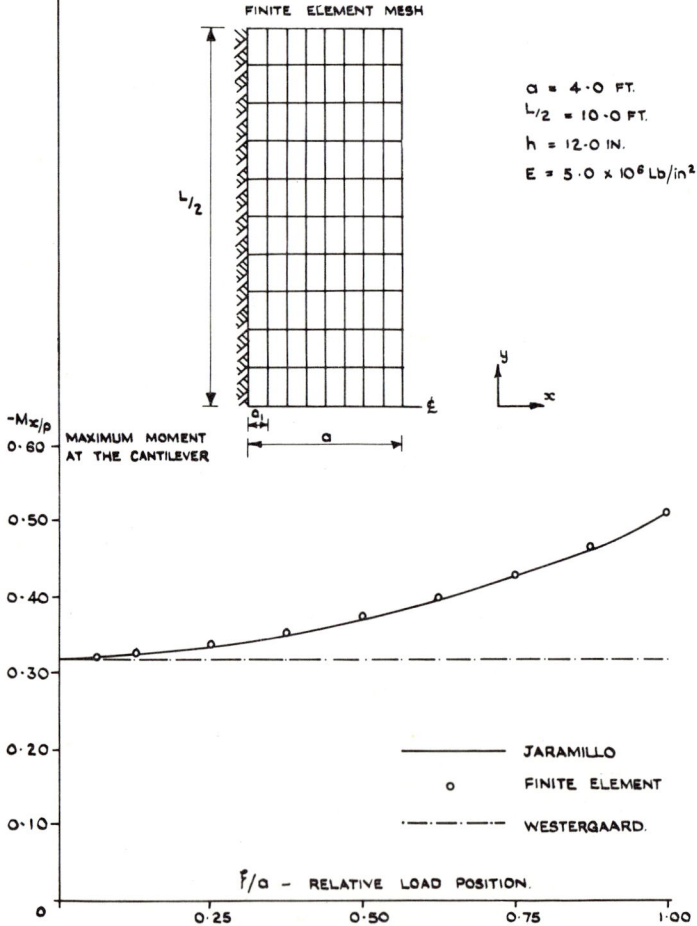

FIG. 3.   COMPARISON OF JARAMILLO AND FINITE ELEMENT FOR THE MAXIMUM MOMENT AT THE CANTILEVER ROOT.

absolute values of the slab length (L), the slab width (a), Young's Modulus (E), and the slab thickness (h) do not affect the results but for completeness are shown in Fig. 3.

The comparison between the finite element results and Jaramillo's solution are shown in Figs. 3 and 4 where Westergaard's solution is also plotted for comparison. Both curves show that the accuracy of the finite element solution is excellent and that it can be used to predict design moments accurately. Fig. 5 shows the variation of the factor A' with relative load position ξ/a where A' is defined by

$$M_{\max_{\text{Jaramillo}}} = \left(-\frac{1}{\pi}\right) \times A' \tag{1a}$$

From Fig. 4 the following remarks can be made about the distribution of hogging moments

(a) the distribution is of symmetric uni-modal form having its peak opposite the load and zero values at $\pm \infty$.

(b) The peak value depends on the load value P, its relative position ξ/a and Poisson's ratio only.

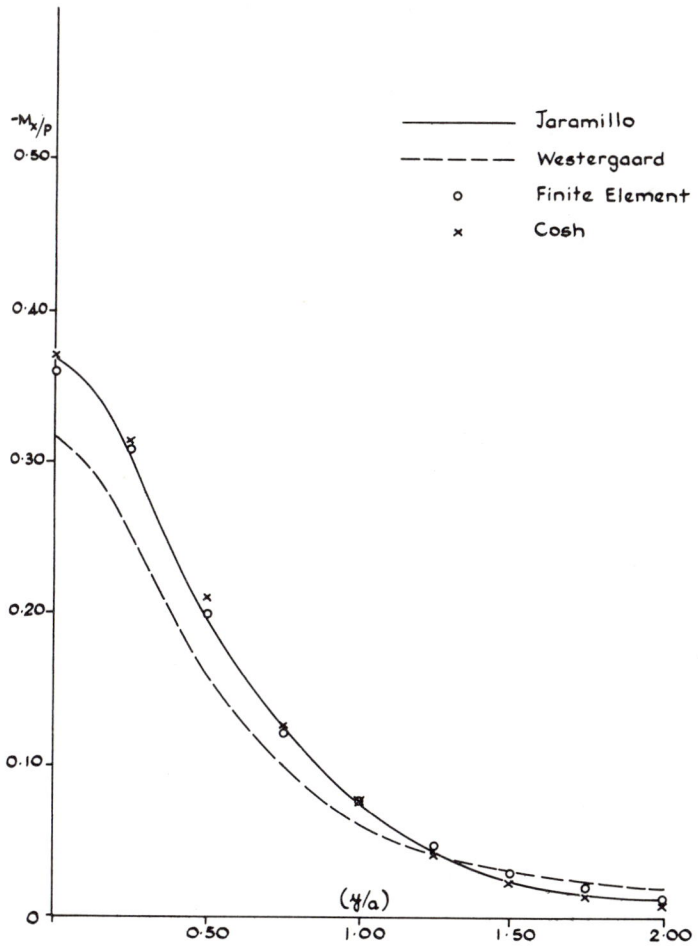

FIG. 4 COMPARISON OF THEORIES FOR THE DISTRIBUTION OF TRANSVERSE BENDING MOMENT ALONG THE CANTILEVER ROOT FOR $\mu = 0.30$, AND $\xi/a = 0.50$

(c) The total area under the curve must equal the statical moment $P \times \xi$.

If the finite element results for the peak are used, then the equation of the curve, which is $a \, \text{fn}\left(\dfrac{y}{a}\right)$, must obey the following boundary conditions

$$
\left.
\begin{array}{l}
\text{for} \qquad \dfrac{\xi}{a}, \ \text{at} \ \dfrac{y}{a} = 0 \quad \text{fn}\left(\dfrac{y}{a}\right) . M_{\text{peak } \xi/a} = F . E_{\text{peak } \xi/a} \\[3mm]
\text{and} \qquad M_{\text{peak } \xi/a} \displaystyle\int_{-\infty}^{+\infty} \text{fn}\left(\dfrac{y}{a}\right) \mathrm{d}y = P \times \xi \\[5mm]
\text{at} \qquad \dfrac{y}{a} = \pm \infty \ \text{fn}\left(\dfrac{y}{a}\right) M_{\text{peak } \xi/a} = 0
\end{array}
\right\}
\tag{2}
$$

The obvious approach is to modify Westergaard's formula, equation (1), to fit the boundary conditions, equation (2). Such a modification takes the form

$$
M_x = -\frac{P}{\pi} A' . \frac{1}{1 + A'^2 \left(\dfrac{y}{\xi}\right)^2}
\tag{3}
$$

where A' may be taken from Fig. 5.

This equation obeys the boundary conditions but was found to give a poor fit to Jaramillo's exact curve and thus another solution was sought.

There are several uni-modal, symmetric, infinite distributions which could be used to fit the boundary conditions of equation (2). The best solution found is a cosh distribution of general form.

$$
M_x = -\frac{P}{\pi} A' . \frac{1}{\cosh\left(4' . \dfrac{y}{a} \dfrac{\xi}{a}\right)}
\tag{4}
$$

where A' is once again taken from Fig. 5.

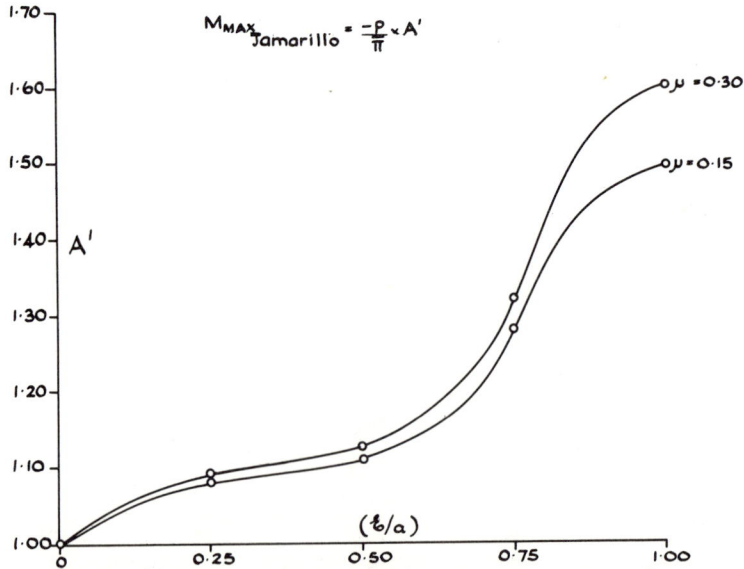

FIG. 5 DEPENDENCE OF A ON THE RELATIVE LOAD POSITION
(ξ/a) — FINITE ELEMENT RESULTS

This curve obeys all the boundary conditions and it can be seen from Fig. 4, where it is plotted and marked as the cosh distribution, that it compares very closely to Jaramillo's results.

It is obvious from Fig. 3 that little error is introduced if, instead of reading A' from the graph, the curve in Fig. 3 be treated as straight line portions between $(0, 0.25)$, $(0.25, 0.50)$, $(0.50, 0.75)$, $(0.75, 1.00)$. Equation (4) can thus be written

$$M_x = -\frac{P}{\pi}\left(A + B\frac{\xi}{a}\right) \cdot \frac{1}{\cosh\left\{\left(A + B\frac{\xi}{a}\right)\frac{y}{a}\Big/\frac{\xi}{a}\right\}} \qquad (5)$$

where A and B depend on ξ/a and μ and are tabulated in Table 2 for $\mu = 0.15$ and 0.30.

Equation (5) has been tested against the finite element solution and Jaramillo's solution over the whole range of ξ/a for $\mu = 0.30$ and against the finite element solution for $\mu = 0.15$ and has been found to be in excellent agreement with both these methods. Thus equation (5) can be safely used to determine the distribution of bending moment along the root of the cantilever due to an arbitrarily placed concentrated load.

(b) Sagging moment under point load

The solutions for the elastic analysis of thin plates predict an infinite value of sagging moment induced under a point load. This is obviously untrue and thick plate theories[10] predict a value which is approximated by the finite element approach for thin plate bending. The finite element results for the variation of this moment with load position is shown in Fig. 6. Although the value of the moment is large (approximately half of the hogging moment at the root) the effects of several point loads can never accumulate as the effects of this sagging moment are extremely local to the load position, Fig. 7(a), The effects, however, cannot be ignored, especially in varying depth slabs where

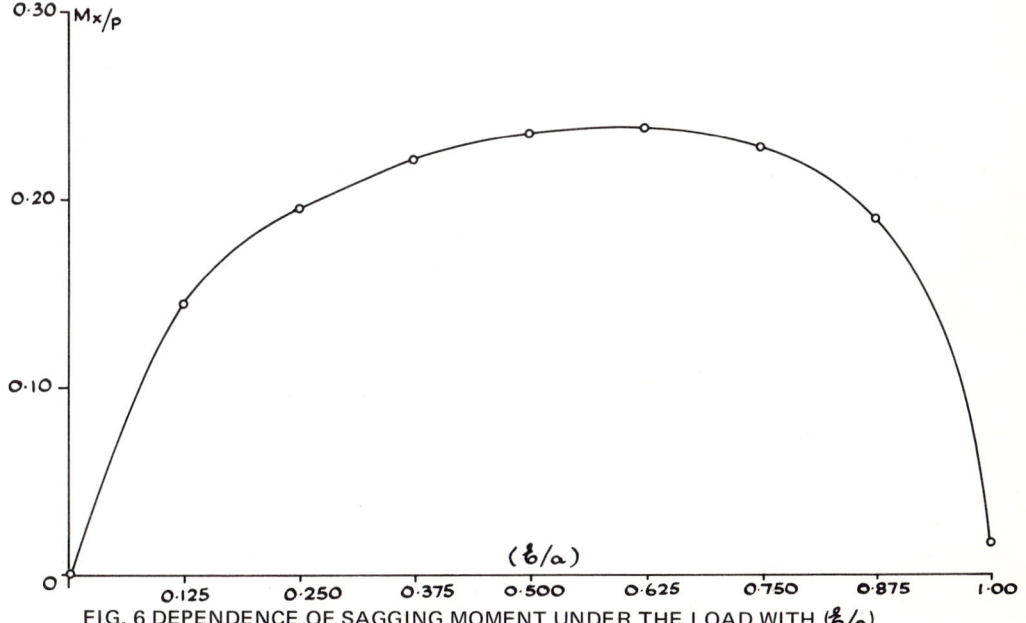

FIG. 6 DEPENDENCE OF SAGGING MOMENT UNDER THE LOAD WITH (ξ/a)

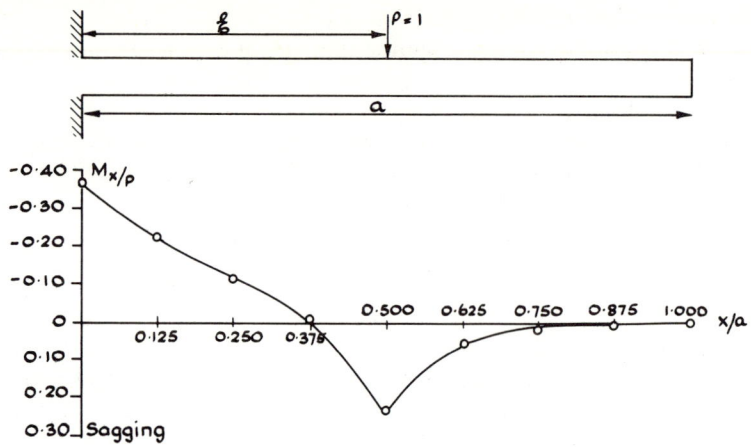

(a) Transverse Distribution of Moment along y/a = 0, for ξ/a = 0.50, μ = 0.15 (Table 3)

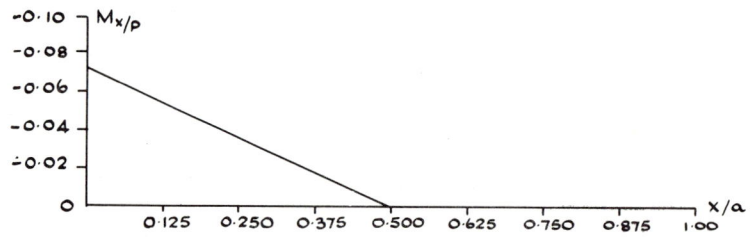

(b) Transverse Distribution of Bending Moment along y/a = 1.0, for ξ/a = 0.50, μ = 0.15

FIG 7. TRANSVERSE DISTRIBUTION OF BENDING MOMENT
DUE TO A UNIT POINT LOAD AT (0.50, 0)

the maximum occurs where the slab is relatively thin. The general form for the distribution of the transverse bending moment along $y = 0$ is as shown in Fig. 7(a), whilst for $y \neq 0$ it is sufficiently accurate to assume the distribution shown in Fig. 7(b), as once again the sagging effect is local. No equation can be found to fit the distribution shown in Fig. 7(a) so the finite element results for various values of relative load position ξ/a are tabulated in Table 3. It is possible from Table 3 to compile the following curves.

(1) The maximum sagging moment in the span due to the standard loading conditions.

(2) The transverse bending envelope diagrams for the standard loading conditions. The authors have compiled both these sets of curves but due to lack of space they are not presented here.

(c) Shear at cantilever root

As previously mentioned little has been published about the shear force distribution along the cantilever root. Jaramillo has not developed the shear equation through his analysis and the only exact solution is for a cantilever plate of constant thickness and of infinite length in both the longitudinal (y) direction and the transverse (x) direction.[10] The solution, derived by Nádai,[14] considers the effect of loads applied close to the support. The assumed deflection function (equation (d) on page 327 of reference (10)) is differentiated to obtain the shear force distribution along the cantilever root as

$$Q_x = \frac{2P}{\pi r} \cos^3 \emptyset \tag{6}$$

where the terms are defined in Fig. 1.

For the case $\emptyset = 0$, i.e. directly opposite the load, the maximum shear is given by

$$Q_{x\,max} = \frac{2P}{\pi \xi} \tag{7}$$

Thus, according to Nádai, the maximum shear force on the cantilever root occurs directly opposite the load and is dependent only on the load position and its value. From equation (7) it can be seen that as ξ tends to zero, $Q_{x\,max}$ tends to infinity. Practically this is impossible and for loads applied close to the cantilever root, the correct approach is to treat them as loads distributed over an area (patch loads).

Timoshenko[10] considers such a patch load placed to produce maximum shear as shown in Fig. 1. By integrating (6) over the limits 0 to u, $-v/2$ to $+v/2$, he obtains the maximum shear force at 0 as

$$(Q_x)_{max} = \alpha \frac{P}{v} \tag{8}$$

where α depends on the ratio v/u (tabulated values are given in reference (10) p. 334).

Timoshenko recommends that equation (8) may be used to calculate the maximum shear force on cantilevers having finite dimensions, and that the effects of any other wheel loads are small enough to be ignored. The authors cannot agree with either of these recommendations and the differences are discussed later.

The alternative methods for determining the shear are the approximate solutions offered by the grillage analysis or the finite element analysis using plate bending elements. Of these two methods, the finite element analysis is usually considered to yield more accurate solutions for the plate bending problem, but the results for shear need careful interpretation.

In the finite element method, the assumed deflection function of the element is used in conjunction with the classical elasticity equations for the bending of thin plates[10] to determine the elemental forces from the nodal deflections. In particular the equation for the shear along $X = 0$ is given by

$$Q_x = \frac{\partial M_x}{\partial x} + \frac{\partial M_{yx}}{\partial y} \tag{9}$$

where the terms are defined in reference (10) as $M_x = -D\left(\frac{\partial^2 w}{\partial x^2} + \mu \frac{\partial^2 w}{\partial y^2}\right)$ and $M_{yx} = -\frac{D(1-\mu)}{2} \frac{\partial^2 w}{\partial x \partial y}$. The boundary condition of the cantilever root is such that the terms $\frac{\partial^2 w}{\partial y^2}$ and $\frac{\partial^2 w}{\partial x \partial y}$ are both zero therefore equation (9) becomes

$$Q_x = \frac{\partial M_x}{\partial x} \tag{10}$$

However, in Zienkiewicz's rectangular plate bending finite element it is not possible to ensure that the term $\frac{\partial^2 w}{\partial x \partial y}$ is generally zero, and hence the finite element solution for the shear distribution along the root using this particular element is incorrect.

A rectangular plate bending element derived by Hansteen[11] has the term $\dfrac{\partial^2 w}{\partial x \partial y}$ as a nodal degree of freedom but the physical interpretation of this term is obscure. Suppressing this term along the root yields incorrect results for shear while results obtained with this degree of freedom unsuppressed are difficult to interpret.

The determination of shear along the boundary is not, therefore, straightforward and the authors have found the most consistent approach is to calculate $\dfrac{\partial M_x}{\partial x}$ from the nodal average output using Zienkiewicz's rectangular plate bending element.

The shear distribution along the root has been calculated in this manner for various load positions and the results are presented in Table 4. As the distribution of M_x is linear over each element in the x direction $\dfrac{\partial M_x}{\partial x}$ is given by

$$\frac{\partial M_x}{\partial x} = \frac{M_{x1} - M_{x2}}{a_1} \tag{11}$$

(see Fig. 3).

It is evident from this equation, that since a_1 is dependent on cantilever length, so is the shear, and values in Table 4 must be divided by cantilever length (in feet).

For the position $y = 0$, i.e. directly opposite the load, the shear force Q_x tends to infinity as the relative load position tends to zero (see Table 4). Analytically this is correct but physically it is impossible.

Leonhardt in his reports on the Stuttgard shear tests[12] has indicated that as the load approaches the root the load is carried by strut and tie action, Fig. 8(a), with shear becoming a secondary load carrying medium, and has presented a reduction factor, dependent on the load position, to be applied to the analytical shear force in order to calculate the actual shear force to be carried by shear action. This is shown in Fig. 8(b). While the Stuttgart shear tests were primarily concerned with beams,

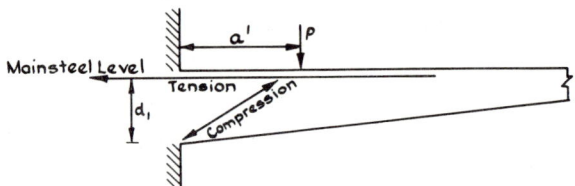

(a) Direct Forces Which Carry the Load when the Load, p, is close to the Root.

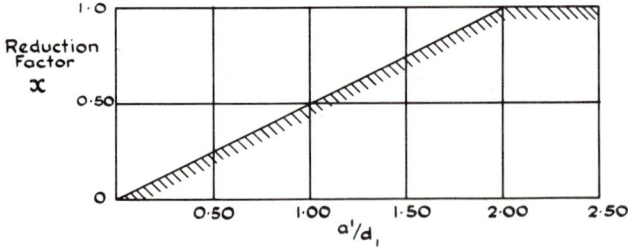

(b) Reduction Factor to be applied to the Shear Force at the Cantilever Root due to Shear Reduction by Direct Forces (Leonhardt)

FIG. 8   LEONHARDT'S SHEAR REDUCTION THEORY

Leonhardt recommends that the results are also applicable to short cantilever slabs.

A second approach to this problem is to consider the wheel load as a patch load distributed over an area defined by the 45° dispersion theory (assuming wheel dimensions given by Henderson).[13] This may then either be used in a finite element analysis or used to formulate a reduction factor to be applied to the point load to calculate the shear to be carried by pure shear action.

The first approach requires a very fine mesh division so that the loaded area is exactly defined and, as mesh grading is not possible with rectangular elements, this produces extremely large meshes. This approach was only used to verify the second approach.

With reference to Fig. 9, it can be seen that as the load approaches the root, part of the loaded area overlaps the support and a further area (hatched) is considered to be carried by direct forces. If this area is ignored then the following modifications apply:

(1) The total load on the cantilever is reduced by an amount RF where

$$RF = (7 \cdot 5 + f + EPS)/(15 \cdot 0 + 2f + d_1) \tag{12}$$

Provided $RF \leqslant 1 \cdot 0$, i.e. provided $EPS \leqslant (d_1 + f + 7 \cdot 5)$, see Fig. 9.

(2) The equivalent centroid of the load is modified to ξ_m where

$$\xi_m = (EPS + 7 \cdot 5 + f + d_1)/(2 \cdot 0) \tag{13}$$

Provided $EPS \leqslant (7\frac{1}{2} + f + d_1)$.

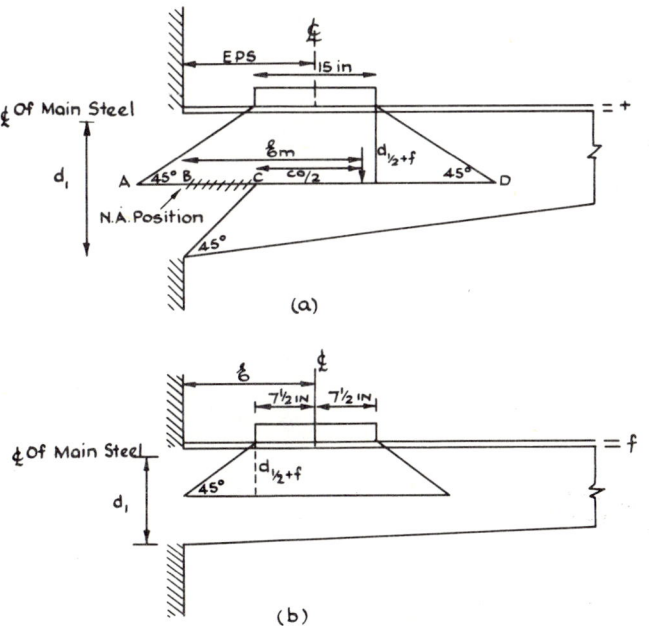

FIG. 9 REDUCTION OF LOAD CARRIED AND MODIFICATION TO ₵ OF LOAD ASSUMING BC IS CARRIED BY DIRECT COMPRESSION, AND POSITION OF WHEEL FOR MAXIMUM SHEAR.

In Fig. 10 a comparison is made between the three theories mentioned above. It can be seen that the approximate approach presented here compares well with the finite element results for the patch loads, Leonhardt's theory producing lower values. These curves also show that the maximum shear force at the cantilever root is produced by a wheel positioned as shown in Fig. 9 and not as Timoshenko predicts, by a wheel as close to the support as possible (Fig. 1).

Figs. 11 and 12 compare Nádai's solution with the finite element solution. Fig. 11 compares Nádai's solution for the quantity (Q_x) x cantilever length, with the finite element solution for this quantity. (Nádai's solution for $Q_{x_{max}} = \dfrac{2P}{\pi\xi}$ is independent of the cantilever length (a), but can be written as $\dfrac{2P}{\pi\xi/a} = (Q_{x_{max}})a$ which is the quantity used in Fig. 11). It can be seen from Fig. 11 that Nádai's solution tends to infinity more rapidly than the finite element solution, the two curves tending to the same asymptote as the relative load position ξ/a increases.

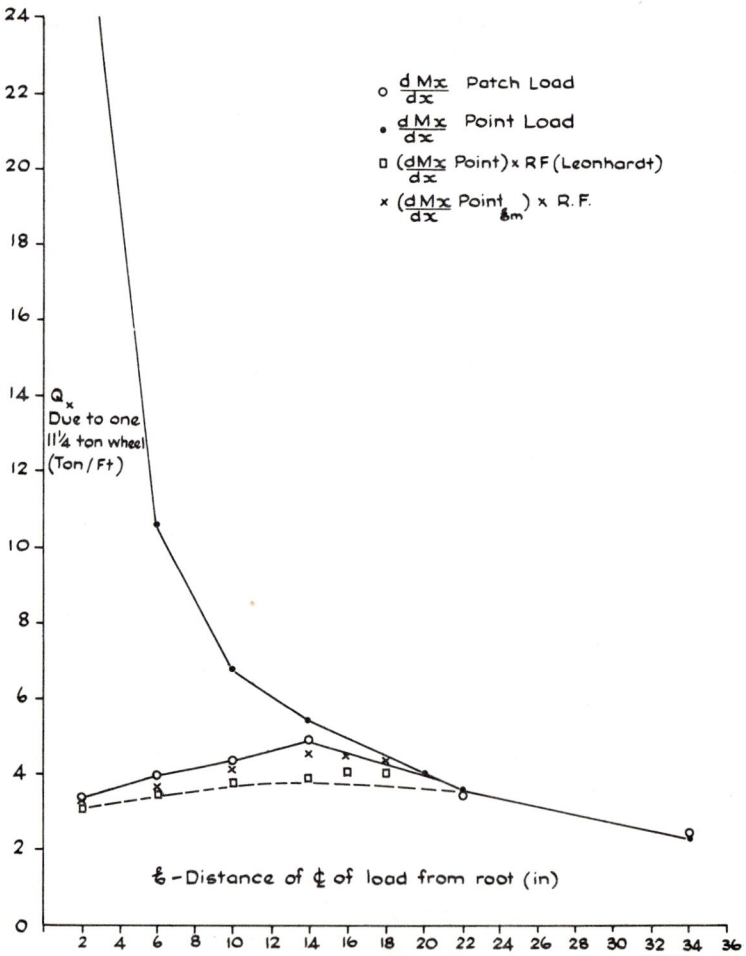

FIG. 10 COMPARISON OF SHEAR THEORIES FOR A LOAD CLOSE TO
THE ROOT FOR $d_1 = 10$ in. AND A CANTILEVER LENGTH OF 8FT.

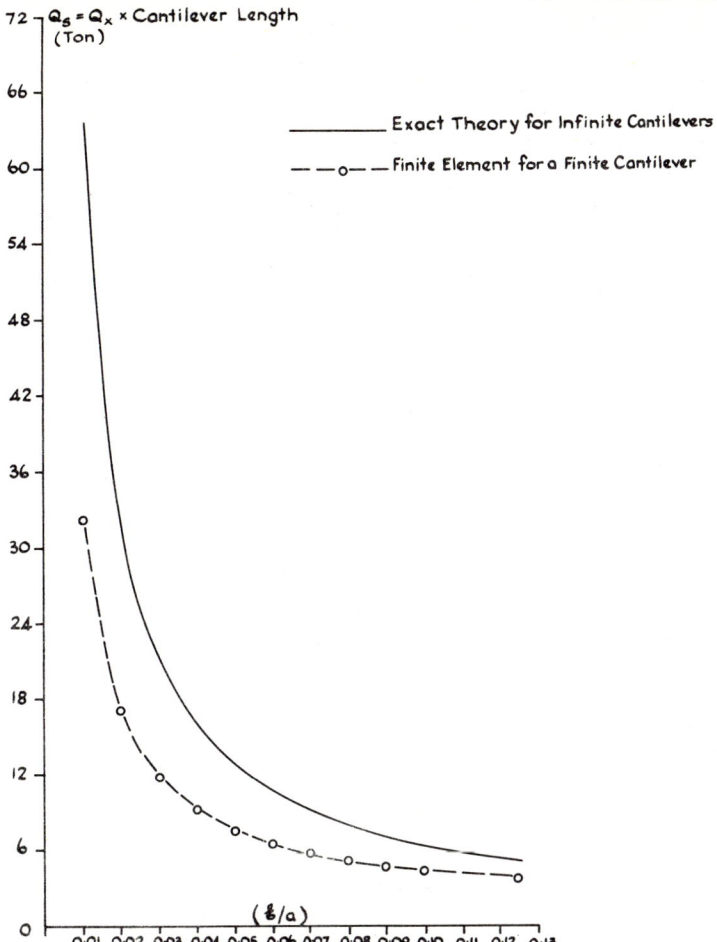

FIG.11 COMPARISON OF INFLUENCE LINES FOR SHEAR AT O
DUE TO A SINGLE POINT LOAD

Fig. 12 compares the maximum shear produced at the cantilever root due to a patch load placed to produce maximum shear according to Nádai's theory (Fig. 1) for differing values of u, v. The three cases shown are for the following conditions.

(1) $u = 15$ in, $v = 3$ in, area of load taken as a wheel of dimensions given by Henderson.[13]

(2) $u = 28$ in, $v = 16$ in, wheel placed as shown in Fig. 9(b) for $f = 1\frac{1}{2}$ in, $d_1 = 10$ in.

(3) $u = 44$ in, $v = 32$ in, wheel placed as shown in Fig. 9(b) for $f = 1\frac{1}{2}$ in, $d_1 = 26$ in.

Generally the results for cases (2), (3) above compare favourably (case (1) is not really a sensible condition). However, as the finite element solution is dependent on the cantilever length, while Nádai's solution is not, the authors feel that the recommendation that Nádai's solution may be used on cantilevers of finite dimensions must be treated with caution, while the recommendation that the effects of other wheel loads may be ignored will, in the authors' view, lead to a section which will be underdesigned for shear. This is shown in the example given to illustrate the design procedures recommended in this paper (Appendix A).

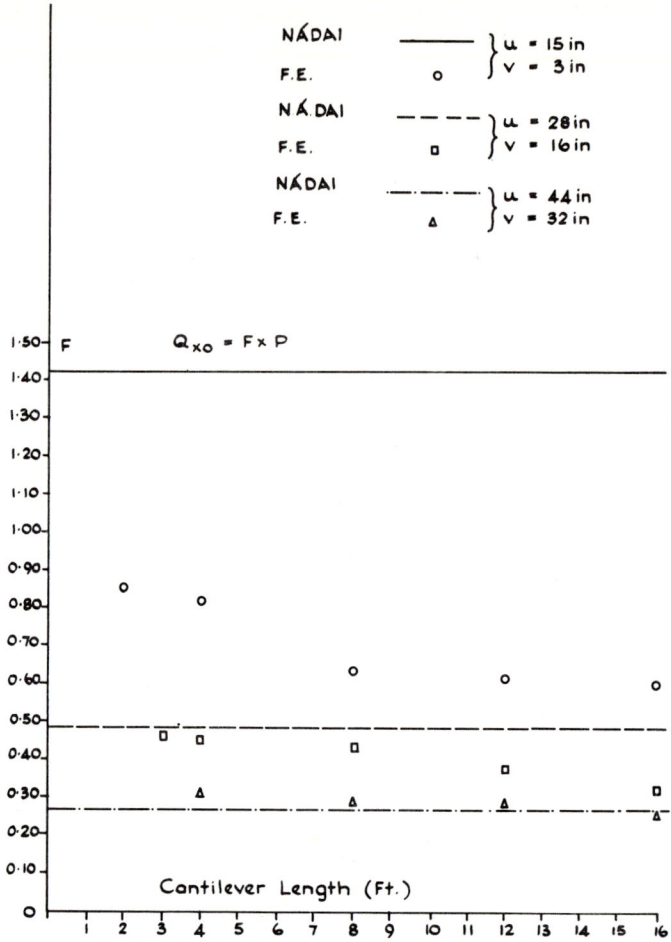

FIG. 12 COMPARISON OF MAXIMUM SHEAR AT O DUE TO
PATCH LOADS OF VARYING DIMENSIONS PLACED
AS SHOWN IN FIGURE I.

Varying Section Cantilevers with Edge Beams

As mentioned previously, Lee has produced curves for the distribution of the bending moment along the cantilever root for a specific example using a grillage analysis, and recommends their use generally. The curves are reproduced in Fig. 13, and the interesting point is that the peak values all lie below $-1/\pi$.

Considering the effect of the edge beam and varying section the following conclusions can be drawn. The effect of the edge beam is to increase the effective length of the cantilever to be used in the analysis, while the varying section tends to reduce this effective length. However, unless the load is applied directly to an edge beam of extremely large stiffness, the peak value cannot fall below the theoretical minimum of $-P/\pi$. The authors have analysed Lee's example using finite element methods, find that this is in fact the case (Fig. 13), and therefore conclude that Lee's results will produce an underdesigned section.

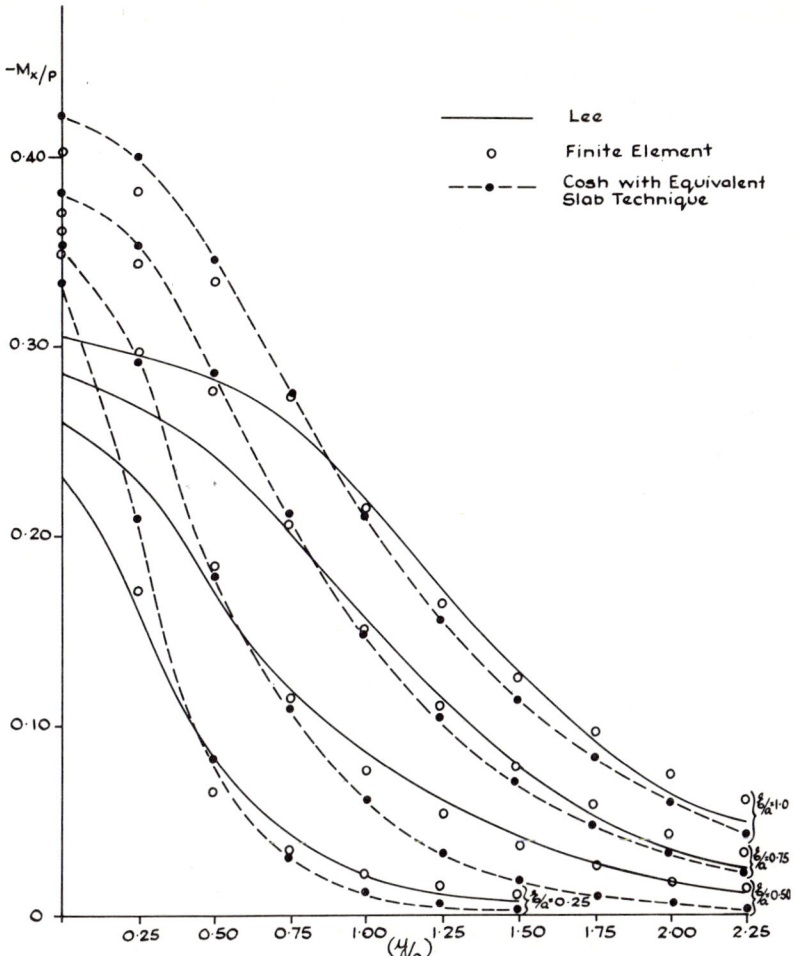

FIG. 13 COMPARISON OF LEE AND FINITE ELEMENT FOR THE
DISTRIBUTION OF MOMENT ALONG THE ROOT FOR $\mu=0.25$

Generally the authors feel that the production of curves for this type of section is not feasible, but that the finite element method using rectangular elements will produce accurate results for all cases. The authors have found that, in most practical cases, the effects of the varying section are small compared to the edge beam effects, and approach (1), which leads to a slight overdesign, is suitable for the majority of cases. The only exception is the case of a concentrated load applied to an extremely stiff edge beam, in which case Reismann and Cheng's solution should be used.

The sagging moment in the span can similarly be evaluated by one of the three suggested approaches. The authors found that the first approach is by far the simplest and produces results which are conservative. The same arguments apply to shear at cantilever root.

Conclusions

The authors conclude that for semi-infinite cantilever slabs of constant width and thickness, the methods of analysis derived in this paper for the moment distribution along the root, the shear distribution along the root and the transverse distribution of the transverse moment, will provide accurate results, while for cantilevers of varying section attached to reinforcing edge beams, the equivalent slab technique described coupled with the procedures defined herein will yield approximate and safe solutions.

Acknowledgement

This work was carried out at the Department of Civil Engineering at Liverpool University and forms part of a research project into the design and analysis of spine beam bridge decks. The authors are indebted to several consulting engineers and bridge design offices whose constant interchange of ideas with the authors is an essential ingredient of this work.

Appendix A

Design curves for 45 units of *HB* load and an example of their use

Example

The cantilever slab shown in Fig. A1(*a*) is to be designed to carry *HA* loading and to be checked for $37\frac{1}{2}$ units of *HB* load.

The analysis of the *HA* loading is straightforward and this example illustrates the analysis for the *HB* load using the design curves presented here.

Design curves shown in Figs. A2 – A4 have been produced, for convenience, for the *HB* train of loads and are based on the authors' formulae derived in the paper. These curves have been automatically plotted on a Calcamp 563 drum plotter operating from the computer. The following explanatory notes might be useful.

(1) Fig. A2 was derived on the basis of the cosh formula assuming that the maximum moment at the cantilever root is produced when the load is as eccentric as possible.

(2) Fig. A3 was derived from Table 3, assuming that the maximum sagging moment at any point occurs when an extreme wheel is over the point.

(3) Fig. A4 was derived from Table 4, assuming the maximum shear due to the *HB* vehicle is produced when the vehicle is positioned to produce the maximum shear due to the innermost wheel.

(*a*) *Effective length of the cantilever*

Average depth of the cantilever $= \dfrac{12 + 8}{2} = 10$ in

Area of the edge beam $= 18 \times 12 + 30 \times 18 = 756$ in^2

∴ Effective cantilever length $= 132 + \dfrac{756}{10} = 207 \cdot 6$ in $=$ 17·3 ft

(*b*) *Maximum hogging moment at the cantilever root*

This occurs at 0, Fig. A1(*b*) with the wheels positioned as shown in Fig. A1(*b*) (see reference (2)).

Thus the extreme wheel position $= 8 \cdot 5$ ft

therefore $\xi/a = \dfrac{8 \cdot 5}{17 \cdot 3} = \cdot 491$ (using modified cantilever length)

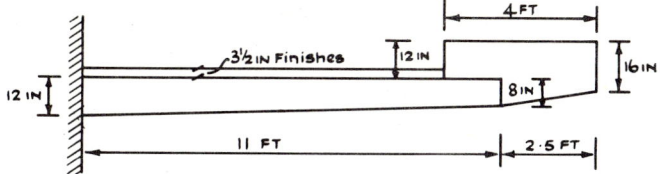

(a) Cross Sectional Dimensions

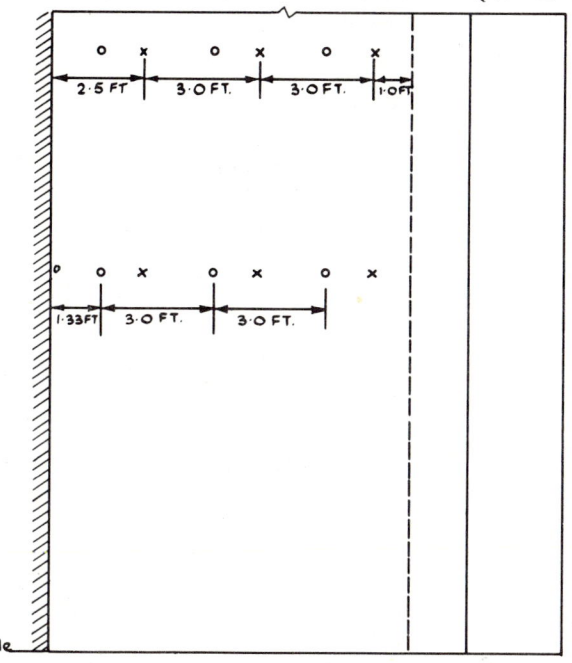

(b) Wheel Positions for (i) Maximum Hogging Moment at 0 (Marked x)
(ii) Maximum Shear at 0 (Marked O)

FIG. A1 DESIGN EXAMPLE AND LOADING.

From Fig. A2 (by interpolation) for $a = 11/\cdot3$ ft, $\xi/a = 0\cdot49$, $M_x = -18\cdot5 \times \dfrac{37\cdot5}{45\cdot0}$

$$= -15\cdot40 \text{ ton ft/ft}$$

For this wheel configuration the corresponding values given by Westergaard and Lee are

Westergaard $= -13\cdot89$ ton ft/ft (10% under)
Lee $= -13\cdot26$ ton ft/ft (14% under)

(c) *Maximum sagging moment in the cantilever span*

As the load has a maximum eccentricity of 0·491 (see (b) above), the maximum sagging moment occurs at $x/a = 0\cdot491$. From Fig. A3, for $a = 17\cdot3$ ft, $x/a = 0\cdot49$,

Maximum sagging moment in the span $= 3\cdot03 \times \dfrac{37\cdot5}{45}$

$$= 2\cdot52 \text{ ton ft/ft}$$

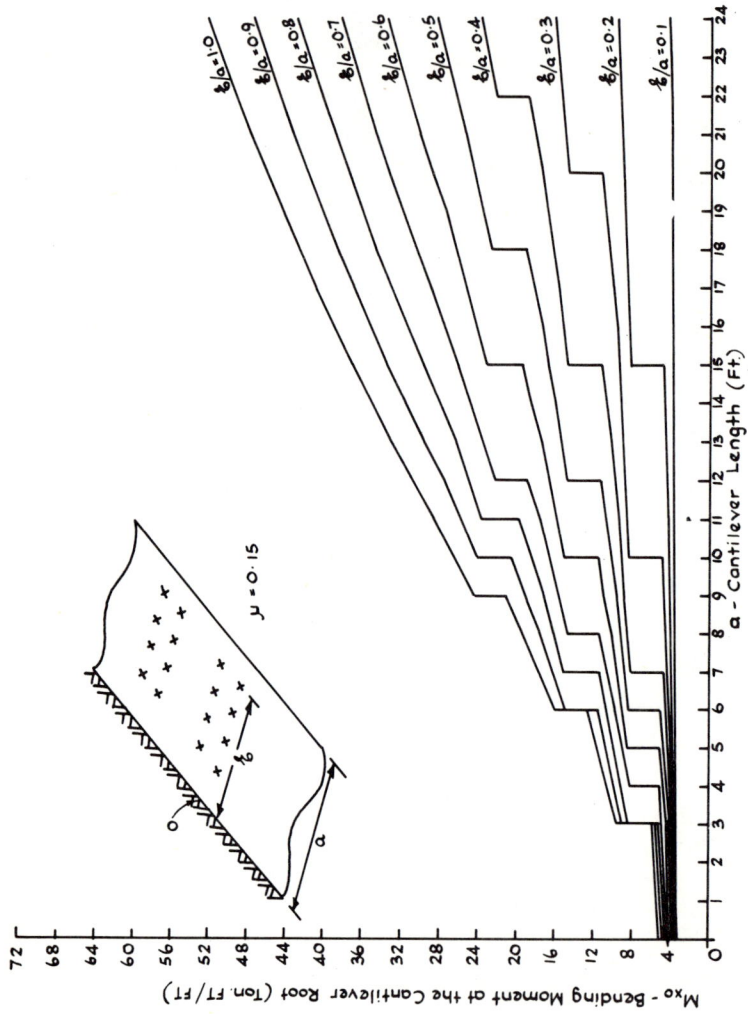

FIG. A2 TRANSVERSE BENDING MOMENT, M_{xo}, AT THE
CANTILEVER ROOT DUE TO 45 UNITS OF HB LOAD

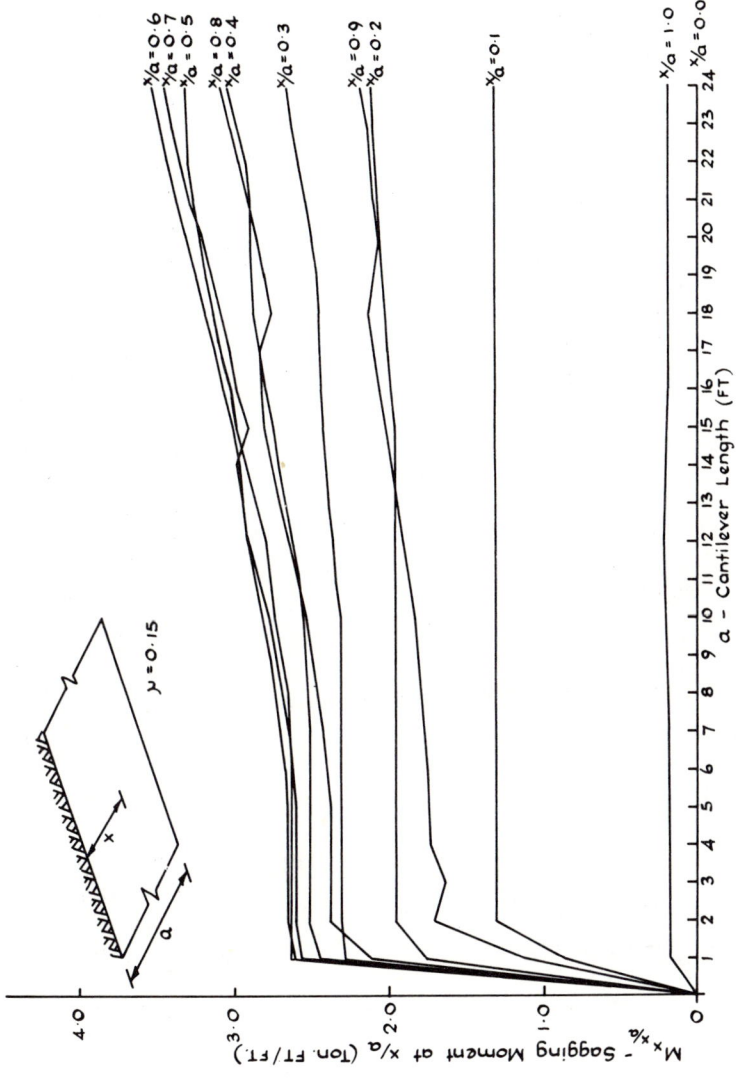

FIG. A3 MAXIMUM TRANSVERSE SAGGING MOMENT AT x/a
DUE TO 45 UNITS OF H.B LOAD

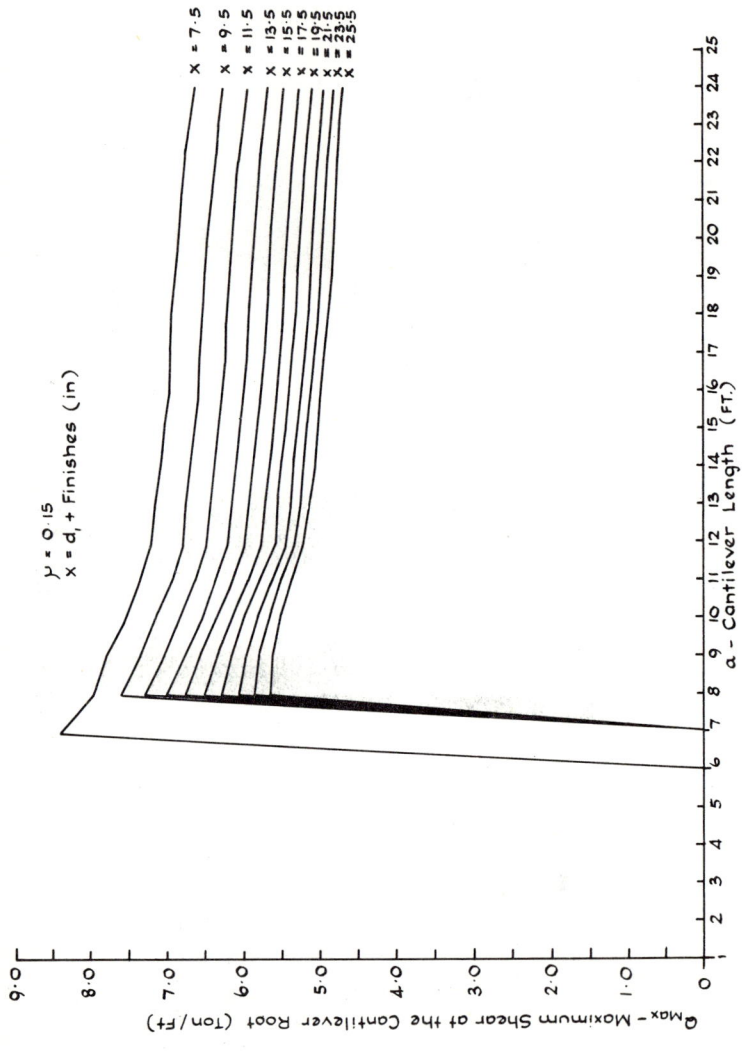

FIG A4 MAXIMUM SHEAR AT THE CANTILEVER ROOT
DUE TO 12 × 11¼ TON WHEEL (s)

(d) Maximum shear force at the cantilever root

This occurs at 0 for the wheel positions as shown in Fig. A1(b), i.e. 12 wheels on the cantilever, with $d_1 + f = 10\frac{1}{2} + 3\frac{1}{2} = 14$ in. From Fig. A4, for $a = 17\cdot3$ ft, $(d_1 + f) = 14$ in.

$$Q_{max} = 5\cdot85 \times \frac{37\cdot5}{45} = 4\cdot87 \text{ ton/ft}$$

The equivalent value given using Timoshenko's recommendations is

$$Q_{max} = 3\cdot81 \text{ ton/ft} \quad (22\% \text{ under})$$

TABLE 1

Categorisation of Published Solutions

Solution / Section Type	Exact	Approximate
(1) Constant Thickness, No Edge Beam 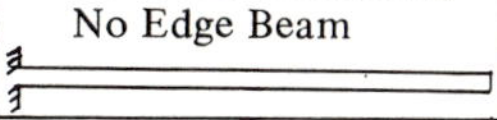	Jaramillo[3]	Westergaard[4]
(2) Constant Thickness, Edge Beam 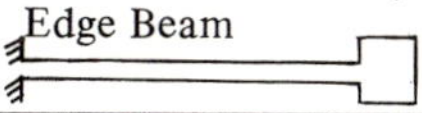	Reismann & Cheng[5]	Equivalent Slab Technique with Jaramillo
(3) Varying Thickness, Edge Beam 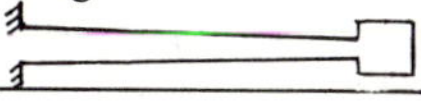		Lee[6]

TABLE 2

A and *B* values for use with equation 5 (see also Fig. 5)

	$\mu = 0\cdot15$		$\mu = 0\cdot30$	
	A	*B*	*A*	*B*
$0 \leqslant \frac{\xi}{a} \leqslant 0\cdot25$	1·0	0·3155	1·0000	0·1760
$0\cdot25 \leqslant \frac{\xi}{a} \leqslant 0\cdot50$	1·0551	0·1081	0·9240	0·4780
$0\cdot50 \leqslant \frac{\xi}{a} \leqslant 0\cdot75$	0·7641	0·6899	0·7960	0·7270
$0\cdot75 \leqslant \frac{\xi}{a} \leqslant 1\cdot00$	0·6416	0·8534	0·5810	1·0200

TABLE 3

Transverse Distribution Coefficients

$\mu = 0.15$

$\dfrac{M_X}{P}$ at $\dfrac{\xi}{a}$ Load Pos.	0	0·125	0·250	0·395	0·500	0·625	0·750	0·825	1·000
0	−0·3183	0	0	0	0	0	0	0	0
0·125	−0·3206	+0·1450	+0·0185	+0·0053	+0·0020	+0·0002	0	−0·0003	0
0·250	−0·3398	−0·0942	+0·1948	+0·0377	+0·0138	+0·0056	+0·0016	−0·0002	+0·0001
0·375	−0·3503	−0·1732	−0·0371	+0·2210	+0·0503	+0·0192	+0·0069	−0·0010	+0·0003
0·500	−0·3676	−0·2253	−0·1185	−0·0095	+0·2346	+0·0556	+0·0192	+0·0042	+0·0007
0·625	−0·3905	−0·2687	−0·1758	−0·0925	−0·0026	+0·2376	+0·0520	+0·0119	+0·0016
0·750	−0·4183	−0·3093	−0·2252	−0·1533	−0·0839	−0·0009	+0·2221	+0·0350	+0·0027
0·875	−0·4505	−0·3497	−0·2719	−0·2075	−0·1502	−0·0937	−0·0223	+0·1891	−0·0033
1·000	−0·4865	−0·3912	−0·3185	−0·2604	−0·2127	−0·1728	−0·1396	−0·1090	+0·0165

(See Figs. 3 and 7)

TABLE 4

Distribution of Shear along the Cantilever Roof for μ = 0·15

− Shear = Value from Table/Cantilever Length

Load Pos. \ Shear/P at y/a	0	0·25	0·50	0·75	1·00	1·25	1·50	1·75	2·00	2·25	2·50
0	(1)	0	0	0	0	0	0	0	0	0	0
0·125	3·7255	0·1250	0·0449	0·0098	0·0023	−0·0002	−0·0007	−0·0006	−0·0004	−0·0002	−0·0002
0·250	1·9653	0·8846	0·0972	0·0363	0·0117	0·0016	−0·0015	−0·0017	−0·0009	−0·0006	−0·0006
0·375	1·4161	0·9067	0·2871	0·0869	0·0292	0·0069	−0·0009	−0·0022	−0·0010	−0·0005	−0·0005
0·500	1·1383	0·8491	0·3936	0·1522	0·0547	0·0160	0·0017	−0·0015	−0·0003	−0·0003	−0·0003
0·625	0·9748	0·7891	0·4470	0·2065	0·0835	0·0282	0·0064	−0·0005	0·0015	0·0022	0·0022
0·750	0·8723	0·7406	0·4715	0·2455	0·1099	0·0418	0·0127	0·0040	0·0044	0·0049	0·0049
0·875	0·8065	0·7041	0·4806	0·2715	0·1321	0·0554	0·0200	0·0083	0·0080	0·0085	0·0085
1·000	0·7620	0·6765	0·4829	0·2892	0·1501	0·0679	0·0274	0·0130	0·0120	0·0123	0·0123

(1) infinity for the range $0 \to 0.125$ use Table 3 and the formula $Q = \dfrac{dM_X}{dx} = \dfrac{M_1 - M_2}{a_1}$

REFERENCES

1. British Standards, BS 153. *Steel girder bridges. Part 3A: Loads.* 1954.
2. G.B. Ministry of Transport. *Standard highway loadings.* . . . Memorandum No. 771, 1961.
3. Jaramillo, T. J. Deflections and moments due to a concentrated load on a cantilever plate of infinite length. *J. appl. Mech.*, Vol. 17, No. 1, March 1950, pp. 67–72.
4. Westergaard, H. M. Computations of stresses in bridge slabs due to wheel loads. *Public Roads*, Vol. 11, 1930.
5. Reismann, H. and Cheng, S. H. The edge reinforced cantilever plate strip. *Publs int. Ass. Bridge struct. Engng*, Vol. 30, 1970, pp. 149–162.
6. Lee, D. J. Huntley's Point overpass. *Struct. Concr.*, Vol. 2, No. 12, Nov./Dec. 1965, pp. 521–534.
7. Somerville, G., Roll, F. and Caldwell, J. A. D. Tests on a one-twelfth scale model of the Mancunian Way. *Cement and Concrete Association Technical Reports*, No. TRA 394, 1965.
8. Sawko, F. and Cope, R. J. The use of finite elements for the analysis of right bridge decks. International symposium on the use of electronic digital computers in structural engineering (University of Newcastle-upon-Tyne, July 1966).
9. Zienkiewicz, O. C. *The finite element method in structural and continuum mechanics.* New York, McGraw-Hill, 1967.
10. Timoshenko, S. P. and Woinowsky-Krieger, S. *Theory of plates and shells.* 2nd ed. New York, McGraw-Hill, 1959.
11. Hansteen, H. Finite element displacement analysis of plate bending based on rectangular elements. International symposium on the use of electronic digital computers in structural engineering (University of Newcastle-upon-Tyne, July 1966).
12. Leonhardt, F. Reducing the shear reinforcement in reinforced concrete beams and slabs. *Mag. Concr. Res.*, Vol. 7, No. 53, Dec. 1965, pp. 187–198.
13. Henderson, W. British highway bridge loading. *Proc. Instn civ. Engrs*, Vol. 3, Part 2, No. 2, June 1954, pp. 325-373.
14. Nádai, A. *Elastische Platten.* Berlin, 1925. p. 203.

FINITE ELEMENT ANALYSIS OF CURVED SLAB BRIDGES WITH SPECIAL REFERENCE TO LOCAL STRESSES

P. T. K. LIM and K. R. MOFFATT

Imperial College of Science and Technology

SYNOPSIS

This paper describes the application of the finite element method to the analysis of continuous slab bridges which are curved in plan. Model test results for two bridges of this type are used as a basis for comparison with the theoretical results.

Attention is drawn to a quadrilateral flexural element which has three degrees of freedom at each node. It is shown that although a relatively coarse assemblage of the basic triangular elements gives good agreement with experimental results for displacements and reactions, quadrilateral elements must be used in rapidly varying stress fields in order to obtain satisfactory stresses with the same coarse mesh.

This paper also shows how an estimate of the local stresses in the region of a column support or a wheel load may be made without recourse to a three dimensional analysis. The technique used involves establishing a relationship between patch size and element size (based on classical two and three dimensional solutions), by means of which a mesh can be selected to give correct local stresses.

Notation

a	radius of plate
$[C]$	matrix relating $\{\delta\}$ to $\{\alpha\}$
c	radius of effective loaded area
c_x, c_y	radii of effective loaded area in respect of M_x and M_y
c'	radius of actual loaded area
$[D]$	rigidity matrix
D_x, D_y	flexural rigidities of plate in x and y directions
D_{xy}	twisting rigidity of plate
D_1	quantity coupling D_x and D_y
h	thickness of plate
J	determinant of the Jacobian matrix
$[k]$	element stiffness matrix
M_x, M_y	bending moments per unit length of sections of a plate perpendicular to x and y axes
M_{xy}	twisting moment per unit length of section of a plate perpendicular to x axis
$[Q]$	matrix relating $\{\epsilon\}$ to $\{\alpha\}$
$[S]$	element stress matrix

u, v	dimensions of effective rectangular loaded area in x and y directions
w	component of displacement in z direction
x, y, z	rectangular cartesian coordinates (right-handed system of coordinates)
$\{\alpha\}$	vector of α_i coefficients
α_i	displacement function coefficient
δ	mesh dimension
δ_x, δ_y	mesh dimensions in x and y directions
$\{\delta\}$	vector of nodal displacements
$\{\epsilon\}$	strain or curvature vector
ϵ_x, ϵ_y	normal components of strain in x and y directions
θ_x, θ_y	components of rotations about x and y axes (positive direction determined by right-hand screw rule)
v	Poisson's ratio
ξ, η	generalised cartesian coordinates
$\{\sigma\}$	vector of stress resultants
σ_x, σ_y	normal components of stress parallel to x and y axes

1. INTRODUCTION

In the finite element analysis of slab bridges with arbitrary boundaries, triangular flexural elements having three degrees of freedom at each node are commonly used. In this paper, attention is drawn to a quadrilateral element which is also suitable for the analysis of this type of bridge. Since this element also has three degrees of freedom at each node, it is interesting to make a comparison between the results obtained for slab bridges using both elements.

In a recent publication,[1] Zienkiewicz and his collaborators observed that the overall behaviour of slab bridges could be adequately described by a two-dimensional flexural solution but that the stresses at a column support could only be found satisfactorily from a costly three-dimensional analysis. It is shown in this paper that an estimate of the local stresses at a column support or under a wheel load may be made on a two-dimensional basis if the reactive or applied is assumed to be uniformly distributed over a finite area.

Model test results obtained for two continuous curved slab bridges on the proposed Gateshead Western Bypass are used as a basis for comparison between theory and experiment. Since the cross-sectional shape, superelevation and vertical curvature of the prototype decks were reproduced to scale in the models, the validity of the assumptions of classical thin plate theory for practical slab bridges could also be investigated.

2. TRIANGULAR AND QUADRILATERAL ELEMENTS

2.1. Displacement Functions

The triangular flexural element used herein has been described by Bazeley *et al.*,[2] while the quadrilateral element is a generalisation of the rectangular element described by Zienkiewicz and Cheung.[3] The displacement function of this rectangular element has been shown to have good convergence characteristics and it is appropriate, therefore, to apply a displacement function of the same form to a general quadrilateral. The formulation of the quadrilateral element stiffness matrix is given in the Appendix.

The displacement functions of both the triangular element and the quadrilateral element satisfy the condition of deflection continuity along the element interfaces, but not the condition of slope continuity, except at the nodes, where continuity of both deflection and slope is imposed.

2.2. Conforming and Non-Conforming Quadrilateral Elements

In connection with continuity conditions between elements, mention should be made of the quadrilateral element described by Veubeke[4] for which the slope continuity requirement is also satisfied. With conforming elements monotonic convergence of the solution to a lower bound on deflection is ensured, although the violation of slope continuity, in the case of non-conforming elements, does not necessarily prevent convergence of deflections to the correct values when the mesh size is reduced. Numerical results presented by Veubeke and Sander[5] indicate that their conforming quadrilateral element has a better convergence rate than the non-conforming rectangular element with respect to the total number of degrees of freedom.

A practical shortcoming of this conforming quadrilateral element is that it requires the introduction of four mid-point nodes at which only the normal slope is specified. Mid-point nodes are inefficient as they are at most common to only two elements. They also result in a disproportionate increase in the band-width of the overall stiffness matrix of the structure, which in turn causes an even larger increase in computer time and core storage. Furthermore, this element may not be readily incorporated into computer programmes which are developed for elements having the same number of degrees of freedom at each node.

One disadvantage of the non-conforming quadrilateral element is that an explicit form of the stiffness matrix cannot be obtained. To obtain the stiffness matrix, it is necessary to resort to numerical integration which is now an established technique in finite element work.[6] It should also be mentioned that computational tests have shown that the accuracy of the results obtained using the non-conforming quadrilateral element deteriorates as the angle between any two opposite sides increases above 30° However, the subdivision into elements which is necessary for representing the behaviour of practical curved bridges involves only moderately distorted quadrilaterals.

2.3. Evaluation of Moments

A significant difference between the triangular element and the quadrilateral element is that the former can only represent the average moment values within its area, while the latter (by virtue of its larger total number of degrees of freedom) attempts to represent its internal moments in a linear manner. In the computer programme, the moment values have therefore initially been evaluated at the centroid of each triangular element, but at all four nodes of each quadrilateral element. Each nodal value of moment is then obtained by averaging the centroidal values of the triangular elements or the nodal values of the quadrilateral elements meeting at the node.

In the case of triangular elements, this averaging process will cause the nodal values obtained along the edges of the slab to represent the values pertaining to some interior region. Similarly the moment value at a node at which a column reaction or a wheel load acts, will represent the value at a point some distance from the node and not one directly at the node. A fair estimate of the moment values at such nodes may be obtained from the moment values in the surrounding elements by an extrapolation procedure. By using extrapolation formulae instead of graphical methods, the extrapolated value will only be dependent on the order n of the polynomial curve used to approximate the bending moment diagram. The average value of each pair of adjacent elements lying on either side of the extrapolation line is assigned to the midpoint of their common side, and the curve is chosen to pass through these values.

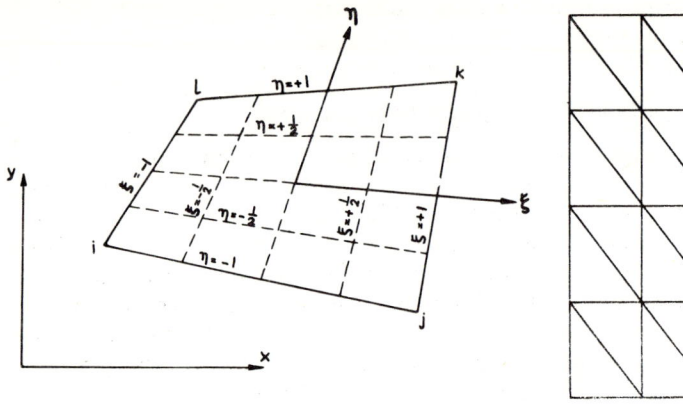

FIG. A.   GENERALIZED CO-ORDINATE SYSTEM
FOR QUADRILATERAL ELEMENT.

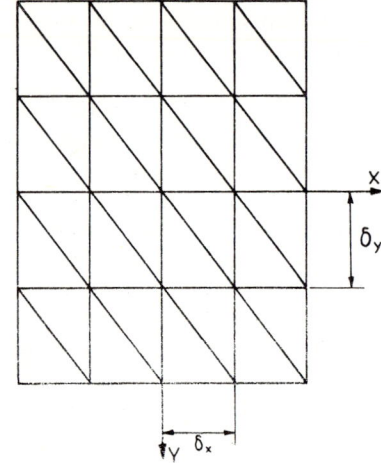

FIG. 1.   TYPICAL SUBDIVISION OF
PLATE INTO TRIANGULAR
ELEMENTS (UNIFORM MESH)

In the case of a uniform division into elements of side δ (Fig. 1), the formula for the extrapolated value may be conveniently written in the form

$$M_0 = \frac{1}{2}(3M_{\frac{1}{2}} - M_{1\frac{1}{2}}) \qquad \text{for } n = 1$$

$$M_0 = \frac{1}{8}(15M_{\frac{1}{2}} - 10M_{1\frac{1}{2}} + 3M_{2\frac{1}{2}}) \qquad \text{for } n = 2$$

$$M_0 = \frac{1}{16}(35M_{\frac{1}{2}} - 35M_{1\frac{1}{2}} + 21M_{2\frac{1}{2}} - 5M_{3\frac{1}{2}}) \qquad \text{for } n = 3$$

$$M_0 = \frac{1}{128}(315M_{\frac{1}{2}} - 420M_{1\frac{1}{2}} + 378M_{2\frac{1}{2}} - 180M_{3\frac{1}{2}} + 35M_{4\frac{1}{2}}) \quad \text{for } n = 4$$

(1)

where M_i denotes the moment corresponding to a point on the extrapolation line at a distance $i\delta$ from the node in question.

3. LOCAL STRESS ANALYSIS

3.1. Definition of Effective Areas

When the dimensions of a column bearing area or wheel contact area are of the same order as the slab thickness, it is usually assumed in a finite element analysis based on classical thin plate theory that the column reaction or the wheel load acts at a single nodal point. This assumption was in fact made in the analysis of the two bridges considered in this paper. Such a treatment will lead to values of local moments which tend to infinity (in accordance with thin plate theory) as the mesh size is made progressively smaller. In dealing with practical meshes, therefore, the question arises as to whether a finite element solution for a load applied at a nodal point will yield meaningful answers for the moments due to a patch load.

For a study of this problem, it is convenient to introduce the notion of an effective area of a patch load acting on the surface of a slab. The effective area is defined here as the loaded area at the middle plane of the slab such that the thin plate solution for this loaded area gives the same maximum tensile stress as a three-dimensional solution for the actual patch load.

The notion of an effective area of a point load in a finite element analysis is also introduced here. This effective area is defined as the area over which the point load may be assumed to be uniformly distributed such that the bending moment at the centre of the patch, according to thin plate theory, is equal to that produced by the same load in the finite element solution. A comparison between the effective areas of a patch load and the equivalent point load should provide a basis for an estimate of the theoretical local moment.

3.2. Effective Area of Patch Load

The first step is to determine the effective area of the column reaction or the wheel load at the middle plane of the slab. The problem of stress distribution near the loaded area is a problem of elasticity in three dimensions. However, if it is assumed that the load is uniformly applied over a circular area, the problem is reduced to an axi-symmetrical problem. The stress distribution within a small distance of the centre of the loaded area is substantially the same as that near the centre of a centrally loaded circular plate. The latter problem has been considered by Woinowsky-Krieger,[7] and his solution is compared here with that given by thin plate theory.

The result of this comparison is summarised in Table 1, which relates the radius c' of the loaded area to an effective radius c based on thin plate theory in respect of the maximum tensile stress under the load for the case of a clamped circular plate of height h and radius a. (Although the compressive stress under the load may be much larger than the tensile stress in the case of a strong concentration of the load, this highly localised stress has less significance than the tensile stress as far as the design of the slab is concerned.) The table shows that the value of the ratio c/c' is practically equal to unity for values of $2c'/h$ greater than one, and that it rapidly approaches infinity as $2c'/h$ approaches zero. This ratio is almost independent of the ratio $h/2a$ for the range in which shear deformation of the plate may be neglected.

3.3. Circular and Rectangular Effective Areas

In practice, the contact area of a column bearing or a wheel will not be circular in shape. Although Table 1 is only strictly applicable to axi-symmetrical problems, it should, nevertheless, provide a reasonable guide to the effective area of, say, a rectangular patch load.

In order that this effective rectangular area may be compared with the effective area of a point load, which will be assumed circular, the equivalent radii c_x, c_y of the rectangular area in respect of the components of moment M_x, M_y at the centre of the area should first be found from the following equations:

$$c_x = \frac{d}{2} e^{\frac{1}{2}\left(\phi - \frac{1-v}{1+v} \psi - 2\right)}$$

(2)

$$c_y = \frac{d}{2} e^{\frac{1}{2}\left(\phi - \frac{1-v}{1+v} \psi - 2\right)}$$

(3)

where $\phi = k \arctan \dfrac{1}{k} + \dfrac{1}{k} \arctan k$

$\psi = k \arctan \dfrac{1}{k} - \dfrac{1}{k} \arctan k$

$k = \dfrac{v}{u}$

$d = \sqrt{u^2 + v^2}$

u, v = dimensions of rectangular area in x and y directions
υ = Poisson's ratio

In the case of a square area, the above equations reduce to

$$c = \frac{u}{\sqrt{2}}\, e^{\pi/4 - 1} = 0 \cdot 57u \qquad (4)$$

as given in Ref. [8], p. 162 to which reference should be made for the basis of Eqs. (2) and (3). Several values of the factors ϕ and ψ are also given in Table 26 of the same reference.

3.4. Effective Area of Equivalent Point Load

The next step is to find the effective radii c_x, c_y of a load acting at a nodal point in respect of the bending moment components M_x, M_y under the load. In order to study the variation of c_x, c_y with the mesh dimensions δ_x, δ_y, results have been obtained for various simply supported rectangular plates of different side ratios and subject to a central point load using a 4 by 4, 8 by 8, and 16 by 16 division into either triangular or rectangular elements. The values of c_x, c_y are obtained by comparing the finite element results with the classical thin plate solution for a load uniformly distributed over a circular area (Ref. [8], p. 143). The values of the ratios c_x/δ_x, c_y/δ_y are given in Table 2 for the case when the moment under the load is obtained by nodal averaging. The table shows that the ratios are practically independent of the mesh division, and this suggests that the results are of general application and not limited to the particular case considered. However, the values presented are for right elements only and will, in general, be different for any other shape of the same type of element, or for elements of orthotropic material, or for elements based on other stiffness formulations. In the special case of elements having equal height and base, it is found that c is approximately equal to δ for triangular elements and equal to $\dfrac{1}{5}\delta$ for rectangular elements.

When the central moment in the square plate is obtained by extrapolation (in the case of triangular elements), it is found that the ratio c/δ is also practically independent of the mesh division. However, it is dependent on the direction from which the extrapolation is made and on the order of the polynomial curve used to approximate the bending moment diagram. The results are summarized in Table 3, which shows that the ratio c/δ can take values between $0 \cdot 39$ and $0 \cdot 69$ for the case considered.

The meshes which are used in practice will not always be uniform. In order to study the effect of a grading of mesh size on the effective area of a point load, the square plate is subdivided in the ratios 6:5:4:3:3:4:5:6 in both directions and again in the ratios 4:3:2:1:1:2:3:4. When the central moment is obtained by nodal averaging, it is found that the ratio c/δ takes the values $1 \cdot 078$ and $0 \cdot 986$ respectively in the case of triangular elements (δ should now be defined as the mesh dimension in the immediate vicinity of the load). Thus even when adjacent mesh dimensions in the load vicinity are in the ratio 1:2, the value of the ratio c/δ deviates from that for the uniform mesh by only 10 per cent. The deviation is less than 6 per cent in the case of rectangular elements, the corresponding values of the ratio c/δ being 0·209 and 0·218.

4. MODEL ANALYSIS

4.1. Description of Models

The model details of the Lobley Hill South Overbridge and the Consett North Overbridge are given in Figs 2 and 3 respectively. A general view of the former model under test conditions is shown in Fig. 4.

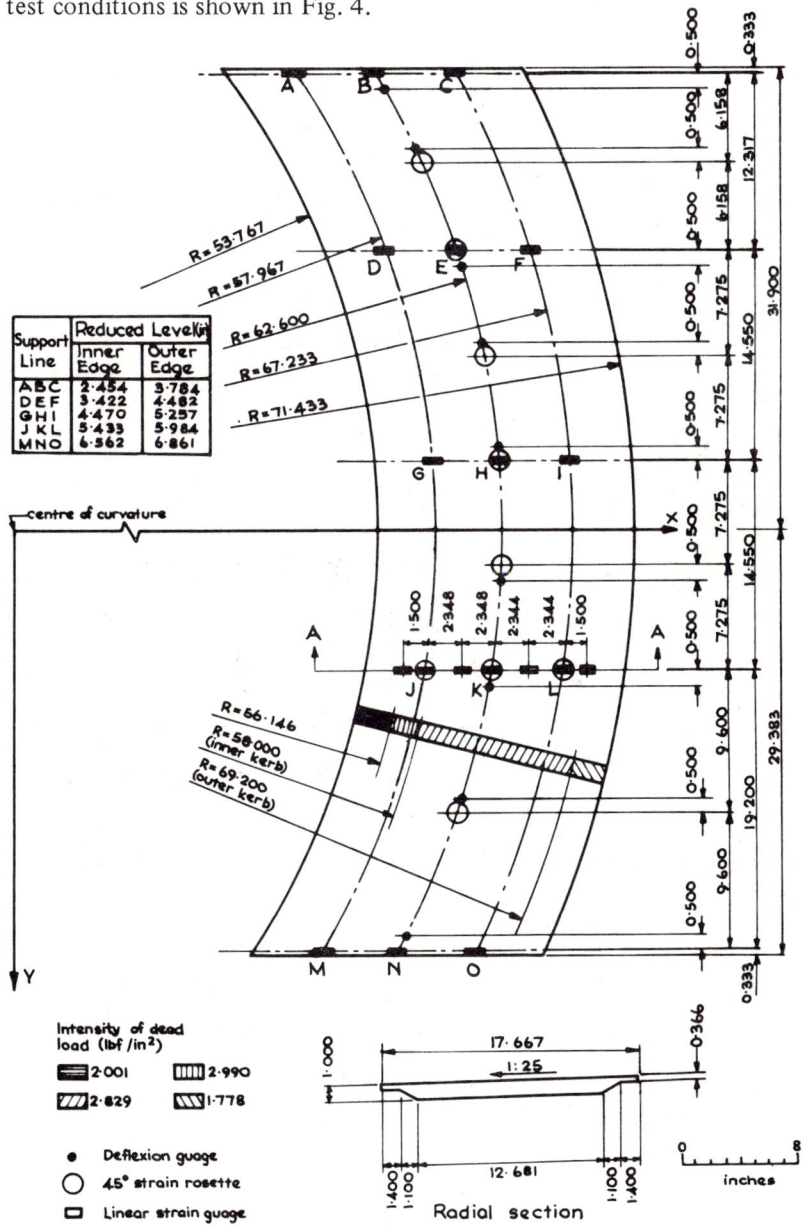

Support Line	Reduced Level	
	Inner Edge	Outer Edge
A B C	2·454	3·784
D E F	3·422	4·482
G H I	4·470	5·257
J K L	5·433	5·984
M N O	6·562	6·861

Intensity of dead load (lbf/in²)

▰ 2·001	▥ 2·990
▨ 2·829	◩ 1·778

• Deflexion guage
○ 45° strain rosette
□ Linear strain guage

Radial section

FIG. 2   LOBLEY HILL SOUTH OVERBRIDGE. DETAILS OF MODEL

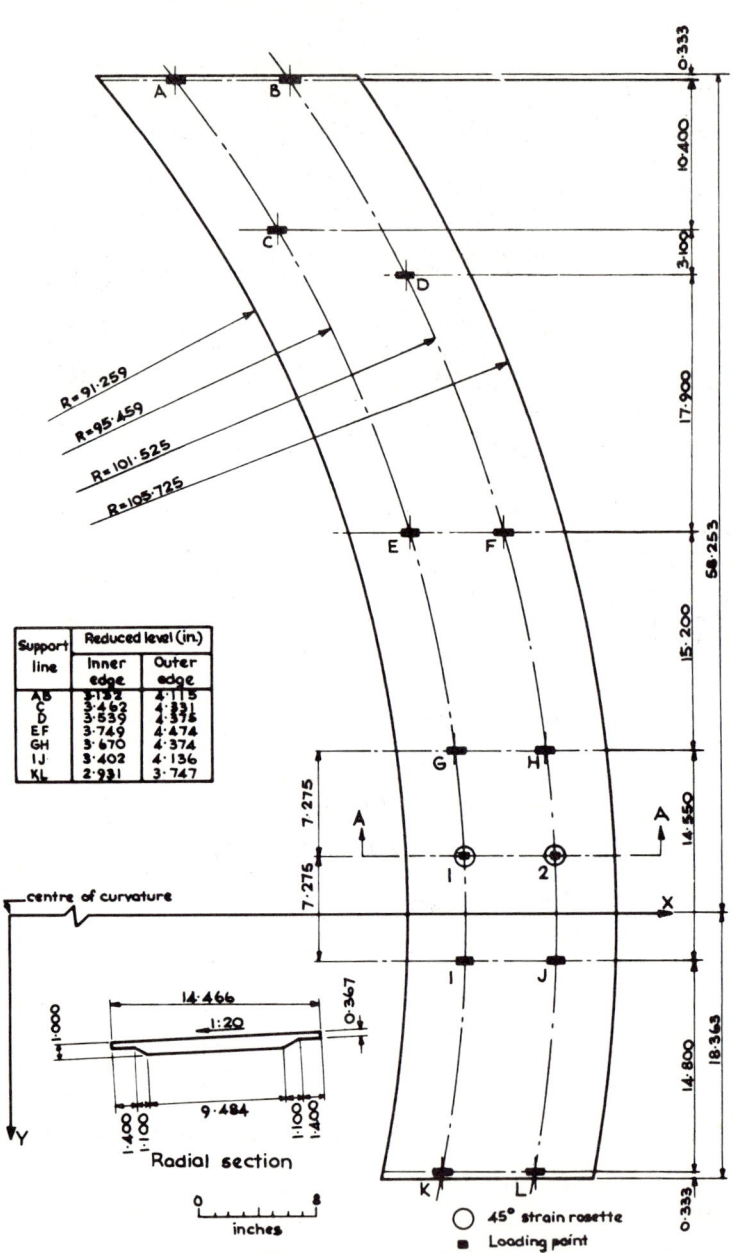

| Support | Reduced level (in.) | |
line	Inner edge	Outer edge
AB	3·132	4·115
C	3·462	4·331
D	3·539	4·375
EF	3·749	4·474
GH	3·670	4·374
IJ	3·402	4·136
KL	2·931	3·747

R = 91·259
R = 95·459
R = 101·525
R = 105·725

centre of curvature

Radial section

○ 45° strain rosette
■ Loading point

FIG. 3   CONSETT NORTH OVERBRIDGE.
         DETAILS OF MODEL

FIG. 4 LOBLEY HILL SOUTH OVERBRIDGE
MODEL UNDER TEST.

The model material was a mixture of 1 part by weight of epoxy resin and 7 parts of sand. This material was chosen because it has a value of Poisson's ratio close to that of concrete, and because the creep of this material is small over the period of application of the load. Also accurate castings of the slabs to 1/30th scale may be readily made from it. The values of the modulus of elasticity and Poisson's ratio for the first model was $2 \cdot 35 \times 10^6$ lbf/in^2 and $0 \cdot 22$, and for the second model was $2 \cdot 90 \times 10^6$ lbf/in^2 and $0 \cdot 20$.

The bridge decks were supported on miniature laminated rubber bearings and mounted on aluminium columns which also acted as reaction dynamometers. The compressive stiffnesses of both the bearings and columns were scaled from those of the prototypes, and their combined stiffnesses are given in Table 4.

The models were instrumented with dial gauges, linear strain gauges of 10 mm gauge length and 45° strain rosettes of the same gauge length. The positions of the gauges discussed in this paper are shown in Figs 2 and 3.

4.2. Loading Conditions

For the sake of brevity, only the results of some of the tests that have been carried out on the two models are presented here; further results are given in Ref. [9].

The condition of dead load is considered for the Lobley Hill South Overbridge model. The actual dead load acting on the prototype was simulated by appropriate weights attached to the underside of the model at approximately 2½ in centres. The positions and magnitudes of these loads are given in Ref. [9], while the equivalent intensities of the load are shown in Fig. 2.

The two positions of an isolated wheel load of 243·25 lbf acting on the Consett North Overbridge model are shown in Fig. 3. The load was applied through a steel block which was mounted on a rubber pad having a contact area of 0·50 in square. This area is intended to correspond to a 12 in by 12 in tyre contact area together with provision for the dispersion of the wheel load at an angle of 45° through the thickness of the non-structural surfacing.

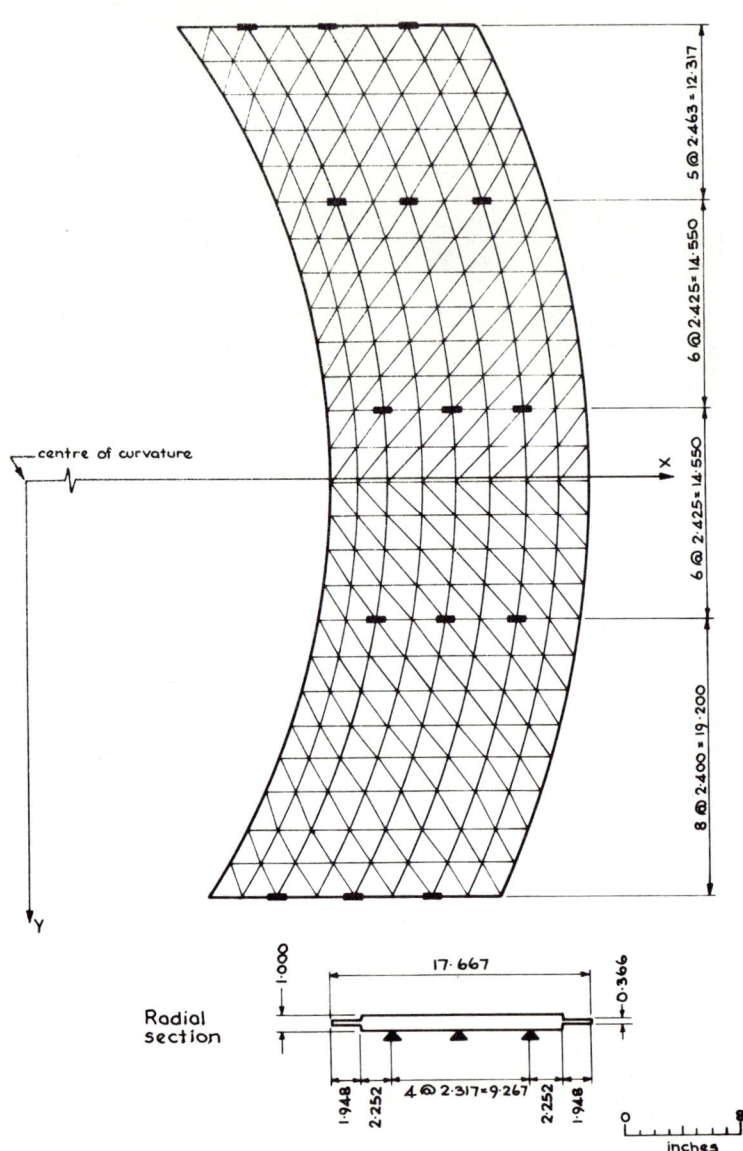

FIG. 5   LOBLEY HILL SOUTH OVERBRIDGE.
FINITE ELEMENT IDEALIZATION (8 × 25 MESH)

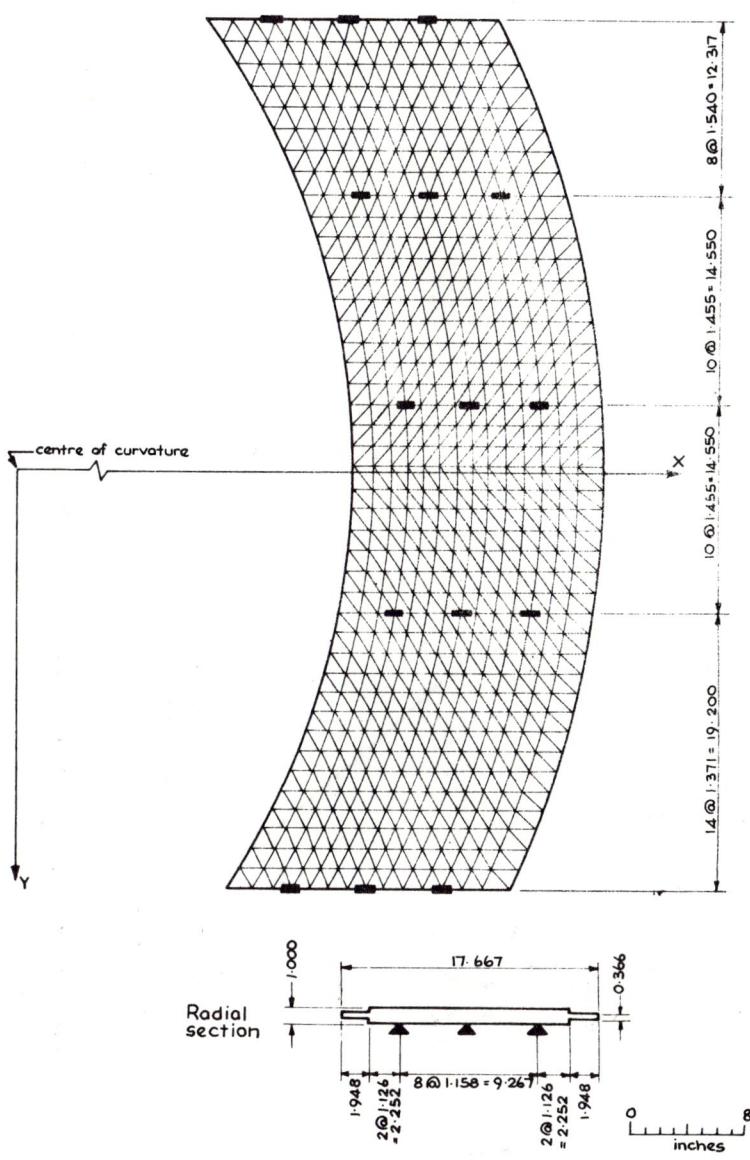

FIG. 6   LOBLEY HILL SOUTH OVERBRIDGE
FINITE ELEMENT IDEALIZATION (14 × 42 MESH)

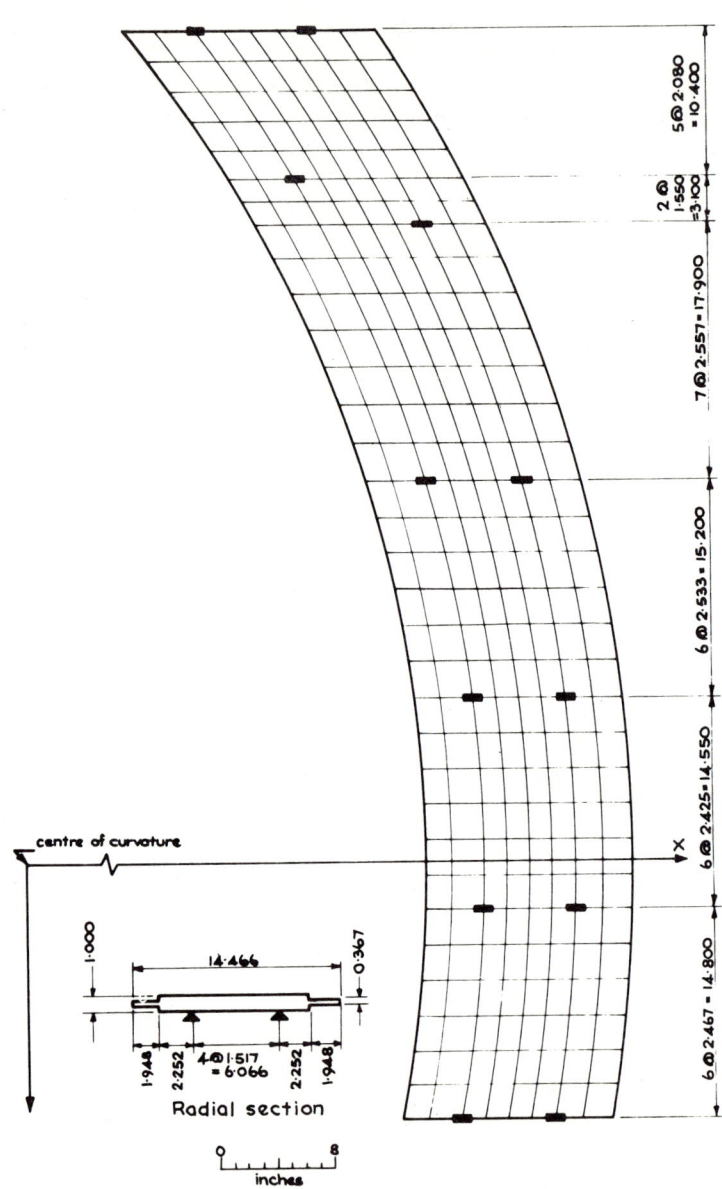

FIG. 7 CONSETT NORTH OVERBRIDGE.
FINITE ELEMENT IDEALIZATION (8 × 30 MESH)

4.3. Finite Element Idealisations

In order that a two-dimensional analysis of each model could be made, the following approximations were introduced:

(i) The idealised deck has a middle plane of symmetry, and a cross-sectional area equal to that of the actual deck.

(ii) The superelevation and vertical curvature of the deck could be disregarded. The effects of these assumptions on the results are shown to be small (generally less than 5 per cent) in Ref.[9], which presents the results of a 'shell' analysis including in-plane action due to the asymmetry of the cross-section.

Two subdivisions into triangular elements were used in the analysis of the Lobley Hill South Overbridge model, the coarse division being shown in Fig. 5 and the fine subdivision in Fig. 6. A further subdivision into quadrilateral elements is not shown as the positions of the nodal points were identical to those for the coarse subdivision into triangular elements. Only one subdivision into quadrilateral elements, as shown in Fig. 7, was used for the analysis of the Consett North Overbridge model. Both slabs were assumed to be supported on elastic springs, the stiffnesses of which are given in Table 4.

In the computer programme, any point load acting within a triangular element is replaced by a statically equivalent system of nodal forces. In order that the distribution of load to the nodes would not be affected by the type of element being used, each quadrilateral element was represented for load distribution purposes by its constituent triangles. It is shown in Ref.[9] that the results obtained using this statical distribution of load differ by only 1·5 per cent from those obtained using a distribution consistent with the formulation of the element stiffness matrix.[3]

5. COMPARISON OF RESULTS

5.1. Deflections

The central longitudinal deflection profile for the Lobley Hill Overbridge model under dead load is shown in Fig. 8. The results of the three different mesh idealisations plot on the same curve, which compares well with the experimental values.

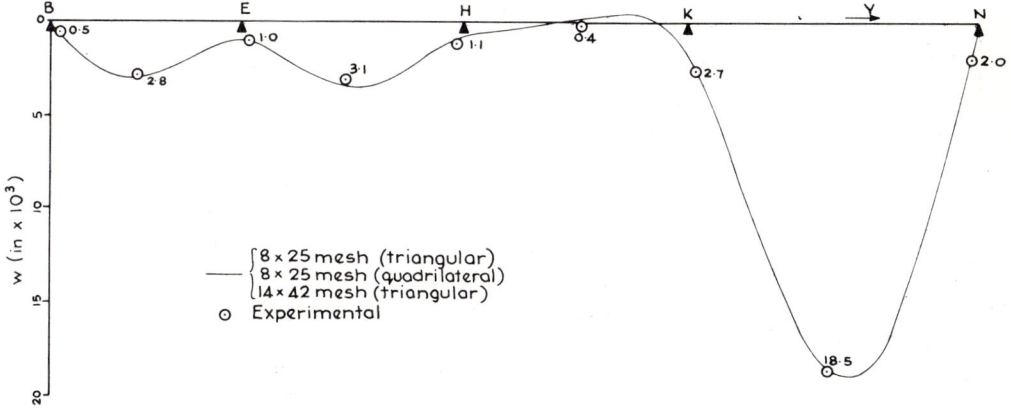

FIG. 8 LOBLEY HILL SOUTH OVERBRIDGE.
CENTRAL LONGITUDINAL DEFLEXION PROFILE
FOR MODEL UNDER DEAD LOAD

5.2. Reactions

The column reactions for the Lobley Hill South Overbridge model obtained using the three different mesh idealisations are in close agreement, as shown in Fig. 9, the maximum difference between the results being only 1 per cent of the load in the most heavily loaded column. The difference between the theoretical and experimental values of the load in each column is within 9 per cent of the maximum theoretical value. Some of this discrepancy may be attributed to the lack of accuracy in the values obtained for the spring constants of the individual supports.

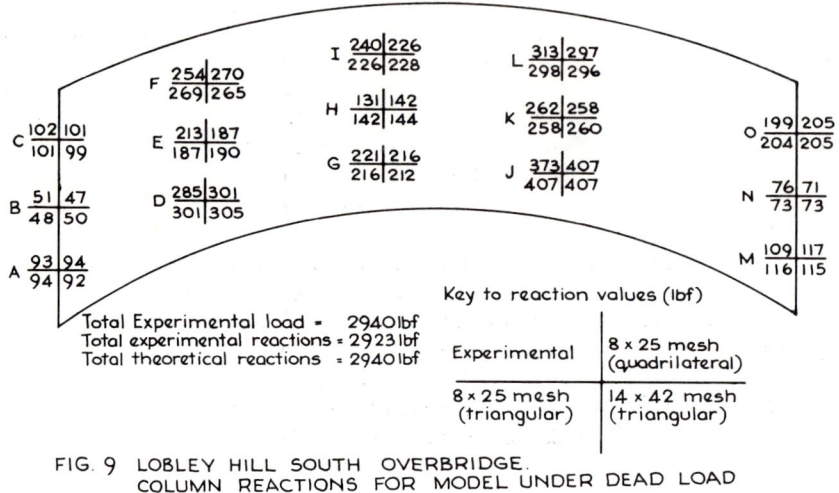

FIG. 9   LOBLEY HILL SOUTH OVERBRIDGE.
COLUMN REACTIONS FOR MODEL UNDER DEAD LOAD

5.3. Moments

The distribution of longitudinal bending moments along the centre line of the Lobley Hill South Overbridge model and across the inner row of columns supporting the longer end span is shown in Figs. 10 and 11(*a*). The distribution of transverse moments and strains across the same section may be seen from Figs 11(*b*) and 11(*c*). Superimposed on the plots of moment values (which are obtained by nodal averaging) are the values of support moments extrapolated from the results of each subdivision into triangular elements. The extrapolated value is based on a cubic variation of the moment component along the circumferential direction, and has been obtained assuming a uniform mesh (Eq. 1). The experimental values shown in Figs 10 and 11 are given by the strains on the top surface of the slab only, since the strains on the bottom surface were not measured for the points shown.

Except near the supports, the results of the three subdivisions into elements are in close agreement. The theoretical and experimental values of the moments also show good agreement at mid-span. The accuracy of the support moments can only be discussed with respect to effective areas, to which reference will be made in Section 6.1. It suffices to say at this stage that the results of the coarse subdivision into triangular elements cannot reproduce the fluctuations in the strain field near the supports, as can be clearly seen from Fig. 11(*c*).

The distribution of both longitudinal and transverse bending moments across Section AA of the Consett North Overbridge model are shown in Fig. 12 for the two load cases considered. The experimental values of moments are given by the strains on the bottom surface. The results, which show good agreement between theory and experiment, will be discussed in Section 6.2.

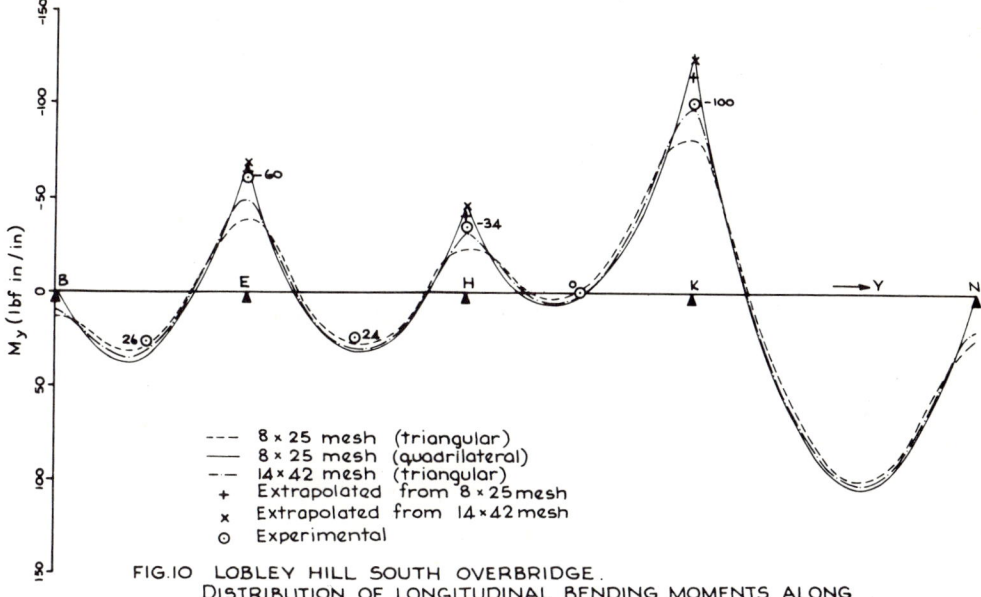

FIG.10   LOBLEY HILL SOUTH OVERBRIDGE.
DISTRIBUTION OF LONGITUDINAL BENDING MOMENTS ALONG
CENTRE LINE FOR MODEL UNDER DEAD LOAD.

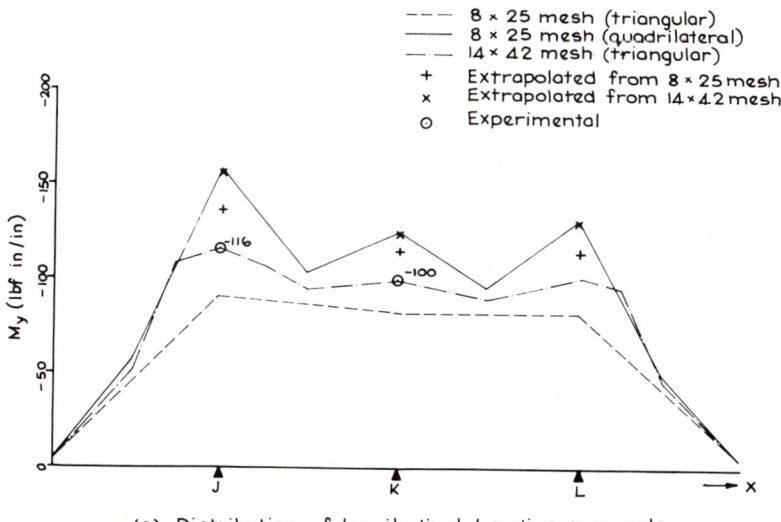

(a) Distribution of longitudinal bending moments.

FIG. 11

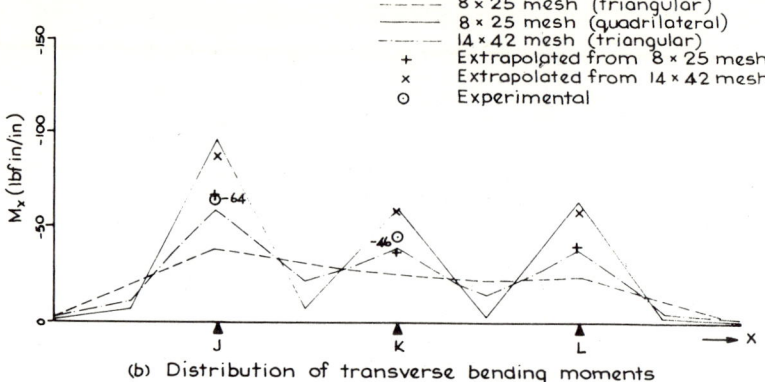

(b) Distribution of transverse bending moments

FIG.II

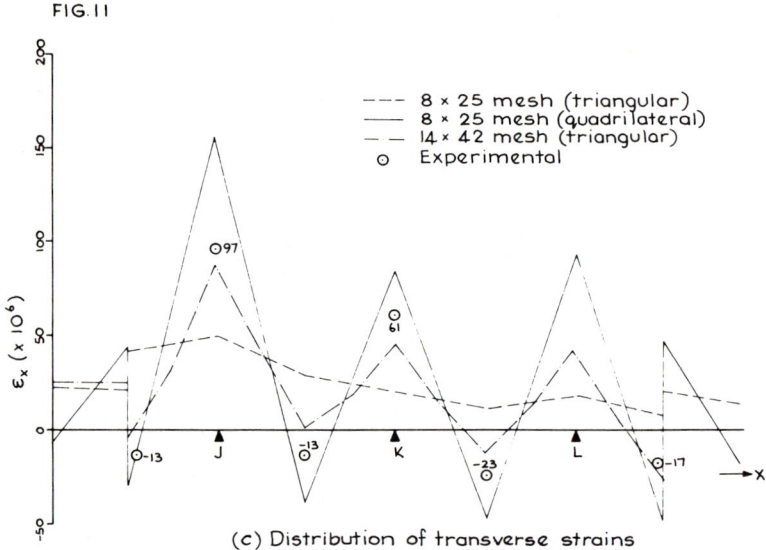

(c) Distribution of transverse strains

FIG. II   LOBLEY HILL SOUTH OVERBRIDGE.
DISTRIBUTION OF BENDING MOMENTS AND STRAINS
ACROSS SECTION AA FOR MODEL UNDER DEAD LOAD.

6. ACCURACY OF LOCAL STRESSES

6.1. Moments at Column Supports

The accuracy of the theoretical values of support moments obtained for the Lobley Hill South Overbridge model will now be examined with respect to effective areas (Section 3). According to Table 1, each of the bearing blocks (the contact area of which measured 1·25 in by 0·50 in, the larger dimension being in the transverse direction x of the deck) has an effective area of 1·19 in by 0·75 in at the middle plane of the slab. This effective rectangular area is, in turn, equivalent to an effective circular area of radius

$$c_x = 0\cdot52 \text{ in}$$
$$c_y = 0\cdot59 \text{ in}$$

in respect of M_x and M_y respectively.

For the elements shown in Figs 5 and 6, it is reasonable to assume that the effective area of a point load at a typical column K may be given by the results obtained for square meshes. The average value of the mesh dimension in the region K, based on the average area of the elements meeting there, is 2·36 in for the coarse mesh, and 1·28 in for the fine mesh.

It follows from Table 2 that the effective radius c of a point load at K, in respect of moment values obtained by nodal averaging, takes the values

$$c_x = c_y = 1.09\delta = 2.58 \text{ in for the coarse triangular mesh}$$
$$c_x = c_y = 1.09\delta = 1.39 \text{ in for the fine triangular mesh}$$
$$c_x = c_y = 0.21\delta = 0.50 \text{ in for the quadrilateral mesh}$$

Since the elements meeting at the other columns were of similar size, it will be further assumed in this discussion that the above values of c apply to all the columns.

Similarly from Table 3, the values of c in respect of moment values obtained by extrapolation are:

$$c_x = 0.58\delta = 1.37 \text{ in}$$
$$c_y = 0.45\delta = 1.06 \text{ in}$$

for the coarse mesh, and

$$c_x = 0.58\delta = 0.74 \text{ in}$$
$$c_y = 0.45\delta = 0.58 \text{ in}$$

for the fine mesh.

In accordance with these values of effective areas of the distributed bearing load and the equivalent point load, the average values of moments as given by the results of both subdivisions into triangular elements should be below the experimental values, with those of the fine subdivision showing a closer agreement. Also, the moment values as given by the results of the subdivision into quadrilateral elements should be above the experimental values. Reference to Figs 10 and 11, which depict the distribution of longitudinal moments M_y and transverse moments M_x, shows this to be the case.

It can also be seen from the same diagrams that the extrapolated values of longitudinal moments obtained from the results of the coarse triangular mesh are above the experimental values (except at column E where the two values are very close),

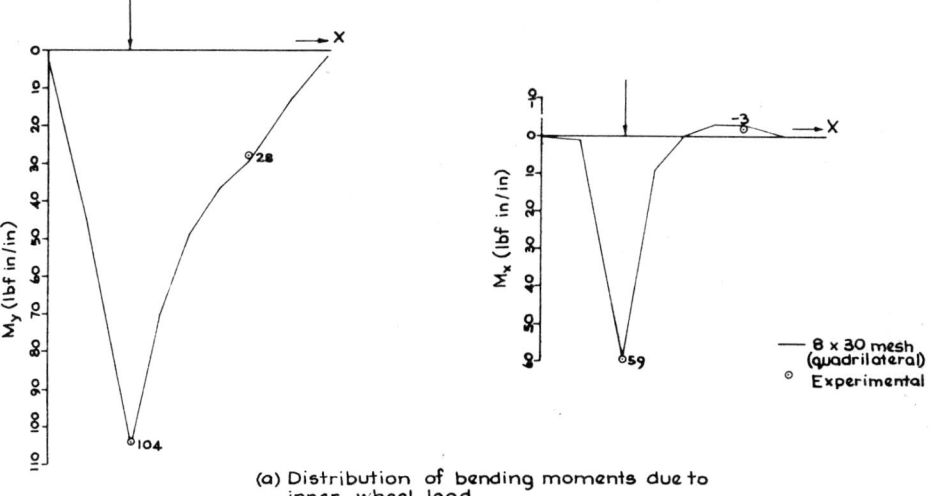

(a) Distribution of bending moments due to inner wheel load

FIG. 12

although the theoretical moment is due to a larger effective loaded area and should therefore be below the experimental value. The extrapolated value of transverse moment at K is on the right side of the experimental value, but not that at J. In the case of the fine mesh, the extrapolated values of both moment components are above the experimental values, although the theoretical value of transverse moment should be below the experimental value.

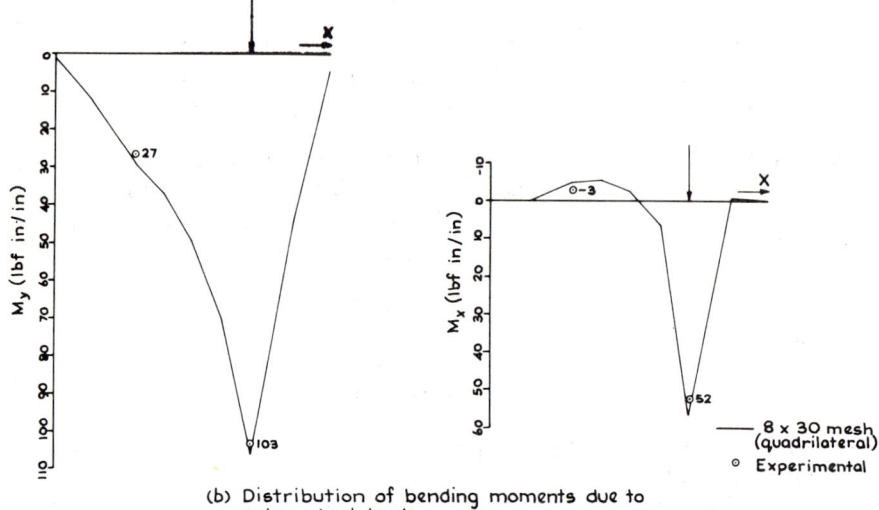

(b) Distribution of bending moments due to
outer wheel load

FIG.12   CONSETT NORTH OVERBRIDGE.
DISTRIBUTION OF BENDING MOMENTS ACROSS
SECTION AA FOR MODEL UNDER WHEEL LOADS

Figure 13 shows the variation of each moment component with the effective radius at K. The smooth plots obtained may be taken as an indication of the validity of the theoretical approach, while the difference between the theoretical and experimental values corresponding to the same effective radius accounts for some of the inconsistencies noted above.

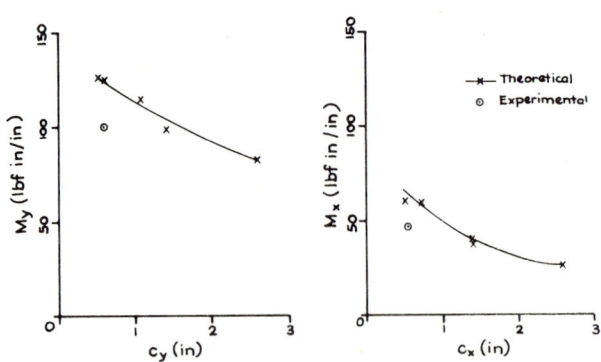

FIG.13 LOBLEY HILL SOUTH OVERBRIDGE
VARIATION OF BENDING MOMENTS
WITH EFFECTIVE RADII AT COLUMN K

6.2. Moments Under Wheel Loads

A similar assessment of the accuracy of the theoretical moment values under wheel loads will now be made for the Consett North Overbridge model. The effective area of the actual square loading pad of side 0·50 in should first be estimated. From Table 1, this pad has an effective area approximately equal to a square of side $u = 0·75$ in which, according to Eq. 4, is equivalent to a circle of radius

$$c = 0·57u = 0·43 \text{ in}$$

in respect of both M_x and M_y.

For the mesh shown in Fig. 7, the aspect ratio of the elements will have to be taken into account, although the assumption will still be made that the effective area of the equivalent point load may be given by the results obtained for right elements. At both loading points, the average mesh dimension in the x direction $\delta_x = 1·89$ in and that in the y direction $\delta_y = 2·42$ in. It follows from Table 2 that the effective radius takes the values

$$c_x = 0·224\delta_x = 0·42 \text{ in}$$
$$c_y = 0·196\delta_y = 0·48 \text{ in}$$

Thus the effective area of the equivalent point load is very close to that of the patch load, especially in respect of transverse moments. The theoretical and experimental values of moments at the two loading points should therefore agree well, and that this is the case may be seen from Fig. 12.

It is noted that there is a closer agreement between theory and experiment for the moment under a wheel load than for the moment at a column support. This result is not unexpected since the magnitude of the wheel load is fixed whereas the magnitude of the reactive load is dependent on the compressive stiffnesses of the rubber bearings and column supports, and any error in the estimation of these values will be reflected in the magnitude of the reactive load (see Section 5.2).

6.3. External and Internal Moments at Support Section

A further assessment of the accuracy of the finite element solution may be obtained by considering the moment equilibrium of, say, the longer end span of the Lobley Hill South Overbridge model. The resultant moment of the external forces about Section AA and the internal moment represented by the area under the curve of the distribution of longitudinal moment across the same section, are given in Table 5.

It may be seen from the above table that agreement to within 1 per cent has been obtained using quadrilateral elements. The results of the coarse triangular mesh show a lack of equilibrium of 30 per cent when the values of support moments are obtained by nodal averaging. This decreases to 14 per cent when the values of support moments are obtained by extrapolation. The corresponding figures for the fine triangular mesh are 15 per cent and 8 per cent. The excellent agreement shown by the quadrilateral elements is to be expected, since the linear distribution of moments within each element is of a type that gives the 'best fit' to the actual moment distribution.[3]

The experimental value of the average external moment at Section AA is 104 lbf in/in, relative to which the experimental value of peak moment at each column support (Fig. 11(a)) appears rather low. This partly supports the observation made in Section 6.1 that the experimental values of support moments are on the low side.

7. CONCLUSIONS

Even with the use of a relatively coarse assemblage of the lowest order flexural elements (triangles), good agreement with experimental results has been obtained for deflections, mid-span moments and support reactions. The maximum difference between the experimental and theoretical values, expressed as a percentage of the maximum theoretical value, is 4 per cent for deflections and mid-span moments, and 9 per cent for support reactions. However, the stress field within each element is effectively constant and cannot reproduce the fluctuations in the stress field near the supports, where there were only two elements between adjacent columns. The results of a fine subdivision into triangular elements show the fluctuations in the stress field but at the expense of a 4½ times increase in computer time. (A decrease in mesh size in the support regions only should give similar results without requiring such a large increase in computer time, although additional time would be required for data preparation.)

The results of a coarse subdivision into quadrilateral elements also show the fluctuations in the stress field near the supports. In this case, there was no increase in computer time since the evaluation of the stiffness matrix of each quadrilateral element, for which a numerical integration technique is necessary, requires approximately only as much computer time as that required for the evaluation of the stiffness matrices of the two triangular elements which it replaces.

Quadrilateral elements also give a direct value for the moment at a column support or under a point load whereas for triangular elements this moment has to be obtained by extrapolation. Even when the local moments are obtained by extrapolation, the results of the coarse subdivision into triangular elements show a difference of 14 per cent between the external moment and the internal moment at the support section considered. The corresponding moment values obtained using quadrilateral elements agree to within 1 per cent.

The actual stress distribution in the vicinity of a column support or a wheel load may only be obtained from a three-dimensional analysis. However, a close estimate of the local stress values may be made from an analysis based on classical thin plate theory by choosing the element size in the vicinity of the load to be such that the effective area of the equivalent point load (according to Table 2 or 3) is approximately equal to that of the reactive or applied patch load at the middle plane of the slab (according to Table 1). Examples on the application of this method are given in Section 6. Apart from a few inconsistencies, the agreement between theoretical and experimental result is good, in spite of the approximate nature of the suggested procedure. From the practical point of view, it is significant that a subdivision into elements which is necessary for representing the overall behaviour of actual bridges is also sufficient for estimating the magnitudes of the local stresses.

The errors involved in neglecting the effects due to the asymmetry of the deck cross-section, and those due to the superelevation and vertical curvature of the deck are noted to be small, in the examples considered.

ACKNOWLEDGEMENTS

The experiments were conducted for the Department of the Environment in association with Ove Arup & Partners, consulting engineers for the Gateshead Western Bypass. The work on local stresses is part of a further investigation for the Department of the Environment. The results were obtained using a computer programme developed with financial assistance from Freeman Fox & Partners, Ove Arup & Partners, and The British Ship Research Association.

The work described in this paper constitutes part of a continuing research programme being carried out under the supervision of Dr J. C. Chapman at Imperial College, London, into the finite element and model analysis of bridge structures. A special acknowledgement is due to Mrs J. E. Slatford who assisted in the running of the computer programme.

APPENDIX

Stiffness Matrix of Quadrilateral Flexural Element

The stiffness matrix of the quadrilateral flexural element described herein is based on a generalisation of the displacement function chosen by Zienkiewicz and Cheung[3] for the rectangular element. In order to satisfy the compatibility condition of a cubic variation of deflection along the boundaries of the element, it is necessary to express the polynomial in terms of the generalised coordinates ξ and η illustrated in Fig. A. In these new coordinates, the displacement function takes the form

$$w = \alpha_1 + \alpha_2\xi + \alpha_3\eta + \alpha_4\xi^2 + \alpha_5\xi\eta + \alpha_6\eta^2 + \alpha_7\xi^3 + \alpha_8\xi^2\eta + $$
$$\alpha_9\xi\eta^2 + \alpha_{10}\eta^3 + \alpha_{11}\xi^3\eta + \alpha_{12}\xi\eta^3$$

where α_1 to α_{12} are arbitrary coefficients.

The coordinates in the two systems are related by the following expressions:

$$x = \frac{1}{4}\{(1-\xi)(1-\eta)x_i + (1+\xi)(1-\eta)x_j + (1+\xi)(1+\eta)x_k + (1-\xi)(1+\eta)x_l\}$$

$$y = \frac{1}{4}\{(1-\xi)(1-\eta)y_i + (1+\xi)(1-\eta)y_j + (1+\xi)(1+\eta)y_k + (1-\xi)(1+\eta)y_l\}$$

Since it is difficult, if not impossible, to express w in terms of x and y, the formulation will proceed in terms of the generalised coordinates.

The derivation of the element stiffness matrix $[k]$ proceeds in the standard manner (Ref.[10], p. 94):

$$[k] = ([C]^{-1})^t \left(\iint [Q]^t [D] [Q]\, dxdy\right) [C]^{-1}$$

The matrices appearing in the above equation are defined in the following paragraphs.

The matrix $[C]$ relates the element nodal displacements $\{\delta\}$ to the coefficients $\{\alpha\}$:

$$\{\delta\} = [C]\{\alpha\}$$

where

$$\{\delta\} = \begin{Bmatrix} w_1 \\ \theta_{x_1} \\ \theta_{y_1} \\ \cdot \\ \cdot \\ \cdot \\ \theta_{y_4} \end{Bmatrix}$$

The rotations $\theta_x = \dfrac{\partial w}{\partial y}$ and $\theta_y = -\dfrac{\partial w}{\partial x}$ may be obtained from the chain rule for partial derivatives:

$$\frac{\partial w}{\partial \xi} = \frac{\partial w}{\partial x}\frac{\partial x}{\partial \xi} + \frac{\partial w}{\partial y}\frac{\partial y}{\partial \xi}$$

$$\frac{\partial w}{\partial \eta} = \frac{\partial w}{\partial x}\frac{\partial x}{\partial \eta} + \frac{\partial w}{\partial y}\frac{\partial y}{\partial \eta}$$

Solving the above equations for $\dfrac{\partial w}{\partial x}$ and $\dfrac{\partial w}{\partial y}$ we obtain

$$\frac{\partial w}{\partial x} = -\frac{1}{J}\left(\frac{\partial w}{\partial \eta}\frac{\partial y}{\partial \xi} - \frac{\partial w}{\partial \xi}\frac{\partial y}{\partial \eta}\right)$$

$$\frac{\partial w}{\partial y} = \frac{1}{J}\left(\frac{\partial w}{\partial \eta}\frac{\partial x}{\partial \xi} - \frac{\partial w}{\partial \xi}\frac{\partial x}{\partial \eta}\right)$$

where J is the determinant of the Jacobian matrix

$$\begin{bmatrix} \dfrac{\partial x}{\partial \xi} & \dfrac{\partial y}{\partial \xi} \\[2mm] \dfrac{\partial x}{\partial \eta} & \dfrac{\partial y}{\partial \eta} \end{bmatrix}$$

It is now possible to obtain the matrix $[C]$ and hence its inverse, for which a numerical procedure is necessary.

The rigidity matrix $[D]$ relates the internal moments $\{\sigma\}$ to the curvatures $\{\epsilon\}$ at any point within the element:

$$\{\sigma\} = [D]\{\epsilon\}$$

For an orthotropic plate with principal directions of orthotropy coinciding with the x and y axes this relationship is of the standard form:

$$\left\{ \begin{array}{c} M_x \\[4mm] M_y \\[4mm] M_{xy} \end{array} \right\} = \begin{bmatrix} D_x & D_1 & 0 \\[3mm] D_1 & D_y & 0 \\[3mm] 0 & 0 & D_{xy} \end{bmatrix} \left\{ \begin{array}{c} -\dfrac{\partial^2 w}{\partial x^2} \\[4mm] -\dfrac{\partial^2 w}{\partial y^2} \\[4mm] 2\dfrac{\partial^2 w}{\partial x \partial y} \end{array} \right\}$$

The matrix $[Q]$ relates the curvatures $\{\epsilon\}$ at any point within the element to the coefficients $\{\alpha\}$:

$$\{\epsilon\} = [Q]\{\alpha\}$$

Following the above procedure for rotations, we obtain expressions for $\dfrac{\partial^2 w}{\partial x^2}$, $\dfrac{\partial^2 w}{\partial y^2}$ and $\dfrac{\partial^2 w}{\partial x \partial y}$ in terms of $\dfrac{\partial^2 w}{\partial \xi^2}$, $\dfrac{\partial^2 w}{\partial \eta^2}$ and $\dfrac{\partial^2 w}{\partial \xi \partial \eta}$:

$$\frac{\partial^2 w}{\partial x^2} = \frac{-A\left(\dfrac{\partial y}{\partial \xi}\right)^2 + 2C\dfrac{\partial y}{\partial \xi}\dfrac{\partial y}{\partial \eta} - B\left(\dfrac{\partial y}{\partial \eta}\right)^2}{J^2}$$

$$\frac{\partial^2 w}{\partial y^2} = \frac{-A\left(\dfrac{\partial x}{\partial \xi}\right)^2 + 2C\dfrac{\partial x}{\partial \xi}\dfrac{\partial x}{\partial \eta} - B\left(\dfrac{\partial x}{\partial \eta}\right)^2}{J^2}$$

$$\frac{\partial^2 w}{\partial x \partial y} = \frac{-A\dfrac{\partial x}{\partial \xi}\dfrac{\partial y}{\partial \xi} + C\left(\dfrac{\partial x}{\partial \xi}\dfrac{\partial y}{\partial \eta} + \dfrac{\partial x}{\partial \eta}\dfrac{\partial y}{\partial \xi}\right) - B\dfrac{\partial x}{\partial \eta}\dfrac{\partial y}{\partial \eta}}{-J^2}$$

where

$$A = \frac{\partial w}{\partial x}\frac{\partial^2 x}{\partial \eta^2} + \frac{\partial w}{\partial y}\frac{\partial^2 y}{\partial \eta^2} - \frac{\partial^2 w}{\partial \eta^2}$$

$$B = \frac{\partial w}{\partial x}\frac{\partial^2 x}{\partial \xi^2} + \frac{\partial w}{\partial y}\frac{\partial^2 y}{\partial \xi^2} - \frac{\partial^2 w}{\partial \xi^2}$$

$$C = \frac{\partial w}{\partial x}\frac{\partial^2 x}{\partial \xi \partial \eta} + \frac{\partial w}{\partial y}\frac{\partial^2 y}{\partial \xi \partial \eta} - \frac{\partial^2 w}{\partial \xi \partial \eta}$$

The equations for the derivation of $[Q]$ are now available.

Before the integration may be carried out, the element of area $dxdy$ must also be expressed in terms of the coordinates ξ and η:

$$dxdy = J\,d\xi d\eta$$

and the limits of integration changed to -1 and $+1$ in both variables. The actual integration of the various expressions is achieved by numerical methods as explicit evaluation of the integrals appears impossible. The numerical technique involves a double application of the Gaussian quadrature formulae.[11]

In the case of a rectangular or parallelogrammic element, it is only necessary to have three pivotal points in each direction in order to obtain a stiffness matrix identical to that obtained by explicit integration. However, as the shape of the element deviates from that of a parallelogram, more pivotal points are required for the result of the integration to converge. For the shape of elements used in this work five pivotal points in each direction are found to be sufficient whereas seven points are found to be necessary for the most extreme shape likely to be encountered in a practical mesh.

Finally the element stress matrix $[S]$ can be obtained from the expression

$$[S] = [D]\,[Q]\,[C]^{-1}$$

Table I Effective area of patch load on clamped circular plate ($y = 0.3$) in respect of maximum tensile stress at bottom of plate

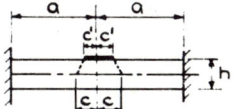

$\dfrac{h}{2a}$		c/c' for $2c'/h =$							
	0	0·25	0·50	0·75	1·00	1·25	1·50	1·75	2·00
0·05	∞	2·77	1·53	1·16	1·01	0·94	0·92	0·93	0·94
0·10	∞	2·70	1·49	1·14	1·00	0·95	0·93	0·93	0·94
0·20	∞	2·74	1·51	1·16	1·01	0·95	0·93	0·92	0·91

Table 2. Effective area of point load on simply supported rectangular plate ($y = 0.3$) in respect of moment values obtained by nodal averaging

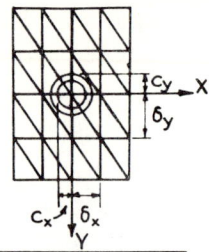

Type of element	Mesh	c_x/δ_x in respect of M_x for $\delta_y/\delta_x =$					
		1.0	1.2	1.4	1.6	1.8	2.0
Triangular	(4×4)	1.075	1.171	1.262	1.350	1.434	1.516
	(8×8)	1.091	1.176	1.261	1.342	1.423	1.503
	(16×16)	1.094					1.501
Rectangular	(4×4)	0.205	0.214	0.226	0.240	0.256	0.272
	(8×8)	0.206	0.217	0.230	0.245	0.261	0.277
	(16×16)	0.206	0.218	0.231	0.246	0.262	0.278

Type of element	Mesh	c_y/δ_y in respect of M_y for $\delta_y/\delta_x =$					
		1.0	1.2	1.4	1.6	1.8	2.0
Triangular	(4×4)	1.075	0.989	0.925	0.874	0.832	0.795
	(8×8)	1.091	1.016	0.962	0.920	0.886	0.858
	(16×16)	1.094					0.878
Rectangular	(4×4)	0.205	0.198	0.196	0.194	0.194	0.194
	(8×8)	0.206	0.198	0.194	0.191	0.190	0.189
	(16×16)	0.206	0.198	0.193	0.190	0.188	0.187

Table 3. Effective area of point load on simply supported square plate ($y = 0.3$) in respect of moment values obtained by extrapolation

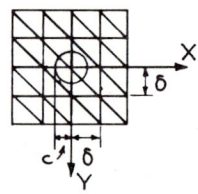

Mesh	c/δ in respect of $\begin{cases} M_x \text{ along } y=0 \\ M_y \text{ along } x=0 \end{cases}$ for order of polynomial $n =$			
	1	2	3	4
(4×4)	0.69	–	–	–
(8×8)	0.69	0.53	0.44	–
(16×16)	0.68	0.53	0.45	0.39

Mesh	c/δ in respect of $\begin{cases} M_x \text{ along } x=0 \\ M_y \text{ along } y=0 \end{cases}$ for order of polynomial $n =$			
	1	2	3	4
(4×4)	0.64	–	–	–
(8×8)	0.68	0.61	0.58	–
(16×16)	0.69	0.62	0.58	0.56

Table 4. Lobley Hill South and Consett North Overbridges. Elastic stiffnesses of column supports.

Lobley Hill South Overbridge	
Column	Elastic stiffness (lbf/in x 10⁻³)
A	193
B	192
C	191
D	192
E	190
F	187
G	184
H	182
I	180
J	177
K	175
L	174
M	194
N	193
O	193

Consett North Overbridge	
Column	Elastic stiffness (lbf/in x 10⁻³)
A	202
B	200
C	198
D	191
E	192
F	190
G	193
H	190
I	195
J	193
K	203
L	202

Table 5. Lobley Hill South Overbridge. External and Internal longitudinal bending moments at section A A for model under dead load.

Type of element	Mesh	M_e (lbf in)	M_i (lbf in)	M_i' (lbf in)	$\frac{M_i - M_e}{M_e} \times 100$	$\frac{M_i' - M_e}{M_e} \times 100$
Triangular	(8 × 25)	1684	1186	1441	−30	−14
Quadrilateral	(8 × 25)	1685	1670	−	−1	−
Triangular	(14 × 42)	1707	1450	1562	−15	−8

M_e = external moment

M_i = internal moment in respect of support moment values obtained by nodal averaging.

M_i' = internal moment in respect of support moment values obtained by extrapolation.

REFERENCES

1. Cheung, Y. K., King, I. P. and Zienkiewicz, O. C. Slab bridges with arbitrary shape and support conditions: a general method of analysis based on finite elements. *Proc. Instn civ. Engrs*, Vol. 40, May 1968, pp. 9–36.
2. Bazeley, G. P., Cheung, Y. K., Irons, B. M. and Zienkiewicz, O. C. Triangular elements in bending – conforming and non-conforming solutions. *Proceedings of the Conference on Matrix Methods in Structural Mechanics (Wright-Patterson Air Force Base, Ohio, 1965)*, 1967.
3. Zienkiewicz, O. C. and Cheung, Y. K. The finite element method for analysis of elastic isotropic and orthotropic slabs. *Proc. Instn civ. Engrs*, Vol. 28, Aug. 1964, pp. 471–488.
4. Veubeke, B. F. de. A conforming finite element for plate bending. *Int. J. Solids & Struct.*, Vol. 4, No. 1, Jan. 1968, pp. 95–108.
5. Veubeke, B. F. de and Sander, G. An equilibrium model for plate bending. *Int. J. Solids & Struct.*, Vol. 4, No. 4, April 1968, pp. 447–468.
6. Ergatoudis, I., Irons, B. M. and Zienkiewicz, O. C. Curved, isoparametric, 'quadrilateral' elements for finite element analysis. *Int. J. Solids & Struct.*, Vol. 4, No. 1, Jan. 1968, pp. 31–42.
7. Woinowsky-Krieger, S. Der Spannungszustand in dicken elastischen Platten. *Ing.-Arch.*, Vol. 4, 1933, pp. 305–331.
8. Timoshenko, S. P. and Woinowsky-Krieger, S. *Theory of plates and shells*. 2nd ed. New York, McGraw-Hill, 1959.
9. Lim, P. T. K. and Moffatt, K. R. Finite element analysis of two curved slab bridge models with special reference to moments at column supports and under wheel loads. Report commissioned by the Department of the Environment, 1970.
10. Zienkiewicz, O. C. *The finite element method in structural and continuum mechanics.* New York, McGraw-Hill, 1967.
11. Kopal, Z. *Numerical analysis.* 2nd ed. London, Chapman and Hall, 1961.

DEVELOPMENTS OF THE FINITE STRIP METHOD IN THE ANALYSIS OF BRIDGE DECKS

YEW-CHAYE LOO and A. R. CUSENS
University of Dundee

SYNOPSIS

This paper summarises recent developments of the finite strip method to facilitate analysis of slab and cellular bridge decks. This work has been carried out as part of a bridge research project sponsored by the Construction Industry Research and Information Association.

The principal developments described are:

(a) a direct approach for the formulation of force matrices
(b) use of a fifth order polynomial as the transverse displacement function
(c) solutions for plate and cellular decks with intermediate discrete columns
(d) use of an auxiliary nodal line technique for slab and cellular decks.

INTRODUCTION

In the analysis of slab and box-beam structures, a numerical technique of solution which assumes the structure to be an assembly of narrow longitudinal strips joined at their edges (nodal lines) has been used by several engineers. Hellan[1] formulated a solution for rectangular isotropic plates simply by direct considerations of continuity and equilibrium at nodal lines. One-way slab behaviour was assumed in the transverse direction of the strip and span-wise behaviour was defined by the characteristic functions of a freely vibrating beam.

A more sophisticated approach for the analysis of rectangular orthotropic plates was suggested by Cheung.[2,3] Known as the finite strip method, this approach is similar in principle to the finite element procedure.[4] The assumed displacement function for a strip in bending consists of a one-way slab function (a third order polynomial) across the width of the strip with a characteristic function defining the longitudinal behaviour. Combining this bending function with first order in-plane displacement functions Cheung has extended the method to deal with simply-supported folded plate structures.[5,6]

Independently, Powell and Ogden[7] used the same bending function and developed a computer program to analyse simply-supported orthotropic steel plate bridge decks.

Recently, the finite strip method has also been used by Willam and Scordelis[8] to deal with simply-supported orthotropic folded plate structures. Bending and in-plane displacement functions identical to those used by Cheung[5] were assumed and the effect of eccentric stiffeners was also included in the analysis.

A third order polynomial function has been incorporated by all these investigators for the analysis of plate bending. This leads to a roughly linear variation of transverse moment across the width of the strip. In the analysis of box structures, linear in-plane displacement functions have been used in conjunction with the third order bending function. Thus a roughly linear variation of transverse moment is accompanied by a virtually constant transverse in-plane stress across the width of a strip. In consequence with this approach a large number of strips has to be used where sharp changes of stress occur.

Force matrices were obtained by means of expansion of the applied loads in the form of a series function similar to that used to prescribe the two opposite end boundary conditions. In this way, matrices for uniformly distributed load, patch load and line load were obtained. A concentrated load situated between strip boundaries was simulated by a line load uniformly distributed across the width of the strip. In other words, the transverse position of the concentrated or patch load inside the strip was assumed to be irrelevant. Very narrow strips were necessary to obtain reliable results for bridge decks under standard vehicle loading and hence, the division pattern of the idealized structures depended on the position of the wheel loads. This has been found to make data preparation cumbersome especially in the analysis of box-beam structures and also tends to increase solution time.

REFINED FINITE STRIP TECHNIQUE FOR ORTHOTROPIC PLATES

A fifth order displacement interpolation function is used by the authors to formulate a refined finite strip solution[9] for simply-supported orthotropic plates on an elastic foundation. The function ensures compatibility of slope and curvature at strip boundaries and gives a cubic polynomial variation for transverse moment across the width of a strip. A direct approach for the formulation of force matrices has also been developed.[9] Unlike the series method used by previous investigators, the direct approach takes into consideration the true nature of the applied forces, which may be in the form of concentrated or linearly distributed loads or concentrated applied moments acting at any position on the strip.

Due to the flexibility of both the fifth order function and the direct approach of evaluating load matrices, the refined method requires a very small number of strips to give reliable results. In some practical cases, only two strips were used and the results are extremely accurate.

In order to demonstrate the improvements given by the refined fifth order solutions, a bridge unit with two intermediate columns under uniform distributed load was analysed by both the fifth and third order methods. Eight strips were used for the refined solution and results were compared with those given by the conventional method using 8 and 20 strips respectively. Column reactions were assumed to be concentrated forces and 19 harmonics (10 terms) were retained for all cases. It is found that the third order solution gives quite favourable results of deflection and longitudinal moments even with only 8 strips. However, Fig. 1 shows that the third order analysis required 20 strips (12 strips at 0·0625 plus 8 strips at 0·03125 near columns) to give transverse moments identical to those of the refined solution.

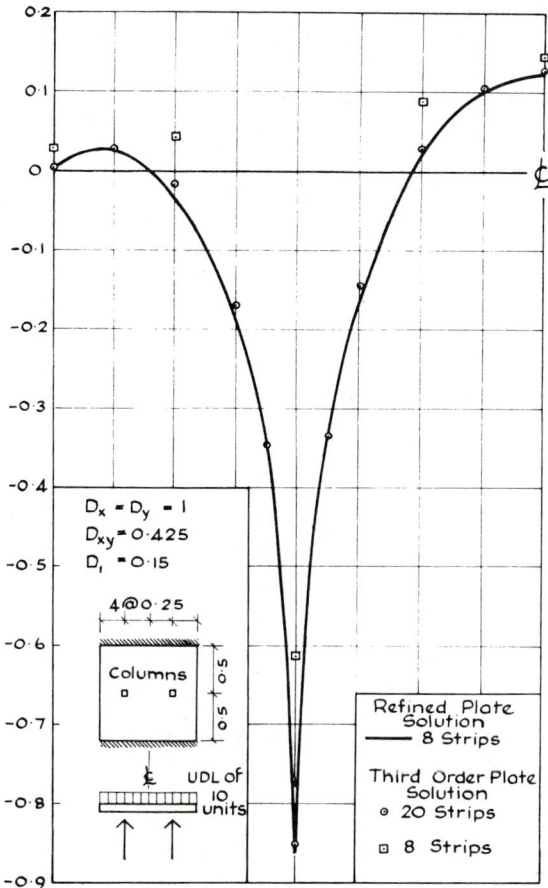

FIG. I   TRANSVERSE MOMENT PROFILES AT LINE
OF COLUMNS FOR A BRIDGE DECK WITH TWO
INTERMEDIATE COLUMN SUPPORTS (19 HARMONICS)

AUXILIARY NODAL LINES AND NODAL LINE CONTINUITIES

In the conventional finite strip approach to plate bending analysis, a third order polynomial function is assumed to represent the deflection profile across the width of a strip. If such a function is incorporated,[2] the unknown parameter can be conveniently chosen as the deflection and slope amplitudes at the left and right nodal lines (see Fig. 2(a)), i.e.

$$\left\{w_m^b\right\} = \left[w_{im}, \left(\frac{\partial w}{\partial x}\right)_{im}, w_{jm}, \left(\frac{\partial w}{\partial x}\right)_{jm}\right]^T \tag{1}$$

where b stands for bending only, $w_{im}, \left(\frac{\partial w}{\partial x}\right)_{im}, w_{jm}$ and $\left(\frac{\partial w}{\partial x}\right)_{jm}$ respectively, are the deflection and slope amplitudes at boundaries i and j of the strip, and m indicates the mth harmonic.

Similarly, in the case of the analysis of box sections, linear in-plane displacement functions across the width of the strip are used in conjunction with the third order function for bending. In this way, the stiffness approach can be employed and formulation for the box structure will be possible. Then, the nodal line deformation amplitude vector is given as:

$$\{w_m\} = \left[u_{im}, v_{im}, w_{im}, \left(\frac{\partial w}{\partial x}\right)_{im}, u_{jm}, v_{jm}, w_{jm}, \left(\frac{\partial w}{\partial x}\right)_{jm} \right]^T \qquad (2)$$

in which u_{im}, v_{im}, u_{jm} and v_{jm} respectively are the in-plane displacement amplitudes in the transverse and longitudinal directions at boundaries i and j of the strip (see Fig. 2(b)).

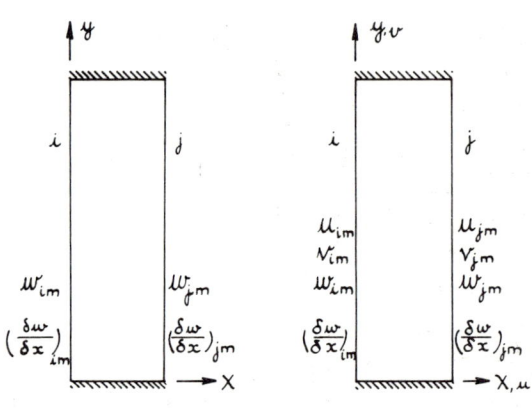

a) Third Order Bending Strip   b) Third Order Strip in
                                      Box-Structures

FIG. 2   CONVENTIONAL FINITE STRIPS

However, due to the low order of the assumed interpolation functions, a large number of strips is necessary to give accurate results. Thus, it would be preferable to use higher order displacement functions in the formulation so that greater accuracy can be obtained using a comparatively smaller number of strips. Two independent procedures are possible to achieve better solutions:

(1) by establishing higher order nodal line compatibilities
(2) by introducing auxiliary nodal lines between the conventional boundaries i and j.

If an Nth order polynomial function is to be incorporated using the first procedure, the associated unknown parameters in a strip (see Fig. 3(a)) can be written as,

$$\{w_m^b\} = \left[w_{im}, \left(\frac{\partial w}{\partial x}\right)_{im}, \left(\frac{\partial^2 w}{\partial x^2}\right)_{im} \cdots, \left(\frac{\partial^k w}{\partial x^k}\right)_{im}, \right.$$

$$\left. w_{jm}, \left(\frac{\partial w}{\partial x}\right)_{jm}, \cdots, \left(\frac{\partial^k w}{\partial x^k}\right)_{jm} \right]^T \qquad (3)$$

where $k = \dfrac{(N+1)}{2}$.

Analysis of a plate using these finite strips will ensure compatibility of higher order derivatives of deformation (curvature etc), which in turn leads to continuity of moment, shear etc at strip boundaries. However, due to the additional compatibilities established at nodal lines, the method ceases to be a true stiffness approach[10] and it is inconvenient to adopt such a function in the analysis of bending moments in box-beam structures.

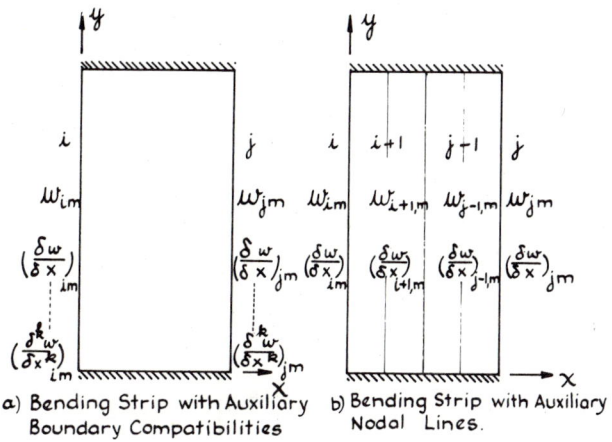

a) Bending Strip with Auxiliary   b) Bending Strip with Auxiliary
Boundary Compatibilities   Nodal Lines.

FIG. 3 HIGHER ORDER FINITE STRIPS

To use the second approach which incorporates auxiliary nodal lines in a strip, the deformation amplitude vector for a strip in bending (Fig. 3(b)) may be written as,

$$\{w^b_m\} = \left[w_{im}, \left(\frac{\partial w}{\partial x}\right)_{im}, w_{i+1,m}, \left(\frac{\partial w}{\partial x}\right)_{i+1,m}, \cdots, \right.$$

$$\left. \left(\frac{\partial w}{\partial x}\right)_{j-1,m}, w_{jm}, \left(\frac{\partial w}{\partial x}\right)_{jm} \right]^T \qquad (4)$$

An Nth order function requires a total of k nodal lines in one strip, where

$$k = \frac{(N+1)}{2}$$

In spite of the fact that compatibility of higher order derivatives is not guaranteed by using auxiliary nodal lines, the solution leads to higher order expressions for displacement and stress. At the same time the analysis preserves the basic requirement of a direct stiffness approach and the strip properties thus developed are readily employable in the analysis of box structures.

The technique of using auxiliary nodes is already established in finite element analysis and has been reviewed by Zienkiewicz.[11]

THE STIFFNESS AND FORCE MATRICES OF A STRIP WITH ANL IN PLATE STRUCTURES

The stiffness and force matrices of a finite strip in bending with one additional nodal line continuity (curvature) have already been developed.[9] The procedure of

incorporating auxiliary nodal lines (ANL) is given briefly here. The displacement interpolation function for a simple-supported strip with a total of k nodal lines as shown
in Fig. 3(b) can be written as,

$$w^b = \sum_{m=1}^{r} \left[C_1^b, C_2^b, \ldots, C_{2k-1}^b, C_{2k}^b \right] \sin k_m y
\begin{bmatrix} w_{im} \\ \theta_{im} \\ \vdots \\ \vdots \\ w_{jm} \\ \theta_{jm} \end{bmatrix}$$

$$= \sum_{m=1}^{r} \left[C^b \right] \sin k_m y \left\{ w_m^b \right\} \tag{5}$$

in which, C_h^b is a function of x only, and $\theta = \dfrac{\partial w}{\partial x}$, and $k_m = \dfrac{m\pi}{a}$.

In an orthotropic strip, the curvature and moment vectors are respectively,

$$\left\{ \phi \right\} = \sum_{m=1}^{r} \left[B_m^b \right] \left\{ w_m^b \right\} \tag{6}$$

$$\left\{ M \right\} = \sum_{m=1}^{r} \left[D^b \right] \left[B_m^b \right] \left\{ w_m^b \right\} \tag{7}$$

where, $\left[B_m^b \right]$ is the coefficient matrix for the bending curvature and $\left[D^b \right]$ is the bending rigidity matrix. The total potential energy of a strip under load function $q(x,y)$
can be expressed as,

$$U^b = \tfrac{1}{2} \int_0^a \int_0^b \left\{ M \right\}^T \left\{ \phi \right\} dxdy - \int_0^a \int_0^b q(x,y) \, w^b \, dxdy \tag{8}$$

Substituting equations (5), (6) and (7) into (8) and minimizing the resulting expression with respect to all the deformation amplitudes leads to,[9]

$$\sum_{m=1}^{r} \int_0^a \int_0^b \left[B_m^b \right]^T \left[D^b \right] \left[B_m^b \right] dxdy \left\{ w_m^b \right\}$$

$$= \sum_{m=1}^{r} \int_0^a \int_0^b \left[C^b \right]^T q \sin k_m y \, dxdy \tag{9}$$

For a particular harmonic,

$$\left[S_m^b \right] \left\{ w_m^b \right\} = \left\{ F_m^b \right\} \tag{10}$$

where $\left[S_m^b \right]$ and $\left\{ F_m^b \right\}$ are the bending stiffness and force matrices for the mth
harmonic. The force matrices can now be evaluated by direct integration of the right-
hand side of equation (10) without expanding the load function q into a Fourier
series. The evaluation of concentrated, patch and uniformly distributed load matrices
is given in Appendix II.

THE STIFFNESS AND FORCE MATRICES OF A STRIP WITH ANL IN BOX-BEAM STRUCTURES

Due to the assumption that no interaction takes place between the bending and in-
plane actions, the bending and in-plane strip properties can be determined separately
and then coupled together using an appropriate direction cosine matrix. The bending stiffness and force matrices developed in the previous section are readily adaptable
to box structures.

In-plane Strip Properties

The in-plane displacement interpolation function for a simply-supported strip (Fig. 4(b)) with k nodal lines is written as,

$$w^p = \begin{bmatrix} u \\ v \end{bmatrix} = \sum_{m=1}^{r} \left[\begin{bmatrix} C_{1m} \end{bmatrix}, \begin{bmatrix} C_{2m} \end{bmatrix}, \ldots, \begin{bmatrix} C_{km} \end{bmatrix} \right] \begin{bmatrix} \begin{bmatrix} u_{im} \\ v_{im} \end{bmatrix} \\ \vdots \\ \vdots \\ \begin{bmatrix} u_{jm} \\ v_{jm} \end{bmatrix} \end{bmatrix} \tag{11}$$

where

$$\begin{bmatrix} C_{hm} \end{bmatrix} = \begin{bmatrix} C_{hu} \sin k_m y & 0 \\ 0 & C_{hv} \cos k_m y \end{bmatrix} \tag{12}$$

in which C_{hu} and C_{hv} are functions of x only. Using this function and following a similar procedure as outlined previously for the plate formulation, we obtained

$$\begin{bmatrix} S_m^p \end{bmatrix} \begin{Bmatrix} w_m^p \end{Bmatrix} = \begin{Bmatrix} F_m^p \end{Bmatrix} \tag{13}$$

where $\begin{bmatrix} S_m^p \end{bmatrix}$ and $\begin{Bmatrix} F_m^p \end{Bmatrix}$ are the in-plane stiffness and force matrices for the mth harmonic.

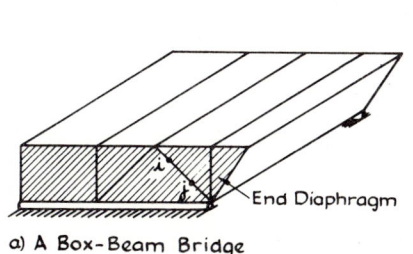

a) A Box-Beam Bridge

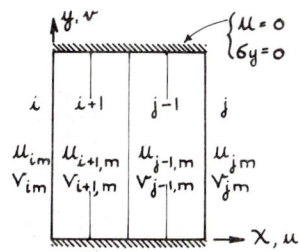

b) In-plane Displacement Amplitudes

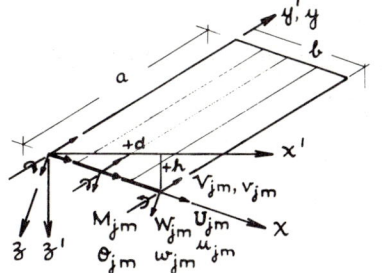

c) Local Force and Displacement Systems

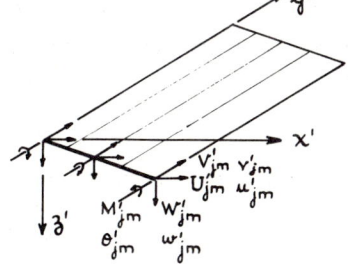

d) Global Force and Displacement Systems

FIG. 4 A REFINED FINITE STRIP IN BOX-BEAM BRIDGE DECKS.

Direction Cosine Matrix

The direction cosine matrix which will transfer the local force system,

$$\{F_m\} = \Big[U_{im}, V_{im}, W_{im}, M_{im}, U_{i+1,m}, \cdots$$
$$\cdots, M_{j-1,m}, U_{jm}, V_{jm}, W_{jm}, M_{jm} \Big]^T \tag{14}$$

to the global system

$$\{F'_m\} = \Big[U'_{im}, V'_{im}, W'_{im}, M'_{im}, U'_{i+1,m}, \cdots$$
$$\cdots, M'_{j-1,m}, U'_{jm}, V'_{jm}, W'_{jm}, M'_{jm} \Big]^T \tag{15}$$

and thus couples the in-plane and bending actions, can be written as,

$$[R] = \begin{bmatrix} [r_i] & [0] & [0] & \cdots & \cdot \\ [0] & [r_{i+1}] & [0] & \cdots & \cdot \\ [0] & [0] & & \cdots & \cdot \\ \cdot & \cdot & & \cdots & [r_j] \end{bmatrix} \tag{16}$$

where

$$[r_i] = [r_{i+1}] = \ldots = [r_j] = \begin{bmatrix} \dfrac{d}{b} & 0 & \dfrac{-h}{b} & 0 \\ 0 & 1 & 0 & 0 \\ \dfrac{h}{b} & 0 & \dfrac{d}{b} & 0 \\ 0 & 0 & 0 & 1 \end{bmatrix} \tag{17}$$

Therefore

$$\{F_m\} = [R]^T \{F'_m\} \tag{18}$$

and

$$\{w_m\} = [R]^T \{w'_m\} \tag{19}$$

where

$$\{w_m\} = \Big[u_{im}, v_{im}, w_{im}, \theta_{im}, u_{i+1,m}, \cdots$$
$$\cdots, \theta_{j-1,m}, u_{jm}, v_{jm}, w_{jm}, \theta_{jm} \Big]^T \tag{20}$$

is the local displacement amplitude vector and,

$$\{w'_m\} = \Big[u'_{im}, v'_{im}, w'_{im}, \theta'_{im}, u'_{i+1,m}, \cdots$$
$$\cdots, \theta'_{j-1,m}, u'_{jm}, v'_{jm}, w'_{jm}, \theta_{jm} \Big]^T \tag{21}$$

is the global system. The two systems of forces and displacements are shown in Fig. 4. But,

$$[S_m]\{w_m\} = \{F_m\} \tag{22}$$

Substituting equations (18) and (19) to (22) gives,

$$[S'_m]\{w'_m\} = \{F'_m\} \tag{23}$$

where

$$[S'_m] = [R][S_m][R]^T \tag{24}$$

is the global stiffness matrix.

Using equation (23), the normal procedure of the stiffness approach can be used to assemble the stiffness and force matrices for all the strips in the idealized structure and solve for the overall unknown deformation amplitude vector. Equation (19) allows $\{w'_m\}$ to be transferred back to the local coordinate system. The local displacements and internal stresses can then be obtained accordingly. The final results are the sum of values from each harmonic.

SOLUTION OF BRIDGE DECKS WITH INTERMEDIATE COLUMNS BY THE FLEXIBILITY APPROACH

With the properties of the strips in plate and folded plate structures established we can now proceed to form a solution for simply-supported plate and box structures with intermediate column supports. For a column-supported structure, the released structure is first analysed (by the finite strip procedure) and the static deflection vector $\{\Delta\}$ is obtained. The unit column reactions are then applied to the structure separately, and the finite strip procedure is again applied. The deflections due to these unit reactions form the so-called flexibility matrix $[F]$, where,

$$[F] = \begin{bmatrix} \delta_{11} & \delta_{12} & \cdot & \cdot & \delta_{1n} \\ \delta_{21} & \delta_{22} & \cdot & \cdot & \delta_{2n} \\ \cdot & & \cdot & \cdot & \cdot \\ \cdot & & & \cdot & \cdot \\ \delta_{n1} & \delta_{n2} & \cdot & \cdot & \delta_{nn} \end{bmatrix} \tag{25}$$

in which δ_{ij} is the deflection at point i due to the unit reaction at j. The effects due to elastic column-shortening and foundation settlement at column positions can also be included in the flexibility matrix, which now becomes:

$$[F] = \begin{bmatrix} (\delta_{11} + \delta'_1) & \delta_{12} & \cdot & \cdot & \delta_{1n} \\ \delta_{21} & (\delta_{22} + \delta'_2) & \cdot & \cdot & \delta_{2n} \\ \cdot & & \cdot & \cdot & \cdot \\ \cdot & & \cdot & \cdot & \cdot \\ \delta_{n1} & \delta_{n2} & \cdot & \cdot & (\delta_{nn} + \delta'_n) \end{bmatrix} \tag{26}$$

where
$$\delta'_i = \delta_{ci} + \delta_{si} \tag{27}$$

in which δ_{ci} is the unit elastic shortening of column i and, δ_{si}, the unit elastic foundation settlement. By considering the compatibility requirement at column positions we can write,

$$[F]\{R'\} + \{\Delta\} = \{\Delta'\} \tag{28}$$

where
$$\{R'\} = \begin{bmatrix} R_1, R_2, \ldots, R_n \end{bmatrix}^T$$

is the column reaction vector,
$$\{\Delta\} = \begin{bmatrix} \Delta_1, \Delta_2, \ldots, \Delta_n \end{bmatrix}^T$$

is the static deflection vector due to applied load $q(x, y)$ and

$$\{\Delta'\} = \left[\Delta_1', \Delta_2', \ldots, \Delta_n'\right]^T$$

is the differential column settlement vector.

Solving equation (28) for the column reactions, the deflections and internal stresses can be calculated by adding results due to applied load $q(x, y)$ and to the individual column reactions on the released structure.

COMPUTER PROGRAMS

Using the finite strip approaches described above, efficient solutions may be determined for a wide range of bridge deck structures. Two computer programs[12] have been developed by the authors specifically for the analysis of slab and cellular bridge decks with discrete intermediate supports.

The first program COSPEQ is based on the refined 'fifth order' plate solution[9] and has been especially developed to give a quick and accurate solution for deflections and moments in orthotropic slab bridge decks with elastic intermediate columns. Provision is also made for the computation of shearing forces (see Appendix I).

The second program COSBOB, for box structures, is based primarily on the conventional 'third order' box solution but employs a flexibility approach to solve for redundant columns. In both programs an improvement is made by introducing the directly evaluated force matrices to give a more accurate representation of concentrated and patch load.

A third program BOBANL[12] employs the Auxiliary Nodal Line technique for the solution of simply supported plate and box structures.

COMPARISON OF RESULTS

Numerical examples are given in this section to illustrate the efficiency of the ANL techniques and the two computer programs COSPEQ and COSBOB developed respectively for the solution of indeterminate slab and box section bridges.

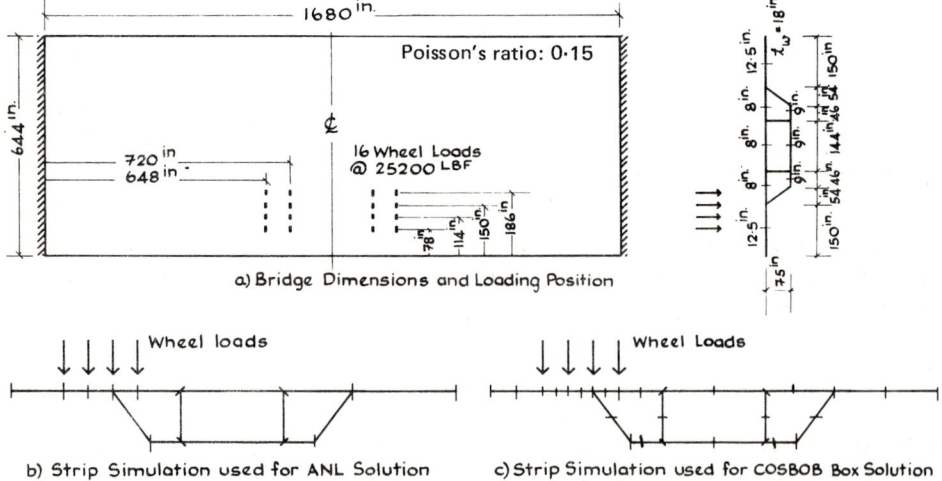

a) Bridge Dimensions and Loading Position

b) Strip Simulation used for ANL Solution

c) Strip Simulation used for COSBOB Box Solution

FIG. 5 SIMPLY-SUPPORTED THREE CELL SPINE BOX BRIDGE AND THE FINITE STRIP SIMULATIONS

Comparison between the ANL Technique and the Conventional Finite Strip Solution for a Simply-Supported Box Bridge

A simply-supported three-cell spine box bridge under 16 wheel loads is analysed using the ANL technique (program BOBANL) and the 'third order' solution (program COSBOB). The bridge dimensions and the loading position are given in Fig. 5(a) and the finite strip simulations used for the two analyses are shown in Figs 5(b) and 5(c). Wheel loads were assumed to be point forces and 15 harmonics (8 terms) were used for the two analyses. Good agreement was found between the two analyses for deflection, plate moments and longitudinal in-plane stresses. An improved solution was given for transverse in-plane stresses by the ANL solution as can be seen in Fig. 6. Similar improvement can also be expected in the case of transverse plate shear distribution.

Comparison between COSPEQ and the Finite Element Program Suite FESS

To demonstrate the efficiency of the computer program COSPEQ in analysing rectangular indeterminate slab bridge decks, comparison is made with the finite element package FESS.[13] FESS is a suite of computer programs developed by Zienkiewicz *et al.* for the data processing, analysis and design of slab bridge deck structures.

The analysis is based on the finite element procedure using straight beam and 'third order' triangular plate elements, which enables the suite to cover a wide range of cases of plane slab structure. In this comparison, a square isotropic bridge with three intermediate column supports under partially distributed load was analysed. For the finite strip approach, the deck was divided into eight equal strips and 15 harmonics were used. In the finite element analysis, 8 x 8, 16 x 16 and 18 x 24 meshes were tried to simulate the behaviour of the deck. The three mesh patterns and the strip idealization of the deck are shown in Fig. 7. Column reactions were assumed to be concentrated forces in all analyses. The comparisons of deflections and transverse and longitudinal moments at the line of the columns are shown in Fig. 8. The values from the finite element analyses were the average nodal values. Table 1 compares the computer time used by each of the four analyses. It can be seen from Fig. 8 that, while no significant difference in deflection profiles can be found in Fig. 8, the moment values given by FESS only converge gradually toward the refined finite strip solution as the mesh becomes very fine. An inspection of Table 1 leads to the conclusion that within its field of application, the finite strip method, as represented by program COSPEQ appears to hold distinct advantages over the finite element method as represented by program FESS.

Approximate Analysis of Multi-span Box Bridges

As a part of the expanding programme of highway construction an increasing number of multi-span box bridges is being built. In this type of bridge, diaphragms are normally provided at intermediate support sections to ensure stability and to reduce probable stress concentrations due to the supporting columns. In the case of spine beam construction, diaphragms are usually filled within the cells of the support sections leaving the cantilever overhangs unstiffened.

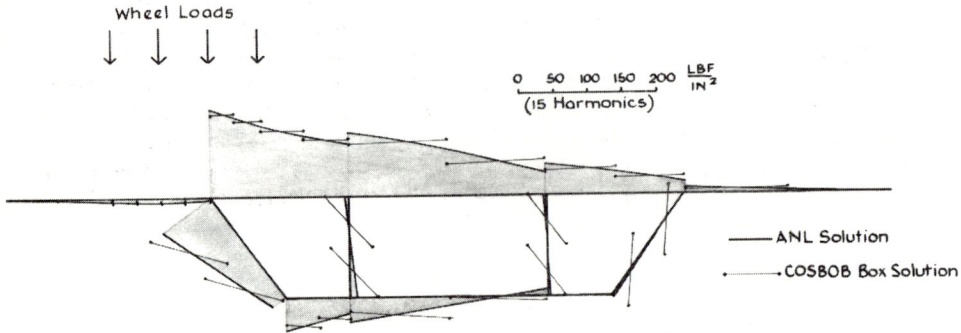

FIG. 6 TRANSVERSE IN-PLANE STRESS PROFILES FOR THE THREE CELL SPINE BOX BRIDGE UNDER SIXTEEN WHEEL LOADS (AT SECTION UNDER THE SECOND AXLE.)

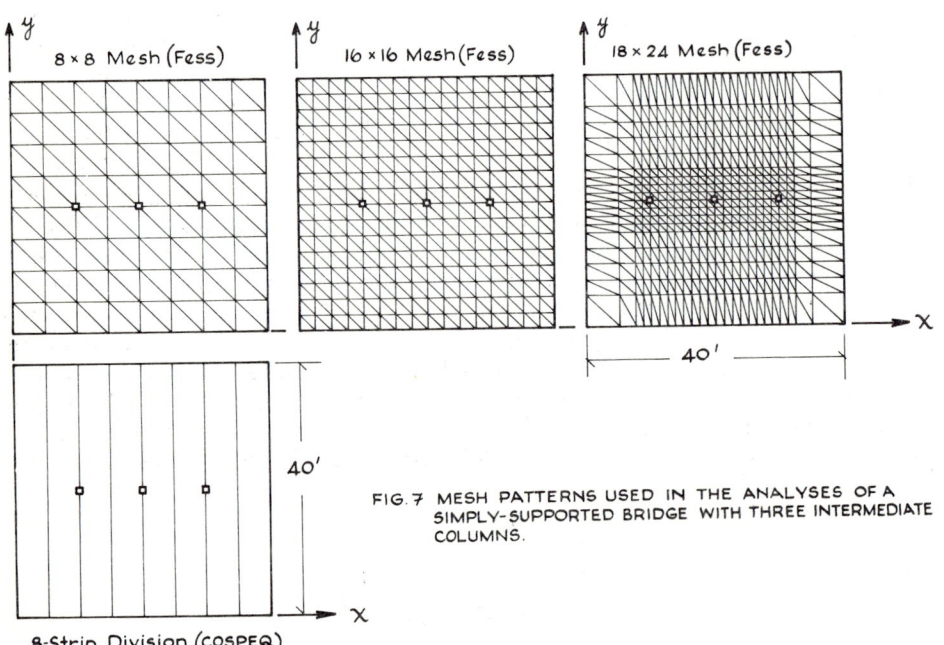

FIG. 7 MESH PATTERNS USED IN THE ANALYSES OF A SIMPLY-SUPPORTED BRIDGE WITH THREE INTERMEDIATE COLUMNS.

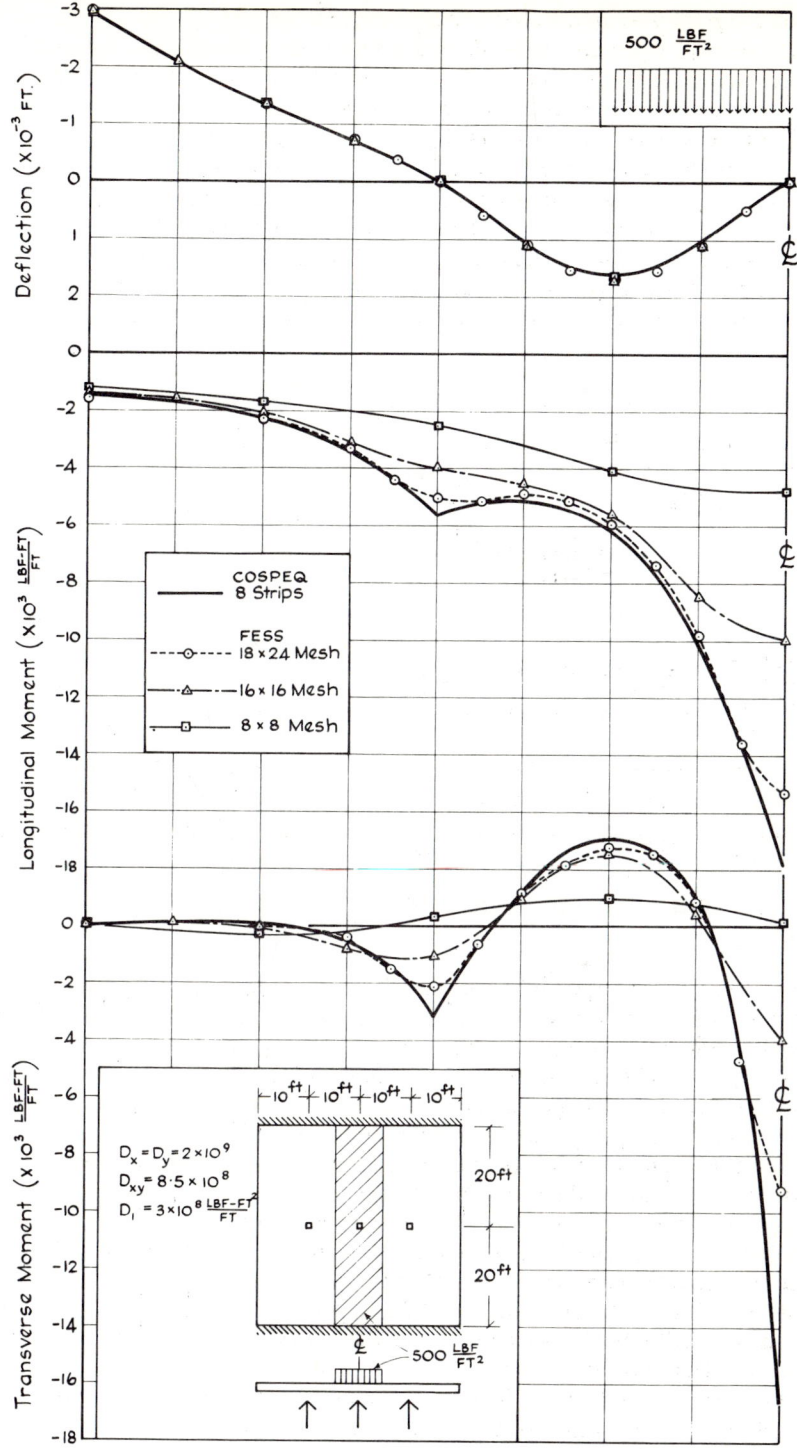

FIG. 8 DISTRIBUTION OF DEFLECTIONS AND LONGITUDINAL AND
TRANSVERSE MOMENTS AT LINE OF COLUMNS FOR A
SIMPLY-SUPPORTED BRIDGE WITH THREE INTERMEDIATE
COLUMNS.

Fig. 9(a) shows a two-span three-cell box-bridge with no overhangs but with rigid diaphragms provided at all support sections. The bridge was first analysed by Scordelis[14] to study the efficiency of the computer programs he developed using folded plate theory, finite element and finite segment approaches.

In the folded plate analysis, the elasticity theory is employed in conjunction with a flexibility approach. The rigid diaphragms at intermediate support sections are simulated by point restraints. In the plane of the diaphragms, rotations and translations are prevented at joints of the constituent plates and in addition, translations are also restricted at third points of individual constituent plates. The corresponding redundant forces at plate joints are assumed to be concentrated where those at third

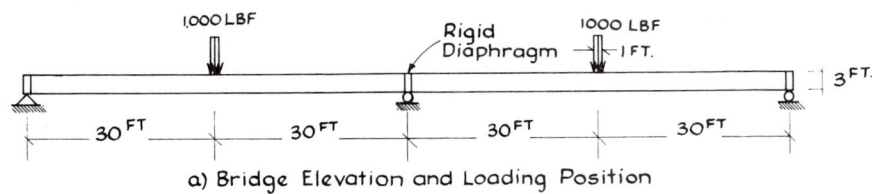

a) Bridge Elevation and Loading Position

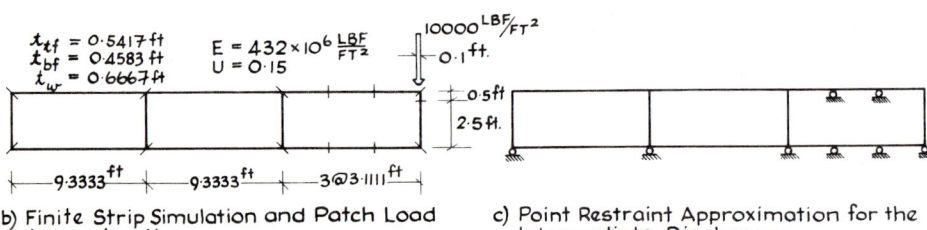

b) Finite Strip Simulation and Patch Load   c) Point Restraint Approximation for the
   Approximation   Intermediate Diaphragm

FIG. 9   TWO-SPAN CONTINUOUS BOX-BRIDGE

points of the plates are taken to be triangularly distributed with centroids of the triangles coincident with the restraint points. All the redundants are assumed to be distributed over one foot length in the direction of the span.

The two span box bridge has also been analysed using the program COSBOB in which the intermediate rigid diaphragms are approximated by vertical point restraints only. The concentrated line load is simulated by a narrow patch load. The finite strip simulation of the bridge and the loading approximation are given in Fig. 9(b). The longitudinal distribution of longitudinal in-plane stresses near the box corner is plotted in Fig. 10, and Fig. 11 shows respectively the longitudinal in-plane stress profiles at the intermediate support section and the loaded section. Ninety-nine harmonics (50 terms) were used in the solution and all results are compared with the folded plate analysis.[14] It can be seen that fairly good results are given by the computer program COSBOB with a simulation of the rigid diaphragm support with 8 rigid columns. The finite strip solution appears to underestimate the longitudinal stresses under the applied loads. However this may be easily remedied by treating the applied line loads as concentrated forces and/or using narrower strips in the vicinity of the applied loads. Improved results especially near the intermediate supports can be obtained by using more closely spaced columns. Although approximate assumptions are involved in simulating rigid diaphragms when using COSBOB, the method provides an economical analysis without undue sacrifice of accuracy.

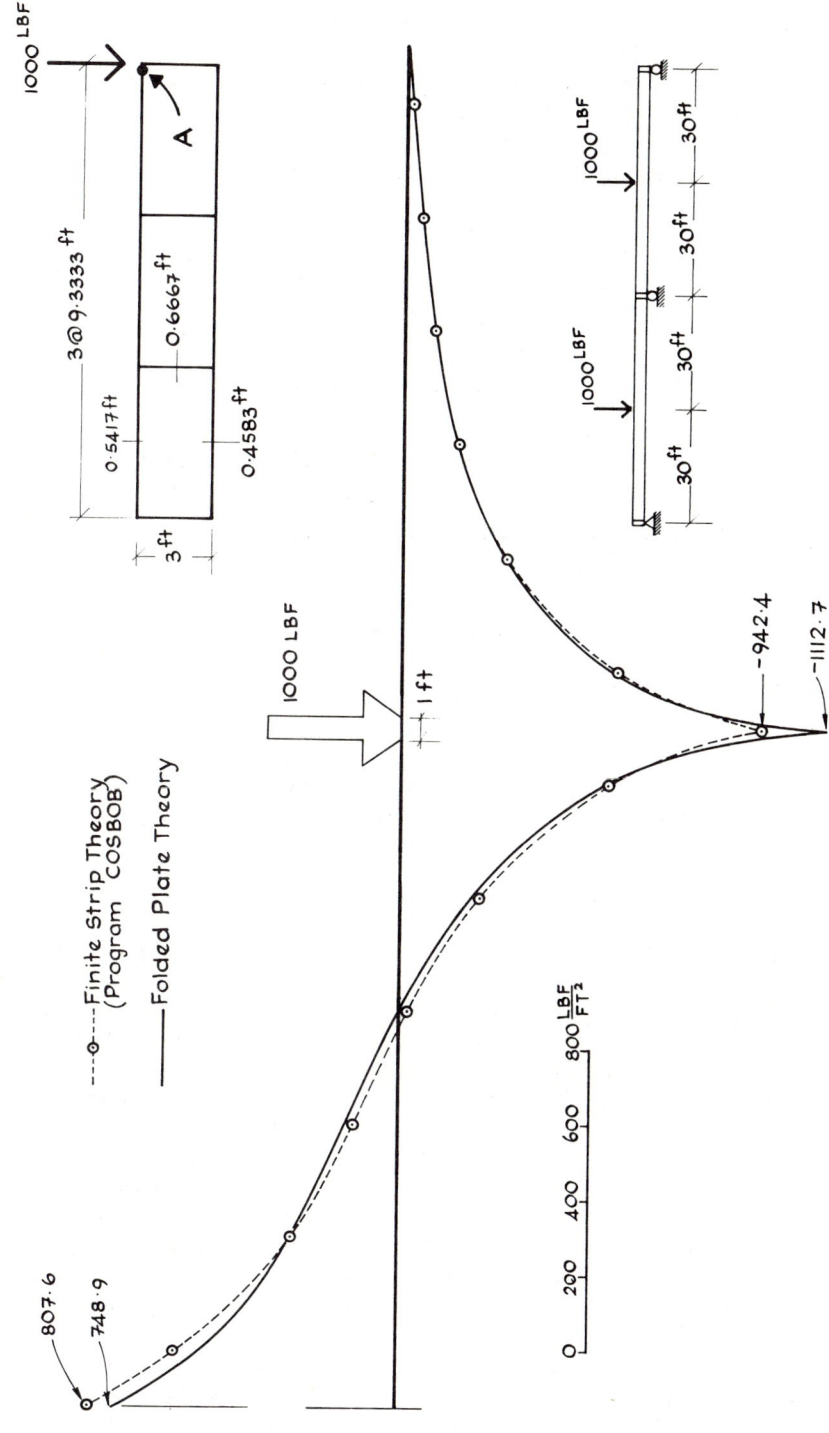

FIG. IO LONGITUDINAL DISTRIBUTION OF LONGITUDINAL IN-PLANE STRESSES (AT POINT A) FOR TWO-SPAN CONTINUOUS BOX BRIDGE.

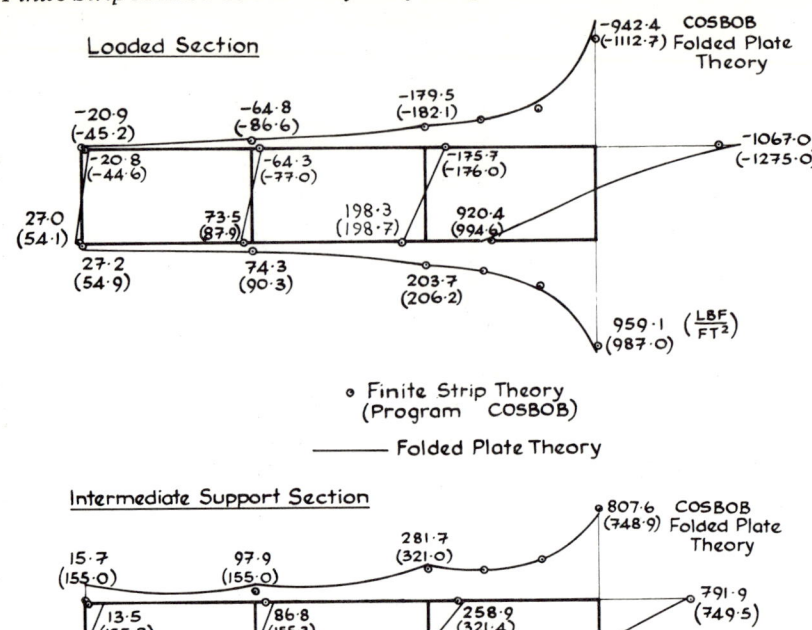

FIG. 11   LONGITUDINAL IN-PLANE STRESS PROFILES FOR
TWO-SPAN CONTINUOUS BOX BRIDGE.

CONCLUSIONS

New developments of the finite strip approach are presented for the analysis of plate and box bridge decks.

The ANL (Auxiliary Nodal Line) technique enables higher order displacement interpolation functions to be incorporated in the finite strip formulation for box structures. When compared with the conventional procedure, the new approach which is based on a strip with one auxiliary nodal line, gives an improved transverse in-plane stress distribution even with a reduced number of strips.

The finite strip method is applied to indeterminate bridge deck structures by use of the flexibility approach. Two computer programs COSPEQ and COSBOB have been developed respectively to give efficient analysis of simply-supported slab and box-type bridge deck structures with elastic intermediate column supports.

Comparison with the finite element program suite FESS[13] shows that the computer program COSPEQ provides an economical and accurate solution for rectangular slab type bridge deck problems.

It has been shown that the computer program COSBOB which was originally developed for simply-supported box bridges with discrete intermediate supports, can also be used for the analysis of multi-span box bridges with rigid diaphragms at support sections.

TABLE 1

Computer Time used by COSPEQ and FESS for the Analysis of an
Isotropic Square Plate in an ICL 4130 Computer
using T30C System

Details	COSPEQ		FESS					
	8 – STRIPS		8 × 8 MESH		16 × 16 MESH		18 × 24 MESH	
	min.	*sec.*	*min.*	*sec.*	*min.*	*sec.*	*min.*	*sec.*
Data checking (FECK)	–	–	–	10	–	12	–	20
Data generating (JPUT)	–	–	1	43	5	38	9	34
Analysis	–	48*	5	6	25	26	47	09
Total	–	48	6	59	31	16	57	3

*Two load cases and 15 harmonics (15 terms)

APPENDIX I

THE EVALUATION OF LATERAL PLATE SHEARS

The transverse and longitudinal plate shears as defined by Timoshenko and
Woinowsky-Krieger[15] can be written as,

$$
\begin{bmatrix} Q_x \\ Q_y \end{bmatrix} = -\begin{bmatrix} D_x & O & H & O \\ O & D_y & O & H \end{bmatrix} \begin{bmatrix} \dfrac{\partial^3 w}{\partial x^3} \\ \dfrac{\partial^3 w}{\partial y^3} \\ \dfrac{\partial^3 w}{\partial x \partial y^2} \\ \dfrac{\partial^3 w}{\partial x^2 \partial y} \end{bmatrix}
\tag{I--1}
$$

where

$$H = 2D_{xy} + D_1$$

If we write,

$$-\frac{\partial^2 w}{\partial x^2} = \sum_{m=1}^{r} \left[B_{1m}^b \right] \left\{ w_m^b \right\}
\tag{I--2}$$

and

$$-\frac{\partial^3 w}{\partial y^2} = \sum_{m=1}^{r} \left[B_{2m}^b \right] \left\{ w_m^b \right\}$$

then

$$
\begin{bmatrix} Q_x \\ Q_y \end{bmatrix} = \sum_{m=1}^{r} \begin{bmatrix} D_x & O & H & O \\ O & D_y & O & H \end{bmatrix} \begin{bmatrix} \dfrac{\partial}{\partial x}\left[B_{1m}^b \right] \\ \dfrac{\partial}{\partial y}\left[B_{2m}^b \right] \\ \dfrac{\partial}{\partial x}\left[B_{2m}^b \right] \\ \dfrac{\partial}{\partial y}\left[B_{1m}^b \right] \end{bmatrix} \left\{ w_m^b \right\}
\tag{I--3}
$$

Thus, after the unknown deformation amplitude vector has been determined, the corresponding shear forces in a strip can be obtained using equation (I–3). As usual, the final results are the sum of individual harmonics. However, it is implied in equation (I–1) that for a given condition, the longitudinal shears Q_y will be comparatively more accurate than the transverse values.

APPENDIX II

FORCE MATRICES DUE TO SOME FORMS OF PLATE LOADING

The force matrix due to lateral load of any form can be written as,[9]

$$\{F_m^b\} = \int_0^a \int_0^b [C^b]^T q(x,y) \sin k_m y \, dxdy \qquad (\text{II}-1)$$

For a strip with one auxiliary nodal line positioned at centre of the two conventional boundaries, $[C^b]^T$ is given as,

$$[C^b]^T = \begin{bmatrix} 1 - \dfrac{23x^2}{b^2} + \dfrac{66x^3}{b^3} - \dfrac{68x^4}{b^4} + \dfrac{24x^5}{b^5} \\[2mm] x - \dfrac{6x^2}{b} + \dfrac{13x^3}{b^2} - \dfrac{12x^4}{b^3} + \dfrac{4x^5}{b^4} \\[2mm] \dfrac{16x^2}{b^2} - \dfrac{32x^3}{b^3} + \dfrac{16x^4}{b^4} \\[2mm] -\dfrac{8x^2}{b} + \dfrac{32x^3}{b^2} - \dfrac{40x^4}{b^3} + \dfrac{16x^5}{b^4} \\[2mm] \dfrac{7x^2}{b^2} - \dfrac{34x^3}{b^3} + \dfrac{52x^4}{b^4} - \dfrac{24x^5}{b^5} \\[2mm] -\dfrac{x^2}{b} + \dfrac{5x^3}{b^2} - \dfrac{8x^4}{b^3} + \dfrac{4x^5}{b^4} \end{bmatrix} \qquad (\text{II}-2)$$

CONCENTRATED LOAD

For a concentrated load P_0 acting at x_0 and y_0 on the strip as shown in Fig. 12, the force matrix can be obtained by direct substitution of local coordinates x_0 and y_0 into equation II–1 which yields,

$$\{F_m^b\} = P_0 [C^b]^T_{x=x_0} \sin k_m y_0 \qquad (\text{II}-3)$$

If the load is at boundary i, the force matrix reduces to,

$$\{F_m^b\} = P_0 \sin k_m y_0 \begin{bmatrix} 1 \\ 0 \\ 0 \\ 0 \\ 0 \\ 0 \end{bmatrix} \qquad (\text{II}-4)$$

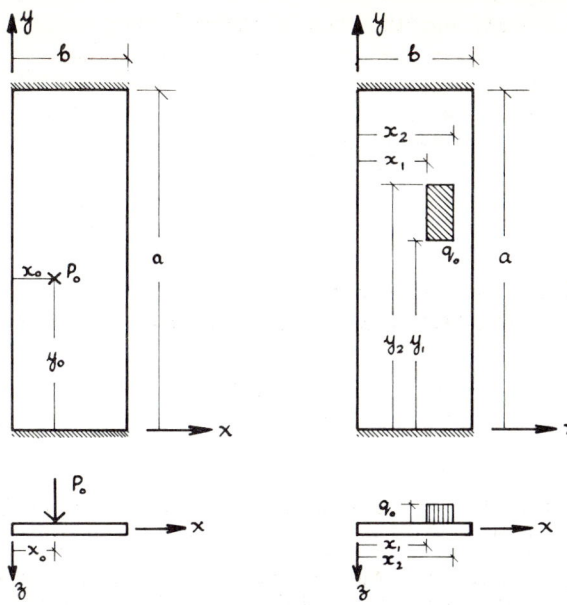

FIG.12   CONCENTRATED AND PATCH LOADING

UNIFORM PATCH LOAD

Fig. 12 shows a patch load of intensity q_0 over a small area at an arbitrary position of the strip. The corresponding force matrix can be obtained by integrating equation (II–1) over the loaded area, which gives

$$\left\{F_m^b\right\} = \frac{q_0}{k_m} \left(\cos k_m y_1 - \cos k_m y_2\right) \begin{bmatrix} \overline{C1} \\ \overline{C2} \\ \overline{C3} \\ \overline{C4} \\ \overline{C5} \\ \overline{C6} \end{bmatrix} \qquad \text{(II–5)}$$

where

$$\overline{C1} = x_{21}^1 - \frac{23}{3b^2} x_{21}^3 + \frac{33}{2b^3} x_{21}^4 - \frac{68}{5b^4} x_{21}^5 + \frac{4}{b^5} x_{21}^6$$

$$\overline{C2} = \frac{1}{2} x_{21}^2 - \frac{2}{b} x_{21}^3 + \frac{13}{4b^2} x_{21}^4 - \frac{12}{5b^3} x_{21}^5 + \frac{2}{3b^4} x_{21}^6$$

$$\overline{C3} = \frac{16}{3b^2} x_{21}^3 - \frac{8}{b^3} x_{21}^4 + \frac{16}{5b^4} x_{21}^5$$

$$\overline{C4} = -\frac{8}{3b} x_{21}^3 + \frac{8}{b^2} x_{21}^4 - \frac{8}{b^3} x_{21}^5 + \frac{8}{3b^4} x_{21}^6$$

$$\overline{C5} = \frac{7}{3b^2} x_{21}^3 - \frac{17}{2b^3} x_{21}^4 + \frac{52}{5b^4} x_{21}^5 - \frac{4}{b^5} x_{21}^6$$

$$\overline{C6} = \frac{-1}{3b} x_{21}^3 + \frac{5}{4b^2} x_{21}^4 - \frac{8}{5b^3} x_{21}^5 + \frac{2}{3b^4} x_{21}^6$$

in which

$$x_{21}^n = x_2^n - x_1^n$$

If the patch area covers the whole of the strip (UDL) equation (II−5) becomes,

$$\left\{F_m^b\right\} = \frac{q_0}{k_m}\left[1-(-1)^m\right]\begin{bmatrix} 7b/30 \\ b^2/60 \\ 8b/15 \\ 0 \\ 7b/30 \\ -b^2/60 \end{bmatrix}$$

In a similar way, the force matrices due to other form of loading can be readily obtained. If several loads are acting on the same strip, the effect is additive.

REFERENCES

1. Hellan, K. Application of a numerical procedure to the analysis of thin rectangular plates of variable thickness. *Acta polytech. scand. Civil engineering and building construction series*, Ci 16,.1963.
2. Cheung, Y. K, The finite strip method in the analysis of elastic plates with two opposite simply supported ends. *Proc. Instn civ. Engrs*, Vol. 40, No. 1, May 1968, pp. 1−7.
3. Cheung, Y. K. Orthotropic right bridges by the finite strip method. Proceedings of the Second International Symposium on Concrete Bridge Design, Chicago, April 1969. (To be published by the American Concrete Institute.)
4. Zienkiewicz, O. C. *The finite element method in structural and continuum mechanics*. New York, McGraw-Hill, 1967.
5. Cheung, Y. K. Analysis of box girder bridges by the finite strip method. Proceedings of the Second International Symposium on Concrete Bridge Design, Chicago, April 1969. (To be published by the American Concrete Institute.)
6. Cheung, Y. K. Folded plate structures by finite strip method. *J. struct. Div. Am. Soc. civ. Engrs*, Vol. 95, ST12, Dec. 1969, pp. 2963−2979.
7. Powell, G. H. and Ogden, D. W. Analysis of orthotropic steel plate bridge decks. *J. struct. Div. Am. Soc. civ. Engrs*, Vol. 95, ST5, May 1969, pp. 909−922.
8. Willam, K. J. and Scordelis, A. C. *Analysis of orthotropic folded plates with eccentric stiffeners*. University of California, Berkeley, Structural Engineering and Structural Mechanics Report No. 70−72, Feb. 1970.
9. Loo, Y. C. and Cusens, A. R. A refined finite strip method for the analysis of orthotropic plates. *Proc. Instn civ. Engrs*, Paper 7340, Vol. 48, Jan. 1971, pp. 85−91.
10. Loo, Y. C. and Cusens, A. R. Discussion on 'Folded plate structures by finite strip method'. *J. struct. Div. Am. Soc. civ. Engrs*, Vol. 96, ST8, Aug. 1970, pp. 1848−1851.
11. Zienkiewicz, O. C. Element shape functions − some general families. Course notes on Recent Developments of Finite Element Methods. University of Wales, Swansea, Jan. 1970.
12. Loo, Y. C. and Cusens, A. R. *Finite strip programs for slab and cellular bridge decks*. Proceedings of the Symposium on Bridge Programs, London, Jan. 1971 (to be published).
13. G.B. Ministry of Transport & Construction Industry Research and Information Association. Finite element package for reinforced concrete slab bridge decks, BECP/1 July 1969.
14. Scordelis, A. C. *Analysis of continuous box girder bridges*. University of California, Berkeley, Structural Engineering and Structural Mechanics Report No. SESM-67-25, Nov. 1967.
15. Timoshenko, S. P. and Woinowsky-Krieger, S. *Theory of plates and shells*. 2nd ed. New York, McGraw-Hill, 1959.

BEHAVIOUR OF SLABS NEAR COLUMN, PIER AND ABUTMENT SUPPORTS

J. D. DAVIES and E. HINTON

University College of Swansea

SYNOPSIS

The paper describes the results of analytical and experimental studies of the behaviour of concrete slabs near various types of supports commonly used in modern bridge construction. Three types of supports are considered:

I. Column Supports. The stress distribution in slabs near isolated column supports, assuming elastic behaviour, is studied for a number of column/slab interaction problems.
II. Pier Supports. Tests on a one-third scale reinforced concrete model of a typical slab/pier connection loaded to failure are described.
III. Abutment Supports. The distribution of bearing reactions at the abutment supports of a simply supported 45° skew slab with upstand edge beams is investigated and a simple method of checking the effects of elastic reactions is suggested.

   The basis of the analytical studies is the finite element method, plate or shell elements being used for overall analysis and isoparametric elements for local analysis.

INTRODUCTION

Before undertaking the analysis of any structure, it is necessary to make a number of assumptions regarding the support conditions in order to obtain a structural idealization that will enable the problem to be studied by the methods of analysis available. Although it is often sufficient, in terms of overall behaviour, to represent supports by points or lines, a more detailed study of local behaviour requires that the finite sizes and properties of real supports must be taken into account. Even the simply supported beam problem becomes complex if information regarding stress distribution near supports (or under concentrated loads) is required in addition to the overall bending and shear actions.

   In modern bridge construction, the bridge deck is frequently supported by a variety of methods. Many techniques are available for the overall analysis of slabs with simply defined support conditions but little data is available to assist the designer in the study of slabs adjacent to supports. This lack of information is largely due to the large number of geometrical configurations possible for the support and the many factors influencing the degree of slab/support interaction. Thus, it is virtually impossible to give simplified 'rules-of-thumb' procedures that will be appropriate for all support conditions. This paper describes studies on examples of three types of support.

I. COLUMN SUPPORTS

Influence of Slab Geometry

When slabs are supported on isolated columns one of the important design criteria is the shear in the slab due to the column reaction. Frequently, in order to keep the slab shearing stresses within acceptable limits, the slab is thickened around the column support area. Apart from reducing the shearing stresses that would arise in a slab of uniform thickness, the thickening must also have an influence (not always beneficial) on the magnitude and distribution of the direct stresses in the slab adjacent to the area of support.

For elastic behaviour, the thin plate theory assumption of plane sections remaining plane is not valid and it is necessary to use thick slab or full three dimensional stress analysis methods to study the behaviour around the column head. The step from 2D to 3D behaviour is a big one, both in terms of the physical understanding and of the size of computations necessary to obtain a solution. Fortunately, for internal columns, the behaviour of the slab under symmetric loading conditions is approximately radially symmetric in character. Under these conditions it is permissible to replace the full 3D thick slab problem by a 2D elasticity problem on any typical radial section. This method has been used to study the stress distributions adjacent to column supports for slabs with three section geometries:

 (*a*) Slab of uniform thickness,
 (*b*) Slab with rectangular drop panel,
 (*c*) Slab with splayed drop panel.

(*a*) Slab of uniform thickness

Fig. 1(*a*) shows an 11 inch thick slab of size 10 ft 6 in square centrally supported on an encased steel column and line loaded around the perimeter by a number of bars

FIG. 1. (a) UNIFORM THICKNESS SLAB LINE LOADED AROUND THE PERIMETER AND SUPPORTED ON A CENTRAL COLUMN.

FIG. 1. (b) SLAB WITH RECTANGULAR DROP PANEL LINE LOADED AROUND THE PERIMETER AND SUPPORTED ON A CENTRAL COLUMN.

FIG. 1. (c) SLAB WITH SPLAYED DROP PANEL LINE LOADED AROUND THE PERIMETER AND SUPPORTED ON A CENTRAL COLUMN.

connected to hydraulic jacks under the laboratory floor. This arrangement was intended to simulate the conditions arising in an extensive slab supported on equally spaced columns at 25 ft centres in two directions. The portion of slab tested corresponded, approximately, to the lines of contraflexure that would develop under uniform loading conditions.

Fig. 2 shows the principal moment vector plot for a typical quarter plan of a segment of the slab system. These moments were determined from a finite element plate bending program treating the columns as point supports. The vectors represent the magnitude, direction and nature of the principal moments at the selected node points — single lines indicate hogging and the double lines sagging moments. It will

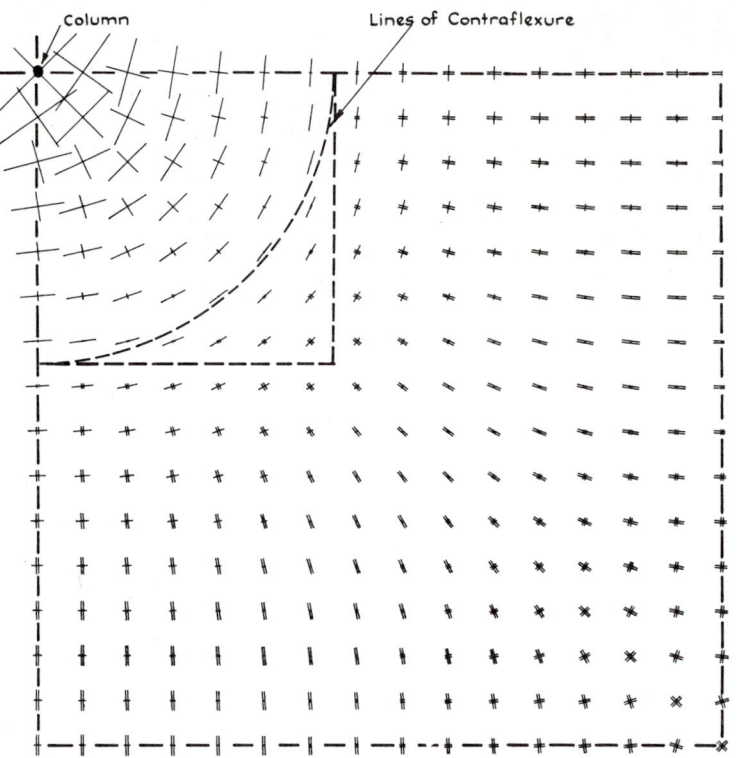

FIG. 2 PRINCIPAL MOMENT VECTOR PLOT FOR A TYPICAL QUARTER SEGMENT OF A SLAB SYSTEM SUPPORTED ON A GRID OF EQUALLY SPACED COLUMNS.

be noted that the slab exhibits radially symmetric behaviour for some distance around the column support point and the line of contraflexure lies on a circle of diameter about 0·44 of the distance between adjacent lines of columns and is represented by the dashed line (the two dashed straight lines show the loading lines chosen, for the convenience of the jacking system available, to test the structure up to failure).

An axisymmetric solution program using isoparametric cubic elements (12 nodes per element, 4 at the corners and 2 at the one-third points of the sides) was used to investigate the distribution of stresses across a typical section of the slab and taking into account the finite size of the column support. In this case (and for (*b*) and (*c*) to follow) it was assumed that the column provided positional restraint in the vertical direction, that is, the boundary condition was specified in terms of displacement restraint and the effects of slab/column continuity were ignored. The effect of this latter effect on the stress distribution at the slab/column interface will be considered in the next section.

The principal stress vector plot on the vertical section of the slab near the column is shown in Fig. 3(a), the single lines indicating compression and the double lines tension. These stresses are due to a line load applied at the circle of contraflexure referred to above. The stresses are shown at the Gaussian integration points of the elements (which are numerically integrated) and not at the nodes lying on the boundaries of the elements.

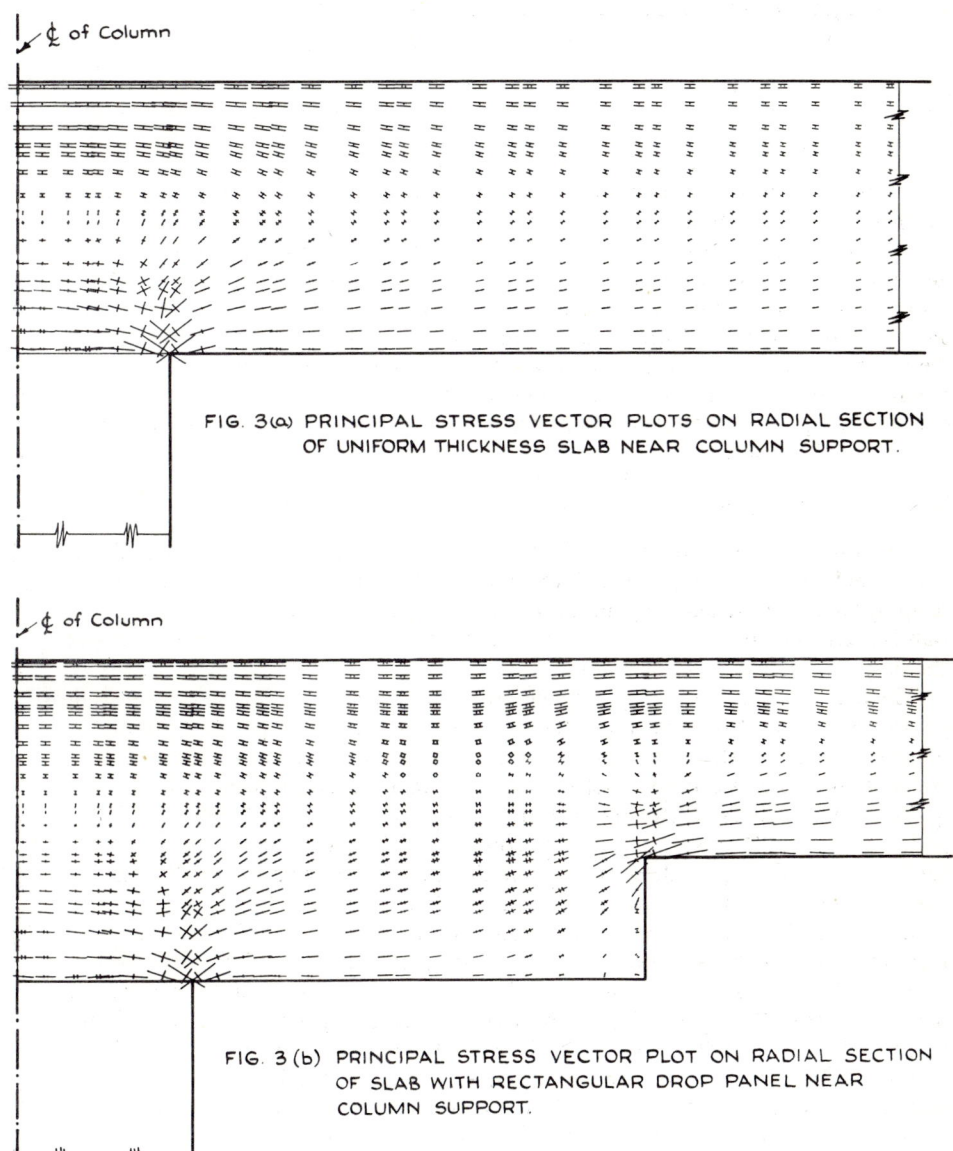

FIG. 3(a) PRINCIPAL STRESS VECTOR PLOTS ON RADIAL SECTION OF UNIFORM THICKNESS SLAB NEAR COLUMN SUPPORT.

FIG. 3(b) PRINCIPAL STRESS VECTOR PLOT ON RADIAL SECTION OF SLAB WITH RECTANGULAR DROP PANEL NEAR COLUMN SUPPORT.

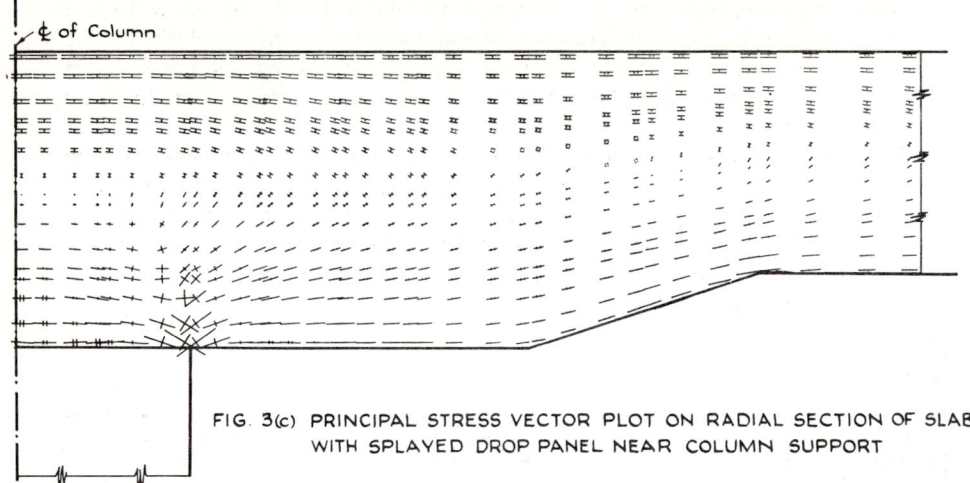

FIG. 3(c) PRINCIPAL STRESS VECTOR PLOT ON RADIAL SECTION OF SLAB
WITH SPLAYED DROP PANEL NEAR COLUMN SUPPORT

The most obvious feature of the stress plot is the high concentration of compressive stresses in the slab near the perimeter of the column support. Between the column perimeter and the centre point there is an area of tensile stress in the vertical direction. Also, as is to be expected, the distribution of direct stresses on the section above the column perimeter is not linear and there are no shear stresses (indicated by the zero inclination of the stress vectors) on the section above the centre point of the column. Within a short distance away from the column perimeter the behaviour rapidly reverts to the 'plane-sections remain plane' characteristic and the principal stresses are equal and opposite and inclined at 45° (maximum shear stress) at the mid-depth of the slab.

(b) Slab with rectangular drop panel

Fig. 1(b) shows an 8 in thick slab of size 10 ft 6 in square centrally supported, via a drop panel, on an encased steel column, and line loaded around the perimeter. The drop panel had an octagonal plan shape but a rectangular section shape where the slab is increased from 8 to 13 in over a distance of 25 in from the perimeter of the drop to the centre of the column.

The overall plate bending analysis described for slab (a) above was repeated allowing for the different thicknesses of the slab elements, and treating the columns as point supports, but there were only minor changes in the relative position of the circle of contraflexure as compared with the uniform thickness slab.

The principal stress vector plot on the vertical section of the slab near the column is shown in Fig. 3(b) assuming that the column provided positional restrain in the vertical direction. Again, these stresses are due to a line load applied at the circle of contraflexure determined from the overall bending analysis.

In this case there are two areas of stress concentration one, as for (a), near the perimeter of the column support and the other at the junction of the slab and the drop panel. It was observed in the test on this slab that cracking of the concrete occurred first in the top of the slab, over the sudden change of section, indicating that the stress concentration effect influences the tensile as well as the compressive stresses at this position. The portion of the drop panel to the left of the change of section is very lightly stressed and could be omitted without detriment to the general stress pattern, that is, this is the obvious position to introduce a splay to effect a gradual change of section.

(c) Slab with splayed drop panel

Fig. 1(*c*) shows a 9 in thick slab of size 10 ft 6 in square centrally supported, via a drop panel, on an encased steel column, and line loaded around the perimeter. The drop panel had an octagonal plan shape but a splayed section shape where the slab is increased linearly from 9 to 12 in over a length of 9 in and maintains this thickness over a further 20 in to the column centre line.

The overall bending analysis was repeated for the modified slab thicknesses but again only minor differences in the relative position of the circle of contraflexure were noted compared with those for (*a*) and (*b*) for the same spacing of the column grid lines.

Using the same loading and column boundary conditions as before, the principal stress plot on the vertical section of the slab is shown in Fig. 3(*c*). The stress distributions show a better 'flow' pattern than that obtaining for the rectangular drop panel shown in Fig. 3(*b*). The concentration of compressive stresses near the column perimeter remain but the ultimate load test indicated that the splayed configuration provided a well balanced design with the soffit concrete at the start of the splay and adjacent to the column perimeter crushing at the point of incipient failure due to punching shear around the column head.

Influence of Boundary Conditions at Slab/Column Interface

The stress plots shown in Fig. 3 are those due to the boundary condition at the slab/column interface being defined as zero vertical displacements. This is equivalent to supporting the slab on a number of rigid rollers permitting free radial displacement at the connection.

In order to study the effects of other boundary conditions on the stress distributions, the uniform thickness slab (no drop panels) was analysed by the same axisymmetric isoparametric element program for three support conditions designated as follows:

(*a*) 'rigid' support — this is identical to that used in Section 3.
(*b*) 'plastic' support — the reaction from the column uniformly distributed across the interface.
(*c*) 'elastic' support — elements were added to include the top portion of the column to represent the full continuity of the connection.

These boundary conditions are shown diagrammatically in Fig. 4 and the graph shows the distribution of vertical stresses across the radius at the interface. The rigid support produces a region of tensile vertical stress but reaches the same order of stress as the 'elastic' stress at the column perimeter when the stress value is about double the average stress value obtained from the plastic support. This result adds weight to modern flat slab design methods in which the column reaction is assumed concentrated around the column perimeter.

The vector stress plots on a radial section for the three support conditions are shown in Fig. 5. Fig. 5(*a*) is identical to Fig. 3(*a*). Fig. 5(*b*) shows no stress concentration effects around the column perimeter because the reactive stresses have been specified as uniformly distributed. The stress distribution within the column for the elastic support (Fig. 5(*c*)) rapidly changes from the non-linear pattern of the interface to a uniform uniaxial state a short distance down the column. Of course, the behaviour could be anticipated from a consideration of the problem in terms of St. Venant's principle.

Apart from affecting the stress distribution around the column head these different boundary conditions can also have a significant effect on the magnitude of

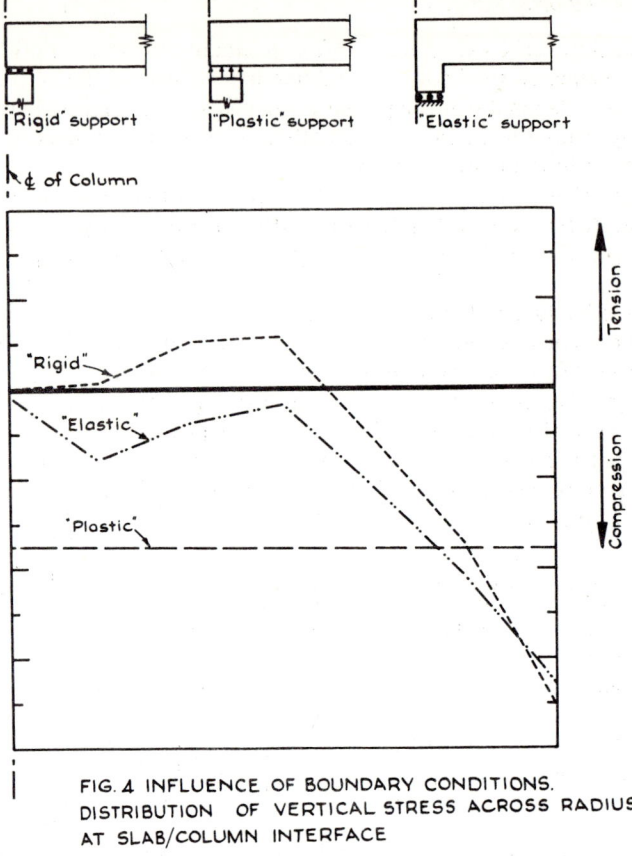

FIG. 4 INFLUENCE OF BOUNDARY CONDITIONS.
DISTRIBUTION OF VERTICAL STRESS ACROSS RADIUS
AT SLAB/COLUMN INTERFACE

the overall slab deflections. In particular, if mid-span deflections in a flat slab system are calculated assuming the columns to provide point supports these values can be as high as twice the values obtained if the finite size of the columns are included in the analysis.

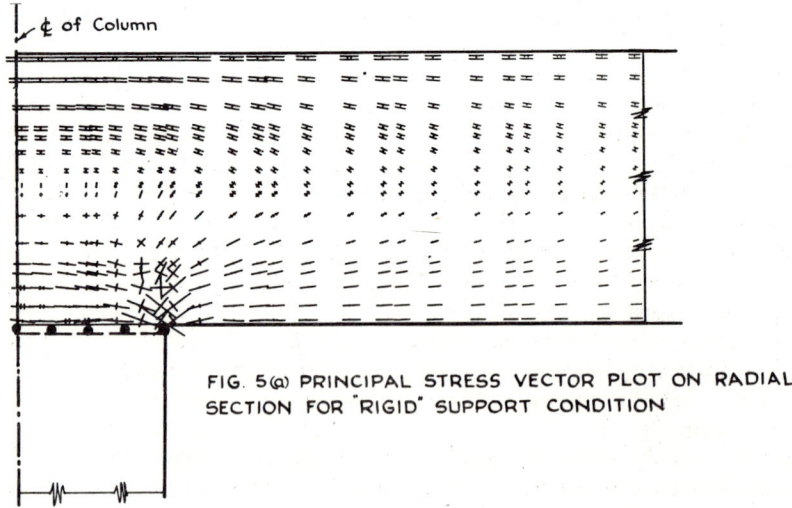

FIG. 5(a) PRINCIPAL STRESS VECTOR PLOT ON RADIAL
SECTION FOR "RIGID" SUPPORT CONDITION

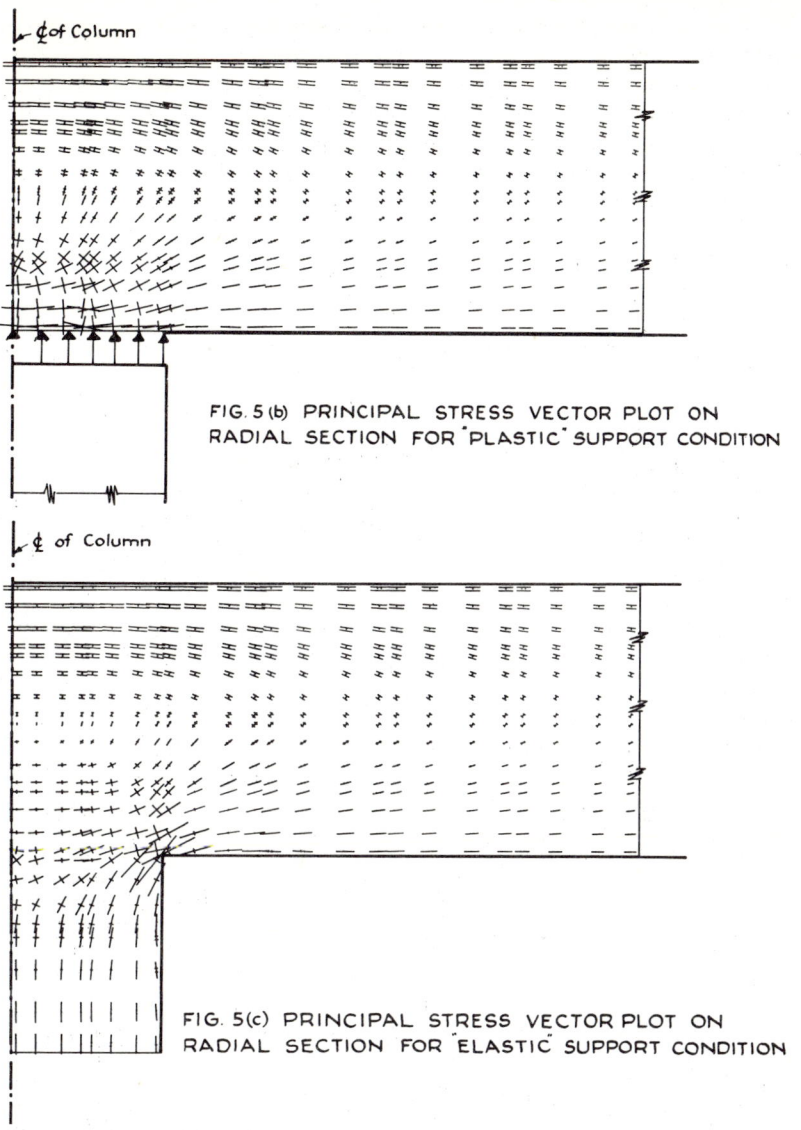

FIG. 5 (b) PRINCIPAL STRESS VECTOR PLOT ON
RADIAL SECTION FOR 'PLASTIC' SUPPORT CONDITION

FIG. 5(c) PRINCIPAL STRESS VECTOR PLOT ON
RADIAL SECTION FOR 'ELASTIC' SUPPORT CONDITION

Influence of Vertical Prestressing Forces

If, for a given uniform thickness slab, the shear stresses around the support point are
excessive, there are generally three methods of improving the situation:

   (i) the total slab thickness is increased,
   (ii) drop panels are added around the column head,
or (iii) shear reinforcement is included to carry the 'excess' shear.

   This section describes the investigation of a fourth possibility: that of providing
vertical prestressing forces to modify the stress distribution in order to reduce the
magnitude of the principal tensile stresses causing shear at the mid-depth of the slab.
This method has been adopted where, for reasons of head room restriction, appearance and economic reasons, the procedures under (i), (ii) and (iii), were not considered suitable.

Fig. 6 shows the part plan of a metre thick slab supported on a column of 0·6 m diameter. The preliminary analysis showed that the shear stresses in the slab would be in excess of those permitted by the design code of slabs with no shear reinforcement. A proposal to prestress the slab by short vertical tendons on concentric circles around the column head was investigated. To simplify the analysis, the individual tendon forces were replaced by statically equivalent circular line loads applying pinching forces to the upper and lower surfaces of the slab at discrete radial

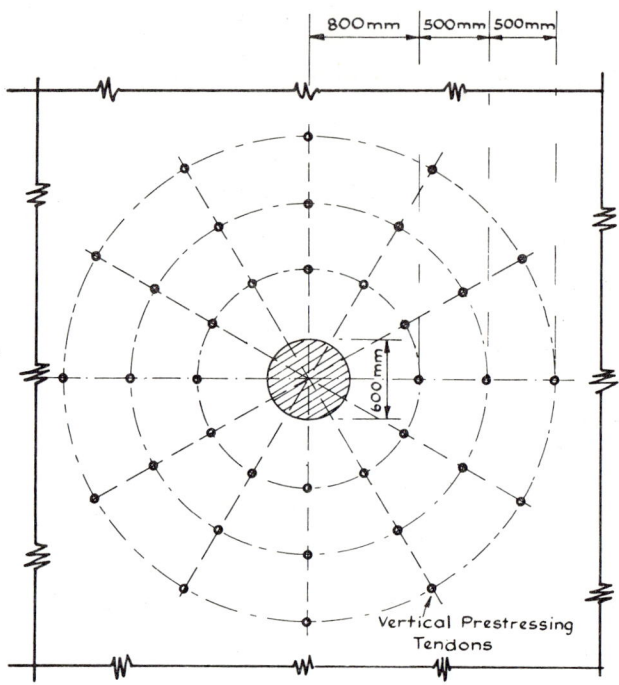

FIG. 6 PART PLAN OF SLAB SUPPORTED ON A COLUMN
AND SUBJECTED TO VERTICAL PRESTRESSING FORCES

intervals. Thus, the problem is once again axisymmetric in character, and the slab was assumed supported on rigid bearings providing vertical restraint only (similar to the 'rigid' support of the previous section).

Fig. 7(*a*) shows the principal stress vector plot for the radial section of the slab adjacent to the column due to the effects of prestress only. This causes high compressive stresses under the prestress application points and these change to horizontal tensile stresses in the body of the slab (it is of interest to note that this behaviour is roughly similar to the stress distribution of horizontal stress across the vertical diametric plane in cylinder splitting tests).

Figs 7(*b*) and 7(*c*) show, to a smaller scale, the principal stress plots on the section due to (*b*) load only and (*c*) load plus prestress. It will be seen that the effect of the prestress is to reduce the principal tensile stresses at mid-depth of the slab along the 45° shear plane (indicated by a dashed line) to about two-thirds of the values developed under load only. This example has been included to demonstrate the types of complex problems that can be studied by the finite element method.

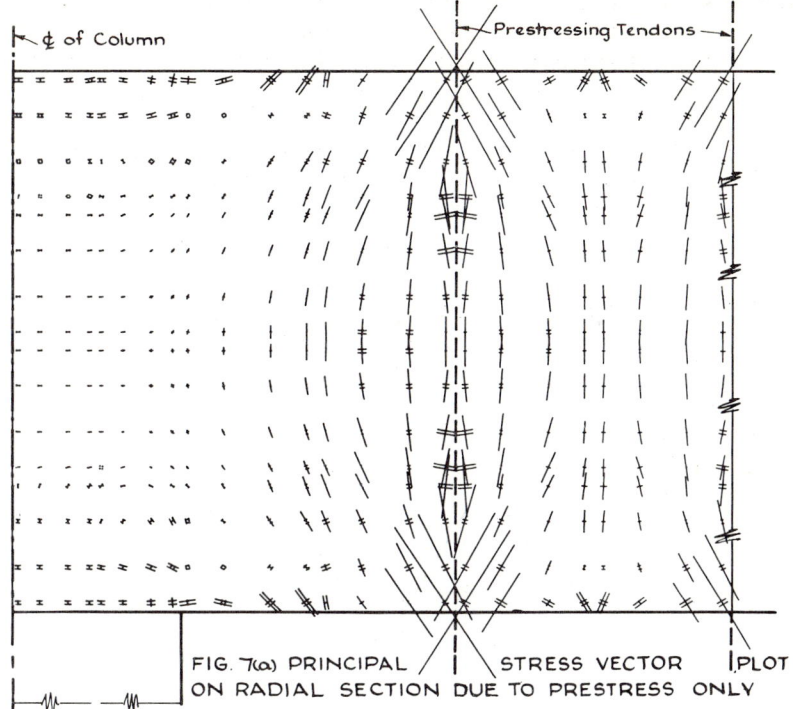

FIG. 7(a) PRINCIPAL STRESS VECTOR PLOT ON RADIAL SECTION DUE TO PRESTRESS ONLY

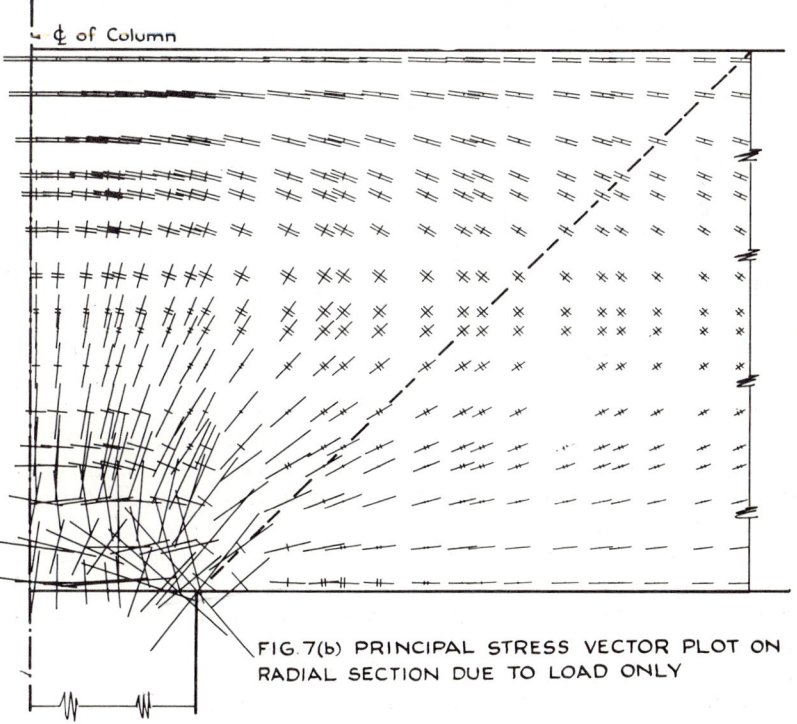

FIG. 7(b) PRINCIPAL STRESS VECTOR PLOT ON RADIAL SECTION DUE TO LOAD ONLY

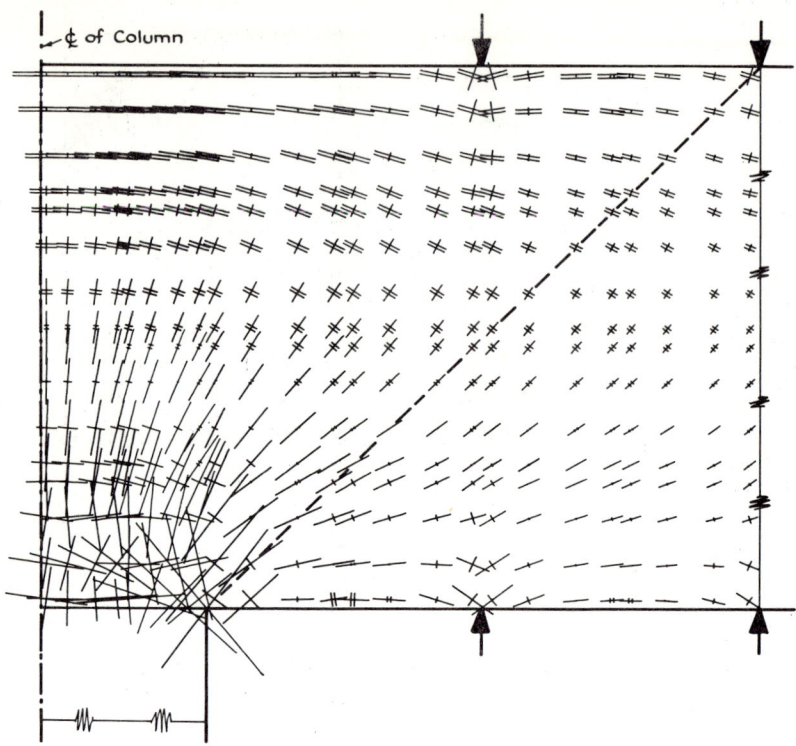

FIG. 7 (c) PRINCIPAL STRESS VECTOR PLOT ON RADIAL
SECTION DUE TO LOAD PLUS PRESTRESS

II. PIER SUPPORTS

Model and Test Arrangements

When one of the section dimensions of a column is large compared with the other it becomes a pier. Piers are frequently used as bridge deck supports and the soffit may be supported on a number of bearing pads fixed at the head of the pier or the pier may be monolithically connected into the slab by continuity reinforcement at the slab/pier junction. This section described tests on the latter type of connection.

Fig. 8(a) shows the general test arrangement for a one-third scale model for a typical slab/pier connection used in an extensive multi-span bridge scheme. The purpose of the test program was to study the behaviour of the connection under-working load conditions for a number of dead load/H.A. loading/H.B. vehicle combinations and then, for a particular loading condition, test the model to failure.

The prototype structure design had been based on the results of the finite element bridge analysis package prepared for the Ministry of Transport. Because of the dominant character of the self-weight loading, the lines of contraflexure in the continuous deck did not vary appreciably for all loading conditions considered. Thus, the length of deck incorporated for the model study was that bounded by these nominal 'no-moment' lines as obtained from the computer analysis.

FIG. 8(a) GENERAL TEST ARRANGEMENT FOR ONE-THIRD SCALE MODEL OF DECK/PIER CONNECTION.

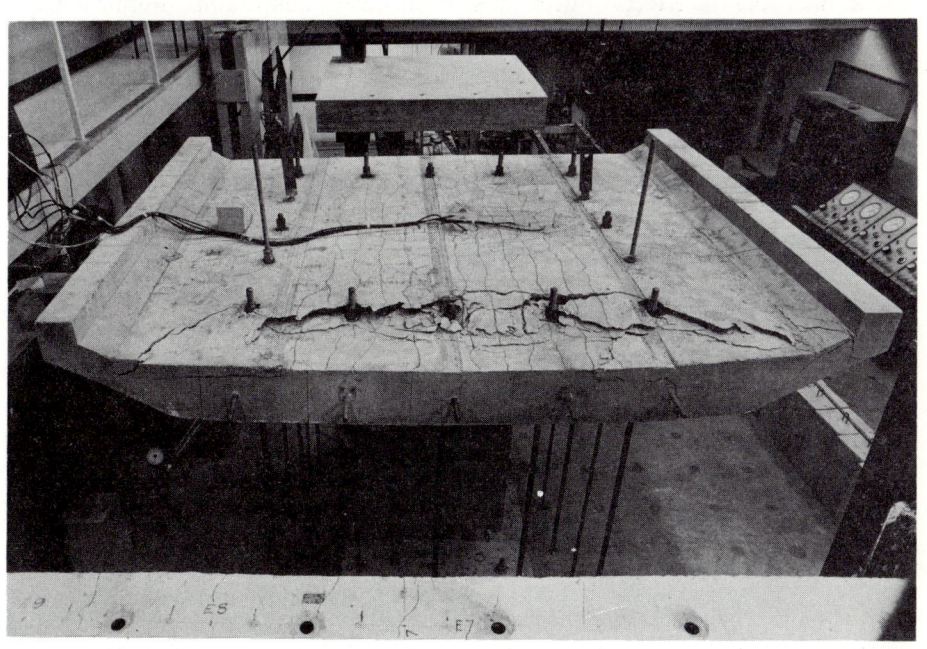

FIG. 8(b) MODE OF FAILURE UNDER ULTIMATE LOAD TEST.

FIG. 8(c) TYPICAL SHEAR FAILURE IN SLAB AROUND PIER HEAD DURING
DESTRUCTION OF MODEL.

Working Load Behaviour

A large number of tests was conducted under the simulated prototype loading conditions, including the effects of deck movements due to temperature causing longitudinal horizontal forces at the top of the pier. However, for brevity, it is only
possible to give some results for a single load case in this section. This is the load
case under which the model was symmetrically loaded, longitudinally and transversely (except for pier strains — see Fig. 9(b)).

The surface and soffit of the slab, and the faces of the pier, were covered with
points at which strains were measured with a Demec demountable gauge. ERS gauges
were attached to a few of the reinforcing bars and vibrating wires were embedded in
the concrete above the pier head. Also, deflections were measured around the perimeter of the slab.

Fig. 9(a) shows the slab strains in the longitudinal direction: line *A* was near the
centre line, line *C* near the edge at the start of the splay and line *B* was at an intermediate position. The top surface strains are larger than the soffit strains because of
the cracking caused by the bending tensile stresses. Some of the largest strains
occurred in the upstand curb on the transverse centre line of the pier. Fig. 9(b)
shows the surface strains near the top, centre and bottom of the pier, these illustrate
the effects of small eccentricities in the loading system.

Ultimate Load Behaviour

The symmetrical loading pattern was also used in the ultimate load test but, due to
small eccentricities in the loading system and distortions in the pier, failure was
initiated on the 'biased' half of the slab.

Fig. 8(b) shows the pattern of cracking and the roughly conical failure mode. The
failure plane was inclined at about 30° to the horizontal and started from the junction of the slab with the pier face (see left hand side of Fig. 8(a)). The model test

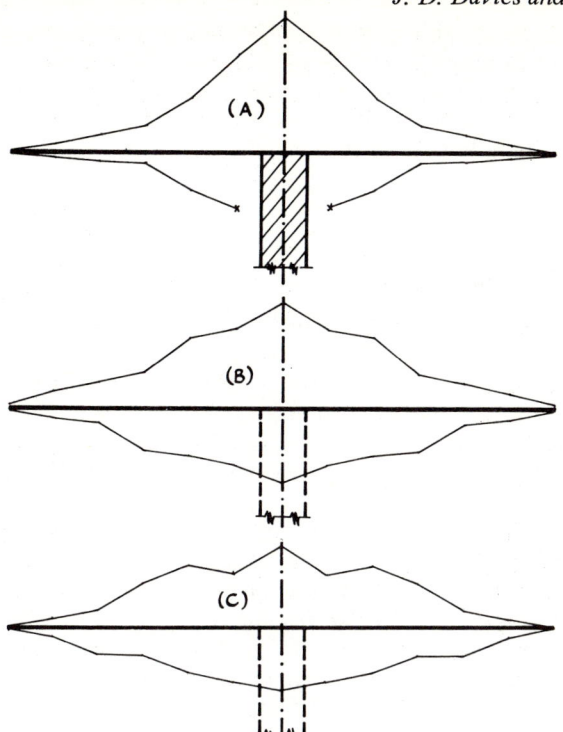

FIG. 9. (a) DISTRIBUTION OF CONCRETE STRAINS IN LONGITUDINAL DIRECTION AT SURFACE AND SOFFIT OF DECK SLAB (SYMMETRIC LOADING).

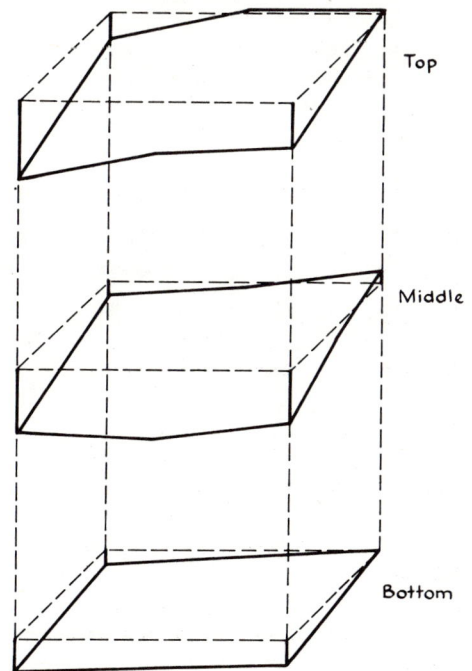

FIG. 9. (b) DISTRIBUTION OF CONCRETE STRAINS IN VERTICAL DIRECTION ON FACE OF PIER (ECCENTRIC LOADING).

was conducted one year after casting and the failure load showed a considerable reserve of strength compared with the specified working loads (a load factor in excess of 4).

Fig. 8(c) shows a typical 45° shear failure around the pier head observed during the subsequent demolition of the model.

III. ABUTMENT SUPPORTS

Skew Plate with Upstand Edge Beams

In the analysis of slabs supported on abutments, a complication arises due to the elasticity of the bearing pads usually interposed between the soffit of the slab and the surface of the abutment. Many finite element programs are available for the analysis of slabs resting on elastic supports and may be employed if the force/deformation behaviour of the bearings is known. When the force/deformation characteristic is not known, it is convenient to obtain data for two extreme conditions.

(a) Non-sinking supports – this is simply obtained by specifying zero vertical displacements at the support points and carrying out the analysis for all the loading cases to be considered for the problem. In effect this 'bound' makes the supports rigid and the slab flexible.

(b) Sinking supports – this other extreme 'bound' is obtained by making the slab initially rigid and the support reactions may be determined, for any particular loading case, from statical conditions only considering the rigid body movements of the slab. The reactions determined in this way may then be considered as loads and applied to the slab (with its correct section properties) and a conventional analysis performed for each of the loading cases considered in (a).

The behaviour of a slab on elastic supports must then lie between the bounds specified by (a) and (b). It is not suggested that this technique should be employed in the detail design of slab structures but it does provide the designer with a tool for

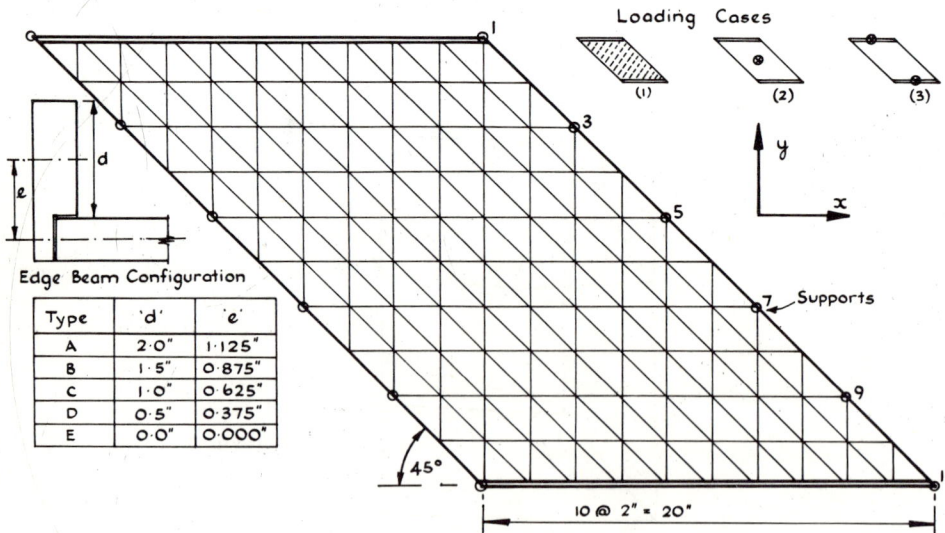

FIG. 10. SKEW PLATE WITH UPSTAND EDGE BEAMS.

assessing the possible effects of support flexibility during the preliminary stages when deciding the spacing of bearings and the thicknesses of slabs.

In order to demonstrate the application of the method, an example has been chosen which introduces two further effects — that of skew and the inclusion of up-stand edge beams. Fig. 10 shows the plan of a 45° skew model plate of span 20 in and width 20 in supported on six equally spaced bearings along each long edge and having upstand edge beams along each short edge. The slab was made from ½ in thick perspex sheet and the edge beams from ½ in wide perspex bars monolithically connected to the edges of the slab. In order to determine the effects of edge beam geometry on the behaviour of the slab and the pattern of bearing reactions, the height of the edge beam above the slab was varied in five steps from 2 in to zero. These are designated A, B, C, D and E and each of these configurations was investigated for three loading cases:

(1) a total uniformly distributed load P,
(2) a central concentrated load P,
(3) two concentrated loads $P/2$ applied at the centres of the two edge beams.

The plan also shows the triangular finite element mesh used in the analysis. Because of the eccentricity of the edge beams, membrane forces as well as bending forces will be developed in the slab/beam system. A five degree of freedom shell program, allowing for eccentrically connected beam elements, was used as the basis of the analytical investigation. The distance between the centroid of the slab and the centroid of the edge beam is designated 'e' and is shown for each of the configurations A, B, C, D and E.

Distribution of Reactions for Non-sinking Supports

From the symmetry of the shape, supports and loading conditions it will be obvious that the sum of the bearing reactions along each supported edge must be $0.5P$. The individual bearing reactions for each configuration for the three loading conditions are given in Tables 1, 2 and 3 respectively. The last row of the tables shows the position of the resultant of the reactions as a ratio of the length of the supported side measured from node 1.

Looking at Table 1 (uniformly distributed load) node 1 is the most highly loaded and its share of the total load increases as the stiffness of the edge beam decreases.

With the concentrated load placed at the centre (Table 2) the most heavily loaded support is at node 3 and its share decreases as the stiffness of the edge beam decreases. Node 3, and its equivalent on the opposite edge, lies on the shortest distance between two supports, that is, the central load is largely carried by the bending action of the slab spanning across this line.

Table 3 shows the distributions for the loads applied at the centres of the edge beams. For large edge beam stiffnesses the reactions at 1 and 11 are of the same order but 1 increases and 11 decreases as sizes are diminished. The negative (uplift) reactions at node 3 will be noted.

For all configurations for the three loading cases the reactions of node 5 show little variation. Also, decreasing the edge beam stiffness results in the resultant of the support reactions moving towards the corner node 1.

Distribution of Reactions for Sinking Supports

In this case the slab is assumed rigid and the supports very flexible. Under symmetric loading the slab will move uniformly downwards causing the same deformation at each support spring. Thus, if the supports are of equal stiffness, they will have equal reactions, that is, the total reaction of $0.5P$ per side (for each loading case) will be provided by a reaction of $0.083P$ at each of the 6 supports. These are listed in the last column of the Tables.

A finite element analysis was carried out for configuration A and E using these reactions as loads. To prevent rigid body movements it was necessary to 'earth' the structure by preventing movements corresponding to the 5 degrees of freedom at any particular node (otherwise, the structure stiffness matrix is singular and cannot be inverted). For convenience, all movements were prescribed zero at the centre node of the slab. Fig. 11 shows the distribution of the M_x moments across the mid-span centre line for loading case (1).

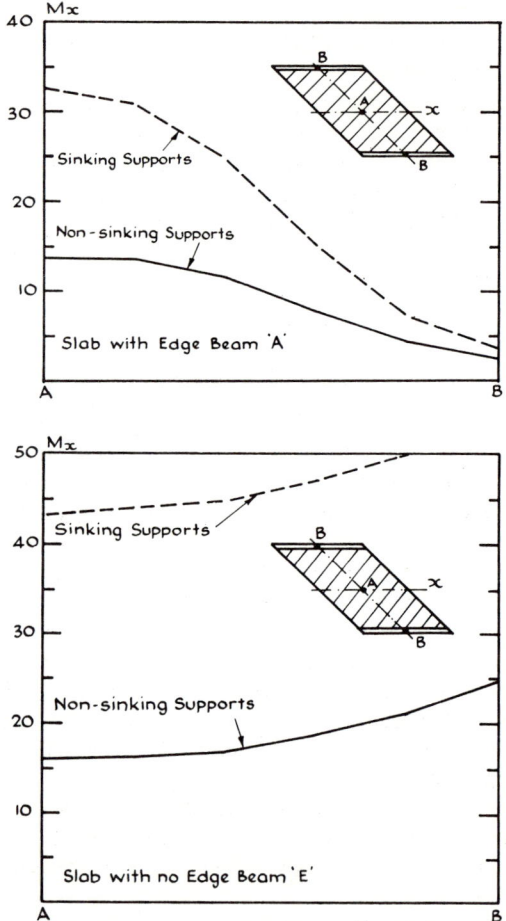

FIG.11 DISTRIBUTION OF M_x MOMENTS ACROSS MID-SPAN CENTRE LINE OF SKEW PLATE FOR LIMITING SUPPORT CONDITIONS

These moments are the bounding values for the two support conditions described above. For most practical bearing characteristics, the behaviour would be expected to be nearer that for non-sinking supports than that predicted by the results for very flexible bearings.

General Conclusions

The paper has described studies of the behaviour of slabs near three types of support.

I. Column supports

For elastic behaviour, geometry of the slab near column supports does not have a marked influence on overall behaviour but has a considerable effect on the load stress patterns. Splayed drop panels promote good stress 'flows' but rectangular drop panels cause stress concentrations at the change of section and, in particular, can cause wide tensile cracks even at working load conditions.

Rigid and elastic support conditions cause the column reaction to be concentrated around the perimeter.

Vertical prestressing of slabs around column heads reduces the principal tensile stresses at the mid-depth of the slab.

II. Pier supports

Monolith slab/pier connections do not significantly effect the flexural behaviour of the slab. The failure mechanism for the particular model displayed an inclined mode associated with bending/shear interaction.

III. Abutment supports

The magnitude and distribution of bearing reactions for a slab/beam system is influenced by the relative geometric configuration. The solution for elastic supports will be bounded by the solutions for non-sinking and sinking supports. Fuller details of these studies may be obtained from the sources given under references.

Acknowledgements

The examples described in this paper have been selected from a number of separate studies.

The work described in section, "Influence of Slab Geometry", is based on the results of studies of the behaviour of reinforced concrete slabs connected, via canti-lever stubs, to structural steel columns and carried out for Messrs Bingham, Blades and Partners, Consulting Engineers.

The effects of vertical prestressing were carried out for Messrs Mott, Hay and Anderson, Consulting Engineers.

The slab/pier model test was conducted for Messrs R. Travers, Morgan and Partners, Consulting Engineers.

Much of the finite element work was based on programs developed (or being developed) on behalf of the Ministry of Transport.

The authors gratefully acknowledge the co-operation of the above and, in particu-lar, wish to thank their colleagues at the University of Wales, Swansea, for various types or assistance over many years.

TABLE 1

Support reactions at abutments of skew plate
Loading (1): total uniformly distributed load $P = 1$

Support Node	Non-sinking Supports Edge beam configurations					Sinking Supports All configurations
	A	B	C	D	E	
1	0·123	0·126	0·131	0·142	0·163	0·083
3	0·111	0·109	0·102	0·089	0·066	0·083
5	0·091	0·091	0·092	0·093	0·093	0·083
7	0·071	0·072	0·073	0·077	0·083	0·083
9	0·042	0·042	0·046	0·056	0·078	0·083
11	0·062	0·060	0·056	0·043	0·017	0·083
Relative position of resultant from node 1	0·394	0·391	0·388	0·378	0·359	0·500

TABLE 2

Support reactions of abutments of skew plate
Loading (2): central concentrated load $P = 1$

Support Node	Non-sinking Supports Edge beam configurations					Sinking Supports All configurations
	A	B	C	D	E	
1	0·048	0·049	0·053	0·061	0·081	0·083
3	0·274	0·273	0·269	0·259	0·237	0·083
5	0·147	0·148	0·149	0·151	0·157	0·083
7	0·016	0·016	0·018	0·021	0·028	0·083
9	0·005	0·004	0·002	0·002	0·000	0·083
11	0·010	0·010	0·009	0·006	−0·003	0·083
Relative position of resultant from node 1	0·274	0·273	0·270	0·265	0·248	0·500

TABLE 3

Support reactions of abutments of skew plate

Loading (3): two concentrated loads $P/2 = 0.5$ applied at centres of edge beams

Support Node	Non-sinking Supports Edge beam configurations					Sinking Supports All configurations
	A	B	C	D	E	
1	0·268	0·273	0·291	0·326	0·374	0·083
3	−0·021	−0·027	−0·047	−0·088	−0·144	0·083
5	0·001	0·001	0·001	0·001	−0·014	0·083
7	0·004	0·006	0·010	0·018	0·038	0·083
9	0·019	0·025	0·046	0·094	0·204	0·083
11	0·229	0·222	0·199	0·149	0·042	0·083
Relative position of resultant from node 1	0·486	0·481	0·466	0·436	0·387	0·500

REFERENCES

General
1. G.B. Ministry of Transport. *Finite element analysis of slab bridges*. BECP/1, 1969.

Column supports
2. *Tests on concrete slab/steel column connections*. University of Wales, Swansea, Research Report C/R/136/71, 1971.

Pier supports
3. *Tests on a 1/3 scale slab/pier connection*. University of Wales, Swansea, Research Report C/R/137/71, 1971.

Abutment supports
4. *Behaviour of a skew plate with edge beams*. University of Wales, Swansea, Research Report C/R/138/71, 1971

PUNCHING OF BRIDGE SLABS

HENRIK NYLANDER and HÅKAN SUNDQUIST

The Royal Institute of Technology, Stockholm

INTRODUCTION

This study forms part of an investigation which is in progress at the Department of Building Statics and Structural Engineering, The Royal Institute of Technology, Stockholm. The purpose of the investigation is to provide a basis for the design regarding punching of bridge slabs carried on columns. Bridge slabs carried on circular columns are very commonly used in Sweden.

The investigations are partly financed by Statens Vägverk (The Swedish National Road Administration).

Punching of Slabs has been studied very thoroughly at the department in earlier investigations. Test results and theoretical investigations are presented in Kinnunen — Nylander,[1] Kinnunen,[2] Andersson,[3] and in "Proposed Regulations for Design of Concrete Flat Slabs and Flat Plates. Comments."[4]

Outline of Tests

Ten different slabs were subjected to investigation in laboratory tests. Five of these slabs were provided with shear reinforcement. The flexural reinforcement in the direction of the largest bending moment and the dimensions of the slabs were equal for all slabs tested.

The aim of the tests was to investigate the influence on the punching load of deviation of the moment distribution from polar symmetry, variation of the amount of flexural reinforcement in the direction of the least bending moment and of shear reinforcement consisting of vertical stirrups. The reinforcement was designed so that punching would occur before yielding of the flexural reinforcement.

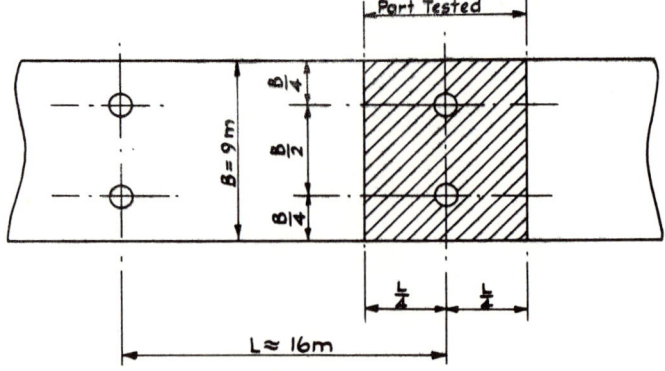

FIG. 1. PLAN VIEW OF A REPRESENTATIVE BRIDGE CONSTRUCTION

Details of Tests

The test specimens are supposed to represent the part at the support of a bridge with a span of about 16 m, width 7–9 m and thickness 60–80 cm. The bridge is resting on two columns at each support according to Fig. 1.

The columns are supposed to be about 6 m high and have a diameter equal to the slab thickness.

The test specimens were made with a scale factor of about 1:5 to 1:7.

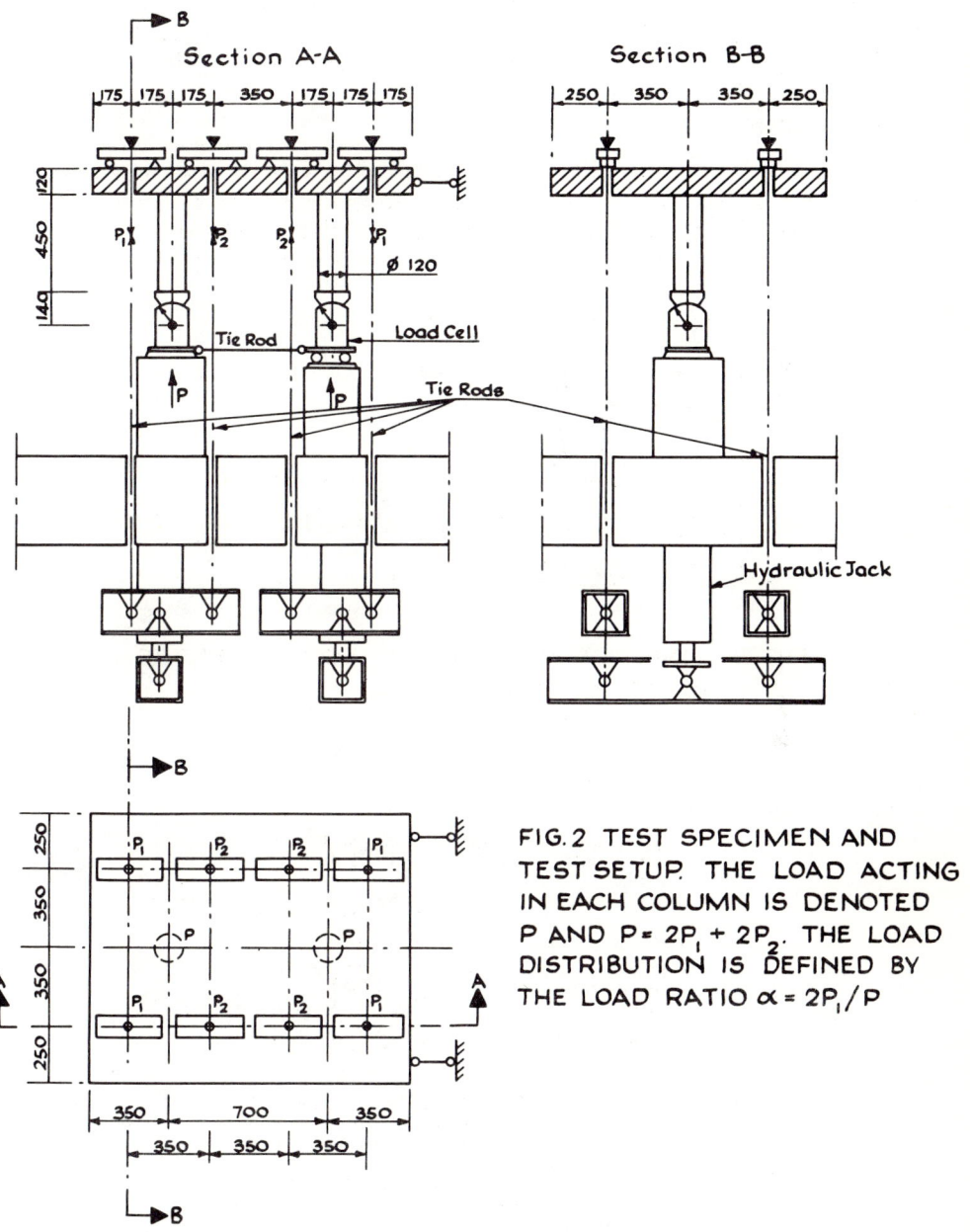

FIG. 2 TEST SPECIMEN AND TEST SETUP. THE LOAD ACTING IN EACH COLUMN IS DENOTED P AND P = 2P$_1$ + 2P$_2$. THE LOAD DISTRIBUTION IS DEFINED BY THE LOAD RATIO α = 2P$_1$/P

The actual load acting on the bridge was in the tests replaced by symmetrical line loads perpendicular to the longitudinal direction of the bridge. The ratio between the point loads forming the line loads was varied in the different tests in order to simulate different distributions of traffic loads. The testing arrangement is shown in Fig. 2. The slabs with load distribution according to the load ratio $\alpha = 2P_1/P = 0.5$ and 0.25 (notations are shown in Fig. 2) were tested with two different amounts of flexural reinforcement in the direction of the least bending moment.

For each alternative of load distribution and distribution of flexural reinforcement two slabs were tested, one slab without and one slab with shear reinforcement. The reinforcement bars used were made of Swedish standard kam steel KS 60 ribbed bars of diameter 8 mm (ϕ 8) and 10 mm (ϕ 10). The yield point stress for the bars used was 6900–7400 kp/cm^2. A list of the different slabs is presented in Table 1.

Testing Procedure

The testing arrangements are shown in Fig. 2. During the tests, the slabs were loaded in several stages at intervals of 10 minutes between two consecutive load steps. The load was increased at a rate of 1.5 tons per column and step until failure.

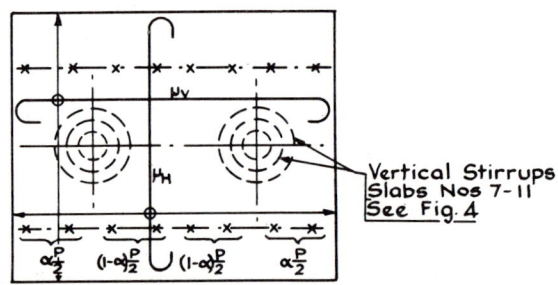

FIG.3 REINFORCEMENT LAYOUT FOR THE SLABS
LISTED IN TABLE I

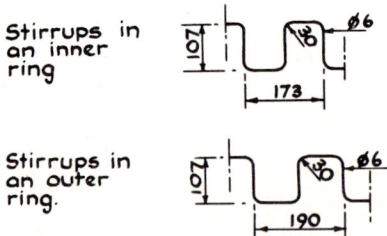

FIG. 4. THE SHEAR REINFORCEMENT FOR SLABS
NOS. 7-11 CONSISTED OF VERTICAL
STIRRUPS ϕ6 IN TWO RINGS CONCENTRI-
CALLY LOCATED AROUND THE COLUMN.
(THE FLEXURAL REINFORCEMENT BARS
WERE THREADED THROUGH THE UPPER
SHANKS OF THE STIRRUPS).

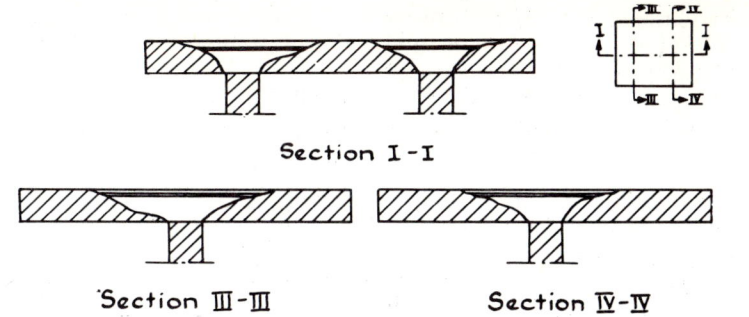

Section I-I

Section Ⅲ-Ⅲ   Section Ⅳ-Ⅳ

FIG. 5. THE SHEAR CRACK OBSERVED AT FAILURE FOR SLAB No. 4

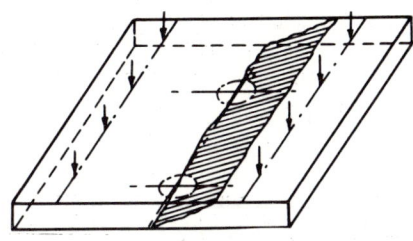

FIG. 6. PRINCIPAL OUTLOOK OF SHEAR CRACK FOR THE SLABS PROVIDED WITH SHEAR REINFORCEMENT.

Test Results

In this short report only test results concerning the ultimate load will be described. More details concerning the tests are to be published.

In Table 2 the principal test results are presented.

The type of failure of the slabs without shear reinforcement was almost the same as for the circular slabs described in ref. [1], i.e. the shear crack observed at failure had the form of a flat cone with almost equal slope in the different directions. Fig. 5 shows a typical view of the slab portion outside the shear crack.

The failure of the slabs with shear reinforcement was almost the same as if the slab had been supported along a line through the heads of the columns, i.e. the slab behaved as a short beam supported in the middle and loaded by two equal symmetrically placed loads, see Fig. 6.

Due to the concentration of stresses to the part of the slabs close to the columns the mean shearing stress at failure τ_{mean} was rather low. Tests made on plate strips with the same shear span, the same percentage of reinforcement and the same concrete mix as for the slabs, gave mean values of shearing stress at failure about 15 to 20 per cent higher than τ_{mean} for the slabs.

A comparison between the different slabs in the test gives rise to the following conclusions.

(a) The ultimate load of the slabs with shear reinforcement was about 35 to 45 per cent greater than for the slabs without shear reinforcement.

(b) The additional flexural reinforcement in the direction of least moment in slabs 3, 6, 8 and 11 did not increase the ultimate load compared with the corresponding slabs without this additional reinforcement.

(c) The ultimate load of the slabs with non-uniform load distribution was about 10 per cent lower than the ultimate load of the slabs tested with uniform load distribution.

The two alternatives for non-uniform load distribution gave approximately the same ultimate load. This is remarkable considering that the measured and calculated load eccentricity in the upper part of the columns was of a quite different magnitude for the slabs with load ratio $\alpha = 3/4$ $(e/d \approx 0.3)$ than for the slabs with load ratio $\alpha = 1/4$ $(e/d < 0.05)$ (e is the load eccentricity and d the diameter of the columns.)

Continued Tests

Continued tests are planned. Slabs with a lower amount of flexural reinforcement and greater thickness (24 cm) will be studied. Because of the favourable effect of shear reinforcement observed in the tests, an attempt will be made to develop appropriate shear reinforcement for bridge slabs.

TABLE 1

Slab No.	Load ratio $\alpha = \dfrac{2P_1}{P}$	Mean value of effective thickness h (cm)	Reinforcement in the longitudinal direction μH	Reinforcement in the transverse direction μV	Shear reinforcement
2	1/2	9·75	ϕ10c70, 1·10%	ϕ 8c100, 0·54%	–
3	1/2	9·60	ϕ10c70, 1·11%	ϕ10c70, 1·21%	–
4	3/4	9·65	ϕ10c70, 1·11%	ϕ 8c64, 0·85%	–
5	1/4	9·75	ϕ10c70, 1·10%	ϕ 8c200, 0·27%	–
6	1/4	9·80	ϕ10c70, 1·09%	ϕ10c70, 1·21%	–
7	1/2	9·75	ϕ10c70, 1·10%	ϕ 8c100, 0·54%	20ϕ6, 5·66 cm^2
8	1/2	9·70	ϕ10c70, 1·10%	ϕ10c70, 1·21%	,,
9	3/4	9·85	ϕ10c70, 1·09%	ϕ 8c64, 0·85%	,,
10	1/4	9·85	ϕ10c70, 1·09%	ϕ 8c200, 0·27%	,,
11	1/4	9·80	ϕ10c70, 1·09%	ϕ10c70, 1·21%	,,

TABLE 2
Test results

Slab No.	Load ratio α	σcube (1)	μV (%)	μH (%)	P_{su} (2)	Type of failure (3)	P_u (4)	τ_{nom} (5)	$\dfrac{\tau_{nom}}{\sigma_{tens}}$ (6)	τ_{mean} (7)	f (8)
2	1/2	271±22	1·10	0·54	–	1+2	18·6	27·9	1·33	–	–
3	1/2	266±26	1·11	1·21	–	2	18·7	28·7	1·38	–	–
4	3/4	272±28	1·11	0·85	–	2	16·3	25·6	1·21	–	–
5	1/4	255±26	1·10	0·27	–	1	15·4	23·2	1·15	–	–
6	1/4	272±24	1·03	1·21	–	1	16·9	25·1	1·19	–	–
7	1/2	268±27	1·10	0·54	29·3	B	25·3	37·9	1·81	17·7	1·36
8	1/2	272±21	1·10	1·21	28·6	B	26·1	39·5	1·87	18·3	1·36
9	3/4	280±12	1·09	0·85	28·4	B+2	24·1	35·6	1·66	16·7	1·37
10	1/4	260±11	1·09	0·27	28·9	B	23·0	34·0	1·66	16·0	1·44
11	1/4	293±17	1·09	1·21	28·5	B	24·5	36·4	1·64	17·1	1·38

Notations

(1) Mean value and standard deviation for eight $15 \cdot 15$ cm^2 cubes.
(2) P_{su} is the vertical force in the shear reinforcement calculated under the assumption that the stress in the shear reinforcement is equal to the yield point stress.
(3) Failure type "1" or "2" refers to failure due to punching of the slab at column 1 or 2 respectively. Failure type B refers to beam type shear failure over the whole width of the slab.
(4) Ultimate load measured in metric tons.
(5) Nominal shear stress at failure $\tau_{nom} = \dfrac{P}{\pi h(d+h)}$.
(6) $\sigma_{tens} = 7 \cdot 5 + \sigma_{cube}/20$, from the cube strength calculated tensile strength of concrete.
(7) Mean value of shearing stress at failure measured over the whole plate width 140 cm.
(8) Value of τ_{nom}/σ_{tens} for slabs with shear reinforcement divided by τ_{nom}/σ_{tens} for the corresponding slab without shear reinforcement.

REFERENCES

1. Kinnunen, S. and Nylander, H. Punching of concrete slabs without shear reinforcement. *K. tek. Högsk. Handl.*, No. 158, 1960.
2. Kinnunen, S. Punching of concrete slabs with two-way reinforcement. . . . *K. tek. Högsk. Handl.*, No. 198, 1963.
3. Andersson, J. L. Punching of concrete slabs with shear reinforcement. *K. tek. Högsk. Handl.*, No. 212, 1963.
4. Statens Betongkommitté (Swedish State Concrete Committee). *Förslag till bestämmelser för dimensionering av betongplattor på pelare, jämte utdrag ur kommentarer* (Proposed regulations for design of concrete flat slabs and flat plates. Comments.). Stockholm, Svensk Byggtjänst, K 1, 1964.

RESEARCH ON SLAB TYPE AND SPINE BEAM BRIDGES

R. E. ROWE and G. SOMERVILLE

Cement & Concrete Association

SYNOPSIS

The scope of the work of the Cement and Concrete Association on slab type and spine beam bridges is described. The essential aspects of the determination of the parameters governing the behaviour of the bridges are discussed and, where possible, recommendations are made on parameter evaluation in design. The primary object of the work, which is supported by the Construction Industry Research & Information Association, is the evolution of simplified design procedures.

INTRODUCTION

The interest in slab type and spine beam structure has increased considerably in recent years as a result of the needs of urban motorways and has been accompanied by a proliferation in analytical techniques, mostly associated with computers, available for these classes of structure. Unfortunately, the available experimental data on the actual behaviour of slab and spine beam type structures are limited. Hence the relative accuracy of different types of analysis in predicting the actual behaviour is not known nor can any reasonable recommendation be made in relation to acceptable simplifications in the mathematical models utilised in the computer approaches.

There was a considerable need, therefore, for experimental research to provide the necessary data to enable the analytical techniques to be assessed; the aspects the authors are concerned with relate specifically to structures built in reinforced and prestressed concrete. The experimental work falls essentially into the category of parameter evaluation for cracked reinforced concrete or jointed precast prestressed concrete structures, relevant to appropriate mathematical models. An important aspect, which must be emphasised, is the need to make recommendations which enable the designer to adopt the simplest possible design approach consistent with the importance of the structure and its specified performance criteria.

Over the past three years an extensive programme of research has been carried out by the Cement and Concrete Association with support from the Construction Industry Research and Information Association in an attempt to produce definitive recommendations. This research has not yet reached the final reporting stage and this paper is intended to discuss some of the major aspects of research, to indicate some of the experimental findings and to consider those areas in which definite recommendations can be made.

SLAB TYPE BRIDGES

In any consideration of slab type bridges the essential problems to be studied are:

  (i) The definition of the basic parameters affecting the actual behaviour of the bridge; these may be defined as the flexural and torsional parameters and are related to the method of construction;

 (ii) the boundary support conditions, i.e. simple, continuous, line or discrete;

(iii) the loading and its history where this may have a significant effect on the the response of the bridge.

The investigations carried out have been concerned with the conventional HA & HB loading of the Department of the Environment (Ministry of Transport); the support conditions have been predominantly those of simple line supports; and the forms of construction conventional *in-situ* reinforced concrete and composite precast prestressed concrete with *in-situ* reinforced concrete.

COMPOSITE SLAB BRIDGE

The form of construction adopted for this bridge was typical of that currently employed to produce fully composite slabs in the span range up to about 16 m but aimed at producing a voided slab, hence extending the span range up to 29 m by virtue of the saving in self-weight. A quarter-scale model of a 22 m span bridge was constructed comprising 18 precast prestressed inverted T-beams, placed side by side, with transverse reinforcing steel passing through holes in the webs at 200 mm centres. *In-situ* concrete was placed between the beams to give an appropriate cover to the transverse steel and finally a doubly reinforced top slab was cast on asbestos-cement sheet as permanent shuttering. The essential details of the individual composite beam are given in Fig. 1 and a general view of the deck in Fig. 2.

Deflection and strain readings were taken throughout the testing enabling the distribution of deflection, longitudinal curvature, and transverse strains on both the reinforcing steel and the concrete. A model HB loading vehicle was applied in a

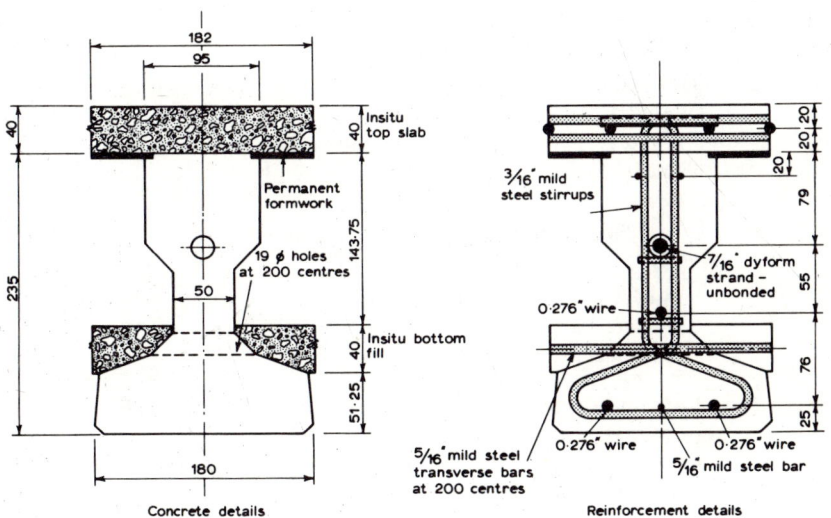

FIG. I   DETAILS OF LONGITUDINAL COMPOSITE BEAM SECTION

FIG. 2

central and eccentric position on the deck and increments of load applied up to the service load. Subsequently the load was increased to 1·25 times the service load and then up to failure. Full details of this investigation will be given by Manton.[1]

For the purposes of this paper, only a limited amount of the experimental data is considered in relation to various possible analytical treatments of this type of deck. For this type of deck, orthotropic plate analysis in the form of the load distribution method is relevant; in applying this method various assumptions may be made. These may be defined as:

Analysis I   — Transverse bottom steel and *in-situ* concrete ignored and deck considered as a T-beam bridge. The flexural and torsional parameters, θ and α respectively, are determined on this basis and are as given in Table 1.

TABLE 1

Relevant flexural and torsional parameters for different assumptions.

Analysis	I	II	III	IV
Flexural parameter	1·20	0·39	0·59	0·68
Torsional parameter	0·41	0·36	0·71	0·51

Analysis II — Bottom *in-situ* concrete considered for derivation of θ and torsional properties are based on thin-wall box theory. Effective thickness of bottom flange *in-situ* concrete taken as that to bottom of the transverse steel. Relevant parameters are given in Table 1.

Analysis III — Longitudinal flexural stiffness as for Analysis II. Transverse flexural stiffness based on top slab and lower transverse steel with a limited amount of 'tension stiffening' associated with each reinforcing bar.

Torsional parameter based upon median perimeter of box in both longitudinal and transverse directions and a mean value adopted.

Analysis IV — Full longitudinal section properties used, including transformed steel sections. Transverse section based on top slab and all steel. Torsional properties based on infinitely wide thin wall section, using only *in-situ* concrete, i.e. two coupled laminae.

The flexural and torsional parameters derived on these various assumptions are given in Table 1. A comparison between the measured distribution coefficients for longitudinal moments and those assessed from the various analyses is given in Figs 3 and 4 for the eccentric and central loading positions of the HB vehicle. From this comparison, it is apparent that if the assumptions made in Analysis III are adopted a satisfactory assessment of the longitudinal moments is obtained. In addition, it was found that the transverse moments, and stress were predicted with satisfactory accuracy. As a result of the tests on this model bridge, the Department of the Environment (Ministry of Transport) agreed to this basis for the analysis and design of decks incorporating the standard MoT/C&CA 'M' Beams.[2]

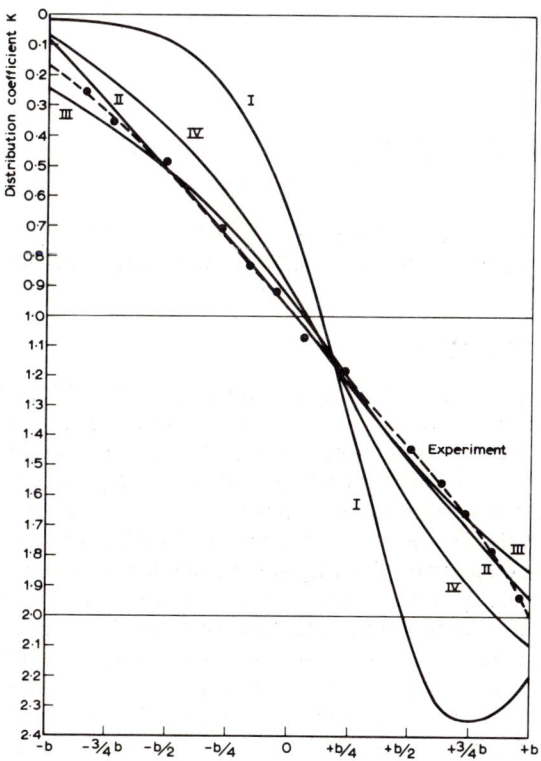

FIG 3 DISTRIBUTION COEFFICIENTS
FOR LONGITUDINAL
MOMENTS-ECCENTRIC LOADING.

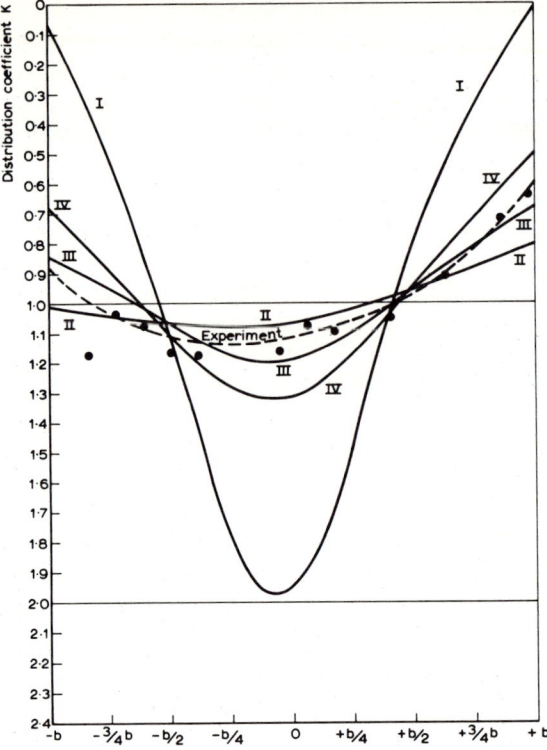

FIG. 4 DISTRIBUTION COEFFICIENTS FOR
LONGITUDINAL MOMENTS - CENTRAL LOADING.

IN SITU R.C. VOIDED SLAB

The second model slab had the same plan dimensions as the deck previously discussed
and effectively the same construction depth; it was intended to be a quarter-scale
model of an actual bridge. The voids in the deck were circular with a diameter of
0·78 times the overall depth of the slab and at centres equivalent to the overall depth.

In this investigation, deflections, strains and reactions are being recorded under a
simulated HB vehicle. To ensure that the effect of traffic is accounted for correctly, a
comprehensive loading history is being applied and the response of the deck defined in
relation to it. Obviously this test will cover all aspects of the behaviour of reinforced
concrete voided slabs up to failure; ancillary work on plastic models is being carried
out to supplement the findings from the point of view of the elastic response of such
structures. Four small scale slabs, having circular voids with diameters of 0·47, 0·67
and 0·8 of the overall depth with corresponding spacings of 0·72, 0·9 and 1·08 times
the overall depth are being tested under point loads and also under pure torsion so
that all the relevant parameters can be defined.

SIGNIFICANCE OF SKEW

The work described so far has been concerned with right bridges and needs to be
extended to cover the skew case. It was considered desirable to start with the solid

slab, treating the efficiency of reinforcing, or prestressing, from a number of different viewpoints namely:

   (i) efficiency of reinforcing for an elastic moment field;
  (ii) efficiency of reinforcing for adequate crack control in reinforced concrete;
 (iii) efficiency of reinforcing for desired ultimate strength.

The first aspect has been dealt with by Clark;[3] in effect the investigation was concerned solely with the amount of reinforcement required for a defined moment field (obtained experimentally) when different arrangements of the longitudinal and transverse steel were adopted. From this study it was clear that orthogonal reinforcement was the most efficient and, of the possible arrangements, that with steel parallel and perpendicular to the abutments was better except for slabs with an angle of skew greater than 30° and an aspect ratio (skew width/skew span) less than 1·0. The use of orthogonal reinforcement, as opposed to the conventional arrangement parallel to both supports and free edges, results not in small savings in steel but very considerable savings which may, in certain cases, be up to 90%. Obviously therefore when the quantities of steel involved are so vastly different, a careful appraisal should be made of the true costs of fixing the reinforcement in the relevant directions so that overall cost of construction is minimised.

The second aspect is being studied for both right and skew slabs on micro-concrete models appropriately reinforced. Associated with this work has been some fundamental research to establish the similitude with regard to cracking in reinforced concrete slabs constructed to different scales; a report on this investigation has been prepared by Clark.[4]

The third, and final, aspect has been covered adequately, at least for solid slabs, in the past work of the Association;[5] the data from both the composite and the *in-situ* slabs, already discussed, should enable the yield line method to be adapted for these construction techniques.

INTERIM RECOMMENDATIONS ON ANALYTICAL TECHNIQUES

Associated with the experimental work described above, various analytical procedures are being considered as tools in predicting the behaviour of slab type bridges. Attention has so far been focussed on conventional orthotropic plate theory (manual) and grid-analysis (computer). Certain recommendations, of an interim nature, have been reported by West;[6] these are concerned with the selection of the longitudinal and transverse beams to represent the actual slab structure and the definition of the stiffnesses associated with them. Subsequently, when all the detailed analysis of the results has taken place, it should be possible to cover all the relevant limit states and hence provide guidance for future bridge codes.

It should be emphasised that, while the interim results described briefly in this paper relate to particular forms of construction or to particular methods of analysis, the aim of the entire programme of research is to make recommendations that are general to all forms of slab construction. In the case of parameter evaluation for example, it is hoped that methods can be evolved that are applicable to all types of section, whether these are solid or hollow, reinforced or prestressed, constructed entirely *in situ* or are partly precast; ideally the methods should also be capable of being used in any type of analysis.

SPINE BEAM BRIDGES

The analysis and design of slab bridges have evolved slowly over a substantial period of time; only a limited number of analytical methods has been used but these have generally been developed and simplified for rapid use in the design office by the production of design aids as the result of extensive experimental work over a wide range of bridge types. As has been mentioned in the first part of this paper, the major problem at the present time is the determination of realistic values for the stiffness parameters associated with modern types of bridge cross-section to be used in these design methods. With spine beam bridges on the other hand, there is a very wide range of analytical methods available, as has been demonstrated by Maisel[7] in his review of literature related to thin-walled beams; these methods have not been

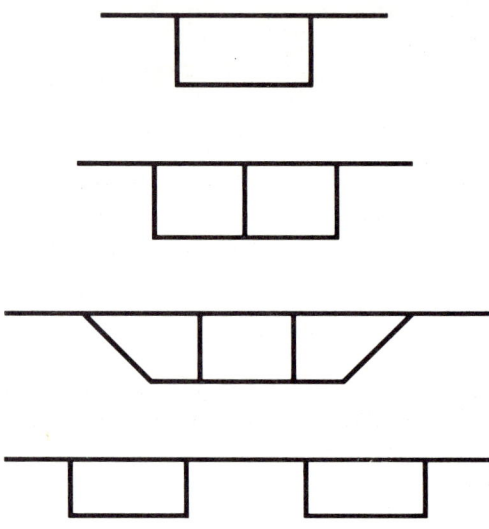

FIG. 5 TYPICAL CROSS-SECTIONS FOR SPINE-BEAM BRIDGES.

developed or simplified to anything like the same extent largely because experimental data, for comparison purposes, are limited to that obtained from a few model tests on individual bridges.

The essential difference between spine beam bridges and other bridge superstructures is due mainly to differences in plan geometry. Spine beam bridges as built in practice may be defined as structural members whose breadth and depth are small in relation to their length and which are therefore subjected mainly to longitudinal bending, transverse shear and torsion. They are generally stiff members whose cross-section consists of a hollow box beam having one or more cells, with or without cantilevers; typical cross-sections of bridges of this type which have been built in this country are shown in Fig. 5 — transverse diaphragms are normally provided only over the supports. The bridges are generally prestressed longitudinally and reinforced transversely unless they are exceptionally wide when they may be transversely stressed. This form of construction is most commonly used in viaducts where, depending on site conditions, they may be built *in situ* or using segmental construction and may be either continuous or simply supported.

The proportion of the bridges built so far have been such that it has been considered reasonable to treat them as beams in designing the longitudinal prestress and in providing the secondary reinforcement to resist vertical shear and torsional shear stress; in the latter case the cross-section is treated as being thin-walled. Transverse strength in the cantilevers and in the flanges of the box is provided on the basis of an analysis for local effects such as that due to Westergaard;[8] if the box is multi-cell in nature, it is the usual practice to idealise the cross-section as a plane frame with prismatic members in order to calculate the transverse effects — this technique has been shown to give reasonable co-relation with test results.[9] The behaviour of a number of bridges designed in this way has been studied using models;[10,11,12,13] generally the tests have been carried through to failure.

The techniques outlined briefly above have been used successfully to design a number of spine beam bridges in this country, but there is still considerable scope for improvement; design methods are still at an early stage of development. The types of section used are determined by trial and error and sufficient experience has not yet been developed for designers to be able to optimise the design to meet a particular set of circumstances. If better use is to be made of our improving material technology

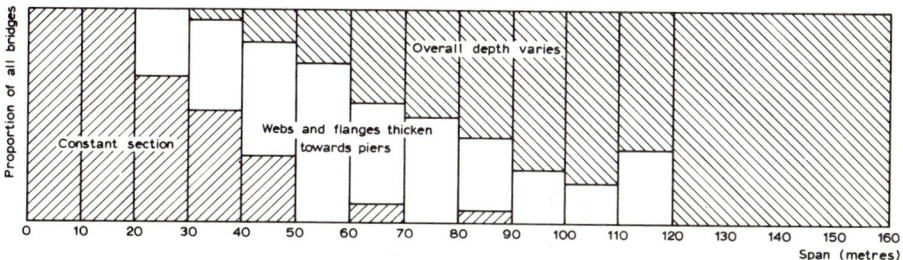

FIG. 6 HISTOGRAM FROM SURVEY OF GEOMETRY OF 150 SPINE BEAM BRIDGES, SHOWING HOW THE TYPE OF LONGITUDINAL GEOMETRY VARIES WITH SPAN.

and construction techniques in the future then it is necessary to know more — more about behaviour generally and, in particular, more about the limiting conditions beyond which simple design methods are no longer sufficiently accurate. As part of an investigation into the behaviour of all types of section used in elevated road structures, the Cement and Concrete Association has, with support from CIRIA, spent some considerable time over the past three years in systematically studying the behaviour of, and developing analytical and design methods for, spine beam bridges generally, to add to the knowledge previously obtained from model tests on individual structures.

The first problem that faces the designer is the initial proportioning of his bridge once he has determined the span and overall width. An examination has been made of the geometry of more than 150 spine beam bridges that have been built throughout the world over the past ten years or so to establish some rules for selecting a particular type of section and for proportioning the members of that section. There is of course considerable scatter in the data due partly to differences in design requirements in different countries, but it has proved possible to derive some general empirical rules. Figure 6 shows how the type of longitudinal section depends on the span. A constant section appears to be economical up to 50 m span; a constant external section with thickening of the webs and/or bottom flange appears to be a viable choice for the span range 40—80 m; while variations in overall depth are apparently most suited to spans exceeding 60 m. In a similar way, by taking span as the key variable, it has proved possible to produce plots enabling an assessment to be made for web thickness and spacing and for the thickness of the top and bottom flanges. A report has been prepared on this work and its publication during 1971 will provide designers with a more rational basis for selecting the initial proportions for their structure.

Having made a reasonable estimate of the proportions of his bridge, the designer must now carry out a more detailed analysis to check that the section he has chosen is satisfactory and to provide the strength to resist the various load effects. In general it will be necessary to consider a number of load cases and therefore the designer would like to use the simplest, quickest and most economical method of analysis consistent with obtaining a sufficiently accurate representation of how his bridge would behave under load.

Before considering what analytical methods are available, it is perhaps worth pausing to reflect on what load effects must be calculated and what types of structural action are peculiar to spine beam construction. In addition to assessing the load

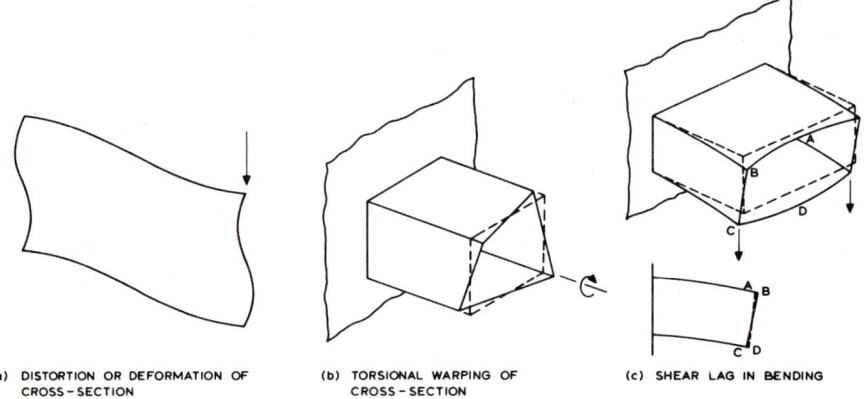

| (a) DISTORTION OR DEFORMATION OF CROSS–SECTION | (b) TORSIONAL WARPING OF CROSS–SECTION | (c) SHEAR LAG IN BENDING |

FIG. 7 TYPES OF STRUCTURAL ACTION WHICH MAY REQUIRE SPECIAL CONSIDERATION IN BOX BEAM BRIDGES.

effects inherent in simple beam theory, one or more of the types of structural action illustrated in Fig. 7 may require consideration under certain circumstances. The various factors shown in the Figure are:

(*a*) distortion or deformation of the cross-section, arising from in-plane displacement of points on the cross-section; this type of action can lead to distortional warping. Resistance to distortion is provided either by transverse diaphragms or, more normally for concrete, by increasing the bending strength of the walls of the box (see part (*a*) of the Figure).

(*b*) warping of the cross-section due to torsion, corresponding to out-of-plane or axial displacements of points on the cross-section, causing plane sections not to remain plane (part (*b*) of the Figure); in this case, no distortion occurs.

(*c*) shear lag, which is the effect of shear deformation in redistributing the bending stresses in a box beam; it permits a structure to resist higher ultimate bending moments than are calculated by simple flexure theory. When a box beam is subjected to torsion, a cross-section tends to warp from its original plane. If one end is restrained against warping, axial stresses are introduced and the shear flows are redistributed near the fixed end; this is also an effect of shear deformation and is sometimes called a shear lag effect.

There is a need, then, to develop methods of analysis of varying complexity which are capable of dealing with all possible types of structure, boundary and support conditions that can occur in practice. It is important to establish the limitations in use of any particular analysis and perhaps even more important to define the conditions under which any of the structural actions mentioned above become critical and require consideration in design; for reasons of economy, the ideal situation would be to be able to choose an analytical method which only just gave sufficiently accurate answers, since experience has shown that the costs of current all-embracing analytical methods can be very high. To arrive at this ideal situation it is essential to make comparisons between the results obtained from the various analytical methods and those obtained from tests on both models and full size bridges.

In his comprehensive review of literature related to the analysis and design of thin-walled beams, Maisel[7] showed that a number of analytical methods existed, most of them being based on elastic theory; many of these are complex and somewhat academic. Exceptions to this are the methods of Knittel[14] and Richmond[15] which are capable of dealing with cross-section distortions; the distortional component of warping stress is dealt with in the work of Kristek.[16] The problem of torsional warping is perhaps best dealt with by Heilig[17] and Vlasov.[18] Heilig's paper is particularly valuable since he goes beyond the development of an analytical procedure to indicate what are the geometrical conditions under which this effect warrants consideration in design. Apart from the methods mentioned above, most of the work published in English is concerned with the folded plate or finite element approaches. For multi-span multi-cell spine beams, these methods lead to high computer costs for analysis and the attention of the Cement and Concrete Association in this field has therefore been concentrated on methods in which an adequate representation of the structural behaviour is combined with a method of calculation requiring small computer capacity or in some cases only a slide rule. In this respect, the book by Dabrowski[19] warrants particular mention since he considers both curved and straight continuous box beams of up to three spans and provides extensive tables for use in design.

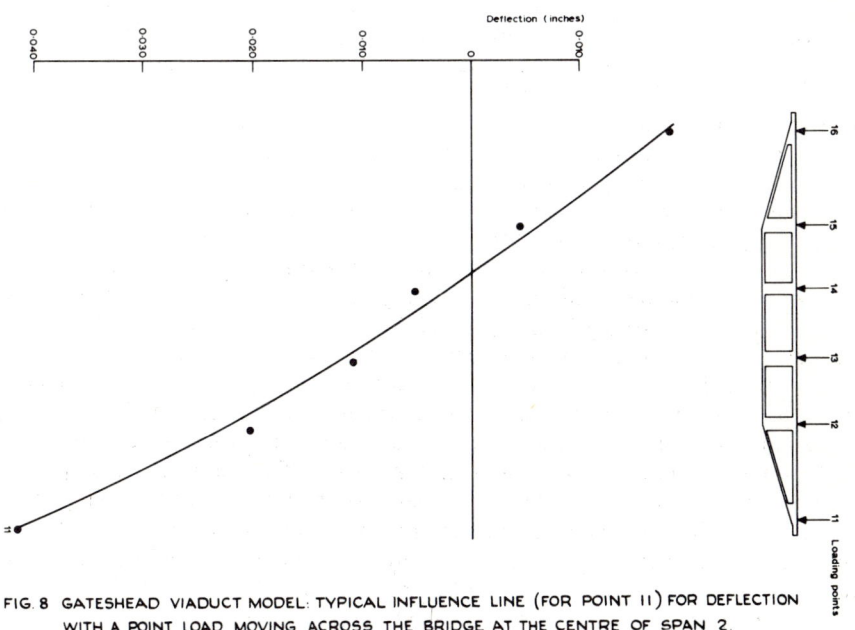

FIG. 8 GATESHEAD VIADUCT MODEL: TYPICAL INFLUENCE LINE (FOR POINT 11) FOR DEFLECTION WITH A POINT LOAD MOVING ACROSS THE BRIDGE AT THE CENTRE OF SPAN 2.

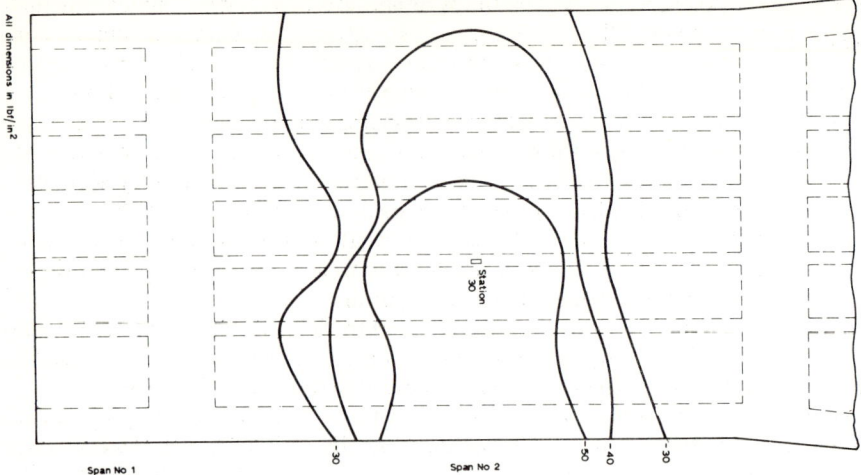

FIG. 9 GATESHEAD VIADUCT MODEL: TYPICAL INFLUENCE CONTOURS FOR STRESS (AT STATION 30 IN SPAN 2) DUE TO A POINT LOAD OF 40 lbf

In developing simplified design methods, it is necessary to know that a particular method does in fact give an adequate representation of the structural behaviour, i.e. it is capable of dealing with the various structural actions mentioned previously should these be critical for the bridge geometry being considered. To do this, it is necessary to make comparisons with experimental data. In completing its development work, the Cement and Concrete Association is making use of the extensive experimental data obtained previously from tests on both models and full size structures. Some of this data is extremely comprehensive; Figs 8 and 9 show typical data obtained from the Gateshead viaduct model,[13] where strains were measured at 200 locations and 450 loading cases were considered. Some recent comparisons have shown that it is important that much of this experimental work should be carried out on concrete or micro-concrete structures so that the possible influence of cracking — particularly on torsional stiffness — can be properly assessed. The work completed so far has indicated that some types of structural action could possible be dealt with in design by introducing simple empirical design rules; shear lag is one problem which could be amenable to this type of solution.

Experimental work on realistic concrete models is also required for other reasons. In his review, Maisel showed that little information was available on the ultimate strength of box beam structures apart from that obtained from individual model tests; with the introduction of limit state design, it is essential that this omission be rectified so that both strength and serviceability criteria can be adequately dealt with in design. With this end in view, the Cement and Concrete Association, again with support from CIRIA, is currently carrying out a series of tests to study the behaviour of prestressed concrete box beams under combined bending, shear and torsion. The test specimens have a span/depth ratio of 20 and are 184 mm deep x 352 mm wide in section with 33 mm thick webs and 22 mm flanges. The main variables are the ratio of bending moment to torque, the shear span, and the quantity of secondary reinforcement. The results obtained from this test series, when considred in conjunction with previous research on shear in prestressed concrete and with recent data from Stuttgart on pure torsion, should lead to a rational design method for providing strength in box beams subjected to combined stress systems.

CONCLUSIONS

In a paper which, of necessity, is both general and brief, it is perhaps impertinent to draw conclusions. However, it does seem relevant to the authors that certain concluding statements should be made which obtain for both types of bridge structure considered.

1. In preliminary design and proportioning which, after all, is the most important part of the design process, the need for simplified techniques is self evident. These should have a defined accuracy, range of application and must be supported by adequate experimental backing. These considerations are of particular importance in the selection of the transverse cross-section, whether of voided slabs or spine beams, since it is desirable to standardise these for various span ranges and widths of deck after carrying out any relevant optimising procedures.

2. Once preliminary proportioning has been achieved, detailed analytical treatment is necessary. Here the question of design time and cost, the definition of the significant data required and the categorisation of appropriate techniques arises. Advances in analysis have been considerable and, for spine beam bridges, there is a plethora of papers as Maisel[7] showed. Associated with the various approaches are computer programs for which the running costs and data preparation costs may be substantial. Only now is the experimental data becoming available which will allow the appropriate guidance to be given to designers on what is the 'best buy' in any given circumstances. It must also be concluded that, so far, perhaps too much attention has been devoted to purely elastic considerations.

3. The introduction of limit state concepts in bridge design is officially accepted and its implications will become more important in the future. Different approaches, therefore, become possible which could greatly simplify design by making the process one of providing the required strength and detailing for serviceability, at least as regards reinforced and prestressed concrete bridges. The experimental data derived from investigations of the type reported here form a vital prerequisite for such future developments.

REFERENCES

1. Manton, B. H. Tests of a quarter-scale model of a voided slab deck incorporating precast, prestressed inverted T-beams. Research Report to be published by Cement and Concrete Association.
2. Manton, B. H. and Wilson, C. M.o.T./C.&C.A. Standard beam sections for concrete bridges: M beams. Cement and Concrete Association, Publication 32.02.
3. Clark, L. A. The provision of reinforcement in simply supported skew bridge slabs in accordance with elastic moment fields. *Cement and Concrete Association, Technical Reports*, No. 42.450, Nov. 1970.
4. Clark, L. A. Crack similitude in 1/3·7 scale models of slabs spanning one-way. Technical Report to be published by Cement and Concrete Association.
5. Granholm, C. A. and Rowe, R. E. The ultimate load of simply supported skew slab bridges. *Res. Rep. Cem. Concr. Ass.*, No. 12, June 1961.
6. West, R. Recommendations on the use of grillage analysis for slab and pseudo-slab decks. Cement and Concrete Association, Interim Technical Note No. 1.
7. Maisel, B. I. Review of literature related to the analysis and design of thin-walled beams. *Cement and Concrete Association, Technical Reports*, No. TRA/440, July 1970.
8. Westergaard, H. M. Computations of stresses in bridge slabs due to wheel loads. *Public Roads*, Vol. 11, No. 1, March 1930.
9. Lee, D. J. The design of Section Five. *Struct. Engr*, Vol. 48, No. 3, March 1970, pp.109–120.
10. Somerville, G., Roll, F. and Caldwell, J. A. D. Tests on a one twelfth scale model of the Mancunian Way. *Cement and Concrete Association, Technical Reports*, No. TRA/394, Dec. 1965.

11. Swann, R. A. The construction and testing of a 1/16th scale model of the prestressed concrete superstructure of Section 5 — Western Avenue Extension. *Cement and Concrete Association, Technical Reports*, No. TRA/441, July 1970.

12. Taylor, H. P. J. and Clements, S. W. Tests on a 1/9th scale model of a transverse section of the West Gate Bridge, Melbourne. *Cement and Concrete Association, Technical Reports*, No. TRA/425, June 1969.

13. Evans, D. J. Tests on a 1/48th scale model of Gateshead Viaduct Cement and Concrete Association, Report No. DN/3008. To be published March 1971.

14. Knittel, G. and Worch, G. Zur Berechnung des dünnwandigen Kastenträgers mit gleichbleibendem symmetrischen Querschnitt. *Beton Stahlbetonb.*, Vol. 60, No. 9, Sept. 1965, pp. 205–211. (Analysis of the thin-walled box girder of constant symmetrical cross-section. Cement and Concrete Association, Library Translation.)

15. Richmond, B. Twisting of thin-walled box girders. *Proc. Instn civ. Engrs*, Vol. 33, April 1966, pp. 659–675.

16. Kristek, V. Tapered box girders of deformable cross section. *J. struct. Div. Am. Soc. civ. Engrs*, Vol. 96, ST8, Aug. 1970, pp. 1761–1793.

17. Heilig, R. Beitrag zur Theorie der Kastenträger beliebiger Querschnittsform. *Stahlbau*, Vol. 30, No. 11, Nov. 1961, pp. 333–349.

18. Vlasov, V. Z. *Thin-walled elastic beams*; 2nd ed., trans. from the Russian by Y. Schectman. National Science Foundation, Washington, D.C., 1961.

19. Dabrowski, R. *Gekrummte dünnwandige Träger, Theorie und Berechnung.* Berlin, Springer-Verlag, 1968. (Translation to be published by the Cement and Concrete Association.)

PRESTRESSED CONCRETE SLAB BRIDGES

R. G. ANDERSON and M. R. DOUGLAS

R. Travers Morgan & Partners

SYNOPSIS

The paper describes a finite element program using triangular elements for the analysis of bending and membrane stresses in slab bridge decks. The suite of programs can be used to prepare the mesh, and to generate the nodal loads for dead and highway loadings. From data concerning the prestressing forces, position and shape of each cable, equivalent inplane and out of plane nodal loads are generated for each mesh point.

When the finite element analysis has been completed any factored combination of applied loading and prestressing force may be processed and the resulting stress history of the top and bottom fibres printed out as the end product. Where the principal stresses exceed the allowable stress quoted in the input these are highlighted. A worked example is given.

1. INTRODUCTION

The slab bridge, is an attractive proposition to the highway designer. It is shallow in depth and thus keeps the cost of adjacent embankments and road works to the minimum. It can be made to any plan shape, and the supports underneath may be randomly placed. Slab bridges are relatively easy to construct and this is usually reflected in the construction costs.

To the designer, the degrees of redundancy have until recently made a continuous, curved or irregular shaped slab bridge a very complex analytical problem.

Today, provided we are designing within an elastic range a number of methods of solution are available. The authors have found the finite element method to be the most versatile for use in the office. The triangular non-conforming plate bending element by Professor O. C. Zienkiewicz,[1,2] together with a simple grillage beam element have been incorporated into a suite of programs for the analysis and design of reinforced concrete slab bridge structures. This suite has been issued by the Ministry of Transport as program No. BECP/1[3] in May 1969 and is being widely used for the design of slab bridges.

For reinforced concrete slab designs a plate bending analysis gives the principal moments acting in specific directions. If sufficient top or bottom reinforcing steel is placed in two directions, not necessarily coinciding with the principal moments, a satisfactory design of the structure can be obtained. A method of proportioning reinforcement has been expounded by Wood[4] and Armer.[5]

Prestress slabs are not a common feature in England, but are used in cases where headroom is at a premium, i.e. renewing railway over bridges where overhead electrification is being installed. Where an elevated road may have spans of medium size using reinforced slabs, if the span exceeds about 100 ft a prestressed slab provides a means of maintaining a shallow construction depth.

The criteria for prestressed concrete differ considerably from those of reinforced concrete. By providing more than the minimum reinforcement an adequate structure may be attained; this is not so with prestressed concrete.

The principal stresses, assuming elastic behaviour under working loads, may be used to proportion the structure and to calculate the prestressing forces required. However, most codes of practice require that all prestressed members should be checked by ultimate strength theory for compliance with specific load factors to ensure that there is an adequate factor of safety.

Such a 'load factor' method is really only adequate for simple flexural members; as the plastic redistribution effects which arise from continuity are not considered.

Limit state design takes account of the inelastic force redistributions which occur prior to collapse, and permits the calculation of a true collapse load. In a slab structure the formation of definite yield lines will be the case of collapse. (See Granholm and Rowe.[6]) This cannot be taken into account in an elastic analysis.

The authors have assumed that the safety and serviceability of a prestressed slab will be adequate if the 'load factor' approach is applied to a unit width at each node in a finite element analysis. The true formation of definite yield lines only takes place after redistribution of forces which should lead to a greater collapse load.

2. ELASTIC DESIGN

The criteria for elastic design are the principal fibre stresses in the top and bottom surfaces of the slab. Both bending and membrane stresses must be considered, this can be done by separating the two effects due to prestressing into the out of plane bending effects due to the cable eccentricities and the inplane membrane effects due to the inplane action, and dependent on the cable plan profile. Both of these stress fields depend upon the transfer of the cable forces to the structure, and the various losses which occur due to friction, creep, etc. The structure can be represented by one finite element configuration and analysed quite separately for the two actions. The stresses in the top and bottom of the slab may be obtained by vector addition of the stresses due to inplane and out of plane forces.

It is not practical to place the prestressing cables along the lines of principal stress, and thus with a prestress slab it is strictly not possible to ensure, as is normally done in prestressed beam design, that principal tensile stresses do not occur. This becomes particularly true of skew slabs, with concentrated loads. We would emphasise, therefore, that a prestressed slab, cannot, in general be designed for zero tensile stress unless it is a right, simply-supported, slab. Where the vector addition of the stresses in the top and bottom surfaces show residual tensile forces then additional reinforcement will be needed.

It is common practice to precast prestress units, and to incorporate these into a slab bridge with insitu concrete between the units and with transverse continuity provided by ordinary reinforcement. Similarly in square or parallelogram decks, it will be sufficient to provide longitudinal prestress only with adequate transverse and longitudinal reinforcement to take account of the σy and τxy stresses remaining.

For reinforcement placed in the x and y directions only the stresses to be resisted are:

$$\overline{\sigma x} = \sigma x + k \left| \gamma xy \right|$$

$$\overline{\sigma y} = \sigma y + \frac{1}{k} \left| \gamma xy \right|$$

which for the case of skew reinforcement are:

$$\overline{\sigma x} = \sigma x - 2\tau xy \cot \alpha + \sigma y \cot^2 \alpha + k \left| \frac{\sigma y \cot \alpha - \tau xy}{\sin \alpha} \right|$$

$$\overline{\sigma \alpha} = \frac{1}{\sin^2 \alpha} \left\{ \sigma y + \frac{1}{k} \left| \sigma y \cot \alpha - \tau xy \right| \right\}$$

where the reinforcement is placed parallel to the x axis and at an angle α to the x axis in an anticlockwise sense.

Under working loads any flexural hairline cracks that occur will close up, and we have in effect a reinforced concrete slab in which the longitudinal reinforcement has been replaced by prestressing tendons allowing a much thinner slab.

This principle may be extended to voided slabs with the prestressing tendons located in the concrete ribs. With oddly shaped slabs the voids, and stressing tendons are placed fan wise. According to Massonnet and Bares[7] a cellular type slab with rectangular voids at regular spacing is theoretically isotropic and with circular or elliptical voids very nearly isotropic. The orthotropic effect swings the principal moment towards the direction of the voids.

When permanent cracks may occur in skew or oddly shaped slabs, subjected to a corrosive atmosphere, transverse prestressing will be needed.

Even with two way prestressing it may still be necessary to provide additional steel reinforcement to deal with local tensile stresses.

3. ULTIMATE LOAD CHECK

At ultimate load we need not consider the inplane effect, indeed the slab may be considered as a reinforced slab, taking into account the prestressing tendons as simple reinforcement. As the bond of the tendon cannot be taken as perfect only a percentage of the yield strength of the tendons should be allowed.

Where only longitudinal prestressing is used it may be assumed that transverse tensile cracks occur as the load increases above the working range. This results in a change in the orthotropy of the slab so that the principal moment lies approximately in the direction of the prestressing tendons. As a result of this cracking the slab is transformed into a set of beams interconnected by the transverse reinforcement. This is analogous to a precast beam/slab situation and can be solved by following the code of practice rules for beams.

Where transverse prestressing is used the slab should be considered node by node to verify that the principal moments under the loading requirements of the code can be duly taken by the slab. It is probable that in such a case the 'Ultimate Load' principal moments will lie in a different direction from those at working load, and it will often be necessary to add additional steel reinforcement.

4. PRESTRESSED CONCRETE SLAB COMPUTER PROGRAM

This suite of programs has been developed from the Ministry of Transport reinforced concrete slab program suite BECP/1 which was prepared and verified by R. Travers Morgan and Partners in conjunction with Professor O. C. Zienkiewicz of University

College, Swansea, where the basic finite element solution system used was developed.[8]

The programs included in the Prestressed Concrete suite of programs are shown in Fig. 2. The function of each program is described below:

FECK 2   Checks the manually prepared data and highlights incompatible data.

IPUT 2   Enables the engineer to develope a suitable finite element mesh with specific reference to the appropriate lane, knife edge and abnormal vehicle loadings specified by BS 153. Regular meshes for all but the most complex geometric shapes can be generated automatically by IPUT 2. Some typical geometric shapes

FIG. I PLAN SHAPES OF STRUCTURES FOR WHICH
MESHES MAY BE AUTOMATICALLY GENERATED

for which meshes may be generated are shown in Fig. 1. The loads specified by BS 153 are readily applied to the structure via IPUT 2. IPUT 2 produces a complete file of the appropriate finite element data for SAADS.

PPUT   The equivalent discrete loading systems (out of plane and inplane) due to the prestressing of a bridge deck are generated with reference to the finite element mesh over the deck by the program PPUT. Individual cables may be processed or all the cables in a plan strip related to the mesh size may be treated as a group force. Data concerning the prestressing forces, losses, eccentricities, position and plan shape of the cable must be given to PPUT for each cable or cable group. The amount of data required depends upon the finite element mesh previously generated for the structure, the number of discrete loads required to represent the cable, and the cable profile. For a parabolic cable profile only the eccentricities of the ends of each parabola and the dip of the

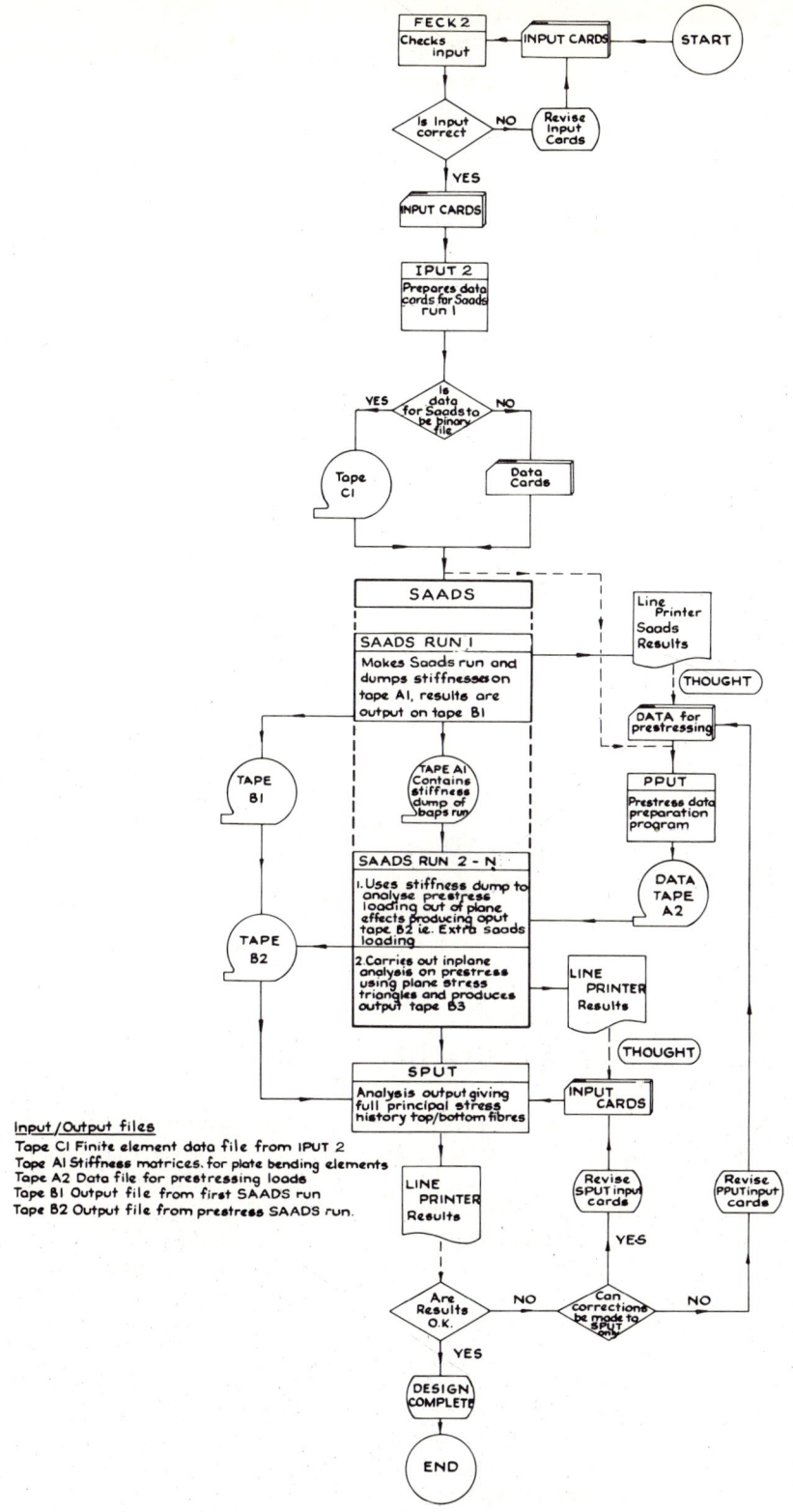

Input/Output files

Tape CI Finite element data file from IPUT 2
Tape AI Stiffness matrices for plate bending elements
Tape A2 Data file for prestressing loads
Tape BI Output file from first SAADS run
Tape B2 Output file from prestress SAADS run.

FIG. 2 FLOW CHART OF PRESTRESS SLAB BRIDGE
DESIGN SYSTEM.

parabola need be input to define all eccentricities. For a cable straight in plan only the end co-ordinates and one co-ordinate at each internal point need be specified to define the geometry of the cable.

SAADS. This is the finite element analysis program for analysing flat plate and beam structures. The plate is represented by triangular constant thickness elements while the beams are represented by straight constant section prismatic members. The beam members are considered to be symmetrically positioned with their axes lying in the centre plane of the plate structure. Both plate and beam members are illustrated in Fig. 3.

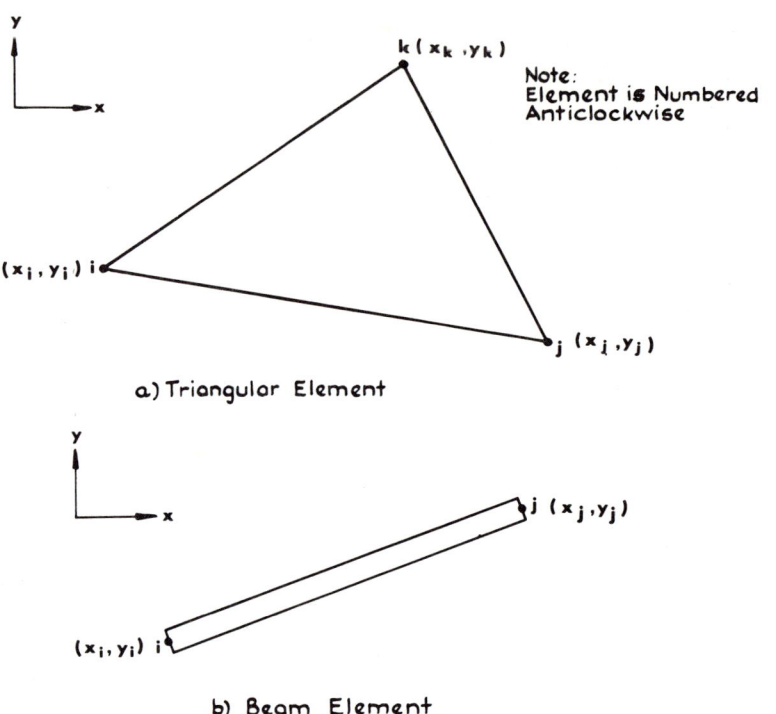

a) Triangular Element

b) Beam Element

FIG. 3 ELEMENTS USED IN ANALYSIS

The program utilises the same shape element with two separate stiffness formulations. Plate bending analysis is carried out using the triangular non-conforming plate element[1] together with a prismatic grillage type beam element. All the out of plane actions are analysed using these elements. The sign convention for nodal displacements and forces together with the moments calculated by the analysis are illustrated in Fig. 4. The membrane analysis of the structure subject to the inplane prestressing loads uses a triangular plane stress element together with a simple prismatic beam element capable of carrying axial loads only. Again the sign conventions are shown in Fig. 4.

Both the inplane and out of plane analyses when used to analyse a prestressed slab use the same finite element mesh configuration which can be prepared by hand or generated by IPUT 2. Either plate bending or plane stress analyses may be carried out by the program SAADS as quite separate analyses. Thus the plane stress analysis of say a deep beam shear wall or diaphragm might readily be accomplished by SAADS. Conversely SAADS can be used together with OPUT for the design and analysis of reinforced concrete slab bridges where membrane loading is not considered.

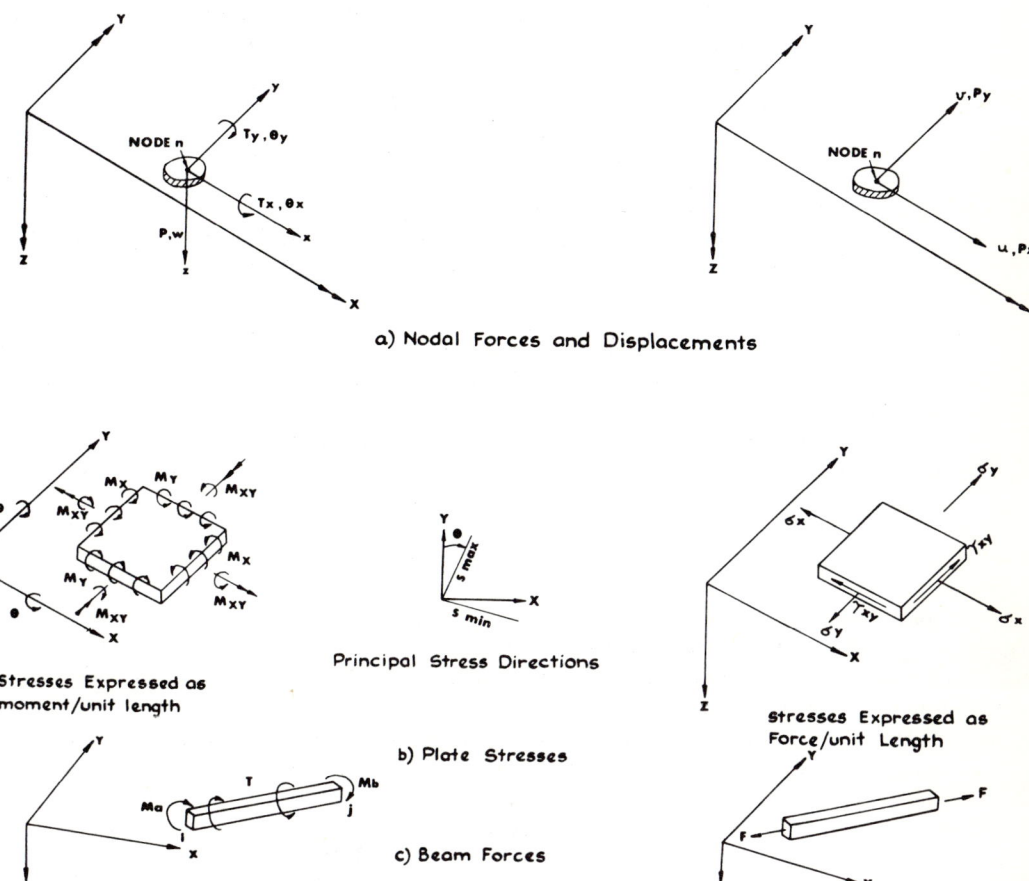

a) Nodal Forces and Displacements

Stresses Expressed as
moment/unit length

Principal Stress Directions

b) Plate Stresses

Stresses Expressed as
Force/unit Length

c) Beam Forces

FIG. 4   SIGN CONVENTIONS FOR PLATE STRUCTURES

SPUT   For Prestressed Concrete, once all the various analyses have been carried out on the structure, it is necessary to determine the complete stress history of the structure throughout its loading cycles. The results of the SAADS analyses have been stored on magnetic tape for this purpose. The program SPUT combines any number of normal loading results with the results of the prestressing analysis. Each loading within an analysis can be factored by any value. The resulting moments and membrane stress fields together with the appropriate reactions due to the combinations can be obtained. These results may be further processed to yield the nodal or element fibre stresses for the top and bottom surfaces of the slab under each combination. The principal fibre stresses and the associated principal angle are also calculated for the combinations and if the values of these principal stresses exceed the allowable principal stresses the relevant stress is highlighted in the output.

OPUT 2 For Reinforced Concrete combines the suitably factored results of chosen load cases using the output file of results created by SAADS. The resulting moment fields can then be output and they are further processed to produce moments of resistance for top and bottom steel in any two specified steel directions. This latter analysis is based on work presented by Wood[4] and allows up to 50 different directions of steel to be specified within a structure.

5. DESIGN PROCEDURE

The geometric shape of the bridge, support conditions, etc., are normally dictated by location, geological conditions, and experience. Finite element techniques are normally found to be too expensive to use for preliminary design comparisons.

Thus, starting with a given slab bridge configuration a suitable finite element mesh is chosen, the number and spacing of elements depending on slab thickness, width of traffic lanes, support conditions and loading cases. The program IPUT 2 is used to generate the mesh and nodal loads for dead and highway loadings.

The finite element analysis program SAADS is used to analyse the structure for the above loading cases for plate bending effects. The results are stored on a tape or disc file for re-use and printed out for examination.

From a visual inspection of these results, and an examination of the practical lay-out of prestressing cables, it is possible (after some experience of the method) to make a first approximation to the prestressing forces required.

Data concerning the prestressing forces, losses, eccentricities, position and plan shape of each cable are prepared from this inspection, and the program PPUT used to generate equivalent inplane and out of plane nodal loads for each finite element mesh point.

The finite element analysis program SAADS is again used to analyse the structure for the prestressing plate bending and membrane effects. The results of this analysis are stored on a tape or disc file for re-use.

All the results stored on file are used in conjunction with the program SPUT to combine the results of any or all of the load cases considered (factored as required), and the prestressing (factored as required).

The results are printed out as a stress history of the top and bottom fibre stresses at each node of the finite element mesh. The principal stresses that exceed the allowable tensile stresses stated in the input data are highlighted.

6. EXAMPLES

6.1. A Two Span Prestressed Beam

The difficulties in carrying out suitable hand calculations on prestressed structures other than simple beams led us to analyse the two span beam shown in Fig. 5 as a check on the method. The beam was considered to have a cross section of width 60 in and depth 30 in with two equal spans of 600 in. The finite element mesh shown in the figure was used to analyse one half of the structure for the three normal loadings of dead load, uniformly distributed live load and point loads at mid spans, and the prestressing shown. The results of the finite element analyses were then processed by SPUT to obtain the longitudinal stresses within the top and bottom plate fibres due to load combinations of:

1. Dead load + Prestress
2. Dead load + Prestress + Uniformly distributed live load
3. Dead load + Prestress + Point loads at mid spans

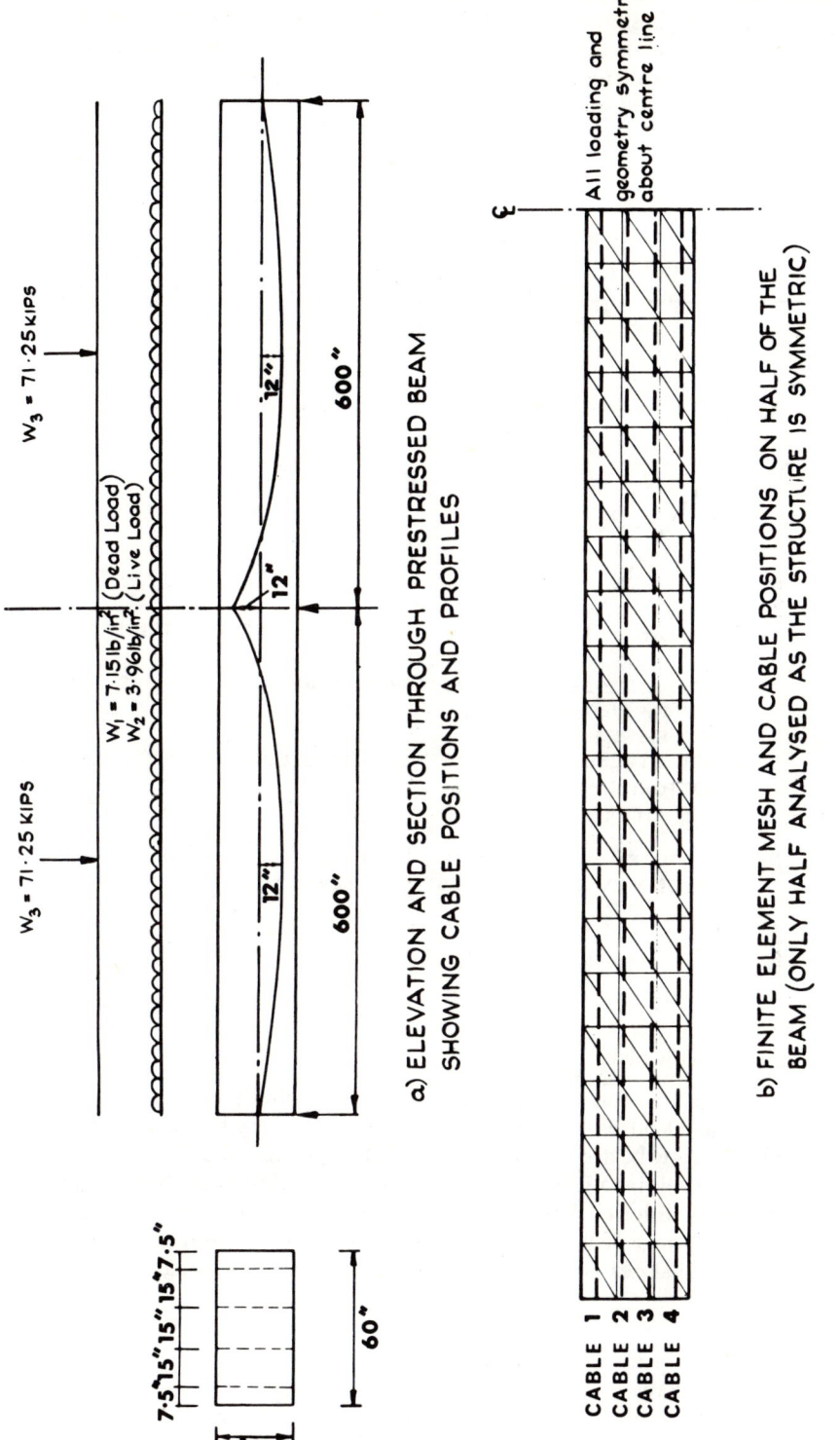

a) ELEVATION AND SECTION THROUGH PRESTRESSED BEAM
SHOWING CABLE POSITIONS AND PROFILES

b) FINITE ELEMENT MESH AND CABLE POSITIONS ON HALF OF THE
BEAM (ONLY HALF ANALYSED AS THE STRUCTURE IS SYMMETRIC)

FIG.5 PRESTRESSING OF TWO SPAN BEAM

The results of these loading combinations were calculated by utilising suitable hand calculations as suggested by Lin.[9] The smooth lines shown in Fig. 6 are the hand calculated values while the points are the values calculated by the programs. It can clearly be seen that for this simple example the computer results are extremely close to the hand calculated values.

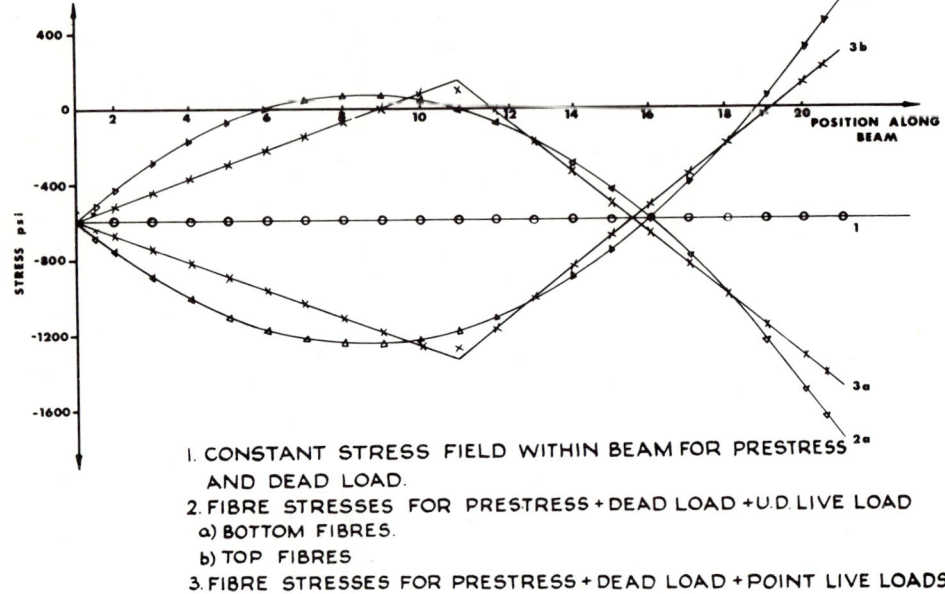

1. CONSTANT STRESS FIELD WITHIN BEAM FOR PRESTRESS AND DEAD LOAD.
2. FIBRE STRESSES FOR PRESTRESS + DEAD LOAD + U.D. LIVE LOAD
   a) BOTTOM FIBRES.
   b) TOP FIBRES
3. FIBRE STRESSES FOR PRESTRESS + DEAD LOAD + POINT LIVE LOADS
   a) BOTTOM FIBRES
   b) TOP FIBRES

NOTE: COMPUTED VALUES ARE INDICATED AS POINTS – SOLID LINES ARE YIELDED BY SIMPLE PRESTRESS BEAM THEORY.

FIG. 6

6.2. A Prestressed Skew Slab

The obvious advantage of this suite of programs is the capability to analyse prestressed concrete slab bridges of complex shape having arbitrary boundary conditions. As an example the skew bridge shown in Figs 7 and 8 was analysed. This skew bridge, used where a slip road joins a four lane motorway, was previously analysed as a reinforced concrete slab bridge of depth 36 in. The reinforcement is shown in Fig. 11. We re-designed this bridge deck as a prestressed slab of depth 20 in for the same live loading cases. Both analyses assumed the supports to be elastic bearings.

Geometric Details.
Span = 53·74 ft
Skew = 135°
Width = 117·5 ft
Depth of Slab = 20 in

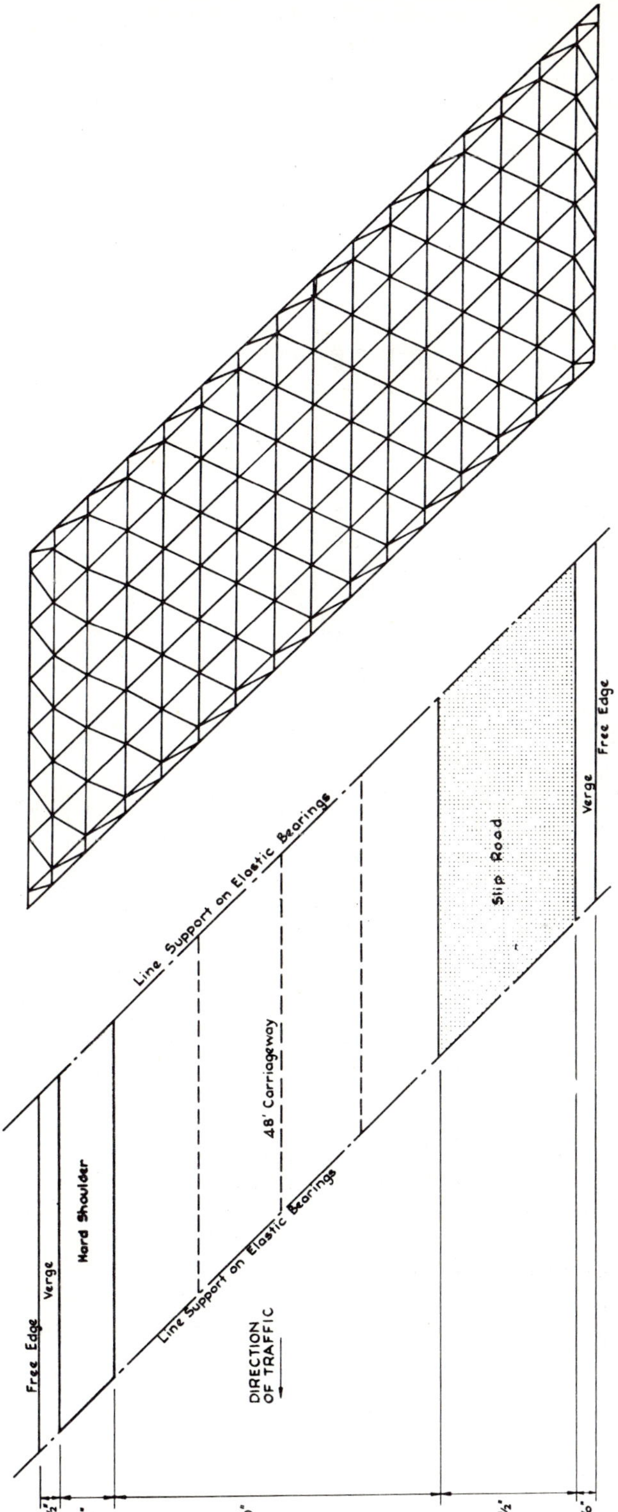

FINITE ELEMENT IDEALISATION OF BRIDGE.

SINGLE SPAN SKEW BRIDGE

FIG. 7

Line Support on Elastic Bearings

Line Support on Elastic Bearings

Free Edge

Verge

Hard Shoulder

48' Carriageway

Slip Road

Verge

Free Edge

DIRECTION
OF TRAFFIC

3'-2½"

8'-3"

48'-0"

20'-1½"

3'-6"

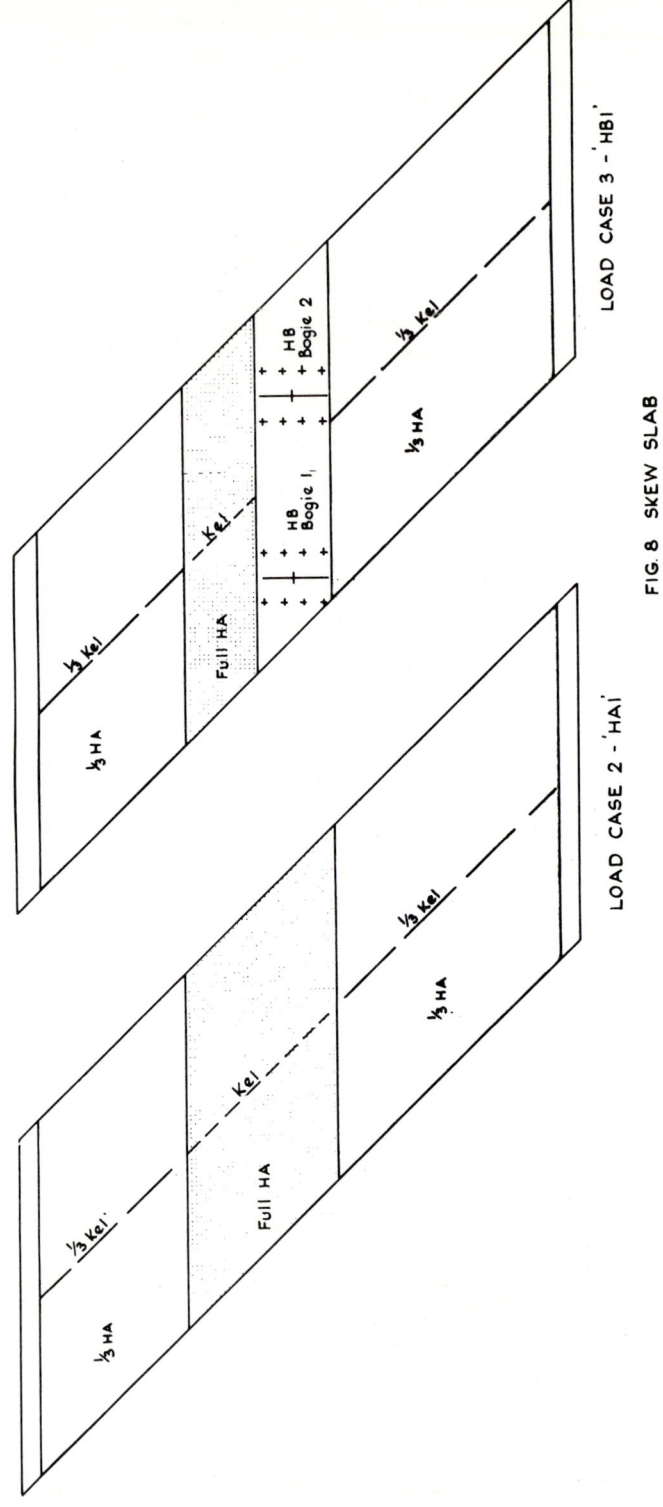

LOAD CASE 3 - 'HB1'

LOAD CASE 2 - 'HA'

FIG. 8   SKEW SLAB

Loadings. The structure was analysed for four typical loadings.

1. Dead load
2. Full H.A. load on two adjacent centre lanes as shown in Fig. 7 (H.A. 1 loading)
3. Full H.A. load on two adjacent edge lanes
4. Abnormal vehicle on the centre lane as shown in Fig. 7 (H.B. 1 loading)

The principal bending moments are shown in Fig. 9 on half the structure for load case (4).

Prestressing. The 12/0·5 in CABCO single strand system was adopted for the prestressing using dyform strand.

$$\text{Design Force per } 12/0\text{·}5 \text{ in cable} = 394\text{·}8 \text{ kips } (1{,}754 \text{ kN})$$

Losses due to friction, etc., taken as 10%
Creep and other losses taken as 10%

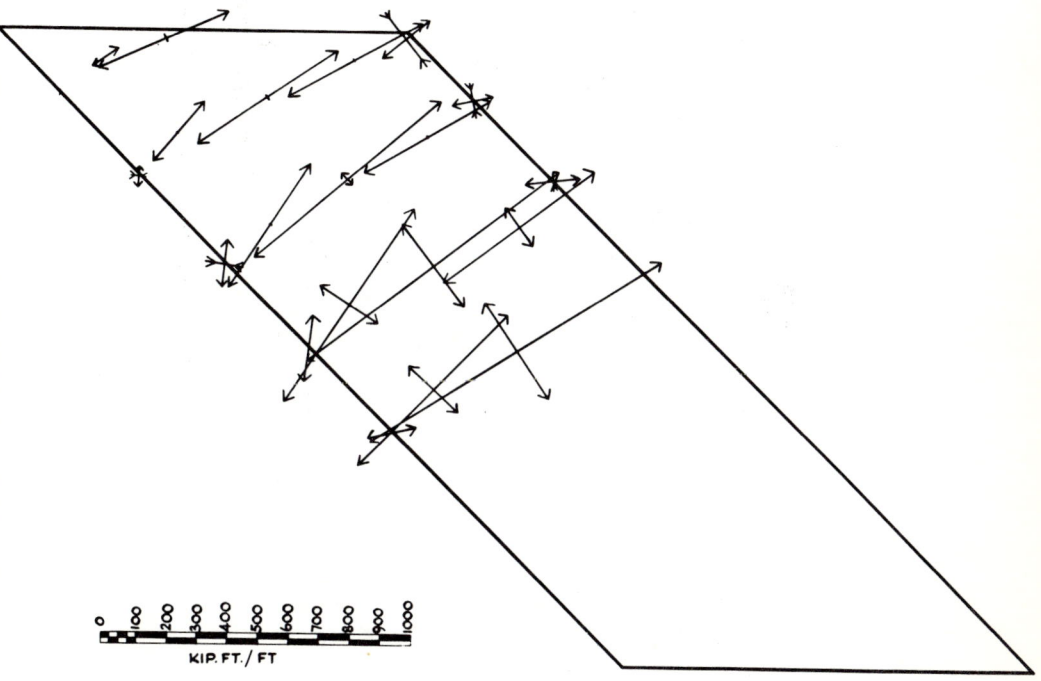

KIP. FT./ FT

FIG. 9 PRINCIPAL MOMENTS – LOAD CASE 4

The prestressing was represented by groups of cables in four regions as shown in Fig. 12. Each region is analysed separately as a prestress load case so that different multiples of the prestressing force may be used to find the correct force.

All cables were assumed parabolic with a maximum eccentricity of 7 in.

The structure was examined for the loading conditions below:

Loading Combinations.
1. Dead load   loading (1)
2. Dead load + H.A. loading (2)
3. Dead load + H.A. loading (3)
4. Dead load + H.B. loading (4)

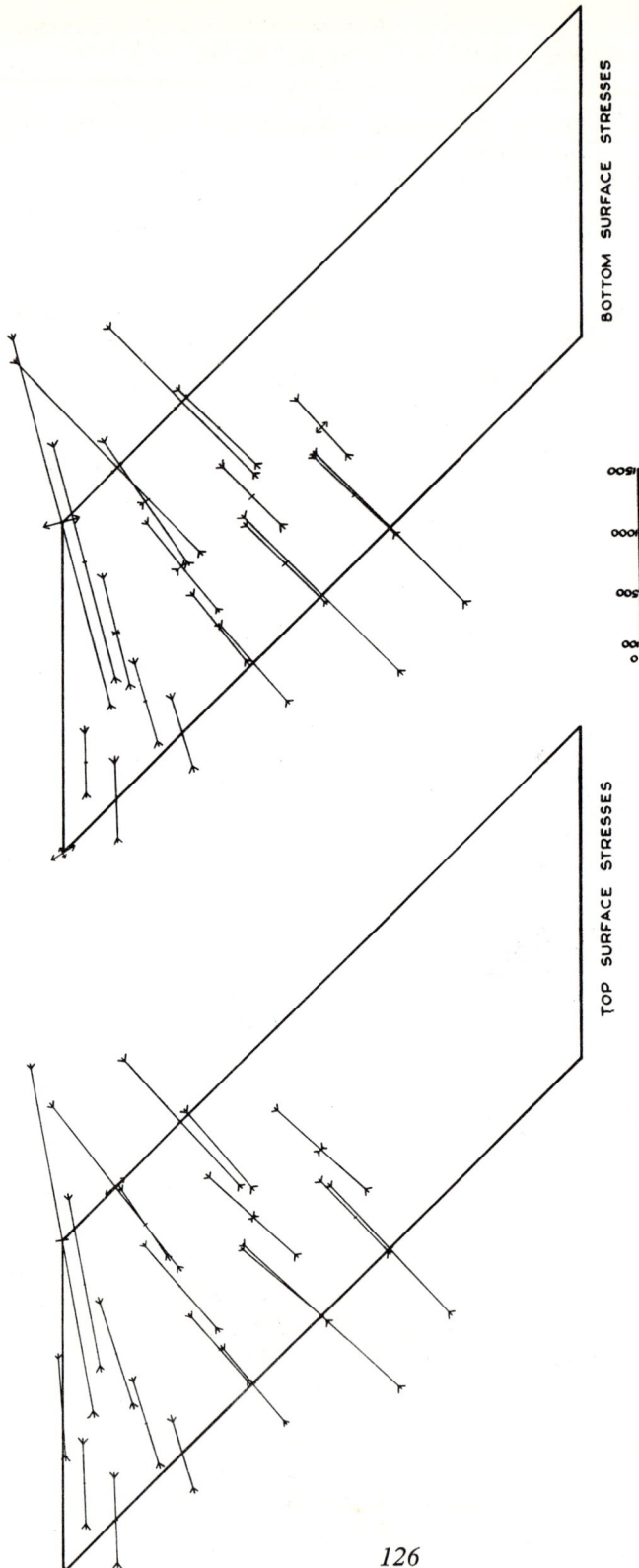

TOP SURFACE STRESSES

BOTTOM SURFACE STRESSES

KIPS / SQ. FT.

FIG. 10 SKEW SLAB PRINCIPAL STRESSES - PRESTRESS PLUS LOAD COMBINATION 2

126

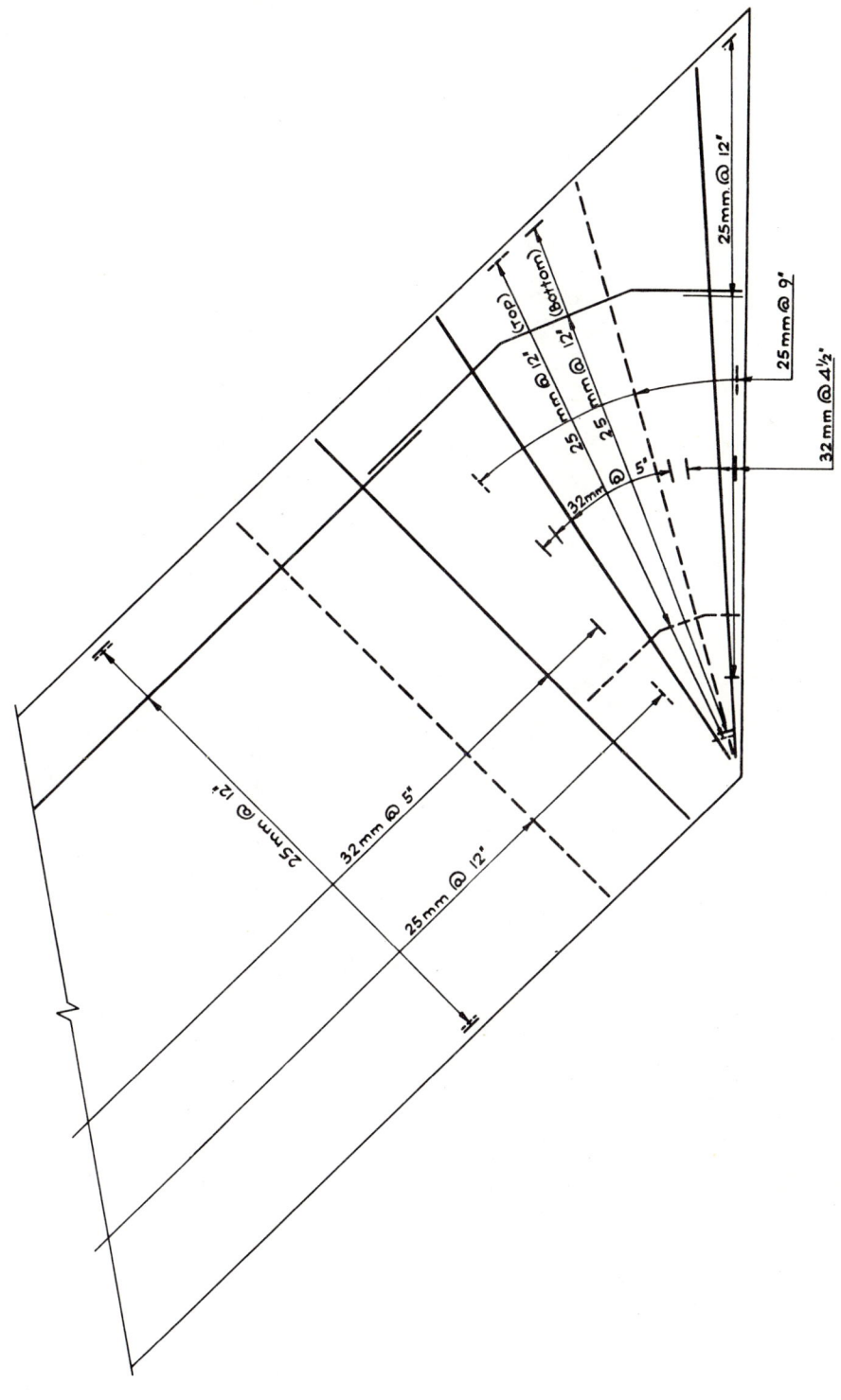

FIG. II SKEW SLAB - TYPICAL REINFORCED CONCRETE DESIGN

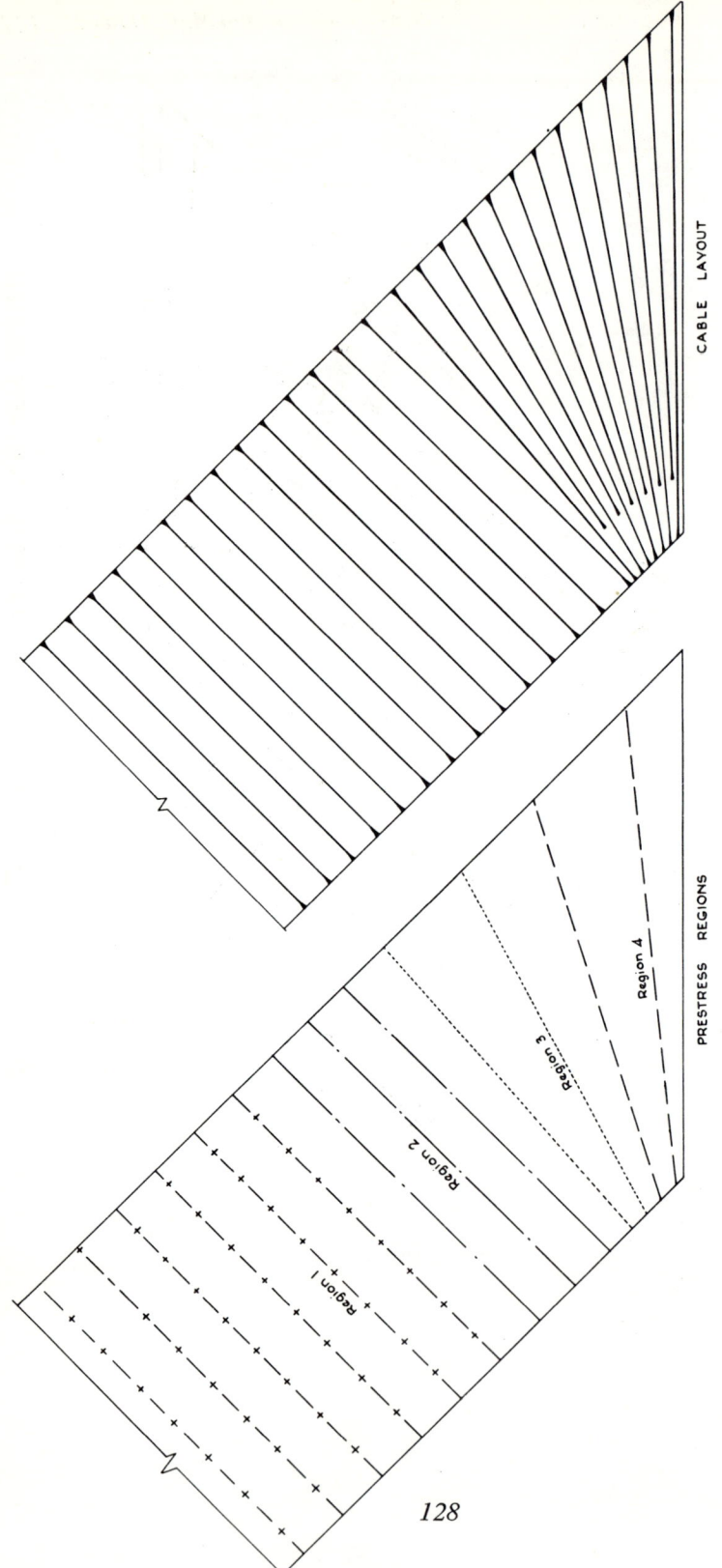

CABLE LAYOUT

PRESTRESS REGIONS

Region 1

Region 2

Region 3

Region 4

128

FIG 12 SKEW SLAB - TYPICAL PRESTRESSED CONCRETE DESIGN

From the results of these loadings a nominal value of prestressing force per region was deducted, and the structure analysed for this nominal force.

The four loading combinations were then combined with a variety of factored prestressing results by using SPUT. The following prestress forces were found to be suitable.

Region 1   722 kips/cable group
Region 2   722 kips/cable group
Region 3   866 kips/cable group
Region 4 1002 kips/cable group

The cables in the acute corner are curtailed to eliminate excessive compression. The selected spacing of cables being

Region 1   12/0·5 at 2 ft 6 in crs
Region 2   12/0·5 at 2 ft 6 in crs
Region 3   12/0·5 at 3 ft 3 in crs tapering to 11 in crs
Region 4   12/0·5 at 4 ft 4 in crs tapering to 7½ in crs

Stress History. Figure 10 shows the stress history on half the structure for the top and bottom surfaces for prestress plus load combination 2. These show a small transverse principal tensile stress, this never exceeds 250 psi for any load case combination. Nominal reinforcement would suffice for crack control.

The maximum compressive stress on the drawing is 2,200 lb per in^2 in the acute corner. This has been reduced in practice by curtailing alternative cables in this region as shown in Fig. 12.

Remarks

The problem was also analysed with a transverse prestress of 64 kips/ft (i.e. 12/0·5 in multistrand cables at 6 ft centres) approximately along the neutral axis.

This completely eliminated all tensile stresses without significantly increasing any of the principal compressive stresses.

A further check on the ultimate strength would be needed for the appropriate strip containing a single cable.

7. CONCLUSIONS

The design of prestressed concrete flat slab bridges can now be carried out easily by using the facilities offered by the suite of programs used in this paper. Although the examples were limited to the analysis of solid slabs the program has been used to analyse voided structures having void ratios of up to 0·6.

Developments

The development of the processing programs, OPUT 2 for reinforced concrete structures and SPUT for prestressed concrete structures, is being extended considerably. OPUT 2 will not only develop moment fields to be resisted in directions of reinforcement, but will calculate the best reinforcement directions and choose suitable reinforcement bars and spacings. SPUT is being extended to cope with the design of reinforcement for mixed prestress/reinforced concrete structures.

The whole suite of programs for the design of reinforcement concrete slab bridges is being written as a subsystem for GENESYS. The form of data input and program flexibility offered by such a system suggests that this method of utilising engineering programs will be the simplest method for such complex suites of programs.

Acknowledgements

The authors wish to make acknowledgement to the Department of the Environment and R. Travers Morgan & Partners for permission to use drawings and other material presented in this paper. Any opinions expressed in the paper are the authors' own.

REFERENCES

1. Zienkiewicz, O. C. *The finite element method in structural and continuum mechanics.* New York, McGraw-Hill, 1967.
2. Bazeley, G. P., Cheung, Y. K., Irons, B. M. and Zienkiewicz, O. C. Triangular elements in plate bending. *Proceedings of the Conference on Matrix Methods in Structural Mechanics, (Wright-Patterson Air Force Base, Ohio, 1965),* 1967.
3. G.B. Ministry of Transport. *Suite of bridge design and analysis programs.* BECP/1, 1968.
4. Wood, R. H. The reinforcement of slabs in accordance with a pre-determined field of moments. *Concrete,* Vol. 2, No. 2, Feb. 1968, pp. 69–76.
5. Armer, G. S. T. Discussion on 'The reinforcement of slabs in accordance with a pre-determined field of moments'. *Concrete,* Vol. 2, No. 8, Aug. 1968, pp. 319–320.
6. Granholm, C. A. and Rowe, R. E. The ultimate load of simply supported skew slab bridges. *Res. Rep. Cem. Concr. Ass.,* No. 12, June 1961.
7. Bareš, R. and Massonnet, C. *Analysis of beam grids and orthotropic plates by the Guyon-Massonnet-Bareš method;* trans. by J. Vaněk. London, Crosby Lockwood, 1968.
8. Cheung, Y. K., King, I. P. and Zienkiewicz, O. C. Slab bridges with arbitrary shape and support conditions. . . . *Proc. Instn civ. Engrs,* Vol. 40, No. 1, May 1968, pp. 9–36.
9. Lin, T. Y. *Design of prestressed concrete structures.* 2nd ed. New York, Wiley, 1964.

CASE STUDIES IN THE USE OF THE BAPS FINITE ELEMENT PROGRAM PACKAGE

K. SRISKANDAN

Department of the Environment

1. The BAPS Finite Element Program Package was made available by the then Ministry of Transport for the analysis of simply supported and continuous skew or curved slab bridge decks of solid or voided construction. The documentation was in two volumes,[1,2] Volume 1 being the User Manual and Volume 2 a report describing the Computer Program, and the verifications carried out. The program package itself was made available in the form of card decks in object form for different makes of computers. This contribution is restricted to experiences in the use of the package in the Department of the Environment's Midland Road Construction Unit on an IBM 360/30 Computer of 64K Bytes with disc operating system.

2. In Midland RCU the BAPS package has been used so far to analyse four different structures which can be grouped into the following categories:—
   (*a*) Simple Supported skew slabs
   (*b*) Two span continuous curved slabs with discrete columns at the Intermediate Supports.

3. SIMPLY SUPPORTED SKEW SLABS

3.1. Simply supported reinforced or prestressed slabs are commonly used for decks of up to about 15 m span. For decks of up to 20° of skew and which are supported along the full width the load distribution analysis that is commonly used is that due to Morice and Little.[3] For over 20° of skew the Influence Surfaces of Rüsch and Hergenroeder[4] or a grillage analysis are amongst some of the methods that have been used in the past. All of these methods suffered from the defect of not being versatile enough for the wide variety of problems one encountered in practice. The Finite Element Method seems to have overcome these problems and the BAPS package is now regarded as an acceptable tool for the analysis of these slabs.

3.2. Of the two skew slabs that were analysed one was solid and the other was of voided construction. As far as the Analysis of these slabs was concerned there was very little difference between the procedures that were adopted and hence only a description of the analysis of the voided slab is given. Where the differences are of importance these are mentioned.

131

3.3. Bankfield Road Bridge:

3.3.1. Bridge description

Bankfield Road Bridge carries a dual two lane highway over Bankfield Road. Although the bridge carries acceleration and deceleration lanes, the crossing of Bankfield Road is sufficiently close to the 'nose' on both sides and hence both edges of the bridge were made parallel to the centre line of the highway. Each carriageway is carried on a separate superstructure and therefore only one half-deck was analysed.

The clear square span of the deck was 46 ft 6 in and the angle of skew was 26° 30'. As the skew span was well over 50 ft it was decided to use an in-situ reinforced concrete voided slab instead of a filler joist slab using PCDG inverted T beams. A deck thickness of 3 ft with 18 in diameter voids at 2 ft 6 in centres was chosen as shown in the cross section of the deck in Fig. 1.

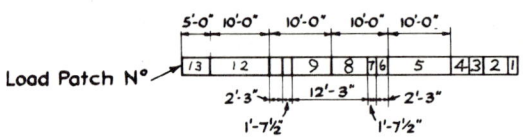

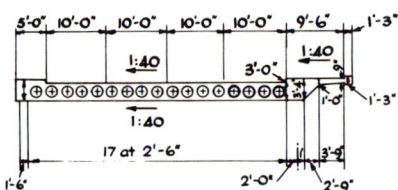

FIG. 1   BANKFIELD ROAD BRIDGE. DECK CROSS SECTION

3.3.2. Mesh division

As stated earlier, the edge of the deck was made parallel to the centre line of the highway and therefore each half deck was a parallelogram in plan. The main mesh lines were therefore chosen parallel to the edges of the deck as shown in Fig. 2. The skew span length of 58 ft was divided up into 10 equal divisions of 5 ft and two of 4 ft each adjoining the abutments as advised in the User Manual. The middle 50 ft coincided with the voided portion of the slab, the end 4 ft at each support being solid.

The verge was divided into 4 mesh lines, each denoting a patch, to take account of the varying material properties and dead load of the deck. The total width of carriageway was 40 ft and this gave four notional 10 ft lanes. Only the two outside notional lanes could be maintained as one load patch, because the BS HB vehicle needs a total width of 12 ft 3 in including the one foot clearance from the kerb.

The 40 ft carriageway had therefore to be split into 8 load patches to enable combinations of HA and HB loading to be applied. The resulting mesh with 208 nodes and 360 elements is shown in Fig. 2.

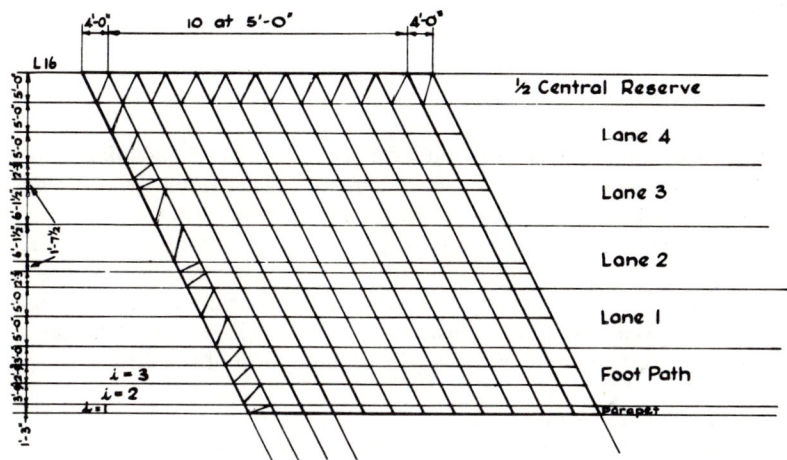

FIG. 2.   BANKFIELD ROAD BRIDGE. FINITE ELEMENT MESH

3.3.3. Material properties and boundary conditions

The deck consists of solid sections (1–4) and voided sections (5–13). The user manual suggests that the material properties for voided orthotropic plates be determined by model tests. Further guidance has however been given on this subject by the Bridges Engineering Division of the Department of the Environment[5] and the various constants were calculated as follows:—

(a) Average values of I_{xx} and I_{yy} were calculated.

(b) $Dx = \dfrac{E I_{xx}}{(1 - \mu^2)}$

(c) $Dy = \dfrac{E I_{yy}}{(1 - \mu^2)}$

(d) $Dxy = 0.75 \dfrac{(1 - \mu)}{2} \sqrt{Dx\, Dy}$

(e) $D_1 = \mu \sqrt{Dx\, Dy}$

The ratio of these constants to the corresponding values for a solid slab of the same thickness are as follows:—

(a) Dx – 0.96
(b) Dy – 0.96
(c) Dxy – 0.72
(d) D_1 – 0.96

Individual bearings were provided between the voids at 2 ft 6 in centres. The initial estimate of the loads on the bearings was made assuming a right deck and using a BS 153 reduction factor for the distribution of the HB load. The type of bearing to be used had to be predetermined and it was assumed that rubber bearings were used. Bearing stiffnesses were calculated from manufacturer's information on the quality of the rubber.

3.3.4. Loading cases

In a right slab bridge of this span it could have been said that HA loading would be critical for longitudinal moments provided adequate edge stiffening was provided. Because of the skew however this is not so apparent and hence the load cases considered had to include HB for maximum longitudinal moment in addition to HB for maximum transverse moments and shear and reactions. Therefore in dividing up the deck into elements provision was made for mesh lines to coincide with load patch boundaries. After inspection it was considered that the following 35 load cases should be analysed to be combined later after examination of the results:

1. Total dead load at entire deck.
2. UDL of 80 lb/ft^2 on patches 2, 3 and 4 (Footway loading).
3-10. HA UDL of 220 lb/ft^2 on each of load patches 5—12 respectively.
11. HB vehicle adjacent to Central Reservation and placed to give maximum Bending Moment at centre span near edge of deck.
12. HB vehicle astride middle two lanes and placed to give maximum Bending Moment at centre span in middle of deck.
13. HB vehicle adjacent to footway and placed to give maximum Bending Moment at centre span near other edge of deck.
14. HB vehicle adjacent to central reservation and placed to give maximum shear at end of span near one edge.
15. HB vehicle astride middle two lanes and placed to give maximum shear at end of span in the middle.
16. HB vehicle adjacent to footway and placed to give maximum shear at end of span near other edge.
17-19. The 4 ton Wheel Load in load patch 2 placed at the centre and 2 ends of the deck respectively.
20-27. HA Knife Edge load placed at centre of span in load patches corresponding to each of the HA UDL cases 3—10 respectively.
28-35. HA Knife Edge load placed at end of span in load patches corresponding to each of the HA UDL cases 3—10 respectively.

In this bridge deck there are only 4 notional lanes and if each of them had been of the order of 12 ft, then they would have been adequate to accommodate the HB vehicle and it would then have been possible to reduce the total number of HA load cases from 24 to 12. Further the HA loading consists of a UDL and a Knife Edge load and of these the program package will automatically calculate the nodal forces for the UDL. The equivalent nodal forces for the Knife Edge load have to be calculated manually and this is a tedious operation. To investigate BM's and shears it is necessary to run separate load cases for UDL and KE loads. The calculation of equivalent nodal forces of HB load cases has also to be performed manually and this in a sense explains the reluctance of the designers to examine more longitudinal positions of the HB vehicle.

3.3.5. Data preparation and computer run

As this was a parallel sided skew bridge it was possible to use the program MIPUT to prepare both the geometry and UDL loading data. For the KEL and HB point load cases the nodal forces had to be calculated and the relevant cards punched by hand and added to the rest of the MIPUT data.

The data was first run through the program FECK. Error messages were printed out stating that overlapping property patches were inadmissable. However, the user manual states that these can be used. Therefore no corrections were made on this account and it was later found that the MIPUT program did in fact accept these overlapping property patches. After other errors had been corrected the data was run through MIPUT and then through MBAPS. Records were not kept at the time the data was being prepared, but it is estimated that the data preparation took one engineer four weeks. Only nodal average forces were output from the Computer.

The estimated run time for MBAPS and the 35 loading cases was 4 hrs 31·5 mins made up of initial compilation 9·5 mins, first loading case 38·5 mins and 6·75 mins, for each subsequent loading case. The actual computer time was 4 hrs 44 mins made up of corresponding figures of 9·00 mins; 33·5 mins and 7·1 mins per subsequent loading case respectively.

3.3.6. Load combination and reinforcement

The output moments from MBAPS were stored on magnetic tape and 15 load combinations were examined. The load combinations are shown in tabular form in Table 1. Combinations 2–11 are for bending and 12–15 are for maximum shears and reactions. The major principal moments under dead load tended to be normal to the abutments. If the slab had been solid in-situ concrete, the two reinforcement directions would have been selected as normal and parallel to the abutments. It was felt however that the presence of voids running parallel to the edges may have some effect on the direction of bending and therefore the directions selected for reinforcement were parallel to the skew and orthogonal to it. The resulting steel arrangement calculated from the results from MOPUT are shown in Fig. 3. The computer time

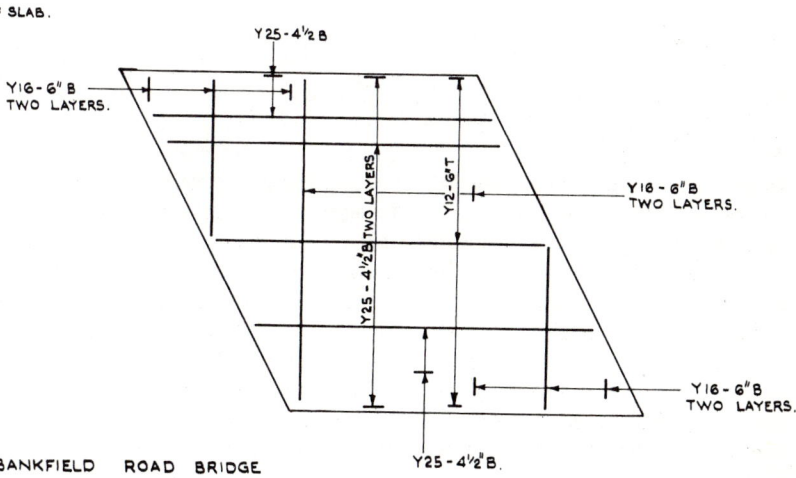

DETAILING KEY.
EXAMPLE Y25-4½"B
Y = HIGH YIELD BARS
25 = BAR DIA.(mm)
4½" = BAR SPACING.
B = BOTTOM OF SLAB.

FIG. 3.   BANKFIELD ROAD BRIDGE
           DECK REINFORCEMENT.

for MOPUT for all 15 combinations was 21·5 minutes and the time taken by one engineer to examine the results and calculate the reinforcement was about three weeks.

3.3.7. In the above example a large number of load cases have been analysed because of the lane width problem mentioned earlier. In a subsequent problem, only 12 load cases were analysed, each load case being a combination giving maximum moments or shears. From the times of the computer runs it is clear that load combinations do not take as much time as load·cases. Therefore it would be useful if some method was devised which would cut down on the number of load cases that have to be considered.

4. CURVED BRIDGES

4.1. Two curved slab bridges have been analysed both of them of similar type. Therefore only the analysis of one bridge, the first to be analysed, will be described in detail. Both bridges have however been analysed by the same design team and therefore the times taken for the second bridge will be compared with those for the first to show that with practice the times can be reduced.

4.2. Rough Heanor Bridge

4.3.1. Bridge description

Rough Heanor bridge is an overbridge carrying a side road over a dual two lane highway. At the crossing the highway is on a curve of 22860·0 m radius while the side road itself is on a sharp curve of 164 m radius and the tangential skew at the point of intersection is 25° 44 ft 17 in. Because of the sharp curvature of the side road, the bridge had to be widened on the inside of the curve for sight line reasons. It was clear therefore that the bridge length had to be kept to the minimum and therefore a two span bridge was adopted with wingwalls parallel to the highway as shown in the plan of the bridge in Fig. 4. As good foundation material was present it was decided to make the bridge continuous with four discrete columns making the intermediate supports. These columns were built into the deck but pinned at the base.

4.3.2. Mesh division

The total width of the deck is 15·58 m consisting of 14·38 m of deck slab 0·6 m thick and two cantilever overhangs each 0·8 m as shown in the X-section in Fig. 4. The cantilever overhangs being small have been left out of the analysis altogether. The carriageway width of 10·19 m was divided into three notional lanes. Longitudinally therefore the deck was divided into five load patches made up of a 0·798 m outside verge, the three lanes totalling 10·19 m and the inside verge of 3·392 m.

On the outside, the edge of the carriageway was very close to the centre of column 1 (Fig. 4). For the purpose of the analysis they were assumed to be coincident. It was therefore possible to divide carriageway lanes 1 and 2 into three equal divisions and have mesh lines coincident with columns 1 and 2. Lane 3 however had to be divided into four divisions as column 3 was situated approximately in the middle of this lane and two divisions looked to be too coarse. The outer verge was divided into three divisions which were not equal as one mesh line had to pass through column 4. In the other direction the mesh lines were maintained at 2 m centres except at the supports. At the Abutments it was attempted to have at least one division which was about a third of the general mesh size and at the columns about three on either side of a size smaller than the diameter plus depth of slab. The mesh that was finally adopted is shown in Fig. 5. It should be noted that the bridge is on a curve and the supports are all parallel to one another and do not lie on radials of the curve.

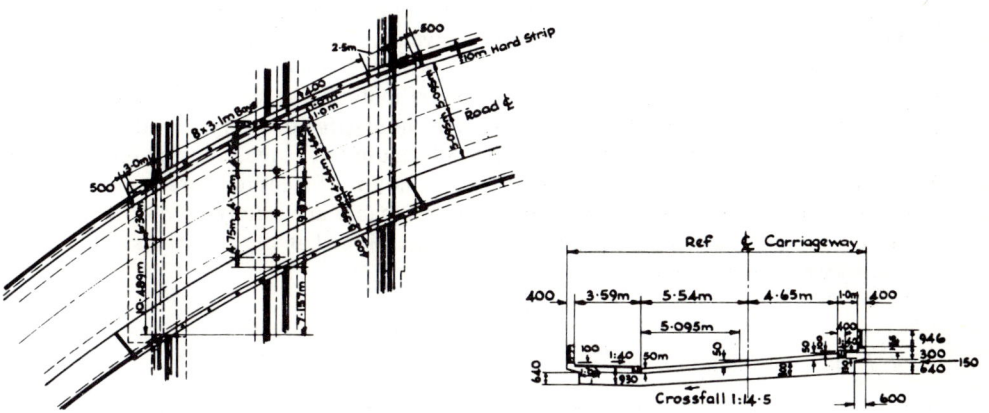

FIG. 4 ROUGH HEANOR BRIDGE. PLAN & DECK CROSS SECTION

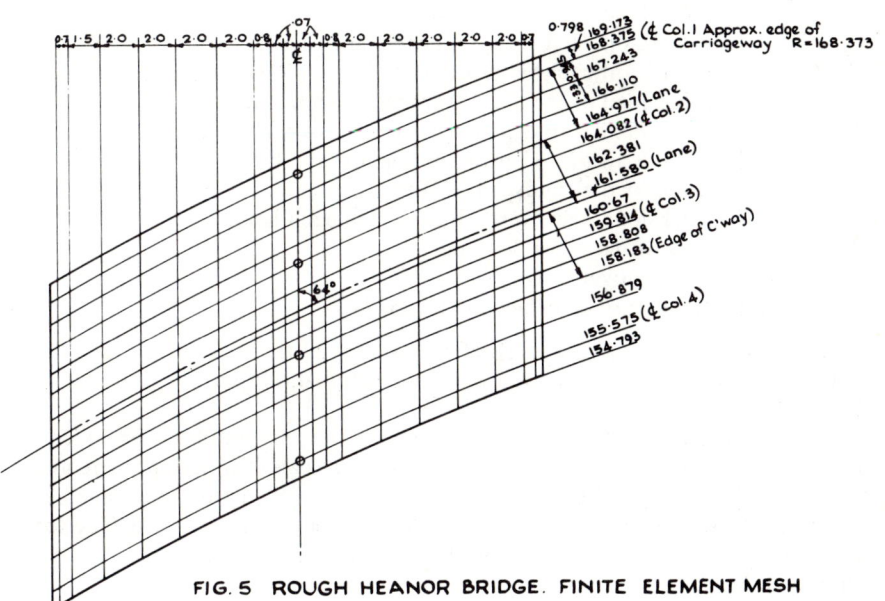

FIG. 5 ROUGH HEANOR BRIDGE. FINITE ELEMENT MESH

4.3.3. Material properties and boundary conditions

The edge cantilevers were ignored for the analysis and therefore the slab that was analysed was of uniform section. The calculation of the material properties was straight forward and was carried out as stated in the User Manual. The deck was assumed to be fixed at one abutment and free at the other, the intermediate columns being flexible enough to allow temperature movements to transfer to the free end. It was assumed that rubber bearings would be provided at each abutment and as in the previous example the stiffnesses had to be estimated for each of the nodes along the centre line of the bearings. The calculation of the elastic boundary conditions at the column supports presented certain difficulties because both areas and the second moments of areas varied considerably with the amount of steel reinforcement that was provided. This again had to be estimated reasonably accurately before the computer run could be executed. As the columns were pinned at the base the stiffnesses k_2 and k_3 were taken to be $3EI/L$.

4.3.4. Loads and loading cases

As this is a continuous bridge a larger number of loading combinations had to be looked at to investigate the effects of one span and both spans being loaded with both HA and HB loading cases. However, as each notional lane was wide enough to accommodate the HB load, the additional problems introduced in the previous example did not arise. In all the following 38 load cases were analysed.

    1. All dead loads including edge loads.
  2–6. Unit UDL on each verge and each lane of East Span.
 7–11. Unit UDL on each verge and each lane of West Span.
12–14. Unit KE load over East Abutment in each lane.
15–17. Unit KE load at midspan of East Span in each lane.
18–20. Unit KE load over Centre Support in each lane.
21–23. Unit KE load at midspan of West Span in each lane.
24–26. Unit KE load over West Abutment in each lane.
27–36. Ten positions of HB loading.
   37. Thermal movement − Contraction.
   38. Settlement of intermediate supports.

Assuming that the columns would deflect by the full thermal movement, the moments produced by contraction were calculated and input as external loads on the whole system.

It was also considered that some degree of settlement should be provided for as the foundations were not on solid rock. However, as all foundations were at the same level, the amount of differential settlement was estimated to be small and a 13 mm differential settlement of intermediate supports was assumed. The force P required to compress the columns by this amount was calculated and input at the column nodes as external forces to the whole system.

4.3.5. Data preparation and computer run

As stated previously the supports are all parallel to one another and therefore the program package could not be used to determine nodal co-ordinates, etc., and the simple procedure of going through FECK- MIPUT to MBAPS could not be adopted. To use the package as it was, it would have been necessary to calculate all nodal co-ordinates and element connectivity by hand. The following steps were adopted in order to eliminate this hand calculation:

1. An existing program to calculate curved bridge geometry was used to generate nodal co-ordinates and numbers.
2. An ancillary program was written to transform results from above program and produce cards in correct fields for MBAPS.
3. Program MIPUT was used to produce element cards only.
4. A plotter program was used to plot nodes and elements and check the mesh geometry.

The rest of the steps taken are the same as in the program package. A flow chart of the full procedure is shown in Fig. 6. The nodal forces equivalent to point loads and knife edge loads had to be calculated manually as in the previous example. As the

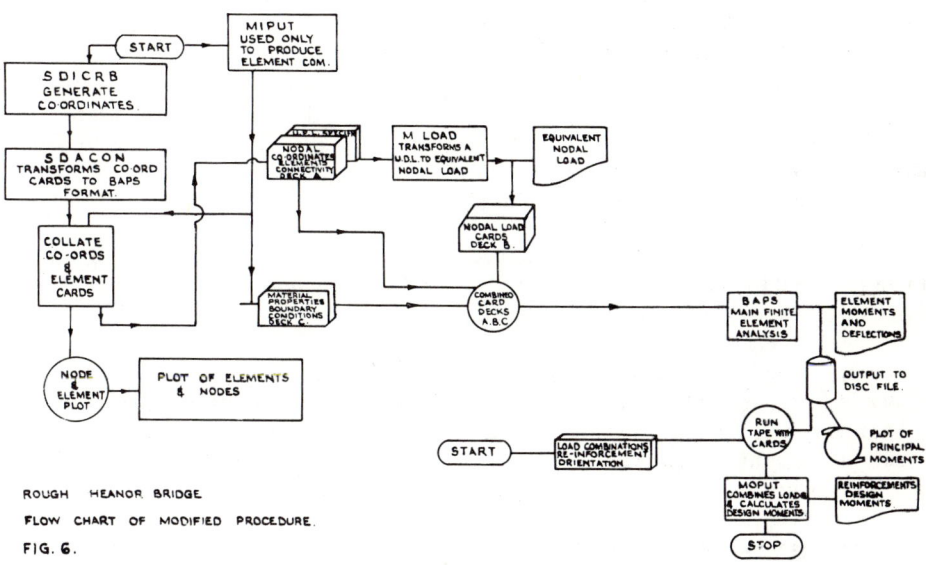

ROUGH HEANOR BRIDGE

FLOW CHART OF MODIFIED PROCEDURE.

FIG. 6.

mesh lines were parallel to the supports and therefore non-radial the knife edge load always cuts through elements and this makes the task of calculating equivalent nodal forces even more tedious. The task of preparing the data from the time of deciding the bridge type is estimated to have taken one engineer four months. Checking of arithmetic of nodal forces, etc., is estimated at a further 2–3 weeks of technician time. The run times on the computer are as follows:

SDICROBOI		2 mins
SDACON		5 mins
MIPUT		6 mins
CJPL		9 mins
MBAPS		402 mins
FECK		1 min
	Total	425 mins
No. of Elements		504
No. of Nodes		285

4.3.6. Load combinations and reinforcement

In all 27 load combinations were carried out to determine maximum positive and negative moments in each span and also shears at all three supports. It was necessary to examine moments and shears in each span because the abutments are parallel to one another and hence the bridge is not symmetrical about the centre support. The combinations that were carried out are shown in Table 2. It can be seen that the multipliers for some of the HA-UDL load cases vary from 9·95 to 10·5. When the multiplier is 10·5 it means that only one span is loaded and therefore the loaded length is smaller than when both spans are loaded in which event the multiplier becomes 9·95. The computer time for the 27 load combinations was 58 mins. The major principal moments due to dead load were again in a direction normal to the supports. The design moments were therefore first calculated normal and parallel to the supports. At the edges of the slab however the steel was placed along the direction of the bridge and it was found that the edge reinforcement and the centre reinforcement combined gave four layers of steel in top and bottom. It was considered that concreting of the edges of the slab would be difficult if not impossible and therefore the design moments were recalculated in directions parallel to and normal to the edges of the slab. Even in this case, at the supports the transverse steel was placed parallel to the supports. However the final steel arrangement was considered to be better and it turned out the total quantity of steel was not substantially different. The inspection of results from MOPUT and preparation of the steel arrangement took one engineer one month.

4.3.7. The number of load cases was reduced to 25 for the second curved bridge and in fact the time for data preparation was halved. The computer time up to the MBAPS run was also reduced to 300 minutes.

5. In carrying out the analysis of these structures it has become apparent that a variety of improvements could be carried out to reduce the amount of time spent by the engineer in preparing the data, examining the results and may be even in actual run time on the computer. The suggested improvements are set out below under these three headings.

5.1. Improvements to Facilitate Input

(*a*) Automatic generation of the mesh for all likely geometries. The second example described is a typical case which could not be handled by the IPUT program.

(*b*) Equivalent nodal forces due to point loads and line loads to be generated automatically, and for these loads to be specified easily. At present this is a laborious manual operation. It is suggested that two pairs of co-ordinates could completely specify the position of an HB vehicle or a knife edge load and giving the load patch within which the load was applied could reduce the amount of computer time to allocate these loads to elements.

(*c*) Additional loading cases to be analysed in a subsequent run without having to input the structure. If this was available it may be possible to reduce the total number of loading cases that were considered and further it would also overcome the reluctance of the designer to examine additional loading cases.

(*d*) Re-analyse the structure with changed boundary conditions with same loads as before or changed loads, but keeping the rest of the structure the same. It has been stated that the stiffness of rubber bearings and columns had to be estimated prior to the run. If the results showed that either of these were inadequate it should be easy enough to estimate the required stiffness and re-analyse the structure.

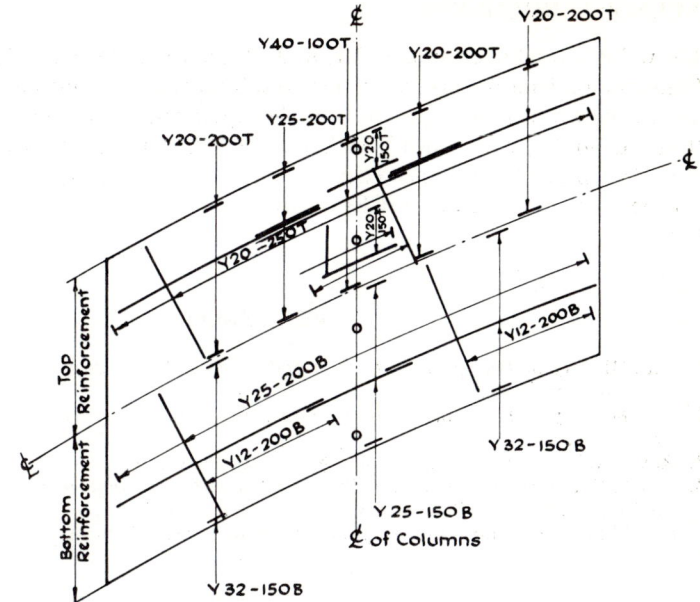

Detailing Key
Example Y40-100 T
Y High Yield Bars
40 = Bar dia (mm)
100=Spacing in (mms)
T = Top of slab

FIG. 7   ROUGH HEANOR BRIDGE. DECK REINFORCEMENT.

5.2. Improvement to Output Facilities

(*a*) A plot of the mesh to check that the input data is correct.

(*b*) Results from a selected area to be output separately, and the rest of the results to be output only optionally. Each load case or combination is usually run to examine either moment or shear in a selected area. Results over this area are therefore of prime importance for that case or combination and the designer should be able to see these without having to wade through reams of results.

(*c*) A vectorial plot of principal moments for any load combination. The designer should have the facility to examine these and decide on the direction of reinforcement before specifying the directions in which he requires his design moments $M*x$ and $M*y$.

(*d*) Automatic calculation of reactions at elastic boundaries for each load combination. MOPUT only combines forces and not displacements. Therefore the process of calculating the reactions on rubber bearings due to combinations of load involves lengthy hand calculations as shown in Table 3.

5.3. Reduction in Computer Time

It can be seen from the load cases that have been considered that in each example there are about 20 HA load cases which account for about 150 minutes of computer time. HA and HB load cases have to be considered as different load cases now because the HB loading is a vehicle and the HA load is a UDL with a knife edge load. If however the HA and HB loading could both be vehicles with same number of axles and wheels all at the same spacing, but the HA wheel loads not being the same as the HB, then separate load cases need not be run for HA and HB. In combining the dead load with HA, or HB or both, different multipliers could be used for HA and the HB loads. It is considered that if such a vehicle could be evolved then the computer times could be more than halved not only for this BAPS program package but for all bridge deck analysis programs.

ACKNOWLEDGEMENTS

The writer would like to thank Mr. R. J. Bridle FICE, Director, Midland Road Construction Unit for his permission to submit this contribution. Thanks are also due to the writer's colleagues in Unit Headquarters, the Warwickshire Sub-Unit (Chief Engineer, Mr. D. Morris MC, TD, BSc, FICE), the Derbyshire Sub-Unit (Chief Engineer, Mr. G. Race BSc, FICE), and the Staffordshire Sub-Unit (Chief Engineer, Mr. R. Musgrave, FICE) for their help in the preparation of this contribution.

REFERENCES

1. G.B. Ministry of Transport. *Suite of bridge design and analysis programs.* BECP/1. 1969. Vol. 1 – User manual.
2. G.B. Ministry of Transport. *Suite of bridge design and analysis programs.* BECP/1. 1969. Vol. 2 – Report.
3. Rowe, R. E. *Concrete bridge design.* New York, Wiley, 1962.
4. Rüsch, H. and Hergenroder, R. *Influence surfaces for moments in skew slabs*; trans. from the German. London, Cement & Concrete Association, 1964.
5. G.B. Ministry of Transport. Bridges Engineering Division Information Document No. 23.

TABLE 1

MULTIPLIERS FOR LOAD COMBINATION

Combination No.	No. of Cases in Combination	D_L 1	F_P 2	HA UDL Load Cases 3	4	5	6	7	8	9	10	HB Load Cases 11	12	13	14	15	16	4 Ton Wheel 17	18	19	HAKEL Load Cases 20	21	22	23	24	25	26	27	HA KEL Load Cases 28	29	30	31	32	33	34	35
1	1	1																																			
2	2	1																1																		
3	2	1																	1																	
4	2	1																		1																
5	18	1	1	1	1	1	1	1	1/3	1/3	1/3										1	1	1	1	1	1/3	1/3	1/3								
6	18	1	1	1/3	1/3	1/3	1	1	1	1	1										1/3	1/3	1/3	1	1	1	1	1								
7	15	1	1	1/3	1/3	1/3	1/3	1/3	1/3	1/3	1/3			1							1/3	1/3	1/3	1/3	1/3	1/3	1/3	1/3								
8	15	1	1	1/3	1/3	1/3	1/3	1/3	1/3	1/3	1/3	1									1/3	1/3	1/3	1/3	1/3	1/3	1/3	1/3								
9	2	1											1																							
10	18	1	1	1/3	1/3	1	1	1	1	1	1		1								1/3	1/3	1	1	1	1	1	1								
11	15	1	1	1/3	1/3	1/3	1/3	1/3	1/3	1/3	1/3										1/3	1/3	1/3	1/3	1/3	1/3	1/3	1/3								
12	15	1	1	1/3	1/3	1/3	1/3	1/3	1/3	1/3	1/3						1																			
13	15	1	1	1/3	1/3	1/3	1/3	1/3	1/3	1/3	1/3				1														1/3	1/3	1/3	1/3	1/3	1/3	1/3	1/3
14	15	1	1	1/3	1/3	1/3	1/3	1/3	1/3	1/3	1/3						1												1/3	1/3	1/3	1/3	1/3	1/3	1/3	1/3
15	15	1	1	1	1	1	1	1	1	1	1																		1/3	1/3	1/3	1/3	1/3	1/3	1/3	1/3

25% overstress allowed when these load cases are combined.

TABLE 2

Combination No.	No. of Cases in Combination	DL	Verges		HA-UDL								HA-KE LOAD				
		1	2	3	4	5	6	7	8	9	10	11	12	13	14	15	16
1		1·0	5·0	5·0	9·95	9·95	3·32	5·0	5·0	9·95	9·95	3·32					
2		1·0		5·0					5·0	10·5	10·5	3·5					
3		1·0	5·0	5·0	9·95	9·95	3·32	5·0	5·0	9·95	9·95	3·32					
4		1·0	5·0	5·0	3·32	9·95	9·95	5·0	5·0	3·32	9·95	9·95					
5		1·0		5·0					5·0	3·5	10·5	10·5					
6		1·0	5·0	5·0	3·32	9·95	9·95	5·0	5·0	3·32	9·95	9·95					
7		1·0		5·0							3·5	3·5					
8		1·0		5·0							3·5	3·5					
9		1·0								3·5		3·5					
10		1·0															
11		1·0	5·0	5·0	3·32	3·32		5·0	5·0	3·32	3·32						
12		1·0	5·0	5·0		3·32	3·32	5·0	5·0		3·32	3·32					
13		1·0			3·5	3·5		5·0								13·3	13·3
14		1·0	5·0			3·5	3·5										13·3
15		1·0	5·0	5·0	9·95	9·95	3·32	5·0	5·0	9·95	9·95	3·32					
16		1·0		5·0					5·0	10·5	10·5	3·5					
17		1·0	5·0	5·0	9·95	9·95	3·32	5·0	5·0	9·95	9·95	3·32					
18		1·0	5·0	5·0	3·32	9·95	9·95	5·0	5·0	3·32	9·95	9·95					
19		1·0		5·0					5·0	3·5	10·5	10·5					
20		1·0	5·0	5·0	3·32	9·95	9·95	5·0	5·0	3·32	9·95	9·95					
21		1·0		5·0							3·5	3·5					
22		1·0		5·0							3·5	3·5					
23		1·0															
24		1·0	5·0	5·0	3·32	3·32		5·0	5·0	3·32	3·32						
25		1·0	5·0	5·0		3·32	3·32	5·0	5·0		3·32	3·32					
26		1·0			3·5	3·5		5·0								13·3	13·3
27		1·0	5·0			3·5	3·5										13·3

LOADS A – No overstress in design.

LOADS A – Any of Loads A + Load C – 25% overs⸱

TABLE 2

MULTIPLIERS FOR LOAD COMBINATIONS

HA-KE LOAD								HB-LOAD										TEMP.	SET.
19	20	21	22	23	24	25	26	27	28	29	30	31	32	33	34	35	36	37	38
		40	40	13·3															
		40	40	13·3															
40	13·3																		
		13·3	40	40															
		13·3	40	40															
40	40																		
			13·3	13·3				1·0											
						13·3	13·3		1·0										
		13·3		13·3						1·0									
											1·0								
13·3		13·3										1·0							
			13·3	13·3									1·0						
															1·0				
																	1·0		
		40	40	13·3														1·0	−1·0
		40	40	13·3														−1·0	1·0
40	13·3																	1·0	−1·0
		13·3	40	40														1·0	−1·0
		13·3	40	40														−1·0	1·0
40	40																	1·0	−1·0
			13·3	13·3				1·0										−1·0	1·0
						13·3	13·3		1·0									−1·0	1·0
											1·0							1·0	1·0
13·3		13·3										1·0						1·0	−1·0
			13·3	13·3											1·0			1·0	−1·0
																	1·0	1·0	1·0
																		1·0	1·0

LOADS B – 25% overstress in design.   LOADS C

Any of Loads B + Load C – 30% overstress in design

TABLE 3

Rough Heanor Bridge
Reaction at East Abutment – Combination 11

Node No.	1	2	7	34	4	5	12	13	Load case		
	1·0	5·0	5·0	1·0	3·5	3·5	13·3	13·3	Load Coefficient		
	Deflections $8(\times 10^6)$								Total	Stiffness $k \times 10^{-3}$	Reaction kN
1	401	403	−7	−77	42	5	34	−3	426	106	45·1
2	393	32	−5	−50	42	1	36	1	449	258	116
3	391	22	−4	−15	42	7	37	7	486	302	147
4	392	13	−3	15	40	16	33	15	523	303	158
5	396	7	−3	45	35	25	25	25	555	271	150
6	398	5	−3	73	30	31	16	31	580	287	167
7	395	2	−2	120	22	35	8	32	614	335	206
8	396		−1	176	15	34	3	25	648	286	185
9	396	−	3	220	11	30	1	17	679	237	161
10	401	−1	11	261	8	26	−	11	717	229	164
11	413	−1	20	290	5	21	−1	6	754	219	165
12	436	−1	36	309	3	17	−1	3	803	285	229
13	507	−2	76	323	−1	10	−1	−	914	351	321
14	637	−3	136	338	−6	4	−1	−1	1104	281	310
15	751	−4	186	360	−10	−	−	−2	1280	106	136

ELASTOPLASTIC ANALYSIS OF REINFORCED CONCRETE SLABS—SOME IMPLICATIONS OF LIMIT STATE DESIGN

LLOYD C. P. YAM and P. C. DAS
Road Research Laboratory

SYNOPSIS

A numerical method for the elastoplastic analysis of plates and slabs is presented. The method is based on the step-by-step solution of linearised partial differential equations. No iteration is required and the method is found to be convergent. Using the 'square' yield criterion, the method is applied to study the behaviour of square isotropic slabs from zero load to collapse. The occurrences of large deflections, excessive strains and collapse are studied. These results are compared with those estimated by a simplified design procedure. The validity and limitations of the latter are discussed.

1. INTRODUCTION

The recent introduction of the concept of limit state design into the draft Unified Code for structural concrete can be regarded as a considerable advance towards a more rational method of design. The evaluation of the probabilities of structures becoming unserviceable is a realistic interpretation of actual structural behaviour. The use of characteristic loads and strengths together with their respective partial safety factors allows for their inherent variability and acceptable low risks of reaching the various limit states. Reference[1] gives an introduction to the limit state analysis of reinforced concrete beams and slabs. The application to the design of highway bridges is given in Reference.[2]

At the time this paper was written, the Unified Code was in the draft form. Improvements will no doubt be incorporated with further research and extensive usage. However, some inherent difficulties seem immediately obvious in view of the use of a simple linear approach to interpret a generally inelastic behaviour.

One way of examining such a simplified approach is to obtain a complete inelastic analysis of the designed structure from zero load to collapse. The various appropriate limit states can then be compared with those estimated by the simplified approach. This study is one of the objectives of the recent development of an analytical method at RRL for the elastoplastic bending of plates and slabs. The purpose of this paper is to illustrate the application of the analytical approach to clarify various design implications.

147

2. REVIEW OF THEORIES OF INELASTIC BENDING OF PLATES

The study of the ultimate strength of rectangular slabs was first published half a century ago in the first issue of the Journal of the Institution of Structural Engineers.[3] The extensive development of the yield-line theory, however, was due to Johansen.[4] The theory enabled upper-bound values of ultimate loads to be obtained by simple calculations. A rigorous foundation for the ultimate load calculation was later provided by the Limit Analysis of Prager and Hodge,[5] who used upper-bound and lower-bound methods to establish 'exact' solutions based on coincidental bounds. Further study and extension of these methods were made by Wood[6] and Jones,[7] whose work has been most useful to both designers and research workers.

By the mid-sixties, when computer application had become the rule rather than the exception in theoretical study, research workers were not so inhibited by computational difficulty and elastoplastic analysis began to emerge. In 1966, Massonnet[8] presented a theory of elastoplastic analysis and suggested an iterative method of solution using finite differences. Cornelis[9] applied the method to analyse a simply supported plate carrying uniform load, assuming an elastic-perfectly plastic material. A similar method was published in America by Bhanmik and Hanley,[10] who were apparently unaware of the work on the other side of the Atlantic. Yet another similar iterative method was published in 1968 by Lin and Ho,[11] who made use of tabulated influence coefficients[12] to reduce the amount of computation. In these three methods, plastic deformations due to each load increment were obtained by iteration. Apart from the large amount of computation, the main drawback these methods have in common is the uncertainty of numerical convergence. This tends to confine the application of the theories to a limited number of expert analysis. Furthermore, there has been little sign of any extension of these methods to examples other than a few simple cases.

A different approach called the 'Discrete Model Analysis' was proposed by Ang and Lopez,[13] who replaced the plate by a grillage of rigid bars connected at deformable nodes, thus confining the iteration to a finite number of plastic nodes. The solutions appear to be reasonable, but it is difficult to appreciate the physical implication of such structural idealisation.

The first application of the Finite Element Method to the elastoplastic bending problem was due to McNiece.[14] However, the violation of the yield criterion within the element casts doubt on the validity of the method. Also, a considerable amount of computer time and storage was required.

Thus, no satisfactory method of elastoplastic analysis is available for a wider application to study the limit state behaviour of slabs: for example, the collapse load and mechanism of a clamped isotropic square slab have not been found.

3. THEORY

A method for the analysis of elastoplastic bending of thin plates is presented below. The stumbling-block as regards numerical convergence is removed by linearising the governing partial differential equation. Inelastic deformations for each load increment are obtained from the linear partial differential equation without involving iteration.

The following formulation is based on the normal assumptions for the bending of thin plates in which membrane actions are neglected. No specific material is referred to, but the material is assumed to be homogeneous and isotropic, to strain-harden isotropically and to obey a definite yield criterion of moments and the associated

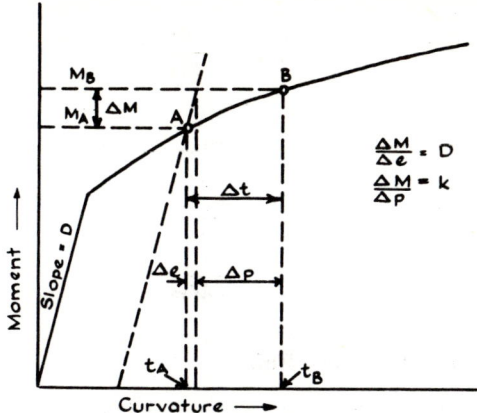

**FIG. I IDEALISED MOMENT – CURVATURE RELATION
FOR UNIAXIAL BENDING**

flow rule. The assumed moment-curvature relation for uniaxial bending is shown in Fig. 1.

The load-increment method of analysis is used, and the governing equations will involve changes of moments and deflections due to the load increment.

3.1. Condition of Equilibrium

The equation of equilibrium for elastic bending of thin plates still holds. In terms of moment changes, it takes the form:

$$\frac{\partial^2 mx}{\partial x^2} + \frac{\partial^2 my}{\partial y^2} - 2\frac{\partial^2 mxy}{\partial x \partial y} = -\Delta q \tag{1}$$

3.2. Yield Criterion

A general yield criterion is given by the function:

$$\emptyset(Mx, My, Mxy) = 0$$

With strain-hardening, an effective plastic moment (Mp) can be introduced which increases in magnitude as plastic flow takes place:

$$f(Mx, My, Mxy, Mp) = 0 \tag{2}$$

The function is such that under the condition of uniaxial bending, the uniaxial bending moment is equal to Mp.

Differentiating (2),

$$-\frac{\partial f}{\partial Mp} \Delta Mp = \frac{\partial f}{\partial Mx} \Delta Mx + \frac{\partial f}{\partial My} \Delta My + \frac{\partial f}{\partial Mxy} \Delta Mxy$$

i.e.

$$F_0 mp = F_1 mx + F_2 my + F_3 mxy \tag{3}$$

where
$$F_0 = -\frac{\partial f}{\partial Mp}$$

$$F_1 = \frac{\partial f}{\partial Mx}$$

$$F_2 = \frac{\partial f}{\partial My} \tag{4}$$

$$F_3 = \frac{\partial f}{\partial Mxy}$$

3.3. Law of Plastic Potential (Flow Rule)

The plastic curvature increments at a point at any instant of loading are given by:

$$\Delta px = \lambda \frac{\partial f}{\partial Mx} = \lambda F_1$$

$$\Delta py = \lambda \frac{\partial f}{\partial My} = \lambda F_2 \tag{5}$$

$$\Delta pxy = \lambda \frac{\partial f}{\partial Mxy} = \lambda F_3$$

where λ is a constant for the point on the plate during the present load increment.

3.4. Moment-Curvature Relations

If a plate in an inelastic state is subject to a small load increment, the change in curvature at a point in the inelastic region may be separated into an elastic component and a plastic component. The total curvature at the end of the load increment is:

$$t = t' + \Delta t \tag{6}$$
$$\Delta t = \Delta e + \Delta p \tag{7}$$
where
    t' = total curvature before load increment
  Δe = change in elastic component
  Δp = change in plastic component.
The linear moment-curvature relations remain valid for the elastic components

$$mx = D(\Delta ex + vey)$$
$$my = D(\Delta ey + v\Delta ex) \tag{8}$$
$$mxy = D(1 - v)\Delta exy$$

The relation between the increments of plastic curvature and plastic moment due to strain-hardening is more complicated. The subject of strain-hardening is treated in great detail in Mendelson's book.[15] The experimental moment-curvature relation for uniaxial bending (Fig. 1) is used. Denoting the curve as an effective moment-effective curvature relation, the desired relation can then be obtained by defining effective moment as Mp in equation (2) and effective plastic curvature by the following equation:

$$\Delta p = \Delta p(\Delta px, \Delta py, \Delta pxy) = A\lambda \tag{9}$$

where A is a function of F_1, F_2 and F_3 (equations 4) and depends on the yield criterion assumed.

The value of k (Fig. 1) will be used to calculate the increase in plastic moment due to plastic flow:

$$\Delta Mp = k\Delta p = kA\lambda \tag{10}$$

3.5. Derivation of Governing Equations

The plate is in equilibrium under a system of loads in excess of the elastic limit. It is assumed that the distributions of plastic moments and curvatures are known. A small load increment is then applied in such a manner that no 'unloading' takes place. A differential equation will be derived which gives the change in deflections (w) due to the load increment.

The changes in total curvatures at a point are:

$$\left. \begin{aligned}
\Delta t_x &= -\frac{\partial^2 w}{\partial x^2} = -d_{20}w \\[2mm]
\Delta t_y &= -\frac{\partial^2 w}{\partial y^2} = -d_2 w \\[2mm]
\Delta t_{xy} &= \frac{\partial^2 w}{\partial x \partial y} = d_{11}w
\end{aligned} \right\} \tag{11}$$

where the d — operator is defined by

$$dmn = \frac{\partial(m+n)}{\partial x^m \partial y^n}$$

The corresponding moment changes at the point are given by equations (8), (7) and (5):

$$\left. \begin{aligned}
mx &= -D[(d_{20} + vd_2)w + \lambda(F_1 + vF_2)] \\
my &= -D[(d_2 + vd_{20})w + \lambda(F_2 + vF_1)] \\
mxy &= D(1-v)(d_{11}w - \lambda F_3)
\end{aligned} \right\} \tag{12}$$

Substituting the above moment changes into equation (3) and noting that $mp = kA\lambda$ (equation (10)), the following equation of elastoplastic deformation is obtained:

$$\lambda = (\alpha d_{20} + \beta d_2 + \gamma d_{11})w \tag{13}$$

where

$$\left. \begin{aligned}
\alpha &= -B(F_1 + vF_2) \\
\beta &= -B(F_2 + vF_1) \\
\gamma &= B(1-v)F_3
\end{aligned} \right\} \tag{14}$$

$$\frac{1}{B} = \frac{kAF_0}{D} + [F_1^2 + F_2^2 + (1-v)F_3^2 + 2vF_1F_2] \tag{15}$$

It should be noted that equation (13) embodies the yield criterion, the flow rule and the complete moment — curvature relation of the plate. This equation of elasto-plastic deformation and the equation of equilibrium are the governing equations for the elastoplastic bending of thin plates.

In order to derive a single partial differential equation consisting of w as the only dependent variable, the equation of equilibrium (equation (1)) is first combined with equations (12) to give:

$$(d_{40} + 2d_{22} + d_4)w = \frac{\Delta q}{D} - [d_{20}(\lambda R) + d_2(\lambda S) + 2d_{11}(\lambda T)] \tag{16}$$

where
$$R = F_1 + vF_2 \\ S = F_2 + vF_1 \\ T = -(1-v)F_3 \quad\rbrace \tag{17}$$

Using differentiation by parts, λ can be eliminated from equations (16) and (13) to give:

$$(c_1 d_{40} + c_2 d_{31} + c_3 d_{22} + c_4 d_{13} + c_5 d_4 \\ + c_6 d_{30} + c_7 d_{21} + c_8 d_{12} + c_9 d_3 \\ + c_{10} d_{20} + c_{11} d_{11} + c_{12} d_2)w = \frac{\Delta q}{D} \tag{18}$$

The formulae for the twelve coefficients (c_1 to c_{12}) are given in the Appendix. Their values are known at each point so that the change in deflections due to the load increment can be obtained by solving equation (18) to satisfy the boundary conditions. It should be noted that the equation is also applicable to the elastic region of the plate, since k is infinitely large there and so λ vanishes (Fig. 1 and equations (13), (14) and (15)). The equation then reduces to the familiar biharmonic form (see also equation (16)).

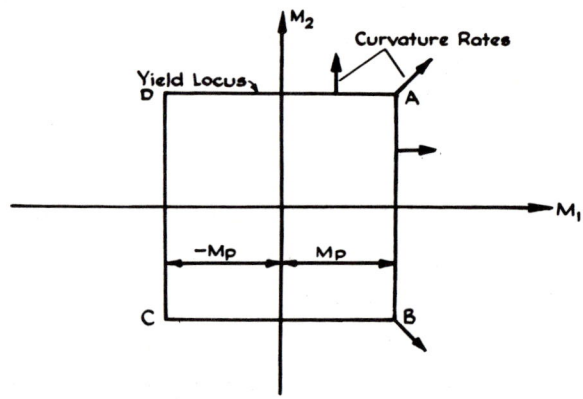

FIG. 2   THE SQUARE YIELD CRITERION FOR ISOTROPIC SLABS

3.6. Method of Solution

Various methods can be used to solve the partial differential equation (18). The application of the Finite Element Method to this equation using the Internal Equilibrium model is dealt with in reference[16]. The method of Finite Difference is used here for illustrative purposes. The square yield criterion is adopted for reinforced concrete slabs. Equations (2) and (4) for various cases are given in Appendix 2.

The procedure of computation is as follows:

3.6.1. **Degree of strain-hardening.** Starting with a loaded plate with known distributions of plastic moments and curvatures, the values of Mp and k at every point in the plastic region are obtained from equation (19) (Appendix 2) and the effective moment — effective curvature relation (Fig. 1). For points in the elastic region, there is no plastic curvature and k is infinitely large.

3.6.2. **Distribution of $\alpha \beta \gamma$ — values.** For every point in the plastic region, the values of F_1, F_2, F_3, A and B are calculated from equations (20), (21), (Appendix 2) and (15). The distributions of $\alpha, \beta, \gamma, R, S$ and T are then calculated from equations (14) and (17).

3.6.3. **Calculation of coefficients.** The twelve coefficients of equation (18) for each point can be calculated from the formulae given in Appendix 1. The partial differentials required are obtained by the method of Finite Differences.

3.6.4. **Solving for deflections.** After calculating the coefficients of the partial differential equation at each point, the solution of w for a given load increment and a given set of boundary conditions is reduced to the solution of a system of linear simultaneous equations.

3.6.5. **Calculation of plastic moments and curvatures.** The effective plastic curvature (Δp) at any point in the plastic region can be calculated from equations (13) and (9). Then, from the effective moment — effective curvature curve, the values of Mp and k are obtained. The moment changes due to the load increment are calculated from equations (12).

3.6.6. **Collapse by incremental loading.** For the next load increment, the same procedure starting from 3.6.1 can be followed. The collapse load is reached when the stiffness matrix of the plastic plate becomes singular. The approach to this condition is usually revealed by the presence of deflections which are very large compared with those at first yield.

4. EXAMPLES

A computer program was written using this method of analysis outlined above, except that equations (13) and (16) were solved instead of (18).

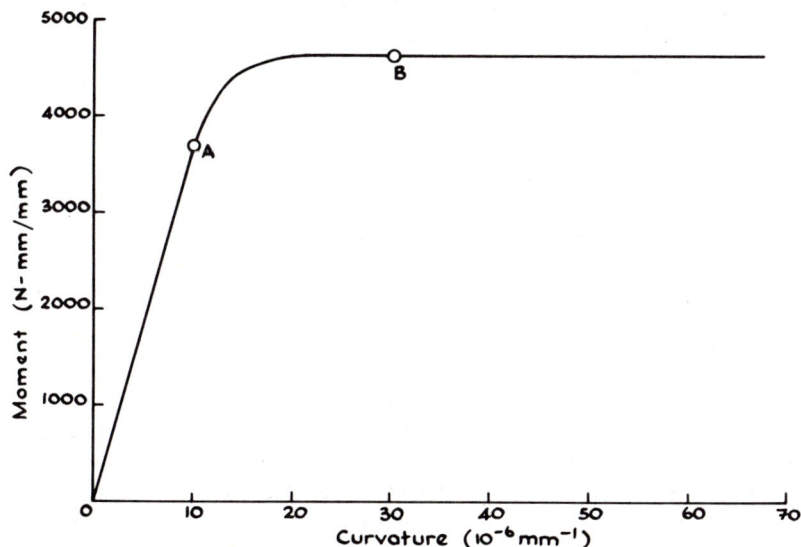

FIG. 3 ASSUMED MOMENT CURVATURE RELATION FOR
WORKED EXAMPLES

Example 1

The first example illustrated below is an isotropic square slab simply supported and loaded uniformly. The slab was divided into a 12 x 12 grid for the method of finite differences. The slab dimensions were taken from one of Muspratt's test slabs.[17] The following values were used in the analysis:

Span $(L) = 3.05$ m
Effective depth of steel = 57 mm
Area of steel = 31.6 mm^2 per bar
Yield strength = 310 N/mm^2
Young's modulus = 185 KN/mm^2
Concrete strength = 28 N/mm^2
Young's modulus for concrete = 25 KN/mm^2
Characteristic load = W_k = 0.0086 N/mm^2

Figure 3 shows the assumed moment-curvature curve for unit width of slab for uniaxial bending. The initial slope of the curve is the flexural rigidity of the uncracked section and the maximum moment is equal to the ultimate moment of the section. The point A will be referred to as the yield point while the term plastic zone will be used to denote the state of bending beyond the point B. It will be assumed that the same curve is valid for hogging (negative) bending.

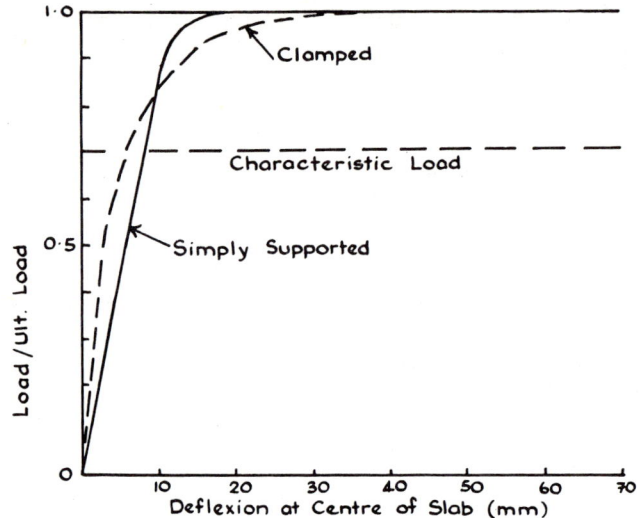

FIG. 4   LOAD - DEFLEXION CURVES

The load-deflection curve is shown in Fig. 4. The elastic deflections were compared with the exact solution and the discrepancy was found to be only 2%. The growth of the plastic zone with increasing load is shown in Fig. 5. At collapse, a large portion of the slab is under positive bending moment equal to the ultimate moment. In addition, the corners are subjected to negative bending moment, reaching the value of the ultimate moment. These zones of ultimate moments are shown in Fig. 6. The distributions of principal moments across a diagonal are plotted in Fig. 7.

Limit state of collapse

Assuming the live load is negligible compared with the dead load, the Unified Code gives a value of 1.4 as the partial safety factor for load. Hence, the design load for collapse is $1.4\ Wk$ (i.e. 0.012 N/mm^2). The exact collapse load for this case is known from the mechanism (Fig. 5) and is $\dfrac{24\ \text{Mult}}{L^2} = 0.012$, assuming that the above given material strengths are characteristic values and that the partial safety factor for strength is unity (for convenience only). In this chosen example the collapse load equals the design load and the slab is therefore just satisfactory with respect to collapse.

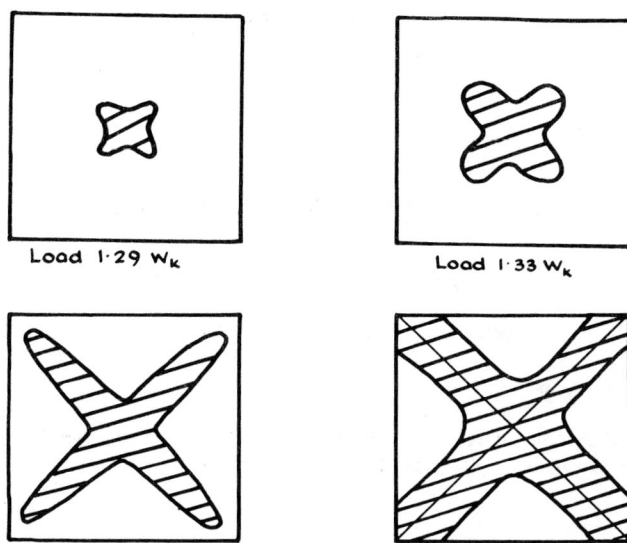

Load 1·29 Wₖ

Load 1·33 Wₖ

Load 1·39 Wₖ

Mechanism 1·4 Wₖ

FIG. 5 GROWTH OF PLASTIC ZONE
(SIMPLY-SUPPORTED SLAB, W_K = 0·0086 N/mm²)

Limit state of local damage

This limit state requires an estimation of steel and concrete strains and the maximum surface crack width with Wk as the design load. Only steel strain is dealt with here. Under this load, the steel strains were estimated to be elastic, with a maximum value of 0·0003 at the centre of the slab, which is much lower than 0·0013 allowed by the Code. The analytical results confirm the validity of the strain calculation, since the slab is elastic under this load.

Limit state of deflection

Neglecting creep and shrinkage, the maximum deflection at a load of W_k is 8 mm, which is below the limit of $L/250$ = 12 mm. This value again coincides with the analytical value.

The example shows that the simple linear estimations are adequate. However, the yield line theory does not give the distributions of bending moments at collapse and does not therefore help in the design of reinforcement. It should also be noted that while the mechanism for a simply supported square slab involves positive yield lines only, it is wrong to ignore the negative bending moments (see Fig. 6).

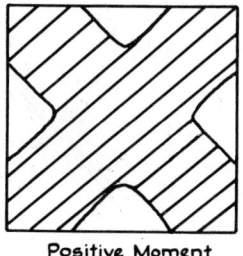

Positive Moment

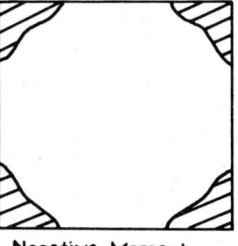

Negative Moment

FIG. 6 ULTIMATE - MOMENT ZONES AT COLLAPSE
(SIMPLY - SUPPORTED SLAB.)

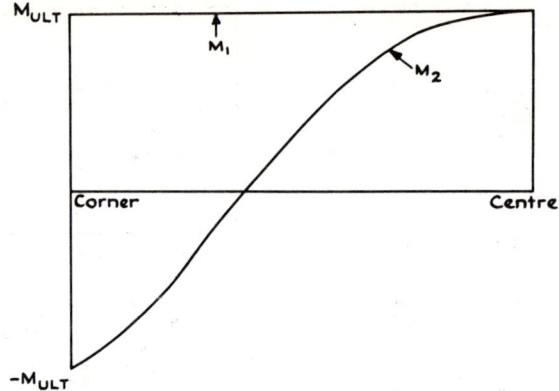

FIG. 7 DISTRIBUTION OF PRINCIPAL MOMENTS ALONG DIAGONAL OF SIMPLY-SUPPORTED SLAB.

Example 2

The same slab was analysed with clamped edges, again assuming the sagging and hogging ultimate moments are equal and that the characteristic load = 0·0147 N/mm². Figure 4 shows the load-deflection curve and Fig. 8 shows the growth of plastic zone and the collapse mechanism. The distributions of principal moments reaching the ultimate values are shown in Fig. 9, and the moments across the diagonal plotted in Fig. 10.

Limit state of collapse

Using the yield line theory, an upper bound $\dfrac{48 \, \text{Mult}}{L^2}$ may be obtained. This is about 15% higher than the collapse load given by this paper. There is no guidance as to how reinforcement should be provided over the slab.

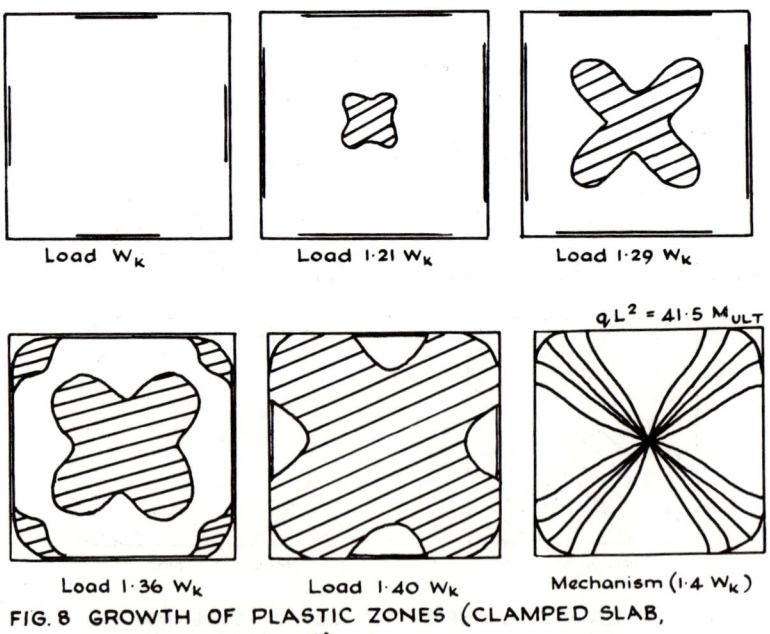

FIG. 8 GROWTH OF PLASTIC ZONES (CLAMPED SLAB,
$W_K = 0.0148$ N/mm²)

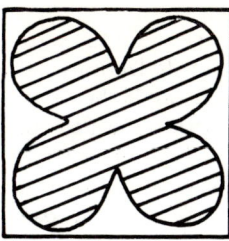 

Positive Moment   Negative Moment

FIG. 9 ULTIMATE - MOMENT ZONES AT COLLAPSE
(CLAMPED SLAB)

Limit states of local damage and deflection

Figure 8 shows that, at the characteristic load ($Wk = 0.0147$) the bending moments at the centres of the edges have reached the ultimate value. The load-deflection curve (Fig. 4) also indicates that the slab is inelastic at this level of load. It follows that the linear approach for estimating strains and deflections is less reliable. For example, the estimated maximum deflection at W_k will be 25% too low while it is impossible to evaluate by the linear approach the region where steel strains exceed the limiting value. This design is no longer governed by the limit state of collapse. On the other hand if the slab were designed against local damage and then checked for the limit state of collapse, it would be found that the reserve strength is rather high for collapse.

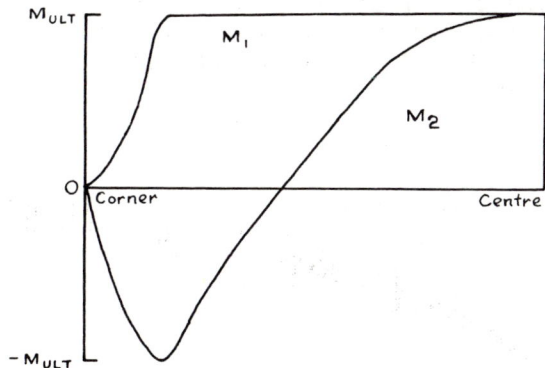

FIG. 10.   DISTRIBUTION OF PRINCIPAL MOMENTS
ALONG DIAGONAL OF CLAMPED SLAB.

5. CONCLUSIONS

This paper illustrates one of the applications of the analytical method. Comments on the Unified Code must be reserved until further study is made subsequent to the publication of the Code. However, two general observations seem worth making at this stage:

(1) The analytical method described above provides a means of studying the behaviour of isotropic slabs loaded from zero load to collapse. Since such existing knowledge is limited only to a few slabs of simple geometry, it is desirable to extend the study to various practical forms of slabs.

(2) The above two examples suggest that it may be difficult to obtain a balanced design for slabs with fixed edges. If the design is based on the state of collapse, strains under the characteristic load may be rather high or deflections are no longer linear to justify the simple design approach. On the other hand, if the design is based on the limit state of local damage, the reserve strength with respect to collapse may be too high to be economical. An optimised design should therefore aim to achieve the minimum safety factors required by the Code for various limit states. This optimisation study may be carried out by the method described in this paper.

6. ACKNOWLEDGEMENTS

The work described forms part of the programme of research of the Bridges Section of the Design Division of the Road Research Laboratory. The authors are grateful to the Section Leader, Mr. W. I. J. Price, for his valuable suggestions and encouragement. This paper is reproduced by permission of the Director of Road Research, Crown copyright and published by permission of Controller of H.M.S.O.

7. NOTATION

D	= flexural rigidity of plate.
k	= rate of change of moment per unit change of plastic curvature under uniaxial bending (Fig. 1).
v	= Poisson's ratio.
Mx, My	= bending moments per unit length of sections of a plate perpendicular to x and y axis, respectively.
Mxy	= twisting moment per unit length of section of a plate perpendicular to x axis.
mx, my, mxy,	changes in Mx, My, Mxy, respectively, due to a load increment.
Mult	= ultimate moment of a section (no strain hardening).
q	= intensity of distributed load.
Qx, Qy	= shearing forces parallel to z axis per unit length of sections of a plate perpendicular to x and y axis respectively.
t	= effective total curvature.
W	= downward deflection of plate.
w	= change in W due to load increment.
Wk	= characteristic load.
x, y, z	= axes of rectangular coordinates.

8. APPENDIX 1

The formulae for the twelve coefficients of the partial differential equation (18) are given below:

$$C_1 = 1 + \alpha R$$
$$C_2 = \gamma R + 2T\alpha$$
$$C_3 = 2 + \beta R + \alpha S + 2\gamma T$$
$$C_4 = \gamma S + 2\beta T$$
$$C_5 = 1 + \beta S$$
$$C_6 = aF + 2\alpha_{10}R + 2\alpha T$$
$$C_7 = \gamma F + \alpha G + 2\gamma_{10}R + 2\alpha_1 S + 2(\alpha_{10} + \gamma_1)T$$
$$C_8 = \beta F + \gamma G + 2\beta_{10}R + 2\gamma_1 S + 2(\beta_1 + \gamma_{10})T$$
$$C_9 = \beta G + 2\beta_1 S + 2\beta_{10}T$$
$$C_{10} = \alpha E + \alpha_{10}F + \alpha_1 G + \alpha_{20}R + \alpha_2 S + 2\alpha_{11}T$$

$$C_{11} = \gamma E + \gamma_{10}F + \gamma_1 G + \gamma_{20}R + \gamma_2 S + 2\gamma_{11}T$$
$$C_{12} = \beta E + \beta_{10}F + \beta_1 G + \beta_{20}R + \beta_2 S + 2\beta_{11}T$$

where,

$$E = R_{20} + S_2 + 2T_{22}$$
$$F = 2R_{10} + 2T_1$$
$$G = 2S_1 + 2T_{10}$$

and the subscripts of $\alpha, \beta, \gamma, R, S$ and T are defined by:

$$Tmn = \frac{\partial(m+n)T}{\partial x^m \partial y^n}$$

9. APPENDIX 2

To obtain the yield conditions and flow rules, let the principal moments be M_1 and M_2, i.e.

$$M_1 = \tfrac{1}{2}(Mx + My) + [\tfrac{1}{4}(Mx - My)^2 + Mxy^2]^{\frac{1}{2}}$$
$$M_2 = \tfrac{1}{2}(Mx + My) - [\tfrac{1}{4}(Mx - My)^2 + Mxy^2]^{\frac{1}{2}}$$

(i) *For yielding due to sagging moment*

The yield condition corresponds to the line AB of the yield locus in Fig. 2. Hence:

$$M_1{}' = Mp$$

Simplifying

$$(Mx + My)Mp - MxMy + Mxy^2 - Mp^2 = 0 \qquad (19a)$$

The above equations take into account the fact that $Mxy = -Mxy$. Strictly speaking, the term Mxy^2 in equation (19a) should be replaced by $\tfrac{1}{2}(Mxy^2 + Myx^2)$. However, this is important only when the equation is differentiated with respect to Mxy to obtain F_3.

By differentiation,

$$\left. \begin{array}{l} F_1 = Mp - My \\ F_2 = Mp - Mx \\ F_3 = Mxy \\ F_0 = 2Mp - (Mx + My) \end{array} \right\} \qquad (20a)$$

Alternatively, the above laws governing plastic curvature increments can be derived from the condition that the principal plastic curvature increment is zero in the M_2 − direction (see arrow perpendicular to AB in Fig. 2). In the M_1 − direction, the increment is taken to be the effective plastic curvature increment Δp. Hence the value of A can be obtained from (5) & (9) as:

$$A = [(Mx - My)^2 + 4Mxy^2]^{\frac{1}{2}} \qquad (21a)$$

(ii) *For yielding due to hogging moment*

The corresponding equations are:

$$(Mx + My)Mp + MxMy - Mxy^2 + Mp^2 = 0 \qquad (19b)$$

$$\left. \begin{array}{l} F_1 = Mp + My \\ F_2 = Mp + Mx \\ F_3 = -Mxy \\ F_0 = -2Mp - (Mx + My) \end{array} \right\} \qquad (20b)$$

$$A = -[(Mx - My)^2 + 4Mxy^2]^{\frac{1}{2}} \qquad (21b)$$

(iii) *For simultaneous yieldings in both principal directions*

These cases correspond to the corners A, B, C and D respectively on the yield locus of Fig. 2. To obtain the flow rules, it is assumed that the principal plastic curvature increments are equal in magnitude in both directions (e.g. the arrow at A is at $45°$ to the M_1 and M_2 axes).

REFERENCES

1. Beckett, D. An introduction to limit state analysis of reinforced concrete beams and slabs. *Build. Sci.*, Vol. 4, No. 1, June 1969, pp. 1–21.
2. Flint, A. R. and Edwards, L. S. Limit state design of highway bridges – a study of the implications with particular reference to certain examples. *Struct. Engr*, Vol. 48, No. 3, March 1970, pp. 93–108.
3. Ingerslev, A. The strength of rectangular slabs. *J. Instn struct. Engrs*, Vol. 1, No. 1, 1923, 3–14.
4. Johansen, K. W. *Yield-line theory*; trans. from the Danish. London, Cement and Concrete Association, 1962.
5. Prager, W. and Hodge, P. G. *Theory of perfectly plastic solids.* New York, Wiley, 1951.
6. Wood, R. H. *Plastic and elastic design of slabs and plates.* London, Thames & Hudson, 1961.
7. Jones, L. L. and Wood, R. H. *Yield-line analysis of slabs.* London, Thames & Hudson, 1967.
8. Massonnet, C. Théorie générale des plaques élasto-plastiques. *Publs int. Ass. Bridge struct. Engng*, Vol. 26, 1966, pp. 289–300.
9. Cornelis, A. Etude à l'aide d'une calculatrice électronique du comportement des dalles en béton armé en phase de fissuration. *Publs int. Ass. Bridge struct. Engng*, Vol. 26, 1966, pp. 301–311.
10. Bhaumik, A. K. and Hanley, J. T. Elasto-plastic plate analysis by finite differences. *J. struct. Div. Am. Soc. civ. Engrs*, Vol. 93, ST5, Oct. 1967, pp. 279–294.
11. Lin, T. H. and Ho, E. Y. Elasto-plastic bending of rectangular plates. *J. Engng Mech. Div. Am. Soc. civ. Engrs*, vol. 94, EM1, Feb. 1968, pp. 199–209.
12. Tuma, J. J., Havner, K. S. and French, S. E. *Analysis of flat plates by algebraic carry-over method.* Publications 118 & 119, Engineering Experiment Station, Oklahoma State University, Stillwater, Okla., 1060–1961.
13. Ang, A. H. S. and Lopez, L. A. Discrete model analysis of elastic-plastic plates. *J. Engng Mech. Div. Am. Soc. civ. Engrs*, Vol. 94, EM1, Feb. 1968, pp. 271–293.
14. McNeice, G. M. and Kemp, K. O. Comparison of finite element and unique limit analysis solutions for certain reinforced concrete slabs. *Proc. Instn civ. Engrs*, Vol. 43, Aug. 1969, pp. 629–640.
15. Mendelson, A. *Plasticity: theory and application.* New York, Macmillan, 1968, pp. 116–129.
16. Yam, L. C. P. A conforming finite element with internal equilibrium for two-dimensional problems. (To be published.)
17. Muspratt, M. A. Destructive tests on rationally designed slabs. *Mag. Concr. Res.*, Vol. 22, No. 70, March 1970, pp. 25–36.

FINITE ELEMENT ANALYSIS OF A BROAD ARCH BRIDGE WITH A SLAB DECK

A. B. SABIR and D. G. ASHWELL

University College, Cardiff

SYNOPSIS

A pilot study is described, of the application of finite element analysis to a bridge consisting of broad circular arch vault supporting a slab deck. The deck rests on the vault at the centre and is either continuous for the whole span, or split into two half-decks. The deck is treated using the rectangular plate element of Zienkiewicz and Cheung,[6] and the vault by a simplified form of Cantin and Clough's[2] cylindrical shell element. The interaction between the deck and vault is taken to consist of vertical forces along the crown of the vault.

The distributions of deflection, thrust and bending moment in the vault are investigated for various positions of a load applied to the deck, and influence surfaces and lines are drawn.

It is suggested that the method used has application to more complicated problems.

NOTATION

a_i	shape function constants
h	thickness of vault
r	mean radius of vault
u, v, w	displacements (see Fig. 2)
x, y, z	co-ordinates (see Fig. 2)
β	half the angle subtended by element
v	Poisson's ratio
ϕ	angular co-ordinate $= y/r$ (see Fig. 2)

INTRODUCTION

The application of finite element methods to cylindrical shells has been discussed by several authors (including Bogner, Fox and Schmit,[1] Cantin and Clough,[2] Sabir and Ashwell[3,4,5]) and several problems have been analysed. The present paper describes a pilot study of the bridge shown in Fig. 1. This consists of a circular arch vault supporting a slab deck. We believe our treatment to be of interest because (i) the arch vault is treated as a cylindrical shell, allowing the effects of unsymmetrical and local loads to be investigated, (ii) the interaction between vault and slab is discussed, both for a continuous full-span slab and for separate half-span slabs, and (iii) the method used can be extended to more complicated arch bridges.

161

Shape Functions for Cylindrical Shells

We shall divide the vault into curved rectangular finite elements. A number of shape functions for such cylindrical elements have been suggested. The simplest is as follows (see Fig. 2 for definition of geometrical symbols).

$$u = a_1 + a_2 x + a_3 y + a_4 xy \tag{1a}$$

$$v = a_5 + a_6 x + a_7 y + a_8 xy \tag{1b}$$

$$w = a_9 + a_{10}x + a_{11}y + a_{12}x^2 + a_{13}xy + a_{14}y^2 + a_{15}x^3$$
$$+ a_{16}x^2 y + a_{17}xy^2 + a_{18}y^3 + a_{19}x^3 y + a_{20}xy^3 \tag{1c}$$

This combines the usual plane stress shape functions for u and v with the rectangular plate bending shape function of Zienkiewicz and Cheung.[6] Five degrees of freedom are considered at each node, namely u, v, w, w_x, w_y (where $w_x = \dfrac{\partial w}{\partial x}$) leading to a 20 x 20 element stiffness matrix given explicitly by Sabir and Ashwell[3] for a shallow element and by Sabir[4,5] if the restriction of shallowness is removed.

Bogner *et al*[1] mention a 24 x 24 element containing the additional terms

$$w = a_{21}x^2 y^2 + a_{22}x^3 y^2 + a_{23}x^2 y^3 + a_{24}x^3 y^3 \tag{2}$$

and including the additional nodal degree of freedom w_{xy}. (Cantin and Clough[2] attribute this element to a thesis by Gallagher[7]). Bogner *et al.*[1] also give a 48 x 48 element in which u, v and w are all expressed by the polynomial form obtained by adding (1c) and (2), and additional nodal degrees of freedom $u_x, u_y, u_{xy}, v_x, v_y, v_{xy}$ are introduced.

Cantin and Clough[2] point out that polynomial forms for u, v, w do not allow rigid body displacements of the element, and removed this limitation by including terms containing trigonometrical functions. Cantin and Clough's shape function is

$$u = a_1 xy + a_2 x + a_3 y + a_4 - a_6 r \sin \phi - a_{20}r(\cos \phi - \cos \beta) \tag{3a}$$

$$v = a_5 xy + a_6 x \cos \phi + a_7 y - a_8 r(1 - \cos \phi \cos \beta)$$
$$- a_{20}x \sin \phi + a_{23} \cos \phi - a_{24} \sin \phi \tag{3b}$$

$$w = (a_9 x^3 y^3) + (a_{10}x^3 y^2) + a_{11}x^3 y + a_{12}x^3 + (a_{13}x^2 y^3) + (a_{14}x^2 y^2)$$
$$+ a_{15}xy + a_{16}x^2 + a_{17}xy^3 + a_{18}xy^2 + a_{19}xy + a_{20}x \cos \phi$$
$$+ a_{21}y^3 + a_{22}y^3 + a_{23} \sin \phi + a_{24} \cos \phi + a_8 r \sin \phi \cos \beta + a_6 x \sin \phi \tag{3c}$$

the nodal degrees of freedom used are $u, v, w, w_x, w_y - (v/r), w_{xy} \cdot w_y - (v/r)$ is used instead of w_y because it equals the rotation of the local tangent plane about the axis of the cylinder. If w_y is used, the 'stiffness matrix' obtained does not give the true generalised forces necessary for expressing the boundary conditions and the equilibrium conditions existing between the arch vault and the other components of the structure.

A partial assessment of the relative effectiveness of these shape functions can be made by applying them to a *linear* arch having negligible width parallel to the axis of curvature. This has been done (Ashwell, Sabir and Roberts[8,9]) and has been found that both the Bogner, Fox and Schmit (48 x 48) and the Cantin and Clough (24 x 24) elements are satisfactory for arches having r/h of the order of 50. (For deep arches, subtending 180°, having thin ribs ($r/h = 320$) the Cantin and Clough element is unsatisfactory).

The shape function used in the present study (in which $r/h = 46.7$) is a simplified form of that of Cantin and Clough (Sabir and Lock[10]). The finite element procedure can be regarded as a process for approximating to a minimum of the potential energy of the structure and loads. In seeking this minimum we must ensure continuity of displacements u, v, w and rotations $w_x, w_y - (v/r)$ at the nodes, because interelement nodal forces do work on these quantities, but imposition of continuity of other (internal) quantities, such as w_{xy}, hinders the approach to the energy minimum and is likely to impair the accuracy of the solution. Bogner *et al.*, for example, state that nodal continuity of u_x, u_y, v_x, v_y should not be imposed when using their (48 x 48) element. A formal process for removing these constraints is given by Ashwell *et al.*[9]; in the present work we adopt the simpler process of ignoring the quantity w_{xy} as a nodal degree of freedom, thus reducing the Cantin and Clough stiffness matrix from 24 x 24 to 20 x 20. This reduction requires a corresponding reduction of the 24 constants a_i in the shape function to 20, and so the four terms in brackets in (3c) are omitted. In this way we also avoid the need to consider the generalised forces corresponding to the generalised nodal displacements w_{xy} when expressing the boundary conditions for the arch vault. When applied to the problems considered by Cantin and Clough, this simplified element (which still satisfies the requirement for rigid-body displacements) gives improved convergence for deflection and almost identical values of the stress resultants (Sabir and Lock[10]).

Problem Considered

The bridge considered is shown in Fig. 1. The dimensions are similar to those of an actual motorway bridge. The vault is hinged along aa and bb, and the deck is simply-supported along cc and dd.

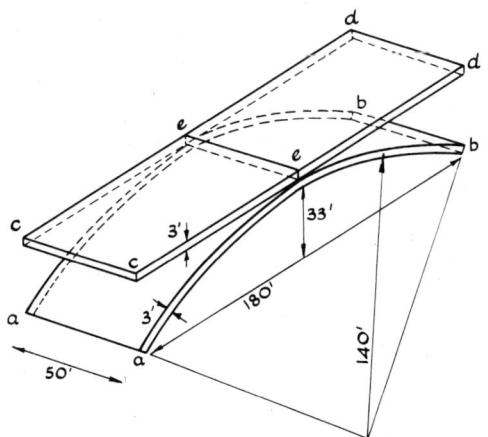

FIG. 1. THE BRIDGE TREATED. THE ARCH VAULT IS HINGED TO FIXED ABUTMENTS ALONG aa AND bb, AND THE DECK IS SIMPLY-SUPPORTED ALONG cc AND dd, AND CAN BE EITHER SPLIT OR CONTINUOUS ALONG ee.

Two conditions are considered along ee. In the first the deck is continuous, and in the second it is split along ee to form two separate half-spans. The deck is considered to rest on the vault so that the only reaction between them is vertical force distributed along ee. While this is a simplification of the conditions of continuity between deck and vault, it is considered to account for their primary interaction, and

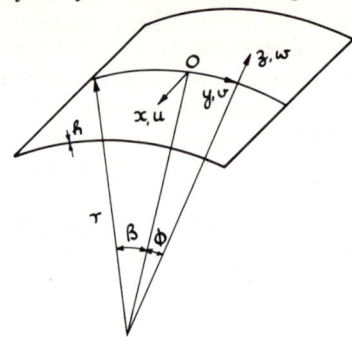

FIG. 2. DEFINITION OF GEOMETRICAL SYMBOLS.
2β IS THE ANGLE SUBTENDED BY AN
ELEMENT.

the method of analysis can be extended to more complete conditions of continuity if it should be thought necessary.

Young's modulus and Poisson's ratio were taken as $3 \cdot 10^6$ lb $-\text{in}^{-2}$ and $0 \cdot 2$.

Method of Analysis

The vault was treated as a cylindrical shell and the deck as a slab with normal loads only. The shape function used for the vault is described in section 2. For the deck the 12 x 12 element of Zienkiewicz and Cheung[6] was used. This has a polynomial for normal displacement similar to (1c), and considers nodal degrees of freedom w, w_x, w_y.

Each half of the vault was divided into 64 equal rectangular elements (i.e. 8 x 8 — a run was done with 8 x 16 but this was found to make very little difference), and each half of the deck was divided into 32 elements (four along the length and eight laterally).

If interaction between deck and vault is restricted to vertical forces at the nine nodes along *ee*, it is found that with these elements the complete structure (with continuous deck) has 999 degrees of freedom. The solution of 999 simultaneous equations can be performed only if their matrix is strongly banded about the principal diagonal. This banding occurs naturally if the structure consists of a single component such as the deck or the vault, for then the structure stiffness matrix can be assembled by starting at one end and proceeding along the structure, at the same time proceeding down the principal diagonal. Since each point in the structure is connected only to neighbouring points, this method of assembly does not lead to non-zero terms remote from the principal diagonal. However, with a structure having several interconnected branches (as in Fig. 1) this procedure is less effective, and in the present work the deck and vault are considered as separate components stitched together along *ee*. The procedure for the continuous deck is as follows.

1. The vault is considered separately. Since each half is divided into 8 x 8 elements, there are nine nodes along *ee* at which the deck will apply vertical forces. The response of the vault to a unit vertical force applied to each node in turn is therefore calculated, and the nodal displacements for half the arch and mid-element stress resultants are printed for each load.

Although there are nine nodes, only five such calculations are necessary from considerations of symmetry. Also, since the vault is symmetrical about *ee*, only half the vault is considered. The nodal reactions between the deck and the vault are called 'stitching forces'.

2. From the information obtained in 1, a 9 x 9 flexibility matrix is formed giving the nine vertical nodal displacements along *ee* in terms of the corresponding vertical nodal forces applied to the vault alone. This flexibility matrix is then inverted to give the corresponding stiffness matrix called the 'stitching matrix'.

3. The stiffness matrix for the deck is calculated, using eight elements in each direction (for the whole deck). Again there are nine nodes along *ee*.

4. The stitching matrix from the vault is added to the stiffness matrix for the deck, so that the terms representing the vertical stiffness at the nodes along *ee* include the stiffnesses of both deck and vault.

5. A unit vertical load is applied to one of the nodes of the deck and the resulting set of simultaneous equations derived from the combined stiffness matrix obtained in 4, is solved for the nodal displacements for the complete deck, which are printed. Mid-element stress resultants (bending moments) are calculated and printed.

6. The nodal vertical displacements along *ee*, obtained in 5, are multiplied by the stitching matrix to obtain the stitching forces applied by the deck to the vault.

7. These values of the stitching forces are used to calculate deflections and stress resultants (thrust and bending moment) in the vault, from the values of these quantities for unit stitching forces printed in 1.

Steps 1 to 6 are performed on the Atlas Computer at Didcot. If steps 5 and 6 are repeated for 20 positions of the unit load on (half) the deck, the complete calculations take about 3 minutes. In this pilot study, step 7 is performed on a desk calculating machine.

A modification is required to this procedure when the split deck is considered, for in this case three components of the bridge are joined along *ee*. The stitching matrix derived from the vault is first added to a stiffness matrix calculated for the half-deck *eedd* (using four elements lengthwise and eight across the width). A unit load is applied in turn to the nodes along *ee*, and the resulting displacements and stress resultants in the deck are printed.

A second stitching matrix is formed (cf. 2 above), relating the vertical nodal forces along *ee* to the corresponding displacements, for the combined structure 'vault + *eedd*'. This second stitching matrix is added to a stiffness matrix calculated for the other half-deck *eecc*. An external load is now applied to *eecc* and the resulting simultaneous equations are solved to obtain displacements, and hence stress resultants in *eecc*. The vertical nodal displacements, thus calculated, along *ee* are then multiplied by the stitching matrix for the vault alone to obtain the stitching forces applied to the vault, and from these its behaviour is investigated from the information printed in step 1 above. Similarly, by multiplying the vertical nodal displacements along *ee* by the second stitching matrix, the behaviour of the half-deck *eedd* can be investigated.

Distribution of Stitching Forces

The main feature of interest in this analysis is the behaviour of the vault for different loads applied to the deck. Thus, although the data obtained included full information concerning deflections and stress resultants for the deck, in making a selection for presentation in this paper, only the vault is considered.

Forces are transmitted to the vault only along *ee*, i.e. as stitching forces, and we first give values of these stitching forces for different load positions. In this way we present a picture of the interaction between deck and vault. In Fig. 3 are shown the nine nodes 1,2, . . . 9 along *ee*, and positions at which we shall consider a unit vertical load to be applied to the deck, i.e. $e_1, e_2, e_3, e_4, f_1, f_2, f_3, f_4, g_1, g_2, g_3$.

We first consider the continuous deck. Table 1 gives values of the nine stitching forces for a load applied at e_1. The first line is for $v = 0.2$. The slightly undulating character of these loads was at first puzzling, so the calculation was repeated for $v = 0$. This (second line of Table 1) gave a much smoother variation of stitching force. Evidently the undulations for $v = 0.2$ were due to the distortion of the cross-section of the arch resulting from the change in its principal curvature (see Ashwell[11] for a discussion of the distortion, produced by change of principal curvature, in the cross-section of a cylindrical shell with free edges). In all remaining calculations a value $v = 0.2$ has been used.

In Table 2 the stitching forces are given for unit loads at e_4, e_3, e_2, e_1. We see how the distribution of the forces changes as the load travels from the abutment to the centre. Also given is the sum of the stitching forces for each load position; this equals the total force applied to the vault.

Tables 3 and 4 give similar results for load positions f_4, f_3, f_2, f_1 and g_4, g_3, g_2, g_1.

Similar results have been obtained for the split deck. To illustrate the order of the differences in stitching force distribution introduced by splitting the deck, Table 5 gives stitching forces for load positions f_4, f_3, f_2, f_1.

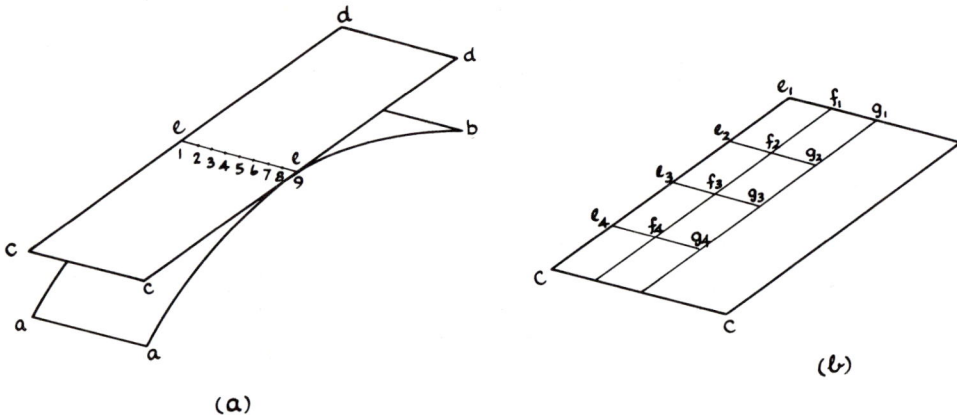

(a)

FIG. 3. (a) THE NINE NODAL POSITIONS ALONG *ee*.
(b) LOADED POINTS ON THE HALF-DECK *ccee*.

Distribution of Deflection, Thrust and Bending Moment

In Figs. 4 to 12 we show distributions of deflection, thrust and bending moment over one half-vault. Since the vault is loaded only along *ee* its behaviour is symmetrical. The object of these figures is to provide a general picture of the behaviour of the vault. They apply to the bridge with the continuous deck; similar figures for the split deck would be only slightly different.

Tables 6, 7 and 8 give the greatest values of the quantities occurring. The stress resultants were calculated at mid-element positions since experience suggests that such values are likely to be more accurate than values calculated at the nodes (Ashwell *et al.*[9]).

Influence Surfaces and Lines

As further examples of the behaviour of the vault, we give an influence surface and influence lines. Fig. 13 shows an influence surface giving the value of the circumferential thrust per unit length in the vault at the point *x* on Fig. 7, plotted against position of a concentrated unit load on the deck. Similar influence surfaces can be plotted for other quantities at other points in the vault and deck.

Fig. 14 shows influence lines for circumferential thrust and bending moment per unit length, at the same point, as a unit load crosses the bridge along the line f_4, f_3, f_2, f_1 (Fig. 7). Lines for both continuous and split decks are given. It is interesting to note that the values of both thrust and bending moment are generally higher for the continuous than for the split deck.

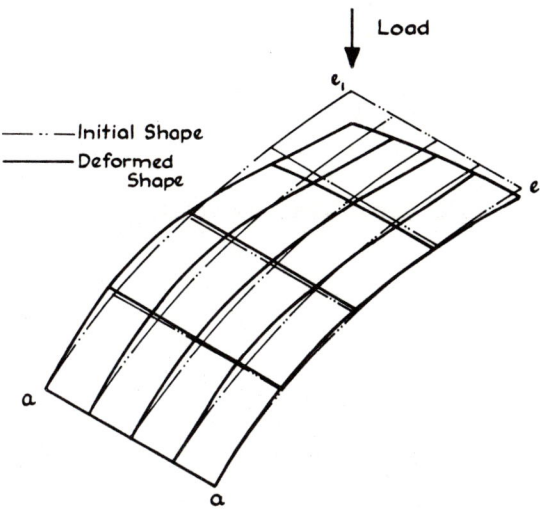

FIG. 4. AN ISOMETRIC PLOT OF THE NORMAL DISPLACE-
MENT FOR HALF THE VAULT, WITH LOAD APPLIED
TO THE BRIDGE AT e_1 THIS IS FOR THE CONTINUOUS
DECK, BUT THE SHAPE FOR THE SPLIT DECK IS NOT
SIGNIFICANTLY DIFFERENT. THE GREATEST VALUE
OCCURRING IS GIVEN IN TABLE 6.

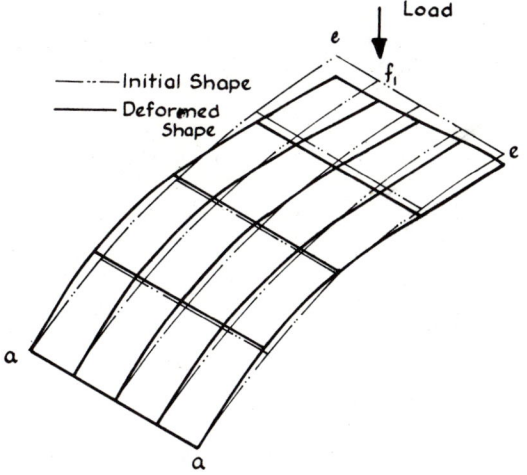

FIG. 5. AS FIG. 4 WITH LOAD APPLIED AT f_1.

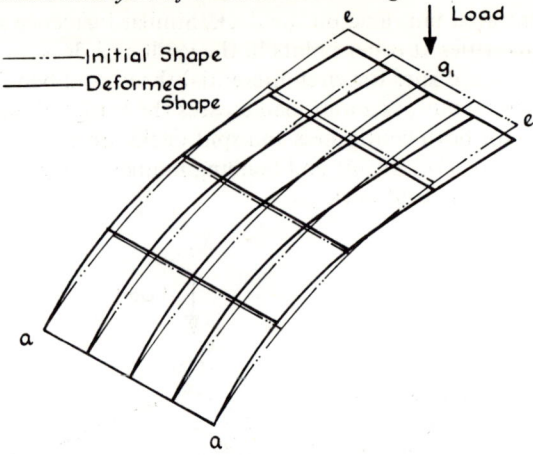

FIG. 6. AS FIG. 4 WITH LOAD APPLIED AT g_1.

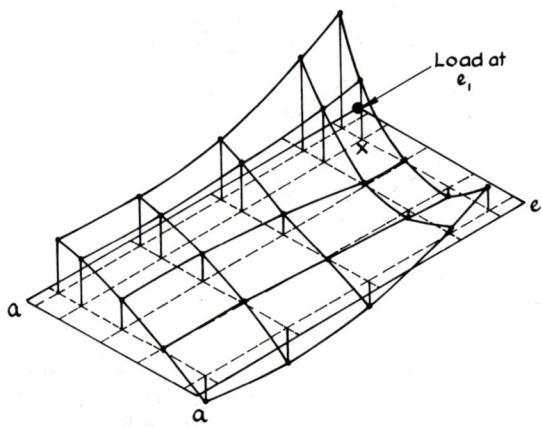

FIG. 7. CIRCUMFERENTIAL THRUST PER UNIT LENGTH TRANSVERSELY CALCULATED AT SELECTED MID-ELEMENT POSITIONS IS THE HALF-VAULT DEVELOPED INTO A RECTANGLE. THE DECK IS CONTINUOUS AND THE LOAD IS APPLIED AT e_1. THE GREATEST VALUE OCCURRING IS GIVEN IN TABLE 7. NOTE THE REGIONS OF TENSION.

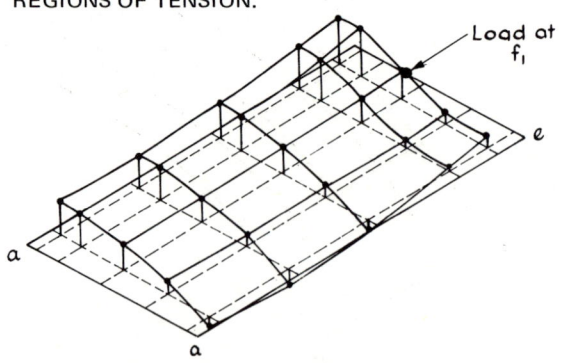

FIG. 8 AS FIG. 7 WITH LOAD APPLIED AT f_1.

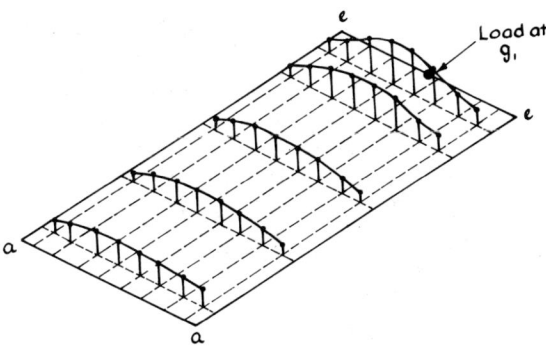

FIG. 9   AS FIG. 7 WITH LOAD APPLIED AT g_1.

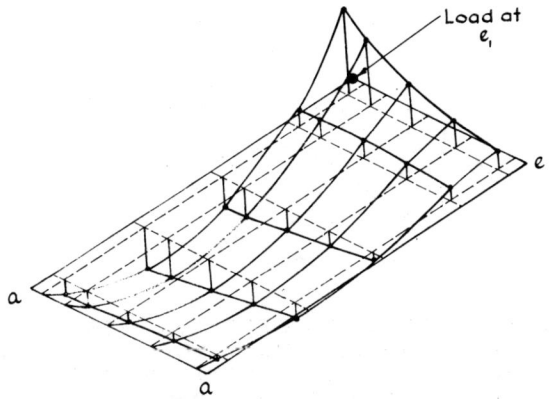

FIG. 10.   CIRCUMFERENTIAL BENDING MOMENT PER UNIT
LENGTH TRANSVERSELY AT SELECTED MID-
ELEMENT POSITIONS. *aaeè* IS THE HALF-VAULT
DEVELOPED INTO A RECTANGLE. THE DECK IS
CONTINUOUS AND THE LOAD IS APPLIED AT e_1
THE GREATEST VALUE OCCURRING IS GIVEN IN
TABLE 8.

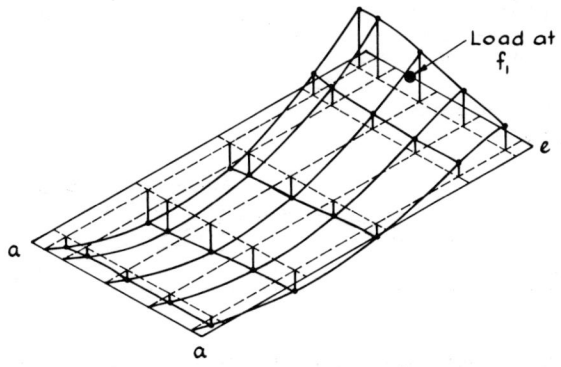

FIG. 11. AS FIG. 10 WITH LOAD AT f_1.

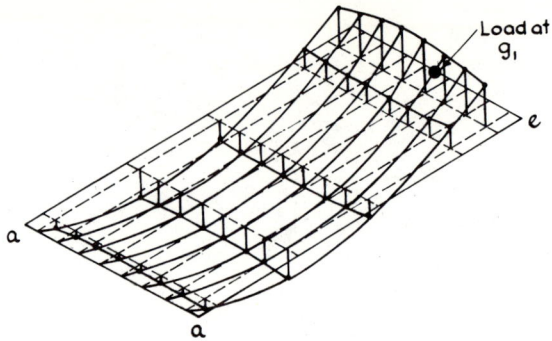

FIG. 12. AS FIG. 10 WITH LOAD AT g_1.

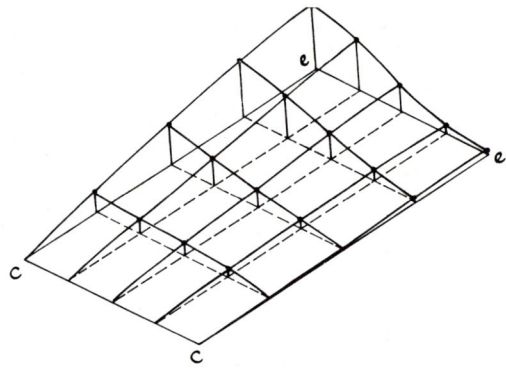

FIG. 13. INFLUENCE SURFACE FOR CIRCUMFERENTIAL
THRUST PER UNIT LENGTH TRANSVERSELY AT THE
POINT x ON FIG. 7. THE THRUST IS PLOTTED AGAINST
THE POSITION OF A POINT LOAD APPLIED TO THE
HALF-DECK *ccee* FOR THE CASE OF THE CONTINUOUS
DECK. THE GREATEST VALUE OCCURRING IS 0·065

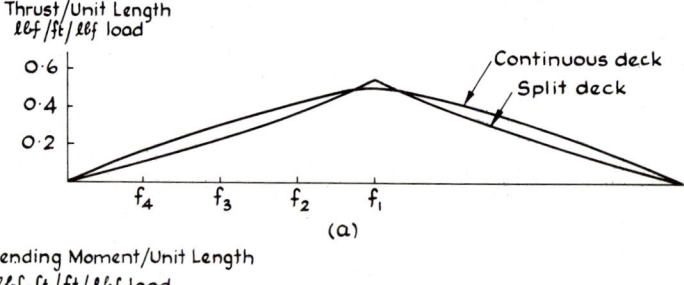

FIG. 14   INFLUENCE LINES FOR CIRCUMFERENTIAL THRUST AND
BENDING MOMENT AT THE POINT x ON FIG. 7.
(a) THRUST.   (b) BENDING MOMENT.

CONCLUSIONS

We have shown how finite element analysis developed for cylindrical shells can be applied to a broad arch bridge. The method used is essentially simple and within the capacity of existing computing facilities. It is capable of extension to other cases including (i) more complex interaction between deck and vault, (ii) vaults of non-uniform thickness, (iii) bridges with columns between deck and vault and (iv) bridges with non-rigid abutments. It should provide the designer with a useful analytical tool.

TABLE 1. Values of stitching forces for a unit load at e_1, and two values of v. Continuous deck.

Position of force	1	2	3	4	5	6	7	8	9
$v = 0\cdot2$	0·5528	0·2601	0·0590	0·0586	0·0324	0·0249	0·0164	0·0304	−0·0608
$v = 0\cdot0$	0·6269	0·1496	0·0790	0·0543	0·0371	0·0266	0·0204	0·0147	−0·0404

TABLE 2. Values of stitching forces for unit loads at $e_4, e_3, e_2, e_1 . v = 0\cdot2$. Continuous deck. Also given are the total forces applied to the vault.

Position of force	1	2	3	4	5	6	7	8	9	TOTAL
load at e_4	·030	·176	·048	·056	·039	·033	·022	·030	−·074	0·360
load at e_3	·089	·328	·091	·098	·063	·050	·032	·041	−·119	0·673
load at e_2	·200	·464	·099	·097	·050	·036	·022	·029	−·103	0·894
load at e_1	·553	·260	·059	·059	·032	·025	·016	·030	−·061	0·973

TABLE 3. As Table 2, for unit loads at f_4, f_3, f_2, f_1

Position of force	1	2	3	4	5	6	7	8	9	TOTAL
load at f_4	·001	·136	·042	·054	·042	·042	·032	·063	−·054	0·358
load at f_3	·015	·242	·090	·094	·082	·073	·056	·101	−·091	0·662
load at f_2	·025	·343	·095	·172	·110	·082	·053	·089	−·084	0·885
load at f_1	·001	·193	·520	·102	·054	·046	·032	·062	−·046	0·964

TABLE 4. As Table 2, for unit loads at g_4, g_3, g_2, g_1

Position of force	1	2	3	4	5	6	7	8	9	TOTAL
load at g_4	−·029	·097	·038	·050	·043	·050	·038	·097	−·029	0·355
load at g_3	−·050	·173	·077	·087	·092	·087	·077	·173	−·050	0·666
load at g_2	−·057	·178	·108	·165	·095	·165	·108	·178	−·057	0·883
load at g_1	−·026	·102	·051	·098	·520	·098	·051	·102	−·026	0·970

TABLE 5. As Table 3, but for split, instead of continuous, deck

Position of force	1	2	3	4	5	6	7	8	9	TOTAL
load at f_4	·056	·065	·023	·029	·023	·023	·021	·029	−·019	0·250
load at f_3	·109	·122	·059	·052	·052	·049	·041	·056	−·040	0·500
load at f_2	·115	·234	·055	·132	·086	·066	·048	·068	−·054	0·750
load at f_1	−·021	·222	·542	·115	·056	·043	·028	·062	−·047	1·000

TABLE 6. Maximum normal deflection of vault, in ft per lbf of load, for different load positions

load position	e_1	f_1	g_1
continuous deck	$1·475 \times 10^{-7}$	$0·929 \times 10^{-7}$	$0·805 \times 10^{-7}$
split deck	$1·510 \times 10^{-7}$	$0·959 \times 10^{-7}$	$0·829 \times 10^{-7}$

TABLE 7. Maximum mid-element circumferential thrust per unit length transversely, in lbf per ft per lbf load

load position	e_1	f_1	g_1
continuous deck	0·121	0·52	0·44
split deck	0·124	0·54	0·54

TABLE 8. Maximum mid-element circumferential bending moment per unit length transversely, in lbf-ft per ft per lbf load

load position	e_1	f_1	g_1
continuous deck	0·281	0·206	0·176
split deck	0·288	0·212	0·181

REFERENCES

1. Bogner, F. K., Fox, R. L. and Schmit, L. A. A cylindrical shell discrete element. *AIAA J.,* vol. 5, no. 4, April 1967, pp. 745–750.
2. Cantin, G. and Clough, R. W. A curved cylindrical-shell, finite element. *AIAA J.,* vol. 6, no. 6, June 1968, pp. 1057–1062.
3. Sabir, A. B. and Ashwell, D. G. A stiffness matrix for shallow shell finite elements. *Int. J. mech. Sci.,* vol. 11, April 1969, pp. 269–279.
4. Sabir, A. B. An extension of the shallow to the non-shallow stiffness matrix for a cylindrical shell finite element. *Int. J. mech. Sci.,* vol. 12, March 1970, pp. 287–292.

5. Sabir, A. B. Contribution to *Seminar on Finite Element Techniques in Structural Mechanics* (*Southampton, April 1970*); edited by Tottenham, H. and Brebbia, C. Southampton U.P., 1970, pp. 317–325.

6. Zienkiewicz, O. C. and Cheung, Y. K. The finite element method for analysis of elastic isotropic and orthotropic slabs. *Proc. Instn civ. Engrs,* vol. 28, August 1964, pp. 471–488.

7. Gallagher, R. H. *The development and evaluation of matrix methods for thin shells structural analysis.* Ph.D. thesis, State University of New York, 1966.

8. Ashwell, D. G. and Sabir, A. B. Limitations of certain curved finite elements applied to arches. Int. J. mech. Sci., in the press.

9. Ashwell, D. G., Sabir, A. B. and Roberts, T. M. Further studies in the application of curved finite elements to circular arches. (To be published.)

10. Sabir, A. B. and Lock, A. C. A curved, cylindrical shell, finite element. (To be published.)

11. Ashwell, D. G. A characteristic type of instability in the large deflexions of elastic plates. *Proc. R. Soc. (A),* vol. 214, 1952, pp. 98–118.

THE LINE SOLUTION TECHNIQUE IN BRIDGE DECK ANALYSIS

H. TOTTENHAM

Department of Civil Engineering,
University of Southampton

INTRODUCTION

The analysis of plate and shell structures by the line solution technique was first discussed by Jenkins and Tottenham[1] in 1962. Applications to shells by Chetty[2] and Desai[3] were of limited direct use owing to the poor computational facilities then available. However, Coull and Kazimi[4] used the technique successfully on a plane stress problem and obtained very useful results even though only a very small computer was available.

In 1966 Coull and Rao[5] showed that the technique could deal with a wide range of problems in the analysis of right isotropic bridge decks. Problems dealt with in detail by Rao[6] include

   (i)   uniformly distributed, block, and concentrated loads
  (ii)   cantilever plates
 (iii)   continuous decks
 (iv)   decks of variable (spanwise) thickness
  (v)   continuous decks with suspended spans.

In all this work the matrix progression method has been used and it is hence very easy to combine the various features.

Das[7] applied the technique to curved isotropic decks with, and without, edge stiffening beams. In his investigation he limited himself to the 'bending' of the edge beams and did not consider the 'in plane' effects. However, comparison with analytical type solutions that he had developed for decks simply supported at the ends with various edge beams and different arrangements of load showed that the method gave reliable results.

Although many of the above authors had stated that in principle the method could also be applied to orthotropic slabs, the first investigation on this topic was carried out by Trockalakis[8]. He considered both isotropic and orthotropic slabs and also laterally articulated slabs such as had been treated by Spindel[9]. The topics treated in detail include simply supported slabs, both rectangular and curved in plan, of isotropic, orthotropic and articulated construction under various forms of loading.

He also investigated rectangular decks continuous over line supports and over individual column supports. He showed that the support moments due to large local loads are much smaller than is usually assumed. Finally, to show the versatility of the approach he considered a curved slab of variable thickness and a two span system in which one slab was rectangular and the second was sectorial.

Since then skew both isotropic and orthotropic slabs, have been tracted as have beam-slab systems considered as a 'bending' problem. Since, however, Shaikh[10] has shown that the equivalent orthotropic plate analysis gives results practically identical with the full bending and in-plane stress analysis of both curved and right decks this topic has not been pursued further.

THE METHOD OF LINE SOLUTIONS

The method of 'line solutions' is a means of reducing a partial differential equation to a set of simultaneous ordinary differential equations. The technique was first suggested by Womersley and Hartree[11], for the mechanical integration of the 'telegraph equation'; they suggest, however, that the method was of limited value due to numerical instability. Unaware of this limitation the Author proposed[12] the technique as being of value in structural mechanics. The great advances in computational machines had, in fact, vastly extended the ranges of direct applicability.

The basic idea is quite simple. If we have some required function $w(x, y)$ which must satisfy some partial differential equation, we assume that along the lines $y = y_k$ ($k = 1 \ldots n$) the solution has the values $w(x, y_k)$, that is a function of x only, which we denote by $w_k(x)$. These are the 'line solutions'. Partial derivatives of $w(x, y)$ with respect to x will, along any line, take the form of complete differentials of $w_k(x)$ with respect to x. Partial derivatives with respect to y can, for any value of x, be replaced by an appropriate 'finite difference' expression and hence for all values of x can be expressed as appropriate difference expressions between the line solutions. For example, if the distance between successive lines y_k is h, then we

can replace $\partial w/\partial y$ along $y = y_k$ by $\dfrac{1}{2h}\left[w_{k+1}(x) - w_k(x)\right]$

Similarly we put

$$\left.\frac{\partial^2 w(x, y)}{\partial x^2}\right|_{y = y_k} = \frac{1}{h^2}\left\{w_{k+1}(x) - 2w_k(x) + w_{k-1}(x)\right\} .$$

and

$$\left.\frac{\partial^4 w(x, y)}{\partial x^4}\right|_{y = y_k} = \frac{1}{h^4}\left\{w_{k+2}(x) - 4w_{k+1}(x) + 6w_k(x) - 4w_{k-1}(x) + w_{k+2}(x)\right\}$$

Mixed derivatives can be expressed in a similar manner, thus

$$\left.\frac{\partial^4 w(x, y)}{\partial x^2 \partial y^2}\right|_{y = y_k} = \left.\frac{\partial^2}{\partial y^2}\left(\frac{\partial^2 w(x, y)}{\partial x^2}\right)\right|_{y = y_k}$$

$$= \frac{1}{h^2}\left\{\frac{d^2 w_{k+1}(x)}{dx^2} - 2\frac{d^2 w_k(x)}{dx^2} + \frac{d^2 w_{k-1}}{dx^2}\right\}$$

If we consider an isotropic plate then the normal displacement $w(x, y)$ must satisfy the partial differential equation

$$\frac{\partial^4 w}{\partial x^4} + 2\frac{\partial^4 w}{\partial x^2 \partial y^2} + \frac{\partial^4 w}{\partial y^4} = \frac{p(x, y)}{D}$$

If we take a set of lines $y = y_k$, distance h apart, we can replace this equation by a set of equations

$$\frac{d^4 w_k}{dx^4} + \frac{2}{h^2}\left(\frac{d^2 w_{k+1}}{dx^2} - 2\frac{d^2 w_k}{dx^2} + \frac{d^2 w_{k-1}}{dx^2}\right)$$

$$+ \frac{1}{h^4}\left(w_{k+2} - 4w_{k+1} + 6w_k - 4w_{k-1} + w_{k-2}\right) = \frac{p(x, y_k)}{D}$$

Equations such as these apply to lines away from an edge of the plate. If an edge of the plate runs along the line y_n, then the equation corresponding to this line would include the 'fictitious line' solutions w_{n+1} and w_{n+2}. These can be eliminated by considering the boundary conditions along this edge.

Thus, if the edge is free the boundary conditions are

$$\left.\frac{\partial^2 w}{\partial y^2} + \mu\frac{\partial^2 w}{\partial x^2}\right|_{y=y_n} = 0 \qquad \left.\frac{\partial^3 w}{\partial y^3} + (2-\mu)\frac{\partial^3 w}{\partial x^2 \partial y}\right|_{y=y_n} = 0$$

Using the appropriate difference expressions we find

$$w_{n+1} = -\mu h^2 \frac{d^2 w_n}{dx^2} + 2w_n - w_{n-1}$$

$$w_{n+2} = (2-\mu) h^4 \frac{d^4 w_n}{dx^4} - 4h^2\frac{d^2 w_{n-1}}{dx^2} + 2(2-\mu) h^2\frac{d^2 w_{n-2}}{dx^2}$$

$$+ 4w_n - 4w_{n-1} + w_{n-2}$$

These expressions can then be substituted back into the equations on lines y_n and y_{n-1}. The other edge can be treated similarly. We thus obtain a set of simultaneous fourth order ordinary differential equations. These can be most readily solved by the matrix progression method.

Matrix progression

In order to apply the matrix progression method the equations are first reduced to a set of first order differential equations. The most suitable way of doing this is by introducing three further sets of functions defined by

$$w_k^1 = \frac{dw_k}{dx}, \quad w_k^2 = \frac{dw_k^1}{dx} = \frac{d^2 w_k}{dx^2}, \quad w_k^3 = \frac{dw_k^2}{dx} = \frac{d^3 w_k}{dx^3}$$

With these a typical equation for a line solution of an isotropic plate is

$$\frac{d^3 w_k^3}{dx} = \frac{d^4 w_k}{dx^4} = \frac{-1}{h^4}\left(w_{k+2} - 4w_{k+1} + 6w_k - 4w_{k-1} + w_{k-2}\right)$$

$$- \frac{2}{h^2}\left(w_{k+1}^2 - 2w_k^2 + w_{k-1}^2\right) + \frac{p(x, y_k)}{D}$$

Treating each equation in this manner we can set up a set of $4n$ equations for the lines $y = y_1$ to $y = y_n$, which can be written in partioned matrix form as:

$$\frac{d}{dx}\begin{bmatrix} W \\ W^1 \\ W^2 \\ W^3 \end{bmatrix} = \begin{bmatrix} 0 & I & 0 & 0 \\ 0 & 0 & I & 0 \\ 0 & 0 & 0 & I \\ A_1 & 0 & A_2 & 0 \end{bmatrix}\begin{bmatrix} W \\ W^1 \\ W^2 \\ W^3 \end{bmatrix} + \frac{1}{D}\begin{bmatrix} 0 \\ 0 \\ 0 \\ P \end{bmatrix}$$

where

$$\underline{W} = \begin{bmatrix} w_1 \\ w_2 \\ \vdots \\ w_n \end{bmatrix} \quad \underline{W}^1 = \begin{bmatrix} w_1^1 \\ w_2^1 \\ \vdots \\ w_n^1 \end{bmatrix} \quad \text{etc.} \ \underline{P} = \begin{bmatrix} p(x, y_1) \\ p(x, y_2) \\ \vdots \\ p(x, y_n) \end{bmatrix}$$

The form of the sub-matrices $\underline{A}_1$ and $\underline{A}_2$ depend upon the form of the plate, these are given in Appendix I for the cases (i) rectangular isotropic, (ii) rectangular orthotropic and (iii) sectorial orthotropic, in all cases with free edges. In all cases we can write the above as

$$\frac{d\underline{W}}{dx} = \underline{A}_0 \underline{W} + \underline{B}_0$$

We can transform these equations so that instead of relating the derivatives of the line solutions they relate the physical quantities or 'actions', which for a rectangular

plate are w, $\qquad \phi = \dfrac{\partial w}{\partial x}$, M_x and V_x.

(Here V_x is the Kirchoff boundary force, the internal shear Q_x could be used but using this quantity would require further modification at the boundaries $x = $ const.) Thus for a typical line

$$w_k = w_k \qquad \phi_k = w_k^1$$

$$M_{xk} = -D\left[w_k^2 + \frac{\mu}{h} \, w_{k+1} - 2w_k + w_{k-1} \right]$$

$$V_{xk} = -D\left[\frac{(2-\mu)}{h^2} (w_{k+1}^1 - 2w_k^1 + w_{k-1}^1) + w_k^3 \right]$$

and on an edge line

$$w_n = w_n, \quad \phi_n = w_n^1 \quad M_{xn} = -D(1 - \mu^2)\, w_n^2$$
$$V_{xn} = -D(1 - \mu^2)\, w_n^3$$

Similar expressions can be derived from other types of decks.
Putting now

$$\underline{Z} = \begin{bmatrix} \underline{W} \\ \underline{\Phi} \\ \underline{M} \\ \underline{V} \end{bmatrix} \quad \text{where } \underline{\Phi} = \begin{bmatrix} \phi_1 \\ \phi_2 \\ \vdots \\ \phi_n \end{bmatrix} \quad \underline{M} = \begin{bmatrix} M_{x1} \\ M_{x2} \\ \vdots \\ M_{xn} \end{bmatrix} \quad \text{etc.}$$

we can write

$$\underline{Z} = \underline{Q}\,\overline{\underline{W}}$$

where $\underline{Q}$ has the partitioned form

$$\underline{Q} = \begin{bmatrix} \underline{I} & \underline{O} & \underline{O} & \underline{O} \\ \underline{O} & \underline{I}_1 & \underline{O} & \underline{O} \\ \underline{Q}_1 & \underline{O} & \underline{I}_2 & \underline{O} \\ \underline{O} & \underline{Q}_2 & \underline{O} & \underline{I}_3 \end{bmatrix}$$

The submatrices are given in the Appendix for the three types of deck mentioned above.

Since Q has only constant values and is non-singular we can write

$$\frac{dZ}{dx} = Q\frac{dW}{dx} = Q\,A_0\,W + Q\,B_0 = (Q\,A_0\,Q^{-1})\,Z + Q\,B_0$$

or

$$\frac{dZ}{dx} = A\,Z + B$$

where

$$A = (Q\,A_0\,Q^{-1}) \qquad B = Q\,B_0$$

The solution of this equation, for loads constant in the x-direction can be written as

$$Z(x) = G(x)\,Z(0) + L(x)$$

where the 'distribution' matrix $G(x)$ is given by

$$G(x) = I + A x + A^2\frac{x^2}{2!} + \ldots A^n\frac{x^n_1}{n!} \ldots$$

the 'loading solution' matrix $L(x)$ is given by

$$L(x) = I x + A\frac{x^2}{2!} + A^2\frac{x^3}{3!} + \ldots A^n\frac{x^{n+1}}{(n+1)!} + B = [G(x) - I]\,A^{-1}\,B$$

and $Z(0)$ is an array containing the values of actions Z at $x = 0$.

At each end of the bridge one half of the set of actions will be specified either explicitly as for example in the case of a 'simple' support for which $w_k = y_{nk} = 0$, or implicitly as in an elastic support where the reaction is related to the displacement by equations of the form $V = S W$. In all cases at the end $x = 0$ we can write

$$Z(0) = K_0\,\bar{Z}_0$$

where K_0 is the 'united restraint matrix' and $\bar{Z}_0$ is the 'reduced actions matrix' containing the unspecified actions. Thus for a simple support

$$Z(0) = \begin{bmatrix} W \\ \Phi \\ M \\ V \end{bmatrix}_0 = \begin{bmatrix} O & O \\ I & O \\ O & O \\ O & I \end{bmatrix} \begin{bmatrix} \Phi \\ V \end{bmatrix}_{x=0} = K_0\,\bar{Z}_0$$

We thus find

$$Z(x) = G(x)\,K_0\,\bar{Z}_0 + L(x)$$

If the load is uniform to the end $x = b$, then at $x = b$

$$Z(b) = G(b)\,K_0\,\bar{Z}_0 + L(b)$$

But here one half of the set of actions is specified so that we can write

$$K_b\,Z(b) = Q$$

For example for a simple support

$$\begin{bmatrix} I & O & O & O \\ O & O & I & O \end{bmatrix} \begin{bmatrix} W \\ \Phi \\ M \\ V \end{bmatrix}_b = \begin{bmatrix} O \\ O \end{bmatrix}$$

The relationships

$$\underline{G}(x_1)\,\underline{G}(x_2) = \underline{G}(x_1 + x_2)$$

$$\underline{L}(x_1 + x_2) = \underline{G}(x_2)\,\underline{L}(x_1) + \underline{L}(x_2) = \underline{G}(x_1)\,\underline{L}(x_2) + \underline{L}(x_1)$$

enable other types of loads to be dealt with readily, details will be found in [11].

Using

$$\underline{K}_b\,\underline{Z}(b) = \underline{K}_b\,[\underline{G}(b)\,\underline{K}_0\,\bar{\underline{Z}}_0 + \underline{L}(b)] = 0$$

we can obtain

$$\bar{\underline{Z}}_0 = -[\underline{K}_b\,\underline{G}(b)\,\underline{K}_0]^{-1}\,\underline{K}_b\,\underline{L}(b)$$

and then by back substitution the values of the actions at any section x = const.

Owing to the fact that the physical quantities are used as the variables, changes in form of the structure present no difficulty. Thus if an 'orthotropic' slab is made solid near the supports it is only necessary to modify the $\underline{A}$ matrix and hence the $\underline{G}(x)$ matrix, and use the appropriate one for the progression over each part of the bridge. Similarly changes from rectangular to sectorial slabs, or changes in radius of curvature, can be included.

Applications

In order to apply this method a programme was prepared in ALGOL 60.* In order to verify both the programme and the method as many cases as could be found in published works were re-analysed and the results compared. Results were also compared with those obtained form another programme giving Levy-type analysis for simply supported slabs. It was found that for reliable results five or more lines should be used. Since the size of the matrices increases in proportion to the number of lines, and the magnitude of the elements in A_1 as the fourth power of the number of lines, an excessive number is not desirable. It is thought by the authors that seven lines will give an accuracy of computation more than sufficient for engineering calculations and at the same time give sufficient flexibility to allow for a realistic representation of lane loading, abnormal load, column supports, etc.

Some idea of the accuracy can be seen from the results of a check made on a continuous deck. A central point load on one span only was considered. An alternative analysis using a single span to represent the effects of a symmetrical and antisymmetrical loading was also made. The results are shown in Table I.

The results shown in Table II give a comparison between the results calculated as above with those given by Timoshenko[12] for a square plate under uniform load (Poisson's ratio = 0.3). It will be noticed that these results, although correct to less than 1% are not as good as those shown in Table I. It has been found that the higher the value of Possion's ratio the more lines required for a given degree of accuracy particularly for results near the edge.

Figs 3 and 4 show the results for the computation of two bridge slabs, shown in Figs 1 and 2 respectively.

*Listings in I.C.T. 1900 hardware representation are available from the Civil Engineering Department, Southampton University.

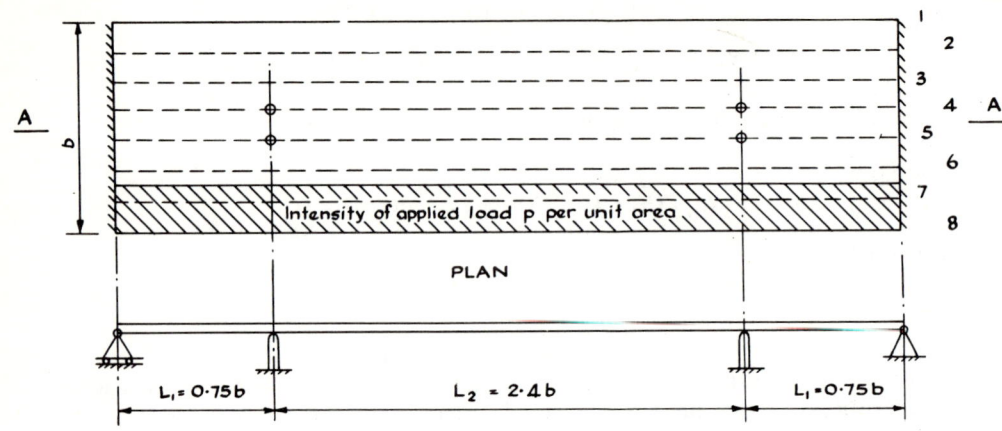

LINES

PLAN

SECTION A-A

FIG. I   ORTHOTROPIC BRIDGE DECK

$$D_x = 5 \cdot 87 \quad D_y = 4 \cdot 75 \quad G_x + G_y = 6 \cdot 60$$

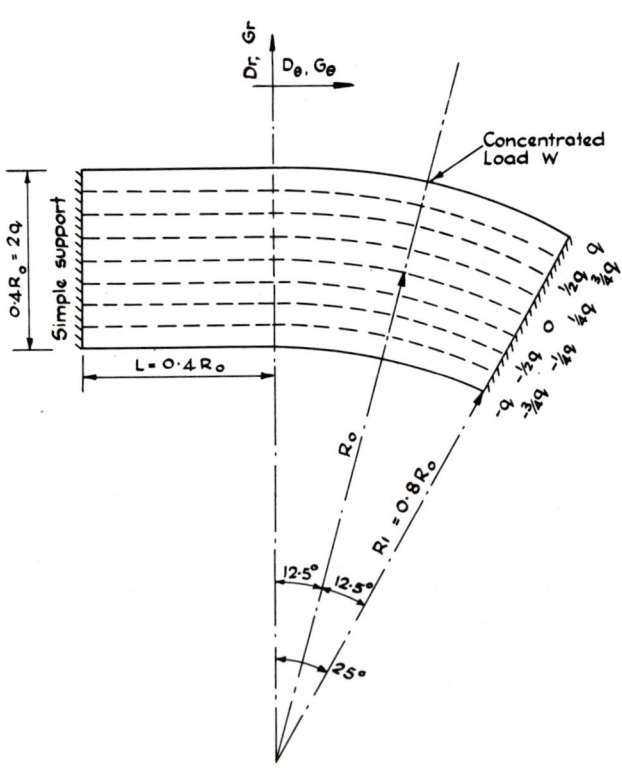

FIG. 2   CONTINUOUS ORTHOTROPIC BRIDGE DECK

$$D_\theta = 2 \cdot 23, \quad D_r = 1 \cdot 00, \quad G_\theta = 0 \cdot 037, \quad G_r = 0 \cdot 013$$

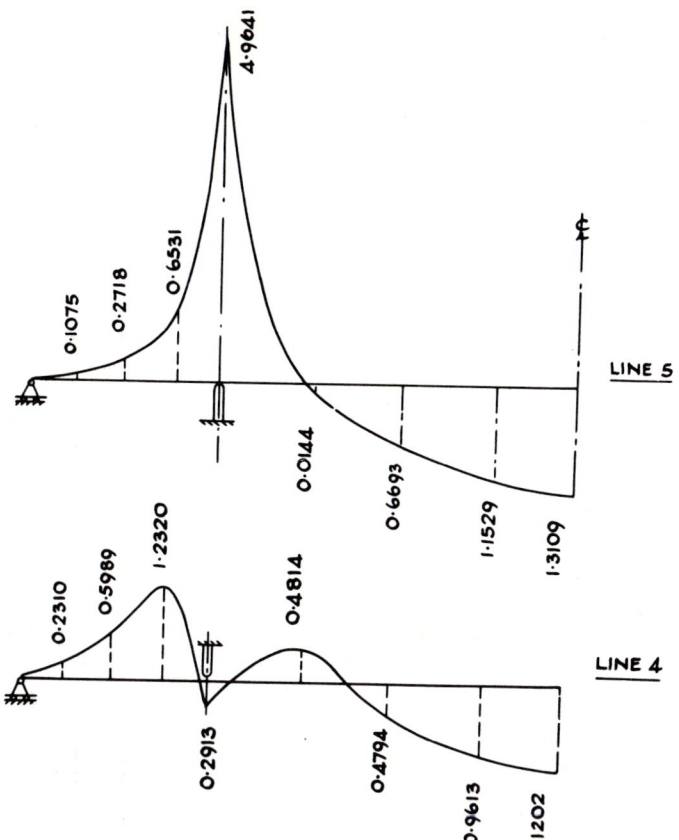

FIG. 3 LONGITUDINAL MOMENT COEFFICENTS

M_x = coef. L_{10}^2, 10^{-2}

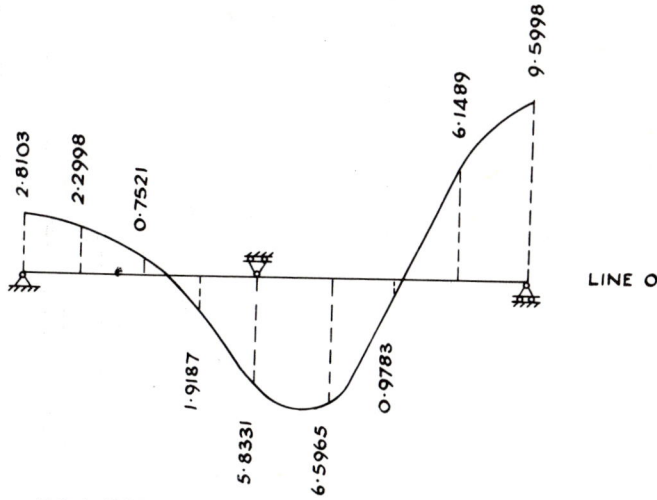

FIG. 4 TORSIONAL MOMENT COEFFICIENTS

SPANWISE VARIATION

$M_{r\theta} + M_{\theta r}$ = coef. $W. 10^{-3}$

APPENDIX

ISOTROPIC RECTANGULAR PLATE

$$\underline{A}_1 = \frac{1}{h^4} \begin{bmatrix} \frac{-2}{1-\mu^2} & \frac{4}{1-\mu^2} & \frac{-2}{1-\mu^2} & & & \\ 2 & -5 & 4 & -1 & & \\ -1 & 4 & -6 & 4 & -1 & \\ & & \cdot & \cdot & \cdot & \cdot & \cdot \\ & & \cdot & \cdot & \cdot & \cdot & \cdot \end{bmatrix}$$

$$\underline{A}_2 = \frac{1}{h^2} \begin{bmatrix} \frac{4(1-\mu)}{1-\mu^2} & \frac{-2(2-\mu)}{1-\mu^2} & & & \\ -(2-\mu) & 4 & -2 & & \\ & -2 & 4 & -2 & \\ & & \cdot & \cdot & \cdot \\ & & \cdot & \cdot & \cdot \end{bmatrix}$$

$$\underline{I}_1 = \underline{I}$$

$$\underline{I}_2 = -D \begin{bmatrix} 1-\mu^2 & & & \\ & 1 & & \\ & & \ddots & \\ & & & \ddots \end{bmatrix} \qquad \underline{I}_3 = -D \begin{bmatrix} (1-\mu)^2 & & & \\ & 1 & & \\ & & \ddots & \\ & & & \ddots \end{bmatrix}$$

$$\underline{Q}_1 = \frac{-D}{h^2} \begin{bmatrix} 0 & 0 & & & \\ \mu & -2\mu & \mu & & \\ & \cdot & \cdot & \cdot & \cdot \\ & \cdot & \cdot & \cdot & \cdot \end{bmatrix}$$

$$Q_2 = \frac{-D}{h^2} \begin{bmatrix} 0 & 0 \\ (2-\mu) & -2(2-\mu) & (2-\mu) \\ & & \ddots & \ddots & \ddots \\ & & & \ddots & \ddots & \ddots \end{bmatrix}$$

ORTHOTROPIC RECTANGULAR PLATE

$$A_1 = \frac{Dy}{Dx}\frac{1}{h^4} \begin{bmatrix} -2 & 4 & -2 \\ 2 & -5 & 4 & -1 \\ -1 & 4 & -6 & 4 & -1 \\ & \ddots & \ddots & \ddots & \ddots & \ddots \end{bmatrix}$$

$$A_2 = \frac{Gx+Gy}{Dx}\frac{1}{h^2} \begin{bmatrix} 2 & -2 \\ -1 & 2 & -1 \\ & -1 & 2 & -1 \\ & & & \ddots & \ddots & \ddots \end{bmatrix}$$

$$\underline{I}_1 = \underline{I}$$

$$\underline{I}_2 = -D_x \cdot \underline{I}$$

$$\underline{I}_3 = -D_x \cdot \underline{I}$$

$$\underline{Q}_1 = \underline{O}$$

$$\underline{Q}_2 = \frac{-Gx+Gy}{h^2} \begin{bmatrix} 0 & 0 \\ 1 & -2 & 1 \\ & \ddots & \ddots & \ddots \end{bmatrix}$$

ORTHOTROPIC CIRCULARLY CURVED PLATE

$$\underline{A}_1 = \frac{Dr.}{D\theta}\frac{1}{h^4}\begin{bmatrix} a_{0,1} & a_{0,2} & a_{0,3} & & & \\ a_{1,1} & a_{1,2} & a_{1,3} & a_{1,4} & & \\ a_{2,1} & a_{2,2} & a_{2,3} & a_{2,4} & a_{2,5} & \\ & \cdot & & \cdot & & \cdot & & \cdot & & \cdot \\ & \cdot & & \cdot & & \cdot & & \cdot & & \cdot & & \cdot \end{bmatrix}$$

$$\underline{A}_2 = \frac{G_\theta + G_r}{D_\theta}\frac{1}{h^2}\begin{bmatrix} b_{0,1} & b_{0,2} & & \\ b_{1,1} & b_{1,2} & b_{1,3} & \\ & \cdot & & \cdot & & \cdot \\ & & \cdot & & \cdot & & \cdot \\ & & & \cdot & & \cdot & & \cdot \end{bmatrix}$$

$$\beta = \frac{D\theta}{G_\theta} + G_r$$

$$b_{0,1} = 2r_0^2 + hr_0(1 - 2\beta) - h^2(3 + 4\beta)$$

$$b_{0,2} = -2r_0^2 - 3hr_0$$

$$b_{1,1} = -r_1^2 + \tfrac{1}{2}hr_1$$

$$b_{1,2} = 2r_1^2 - h^2(2\beta + 1)$$

$$b_{1,3} = -r_1^2 - \tfrac{1}{2}hr_1$$

$$\alpha = D\theta/Dr$$

$$a_{0,1} = -2r_0^4 - 2\alpha h^2 r_0^2 - 3\alpha h^3 r_0$$

$$a_{0,2} = 4r_0^4 + 2\alpha h^2 r_0^2 + 3\alpha h_r^3$$

$$a_{0,3} = -2r_0^4$$

$$a_{1,1} = 2r_1^4 + h^2 r_1^2 - \frac{\alpha}{2}h^3 r_1$$

$$a_{1,2} = -5r_1^4 + hr_1^3 - 2\alpha h^2 r_1^2$$

$$a_{1,3} = 4r_1^4 - 2hr_1^3 + \alpha h^2 r_1^2 + \frac{\alpha}{2} h^3 r_1$$

$$a_{1,4} = -r_1^4 + hr_1^3$$

$$a_{2,1} = -r_2^4 - r_2^3 h$$

$$a_{2,2} = 4r_2^4 + 2hr_2^3 + \alpha h^2 r_2^2 - \frac{\alpha}{2} h^3 r_2$$

$$a_{2,3} = -6r_2^4 - 2\alpha h^2 r_2^2$$

$$a_{2,4} = 4r_2^4 - 2hr_2^3 + \alpha h^2 r_2^2 + \frac{\alpha}{2} h^3 r_2$$

$$a_{2,5} = -r_2^4 + hr_2^3$$

$$\gamma = \frac{D\theta - Gr}{Gr + G\theta}$$

$$\underline{Q}2 = \frac{(G\theta + Gr)}{h^2} \qquad\qquad \delta = \frac{Gr - G\theta}{Gr + G\theta}$$

$$
\begin{bmatrix}
-\gamma \dfrac{h}{r_0^2} - \delta \dfrac{h^2}{r_0^2} & \gamma \dfrac{h}{r_0^2} & & \\[2ex]
-\dfrac{1}{r_1} - \gamma \dfrac{h}{2r_1^2} & \dfrac{2}{r_1} - \dfrac{\delta h^2}{r_1^3} & \dfrac{1}{r_1} + \gamma \dfrac{h}{2r_1^2} & \\[2ex]
& \cdot & \cdot & \cdot \\[2ex]
& \cdot & \cdot & \cdot
\end{bmatrix}
$$

$$
\underline{I}_1 =
\begin{bmatrix}
1/r_0 & & \\
& 1/r_1 & \\
& & \cdot
\end{bmatrix}
$$

$$
\underline{I}_2 = D\theta
\begin{bmatrix}
-1/r_0^2 & & \\
\cdot & & \\
& -1/r_1^2 & \\
& & \cdot
\end{bmatrix}
$$

$$\mathbf{\underline{I}}_3 = D\theta \begin{bmatrix} -1/r_0^3 & & & \\ & -1/r_1^3 & & \\ & & \cdot & \\ & & & \cdot \end{bmatrix}$$

$$\mathbf{\underline{Q}}_1 = \frac{D}{h} \begin{bmatrix} -1/r_0 & 1/r_0 & & \\ -\tfrac{1}{2}r_1 & 0 & \tfrac{1}{2}r_1 & \\ & \cdot & \cdot & \cdot \\ & & \cdot & \cdot & \cdot \end{bmatrix}$$

REFERENCES

1. Jenkins, R. S. and Tottenham, H. The solution of shell problems by the matrix progression method. *Proceedings of the World Conference on Shell Structures. (San Francisco, 1962)*, Washington, National Research Council, 1964.
2. Chetty, S. M. K. *An investigation into linear analysis of hyperbolic paraboloid shells.* Ph.D. thesis, Southampton University, 1962.
3. Desai, J. R. *Approximation solutions in the analysis of arch dams on rigid foundations.* Ph.D. thesis, Southampton University, 1963.
4. Kazimi, S. M. A. and Coull, A. The application of line-solution techniques to the solution of plane-stress problems. *Int. J. Mech. Sci.,* vol. 6, no. 5, 1964, pp. 391–399.
5. Coull, A. and Rao, K. S. Analysis of bridge decks by line solution techniques. International symposium on the use of electronic digital computers in structural engineering (University of Newcastle-upon-Tyne, 1966).
6. Rao, K. S. *The application of line solution to plate problems, with particular reference to bridge decks.* Ph.D. thesis, Southampton University, 1966.
7. Das, P. C. *Analysis of curved bridge decks with concentrated loads.* Ph.D thesis, Southampton University, 1968.
8. Trockalakis, T. M. *The analysis of bridge decks by the line solution matrix progression technique.* Ph.D thesis, Southampton University, 1968.
9. Spindel, J. E. *A study of bridge slabs having no transverse flexural stiffness.* Ph.D. thesis, London University, 1961.
10. Shaikh, A. H. *An investigation into the elastic behaviour of right and curved bridge-decks.* Ph.D. thesis, Southampton University, 1969.
11. Tottenham, H. The matrix progression method in structural analysis: Chapter 7 in Rydzewski, J. R. *Introduction to structural problems in nuclear reactor engineering.* London, Pergamon, 1962.
12. Timoshenko, S. P. *Theory of plates and shells.* New York, McGraw-Hill, 1940.

TABLE I

Position		Sym. Load Mom.	Anti-Sym. Load Mom.	Pt. Load Mom.
A	LINE 1	+3·64458	+4·48505	+8·12963
	2	+1·40559	+2·23336	+3·63895
	3	+0·61994	+1·38540	+2·00534
	4	+0·11687	+0·77742	+0·89428
B	1	−0·43643	+1·12370	+6·87270
	2	−0·08344	+1·31151	+1·22807
	3	−0·04944	+0·96954	+0·92009
	4	−0·07903	+0·56420	+0·48518
C	1	−2·92619	0	−2·92619
	2	−2·18342	0	−2·18342
	3	−1·03699	0	−1·03699
	4	−0·00799	0	−0·79954
D	1	−0·43643	−1·12370	−1·56013
	2	−0·08344	−1·31151	−1·39995
	3	−0·04944	−0·96954	−1·01899
	4	−0·07903	−0·65420	−0·64323
E	1	+3·64458	−4·48505	−0·84048
	2	+1·40559	−2·23336	−0·82777
	3	+0·61994	−1·38540	−0·76546
	4	+0·11687	−0·77742	−0·66055

(Moment = 6kW 10^{-2}, deflection = $(6kL^2/D) \, 10^{-3}$)

TABLE II

Deflections and bending moments in uniformly loaded square plate with edges $x = 0, x = a$ simply supported and the other two free.

Poisson's Ratio = 0·3

Method	$x = a/2, \quad y = 0$			$x = a/2 \quad y = \pm b/2$	
	$w = \alpha \dfrac{qa^4}{D}$	$Mx = \beta_1 qa^4$	$My = \beta_1 qa^4$	$w = \alpha_2 \dfrac{qa^4}{D}$	$Mx = \beta_2 qa^2$
	α	β_1	β_1	α_2	β_2
Levy-type (Timoshenko)	0·01309	0·1225	0·0271	0·01509	0·1318
Line solution Matrix Progression	0·01309	0·1224	0·0269	0·01498	0·1309

w = deflections Mx = longitudinal moment My = transverse moment.

FREQUENCY ANALYSIS OF CERTAIN SINGLE AND CONTINUOUS SPAN BRIDGES

M. S. CHEUNG and Y. K. CHEUNG
University of Calgary
D. V. REDDY
Indian Institute of Technology

SYNOPSIS

The paper describes the application of the finite strip method to the determination of frequencies and modal eigen-functions for the free vibration of single and continuous spans of constant or varying thickness bridges with isotropic or orthotropic properties. The analysis is based on the engineering theory of plates. Only linearly elastic materials and free, undamped vibration are considered.

The stiffness matrix of a strip is derived by applying the Rayleigh-Ritz approach assuming that the displacement profile along the length is a suitable characteristic beam function series satisfying the end conditions and the profile along the width, a simple cubic polynomial. A consistent type mass matrix is also formed for the same assumed functions. The stiffness and mass matrices of all the strips making up a plate are then assembled to form the eigenvalue matrix. The method is simple and accurate, and all the natural frequencies and corresponding modal shapes can be obtained rapidly from an intermediate or even small size electronic computer.

INTRODUCTION

Among the earliest studies of the dynamic behaviour of single span, constant thickness bridges, mention must be made of the investigations of Inglis,[1] Hillerborg,[2] Biggs and others,[3] Mise and Kunii[4] in which the bridge was assumed to behave like a beam. The assumption is valid only for bridges with long and heavy girders as are usual in railway plate girder bridge practice. Huffington and Hoppmann[5] determined the exact frequencies and modal eigen-functions of rectangular orthotropic plates with 'bridge-type' of boundary conditions by direct solution of the governing differential equation of motion using the Lévy approach. Yamada and Veletsos[6] have studied the vibration of single-span I-beam and slab bridge decks treating the slab as continuous over a series of flexible beams and using the Rayleigh-Ritz method. The free vibration problem idealising the structure as an equivalent orthotropic plate has also been solved by them using the energy method. The two approaches have been compared by Huang and Walker[7] for five and nine beam bridges with edge beams. However, there are very few references to analysis for continuous bridges, especially for the cases of variable thickness and unequal span bridge decks. The finite strip method, which is really a piece-wise Rayleigh-Ritz analysis, is a powerful tool for the solution of such problems. The validity of the assumption of an equivalent ortho-tropic plate is demonstrated by comparison with exact theory.

In the finite strip method, the bridge deck is assumed as an assemblage of many parallel orthotropic strips, each of which may have different orthotropic properties, but constant within the strip. The displacement function chosen for each strip is of the form $\psi(x)$, $Y(y)$ in which $Y(y)$ is a characteristic beam function that satisfies the end conditions of the strip, while $\psi(x)$ is a simple polynomial given in terms of the displacement parameters at the two adjacent nodal lines. Compatibility between adjacent strips is ensured as the function defines the displacements and its first derivatives along the interface uniquely by the common nodal displacement para-meters. By using a function of this type of a two-dimensional problem is reduced to a one-dimensional case similar to a Lévy-Kantorovich type solution. As a consequence, the sizes of the resulting matrices are considerably reduced. It should also be men-tioned that the computer programme for such an approach is comparatively short and requires very little input data for its execution. The method is ideally suitable for programming on a small computer.

Theoretical Formulation

The general equation of motion can be expressed as:

$$[M]\,\{\ddot{x}\} + [C]\,\{\dot{x}\} + [K]\,\{x\} = \{p(t)\} \tag{1}$$

where, $[M]$ is the mass matrix of the system, $[C]\,\{\dot{x}\}$ is damping force matrix, $[K]\,\{x\}$ is the restoring force matrix of the system in which $[K]$ is the system stiff-ness matrix.

Equation 1 gives the equation of motion of a system subject to a disturbing force vector $p(t)$. For free undamped vibration the eigenvalue formulation for Eq. 1 is:

$$\{[K] - \omega^2\,[M]\,\{\delta\} = \{0\} \tag{2}$$

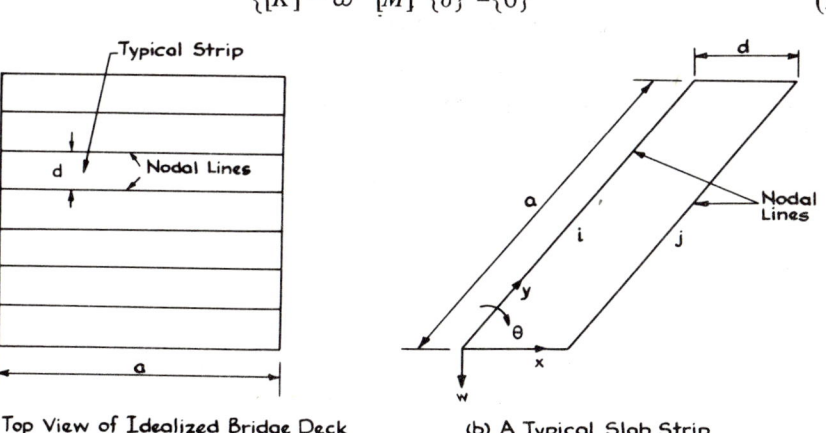

(a) Top View of Idealized Bridge Deck   (b) A Typical Slab Strip

FIG. 1

Strain Energy of Strip

The bridge deck is assumed to be an assemblage of a number of strips, each strip having its individual orthotropic elastic properties (Figs 1(a) and 1(b)). The thickness of each strip can vary in both directions, although for the sake of convenience the transverse variation of thickness is usually approximated by a series of strips with different thicknesses in a step-wise manner.

A suitable displacement function for a typical strip (Fig. 1) that satisfies different end conditions and the compatibility conditions, δ along the boundary lines i and j, can be assumed as:

$$w = [\psi(x)] \cdot Y(y) \tag{3}$$

where $Y(y)$ is a suitable characteristic beam function that satisfies the end conditions and $\psi(x)$ is a simple polynomial with a number of terms that corresponding to the number of degrees of freedom at each nodal line.

The polynomial constants in Eq. 3 can be determined from two boundary conditions:

At $x = 0$,
$$\delta_i = \begin{Bmatrix} w_i \\ \theta_i \end{Bmatrix}$$

and $x = d$,
$$\delta_j = \begin{Bmatrix} w_j \\ \theta_j \end{Bmatrix} \tag{4}$$

Therefore, the nodal displacement parameters, $\{\delta\} = [w_i, \theta_i, w_j, \theta_j]^T$, are related to the multiplying constants, $\{A\}$, as follows:

$$\{\delta\} = [C] \{A\} \tag{5}$$

Substituting Eq. 5 into Eq. 3 and differentiating the displacement function, the curvature of the strip $\left(-\dfrac{\partial^2 w}{\partial x^2} - \dfrac{\partial^2 w}{\partial y^2}, 2\dfrac{\partial^2 w}{\partial x \partial y} \right)$, can be expressed in terms of displacement parameters as follows:

$$\{\chi\} = [B] \{A\} = [B] [C^{-1}] \{\delta\} \tag{6}$$

The moments $[M] = [M_x, M_y, M_{xy}]$ are related to curvatures $\{\chi\}$ as:

$$[M(x,y)] = [D(x,y)] \{\chi(x,y)\} \tag{7}$$

where $[D(x,y)]$ is the flexural rigidity matrix of the strip which is homogeneous and orthotropic. If the directions of orthotropy coincide with x, and y, $[D(x,y)]$ is independent of x and y and can be written as:

$$[D] = \begin{bmatrix} D_x & D_1 & 0 \\ D_1 & D_y & 0 \\ 0 & 0 & D_{xy} \end{bmatrix} \tag{8}$$

where $\quad D_x = \dfrac{E_x h^3}{12(1 - v_x v_y)}, \quad D_y = \dfrac{E_y h^3}{12(1 - v_x v_y)}, \quad D_{xy} = \dfrac{G h^3}{12}$

and $\quad D_1 = \dfrac{v_x E_y (= v_y E_x) \, h^3}{(1 - v_x v_y) \quad 12}$

in which E_x, E_y, G, v_x, v_y are the elastic properties.

The total strain energy of the system for plate bending is given by

$$U = \tfrac{1}{2} \int_0^a \int_0^d \left(M_x \frac{\partial^2 w}{\partial x^2} + M_y \frac{\partial^2 w}{\partial y^2} + 2M_{xy} \frac{\partial^2 w}{\partial x \partial y} \right) dx \, dy \tag{9}$$

Substituting Eqs. 6 and 7 into Eq. 9 and integrating over the mid-surface of the strip gives the total strain energy.

Kinetic Energy of Strip

Neglecting rotary inertia, the kinetic energy, can be written as:

$$T = \int_0^a \int_0^d \int_{-\frac{1}{2}h}^{+\frac{1}{2}h} \frac{1}{2} \rho \left(\frac{\partial w}{\partial t}\right)^2 dx \, dy \, dz$$

$$= \frac{1}{2}\rho h \int_0^a \int_0^d \left(\frac{\partial w}{\partial t}\right)^2 dx \, dy \tag{10}$$

or $$= \frac{1}{2} \omega^2 \rho h \int_0^a \int_0^d (\dot{w})^2 \, dxdy$$

where h is the thickness of the strip and ρ is the mass density.

Substituting Eq. 3 into Eq. 10 and integrating over the mid-surface gives the kinetic energy of the strip.

Strip Stiffness and Mass Matrices

The total potential energy of the strip $\bar{U}$ is:

$$\bar{U} = U - T \tag{11}$$

The stiffness and mass matrices of the strip viz. $[k]$ and the $[m]$ obtained by using the minimum potential energy principle, are given in Tables 1 and 2.

Bridge Decks

The finite strip solution is now applied to the vibration of single and multi-span beam and slab bridge decks with equivalent orthotropic plate properties. In the analyses, two different types of strip layout are used. In the first case, the deck slab is divided into longitudinal strips with simply supported end conditions which can be easily represented by a simple sine series. The orthogonality property of the assumed displacement function (when $m \neq n$, $[K]_{mn} = 0$), reduces the size of the final matrix and thereby the storage requirement. In the second case, the deck slab is divided into transverse strips with free-free end conditions and the characteristic function satisfying such free boundary conditions can be taken as:

$$Y_1(y) = 1,$$

$$Y_2(y) = 1 - \frac{2y}{a}$$

$$Y_m(y) = \sin\frac{\mu_m y}{a} + sh\frac{\mu_m y}{a} - \alpha_m \left(\cos\frac{\mu_m y}{a} + ch\frac{\mu_m y}{a}\right) \tag{12}$$

where $$\alpha_m = \frac{\sin\mu_m - sh\,\mu_m}{\cos\mu_m - ch\,\mu_m}$$

and $$\mu_m = 4{\cdot}7300, \ 7{\cdot}8532, \ 10{\cdot}9960, \ 14{\cdot}1370, \ \ldots \ \frac{2m-3}{2}\pi$$

for $$m = 3, 4, 5 \ldots$$

However, for such a strip the orthogonality property no longer holds and the series terms have to be evaluated by solving a combined set of simultaneous equations. The free-free strip is particularly suitable for the analysis of a continuous bridge with any number of spans of constant or varying thickness and isotropic or orthotropic material properties. The use of the transverse strip eliminates the difficulty encountered in the assumption of a suitable function for a multi-span longitudinal strip.

EXAMPLES

1. The frequency analysis of a number of isotropic and orthotropic simply supported square and rectangular bridge decks will now be presented. The solution for an isotropic slab bridge is compared in Fig. 2 with Odman's[9] solution based on characteristic functions and for the orthotropic case with the Rayleigh-Ritz analysis of Yamada and Veletsos[6] and Huang and Walker.[7] The results are given in Table 3.

2. The next example is that of a two-equal span, isotropic bridge ($\rho h = 1$) for which the results are compared with those in reference [10] for the case of zero Poisson's ratio (Table 4). The use of only four term functions for an 8-strip lay-out gives fairly accurate results.

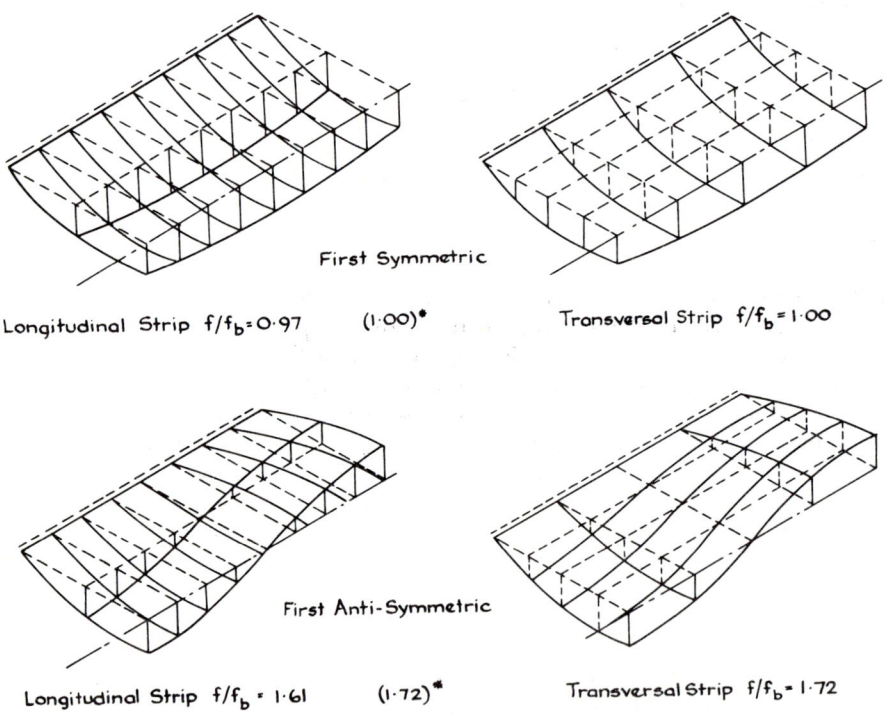

First Symmetric

Longitudinal Strip $f/f_b = 0.97$ $(1.00)^*$ Transversal Strip $f/f_b = 1.00$

First Anti-Symmetric

Longitudinal Strip $f/f_b = 1.61$ $(1.72)^*$ Transversal Strip $f/f_b = 1.72$

FIG. 2(a) TRANSVERSE MODES FOR A SQUARE ISOTROPIC BRIDGE DECK
($m = 1$, $v = 1/6$ VALUES OF F/F_{beam}

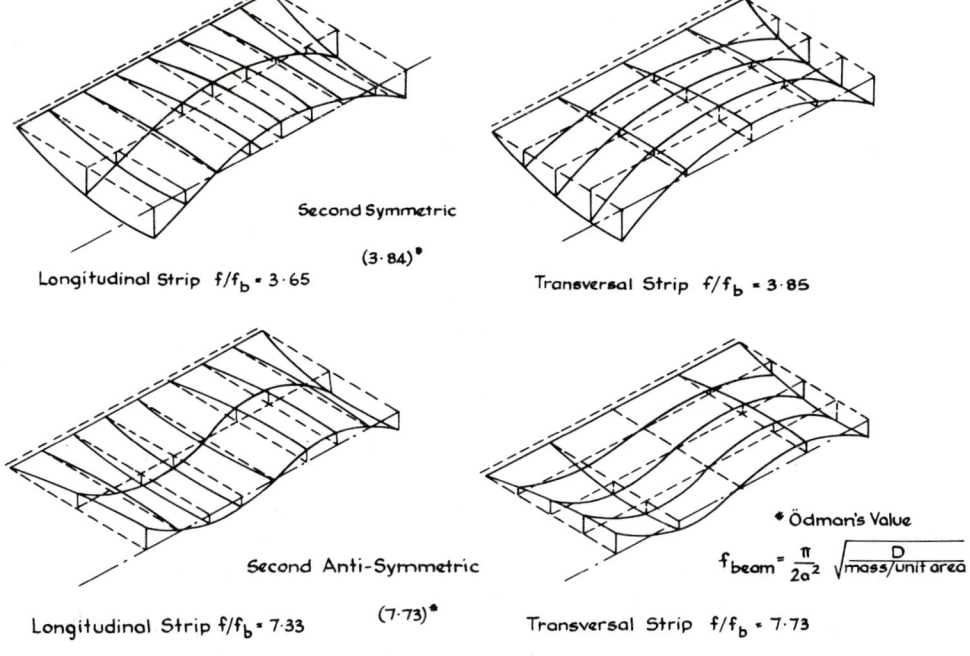

Second Symmetric

$(3.84)^*$

Longitudinal Strip $f/f_b = 3.65$ Transversal Strip $f/f_b = 3.85$

Second Anti-Symmetric

$(7.73)^*$

Longitudinal Strip $f/f_b = 7.33$ Transversal Strip $f/f_b = 7.73$

* Ödman's Value

$$f_{beam} = \frac{\pi}{2a^2} \sqrt{\frac{D}{mass/unit\ area}}$$

FIG. 2(b) TRANSVERSE MODES FOR A SQUARE ISOTROPIC BRIDGE
$(m = 1, v = 1/6)$ VALUES OF F/F_{beam}

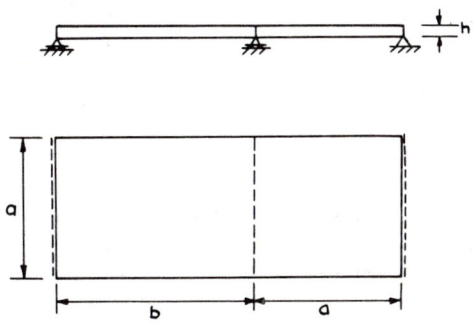

FIG. 3(a) A TWO UNEQUAL SPAN ISOTROPIC CONTINUOUS
BRIDGE

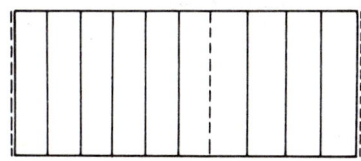

FIG. 3(b) FINITE STRIP IDEALISATION OF THE BRIDGE
DECK

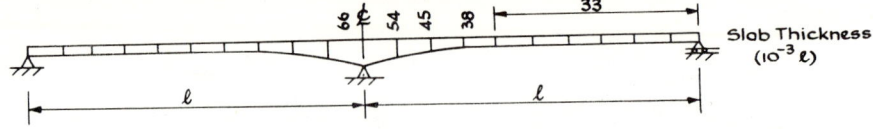

FIG. 4(a) LONGITUDINAL SECTION

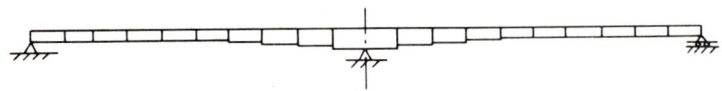

FIG. 4(b) IDEALISED SECTION

Circular Frequencies; 3·40, 6·76, 7·58, 9·30, 14·46

$$\left[\text{Unit:} \quad \frac{1}{10a^2} \quad \sqrt{\frac{H}{\rho}}, \quad H = \frac{E}{12(1-\nu^2)} \right]$$

3. A two-span isotropic bridge deck and orthotropic bridge deck of constant thickness (Fig. 3) and varying span ratios are solved and the first six frequencies are listed in Tables 5 and 6. Frequencies which are not affected by the variation in Poisson's ratio all correspond to beam modes in which the first term of the series is dominant.

4. A three-span bridge is also solved and the results are listed in Table 7. It is obvious that the method can be readily applied to any number of spans. However, as the number of spans increases the size of the matrices increases. The computer time required varies as the cube of the number of equations. Fortunately, from many tests, it was found that only three strips are sufficient for each span to define the modal shapes. Thus, even for the problem of a large number of spans, the finite strip method is still economical compared with any other existing method such as finite element and finite difference formulations. For instance, in the computation of the first nine frequencies for a three-span bridge deck, using nine strips and four terms of the series, the computing time required is only five to six minutes on a medium size computer IBM 360/50.

5. To further demonstrate the versatility of the method, a continuous, variable thickness bridge deck (Fig. 4(a)) is solved by the same programme. The deck is divided into strips in the transverse direction, by averaging the thickness within each strip and the idealised deck slab is shown in Fig. 4(b). The validity of such an idealisation has already been established in earlier investigations.[11,12] The first five circular frequencies are listed in Fig. 4(b).

CONCLUSION

The finite strip method has been shown to be a versatile tool for vibration analysis of single span as well as continuous span bridges. It is particularly effective for the case of continuous bridge analysis and with the advantage of relatively small matrices, compares favourably with finite difference and finite element analyses.

ACKNOWLEDGEMENT

The authors wish to thank the National Research Council of Canada for the financial support of this project.

NOTATION

$[m]$ = mass matrix of a strip
$[k]$ = stiffness matrix of strip
$\{\delta\}$ = vector of nodal displacement
$\{p(t)\}$ = force vector
ω = circular frequency of structure (radians/second)

$\begin{Bmatrix} w_i \\ \theta_i \end{Bmatrix}_m$ = nodal displacement and rotation of ith nodal line for the mth term of the series

ρ = mass per unit volume
h = thickness of strip
d = width of strip
w = deflection function
a = length of bridge deck in y direction
U = strain energy of system
$\overline{U}$ = total potential energy of system
T = kinetic energy of system
Y = characteristic function

TABLE 1   Stiffness matrix for an orthotropic strip

$$[k]_{mn} = \frac{1}{420d^3}$$

$5040\ D_x\ I_1$ $-504\ d^2\ D_1\ I_2$ $-504\ d^2\ D_1\ I_3$ $156\ d^4\ D_y\ I_4$ $2016\ d^2\ D_{xy}\ I_5$	$2520\ d\ D_x\ I_1$ $-462\ d^3\ D_1\ I_2$ $-42\ d^3\ D_1\ I_3$ $22\ d^5\ D_y\ I_4$ $168\ d^3\ D_{xy}\ I_5$	$-5040\ D_x\ I_1$ $504\ d^2\ D_1\ I_2$ $504\ d^2\ D_1\ I_3$ $54\ d^4\ D_y\ I_4$ $-2016\ d^2\ D_{xy}\ I_5$	$2520\ d\ D_x\ I_1$ $-42\ d^3\ D_1\ I_2$ $-42\ d^3\ D_1\ I_3$ $-13\ d^5\ D_y\ I_4$ $168\ d^3\ D_{xy}\ I_5$
$2520\ d\ D_x\ I_1$ $-462\ d^3\ D_1\ I_3$ $-42\ d^3\ D_1\ I_2$ $22\cdot\ d^5\ D_y\ I_4$ $168\ d^3\ D_{xy}\ I_5$	$1680\ d^2\ D_x\ I_1$ $-56\ d^4\ D_1\ I_2$ $-56\ d^4\ D_1\ I_3$ $4\ d^6\ D_y\ I_4$ $224\ d^4\ D_{xy}\ I_5$	$-2520\ d\ D_x\ I_1$ $42\ d^3\ D_1\ I_2$ $42\ d^3\ D_1\ I_3$ $13\ d^5\ D_y\ I_4$ $-168\ d^3\ D_{xy}\ I_5$	$840\ d^2\ D_x\ I_1$ $14\ d^4\ D_1\ I_2$ $14\ d^4\ D_1\ I_3$ $-3\ d^6\ D_y\ I_4$ $-56\ d^4\ D_{xy}\ I_5$
$-5040\ d\ D_x\ I_1$ $504\ d^2\ D_1\ I_3$ $504\ d^2\ D_1\ I_2$ $54\ d^4\ D_y\ I_4$ $-2016\ d^2\ D_{xy}\ I_5$	$-2520\ d\ D_x\ I_1$ $42\ d^3\ D_1\ I_3$ $42\ d^3\ D_1\ I_2$ $13\ d^5\ D_y\ I_4$ $-168\ d^3\ D_{xy}\ I_5$	$5040\ D_x\ I_1$ $-504\ d^2\ D_1\ I_2$ $-504\ d^2\ D_1\ I_3$ $156\ d^4\ D_y\ I_4$ $2016\ d^2\ D_{xy}\ I_5$	$-2520\ d\ D_x\ I_1$ $462\ d^3\ D_1\ I_2$ $42\ d^3\ D_1\ I_3$ $-22\ d^5\ D_y\ I_4$ $-168\ d^3\ D_{xy}\ I_5$
$2520\ d\ D_x\ I_1$ $-42\ d^3\ D_1\ I_3$ $-42\ d^3\ D_1\ I_2$ $-13\ d^5\ D_y\ I_4$ $168\ d^3\ D_{xy}\ I_5$	$840\ d^2\ D_x\ I_1$ $14\ d^4\ D_1\ I_3$ $14\ d^4\ D_1\ I_2$ $-3\ d^6\ D_y\ I_4$ $-56\ d^4\ D_{xy}\ I_5$	$-2520\ d\ D_x\ I_1$ $462\ d^3\ D_1\ I_3$ $42\ d^3\ D_1\ I_2$ $-22\ d^5\ D_y\ I_4$ $-168\ d^3\ D_{xy}\ I_5$	$1680\ d^2\ D_x\ I_1$ $-56\ d^4\ D_1\ I_2$ $-56\ d^4\ D_1\ I_3$ $4\ d^6\ D_y\ I_4$ $224\ d^4\ D_{xy}\ I_5$

$$I_1 = \int_0^a Y_m Y_n\, dy; \quad I_2 = \int_0^a Y''_m Y_n\, dy; \quad I_3 = \int_0^a Y_m Y''_n\, dy; \quad I_4 = \int_0^a Y''_m Y''_n\, dy; \quad I_5 = \int_0^a Y'_m Y'_n\, dy$$

TABLE 2   Consistent mass matrix of the strip

$$[m]_{mn} = \int_0^a \rho h \left[\ \cdots\ \right] dy$$

$\dfrac{78}{210}\, d\, Y_m Y_n$	$\dfrac{11}{210}\, d^2\, Y_m Y_n$	$\dfrac{9}{70}\, d\, Y_m Y_n$	$-\dfrac{13}{420}\, d^2\, Y_m Y_n$
$\dfrac{11}{210}\, d^2\, Y_m Y_n$	$\dfrac{1}{105}\, d^3\, Y_m Y_n$	$\dfrac{13}{420}\, d^2\, Y_m Y_n$	$-\dfrac{3}{420}\, d^2\, Y_m Y_n$
$\dfrac{9}{70}\, d\, Y_m Y_n$	$\dfrac{13}{420}\, d^2\, Y_m Y_n$	$\dfrac{13}{35}\, d^2\, Y_m Y_n$	$-\dfrac{11}{210}\, d^2\, Y_m Y_n$
$-\dfrac{13}{420}\, d^2\, Y_m Y_n$	$-\dfrac{3}{420}\, d^3\, Y_m Y_n$	$-\dfrac{11}{210}\, d^2\, Y_m Y_n$	$\dfrac{1}{105}\, d^3\, Y_m Y_n$

TABLE 3

Transverse modes for orthotropic bridges ($a/b = 1$)

Values of f/f_{beam} where $f_{beam} = \pi/2a^2 \sqrt{\dfrac{D_x}{\text{mass/unit area}}}$

Longitudinal mode $m = 1$

Orthotropic Property	$D_x = 1.0, D_y = 1.0$ $D_1 = 0, D_{xy} = 0.1$			$D_x = 1.0, D_y = 0.19753$ $D_1 = 0, D_{xy} = 0.06666$			$D_x = 1.0, D_y = 0.0625$ $D_1 = 0, D_{xy} = 0.0625$			$D_x = 1.0, D_y = 0.19753$ $D_1 = 0, D_{xy} = 0.2222$			$D_x = 1.0, D_y = 0.01234$ $D_1 = 0, D_{xy} = 0.05555$		
Transverse Mode	Finite Strip		Ref.	Finite Strip		Ref.	Finite Strip		Ref.	Finite Strip		Ref.	Finite Strip		Ref.
	L	T	(6)	L	T	(6)	L	T	(6)	L	T	(6)	L	T	(6)
First Symmetric	0.97	1.00	1.00	0.97	1.00	1.00	0.97	1.00	1.00	0.97	1.00	1.00	0.97	1.00	1.00
First Anti-Symmetric	1.15	1.21		1.10	1.12		1.08	1.13	1.13	1.38	1.41	1.41	1.07	1.11	1.11
Second Symmetric	2.64	2.84		1.70	1.83		1.50	1.60	1.58	2.36	2.53	2.48	1.35	1.46	1.42
Second Anti-Symmetric	3.73	4.00		3.14	3.41		2.28	2.49	2.47	4.05	4.30	4.23	1.80	1.97	1.93

L, Longitudinal Strip   T, Transversal Strip

TABLE 4

Circular frequencies for a continuous bridge of two equal spans

$$a = b_1 = b_2 = 1$$

Frequencies	Transverse Strip	Longitudinal Strip	Reference (10)
ω_1	9·87	9·63	9·86
ω_2	15·43	15·41	
ω_3	17·92	16·83	17·17
ω_4	22·21	22·16	
ω_5	39·65	38·25	39·48

TABLE 5

Frequencies of a continuous isotropic bridge deck
with two unequal spans

$$a = 1, \ \rho h = 1$$

Poisson's Ratio	$\dfrac{b}{a}$	ω_1	ω_2	ω_3	ω_4	ω_5	ω_6
$v = 0$	1·5	5·35	11·74	12·31	19·85	20·50	28·64
	2·0	3·16	8·45	9·87	13·81	17·92	20·98
$v = 0·167$	1·5	5·35	10·98	12·31	18·87	20·50	27·56
	2·0	3·16	7·84	9·87	13·81	16·92	20·04
$v = 0·3$	1·5	5·35	10·30	12·31	18·03	20·50	26·61
	2·0	3·16	7·30	9·87	13·81	16·03	19·22

TABLE 6

Frequencies of a continuous orthotropic bridge with two unequal spans

Property		$\dfrac{b}{a}$	ω_1	ω_2	ω_3	ω_4	ω_5	ω_6
$D_y = 1\cdot0$	$D_{xy} = 0\cdot5$	1·5	5·35	11·74	12·31	19·85	20·50	28·64
		2·0	3·16	8·45	9·87	13·81	17·92	20·98
	$D_{xy} = 0\cdot25$	1·5	5·35	9·19	12·31	16·62	20·50	25·02
		2·0	3·16	6·42	9·87	13·81	14·56	17·87
$D_y = 0\cdot19753$	$D_{xy} = 0\cdot22222$	1·5	5·35	8·73	12·31	16·01	18·23	20·50
		2·0	3·16	6·08	9·87	13·81	13·91	14·93
	$D_{xy} = 0\cdot1111$	1·5	5·36	7·29	12·31	14·34	15·20	20·50
		2·0	3·16	4·88	9·87	12·12	12·90	13·81
$D_y = 0\cdot0625$	$D_{xy} = 0\cdot1250$	1·5	5·35	7·41	12·31	13·27	14·46	20·50
		2·0	3·16	5·00	9·87	10·29	12·26	13·81
	$D_{xy} = 0\cdot0625$	1·5	5·35	6·50	10·89	12·31	13·46	17·78
		2·0	3·16	4·20	8·59	9·87	11·16	13·81

TABLE 7

Frequencies of a continuous isotropic bridge deck
of three equal spans

$\rho h = 1$

Frequency / Span Ratio	ω_1	ω_2	ω_3	ω_4	ω_5
$b_1 = a$ $b_2 = a$	9·87	12·67	17·93	18·52	20·01
$b_1 = a$ $b_2 = 1\frac{1}{2}a$	6·20	11·69	12·49	13·70	19·46
$b_1 = a$ $b_2 = 2a$	3·86	8·92	9·88	12·51	15·45

Multiplier: $\dfrac{1}{a^2}\sqrt{\dfrac{\rho\theta}{\rho h}}$

REFERENCES

1. Inglis, C. E. *A mathematical treatise on vibrations in railway bridges.* London, Cambridge U.P., 1934.
2. Hillerborg, A. Dynamic influences of smoothly running loads on simply-supported girders. *K. tek. Högsk. Avh.,* No. 68, 1951
3. Vibration and stresses in girder bridges. *Bull. Highw. Res. Bd.,* No. 124, 1956.
4. Mise, K. and Kunii, S. A theory for the forced vibrations of a railway bridge under the action of moving loads. *Q. Jl Mech. appl. Math.,* Vol. 9, No. 2, June 1956, pp. 195–206.
5. Huffington, N. J. and Hoppmann, W.H. On the transverse vibrations of rectangular ortho-tropic plates. *J. appl. Mech.,* Vol. 25, 1958, pp. 389–395.
6. Yamada, Y. and Veletsos, A. S. Free vibration of simple span I-beam bridges. Part B. 8th Progress Report, Highway Bridge Impact Investigation, University of Illinois, 1958.
7. Hunag, C. L. and Walker, W. H. Free vibration of simple span I-beam bridges. Part D. 9th Progress Report, Highway Bridge Impact Investigation, University of Illinois, 1958.
8. Anderson, R. G. *A finite element eigenvalue solution system.* Ph.D. thesis, University of Wales, Swansea, 1968.
9. Odman, S. T. A. Studies of boundary value problems – Part II. Characteristic functions of rectangular plates. *Handl. svenska ForskInst. Cem. Betong.,* No. 24, 1955.
10. Warburton, G. B. The vibration of rectangular plates. *Proc. Instn mech. Engrs,* Vol. 168, No. 12, 1954, pp. 371–381.
11. Cheung, Y. K. and Cheung, M. S. Free vibration of orthotropic variable thickness, continuous rectangular plates. *Revue Roum. Sci. tech., Sér. Méc. appl.,* Vol. 15, No. 3, 1970.
12. Cheung, M. S., Cheung, Y. K. and Ghali, A. Analysis of slab and girder bridges by the finite strip method. *Build Sci.,* Vol. 5, No. 2, Oct. 1970, pp. 95–105.

ANALYTICAL SOLUTIONS FOR BOX GIRDER BRIDGES

A. C. SCORDELIS

University of California, Berkeley

SUMMARY

A review of analytical methods and general computer programs which have been developed at the University of California for the analysis of box girder bridges is presented. Several solutions based on two general methods, the direct stiffness harmonic analysis and the finite element method, are described. These solutions may be used to analyse multi-cell box girder bridges which are straight, skew, curved or of arbitrary general geometry and under general loading. Numerical results from three examples involving a straight two span continuous bridge, a single span skew bridge and a single span curved bridge, are given.

INTRODUCTION

General Remarks

In recent years approximately 60% of the concrete bridges (computed on the basis of deck area) built in California have been multi-cell reinforced concrete box girder

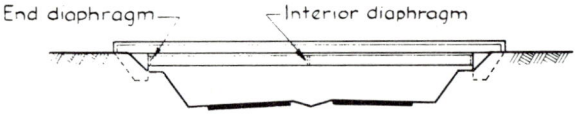

a) Elevation of Typical Simple Span Bridge

b) Elevation of Typical Continuous Bridge

c) Typical Cross-sections.

FIG. I MULTI-CELL BOX GIRDER BRIDGES

200

bridges. These bridges have proven to be economical compared to other types and have found wide usage both as simple span and continuous structures (Fig. 1), primarily in the span ranges between 60 and 100 ft. These cast in place bridges can be constructed to follow any desired alignment in plan so that straight, skew and curved bridges of every shape are common in today's highway system.

For longer spans, with some recently designed up to 300 ft, post-tensioned box girder bridges have been used extensively. In the past few years, this type of bridge has comprised almost half of all prestressed highway structures built in California. One set of current plans calls for a segmentally constructed prestressed box girder bridge with a 450 ft span in Southern California.

Less commonly used in California but often used in other part of the United States, are bridges consisting of individual thin walled steel box girders or precast-prestressed hollow box girders with a composite cast-in-place slab deck (Fig. 2).

The typical cross-section of a concrete box girder bridge (Fig. 1(c)) consists of a top and bottom slab connected monolithically with vertical webs to form a

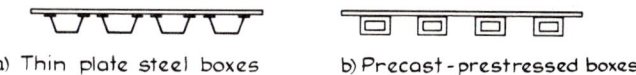

a) Thin plate steel boxes   b) Precast-prestressed boxes

FIG. 2 BRIDGES WITH COMPOSITE SLAB OVER
INDIVIDUAL BOX GIRDERS

cellular structure. Transverse diaphragms are placed at the end and interior support sections and in some cases, additional interior diaphragms are utilised between supports.

In a previous study[1] of over 200 concrete box girder bridges constructed in California, it was found that the majority of the bridges had a depth-span ratio in the ·050 to ·065 range; a top slab thickness of 6 to 7 in; a bottom slab thickness of $5\frac{1}{2}$ in; web thicknesses of 8 in; and cell widths (centre to centre spacing of webs) of between 7 and 9 ft. The number of cells, which is directly related to the overall bridge width, ranged from 2 to more than 16, with a preponderance of bridges having from 3 to 9 cells.

Because of the large use of concrete box girder bridges, it is evident that research directed toward improved design methods is desirable. These design methods should be based on rational analytical and experimental studies which can be used to accurately predict the structural response of the bridge to dead and live loads.

Purpose of Present Paper

A continuing program of research on box girder bridges has been conducted at the University of California at Berkeley since 1965. A systematic plan was initially developed to study successively straight, simple and continuous bridges, skew bridges and curved bridges. For each of these configurations the approach has been to: (1) study the available literature; (2) develop analytical methods and general computer programs; (3) perform experimental studies on elastic models to verify the analytical methods developed if deemed necessary; (4) make analytical parameter studies; (5) test large scale reinforced concrete models or prototypes; and (6) develop recommended design procedures. At present (1971), step (6) has been reached for the straight bridges, step (3) for the skew bridges, and step (2) for the curved bridges.

The purpose of the present paper is to review the analytical methods and general computer programs which have been developed for these bridges and to present numerical results from several examples. Detailed information on this research including computer program listings for many of the programs to be described may be found in a series of published research reports.[1-8]

Analytical Models and Methods

An analytical solution of the true response of a box girder bridge under load is complicated by the usual factors common to other reinforced concrete structural systems. It is a highly indeterminate structure; it is made of two materials, concrete and steel; under increasing load it experiences cracking and thus some redistribution of internal forces; and also the internal forces are time-dependent because of creep and shrinkage in the concrete. Nevertheless, as for other reinforced concrete systems, such as frames, slabs, and shells, it is generally accepted that for design purposes, the distribution of internal forces, moments and displacements in a box girder bridge due to applied loads can be based on an elastic analysis of an uncracked homogeneous concrete system.

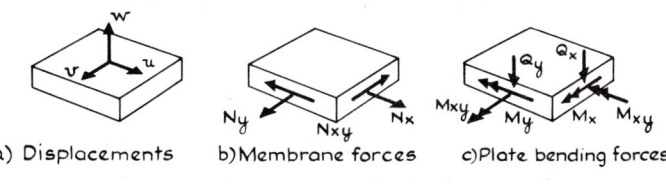

a) Displacements   b) Membrane forces   c) Plate bending forces

FIG 3   DISPLACEMENTS AND INTERNAL FORCES ON
A DIFFERENTIAL ELEMENT

In a complete analysis of multi-cell box girder bridge, all of the internal forces and displacements shown on a typical element in Fig. 3 taken from a deck or web plate of the bridge should be determined. The internal forces N_x, N_y and N_{xy} are termed membrane forces while M_x, M_y, M_{xy}, Q_x, and Q_y are internal forces due to plate bending. In many approximate analyses certain internal forces are assumed to be negligible and are thus taken as zero.

Of prime interest from a design standpoint are those internal forces which determine the reinforcing steel requirements for the bridge. These are N_x for the main longitudinal tension steel; N_{xy} for the diagonal tension steel; and M_y and N_y for the transverse steel.

Many analytical models and methods have been developed for analysing box girder bridges. Among these are approximate methods based on simplified structural behaviour such as the use of an equivalent beam grillage or anisotropic slab to represent the system; exact and approximate methods based on folded plate theory; and numerical solutions based on finite element or finite difference methods. No attempt will be made to review all of these methods here. Instead, a number of solutions and associated computer programs which have been developed at the University of California for straight, skew and curved bridges will be described. These solutions are based on an elastic analysis of an uncracked homogeneous structure and give a complete solution for all of the internal forces, moments and displacements shown in Fig. 3 for any type or position of loading on the bridge. These methods can all be classified under two major solution techniques: (1) direct stiffness harmonic analysis and (2) finite element solutions.

DIRECT STIFFNESS HARMONIC ANALYSIS

General Remarks

This method was originally developed by Jenkins[10] for the analysis of prismatic cylindrical shell systems simply supported at the two ends. DeFries-Skene and Scordelis[11] first utilised the method for the analysis of straight prismatic folded plates. A number of additional studies have also used or extended the procedure to analyse straight multi-cell box girder bridges.[1,2,4,12,13,14,16] More recently (1970) the method has been extended by Meyer and Scordelis[6,8] to the analysis of bridges curved in plan.

For a multi-cell prismatic box girder bridge (Fig. 4), the problem to be solved is the determination of the internal forces and displacements (Fig. 3) in a structural system consisting of an assembly of longitudinal plate elements interconnected at joints along their longitudinal edges and simply supported at the two ends by transverse diaphragms. The known quantities input into the problem include the geometry, dimensions and material properties of the plate elements, the surface and joint loadings and the boundary conditions along the longitudinal joints. Each plate element selected is assumed to extend longitudinally over the entire span and transversely between designated joints on the cross-section (Fig. 5).

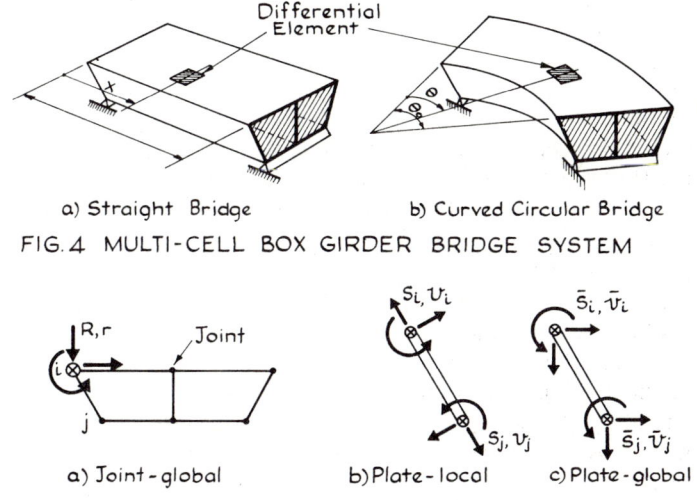

a) Straight Bridge   b) Curved Circular Bridge

FIG. 4 MULTI-CELL BOX GIRDER BRIDGE SYSTEM

a) Joint-global   b) Plate-local   c) Plate-global

FIG. 5 JOINT AND PLATE EDGE FORCES AND
DISPLACEMENTS IN GLOBAL AND LOCAL COORDINATE
SYSTEMS.

Direct Stiffness Method

An analysis for applied loads with any arbitrary longitudinal distribution for straight bridges or any arbitrary circumferential distribution for curved circular bridges may be performed using a harmonic analysis. The applied loads are first resolved into Fourier series components. An analysis is carried out for all loading components of each particular harmonic and then the final results are obtained by summing the results for all harmonics used to represent the load. Once the solution technique, which involves extensive computations, has been developed for a single harmonic, it can be reused for any harmonic, and thus the method is ideally suited to the application of a digital computer.

The analysis for each harmonic load has the advantage that for straight or curved circular bridges such loads will produce displacements of the same variation and vice versa and thus a single characteristic value may be used to describe any force or displacement pattern. For example the displacement pattern for the nth harmonic:

$$r(x) = r_0 \sin \frac{n\pi x}{L} \quad \text{for straight bridges}$$

$$r(\theta) = r_0 \sin \frac{n\pi\theta}{\theta_0} \quad \text{for curved circular bridges}$$

may be described by the single value r_0. L and θ_0 define the span for the straight and circular bridges respectively (Fig. 4). This makes it possible to treat an entire joint as a single nodal point and to operate with single forces and displacements instead of functions. If the conditions of static equilibrium and geometric compatibility are maintained at a nodal point they will automatically be satisfied along the entire longitudinal joint. Thus the three dimensional prismatic bridge problem may be treated as a two dimensional problem in the transverse direction. A direct stiffness method applied to such a system results in a structure stiffness matrix which is extremely well conditioned for solution since the non-zero coefficients are all grouped in a narrow band about the main diagonal.

Each joint has four degrees of freedom; it can displace vertically and horizontally in the plane of the cross-section; it can move longitudinally tangent to the joint; and it can rotate about an axis tangent to the joint. These directions define a 'global coordinate system' for displacements or forces at the joint (Fig. 5(a)).

The direct stiffness method has been described in detail in many publications[1,11,13,20] and thus need only be briefly outlined here by the following steps:

1. Determine the element stiffness matrix for each plate element in the local coordinate system (Fig. 5(b)).
2. Transform the element stiffnesses to a global coordinate system (Fig. 5(c)) and assemble these into the structure stiffness matrix K.
3. Solve the equilibrium equations $R = Kr$, where R represents the applied loads, for the unknown joint displacements r (Fig. 5(a)).
4. Determine the plate element internal forces and displacements by expressions relating these quantities to the joint displacements r.

The basic logic of a general computer program for the direct stiffness harmonic analysis is independent of the method chosen for determining the element stiffness matrices. In the following sections elements which have been used for straight and curved bridges will be described.

Straight Bridges

Formulas based on elasticity theory for the stiffness matrix coefficients defining membrane and plate bending action for isotropic, linearly elastic plates have been summarised for easy usage by DeFries-Skene and Scordelis.[11] They have been utilised to develop the following computer programs. Figures in parenthesis indicate the year the program was first used and figures in brackets indicate references where detailed descriptions of the program and theoretical development may be found.

MULTPL Program (1965) [1]

This program provides a rapid solution for open or cellular folded plate structures having single simple spans. A direct stiffness harmonic analysis is performed in the solution. The elasticity theory is used for the isotropic plate elements. Uniform or partial surface loads as well as line loads and concentrated loads may be applied anywhere on the structure. Up to 100 non-zero terms of the appropriate Fourier series may be used to approximate the loading.

MUPDI Program (1966) [1, 2]

This program can be utilised to analyse open or cellular folded plate structures simply supported at the two ends and having up to four interior rigid diaphragms or supports between the two ends. A direct stiffness harmonic analysis based on the elasticity theory is used for the folded plate system. Compatibility at the interior rigid diaphragms or supports is accomplished by a force method of analysis. Loads and redundant forces may be approximated by up to 100 non-zero terms of the appropriate Fourier series.

MUPDI3 Program (1971)

This program extends the original MUPDI program such that up to twelve interior diaphragms or supports may be used and they need no longer be rigid. Diaphragms may be defined by flexible beams and supports may be defined by two dimensional planar rigid frame bents. Options permit evaluation of internal forces in the diaphragms and bents as well as in the bridge.

Direct application of the elasticity theory to determine the stiffness matrix of plate elements which are not isotropic becomes exceedingly complex and resort must be made to simpler approaches. A theory known as the 'finite strip method' has great potential for use in these cases. The finite strip method may be thought of as a special form of the finite element method. It approximates the behaviour of each plate by an assemblage of longitudinal finite strips for which selected displacement patterns varying as harmonics longitudinally and as polynomials in the transverse direction are assumed to represent the behaviour of the strip in the total structure. With this assumption the displacement at any point in the strip can be expressed in terms of the eight nodal point displacements shown in Fig. 5(*b*). Using successively strain-displacement relationships, a stress-strain law and thence either the principle of virtual displacements or the principle of minimum potential energy, the element stiffness matrix and generalised or consistent loads can be derived for the strip.

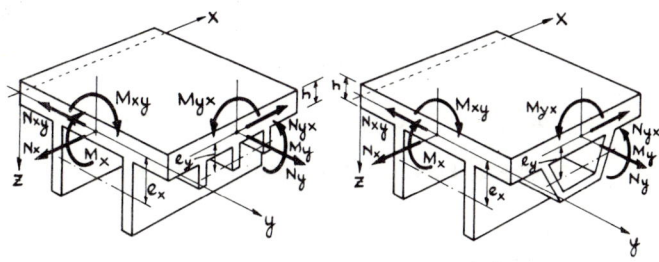

a) Torsionally soft ribs   b) Torsionally stiff ribs

FIG 6   INTERNAL FORCES ON A DIFFERENTIAL ELEMENT
        TAKEN FROM AN ORTHOTROPIC PLATE WITH ECCENTRIC
        RIBS

The stiffness matrix for an orthotropic plate has been developed by Cheung[15, 16] and Powell and Ogden[17] using this technique. Willam and Scordelis[4, 18] have used the same method to derive a stiffness matrix for an orthotropic plate with closely spaced eccentric ribs in both longitudinal and transverse directions. Account is taken of the coupling of the membrane and plate bending action due to the eccentricity of the ribs. Solutions have been obtained for plate elements stiffened by torsionally soft or stiff eccentric ribs (Fig. 6) and the following program is available for the analysis of straight box girder bridges made up of plate elements of this type.

MULSTR Program (1970) [4]

This program is capable of analysing straight prismatic folded plates made up of orthotropic plate elements with eccentric stiffeners. The structure must be simply supported at its two ends. Each plate element is idealised by a number of longitudinal finite strips in which the properties of the longitudinal and transverse stiffeners are distributed uniformly over the area of each strip and are accounted for in the analysis. The finite strip method is used to determine the strip stiffness. The displacement patterns are assumed to vary as harmonics longitudinally. In the transverse direction, a linear variation of the in-plane displacements and a cubic variation of the normal displacements are chosen. A direct stiffness harmonic analysis is used to analyse the assembled structure.

Curved Bridges

The harmonic analysis approach has been successfully applied to circular curved bridges. Cheung[19] has presented solutions for curved bridge decks and Meyer and Scordelis[6, 8] for curved box girder bridges. Both have used the finite strip method to evaluate the element stiffnesses. For the curved box girder bridge (Fig. 4(*b*)), these elements may consist of segments with orthotropic material properties taken from a general cone, a cylindrical shell, or a circular ring plate (Fig. 7). After an extensive study[8] of the use of various shell theories and various assumed transverse displacement functions for the curved finite strips, the following program was developed as the best balance between required computer time and accuracy of results obtained.

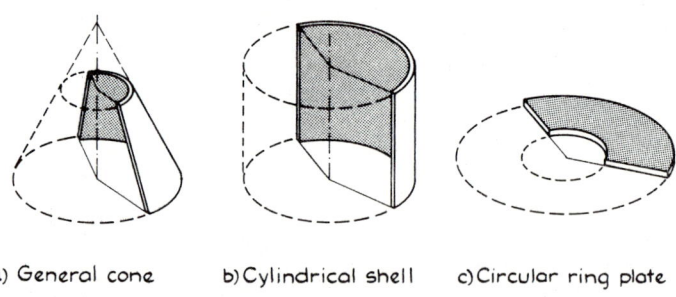

a) General cone   b)Cylindrical shell   c)Circular ring plate

FIG. 7   CURVED BOX GIRDER BRIDGE ELEMENTS.

CURSTR Program (1970) [6]

This program is capable of analysing prismatic folded plate structures which are circular in plan and made up of orthotropic plate elements. The structure must be simply supported by radial diaphragms at its two ends. Each plate element is idealised by a number of circumferential finite strips. The finite strip method is used to determine the strip stiffness. The displacement patterns are assumed to vary as harmonics in a circumferential direction. In the transverse direction, a linear variation of the in-plane displacements and a cubic variation of the normal displacements are chosen. A direct stiffness harmonic analysis is used to analyse the assembled structure.

Skew Bridges

The writer has made a number of unsuccessful attempts to develop a simple harmonic analysis for skew bridges similar to the methods used for straight and curved circular bridges. The problem lies in the fact that for skew plate elements, unlike for straight and curved elements, there is coupling between the harmonics which essentially negates the advantages of this approach over the more general finite element method.

Advantages and Disadvantages of Direct Stiffness Harmonic Analysis

Advantages of this method are:

1. It is well suited for computer programming and can yield a complete and accurate solution in a reasonable amount of computer time.
2. Any desired theory can be used to determine the response of the individual plate elements.
3. Both surface and joint loadings of arbitrary longitudinal variation can be treated.
4. Any combination of displacement and force boundary conditions along the longitudinal joints can be used.

Disadvantages of this method are:

1. It is restricted to prismatic structures which may have interior supports; but must be simply supported at the extreme ends.
2. The material and geometric properties of each plate element making up the cross-section must be constant in the longitudinal direction.

FINITE ELEMENT SOLUTIONS

The finite element method has been described extensively in the literature during the past decade. A comprehensive discussion of the theory and application of the method is given in the book by Zienkiewicz.[20]

In the finite element method the actual continuum is replaced by an assembly of finite elements interconnected at nodal points (Fig. 8). For a general box girder bridge system, the finite elements may consist of two dimensional shell or plate elements and transverse or longitudinal one dimensional frame type elements. Stiffness matrices which approximate the behaviour in the continuum, are developed for the finite elements based on assumed displacement or stress patterns, after which an analysis based on the direct stiffness method may be performed to determine nodal point displacements and thence the internal stresses in the finite elements. It should be recognised that the accuracy obtained is dependent on the assumptions used in deriving the stiffness matrices and on the fineness of mesh used in subdividing the structure. As generally applied, the results obtained closely satisfy compatibility, but not necessarily equilibrium in the continuum until a sufficiently fine mesh is used.

A number of investigators have developed general shell programs which could be used for analysing box girder bridges. However, if available, it is better to use special purpose programs which take advantage of the repetitive and special nature of these structures. These should provide the required accuracy in the results with

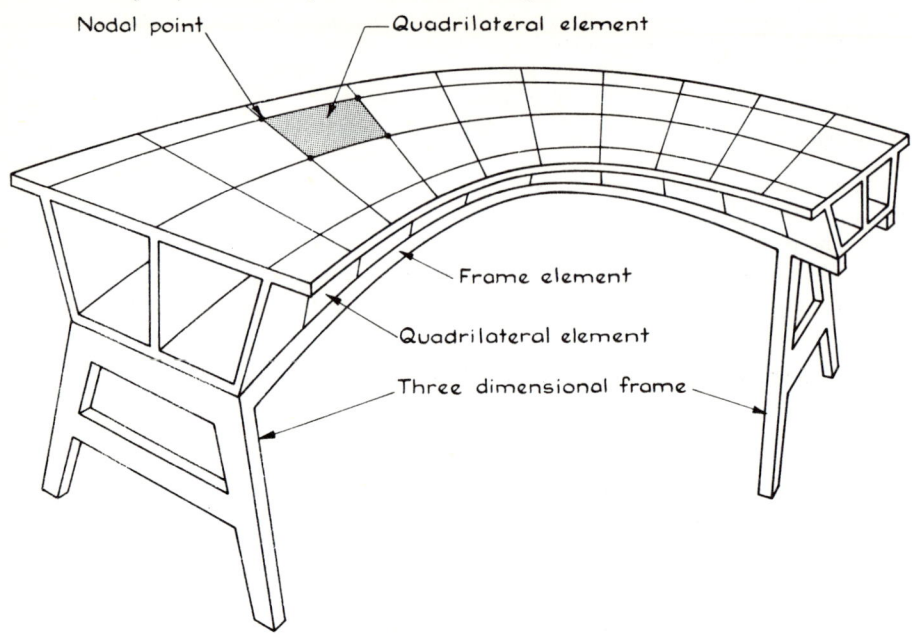

FIG. 8 BOX GIRDER BRIDGE OF GENERAL GEOMETRY

a minimum of the following: (1) required amount of input preparation; (2) execution time and core storage in the computer; and (3) amount of output data reduction necessary for meaningful interpretation. A comprehensive study of twenty different quadrilateral and trapezoidal elements was made by Willam[9] with the above criteria as a basis for selecting optimum elements for the programs to be described.

Bridges with Arbitrary Plan Geometry and Constant Depth[7,9]

The bridge is assumed to be made of top and bottom horizontal deck slabs and vertical webs.

The deck slabs are idealised by quadrilateral elements having a total of 5 degrees of freedom (DOF) per node, 3 translations and 2 rotations (Fig. 9). The in-plane action of the quadrilateral elements is represented by the plane stress mixed model Q8D11 having 2 translational DOF at each external corner node and 3 internal DOF (Fig. 9(*a*)). The mixed model is constructed using separate expansions for the displacement and strain fields. The variations of the *u* and *v* components of the displacement field are approximated by the standard bi-linear expansion for the 8 corner node DOF and by bi-quadratic expansions for 2 of the internal DOF. The third internal DOF is used to enforce a constant shear-strain variation over the entire element, which produces a more flexible and better element. After the element stiffness is formed the 3 internal DOF are eliminated by an internal static condensation process. The quadrilateral plate bending element Q19 (Fig. 9(*b*)) used for the deck slabs has been described in detail by Clough and Felippa.[21] This compatible element is made up of four subtriangles, each of which has 11 DOF associated with full cubic expansions of the *w*-displacement field and an enforced linear variation of the normal slope along one edge. In combining the four sub-elements, a quadrilateral with 19 DOF is obtained. However, the 7 internal DOF are eliminated by static condensation leaving the essential 3 DOF at each corner node, 2 rotations and a translation (Fig. 9(b)).

The vertical webs of the bridge are idealised by special rectangular spar elements having a total of 5 DOF at each corner node, 3 translations and 2 rotations. A

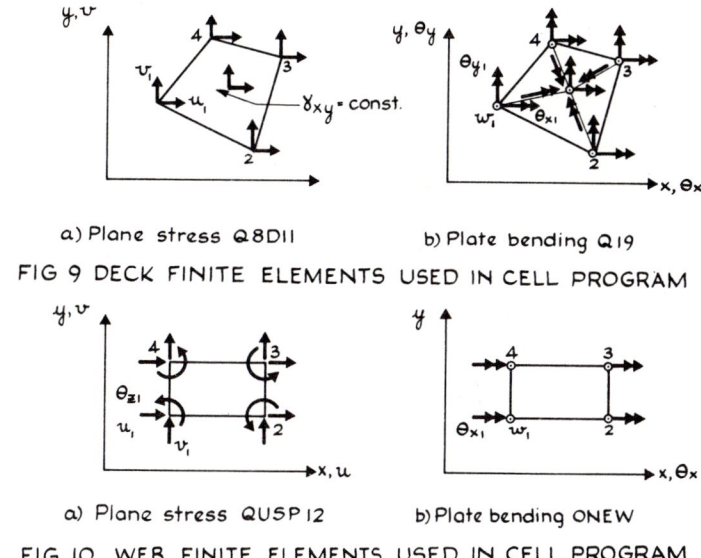

a) Plane stress Q8DII   b) Plate bending Q19

FIG 9 DECK FINITE ELEMENTS USED IN CELL PROGRAM

a) Plane stress QUSP 12   b) Plate bending ONEW

FIG 10 WEB FINITE ELEMENTS USED IN CELL PROGRAM

single element over the entire depth of the bridge can be used to capture the essential behaviour of the web. The in-plane action of these elements is represented by the model QUSP12 (Fig. 10(a)). A bilinear expansion for u and v is associated with the 2 translational DOF at each node and a cubic variation in the x-direction of v is defined by the rotation $\theta_{zi} = \partial v / \partial x$ at each node. The plate bending is represented by a simple one-way bending element ONEW having 2 DOF at each node, a rotation and a translation (Fig. 10(b)).

The above elements were utilised in developing the following program.

CELL (1970) [7]

This program analyses cellular structures of constant depth with arbitrary plan geometry. The structure must be made up of top and bottom decks and vertical webs. Two different finite element types are used to capture the main behaviour of the deck and web components. Orthotropic plate properties and arbitrary loadings and boundary conditions can be treated. Automatic element and coordinate generation options minimize the required input data.

Bridges of General Geometry with Arbitrary Integrated Three-Dimensional Frame[8]

A structure of this type (Fig. 8), is made up of quadrilateral two dimensional elements and one dimensional frame elements, each of which is assumed to have 6 DOF at each node, 3 translations and 3 rotations. Although 5 DOF have often been used for general shell programs, special techniques are required to account for the missing DOF, the rotation normal to the shell or plate surface. The use of 6 DOF for the two dimensional elements eliminates this complication and also makes the integration of the standard 6 DOF per node frame elements into the system a simple matter.

The quadrilateral elements for both the decks and the web may have an arbitrary orientation in space. They are taken as flat plate elements which give the best least squares fit through the actual location of the corner nodes. For box girder bridges of general plan and elevation geometry this assumption should be quite good. The plane stress model used was derived in its original form by Abu Ghazeleh[22] for a rectangular shape and has also been described in detail.[2] Willam[9] rederived the stiffness in skew natural coordinates for a general quadrilateral shape. The special feature of this element, Q12R12, is that in addition to the commonly used 2 translational DOF per node, each node is also assigned a rotational DOF (Fig. 11). This is defined as the average rotation about the element z-axis, i.e. for a rectangular element,

$$\theta_{zi} = \tfrac{1}{2}\left[\left(\frac{\partial v}{\partial x}\right)_i - \left(\frac{\partial u}{\partial y}\right)_i\right]$$

The internal element displacements u and v are assumed to vary linearly with the 2 translational DOF and as beam functions (Fig. 11) with the rotational DOF. The

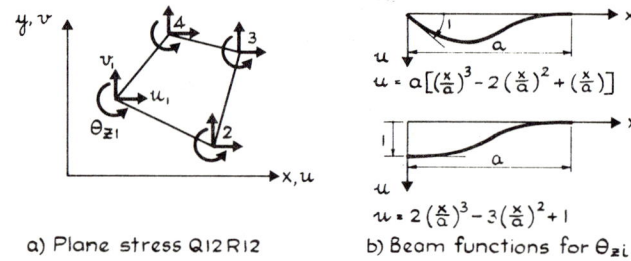

a) Plane stress Q12R12

$u = a\left[\left(\frac{x}{a}\right)^3 - 2\left(\frac{x}{a}\right)^2 + \left(\frac{x}{a}\right)\right]$

$u = 2\left(\frac{x}{a}\right)^3 - 3\left(\frac{x}{a}\right)^2 + 1$

b) Beam functions for θ_{zi}

FIG. 11   PLANE STRESS ELEMENT USED FOR DECK
AND WEB ELEMENTS IN FINPLA2 PROGRAM

assumed nodal rotations introduce small angular discontinuities so that the element is not fully compatible.

The plate bending model used for the quadrilaterals is identical to that described in the preceding section (Fig. 9(*b*)) for the CELL program. The stiffness matrix for the standard one dimensional frame elements can be found in any modern textbook.

The above elements were first incorporated in 1967 into a program called FINPLA for straight prismatic bridges[2,5] which has been used extensively. More recently[8] it has been extended to bridges with general geometry, FINPLA2, and complete checking and documentation of this program is presently in progress.

FINPLA2 Program (1970) [8]

This program can analyse general non-prismatic cellular structures of varying width and depth and may have an integrated three-dimensional frame. The structure is discretised by dividing it longitudinally into a certain number of structure segments by vertical sections, and by subdividing each such segment into finite elements. The structure alignment is described by a longitudinal reference line which may be a straight line, a circular curve or an arbitrary planar string polygon and cross-sections are defined with respect to this line. Orthotropic plate properties and arbitrary loadings and boundary conditions can be treated. Automatic element and coordinate generation options minimise the required input data.

Other Programs

A program called SIMPLA developed in 1967 and described previously in detail[2,14] has given excellent solutions for straight prismatic box girder bridges with arbitrary boundary conditions. It has the advantage that it uses an equilibrium finite element model with 14 DOF and yields a solution which automatically satisfies statics.

Advantages and Disadvantages of Finite Element Method

Advantages of this method are:

1. It is the most general method available and can treat arbitrary loadings, boundary conditions, varying material and dimensional properties and cutouts.
2. One dimensional frame type elements can be readily incorporated as an integral part of the structural system.

Disadvantages of this method are:

1. It requires a greater amount of computer time than a direct stiffness harmonic analysis to obtain a solution of comparable accuracy.
2. A refined mesh size must be used to achieve accurate results in the vicinity of steep stress gradients.
3. Static equilibrium is not automatically satisfied for the displacement models normally used, but is approached as the mesh size is refined. Judgement must be used in selecting an appropriate mesh layout and in interpreting the results.
4. Automatic mesh and load generation schemes need to be incorporated into the computer program used to avoid the large manual input of data otherwise required.

NUMERICAL EXAMPLES

Three numerical examples involving a straight two span continuous bridge, a single span skew bridge and a single span curved bridge are presented. These illustrate results obtained using the analytical methods and computer programs which have been described. Computer execution times indicated were on a CDC 6400 computer. As mentioned earlier, detailed descriptions of these programs may be found in References [1–8].

Example 1 – Straight, Two Span, Four Cell Bridge (Figs 12, 13)

This example presents results for longitudinal stresses N_x and vertical deflections at a midspan section directly under a single 1 kip load for a two span, four cell box girder bridge. Dimensions, loading, modulus of elasticity E, and Poisson's ratio, v, are given in Fig. 12.

A comparison of results is given in Fig. 13 for solutions by MUPDI, which assumes the interior support and the midspan diaphragm in span 1 are completely rigid, and MUPDI3, which accounts for the flexibilities of the interior support single column bent and the midspan diaphragm. No midspan diaphragm exists in span 2. As would be expected, the solution by MUPDI3 gives slightly greater deflections but the values of N_x are hardly altered. A total of 100 harmonics were used in each solution. Computer times required were 203 and 232 seconds respectively, for the MUPDI and MUPDI3 solutions.

Example 2 – Skew, Single Span, Two Cell Bridge (Figs 14, 15)

This example presents results obtained using the finite element program CELL for a single span, two cell box girder bridge on skewed simple supports. Dimensions, loading, material properties and finite element mesh layout are given in Fig. 14. Studies have shown that more accurate results for a given mesh size are obtained when rectangular and triangular elements are used instead of parallelogrammic elements to idealise skewed regions. Hence, the structure's top and bottom decks are discretised by 3 x 3 ft square elements while right triangles adjust the mesh to the skewed boundaries.

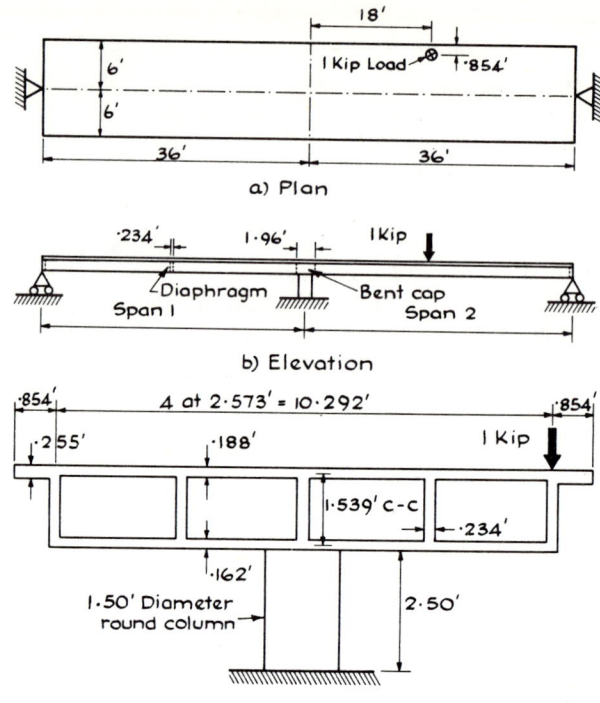

a) Plan

b) Elevation

c) Typical Section
E = 432,000 KSF; ν = 0·15

FIG. 12 EXAMPLE 1 – DIMENSIONS AND LOADING.

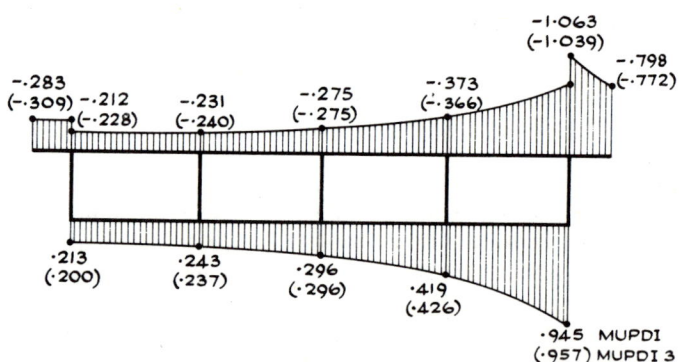

a) Longitudinal stresses N_x (KIP/FT.)
in top and bottom slabs

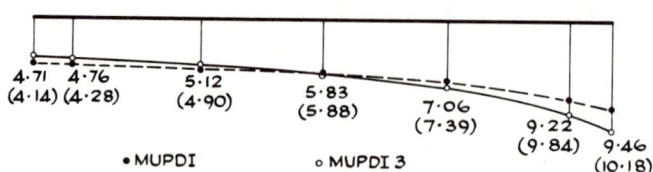

b) Vertical deflections (FT × 10^{-4}) of top slab

FIG. 13 EXAMPLE 1 – LONGITUDINAL STRESSES AND
DEFLECTIONS AT SECTION AT MIDDLE OF SPAN 2

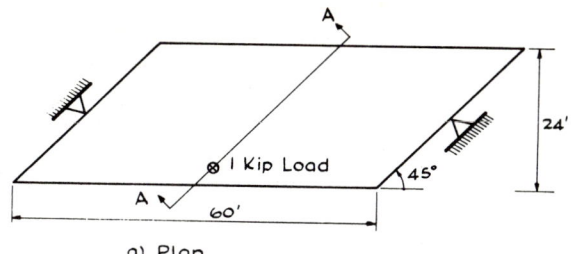

a) Plan

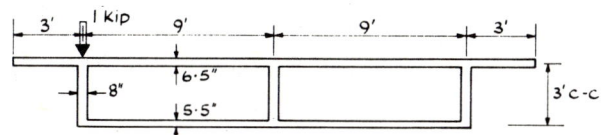

b) Typical cross-section

E = 432,000 KSF; υ = 0·15

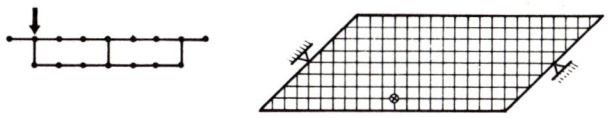

c) Finite element mesh layout

FIG. 14 EXAMPLE 2 – DIMENSIONS, LOADING AND
FINITE ELEMENT MESH LAYOUT

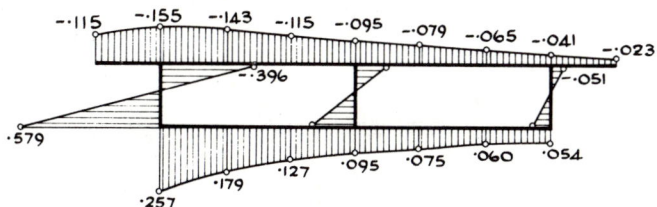

a) Normal stresses N_n (KIP/FT) in top and bottom slabs
and longitudinal stresses N_x (KIP/FT) in webs

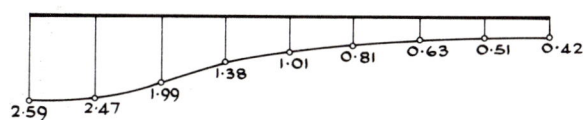

b) Vertical deflections (FT × 10^{-4}) of top slab

FIG. 15 EXAMPLE 2 – STRESSES AND DEFLECTIONS
AT SKEWED MIDSPAN SECTION AA

Figure 15(a) illustrates the distribution of the normal stress resultants N_n at the skewed midspan section AA (Fig. 14(a)) for the decks and the longitudinal stresses N_x for the girders. The total internal moment on this section can be found by numerically integrating the contributions of the above stress resultants and of the plate bending moments which are not shown. These yield values of 9·92 and 0· 0·33 ft kips respectively, or a total of 10·25 ft kip. For purposes of checking the solution, this value can be compared with the total statical moment on the section which equals 0·5 (21·21) = 10·61 ft kips, where the 21·21 ft distance is the normal distance between the skewed support and midspan sections. Figure 15(b) shows the transverse distribution of the vertical deflections at the same midspan skewed section. Computer time required for this solution was 193 seconds.

Example 3 — Curved, Single Span, Two Cell Bridge (Figs 16, 17)

This example presents results for longitudinal stresses N_θ and transverse slab bending moments M_r at the midspan section of a single span, two cell box girder bridge which is circularly curved in plan and simply supported at its two ends on radial diaphragms. Results are compared as obtained by a finite strip harmonic analysis using the CURSTR program with 50 harmonics and by a finite element analysis using the FINPLA2 program. Dimensions, loading, material properties and finite element mesh layout are shown in Fig. 16. Subdivision of the mesh along the longitudinal centre-line of the bridge was 8 at 6, 2 at 2, and 8 at 6 ft. The loading consisted of a standard AASHO truck in position A as shown.

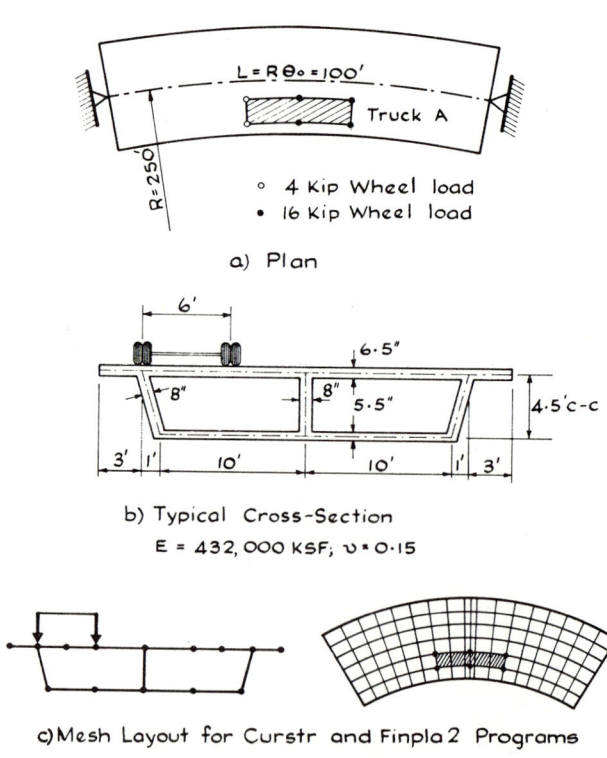

a) Plan

b) Typical Cross-Section
E = 432,000 KSF; υ = 0·15

c) Mesh Layout for Curstr and Finpla 2 Programs

FIG. 16 EXAMPLE 3 – DIMENSIONS, LOADING AND MESH LAYOUT.

Agreement between the two theories for the longitudinal stresses N_θ (Fig. 17(a)) is satisfactory. Differences in tabulated values for the top and bottom deck range from 1—8% and for the web slightly higher being from 4—11%. A more refined mesh size for the FINPLA2 solutions and the use of more harmonics for the CRUSTR solution would yield even closer agreement.

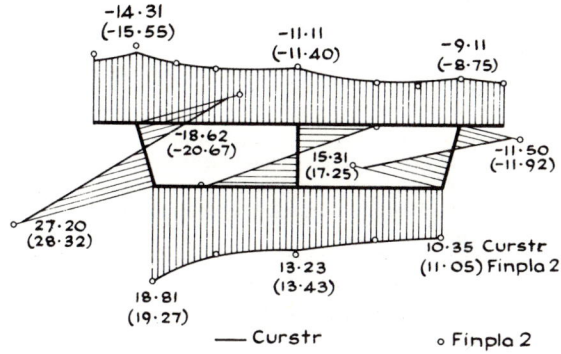

a) Longitudinal Stresses N_θ (KIP/FT)

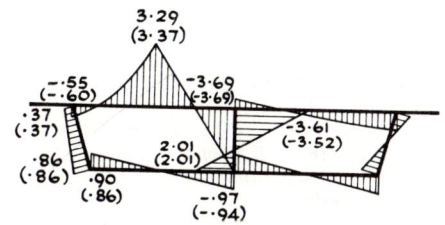

b) Transverse Slab Bending Moments M_r (FT KIP/FT)

FIG. 17   EXAMPLE 3 - LONGITUDINAL STRESSES AND
TRANSVERSE SLAB MOMENTS AT MIDSPAN SECTION

Agreement between the two theories for the transverse slab bending moments M_r (Fig. 17(b)) is excellent, with differences in significant values being less than 3%.

Computer times required were 27 and 300 seconds respectively, for the CURSTR and FINPLA2 solutions.

SUMMARY

A review of a number of available analytical methods and general computer programs for box girder bridges has been presented. These programs make it possible to obtain solutions for bridges which are straight, skew, curved or of arbitrary geometry and under general loading.

ACKNOWLEDGEMENTS

A number of the computer programs described in this paper were developed as part of a continuing research program at the University of California which is being sponsored by the Division of Highways, Department of Public Works, State of California, and the Bureau of Public Roads, Federal Highway Administration, United States Department of Transportation. The opinions, findings and conclusions expressed in this paper are those of the author and not necessarily those of the Division of Highways or the Bureau of Public Roads.

The important contributions of B. N. Abu Ghazeleh, K. S. Lo, C. Meyer, K. J. Willam, D. Ngo and C. S. Lin, all of whom have worked with the author on this research program sometime during the past five years, is gratefully acknowledged.

REFERENCES

1. Scordelis, A. C. *Analysis of simply supported box girder bridges.* University of California, Berkeley, Structural Engineering and Structural Mechanics Report No. SESM 66–17, Oct. 1966.
2. Scordelis, A. C. *Analysis of continuous box girder bridges.* University of California, Berkeley, Structural Engineering and Structural Mechanics Report No. SESM 67–25, Nov. 1967.
3. Scordelis, A. C. and Meyer, C. *Wheel load distribution in concrete box girder bridges.* University of California, Berkeley, Structural Engineering and Structural Mechanics Report No. SESM 69–1, Jan. 1969.
4. William, K. J. and Scordelis, A. C. *Analysis of orthotropic folded plates with eccentric stiffeners.* University of California, Berkeley, Structural Engineering and Structural Mechanics Report No. SESM 7–2, Feb. 1970.
5. Meyer, C. and Scordelis, A. C. *Computer program for prismatic folded plates with plate and beam elements.* University of California, Berkeley, Structural Engineering and Structural Mechanics Report No. SESM 7–3, Feb. 1970.
6. Meyer, C. and Scordelis, A. C. *Analysis of curved folded plate structures.* University of California, Berkeley, Structural Engineering and Structural Mechanics Report No. SESM 70–8, June 1970.
7. William, K. J. and Scordelis, A. C. *Computer program for cellular structures of arbitrary plan geometry.* University of California, Berkeley, Structural Engineering and Structural Mechanics Report No. SESM 70–10, Sept. 1970.
8. Meyer, C. *Analysis and design of curved box girder bridges.* University of California, Berkeley, Structural Engineering and Structural Mechanics, Report No. SESM 70–22, Dec. 1970.
9. Willam, K. J. *Finite element analysis of cellular structures.* Ph.D. thesis, University of California, Berkeley, Dec. 1969.
10. Jenkins, R. S. *Theory and design of cylindrical shell structures.* London, O. N. Arup Group, 1947.
11. DeFries-Skene, A. and Scordelis, A. C. Direct stiffness solution for folded plates. *J. struct. Div. Am. Soc. civ. Engrs*, vol. 90, ST4, August 1964, pp. 15–47.
12. Chu, K. H. and Pinjarkar, S. G. Multiple folded plate structures. *J. struct. Div. Am. Soc. civ. Engrs*, Vol. 92, ST2, April 1966, pp. 297–321.
13. Scordelis, A. C., Davis, R. E. and Lo, K. S. Load distribution in concrete box girder bridges. *First International Symposium on Concrete Bridge Design (Toronto, 1967).* American Concrete Institute Publication SP-23, 1969.
14. Scordelis, A. C. and Davis, R. E. Stresses in continuous concrete box girder bridges. *Second International Symposium on Concrete Bridge Design (Chicago, April 1969).* American Concrete Institute.
15. Cheung, Y. K. *Orthotropic right bridges by the finite strip method.* University of Calgary, Faculty of Engineering, Technical Publications, Oct. 1968.
16. Cheung, Y. K. Analysis of box girder bridges by the finite strip method. *Second International Symposium on Concrete Bridge Design (Chicago, April 1969).* American Concrete Institute.
17. Powell, G. H. and Ogden, D. W. Analysis of orthotropic steel plate bridge decks. *J. struct. Div. Am. Soc. civ. Engrs*, Vol. 95, ST5, May 1969, pp. 909–922.
18. William, K. J. and Scordelis, A. C. Analysis of eccentrically stiffened folded plates. Presented at the Internatinoal Association for Shell Structures Symposium on Folded Plates, Vienna, Sept. 1970.
19. Cheung, Y. K. The analysis of cylindrical orthotropic curved bridge decks. *Publs int. Ass. Bridge struct. Engng*, vol. 29, part 2, 1969.
20. Zienkiewicz, O. C. *The finite element method in structural and continuum mechanics.* New York, McGraw-Hill, 1967.
21. Clough, R. W. and Felippa, C. A. A refined quadrilateral for the analysis of plate bending. *Proceedings of the 2nd Conference on Matrix Methods in Structural Mechanics (Wright-Patterson Air Force Base, Ohio, 1968).*
22. Abu Ghazeleh, B. N. *Analysis of plate type prismatic structures.* Ph.D. thesis, University of California, Berkeley, Jan. 1966.

ANALYSIS OF VARIOUS TYPES OF BRIDGES BY THE FINITE ELEMENT METHOD

J. D. DAVIES, I. J. SOMERVAILLE and O. C. ZIENKIEWICZ
University College of Swansea

SUMMARY

Whilst general programs utilising finite elements are capable of solving the elastic problem in any configuration — and hence any form of a bridge structure — much attention has recently been given to improving the economy of solution.

In this paper attention will be focussed on the type of idealization necessary for the solution and on the elements used therein to achieve, economically, a sufficient engineering accuracy. It is convenient in this context to classify the types of bridges as:

(*a*) slabs (both thick and thin) and grillages
(*b*) slabs with eccentric beams
(*c*) shallow cellular structures
(*d*) deep box structures.

Elastic analysis with small deformation only is considered and thus problems of elastic instability are expressly here excluded. Such problems associated with very thin walled steel structures pose problems of their own beyond the scope of this review.

This paper is illustrated with many examples.

1. INTRODUCTION

The finite element method as a universal tool of solution of structural engineering problems is now so well known and established that no description of it is here necessary.[1] Its early application to problems of plate flexure led to its adoption as a convenient tool in the solution of many bridge problems where its generality gives it a considerable edge over more specialised techniques which on occasion provide equally economical, if narrower, applications.[2,8]

The objective of this present paper is to consider the economy and merits presented by different forms of physical and element idealization of the problem.

In this context it is convenient to categorise bridge structures into four general types:

(*a*) slabs or grillages with concentric beams
(*b*) slabs with eccentric beams
(*c*) shallow cellular structures
(*d*) deep box structures.

Fig. 1 illustrates the four categories which will be discussed separately. Clearly the classification is only general and many types of intermediate character will arise.

Further in each type the members can be thin, if steel construction is used, or of very massive section associated with reinforced concrete. Indeed this subclassification will necessitate on occasion, different treatment.

In all problems we shall be concerned with elastic analysis only. This basic assumption is in accordance with the usually accepted design criteria although on occasion non-linear limit load conditions have to be considered. Although much work is in

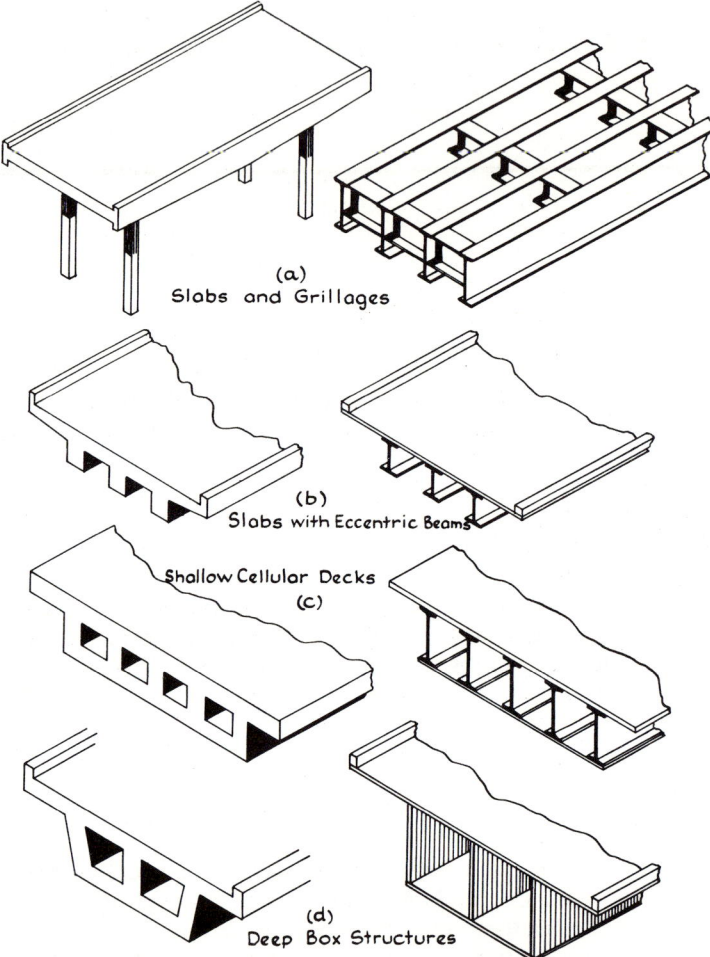

(a)
Slabs and Grillages

(b)
Slabs with Eccentric Beams

Shallow Cellular Decks
(c)

(d)
Deep Box Structures

FIG.I TYPES OF BRIDGE STRUCTURE

progress in the field of non-linear analysis and once again finite element techniques prove to give a convenient tool,[1] the application of such processes is liable to be made only in the final check analysis, the working tool design being considered on a linear basis. For this, and space limitation reasons, the present paper excludes the phenomena of concrete cracking, compressive nonlinearity, buckling behaviour, etc.

When comparing different elastic finite element solutions of a particular type of structure the designer is often faced with a bewildering array of elements and the 'propaganda' about their relative merits. Various methods of comparison and assessment have from time to time been proposed and indeed, in this paper, we shall at all times make comparisons. The following should, however, be always borne in mind.

(i)   One cannot expect more out of the solution than is inherent in the assumptions
       introduced (e.g. if the problem is idealised to that of a thin elastic plate — the
       only standard of assessment is an exact solution based on the same assumption).
(ii)   All properly formulated finite element solutions (or indeed other types) must
       in the limit give the exact solution under the limitations of (i). For a given
       approximation the computational effort (and cost) will doubtless differ between
       alternatives given.
(iii) In the finite element field, comparisons of efficiency of different elements are
       often based on accuracy (and convergence rate) attainable for an idealisation
       using the same number of degrees of freedom in a test case. Such comparisons
       are the most objective as computer times (depending on quality of program-
       ming) are excluded.

A word of warning however, about such comparisons must be given. In the first place,
all attention is focussed on the number of variables in the simultaneous equations —
although much cost can be incurred in the actual element formulation. In the second
place — the time of solution using a standard method will be dependent not only on
the number of unknowns but also on the band width, which may vary with different
elements.

(iv)   Comparisons of the element merits should never be made on computer times
       alone unless a standard form of program is used. This fairly obvious fact is all
       too often ignored.

2. SLAB BRIDGES WITH CONCENTRIC BEAMS (TYPE (a))

Bridges which consist of relatively thin decks supported by columns or walls are
frequently used. For these, obviously, solutions based on plate bending theory are
applicable — and, where beams whose neutral axis is coincident with the middle
plane are used, the conventional beam bending analysis can be easily incorporated.
    If the deflections are small, and if the slab is relatively thin, it is usual to consider
the problem as governed by the classical thin plate theory which specifically neglects
effects of shear deformation.[9]
    If the slab is relatively thick, or if a cellular (sandwich) type is used in construc-
tion, the shear deformation becomes important and both deflections and moment
fields can be substantially influenced by it. For such problems use of a more elaborate
plate theory becomes necessary.

2.1. 'Thin Plate' Theory Solution

The greatest variety of elements applicable to one problem is probably presented in
this category. A complete survey is impracticable here and instead is given elsewhere.[1,10]
    Fig. 2 summarises the most important types proved in practice.[11-25]
    It will be noted that some are based on non-conforming shape functions for which
the usual Rayleigh-Ritz assumptions do not apply. Convergence of these as well as of
all the others have been proved experimentally, as well as theoretically,[15,30] and thus
all are acceptable.
    Any one of the above elements can be used successfully in the problem of slab deck
analysis.
    If only one type of a plate element is to be used, rectangles 1NC, 6C and parallelo-
gram 2NC must be excluded as not allowing the necessary freedom to describe all

	TOTAL DEGREE OF FREEDOM	TYPE	REFERENCE	SHEAR DEFORMATION
	12	1NC	11, 12	NO
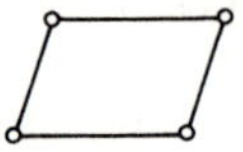	12	2NC	13, 14	NO
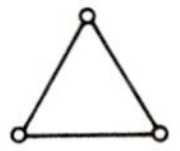	9	3NC 3 C 3 H	15, 2 15, 16 18, 19	NO NO YES
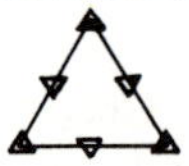	6	4E	17	NO
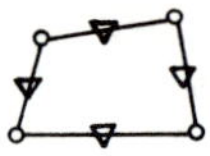	16	5Ca 5Cb	22 10	NO
	16	6C	23, 24	NO
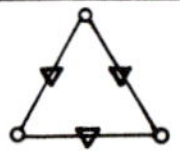	12	7C	15, 16	NO
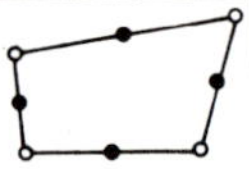	24	8C	20, 21	YES
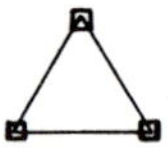	18	9C	25, 26, 27, 28	NO
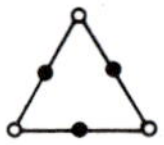	18	10C	29	NO

o $w, \dfrac{\partial w}{\partial x}, \dfrac{\partial w}{\partial y}$ ▽ $\dfrac{\partial w}{\partial n}$

• $w, \dfrac{\partial w}{\partial n}, \dfrac{\partial^2 w}{\partial n \partial s}$ □ $w, \dfrac{\partial w}{\partial x}, \dfrac{\partial w}{\partial y}, \dfrac{\partial^2 w}{\partial x^2}, \dfrac{\partial^2 w}{\partial x \partial y}, \dfrac{\partial^2 w}{\partial y^2}$

Δ w ■ $w, \dfrac{\partial w}{\partial x}, \dfrac{\partial w}{\partial y}, \dfrac{\partial^2 w}{\partial x \partial y}$

DEGREES OF FREEDOM AT A NODE

NC - Displacement type : non-conforming
C -   "   "   : conforming
E - equilibrium   "
H - hybrid   "

FIG.2 SOME SUCCESSFUL PLATE BENDING ELEMENTS

shapes. However, when simple shapes and specialised programs are desired, these elements are found to perform rather better than simple triangles and here element 6C is slightly superior to 1NC.

From more general shapes the choice is more difficult. For reasons of engineering simplicity, ease of program assembly and the extensions which will be discussed later, elements with three degrees of freedom $(w, \frac{\partial w}{\partial x}, \frac{\partial w}{\partial y})$ at nodes are desirable. These permit a simple coupling with beam elements in a standard manner familiar to structural engineers. Performance of some such elements is shown for a specific problem in Fig. 3. With the exception of 3C all elements give acceptable answers even at coarse subdivisions.

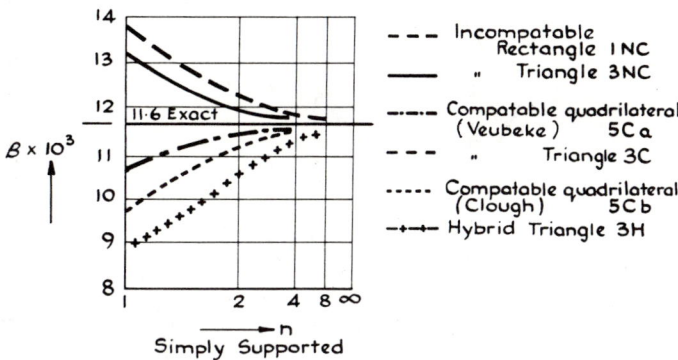

Simply Supported

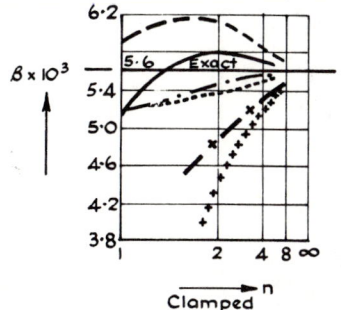

Clamped

FIG. 3 COMPARISON OF VARIOUS FINITE ELEMENT SOLUTIONS FOR A SQUARE PLATE UNDER A CENTRAL LOAD P. n – no. OF DIVISIONS OF HALF-EDGE a $\beta = {}^{w}/_{Pa^2 D}$

The natural choice seems to lie here between 3NC and 3H. In fact the first has been incorporated in the standard bridge programs used by the Ministry of Transport.[31, 32]

Beam elements whose neutral axis lies on (or close to) the middle plane are easily incorporated into such a program which has been widely used in practice. Fig. 4 shows some typical results obtained by an automated plot. Obviously, by omission of plate elements, plane grillage structures are simply dealt with. Similarly, by suitable choice of stiffnesses, orthotropic slabs are easily tackled.

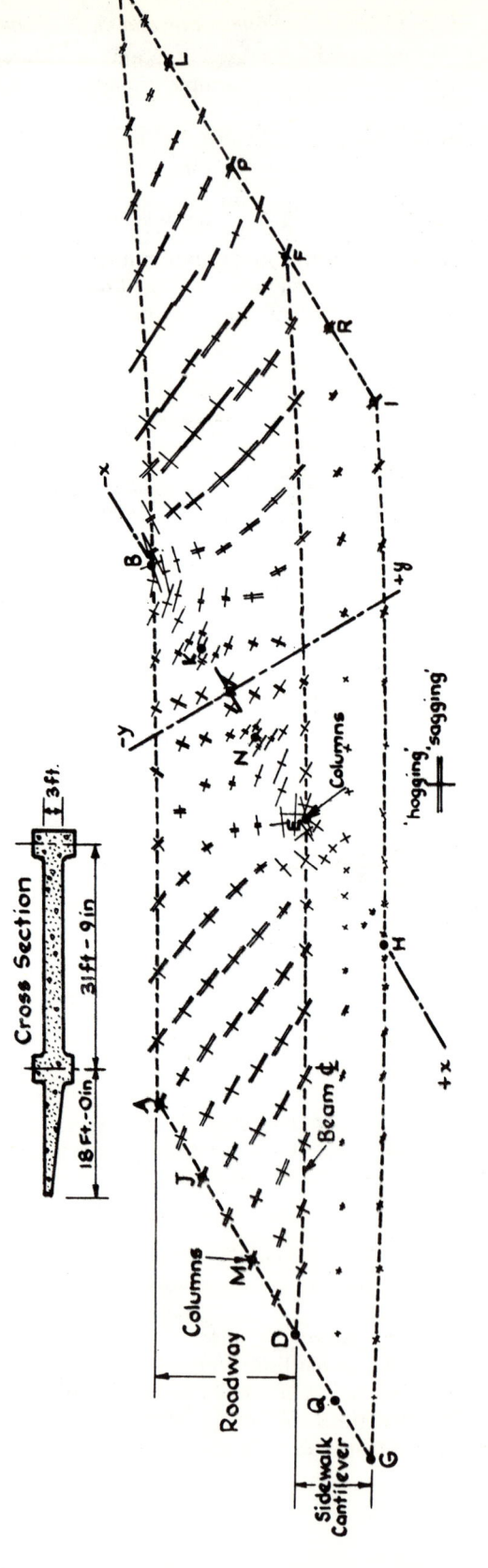

FIG. 4 TWO SPAN BRIDGE : PRINCIPAL MOMENT PLOT.

2.2. Thick Plate Solutions

Where the slab is thick, or when a sandwich construction is used, the importance of shear deformation in the slab is significant. Only two of the elements shown in Fig. 2 have the facility of dealing with this (3H, 8C). The second has been extensively tested in this respect and is easily adapted to sandwich construction while the first can be used effectively only for isotropic slabs. It is anticipated that a very wide use of this element will be made in future programs, as thickness variation, etc., give no problems in its formulation.

2.3. Specialised Programs

If the cross section of the slab is constant and boundary conditions are suitable, then the finite element divisions can be made in the section plane only accompanied by a Fourier expansion in the longitudinal direction. Variable section bridges may be solved by carrying out a numerical integration process along the length of the bridge to obtain

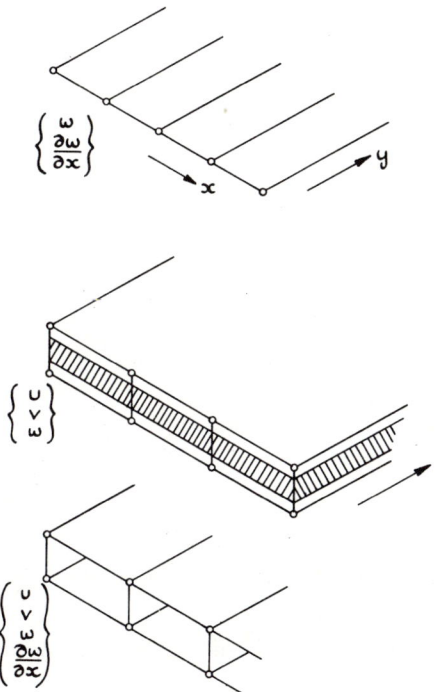

FIG. 5 FINITE ELEMENT SOLUTION WITH THE AID OF FOURIER EXPANSIONS FOR BRIDGES OF CONSTANT SECTION.

section properties for the input data. With the load also expanded in Fourier series a decoupling occurs and the size of the problem requiring solution for each harmonic is reduced by one dimension. Cheung[33,34,35] used this procedure 'finite strip method' in the context of thin slabs and boxes and a recent extension to thick slabs and prismatic solids has been made by Zienkiewicz and Too.[36]

A mention of these processes is made here in passing as they are essentially finite element ones. Their economy, for cases where only a few load harmonics need be considered, is obvious. For complex boundary conditions or concentrated loads the economy is less apparent (Fig. 5).

3. SLABS WITH ECCENTRIC BEAMS (TYPE (b))

This type of structure is the one probably most frequently encountered. As already mentioned if the eccentricity is very small the approach of the previous section is applicable — especially if amended section properties for the beam are used.[37] The apparent alternative of treating the problem as that of a full, three dimensional structure (see Section 5) is always available — but its cost is much greater.

A simple process by which eccentric beams can be treated is to note that if the cross section of the beam is sturdy its deformations are entirely defined by the middle surface displacements of the slab (Fig. 6a). The middle surface displacement at a node of the slab, now including the in-plane components is

$$[\delta_i] = \begin{bmatrix} u_i \\ v_i \\ w_i \\ \theta_{xi} \\ \theta_{yi} \end{bmatrix} \qquad (1)$$

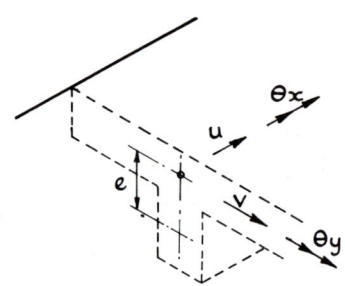

FIG. 6(a) SLAB WITH ECCENTRIC BEAM

This uniquely determines the displacements at the centroid of the beam

$$\left[\delta_i^b\right] = [T]\ [\delta_i] \qquad (2)$$

with the stiffness matrix of a beam under these degrees of freedom $[K'_b]$ known from elementary considerations, the stiffness with respect to the plate co-ordinates can be simply found as

$$[K_b] = [T]^T\ [K'_b]\ [T] \qquad (3)$$

and added in the usual manner.

Thus, if the plate program is written so as to include 'in-plane' degrees of freedom (i.e. incorporating a plane stress matrix) the problem is not complicated by the addition of eccentric beams.

The first successful attempt made on these lines was that by Gustafson and Wright[38] using a simple rectangular and parallelogram plate bending element (1NC) with corresponding plane strain action[1] given by the freedoms associated with some nodes.

Clearly the process can be applied with any plate bending element (noting that if derivatives higher than first order are used as degrees of freedom, these must not be coupled).

The same subdivision as in the simple slab bridge program can be used but now at each node the degrees of freedom are increased by two in plane components. Data preparation, etc., thus follows the identical pattern but now, in addition to bending moments, direct stress components in the beams and in the slab will be considered in the results.

A program on the lines of that described in ref. 31 has been prepared for the Ministry of Transport[39] and extensively tested. In the next subsection several examples of the accuracy attainable are demonstrated.

It is important to realise that only two additional assumptions have been made, namely

(*a*) that the beam is sufficiently stiff not to change its section appreciably during deformation.

(*b*) that the rotational stiffness of the beams about the vertical axis is negligible.

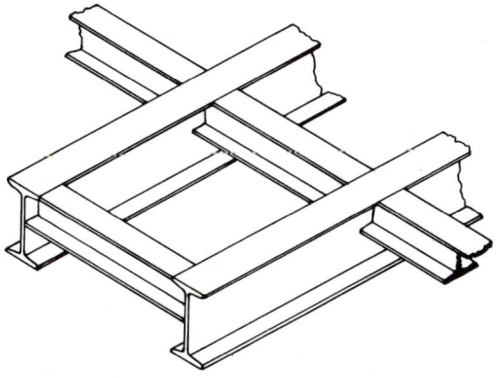

FIG. 6(b) GRILLAGE SYSTEM

Neither of these assumptions are completely true. The first fails in the case of very deep slender beams and the second influences in all cases the in plane slab stresses, and the local torsional stiffness of the bridge about an axis parallel to the beams. Nevertheless, for a wide variety of problems, the approach is successful and it is possible to eliminate the effects of the second assumption by including the sixth degree of freedom in the beam stiffness.

In the following section some examples are given to show the versatility and reasonable accuracy obtained with the assumptions involved.

It is worth remarking that the six degree of freedom version of this program is, obviously, capable of dealing with grillages where the neutral axis of the beams are eccentric, (Fig.6b).This is done simply by omitting the slab elements.

3.1. Some Examples

Example 1. A two span rectangular bridge

Results are available for a reinforced concrete bridge of two spans with longitudinal beams (see sections in Fig. 7) tested by a plexiglass model (1:24 scale) and also a concrete model of half scale,[40] for a unit load at the exterior girder at the centre of one span. The bridge was analysed using a subdivision of 10 transversely and 16 in each span longitudinally. Results of deflection normalised to a percentage of the

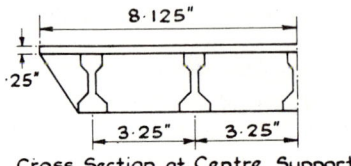

Cross Section at Centre Support

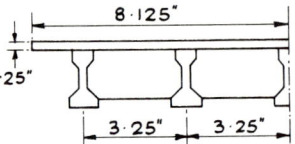

Cross Section at Interior Diaphragm

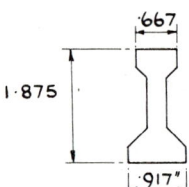

Beam Cross Section

FIG. 7 BRIDGE MODEL CROSS SECTIONS

sum of girder deflection are shown in Table 1, while results of lower flange strain normalised to a percentage of the sum of the lower flange stresses in each girder are given in Table 2.

Example 2

This example concerns a curved two span girder bridge model (41) whose section is shown in Fig. 8. This perspex model spanned 28·05 degrees between supports, with seven transverse intermediate diaphragms in each span as well as over the supports. Fig. 9 shows deflections of the inner and outer girder arising from the application of a 15 lb load on the centreline midway between the two supports, as deduced from the published experimental results, and obtained by finite element analysis. It will be noted that deflections of the outer girder are apparently overestimated while inner girder deflections are underestimated, suggesting that the torsional stiffness of the structure is not completely realised in the finite element solution.

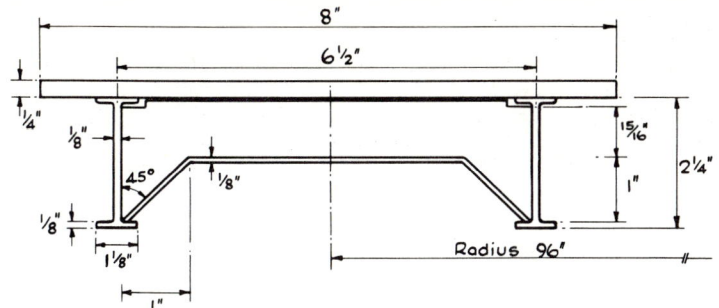

FIG 8 CROSS SECTION OF CURVED BRIDGE MODEL

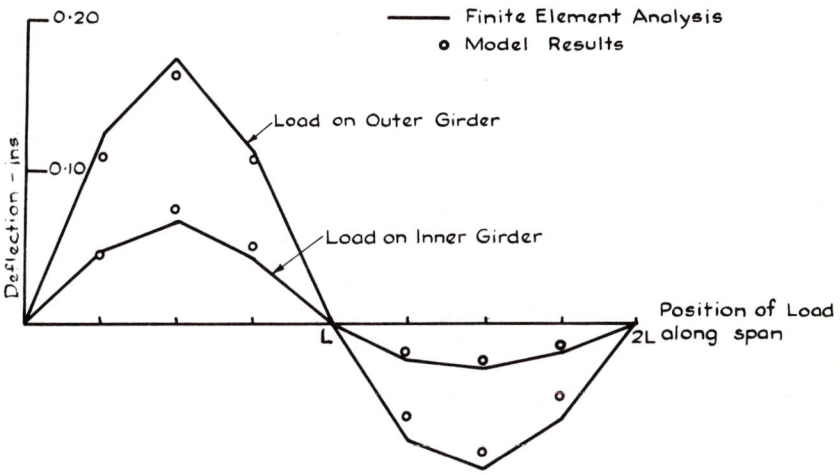

FIG. 9(a) DEFLECTIONS AT CENTRE OF ONE SPAN OF A TWO SPAN
CURVED GIRDER BRIDGE

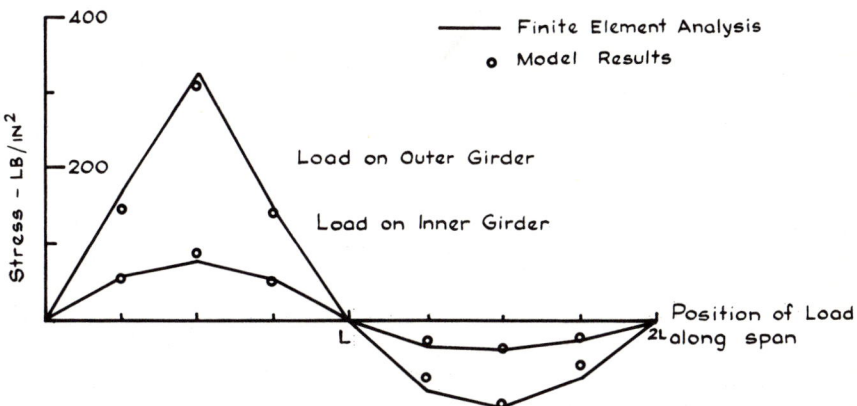

FIG 9(b) STRESSES AT LOWER FLANGE OF OUTER GIRDER IN
MID-SPAN OF FIRST SPAN.

Example 3

A skew slab with eccentric edge beams as shown in Fig. 10 has been constructed in perspex and tested. Results are shown in Fig. 11 for deflections across the skew transverse centreline, compared with values given by finite element analysis using the five degree of freedom program. Agreement will be seen to be very satisfactory.

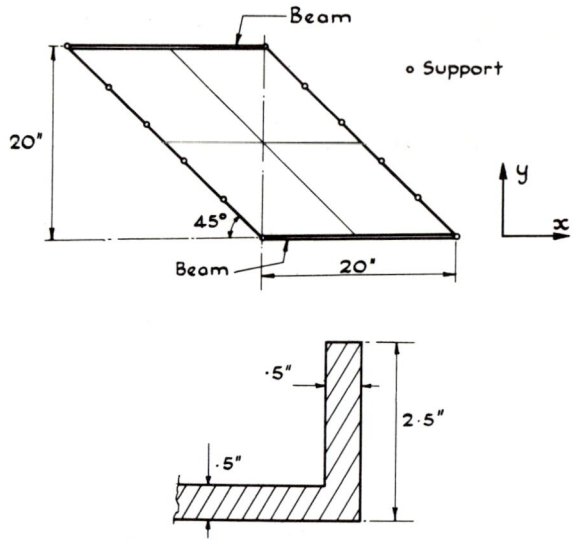

FIG. 10 PLAN OF SKEW PLATE AND DETAIL
OF EDGE BEAM

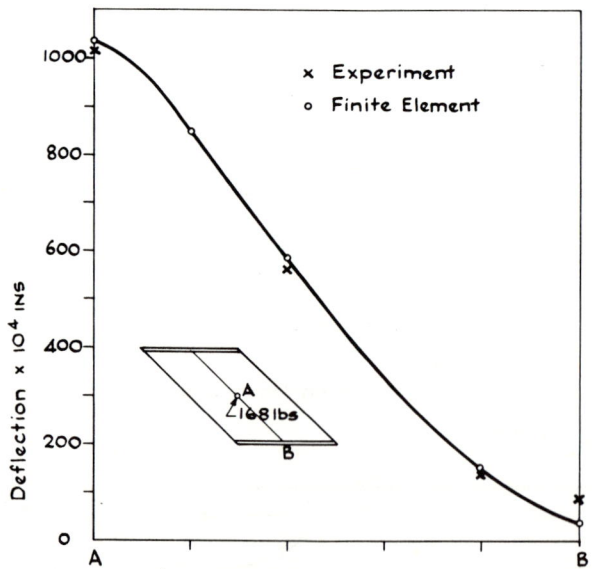

FIG.11 DEFLECTIONS IN SKEW SLAB WITH ECCENTRIC
EDGE BEAMS DUE TO CENTRAL CONCENTRATED LOAD.

4. SHALLOW CELLULAR STRUCTURES (TYPE (c))

For shallow cellular structures it is possible to develop specialised techniques which allow a solution for any plan form or loading to be achieved economically.

The simplest of these is obviously the idealisation of the structure as a corresponding orthotropic plate, as already mentioned. If purely flexural behaviour is admitted (as in the simple thin plate theory) a substantial error can result due to the neglect of shear deformations. It is usually desirable to use here thick, sandwich plate, theory typified by such elements as 8C of Fig. 2. The determination of the effective core moduli needs to be given considerable care in such approximations.

For thicker cellular sections two alternative possibilities arise.

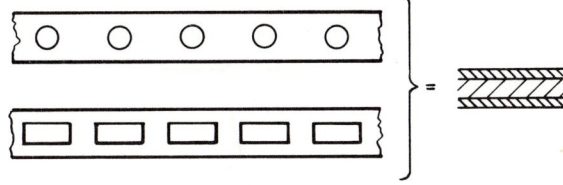

FIG. 12(a) CELLULAR STRUCTURES AS SANDWICH PLATES

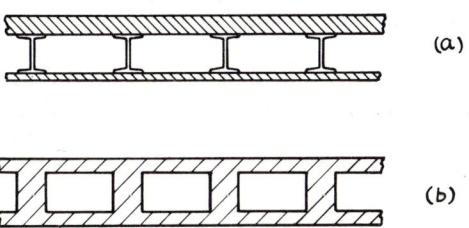

(a)

(b)

FIG. 12(b) SPECIAL TYPES OF CELLULAR CONSTRUCTION

In the case illustrated in Fig. 12a, b where a thick flexural deck is attached to relatively rigid beams in turn connected by a membrane or slab the procedures of the previous section can be used by attaching a membrane stiffness eccentric to the main slab. An appropriate transformation can be made to the main slab co-ordinates noting that the slab displacements *and rotations* jointly transform into one pure membrane displacement of the bottom slab, while the bending displacements of both upper and lower slabs are identical.

This extension of the previous programme is simple and no additional degrees of freedom arise. One limitation is, however, that nodal points can only be placed now where beams exist.

A special case arises in cellular slabs of symmetric section as shown in Fig. 13. If load action is taken to act only in the longitudinal diaphragm planes, the secondary effects of local bending on the deck slabs may be ignored and we have a situation of antisymmetry of u and v displacements. With diaphragms free of warping the lateral displacements at all points in the deck are determined completely by rotation values at corresponding points in the nodal plane, consequently the problem may be formulated in terms of three degrees of freedom at nodes in the nodal plane. In fact the solution for displacements can be obtained by means of a concentric plate and beam program of Type (a), using plate elements whose bending stiffness is equal to the bending stiffness of two equal parallel plate elements.

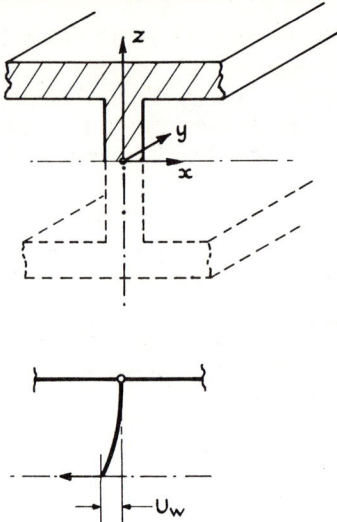

FIG. 13 ANALYSIS OF SYMMETRIC CELLULAR SLABS.

The effects of lateral warping of the longitudinal diaphragms may be included by introducing an additional variable at each node to represent the magnitude of the warping. In this case the lateral displacement at any point in the upper slab plane is related to the corresponding variables in the central plane by

$$u = \theta_y . z + u_w$$

the last term representing the warping displacement. It will be noted that because of the antisymmetry of u there is no warping moment at the neutral plane, and consequently the warping stiffness is easily calculated by elementary considerations.

5. DEEP BOX STRUCTURES (TYPE (d))

This type of bridge, frequently used in large structures, presents a general problem which can be tackled by any generalised shell programs available in finite element methodology.[1] As almost always the sections are flat, the refinement of curved shell elements is unnecessary, and flat elements combining bending and in plane action are used. While in principle no difficulties of analysis in such structures exist, we focus the attention here on economies possible to reduce the obviously much greater computational effort than that necessary in the previous sections.

5.1. Element Type

Any of the plate bending elements may be combined with plane stress action on the same element. It is usual (though not imperative) to associate the in plane degrees of freedom with the same nodes. Some such plane stress elements are illustrated in Fig. 14. In the rectangle/quadrilateral families the suffix r stands for reduced order of integration. It was found recently[21, 42] that by using a reduced order of numerical integration such element 'mutants' give a much improved bending response; a

factor important for such representation. Only one such element in the depth of a
beam can account for the full action (whereas several constant strain elements are
needed to obtain reasonable accuracy).

	TOTAL DEGREES OF FREEDOM	TYPE
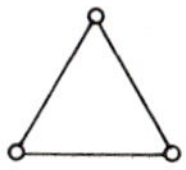	6	T6
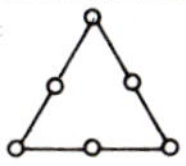	12	T12
	8	R8 R8 r
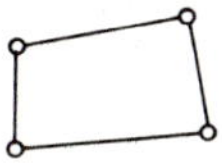	8	I8 I8r
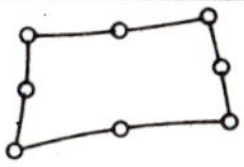	16	I16 I16r

FIG. 14   MEMBRANE ELEMENTS

If bending elements with three flexural degrees of freedom are used in conjunction
with two degrees of freedom for in plane action, the total of five degrees of freedom
is obtained at a node. It is usual to assemble in six degrees of global freedom[1,43]
using either local assembly or a fictitious, sixth, rotational stiffness.[42]

With elements meeting at a finite angle continuity of displacements along the
edge is usually violated even by so called conforming elements, but experience
shows that errors are generally not severe. The possible elements in this context are
very numerous and several have been investigated (Table 3).

The last of these, being capable of representing full shear deformation, is of
particular interest for massive concrete sections.

While in the elements listed six degrees of freedom exist at each node, one simpler
possibility is presented by combining the bending element 4E with the constant
strain triangle T6. Such an element will have three corner nodes each with three degrees
of freedom and mid side nodes with one degree of freedom only. Though the number
of such elements used must be large it is possible that (as nodal degrees of freedom
are reduced) good general results can be obtained economically. This combination
has not yet been tested, but it is being investigated.

5.2. Slender Sections – A Simplification

In the case of thin wall box type bridges with diaphragms, load carrying is achieved
primarily through the action of membrane stresses. Thus if in a typical program of
the preceding section the bending stiffness is omitted and nodal degrees of freedom
reduced to three displacements (u, v, w) only, then a very considerable computational
advantage arises. Such processes necessitate precautions that the structure does not
become a mechanism. In the context of box type panel buildings very successful
results have been obtained by this process.[45] For bridges, in order to produce lateral

stability, Sawko[44] introduces artificial diaphragms whose rigidity approximates to that attained by flexural action, and gets excellent results.

As an indication of how satisfactory such simplification is, the following results for a fairly solid box section with diaphragms shows that quite good results may be obtained. The box is one analysed by Meyer and Scordelis. Its section dimensions are 112 in by 36 in deep with flanges 6·5 in thick at the top and 5·5 in at the bottom. The sides are 8 in thick. The span is 720 in with a load of 1 kip on the centre line over one side. There is a central diaphragm and the end diaphragms are taken as rigid.

The following Table 4 compares the deflections calculated with and without taking bending action into account, with values given by Meyer and Scordelis from a folded plate analysis.

The fact that the membrane only solution is apparently better should not be mis-interpreted. This solution will obviously be the more flexible of the two, and the mesh used was quite coarse, two elements across the flanges and one in the depth of the web. Since the computation time for a membrane solution may conceivably be 1/8th of that of the full solution it is clearly a useful technique for exploratory work.

5.3. Constant Section — Partial Fourier Expansion

The processes described in Section 2.3 are once again applicable to situations where sectional properties do not vary along the bridge axis. The examples given by Cheung[33-35] have demonstrated that in the context of thin wall prismatic sections. Here they give an alternative to the general shell processes just described. For thick wall sections however, inherently *more accurate* results can be obtained by this method[36] in some cases. While shell type programs even utilizing thick shell elements, and thus accounting for shear deformation, give a good approximation, the structure is reduced to a 'middle line' configuration, as shown in Fig. 15. A Fourier type solution by using two dimensional elements in the cross section gives a *full three dimensional solid* solution. Errors due to using linear idealisation of members and various

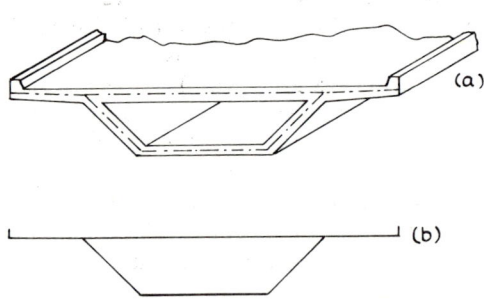

FIG. 15 BOX GIRDER BRIDGE: LINEAR IDEALIZATION

stress concentration are thus revealed. Fig. 16 shows how the real bridge section idealised in Fig. 15 has been represented by two dimensional elements in the cross section, to generate solutions for harmonic loading which may then be used as a reference example against which shell type programs may be tested.

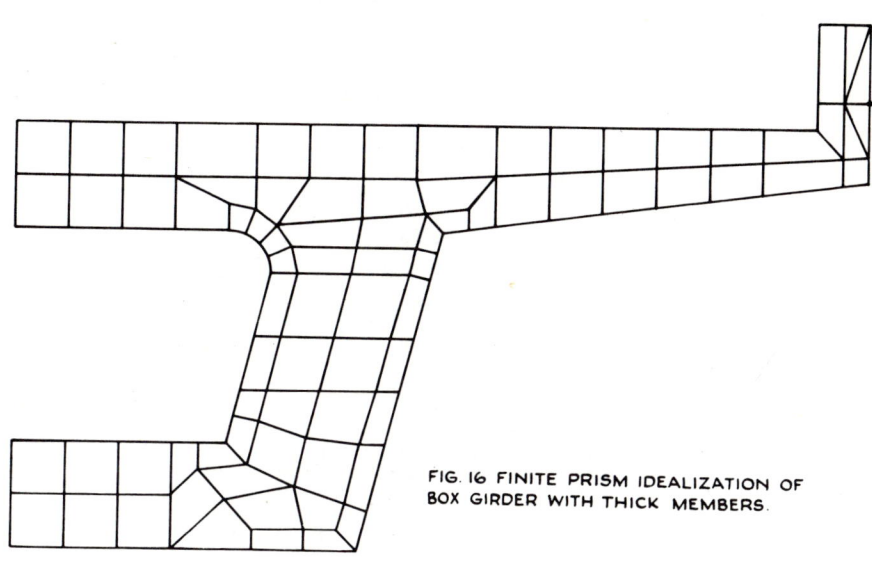

FIG. 16 FINITE PRISM IDEALIZATION OF BOX GIRDER WITH THICK MEMBERS.

TABLE 1

Distribution of Girder Deflections at Middle of Loaded Span

Loaded Girder	Model	% Sum of Girder Deflections				
		A	B	C	D	E
A	½ scale	64·5	27·9	9·1	0·9	−2·4
	1/24 scale	64·8	28·4	8·1	0·8	−2·1
	Finite Element	63·6	28·2	9·1	1·3	−2·4
B	½ scale	28·8	41·1	21·6	7·3	1·2
	1/24 scale	28·5	40·1	22·8	7·9	0·7
	Finite Element	28·1	40·1	22·4	8·1	1·3
C	½ scale	9·7	21·8	37·8	21·6	9·1
	1/24 scale	8·4	23·0	37·5	23·0	8·1
	Finite Element	9·1	22·3	37·3	22·3	9·1

TABLE 2

Distribution of Lower Flange Strain at Middle of Loaded Span

Loaded Girder	Model	% Total Lower Flange Strain on Girder				
		A	B	C	D	E
A	½ scale	75·9	17·7	6·2	0·6	−0·4
	1/24 scale	75·4	21·3	3·7	−0·1	−0·4
	Finite Element	74·3	21·1	4·7	0·2	−0·3
B	½ scale	20·1	60·4	19·5	0·9	−0·9
	1/24 scale	23·2	53·0	18·2	5·3	0·3
	Finite Element	21·1	54·5	18·6	5·1	0·6
C	½ scale	6·5	17·9	51·2	17·9	6·5
	1/24 scale	4·3	20·0	53·0	18·4	4·3
	Finite Element	4·5	18·7	53·5	18·7	4·6

TABLE 3

Membrane and Bending Elements

Type	Bending	in plane	Nodes
A	1NC	R8r or R8	4
B	3NC	T6	3
C	4 × 1NC	18r	4
D	8C	I16	8

TABLE 4

Solution	Midspan Deflections (ft)	
	Unloaded Web	Loaded Web
Folded Plate	·000444	·000493
Finite Element	·000433	·000484
Finite Element (Membrane action only)	·000436	·000490

REFERENCES

1. Zienkiewicz, O. C. The finite element method in engineering science. McGraw-Hill, 1971. (To be published).
2. Cheung, Y. K., King, I. P. and Zienkiewicz, O. C. Slab bridges with arbitrary shape and support conditions: a general method of analysis based on finite elements. *Proc. Instn civ. Engrs*, Vol. 40, May 1968, pp. 9–36.
3. Sawko, F. and Cope, R. J. Analysis of spine beam bridges using finite elements. *Civ. Engng publ. Wks Rev.*, Vol. 65, No. 763, Feb. 1970, pp. 146–147.
4. Meyer, C. and Scordelis, A. C. *Computer program for prismatic folded plates with plate and beam elements*. University of California, Berkeley, Structural Engineering and Structural Mechanics Report No. 70–3, Feb. 1970.
5. Powell, G. H. and Ogden, D. W. Analysis of orthotropic steel plate bridge decks. *J. struct. Div. Am. Soc. civ. Engrs*, Vol. 95, ST5, May 1969, pp. 909–922.
6. Malcolm, D. J. and Redwood, R. G. Shear lag in stiffened box girders. *J. struct. Div. Am. Soc. civ. Engrs*, Vol. 96, ST7, July 1970, pp. 1403–1419.
7. William, K. *Finite element analysis of cellular structures*. Doctoral dissertation, University of California, Berkeley, Dec. 1969.

8. Douglas, M. R. and Lunniss, R. C. An application of the finite element technique in bridge design. *Concrete*, Vol. 4, No. 5, May 1970, pp. 197–200.
9. Timoshenko, S. P. and Woinowsky-Krieger, S. *Theory of plates and shells.* 2nd Ed. New York, McGraw-Hill, 1959.
10. Clough, R. W. and Fellipa, C. A. A refined quadrilateral element for analysis of plate bending. *Proceedings of the 2nd Conference on Matrix Methods in Structural Mechanics (Wright-Patterson Air Force Base, Ohio, 1968).*
11. Zienkiewicz, O. C. and Cheung, Y. K. The finite element method for analysis of elastic isotropic and orthotropic slabs. *Proc. Instn Civ. Engrs*, Vol. 28, Aug. 1964, pp. 471–88.
12. Adini, A. and Clough, R. W. *Analysis of plate bending by the finite element method.* Report to National Science Foundation, U.S.A. G.7337, 1961.
13. Dawe, D. J. Parallelogrammic elements in the solution of rhombic cantilever plate problems. *J. Strain Anal.*, Vol. 1, No. 3, 1966, pp. 223–230.
14. Argyris, J. H. Continua and discontinua. *Proceedings of the Conference on Matrix Methods in Structural Mechanics (Wright-Patterson Air Force Base, Ohio, 1965)*, 1967.
15. Bazeley, G. P., Cheung, Y. K., Irons, B. M. and Zienkiewicz, O. C. Triangular elements in plate bending – conforming and non-conforming solutions. *Proceedings of the Conference on Matrix Methods in Structural Mechanics (Wright–Patterson Air Force Base, Ohio, 1965)*, 1967.
16. Clough, R. W. and Tocher, J. L. Finite element stiffness matrices for analysis of plate bending. *Proceedings of the Conference on Matrix Methods in Structural Mechanics (Wright-Patterson Air Force Base, Ohio, 1965)*, 1967.
17. Morley, L. S. D. On the constant moment plate bending element. J. Strain Anal. (To be published).
18. Severn, R. T. and Taylor, P. R. The finite element method for flexure of slabs when stress distributions are assumed. *Proc. Instn civ. Engrs*, Vol. 34, June 1966, pp. 153–70.
19. Allwood, R. J. and Cornes, G. M. M. A polygonal finite element for plate bending problems using the assumed stress approach. *Int. J. numer. Meth. Eng.*, Vol. 1, No. 2, April/June 1969, pp. 135–49.
20. Ahmad, S., Irons, B. M. and Zienkiewicz, O. C. Analysis of thick and thin shell structures by curved finite elements. *Int. J. numer. Meth. Eng.*, Vol. 2, No. 3, July/Sept. 1970, pp. 419–51.
21. Zienkiewicz, O. C., Taylor, R. L. and Too, J. Reduced integration techniques in general analysis of plates and shells. *Int. J. numer. Meth. Eng.* (To be published).
22. Veubeke, B. F. de. Bending and stretching of plates. *Proceedings of the Conference on Matrix Methods in Structural Mechanics (Wright-Patterson Air Force Base, Ohio, 1965)*, 1967.
23. Bogner, F. K., Fox, R. L. and Schmit, L. A. The generation of inter-element-compatible stiffness and mass matrices by the use of interpolation formulae. *Proceedings of the Conference on Matrix Methods in Structural Mechanics (Wright-Patterson Air Force Base, Ohio, 1965)*, 1967.
24. Butlin, G. A. *The finite element method applied to plate flexure.* Ph.D. thesis, Cambridge University, 1966.
25. Butlin, G. A. and Ford, R. *A compatible plate bending element.* University of Leicester Engineering Department Report 68–115, 1968.
26. Argyris, J. H., Fried, I. and Scharpf, D. W. The TUBA family of plate elements for the matrix displacement method. *Aeronautical Journal*, Vol. 72, Aug. 1968, pp. 701–709.
27. Bell, K. A refined triangular plate bending finite element. *Int. J. numer. Meth. Eng.*, Vol. 1, No. 1, Jan.-March 1969, pp. 101–122.
28. Cowper, G. R., Kosko, E., Lindberg, G. M. and Olson, M. D. Formulation of a new triangular plate bending element. *CASI Trans.*, Vol. 1, No. 2, Sept. 1968, pp. 86–90. (See also N.R.C. Aero Report LR 514, 1968.)
29. Irons, B. M. A conforming quartic triangular element for plate bending. *Int. J. numer. Meth. Eng.*, Vol. 1, No. 1, Jan.-March 1969, pp. 29–45.
30. Walz, J. E., Fulton, R. E. and Cyrus, N. J. Accuracy and convergence of finite element approximation. *Proceedings of the 2nd Conference on Matrix Methods in Structural Mechanics (Wright-Patterson Air Force Base, Ohio, 1968).*
31. University of Wales, Swansea. Bridge and plate systems (Type 1) FESS. Centre for Numerical Methods in Engineering Computer Program Report No. 6, April 1968.
32. R. Travers Morgan & Partners. Report on slab bridge analysis by computer. Ministry of Transport, Oct. 1968.
33. Cheung, Y. K. The finite strip method in the analysis of elastic plates with two opposite simply supported ends. *Proc. Instn civ. Engrs*, Vol. 40, May 1968, pp. 1–7.

34. Cheung, Y. K. Finite strip method of analysis of elastic slabs. *J. Engng Mech. Div. Am. Soc. civ. Engrs*, Vol. 94, EM6, Dec. 1968, pp. 1365–1378.

35. Cheung, Y. K. Folded plate structures by the finite strip method. *J. struct. Div. Am. Soc. civ. Engrs*, Vol. 95, ST12, Dec. 1969, pp. 2963–2979.

36. Zienkiewicz, O. C. and Too, J. L. Finite elements combined with Fourier expansion in analysis of solid prismatic bodies. (To be published).

37. Kretsis, K. Eccentric edge beams on bridge slabs. *Struct. Engr*, Vol. 48, No. 8, Aug. 1970, pp. 315–317.

38. Gustafson, W. C. and Wright, R. N. Analysis of skewed composite girder bridges. *J. struct. Div. Am. Soc. civ. Engrs*, Vol. 94, ST4, April 1968, pp. 919–941.

39. Somervaille, I. J. *Slab and girder bridge program.* University of Wales, Swansea, Computer Report CNME/CR/41, July 1970.

40. Carpenter, J. E. and Magura, D. D. Structural model testing – Load distribution in concrete 1 – Beam bridges. *J. Res. Dev. Labs Portld Cem. Ass.*, Vol. 7, No. 3, Sept. 1965, pp. 32–48.

41. Culver, C. G. and Christiano, P. P. Static model tests of a curved girder bridge. *J. struct. Div. Am. Soc. civ. Engrs*, Vol. 95, ST8, Aug. 1969, pp. 1599–1614.

42. Docherty, H. P., Wilson, E. L. and Taylor, R. L. *Stress analysis of axisymmetric solids utilizing higher order quadrilateral finite elements.* University of California, Berkeley, Structural Engineering and Structural Mechanics Report No. 69–3, Jan. 1969.

43. Zienkiewicz, O. C., Parekh, C. J. and King, I. P. Arch dam analysis by a linear finite element shell solution program. Proceedings of the Symposium on Arch Dams, Institution of Civil Engineers, London, 1968.

44. Sawko, F. and Cope, R. J. Analysis of multi-cell bridges without transverse diaphragms – a finite element approach. *Struct. Engr*, Vol. 47, No. 11, Nov. 1969, pp. 455–460.

45. Zienkiewicz, O. C., Parekh, C. J. and Teply, B. *Analysis of buildings comprised of floors and walls.* University of Wales, Swansea, Computer Report C/R/128/70, 1970.

THE RELATIONSHIP OF BOX BEAM THEORIES
TO BRIDGE DESIGN

B. RICHMOND

G. Maunsell & Partners

SYNOPSIS

Methods of analysing a wide range of box structures have been developed which give a physical understanding of structural behaviour and facilitate a synthetic approach to design.

Several of these methods are described in this paper. Comparisons demonstrating their accuracy for design are given for a spine box structure under twisting loads. In particular the effect of loads applied to cantilever systems is discussed with numerical confirmation of the conclusions. An extension of the approaches to a series of beams interconnected by a plate or cross-girder system is presented.

Mechanisms for representing the essential characteristics of multi-cell box beams without transverse diaphragms are described with comparisons of stresses and reactions for a skew multi-cell box.

Curved box beams are discussed and the effects of thin walls on the primary bending moments mentioned.

Finally the use of finite element methods for solving complex bridge structures is discussed.

The box girder, in single cell to multi-cell forms, has a number of well-known advantages over the comparable structures with open cross-sections on solid slab structures. If full use is made of its structural possibilities it follows that there are correspondingly more design criteria to be satisfied. The use of model testing, full-scale testing and modern methods of stress analysis are all valuable tools for ensuring bridge designs are adequate. But at the earlier stages of design an understanding of the basic mechanisms governing the structures is necessary. This understanding enables the designer to design more rationally and to decide if a particular bridge structure requires a detailed investigation. In economic terms it may well be more advantageous to increase the thickness of a small part of a structure rather than undertake further investigations, but this cannot be done by intuition. Where the detailed investigation is justified the basic mechanisms give a framework to the results and to their use in future designs. The last point, though not a subtle one, has been overlooked in almost all published research work on bridges other than by using the engineering theories of bending and torsion. These are the most important of the basic mechanisms but are not sufficient to differentiate between the different ranges of box beam behaviour and only permit a very limited degree of extrapolation from test results.

The use of computers and, for example, finite element and folded plate approaches has enabled theoretical predictions to be made that agree well with experimental work in the elastic range. These methods can be used without the physical under-standing of structures necessary for what are often thought of as simpler approaches. The user is not likely to recognise in the mass of numbers representing his solution the significant values without physical understanding. Nor will he be able to con-sider stages of behaviour beyond the elastic range.

The following discussion on the basic mechanisms of box girders and their appli-cation is therefore intended to be used in conjunction with the other methods of stress analysis and experimental work and at G. Maunsell & Partners extensive use is made of finite element methods adapted specifically for both concrete and steel box girders representing the geometry of the structure in three dimensions, and of course model testing. But for many structures where relatively quick hand calculations suggest that further investigation is necessary the mechanisms described can be used as the basis of a computer solution which will be more economical than a normal finite element approach. The economy may be as much in the interpretation of the results as in the computing. The term finite element methods is used in this paper to mean the usual plate bending and membrane elements but in a wider sense the methods described here are finite element methods when dealt with as a set of dis-crete members, since deflection or stress modes are assumed in order to calculate the member properties. Seen the other way round, the finite element approach to pris-matic structures of any cross-section, particularly when using rectangular elements, is a generalisation of engineering beam theory.

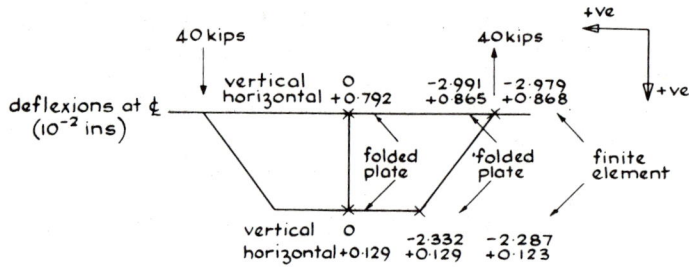

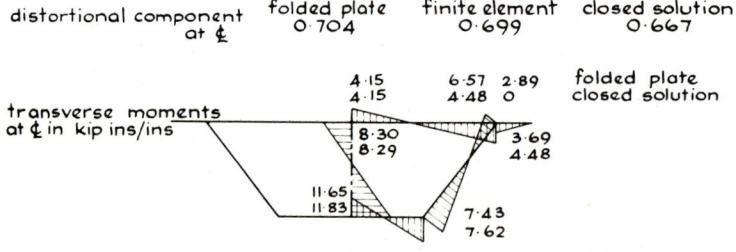

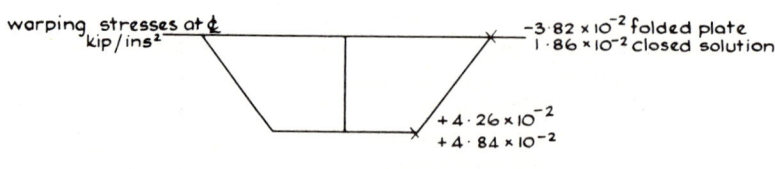

FIG. I

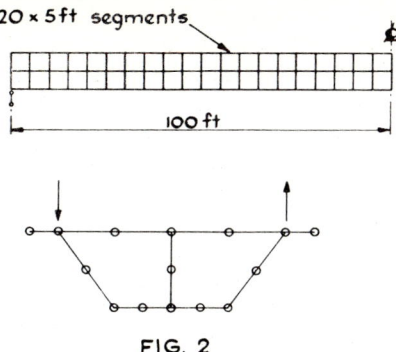

FIG. 2

Single or Double Cell Box Beams under Twisting Loads

The concepts of generalised force and deflection have been used[1,2] to describe the distortional and torsional behaviour of box beams including those of trapezoidal cross-section. The simulation of these effects by an equivalent beam with vertical deflections equal to those of the corners of the box, and direct relationships in shear forces and bending moments permits a more direct appreciation of the problem than the alternative approach based on warping functions. The latter approach can still be linked to an analogous beam but it is more a mathematical than physical simulation. Figure 1 shows the closed solution of the example of the reference[2] but with one set of twisting loads only, compared with the stresses and deflections of a folded plate solution of the elastic equations using 50 terms in the harmonic series together with a finite element solution with the system of elements shown in Fig. 2.

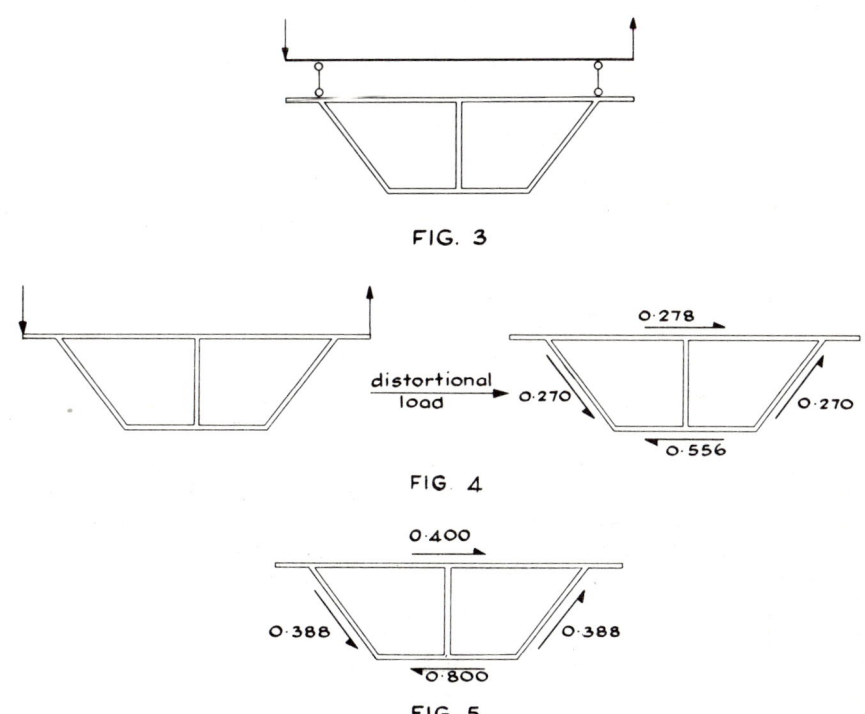

FIG. 3

FIG. 4

FIG. 5

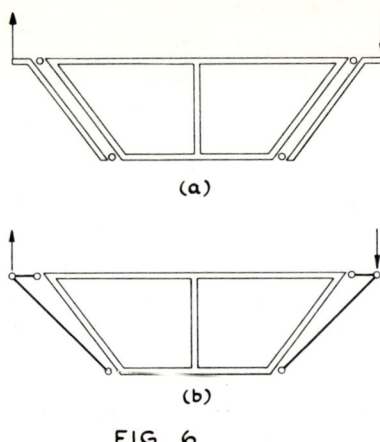

FIG 6

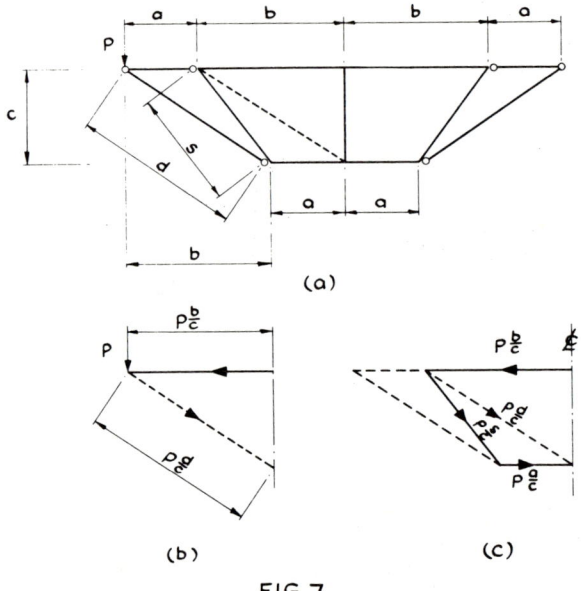

FIG. 7

The only significant difference in longitudinal stresses and transverse moments was in the regions of the local disturbance at point loads. The Karman Fourier integral correction to engineering beam theory gave a correction to the closed solution of the right order. This demonstrated that the closed solution is adequate for design purposes where, as is usually the case, the local effect of wheel loads is provided for by a suitable slab or stiffener system. The difference in transverse bending moments between the folded plate solution and the closed solution and finite element solution is due to local plate bending of the cantilever which is more accurately allowed for by the folded plate solution. It is, however, of little practical significance.

If the twisting loads act at the tips of the cantilevers the torsional load increases and if the cantilevers are supported, as in Fig. 3, the distortional loading and transverse moments on the walls of the box increase in the same proportion. In fact the cantilevers are part of the frame of the whole box and the cantilever fixing moment is taken partly by the web and partly by the top flange. If the corners of the box are assumed supported horizontally and vertically the moment in the frame formed from a slice of the bridge can be calculated and from the shears in the plates the distribution of load into the box is found to be as Fig. 4, for unit twisting loads whereas a unit distortional force derived from twisting loads at the upper corners is

as in Fig. 5, so that the distortional force in Fig. 4 is $\dfrac{0.270}{0.388} = 0.695$.

A 25 term harmonic solution using folded plate theories gave the distortional deflection for the cantilever loaded as being 0.685 of the distortional deflections for the corner loaded case.

If thick webs are used almost all the moment is taken by the web which is equivalent to the mechanism shown in Fig. 6(*a*), which for distortional forces is equivalent to the system of Fig. 6(*b*).

Figure 6(*b*) shows how the load is fed into the box in something approaching a pure torsional distribution.

Figure 7 shows that there is a position that gives forces in the direction of the sides that are as required for pure torsion.

Thus either thick shallow webs or propped cantilevers can reduce distortional bending to negligible levels. The thick web must of course carry all the bending from the cantilever.

The distortional effect of an eccentric load may, therefore, reduce as the load moves outwards from the corners until it becomes zero. Further increases in eccentricity increase the distortion again but in the opposite sense. This phenomenon has been confirmed by model tests on a beam of complex cross-section with a propped cantilever. The null point was very close to the position predicted by Fig. 7(*a*).

Multiple Box Beams

The concepts of an equivalent beam system and generalised forces referred to above can be used for multiple box beams interconnected by a slab or beam system. In systems with cross-girders or where the rotational stiffness ratio of the flange to web of the transverse frame is high enough to give slab to box interaction as in Fig. 3, the substitute system shown in Fig. 8 can be used. *EI* and *GK* are the usual flexural and torsional rigidities of each composite box beam.

I_θ = Second moment of area for distortional bending
$I_m = I - 2I_\theta$ $(2I_\theta + I_m = I)$
I_d = Second moment of area of a beam with equal stiffness to frame stiffness in distortional mode of deflection
I_s = Deck slab second moment of area
I = Large value of second moment of area to restrain middle beam to mean deflection of outer beams.

The above approach is mainly applicable to steel boxes and the effects of internal diaphragms can be allowed for by suitable values of I_d. The substitute grillage can be analysed by a computer but, apart from obtaining numerical results, the substitute system enables the structural mechanisms involved to be understood for the purpose of design.

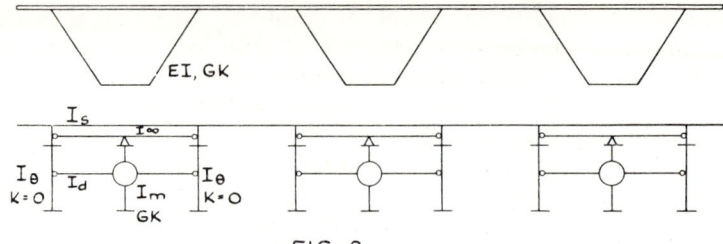

FIG. 8

The frame action of a concrete box system cannot be conveniently simulated in the same way because of the effect on distortional loading already described but even so the above representation is valuable for studying the effect of this additional mechanism. Thinking in these terms shows that it is possible to predict whether distortional behaviour need be allowed for in the analysis of multiple concrete box beams. One way of doing this is to estimate the points of contraflexure in the deck slab and compare them with the null points of the box beam frames. It was pointed out in the case of single box beams that, even if no distortion is caused, the bending moments from the cantilever moment must be carried by the transverse frame of the box. The moments in the slab between the multiple boxes must also be carried by the transverse frame. The difference between one form of frame bending and another may or may not be advantageous. The essential point is, however, that it is possible to design the cross-section to work in whichever way is appropriate.

Detailed solutions of both steel and concrete box systems show that an overall analysis based on simple beam theory is often sufficient provided the design of individual beams allows for distortional behaviour. The mechanisms described enable the degree of approximation permissible to be judged. Folded plate and finite element solutions are valuable in this context. The particular parameters used give reference points to the more general results that can be obtained by the approaches described.

In the opposite sense the writer has found these engineering theories to be very valuable for verifying finite element and folded plate programs. Even when the programs have been thoroughly checked it is still important to examine the results of each problem.

Changes of direction in plan can be allowed for using the equivalent system of beams.

Multi-cell Box Beams without Transverse Diaphragms

The basic mechanisms of multi-cell box beams depend on the shear stiffness of the structure. For structures with a normal shear stiffness in the longitudinal direction and a low shear stiffness transversely the basic mechanisms depend on the form of loading. Anti-symmetrical loading gives a reasonably close approximation to simple torsional behaviour provided allowance is made for the effects of differential bending and transverse bending. Symmetrical bending gives essentially shear lag behaviour with no significant torsional behaviour at all. For arbitrary loading ref.[3] shows that, provided the most significant aspects of the two mechanisms are combined, surprisingly close agreement is obtained with more complete solutions. The results of such 'simplified' solutions are however meaningless unless the mechanism on which they are based is understood.

Skew beams can also be solved by the same approach as shown in the comparisons of Fig. 9, 10, 11 and 12, between a grillage representation of the above mechanisms and the lattice approach described in the next section.

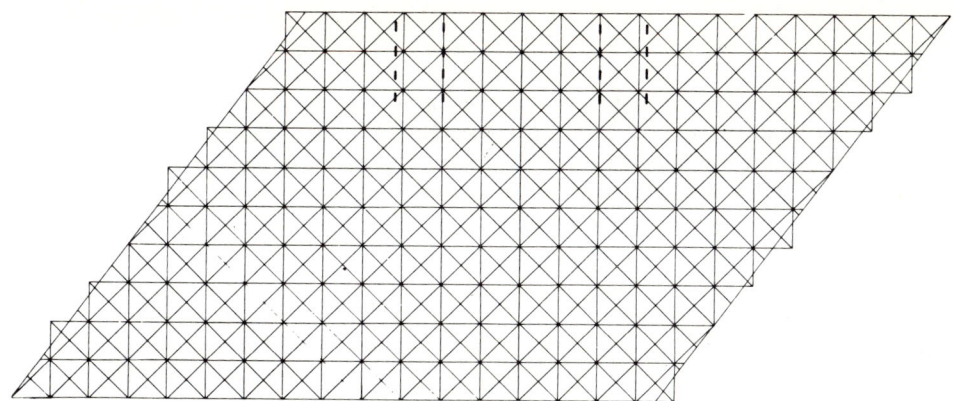

FIG 9 TOP HAT BEAMS   SKEW DECK

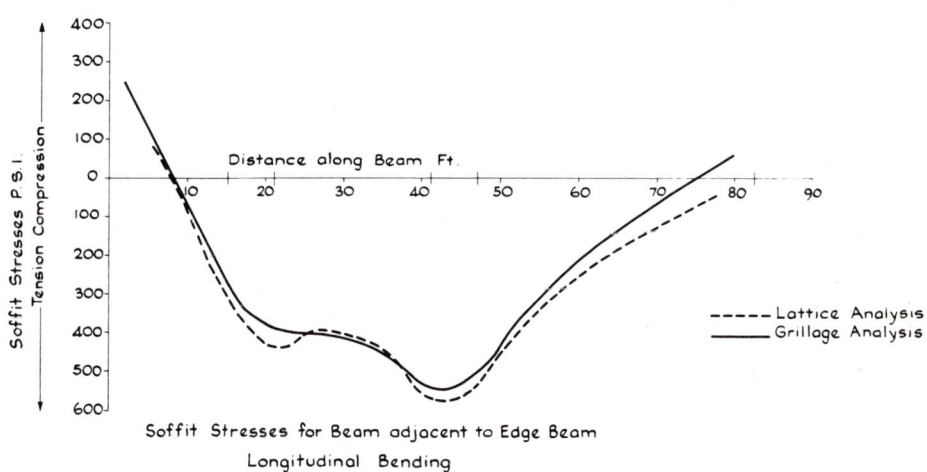

Soffit Stresses for Beam adjacent to Edge Beam

Longitudinal Bending

FIG. 10   TOP HAT BEAMS. SKEW DECK.

Box Beams Curved in Plan

Box beams curved in plan do not differ in essentials from straight box beams apart from the effect of geometry on the statical relationships unless the webs and flanges are very thin. If, however, they are thin the primary bending moments can be significantly increased by the interaction between distortional and simple bending deflections which is not found in straight box beams.

This phenomenon has been confirmed experimentally by Sanchez, ref.[4]. He found reasonable agreement between a curved in plan perspex box beam, continuous over two spans, and the Dabrowski-Wlassov theory of curved beams.

The depth to thickness ratio of the model box sides would be extreme for large structures without diaphragms and in many cases the approaches described earlier can be used.

The usual finite element method can also be used but a useful alternative based on the lattice analogy can transform the curved box of rectangular cross into a system of pin-jointed members for the flanges and beams for the web members. The upper and lower flange members can be represented in pairs as beams with zero shear stiffness. The transformed system can then be solved exactly by a grillage program as described in ref.[3] for multi-cell boxes.

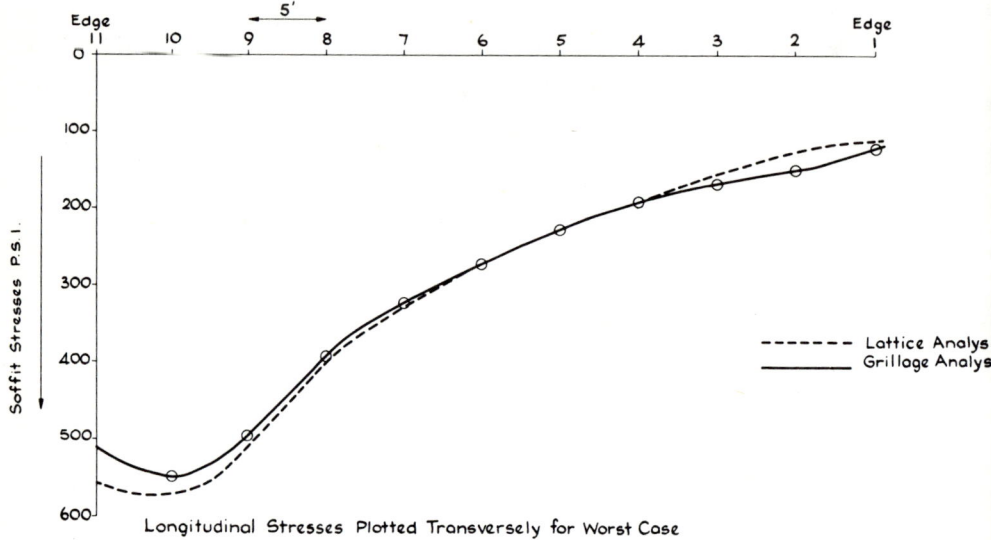

Longitudinal Stresses Plotted Transversely for Worst Case

FIG.11   TOP HAT BEAMS   SKEW DECK.

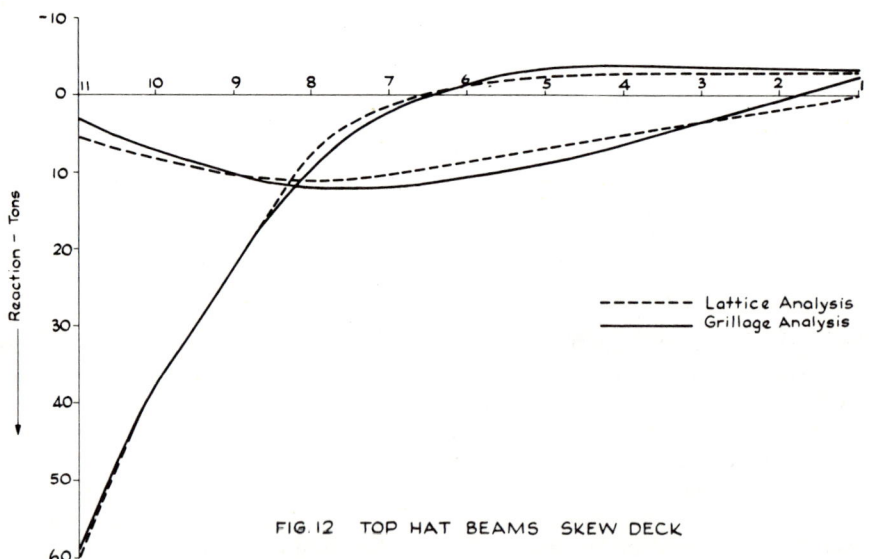

FIG. 12   TOP HAT BEAMS   SKEW DECK

Box Beams of Complex Cross-section

Box beams of complex cross-section or with complications such as cut-outs, anchor-ages of cable-stayed bridges, etc. can be examined using various approaches that have been evolved in the analysis of aircraft structures. Kuhn, ref.[6], describes these in his book and compares them with experimental work. The physical understanding given by these theories and the accuracy with which many complicated systems can be solved by relatively simple calculations makes a study of his work very rewarding.

It is, however, to the finite element approach that we now turn for the most economical solution of complex problems. The length of time required to obtain reliable answers using simple numerical methods can be very great because of the difficulties in producing satisfactory substitute structures. Different substitute struc-tures must be used for various loading cases and different parts of the structure. An understanding of these approaches is still valuable during design and provided a more exact solution is obtained to check the design they can be used quickly.

The finite element approach can be very expensive in data preparation and in evaluation of the results. In order to enable it to be used as a standard design tool G. Maunsell & Partners have produced a program, ref.[5], specifically for thin-walled beams of arbitrary cross-section. By restricting the range of the problem it has been possible to reduce the data input to the same amount as would be required for a two-dimensional frame with the shape of the cross-section. Special diaphragms, cut-outs, etc. increase the data but it is still possible to regard it as a design tool. The output is in a form that is compatible with normal engineering beam calculations.

Finite element methods give approximate answers to stress distributions and to use these methods the designer must know what reliance he can place on the results. In consequence he needs to understand both the structure and the basis of the finite element theory. The essentials of finite element theory should therefore be seen as a basic requirement for all structural engineers engaged in design.

A good example of this type of approximation in the three-dimensional repre-sentation of a box beam occurs at the junction of a thin plated diaphragm and the web and bottom flange of a box of trapezoidal cross-section. The diaphragm acts as the web of a transverse beam with the box flanges acting as transverse flanges to that beam. The support from the diaphragm at the junction will be effective over a considerable part of the elements on either side of the diaphragm web. This enables the flanges to take loads which, unless the elements are very small, they cannot carry in the actual structure. A correction can be made to allow for this error if it is recognised. Alternatively smaller elements can be used adjacent to diaphragms.

REFERENCES

1. Dalton, D. C. and Richmond, B. Twisting of thin-walled box girders of trapezoidal cross-section. *Proc. Instn civ. Engrs*, Vol. 39, Jan. 1968, pp. 61–73.
2. Richmond, B. Trapezoidal boxes with continuous diaphragms. *Proc. Instn civ. Engrs*, Vol. 43, Aug. 1969, pp. 641–650.
3. Hook, D. M. A. and Richmond, B. Precast concrete box beams in cellular bridge decks. *Struct. Engr*, Vol. 48, No. 3, March 1970, pp. 120–128.
4. Moya, V. S. *Twisting of thin walled box beams*. M.Phil. thesis, University of Surrey, 1967.
5. Richmond, B. Thin-walled beams. Bridge program review symposium, Jan. 1970.
6. Kuhn, P. *Stresses in aircraft and shell structures*. New York, McGraw-Hill, 1956.

A FOLDED PLATE APPROACH TO THE ANALYSIS OF BOX GIRDERS

H. R. EVANS and K. C. ROCKEY

Department of Civil and Structural Engineering,
University College, Cardiff

SYNOPSIS

The paper presents a method of analysis for box girders based on the Ordinary Method of analysis of folded plate structures. The method is termed the "Nodal Section Method" and, since it assumes a simplified structural behaviour, it does not make excessive demands on computer time and storage space. Furthermore, the method may be applied to the analysis of a girder in which the cross-section varies along the span. Results given by the method are compared to values obtained from experiments and also to values given by the Finite Element Method, and satisfactory agreement is noted in all cases.

1. INTRODUCTION

The term "folded plate" is usually used to describe a structure such as that shown in Fig. 1a, and this type of structure has been much used for roofing large areas. A box girder such as that shown in Fig. 1b or Fig. 1c may be regarded as a special type of folded plate in that the plates are arranged so as to form a closed section. Methods of analysis originally developed for folded plates may thus be adapted for the analysis of box girders.

A large amount of research effort has been devoted to folded plate analysis and three main methods have been established. These are the Finite Element Method[1, 2] the Elasticity Method[2, 3, 4] and the Ordinary Method[2, 4, 5, 6, 7] details of each of these being given in the listed references.

The Finite Element Method can, of course, be used for box girder analysis and the adaptability of this method is such that it can readily take into account a variety of support and loading conditions. However, it involves very extensive computations which lead to excessive demands on computer time. This makes it unsuitable for use as a tool during a design process in which many analyses of the structure are desirable so as to assess the effect of varying certain parameters. However, this method is invaluable for analysing the designed structure accurately to ensure that it is satisfactory, and many computer programs are available to do this.

The most accurate of the Folded Plate Methods of analysis, the Elasticity Method, has also been applied successfully to box girder analysis by Scordelis[8, 9]. In this method, termed the "Folded Plate Method" by Scordelis, the bending of

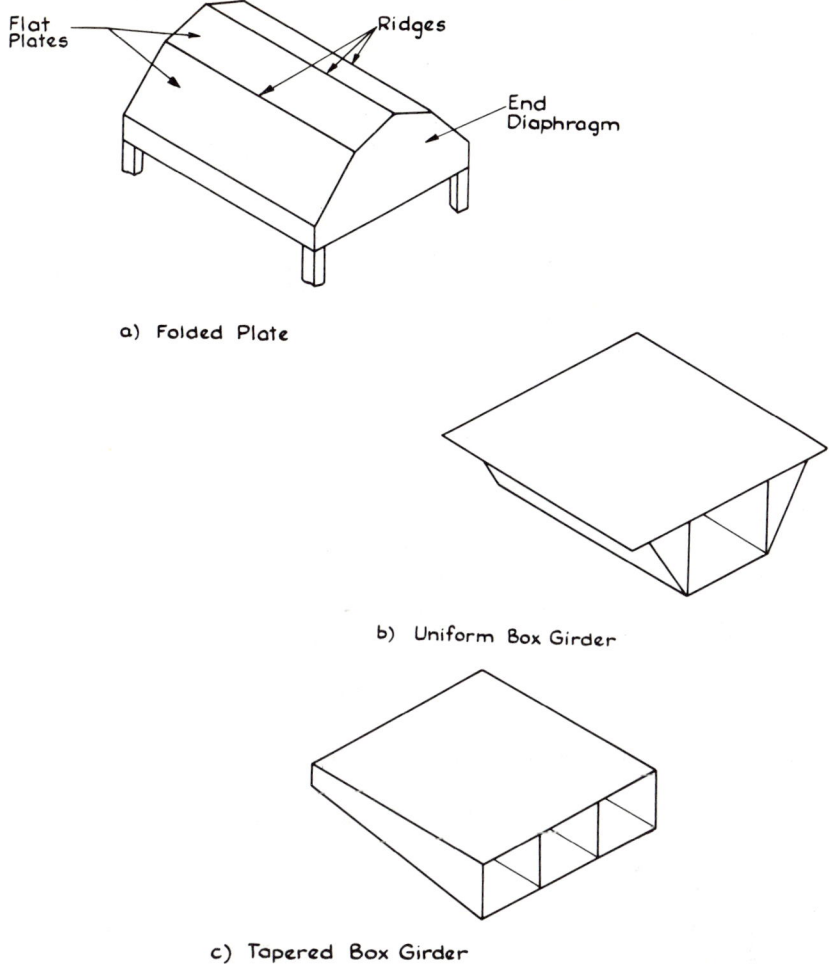

a) Folded Plate

b) Uniform Box Girder

c) Tapered Box Girder

FIG. I - TYPICAL FOLDED PLATES & BOX GIRDERS

each plate normal to its plane is analysed by plate flexure theory and the bending
in the plane is analysed by plane stress theory. These classical theories necessitate
the representation of the external loads by Fourier Series with the result that the
computational effort required is still considerable, although only a small fraction
of that required in the Finite Element Method. Its range of application is, however,
more limited than that of the Finite Element Method and it can only be applied to
box girders consisting of rectangular plates. A girder in which the cross-section varies
along the span, such as that shown in Fig. 1c, cannot therefore be analysed by the
Elasticity Method.

The Ordinary Method is an approximate method of analysis since it assumes a
simplified structural behaviour. This significantly reduces the amount of computation required. It has been shown[10] that the errors introduced by these simplifying
assumptions are small for uniformly loaded folded plates, provided that the length/
width ratio of the component plates exceeds 3. Now for most bridge girders, this
ratio will be much larger than 3 so that the Ordinary Method can be safely applied.

Scordelis[9] has, in fact, successfully adapted the Ordinary Method to the analysis of single-span and continuous box girders. This approach was termed the Finite Segment Method and its formulation was such as to restrict its application to the analysis of box girders containing uniform rectangular plates under loads applied at the ridges only.

Now Johnson and Lee[11] have shown how the Ordinary Method can be applied to the analysis of folded plates containing tapered plates, provided that the plate taper is not excessive. This restriction on taper arises from the fact that each plate must have a length/width ratio greater than 3 before the Ordinary Method can be applied. A conservative interpretation of this condition for a tapered plate requires the length to be more than 3 times the greatest width and this automatically restricts the inclination of one edge of the plate to the other edge to a maximum value of 19°, see Fig. 2. Once again this limitation is not too serious from the

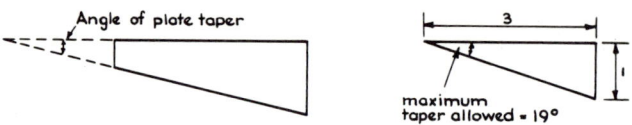

FIG. 2.   DEFINITION OF ANGLE OF PLATE TAPER

point of view of bridge girders since the taper will be considerably less than this limiting value in many practical cases.

It would seem, therefore, that an extension of the Ordinary Method of folded plate analysis, as developed by Johnson and Lee, to the analysis of box girders would be advantageous. It would provide a solution procedure economical of computer time and thus suitable for use as a design tool, and a procedure that would, furthermore, be able to take tapered girders into account. Such an approach is discussed in this present report for simply supported box girders.

2. FOLDED PLATE ANALYSIS

The Ordinary Method of folded plate analysis makes the following basic assumptions:

1. The material of the structure is homogeneous and linearly elastic.
2. The actual deflections are small compared to the structural dimensions.
3. The principle of superposition holds.
4. The connections between plates are fully monolithic.
5. Each supporting end diaphragm is infinitely stiff in its plane and perfectly flexible normal to its plane.

    In general, a component plate will be subjected to in-plane and normal load components and will deflect both in its plane and normal to its plane. The Ordinary Method simplifies the structural action by making two further assumptions.

6. The in-plane action of an individual plate is similar to that of a simple beam spanning between the end diaphragms.
7. The action of each plate normal to its plane is similar to that of a transverse one-way slab strip.

The behaviour of a folded plate may then be considered to consist of the action of a series of transverse one-way slab strips supported elastically at the ridges by a series of interconnected plate beams spanning longitudinally between the end

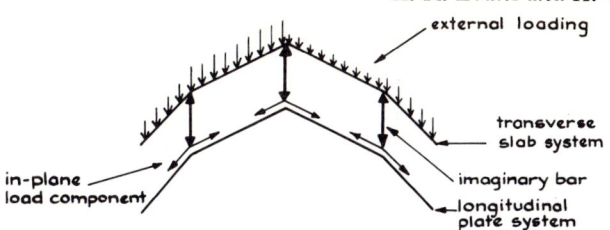

FIG. 3. IDEALISED FOLDED PLATE BEHAVIOUR

diaphragms. The slab strips only transmit shears and moments in the transverse direction, this action being termed the "transverse slab action", while the plate beams only transmit forces in their plane, this action being termed the "longitudinal plate action" of the structure. This idealized behaviour is illustrated in Fig. 3 and it is apparent that the reactions of the slab system must be equal to the ridge loads of the plate system and also that the ridge deflections of the two systems must be compatible.

The transverse slab action is analysed by considering each transverse strip as a continuous beam, the equilibrium of transverse moments and the compatibility of rotations being satisfied at the ridges. In the analysis of the longitudinal plate action, each component plate is analysed as a simply supported beam spanning between the end diaphragms, the connection between the adjacent plates being taken into account by ensuring the equilibrium of longitudinal shear forces and the continuity of longitudinal strains at the ridges.

The analysis is not straightforward because the slab system is ELASTICALLY supported by the plate system. A situation is thus created in which the displacements of the plate system are produced by the slab reactions, while these reactions themselves depend partly on the plate displacements. Several convenient solution techniques have, however, been proposed and the Method of Particular Loadings developed by Yitzhaki[6] was adopted in the present investigation. A matrix formulation of this technique has been presented by Scordelis[7] and this makes the solution procedure suitable for a digital computer.

In this Method of Particular Loadings the structure, under the action of the externally applied loads, is analysed first of all by assuming the slab system to be RIGIDLY supported by the plate system and the displacements thus set up in the plate system by these ridge loads are calculated. These plate displacements lead to relative joint displacements, as defined in Fig. 4, which set up additional 'sway' moments in the transverse slab system. A subsequent correction analysis is then

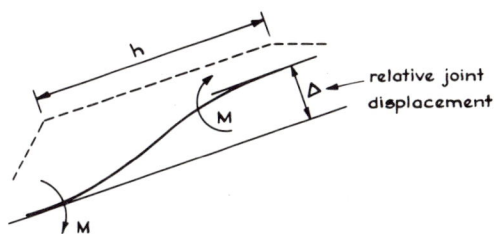

M = Additional 'sway' moments due to relative joint displacements = $\dfrac{6EI\Delta}{h^2}$

FIG. 4.– RELATIVE JOINT DISPLACEMENT OF A TYPICAL PLATE

carried out to take these additional moments into account, and the final results are obtained by superposition of the individual parts of the analysis.

The importance of this correction for relative joint displacements depends on the structural dimensions and on the loading conditions. For a uniformly loaded folded plate structure it has been shown[10] that when the individual plate dimensions resemble those of the component plates of a bridge girder, then the correction is extremely significant. When it is also remembered that bridges are generally subjected to skew loads and that such loads are particularly conducive to large transverse distortions of the cross-section, then it may be concluded that the correction for relative joint displacements will be an essential part of most box girder bridge analyses.

When the Ordinary Method is applied to a 'conventional' folded plate structure, i.e. one in which all the plates are rectangular and the loading on each plate has a similar distribution in the longitudinal direction, then it is sufficient to consider the satisfaction of equilibrium and compatibility conditions between adjacent plates at one point only on their common ridge. If the required conditions are satisfied at this one point then they will be automatically satisfied all along the ridge. Consequently, only one transverse slab strip has to be considered and the longitudinal shear forces and stresses have to be calculated at one cross-section only of each longitudinal plate beam.

In a tapered structure, the cross-section varies along the span so that the analysis cannot now be based on considerations for one typical cross-section. Instead, the structure has to be divided into a number of transverse nodal sections, as shown in Fig. 5, and equilibrium and compatibility conditions at the ridges must be satisfied at each nodal section. Hence, at each nodal section a transverse

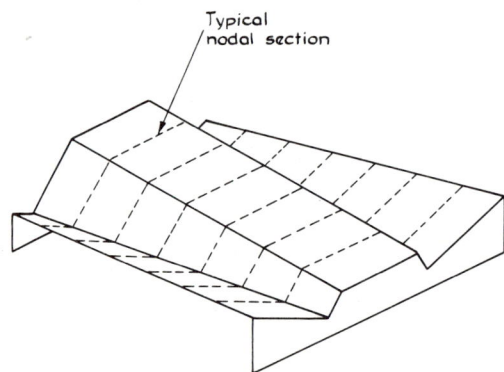

FIG. 5. NODAL SECTIONS ON A TAPERED STRUCTURE

slab strip must be analysed for the action of the external loading pertaining to that particular section and a series of point slab reactions determined for each ridge. These are then applied to the plate system and the longitudinal shear forces and stresses set up in the plate beams at each nodal section are taken into considera- tion. The solution process is thus considerably lengthened for tapered structures but it is of a repetitive nature and is ideally suited for automatic computation. The Nodal Section technique also enables conditions in which the longitudinal distribu- tion of the applied loads differs for the various plates to be readily taken into account.

Obviously, such a Nodal Section analysis is approximate because equilibrium and compatibility conditions between adjacent plates are only satisfied explicitly at a few points along the span. The more nodal sections taken, then the more accurate will be the solution obtained and the rate of convergence of the solution is illustrated later in some specific cases. In the general analysis, a uniform structure becomes a particular case in which all nodal sections have the same dimensions.

3. FURTHER CONSIDERATIONS FOR BOX GIRDER ANALYSIS

Although box girders are basically similar to folded plates there are two possible points of difference that must be taken into account.

The first of these points concerns the number of plates meeting at one ridge. Now it is very rare in folded plate structures for more than two plates to intersect, although most cellular girders will have such an arrangement of plates, as shown in Figs 1b and c. This introduces a difficulty into the proposed method of analysis at the point where the reaction of the slab system at a ridge is resolved into component loads in the planes of the plates meeting at the ridge. If there are more than

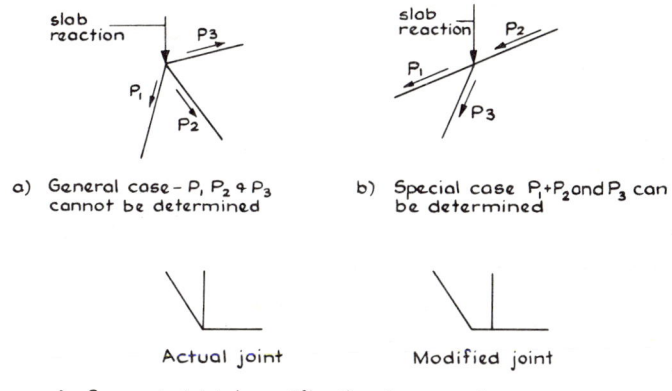

a) General case – P_1, P_2 & P_3 cannot be determined

b) Special case P_1+P_2 and P_3 can be determined

Actual joint

Modified joint

c) Suggested joint modification in general case

FIG. 6 – DIFFICULTIES IN RESOLUTION OF JOINT FORCES

two intersecting plates, as shown in Fig. 6a, then the component load carried by each plate cannot be determined in the general case since only two equilibrium equations are available.

This difficulty has been overcome for the particular case when three plates intersect at a ridge, two of these plates being co-planar, as shown in Fig. 6b. In such a case the two equilibrium equations enable the in-plane load on the non-aligned plate to be determined and also the sum of the in-plane loads on the two co-planar plates, although the proportion of this total load taken by each plate cannot be determined. Because of this, the two co-planar plates are considered to act as one deep beam during the longitudinal plate analysis, this plate beam being connected to adjacent plates not only along its edges, as in the usual case, but also along the appropriate line within its width. The equilibrium of longitudinal shear forces and the continuity of longitudinal stresses is then satisfied as usual at each of these lines of interconnection. The analysis of the transverse slab system is not affected by this modification and a diagrammatic representation of the modified solution procedure is shown in Fig. 7.

The proposed modification will only apply when 3 plates meet at a ridge and 2 of these plates are co-planar. However, many cellular box girders will fall into this category, as shown in Fig. 1c. For other cases, such as that shown in Fig. 1b

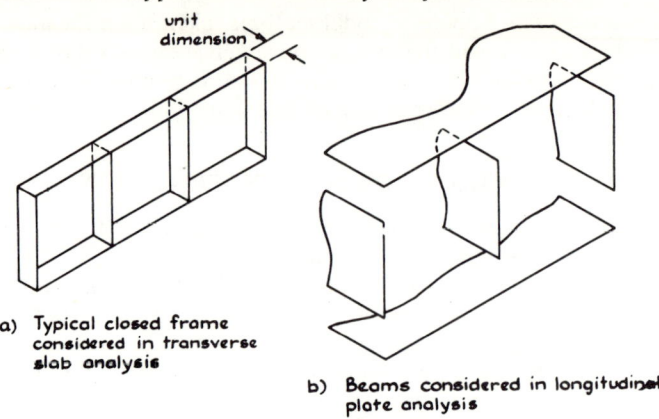

a) Typical closed frame considered in transverse slab analysis

b) Beams considered in longitudinal plate analysis

FIG. 7 – STRUCTURAL IDEALISATION FOR MULTI-CELL GIRDER.

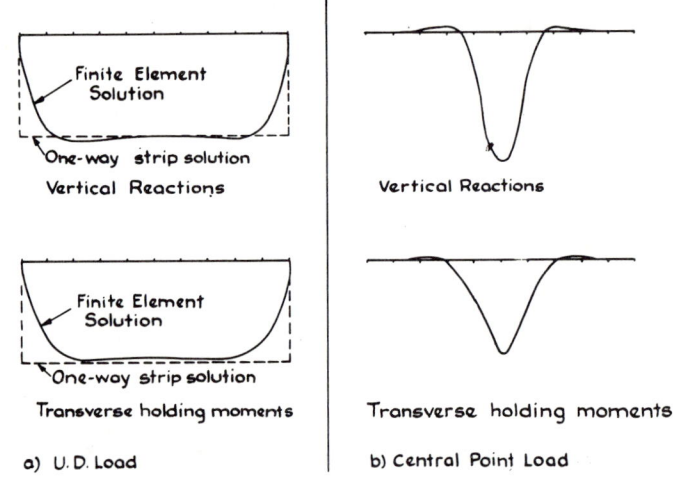

a) U.D. Load

b) Central Point Load

FIG. 8 - EDGE HOLDING FORCES FOR A PLATE IN FLEXURE

where the assumed conditions are not satisfied at the edges of the bottom flange, then it may be possible to assume a slightly modified form of the appropriate joints, as shown in Fig. 6c, so that the analysis can still be applied. In fact, the modified arrangement may represent practical conditions more closely than the assumption of a single intersection point for the plates.

The second additional point that has to be considered for box girders concerns the distribution of the applied loads. The design loads normally considered for folded plate roofs are the self weight and wind and snow loads, all of which have a fairly uniform distribution over the surface. In box girder bridges, however, much more localized loading effects on the top flange, such as the abnormally heavy vehicle, must be taken into account. Under such loads, the analysis of the top flange by taking a transverse one-way strip is rarely justified. This is demonstrated in Fig. 8 where the clamping forces at the longitudinal edge of a plate in flexure are plotted. The length/width ratio of the plate considered is 4·8 and values obtained by taking a one-way strip are compared to those obtained from a Finite Element solution for two different load cases. Whereas the transverse strip analysis is accurate for the uniformly distributed load case, it is clearly not so for the point loading.

To take such loadings into account, it is proposed that a Finite Element plate flexure analysis of the loaded top flange plate be carried out first with the longitudinal edges of the plate completely clamped. The edge holding forces thus calculated are then applied in the reverse direction to the box girder as ridge loads and a complete analysis of the box under these ridge loads carried out by the Nodal Section Method, the transverse holding moments being assumed to be carried by the slab system and the vertical reactions being assumed to be carried by the plate system. This solution is then superimposed onto the initial Finite Element solution so that the fictitious edge holding forces introduced are eliminated.

It is often of interest to determine the bending moments set up in the flange plate in the locality of the applied loads. The initial Finite Element solution will give these moments for any applied loading for the case when the longitudinal edges of the plate are clamped. However, in order to obtain the complete picture, another Finite Element flexure solution of the relevant plate has to be carried out, subsequent to the Nodal Section analysis. In this final step, the ridge deflections calculated by the Nodal Section Method are imposed on to the plate, which is otherwise considered to be unloaded now. Superposition of the two Finite Element solutions will then define the localized effects set up within the loaded plate.

The complete solution procedure is illustrated in Fig. 9. By introducing a

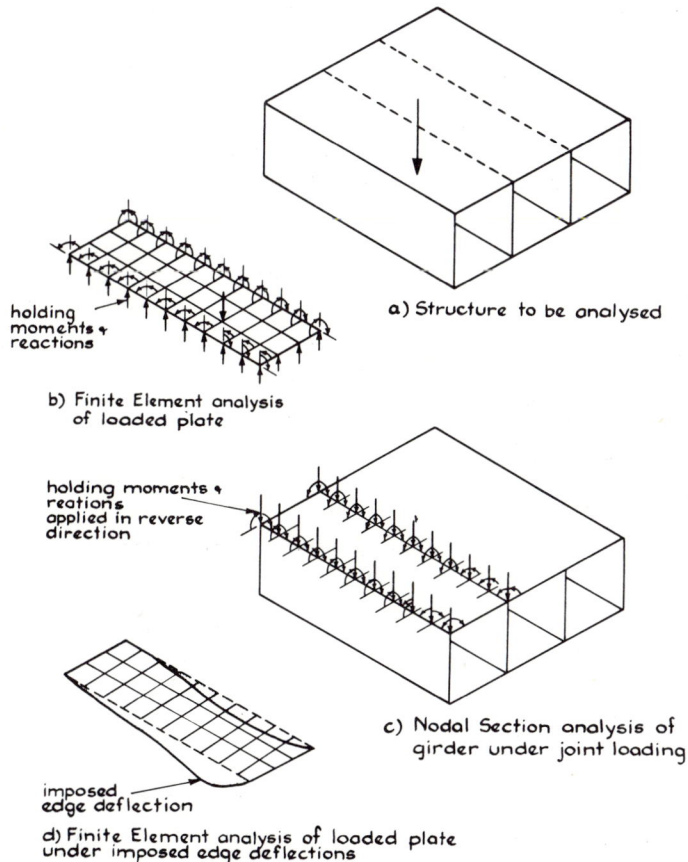

holding moments + reactions

a) Structure to be analysed

b) Finite Element analysis of loaded plate

holding moments + reactions applied in reverse direction

c) Nodal Section analysis of girder under joint loading

imposed edge deflection

d) Finite Element analysis of loaded plate under imposed edge deflections

FIG.9 - SOLUTION PROCEDURE FOR LOCALISED LOADING CONDITIONS.

Finite Element solution in this way the overall solution time will, of course be increased. However, the increased time will still be very much less than that required for a full Finite Element solution of the complete box girder. It must be appreciated that, in the proposed procedure, a Finite Element solution is only required for the plates that are subjected to localized loading conditions, and the number of plates loaded in this manner is usually small compared to the total number of plates in the cross-section. Furthermore, these plates are only analysed for flexure and such an analysis involves the consideration of only 3 degrees of freedom at each node compared to the 6 degrees of freedom that would have to be considered at each node in a full Finite Element analysis of the girder.

4. COMPARISON OF RESULTS

In this section, theoretical results obtained by the Nodal Section Method will be compared to values given by the Finite Element Method and to the results of preliminary model tests.

The models tested are shown in Fig. 10, most of the models being made from mild steel. The end diaphragms for these steel models were made from $3/8$ in thick mild steel plate and freedom of movement of each diaphragm normal to its plane, together with complete rigidity in its plane, was ensured.

In the tests, vertical distributed loads were applied to the top flange of each model. These loads were applied to the steel models by means of a 'whiffle-tree system', the distributed load being simulated by a series of closely spaced point loads. By using such a loading system it was possible to vary the type of loading applied to the models very conveniently.

Only the deflections of the models were measured during these initial tests. These were measured at the mid-span cross-section and the position on the cross-section at

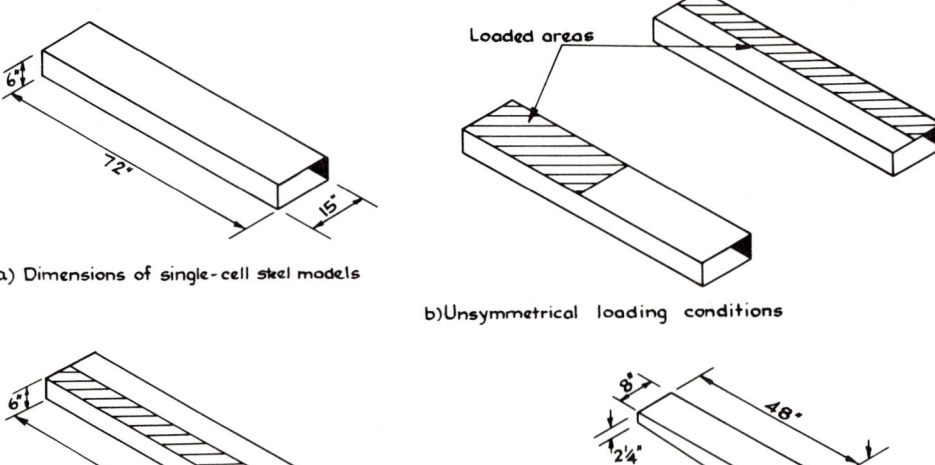

a) Dimensions of single-cell steel models

b) Unsymmetrical loading conditions

c) Double-cell steel box showing one cell loaded.

All plates 0·08 ins thick

d) Tapered single-cell perspex box

All plates 0·25 ins thick

FIG. 10 – DETAILS OF EXPERIMENTAL MODELS

which these measurements were taken are shown in Fig. 11, only centre plate deflections being measured in most cases. A number of loading increments were applied to each model so that its load/deflection characteristics could be determined accurately. In the present comparison, values corresponding to a loading intensity of 1 lb/in² will be considered although very much higher loads than this were applied in some cases.

The first steel model tested had a plate thickness of 0·08 in and the results of the test in which a uniformly distributed load was applied over the complete area of the top flange are given in Table 1. Theoretical results given by the Nodal Section and Finite Element Methods are included in this table, theoretical values of the transverse moments and longitudinal stresses at the ridges being included in addition to the deflection values. Hogging moments and tensile stresses are taken as positive.

Two Finite Element and four Nodal Section solutions were carried out and the finest structural subdivision employed in each method is shown in Fig. 11b. The results obtained using these subdivisions agree reasonably well with the experimental values, the error in the value given by the Nodal Section Method for the maximum deflection of the cross-section (δ_4) being + 6·15% and the corresponding error in

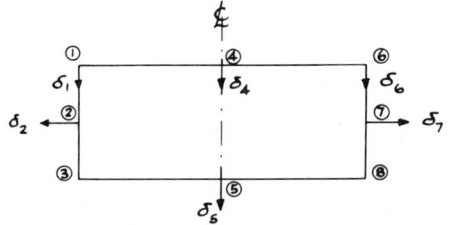

Deflections measured in tests are $\delta_2, \delta_4, \delta_5$ & δ_7

a) Positive directions of deflections compared for single-cell boxes

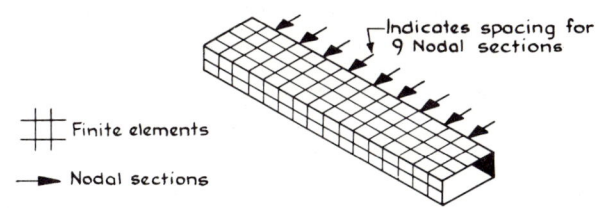

a) Typical structural idealisations

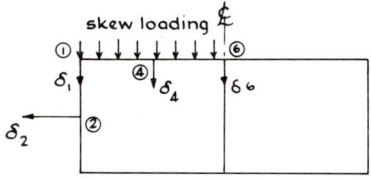

c) Positive directions of deflections compared for double-cell boxes

FIG. 11 – DETAILS OF TABULATED RESULTS

TABLE 1

Uniform single-cell steel box – thin flange – u.d. load

Method		Deflections (ins)				Longit. Stresses (lbs/sq. ins)		Trans. Moments (lbs. ins/inch)	
		δ_1	δ_2	δ_4	δ_5	σ_1	σ_3	M_1	M_3
Nodal Section	3 sections	6·34	20·6	190	−25·9	−112	112	14·8	−1·55
	5 sections	6·80	20·6	190	−26·4	−116	116	14·8	−1·55
	7 sections	6·97	20·6	190	−26·6	−118	118	14·8	−1·55
	9 sections	7·15	20·6	190	−26·7	−118	118	14·8	−1·55
Finite Element	72 elements	7·81	–	158	−16·0	−123	122	11·2	−1·18
	192 elements	8·13	20·1	173	−21·6	−126	125	14·1	−1·47
Experimental		–	18·5	179	−26·0	–	–	–	–
Multiplier		10^{-3}				10		1	

Span/width = 4·8   Width/depth = 2·5   Width/thickness = 188

TABLE 2

Uniform single-cell steel box – thick flange – u.d. load

Method		Deflections (ins)					Longit. Stresses (lbs/sq. ins)		Trans. Moments (lbs. ins/inch)	
		δ_1	δ_2	δ_4	δ_5		σ_1	σ_3	M_1	M_3
Nodal Section	3 sections	4·65	8·75	51·5	−9·41		−61·5	105	6·12	−0·647
	5 sections	4·99	8·75	51·7	−9·12		−63·7	109	6·12	−0·647
	7 sections	5·12	8·75	51·9	−9·01		−64·5	111	6·12	−0·647
	9 sections	5·17	8·75	51·9	−8·90		−64·9	111	6·12	−0·647
Finite Element	192 Elements	6·37	8·29	47·3	−5·93		−71·3	120	5·81	−0·612
Experimental		–	7·44	54·3	−9·90		–	–	–	–
Multiplier				10^{-3}				10	1	

Span/width = 4·8   Width/depth = 2·5   Width/thickness = 94

the Finite Element value being −3·35%. The Finite Element method however significantly underestimates the upward deflection at the centre of the bottom flange (δ_5). This deflection is quite small because the bottom flange is unloaded and the upward movement depends on the transfer of transverse moments from the top flange. The theoretical values of moments and stresses obtained by the two methods are seen to be in close agreement.

The solution obtained by using only 72 finite elements is seen to be very inaccurate, particularly for the transverse moments and consequently for the centre plate deflections. The convergence of the Nodal Section Method on the other hand is rapid but this is partly due to the fact that, in this particular case, the model cross-section and the applied loading does not vary along the span. Also, in this case, an initial plate flexure solution was not required in the Nodal Section Method because the loading was uniformly distributed over the complete flange area and the transverse symmetry of the structure meant that a correction for relative joint displacements was not necessary.

In the second model tested, the top flange thickness was increased to 0·16 in, all other model dimensions remaining the same. The results for a uniformly distributed load test on this model are given in Table 2 together with the results of 4 Nodal Section solutions and one fine-mesh Finite Element solution. The model behaviour is now seen to be considerably different from that observed in the first test, the ratio of the deflection at the centre of the top flange to the ridge deflection (δ_4/δ_1) as predicted theoretically being only just over one-third of the ratio for the first model. However, the good agreement between the Nodal Section, Finite Element and experimental values is maintained and the rapid convergence of the Nodal Section solution is again evident.

Two further tests were carried out on this second model, different loading conditions being considered as shown in Fig. 10b. A uniformly distributed load was first applied over the complete width of the model and over half the longitudinal span thus giving a longitudinally unsymmetrical loading. Secondly, a uniformly distributed load was positioned to cover half the width of the model over the complete longitudinal span thus giving a transversely unsymmetrical loading.

The experimental and theoretical results for both these loading cases are given in Table 3. In the first case a preliminary plate flexure solution was carried out in the Nodal Section analysis since the load was not applied over the complete longitudinal span. Also in the case of the transversely unsymmetrical loading large distortions of the cross-section were set up and a correction for relative joint displacements was carried out. The Finite Element mesh used in the analysis of the longitudinally unsymmetrical loading case was not as refined as in the other cases considered, only 144 elements being taken instead of 192. This was done to save computer time because in this case advantage could not be taken of longitudinal symmetry and the complete structure had to be analysed.

The agreement between theoretical and experimental values is again reasonable although the relatively coarse mesh used in the Finite Element solution does lead to some discrepancies in the longitudinally unsymmetrical loading case. The values given by the Nodal Section Method are satisfactory and show good agreement with the Finite Element values obtained using the fine mesh in the transversely unsymmetrical loading case.

The double-cell steel box shown in Fig. 10c was tested next. A uniformly distributed load was first applied over the complete area of the top flange and secondly the load was applied to one cell only. Finite Element solutions have not been obtained for these models but values obtained by a Nodal Section analysis are compared to experimental values for both loading conditions in Table 4. Once again

TABLE 3
Uniform single-cell steel box – thick flange – unsymmetrical loading

Method		δ_1	δ_2	δ_4	δ_5	δ_6	δ_7	Loading
				Deflections (ins)				
Nodal Section	9 Sections	2·69	4·10	26·7	−4·05	2·69	4·10	u.d. load
Finite Element	144 Elements	3·17	4·15	23·7	−2·99	3·17	4·15	on ½ span.
Experimental		–	3·94	27·0	−4·72	–	3·94	
Nodal Section	9 Sections	2·29	3·52	24·6	−3·82	5·20	4·70	u.d. load
Finite Element	192 Elements	2·28	3·72	23·7	−2·96	4·08	4·58	on ½ width
Experimental		–	3·17	26·1	−4·11	–	4·68	
Multiplier				10^{-3}				

Span/width = 4·8   Width/depth = 2·5   Width/thickenss = 94

TABLE 4

Uniform double-cell steel box

Method		Deflections (ins)				Loading
		δ_1	δ_2	δ_4	δ_6	
Nodal Section	9 Sections	6·62	2·30	16·6	6·62	u.d. load on both cells
Experimental		6·76	2·23	15·4	7·15	
Nodal Section	9 Sections	6·81	3·09	18·3	3·31	u.d. load on one cell
Experimental		6·90	3·44	20·9	3·58	
Multiplier				10^{-3}		

Span/cell width = 9·6   Cell width/depth = 1·25   Cell width/thickness = 94

a correction for relative joint displacements was included in the Nodal Section solution for the skew loading and the agreement between theoretical and experimental values is seen to be reasonable for both loading cases.

Finally a test was carried out on a tapered perspex box having the dimensions shown in Fig. 10d. Loading was applied to this model by placing bags of lead shot on the top flange and end conditions similar to those for the steel models were imposed. Values from five Nodal Section solutions and a coarse mesh Finite Element solution are compared to experimental values in Table 5.

The cross-sectional dimensions of this perspex box differ significantly from those of the steel boxes considered. This is particularly true of the width/thickness

TABLE 5

Tapered single-cell perspex box — u.d. load

Method		Deflections (ins)		
		δ_1	δ_4	δ_5
Nodal Section	3 Sections	84·7	142	76·7
	5 Sections	91·2	149	83·2
	7 Sections	93·5	151	85·5
	9 Sections	94·6	152	86·0
Finite Element	72 Elements	102	133	95·7
Experimental		96·5	144	–
Multiplier		$+10^{-3}$		

Span/width = 6·0   Width/depth varies from 3·6 to 1·5   Width/thickness = 32

ratio which now has a value of 32 compared to values of 188 and 94 for the steel models. As a consequence the structural behaviour differs, far less distortion of the cross-section being observed so that the structure acts more like a beam. For example the centre point of the bottom flange now deflects downwards instead of upwards as for the steel boxes. However, fair agreement is again observed between the theoretical and experimental values and the Nodal Section solution converges quickly. Some discrepancies are observed in the Finite Element values but these are of course anticipated when such a coarse mesh is used.

5. CONCLUSIONS

This initial study has shown that the Nodal Section Method predicts the behaviour of the models tested to a satisfactory degree of accuracy and furthermore, the values given by the Nodal Section Method agree well with values given by Finite Element solutions employing fine meshes.

The range of models considered in the present report is very limited and the Nodal Section Method still has to be tested under rather more exacting and realistic conditions before it can be recommended for general application. Such an investigation is at present under way. The Method can be adapted to consider end conditions other than the simple supports considered in the present report and its extension to girders continuous over intermediate supports, along lines similar to those suggested by Beaufait[12] for folded plates, is also under consideration at present.

The advantage of the Nodal Section Method is that, by assuming a simplified structural behaviour, it greatly reduces the amount of computation required in the analysis. It is estimated that in the typical case of a single-cell box under completely unsymmetrical loading a complete Nodal Section analysis, including a correction for relative joint displacements and an initial and final plate flexure analysis to give localized effects under point loads, could be obtained in about 1/20th of the time required for a full Finite Element solution involving 192 elements. About 75% of the time for the complete Nodal Section solution would be taken up by the two plate flexure analyses. In addition it should be appreciated that the input data required for the Nodal Section Method is very much less than that involved in a Finite Element Study and the solution can be carried out using a relatively small computer. It must be appreciated that the Finite Element solution will provide much more information about the overall behaviour of the structure since displacement and stress values will be calculated at each nodal point. However, many of these values will be irrelevant during the design stage when critical regions only are being checked.

It is suggested, therefore, that the Nodal Section Method is suitable for use as a tool during a design process since it will enable many analyses of the structure to be carried out economically. Once the structure has been so designed, then, if considered necesary, a full Finite Element solution of the designed structure could be carried out to obtain a more detailed picture of the overall structural behaviour.

ACKNOWLEDGEMENTS

The Authors wish to thank the Department of the Environment and R. Travers Morgan & Partners, Consulting Engineers, for partly sponsoring this investigation. Thanks are also due to Dr. K. A. Johns and Messrs. F. R. Hoyes, R. A. Morgan and J. W. Waddell of University College, Cardiff for their assistance during the investigation.

REFERENCES

1. Rockey, K. C. and Evans, H. R. A finite element solution for folded plate structures. In Davies, R. M. *Space Structures*. Oxford, Blackwell Scientific Publications, 1967, pp. 165–188.
2. Evans, H. R. *The analysis of folded plate structures*. Ph.D. thesis, University of Wales, Swansea, 1967.
3. Goldberg, J. E. and Leve, H. L. Theory of prismatic folded plate structures. *Publs int. Ass. Bridge struct. Engng*, vol. 17, 1957, pp. 59–86.
4. De Fries-Skene, A. and Scordelis, A. C. Direct stiffness solution for folded plates. *J. struct, Div. Am. Soc. civ. Engrs*, vol. 90, ST4, August 1964. pp. 15–47.
5. Gaafar, I. Hipped plate analysis, considering joint displacements. *Trans. Am. Soc. civ. Engrs*, vol. 119, 1954, pp. 743–784.
6. Yitzhaki, D. *The design of prismatic and cylindrical shell roofs*. Haifa, Haifa Science Publishers, 1958.
7. Scordelis, A. C. A matrix formulation of the folded plate equations. *J. struct. Div. Am. Soc. civ. Engrs*, vol. 86, ST10, October 1960, pp. 1–22.
8. Scordelis, A. C. *Analysis of simply supported box girder bridges*. University of California, Berkeley, Structural Engineering and Structural Mechanics Report no. SESM 66–17, October 1966.
9. Scordelis, A. C. *Analysis of continuous box girder bridges*. University of California, Berkeley, Structural Engineering and Structural Mechanics Report no. SESM 67–25, November 1967.

10. Evans, H. R. and Rockey, K. C. A critical review of the methods of analysis for folded plate structures. To be published in the Proceedings of the Institution of Civil Engineers — June 1971.
11. Johnson, C. D. and Lee, T. Long nonprismatic folded plate structures. *J. struct. Div Am. Soc. civ. Engrs,* vol. 94, ST6, June 1968, pp. 1457–1484.
12. Beaufait, F. W. Analysis of continuous folded plate surface. *J. struct. Div. Am. Soc. civ. Engrs,* vol. 91, ST6, December 1965, pp. 117–140.

FINITE ELEMENT ANALYSIS OF CURVED BOX GIRDER BRIDGES

P. T. K. LIM, J. T. KILFORD and K. R. MOFFATT

Imperial College of Science and Technology

SYNOPSIS

Existing finite element solutions of box girder bridges employ the basic triangular or rectangular shell elements. These elements do not represent the in-plane bending action of the cell walls efficiently and therefore to obtain accurate results a fine mesh idealization must be used. In this paper an element has been developed which, since it has a biased, beam-like, in-plane displacement field, is suitable for the analysis of box girders. The element presented is of trapezoidal shape for application to right, skewed or curved box girder bridges having constant width and depth. Results of a model test on an actual curved box girder bridge are used to demonstrate the accuracy and efficiency which can be achieved with the proposed element.

Notation

A	area of element
a, b, h	dimensions of element in x, y and z directions
$[C]$	matrix relating $\{\delta\}$ to $\{\alpha\}$
C_x, C_y	axial rigidities of plate in x and y directions
C_{xy}	shear rigidity of plate
C_1	quantity coupling C_x and C_y
$[D]$	rigidity matrix
E	modulus of elasticity in tension and compression
$[k]$	element stiffness matrix
$[\bar{k}]$	generalized co-ordinate element stiffness matrix
M_z	moment about z axis (positive direction determined by right-hand screw rule)
N, n	number of longitudinal mesh divisions (Fig. 6)
$[Q]$	matrix relating $\{\epsilon\}$ to $\{\alpha\}$
$[S]$	element stress matrix
u, v, w	components of displacements in x, y and z directions
x, y, z	rectangular cartesian co-ordinates (right-handed system of co-ordinates)
α	coefficient associated with fictitious rotational stiffness
$\{\alpha\}$	vector of α_i coefficients
α_i	displacement function coefficient
γ_{xy}	shearing strain component in xy-plane
$\{\delta\}$	vector of nodal displacements
$\{\epsilon\}$	strain vector

ϵ_x, ϵ_y normal components of strain in x and y directions

θ_x, θ_y, θ_z components of rotations about x, y and z axes (positive direction determined by right-hand screw rule)

ν Poisson's ratio

ξ, η generalized cartesian co-ordinates

$\{\sigma\}$ vector of stress resultants

σ_x, σ_y normal components of stress in x and y directions

τ_{xy} shearing stress component

Subscripts

$i, j, k, \ldots$ nodal points of element

Superscripts

E extensional component

F flexural component

1. INTRODUCTION

Three-dimensional assemblages of finite elements capable of simulating both extensional and flexural actions have been employed in the analysis of arch dams,[1, 2, 3] folded plate structures,[4] infilled frames,[5] and right box girder bridges.[6, 7] The need for this type of analysis in the case of box girder bridges arises when the structure is of such complexity as to be outside the scope of more rapid methods of analysis such as those based on folded plate theory,[8] thin-walled beam theory,[9, 10] and the beam on elastic foundation analogy.[11] Arbitrary boundaries, discrete random supports, and variations in orthotropy of the cell walls are examples of features not readily amenable to analysis by these simpler methods.

However, the versatility of the finite element method is only achieved at the expense of considerable computer time and core storage. This is because it requires the solution of a very large system of equations even in the case of a single span, single cell box girder bridge. One method of overcoming this problem is to develop more efficient elements for representing the behaviour of box girder bridges, and such an element is described in this paper.

A further aim of the paper is to make generally available a set of model results of an actual curved box girder bridge. Results of the model test are used as a basis for comparison between solutions obtained using the proposed element and other existing elements.

2. SOME SHELL ELEMENTS

2.1 Basic Triangular Element

The lowest order element suitable for the analysis of curved box girder bridges is the basic triangular shell element, the stiffness matrix of which is formed from those of the basic triangular plane stress and plate bending elements. According to this formulation, the element has at each of its nodal points two in-plane degrees of freedom u, v and three out-of-plane degrees of freedom w, θ_x, θ_y but not the in-plane rotation θ_z.

Although a relatively coarse assemblage of such elements gives satisfactory results for arch dams,[1, 2, 3] this is not so for box girder bridges. For such structures a fine mesh division is required, since the imposed in-plane displacement field of the element cannot adequately simulate beam action in the cell walls, which is the predominate effect in the webs.

Also this element can only indicate the average value of each of the extensional and flexural components of stresses within its area. Thus each cell wall must be represented by at least two transverse mesh divisions in order that the stresses at the web-flange junction can be estimated. Even with the use of a fine mesh division, it would be necessary to use an extrapolation procedure in order to obtain the stresses at the web-flange junction, or the local stresses near a support or under a wheel load.

Apart from these limitations which necessitate the use of a fine mesh, the basic triangular shell element has the following theoretical disadvantages:

(i) At the junction of non-coplanar elements the in-plane displacement of one plate, which is assumed to vary linearly between adjacent nodes, will not be compatible with the out-of-plane displacement of the adjoining plate, for which a cubic variation between nodes is assumed. This lack of compatibility, although not very significant in respect of the overall behaviour of the box, can cause considerable local fluctuations of the nodal slopes along the web-flange junction.[12]

(ii) Since θ_z is not considered in the formulation of the stiffness matrix, any cross-coupling between the in-plane rotational stiffness and the out-of-plane bending stiffness of adjoining non-coplanar elements is neglected. Especially in the case of thin plates, the former stiffness could be much larger than the latter, and therefore might have a significant effect on the flexural stresses which are developed.

(iii) Any node within an assemblage of coplanar elements would not have an in-plane rotational stiffness associated with it. As it is convenient in the matrix formulation to retain all six degrees of freedom at every node, a singularity would appear in the overall stiffness matrix of the structure at such a node. This singularity would have to be eliminated prior to the solution of the equations.

The disadvantage mentioned in (iii) can be overcome by adding an arbitrary quantity to the diagonal term of the stiffness matrix corresponding to the singularity.[13] A simpler alternative method suggested in Ref. 3 assumes that an element of elastic modulus E, area A and thickness h has a fictitious rotational stiffness of the form

$$\begin{Bmatrix} M_{z_i} \\ M_{z_j} \\ M_{z_k} \end{Bmatrix} = \alpha EAh \begin{bmatrix} 1 & -0.5 & -0.5 \\ -0.5 & 1 & -0.5 \\ -0.5 & -0.5 & 1 \end{bmatrix} \begin{Bmatrix} \theta_{x_i} \\ \theta_{x_j} \\ \theta_{x_k} \end{Bmatrix}$$

where α is an undetermined coefficient. Suitable values of α for application to box girders are discussed in Section 4.3.

2.2   Some higher order elements

For right box girder bridges, an obvious choice of element is the basic rectangular shell element (Fig. 1a), the stiffness matrix of which is formed from those of the basic plane stress and plate bending elements. In the idealization of the box cross-section, fewer nodes would be required with this element than with the basic triangular element. This is because the displacement functions assumed for the rectangular element are of a form that permits the flexural components of strain

$\epsilon_x^F, \epsilon_y^F$ to vary linearly in both the x and y directions of the element, and the

extensional components of strain $\epsilon_x^E, \epsilon_y^E$ to vary linearly in the y and x directions

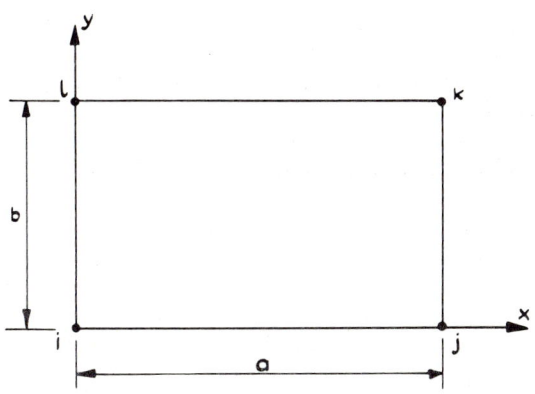

(a) Rectangular element

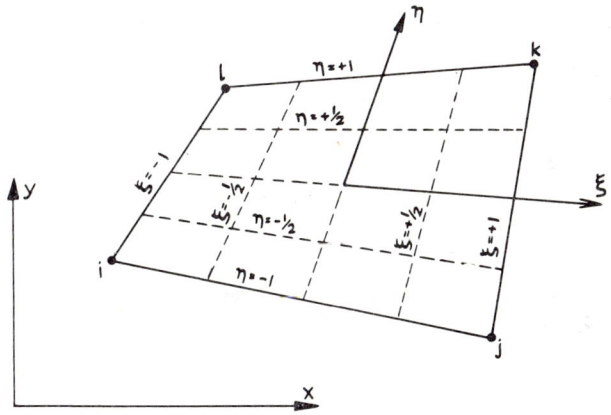

(b) Quadrilateral element and generalized co-ordinate system

FIG. I. RECTANGULAR AND QUADRILATERAL ELEMENTS.

respectively. However, it would still be necessary to find the peak values of the longitudinal extensional strain in the box by an extrapolation procedure since the strain is constant over the length of the element. As in the case of the basic triangular element, this rectangular element would not be compatible along the web-flange junction and does not have a rotational stiffness. However, the improved stress field within the element provides a better solution than that given by the basic triangular element for right box bridges.

To enable curved bridges to be analysed, displacement functions of the same form as those used for the rectangular shell element, but expressed in terms of the generalized co-ordinates ξ and η illustrated in Fig. 1b, could be applied to a general quadrilateral. This extension can be achieved using numerical integration techniques although Allwood[14] has presented an explicit form of the stiffness matrix for the quadrilateral plane stress element. In Ref. 15, numerical integration techniques are used to obtain the stiffness matrix of the quadrilateral plate bending element.

In a recent paper, MacLeod[16] suggested a rectangular plane stress element, which has some in-plane rotational stiffness, for application to shear wall structures. This element has a quadratic variation of displacements along its sides. A shell element, the stiffness matrix of which is formed from those of the MacLeod plane stress element and the basic plate bending element should provide a better solution to that given by the basic rectangular shell element for right box girder bridges. However, it will not be discussed further because the element described in Section 2.3 is more suitable for the present analysis.

The triangular plane stress element presented by Tocher and Hartz[17] overcomes the deficiencies associated with the previously mentioned plane stress elements but at the expense of four additional degrees of freedom $\partial u/\partial x$, $\partial u/\partial y$, $\partial v/\partial x$, $\partial v/\partial y$ per node. It should be possible to assemble the stiffness matrix of this element with that of the basic triangular plate bending element in a global system having, in general, twelve degrees of freedom at each node which would include all the first derivatives of u, v, w. However, this shell element would be limited by the deficiencies of the constituent plate bending element.

2.3   Proposed element

In this section, a new plane stress element is described for use in conjunction with the basic quadrilateral plate bending element.

Consider first the basic rectangular plane stress element (Fig. 1a) the displacement field of which is defined by

$$u = \alpha_1 + \alpha_2 x + \alpha_3 y + \alpha_4 xy \tag{i}$$

$$v = \alpha_5 + \alpha_6 x + \alpha_7 y + \alpha_8 xy \tag{ii}$$

The eight coefficients α_i appearing in the above displacement functions may be expressed in terms of the two displacement components at each of the four nodes of the element.

If, besides the displacement components u, v, an additional derivative of displacement $\partial v/\partial x$ is required to be continuous at the nodes, then this specification of four additional degrees of freedom for the element will permit the introduction of four more terms in the polynomial for v:

$$v = \alpha_5 + \alpha_6 x + \alpha_7 y + \alpha_8 x^2 + \alpha_9 xy + \alpha_{10} x^3 + \alpha_{11} x^2 y + \alpha_{12} x^3 y \tag{iii}$$

It should be noted that the cubic displacement component v along each of the sides *ij* and *kl* is uniquely defined by the nodal values of displacement and slope at each end. Also, the linear variation of v along the sides *il* and *jk* is retained in Equation (iii). Thus this choice of the four additional terms for the polynomial satisfies the condition of displacement continuity along the element interfaces.

An element stiffness matrix based on Equations (i) and (iii) has been derived, and details of the formulation are given in the Appendix. Clearly the biased displacement field imposed on the element limits its range of applicability, at least from an aesthetic viewpoint. However, this displacement field, when orientated in the appropriate direction, permits a close representation of beam action within the element as will be shown in Section 3. In the case of box girder bridges, this special feature can be used to advantage. Furthermore, use of the proposed plane stress element will ensure compatibility with a non-coplanar plate bending element, if they are connected along the sides *ij* of *kl*, and the difficulties caused by neglecting the in-plane rotational stiffness will also be avoided.* As in the case of the basic rectangular element, the longitudinal extensional strain is constant over the length of the element.

By expressing the displacement functions in terms of the generalized coordinates ξ and η, the compatibility condition for continuous displacements between adjacent coplanar or non-coplanar elements would also be satisfied for the case of a trapezium. This generalization of the rectangular element extends its range of applicability considerably, as skew and curved box girder bridges of constant width and depth could then be considered. The procedure for the generalization is the same as that indicated for the quadrilateral plate bending element in Ref.15.

3. APPLICATION TO STRAIGHT AND CURVED BEAMS

3.1 Convergence tests

In order that a comparison between the convergence characteristics of the basic plane stress elements and the proposed element may be made, a straight beam and a curved beam, both built-in at one end and loaded by a transverse force at the free end were considered. These problems were chosen since beam action is predominant in the cell walls of box girders and also since the exact solutions for them have been presented by Timoshenko.[18] The dimensions of the beams are shown in Figs 2a and b, the arc length of the axis of the curved beam being equal to the length of the straight beam.

The tests were carried out for a 1 by 8, 2 by 16, and 4 by 32 division of each beam into elements. It should be noted that the finite element idealization of the curved beam involves a further approximation compared with that of the straight beam in that the straight sides of the elements provide only an approximation to the smooth curve. However, this type of discrepancy diminishes with decreasing mesh size. In the analysis, the proposed element was orientated so that the two opposite edges along which the displacement varies as a cubic were in the longitudinal direction of the beam.

The boundary conditions at the built-in end were chosen to allow free warping, as assumed in the exact theory, and the continuously varying loads over the end sections were distributed statically to the nodal points of the idealized models.

The accuracy of each solution is measured by the percentage error in the deflexion of the axis of the beam, which is calculated as

$$\left(\frac{\text{finite element value} - \text{exact value}}{\text{exact value}}\right) \times 100$$

The variation of this percentage error with mesh size is given in Fig. 2c. The corresponding values obtained for each beam are in such close agreement that they plot on the same curve. It is apparent from the diagram that the proposed element is more effective in representing beam action than the basic elements, even when the comparison is made on the basis of the total number of degrees of freedom. The results also show the expected monotonic convergence of deflexion to a lower bound solution.

---

* In transforming the stiffness matrix of the proposed element from a local to a global coordinate system, prior to the assembly of the overall stiffness matrix of the structure, the derivative $\partial v/\partial x$ is treated as a 'true' in-plane rotation of the element.

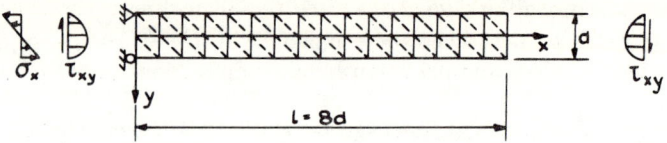

(a) Details of straight beam and subdivision into elements

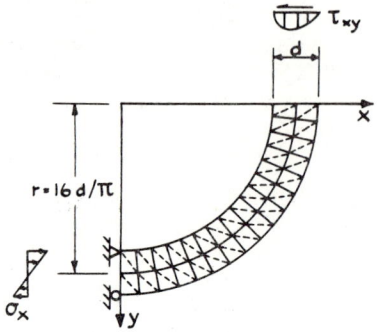

(b) Details of curved beam and subdivision into elements

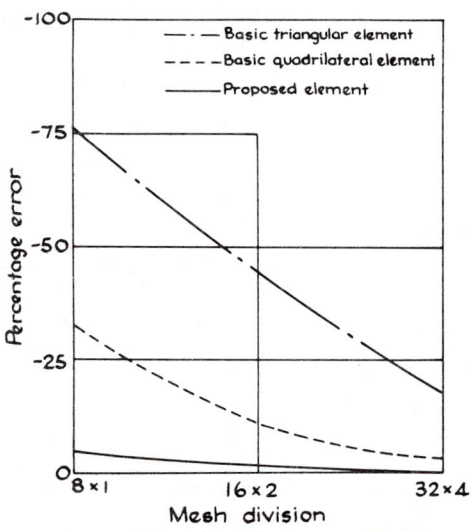

(c) Convergence curves for straight and curved beams

FIG. 2. EFFECT OF MESH SIZE ON DISPLACEMENTS FOR
STRAIGHT AND CURVED BEAMS ($\nu = 0.2$)

3.2   Pure bending of cantilever

The second example considered was the pure bending of a cantilever, for which
Poisson's ratio $\ddot{\nu} = 0$. This was chosen as a further illustration of the capability of
the proposed element for representing beam action. The use of only one element
gives values of deflexions and stresses that are identical to the exact solution for
any value of the depth–span ratio, d/1, of the beam. This result is to be expected
since the exact solution is contained in the polynomial expressions chosen to

describe the displacement field within the element. In the case of the basic rectangular element, a single element representation of a beam having d/1 = 8 gave an error of 99·5 per cent in the maximum deflexion and 33·3 per cent in stresses. Errors of 33·3 per cent were obtained for both deflexions and stresses when the number of elements along the length of the cantilever was increased to 8.

4. APPLICATION TO MODEL OF CURVED BOX GIRDER BRIDGE

4.1 Description of model

The curved box girder bridge which has been used as a basis for comparing the results of the various finite element idealizations is a 1/30th scale elastic model of part of the proposed prestressed concrete slip-road structures for the Stockton Road Interchange at Teesside. The model analysis was performed during the initial stages of design in order to investigate the deformations and the transverse

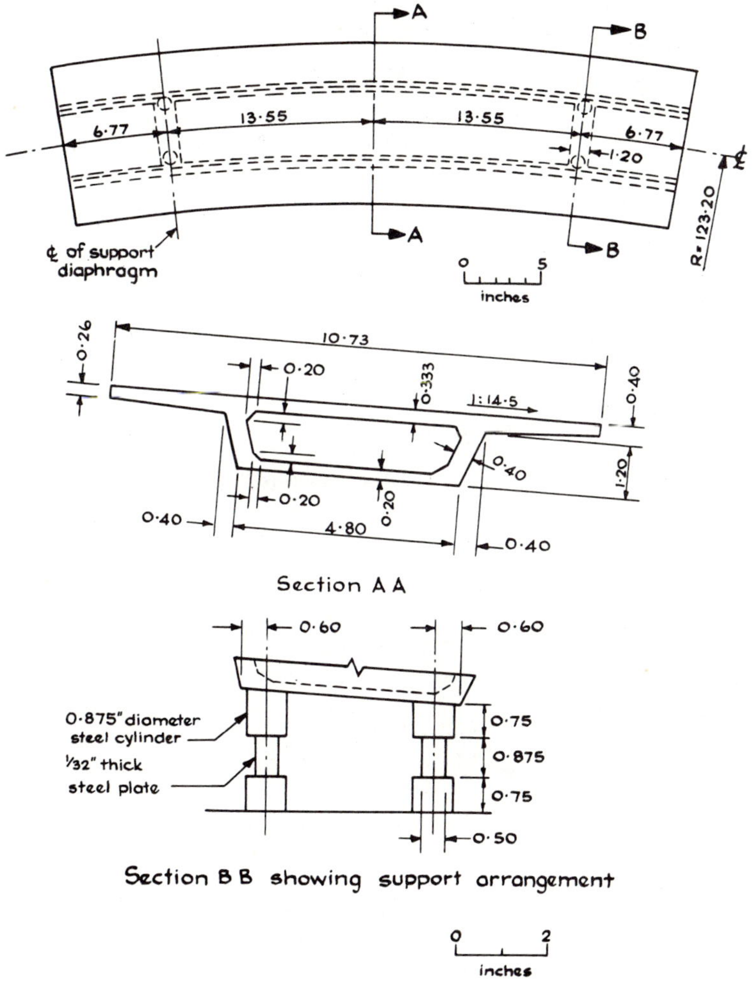

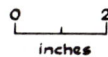

FIG. 3 STOCKTON ROAD INTERCHANGE: SLIP-ROAD STRUCTURE. DETAILS OF MODEL

distribution of stresses across a section of a curved box girder bridge and to check the applicability of the computer program for use in subsequent design modifications.

The model was of a single span, with quarter span cantilevers at each end, which thus enabled the effects of continuity over the supports to be investigated without the expense of constructing a multi-span structure. Rigid diaphragms were incorporated at the support sections. Details of the superstructure and the supports are given in Fig. 3. The model material was an araldite and sand mixture, the values of the modulus of elasticity and Poisson's ratio being 2.75×10^6 lbf/in^2 and 0.23.

Instrumentation consisted of dial gauges at the centre cross-section, together with a line of 5 mm electrical resistance strain gauges at the same section and at a section 1.31 in from the support centre line. Details of the locations of the gauges are given in Fig. 4.

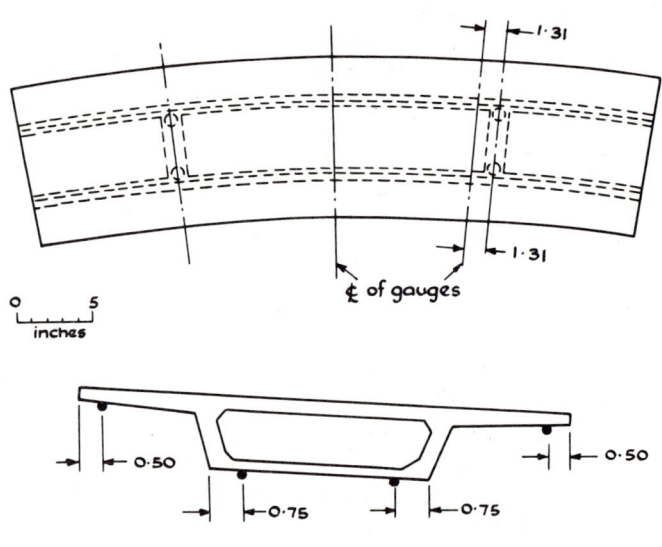

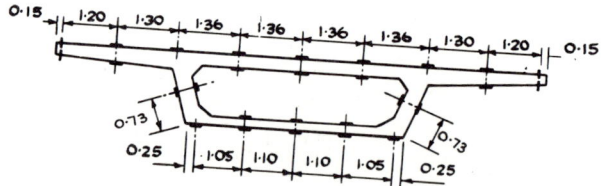

Position of deflexion gauges at centre line

Position of strain gauges at centre line
and near support line

- Deflexion gauge
- 45° strain rosette
I Linear strain gauge

FIG. 4 STOCKTON ROAD INTERCHANGE: SLIP-ROAD STRUCTURE
INSTRUMENTATION OF MODEL.

The model was subjected to both point loads and Type HB loading.[19] Figure 5 shows the application of a point load to the model. The Type HB loading was represented by an articulated framework which distributed the applied load equally on to 16 scaled rubber based steel pads representing the wheels of the vehicle.

FIG. 5 STOCKTON ROAD INTERCHANGE: SLIP-ROAD STRUCTURE.
MODEL SUBJECT TO CENTRAL POINT LOAD

4.2   Finite element idealizations

The idealization of the model (Fig. 6) involved the following approximations:

(i) The flanges, webs and diaphragms extend between mesh lines and hence overlap at the junction of the cell walls.

(ii) The haunches at the web-flange junctions were omitted.

(iii) The tapered side cantilevers were represented by constant thickness elements having the same cross-sectional area.

The effect of these approximations is to cause the second moment of area of the model cross-section to be over-estimated by only 1 per cent. A further approximation was introduced by supporting the box under the webs rather than at the actual positions shown in Fig. 3. The effect of this approximation on the analysis of the model is considered to be negligible.

In the diagrams and the discussion to follow, a mesh designation of the form $n \times p \times q$ is used where n, p, q denote the number of mesh divisions over the length, width and depth of the enclosed box respectively. Figure 6 shows the two cross-sectional arrangements of elements used in the analysis of the model, designated as $n \times 3 \times 1$ mesh and $n \times 3 \times 2$ mesh. Three values of n were considered, namely 6, 12 and 18, and the total number of mesh divisions in the longitudinal direction corresponding to these were 10, 18 and 24 respectively.

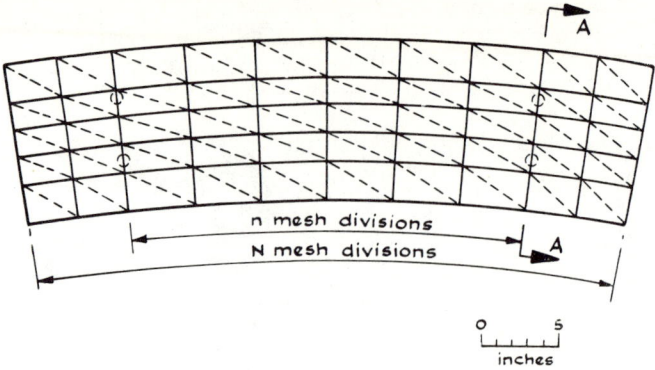

n mesh divisions

N mesh divisions

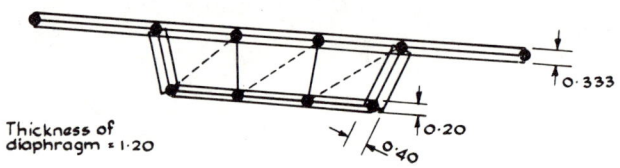

Thickness of
diaphragm = 1·20

Section AA showing section of n × 3 × 1 mesh

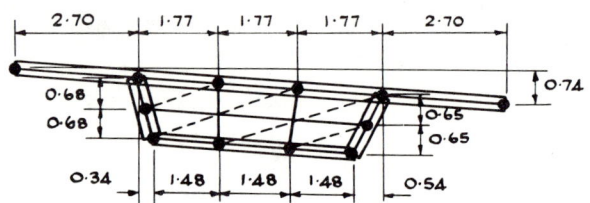

Section AA showing section of n × 3 × 2 mesh

FIG. 6   STOCKTON ROAD INTERCHANGE: SLIP-ROAD STRUCTURE
SUBDIVISION OF MODEL INTO TRIANGULAR OR
QUADRILATERAL ELEMENTS.

4.3   Effect of fictitious rotational stiffness

The technique of assigning a fictitious rotational stiffness to the basic triangular
element, referred to in Section 2.1, can clearly be extended to the case of the
basic quadrilateral element. In order to estimate the effect of variations in the
value of the coefficient α associated with this fictitious stiffness, on the behaviour
of box girders, results have been obtained for the model loaded at mid-span by
a point load over the outer web using a 6 × 3 × 1 mesh division into triangular and
quadrilateral elements.

Figure 7a shows the variation in deflexion at the point of application of the load
with α. It can be seen that for the range of α considered the deflexion varies by

almost 50 per cent for both elements and that for practical box girder bridges a suitable value of α would not be larger than 10^{-3}. This value may be compared with a value of $\alpha = 3 \times 10^{-2}$ which is recommended as being suitable for arch dams.[3] On the CDC 6600 computer, which has an accuracy of 15 significant decimal digits, the solution was unstable for values of α less than 3×10^{-7}.

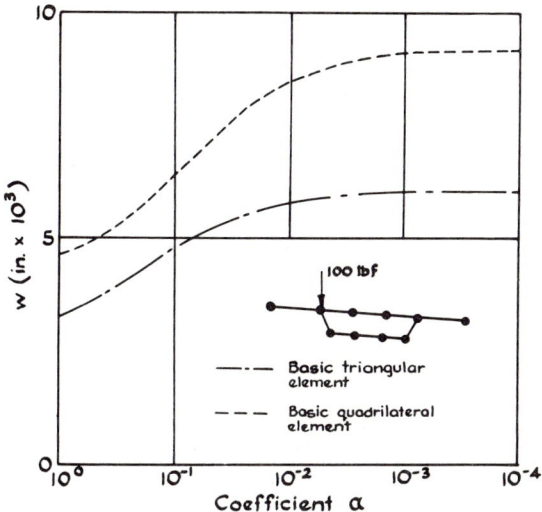

(a) Mid-span deflexion at top of outer web

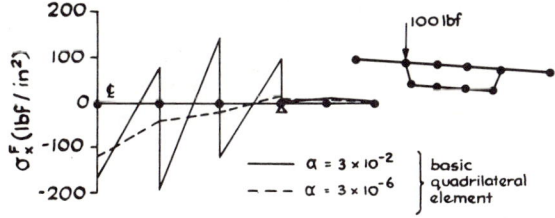

(b) Distribution of longitudinal flexural stresses along longitudinal section of top flange immediately to the right of outer web.

FIG. 7   STOCKTON ROAD INTERCHANGE: SLIP-ROAD STRUCTURE. EFFECT OF FICTITIOUS ROTATIONAL STIFFNESS ON DEFLEXIONS AND STRESSES FOR MODEL UNDER CENRAL POINT LOAD ($6 \times 3 \times 1$ MESH)

Figure 7b shows how the flexural component of longitudinal stress at the web-flange junction varies along the length of the model. The values plotted pertain to the line of elements in the top flange of the box adjacent to the outer web. For $\alpha = 3 \times 10^{-6}$ a smooth plot is obtained, whereas for $\alpha = 3 \times 10^{-2}$ considerable oscillations about the "correct" values are evident, due to the cross-coupling between the fictitious rotational stiffness of the web and the bending stiffness of the flange.

The importance of choosing suitable values of α having been illustrated, all subsequent solutions using the basic shell elements presented herein have been based on a value of $\alpha = 3 \times 10^{-6}$.

4.4   Effect of mesh size

The effect of mesh size on the accuracy of the results is studied with reference to the model for the two basic elements and the proposed element. The values of displacement directly under the point load are plotted in Fig. 8. It is significant that the results obtained using the different elements appear to converge to the same value. The results show that for the proposed element, satisfactory accuracy would be obtained using a 6 x 3 x 1 mesh division. However, for the basic

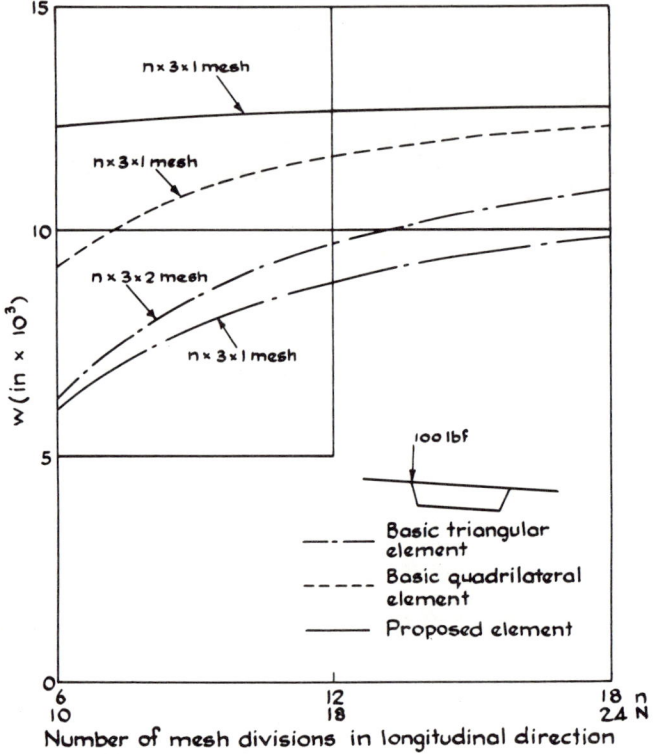

FIG. 8   STOCKTON ROAD INTERCHANGE: SLIP-ROAD STRUCTURE. EFFECT OF MESH SIZE ON DEFLEXION AT TOP OF OUTER WEB FOR MODEL UNDER CENTRAL POINT LOAD.

quadrilateral element, this order of accuracy could only be achieved using an 18 x 3 x 1 mesh division and the value obtained using an 18 x 3 x 2 mesh division into triangular elements is 12 per cent lower. Subsequent solutions presented herein are obtained using a 6 x 3 x 1 mesh division.

4.5   Comparison of displacements

The deformed centre cross-section of the model for the same load case is shown in Fig. 9. The results obtained using the three different elements show approximately the same distorted shape. However, only the results obtained using the proposed element compare well with the experimental values of vertical displacements.

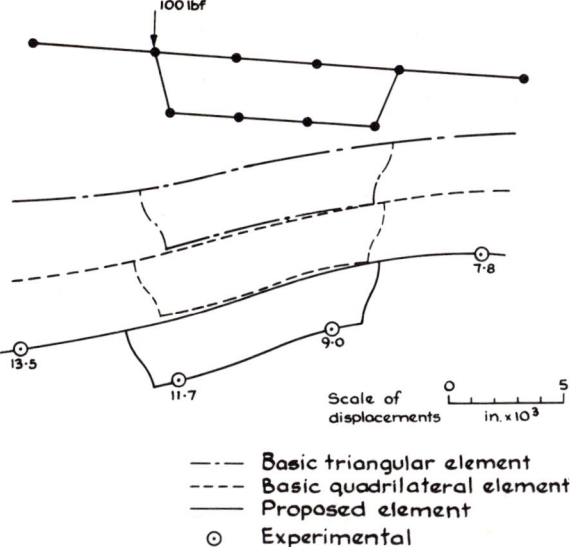

100 lbf

7·8

13·5

11·7

9·0

Scale of displacements
0   5
in. x 10³

—·— Basic triangular element
- - - - Basic quadrilateral element
——— Proposed element
⊙ Experimental

FIG. 9   STOCKTON ROAD INTERCHANGE: SLIP-ROAD STRUCTURE.
COMPARISON OF DEFLEXIONS AT MID-SPAN FOR
MODEL UNDER CENTRAL POINT LOAD (6 × 3 × 1 MESH)

4.6   Comparison of stresses

In this section, a comparison is made between the experimental and theoretical values of stresses in the model under Type HB loading. The two positions of the loading considered and the corresponding experimental results are shown in Figs 10 and 11.

The theoretical results were obtained using the basic quadrilateral element and the proposed element. For the first load case, the analysis was made using the previous 6 × 3 × 1 mesh which has a uniform division along the span. For the second load case, the mesh divisions were adjusted so that the mid-points of the first ring of elements within the span coincided with the gauge line. This was done in order that a direct comparison could be made between the experimental and theoretical results.

In the analysis, each wheel load was assumed to act at a point, and any point load acting within the area of an element was replaced by an equivalent system of forces at the nodes of the element. In order that the load may be distributed by simple statics, a quadrilateral element was divided into two triangles in two distinct ways, and half the load was distributed to the nodes of each of the triangles on which it acts.

In the transverse plots shown in Fig. 12 for the first load case, the values at each node pertaining to a cell wall or a side cantilever were obtained by averaging the appropriate nodal values of each constituent element meeting at that node. The stress values plotted in Fig. 13 for the second load case were obtained by applying a similar averaging process to the stresses at the mid-points of the longitudinal sides of the element. In comparing the theoretical and experimental values of stresses, it should be noted that the representation of the tapered side cantilevers by constant thickness elements will affect the flexural components of stress more than the extensional components.

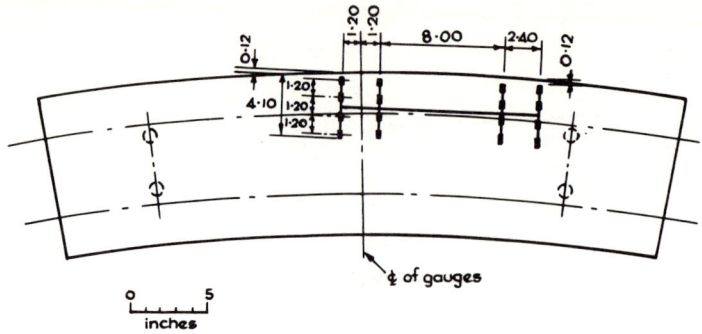

(a) Position of 100 lbf Type HB loading

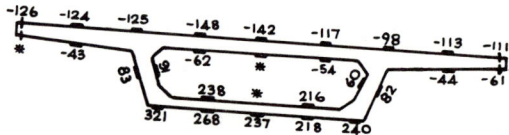

(b) Longitudinal stresses (lbf / in²)

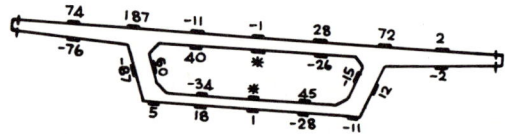

(c) Transverse stresses (lbf / in²)

❋ Bad reading

FIG. 10 STOCKTON ROAD INTERCHANGE: SLIP-ROAD STRUCTURE.
EXPERIMENTAL VALUES OF STRESSES AT CENTRE
SECTION FOR MODEL UNDER TYPE HB LOADING
within mid-span.

The results shown in Figs 12 and 13 can be summarized as follows:

(1) Figure 12a shows the almost constant value of the longitudinal extensional stress over the length of each element. This characteristic of the element should be kept in mind when interpreting the results.

(2) Figures 12b and c show good agreement between the results obtained using the proposed element and the experimental values of longitudinal stresses, except for those on the top surface of the outer cantilever. The results obtained using the basic quadrilateral element are about 30 per cent lower.

(3) Figure 12d shows that both elements give a good representation of the transverse flexural stresses.

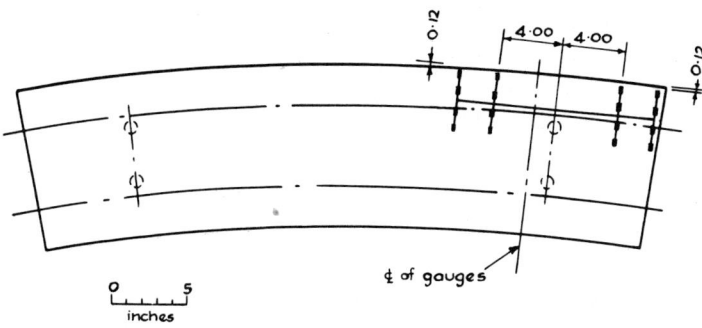

(a) Position of 100 lbf Type HB loading

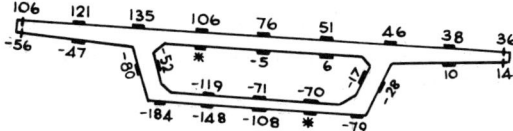

(b) Longitudinal stresses (lbf/in²)

(c) Transverse stresses (lbf/in²)

✳ Bad reading

FIG. 11 STOCKTON ROAD INTERCHANGE: SLIP-ROAD STRUCTURE
EXPERIMENTAL VALUES OF STRESSES NEAR SUPPORT
SECTION FOR MODEL UNDER TYPE HB LOADING
OVER SUPPORT

(4) Figure 13 shows that the proposed element also predicts the longitudinal
stresses near the supports with a good degree of accuracy.

A further assessment of the accuracy of the finite element solution may be
obtained by considering the moment equilibrium of, say, half the span of the
bridge. The results obtained using the basic quadrilateral element show a lack
of equilibrium of 36 per cent between the external and internal moments. The use
of the proposed element results in a lack of equilibrium of 7 per cent, which is
acceptable having regard to the coarseness of the mesh and the assumed constant
state of longitudinal strain along the length of the element.

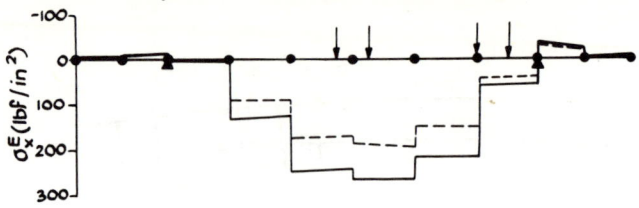

(a)   Distribution of longitudinal extensional stresses
      along bottom of outer web.

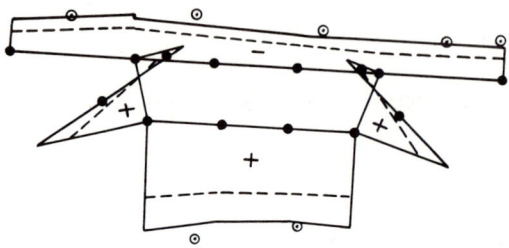

(b) Distribution of longitudinal extensional stresses
    at centre section.

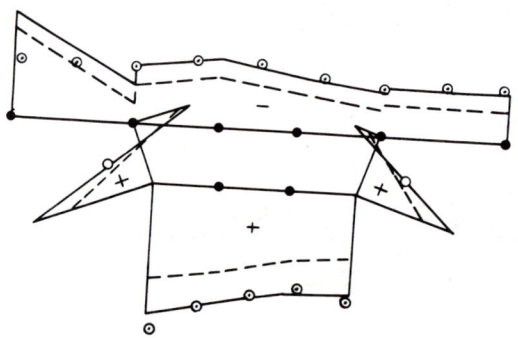

(c) Distribution of longitudinal outer surface stresses
     at centre section

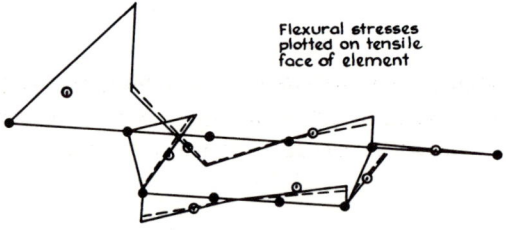

(d) Distribution of transverse flexural stresses
    at centre section

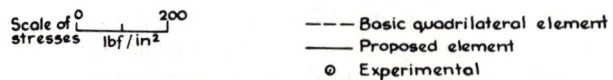

FIG. 12   STOCKTON ROAD INTERCHANGE: SLIP-ROAD STRUCTURE.
           COMPARISON OF STRESSES AT CENTRE SECTION FOR
           MODEL UNDER TYPE HB LOADING WITHIN SPAN
           (6 × 3 × 1 MESH)

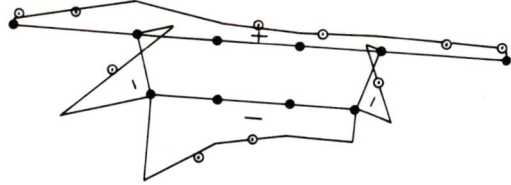

(a) Distribution of longitudinal extensional stresses
     near support section

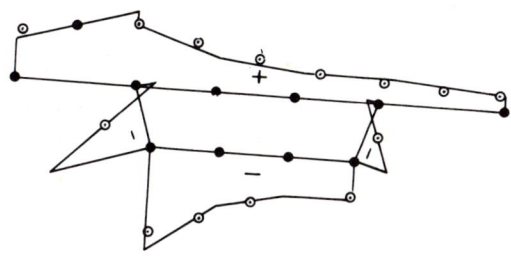

(b) Distribution of longitudinal outer surface stresses
     near support section

Scale of   0   200   ——— Proposed element

stresses   lbf /in²   ⊙ Experimental

FIG. 13   STOCKTON ROAD INTERCHANGE: SLIP-ROAD STRUCTURE.
             COMPARISON OF STRESSES NEAR SUPPORT SECTION
             FOR MODEL UNDER TYPE H8 LOADING OVER
             SUPPORT (6 × 3 × 1 MESH)

5.   CONCLUSIONS

(1) The plane stress element presented in this paper has a biased, beam-like
displacement field and, when used in conjunction with a plate bending element,
is well suited to the analysis of box girder bridges.

(2) The proposed element is of trapezoidal shape for application to right, skewed,
or curved box girder bridges having constant width and depth.

(3) Model test results of an actual curved box girder bridge are used as a basis
for comparison between solutions obtained using the proposed element and the
basic triangular and quadrilateral shell elements.

(4) With only one element over the depth and six elements along the span,
the results obtained using the proposed element compare well with the experimental
values of deflexion under a point load, whereas the results obtained using the
basic elements are up to 50 per cent lower.

(5) With the same mesh division, the results obtained using the basic quadrilateral
element show a lack of equilibrium of 36 per cent between the external and internal

moments at the centre cross-section for Type HB loading. This value decreases to 7 per cent when the proposed element is used.

(6) These improvements in the accuracy of the solution are gained with only a marginal increase in computer cost.

ACKNOWLEDGEMENTS

The experiments were conducted for the Department of the Environment in association with Dobbie, Sandford Fawcett & Partners, consulting engineers for the A19 Teesside Diversion. The computer program employed in the analysis was developed with financial assistance from Freeman Fox & Partners, Ove Arup & Partners, and The British Ship Research Association.

The work described herein constitutes part of a continuing research programme being carried out under the supervision of Dr. J. C. Chapman at Imperial College, London, into the finite element and model analysis of bridge structures. A special acknowledgement is due to Mrs. J. E. Slatford who assisted in the running of the computer program.

REFERENCES

1. Clough, R. W. and Tocher, J. L. Analysis of thin arch dams by the finite element method. *International Symposium on the Theory of Arch Dams (Southampton University, 1964)*; edited by Rydzewski, J. R. Oxford, Pergamon, 1965.
2. Zienkiewicz, O. C. and Cheung, Y. K. Finite element method of analysis for arch dam shells and comparison with finite difference procedures. *International Symposium on the Theory of Arch Dams (Southampton University, 1964)*; edited by Rydzewski, J. R. Oxford, Pergamon, 1965.
3. Zienkiewicz, O. C., Parekh, C. J. and King, I. P. Arch dam analysis by a linear finite element shell solution program. Proceedings of the Symposium on Arch Dams, Institution of Civil Engineers, London, 1968.
4. Rockey, K. C. and Evans, H. R. A finite element solution for folded plate structures. In Davies, R. M. *Space structures.* Oxford, Blackwell Scientific Publications, 1967, pp. 165–188.
5. Majid, K. I. and Williamson, M. Linear analysis of complete structures by computers. *Proc. Instn civ. Engrs,* vol. 38, Oct. 1967, pp. 247–266.
6. Sawko, F. Recent developments in the analysis of steel bridges using electronic computers. *Proceedings of the Conference on Steel Bridges (London, 1968).* London, British Constructional Steelwork Association, 1969.
7. Mehrotra, B. L., Mufti, A. A. and Redwood, R. G. Analysis of three dimensional thin-walled structures. *J. struct. Div. Am. Soc. civ. Engrs,* vol. 95, ST12, Dec. 1969, pp. 2863–2872.
8. Goldberg, J. E. and Leve, H. L. Theory of prismatic folded plate structures. *Publs int. Ass. Bridge struct. Engng,* vol. 17, 1957, pp. 59–86.
9. Resinger, F. *Der dünnwandige Kastenträger . . .* (Thin-walled box girders). Koln, Stahlbau-Verlag, 1959.
10. Richmond, B. Twisting of thin-walled box girders. *Proc. Instn civ. Engrs,* vol. 33, April 1966, pp. 659–675
11. Wright, R. N., Abdel-Samad, S. R. and Robinson, A. R. BEF analogy for analysis of box girders. *J. struct. Div. Am. Soc. civ. Engrs,* vol. 94, ST7, July 1968, pp. 1719–1743.
12. Sawko, F. and Cope, R. J. Discussion on 'Linear analysis of complete structures by computers'. *Proc. Instn civ. Engrs,* vol. 40, June 1968, pp. 205–209.
13. Eisemann, K., Woo, L. and Namyet, S. Space frame analysis by matrices and computer. *J. struct. Div. Am. Soc. civ. Engrs,* vol. 88, ST6, Dec. 1962, pp. 245–277.

APPENDIX

Stiffness matrix of proposed plane stress element

The formulation of the stiffness matrix of the proposed plane stress element will be described here for the limiting case of a rectangular element for which an explicit form of the stiffness matrix is possible. The generalization of this element to the case of a general quadrilateral follows exactly the same procedure as that described in Ref. 15 for a quadrilateral plate bending element and will not be reiterated here.

For ease of reference the polynomial expressions for the two components of displacements and the additional derivative of one displacement component which are prescribed to be continuous throughout the system of elements are given below:

$$u = \alpha_1 + \alpha_2 x + \alpha_3 y + \alpha_4 xy$$

$$v = \alpha_5 + \alpha_6 x + \alpha_7 y + \alpha_8 x^2 + \alpha_9 xy + \alpha_{10} x^3 + \alpha_{11} x^2 y + \alpha_{12} x^3 y \quad \text{(i)}$$

$$\frac{\partial v}{\partial x} = \alpha_6 + 2\alpha_8 x + \alpha_9 y + 3\alpha_{10} x^2 + 2\alpha_{11} xy + 3\alpha_{12} x^2 y$$

The displacement components have been expressed in terms of the coefficients $\{\alpha\}$ for the element shown in Fig. 1a.

Having specified these displacement functions for the element, we may proceed with the derivation of the element stiffness matrix $[k]$ in the standard manner:[20]

$$[k] = ([C]^{-1})^t \left(\iint [Q]^t [D][Q] \, dxdy \right) [C]^{-1} \quad \text{(ii)}$$

The matrices appearing in Eq. (ii) are defined below:

(1) The rigidity matrix $[D]$ relates the stress resultants $\{\sigma\}$ to the strain components $\{\epsilon\}$:

$$\{\sigma\} = [D] \{\epsilon\} \quad \text{(iii)}$$

For an orthotropic plate, Equation (iii) takes the form

$$\begin{Bmatrix} \sigma_x h \\ \sigma_y h \\ \tau_{xy} h \end{Bmatrix} = \begin{bmatrix} C_x & C_1 & 0 \\ C_1 & C_y & 0 \\ 0 & 0 & C_{xy} \end{bmatrix} \begin{Bmatrix} \epsilon_x \\ \epsilon_y \\ \gamma_{xy} \end{Bmatrix} \quad \text{(iv)}$$

For an isotropic material the rigidity matrix $[D]$ reduces to the form

$$\frac{Eh}{1-v^2} \begin{bmatrix} 1 & v & 0 \\ v & 1 & 0 \\ 0 & 0 & \frac{1-v}{2} \end{bmatrix}$$

(2) The matrix $[Q]$ relates the strain components $\{\epsilon\}$ to the coefficients $\{\alpha\}$:

$$\{\epsilon\} = [Q] \{\alpha\} \quad \text{(v)}$$

By virtue of the prescribed displacement functions, Equation (v) takes the form

$$
\left\{
\begin{array}{c}
\dfrac{\partial u}{\partial x} \\[2mm]
\dfrac{\partial v}{\partial y} \\[2mm]
\dfrac{\partial u}{\partial y} + \dfrac{\partial v}{\partial x}
\end{array}
\right\}
=
\begin{bmatrix}
0 & 1 & 0 & y & 0 & 0 & 0 & 0 & 0 & 0 & 0 & 0 \\
0 & 0 & 0 & 0 & 0 & 0 & 1 & 0 & x & 0 & x^2 & x^3 \\
0 & 0 & 1 & x & 0 & 1 & 0 & 2x & y & 3x^2 & 2xy & 3x^2y
\end{bmatrix}
\left\{
\begin{array}{c}
\alpha_1 \\
\vdots \\
\alpha_{12}
\end{array}
\right\}
\tag{vi}
$$

(3) The matrix $[C]$ relates the nodal displacements $\{\delta\}$ of the element to the coefficients $\{\alpha\}$:

$$\{\delta\} = [C]\,\{\alpha\} \tag{vii}$$

which may be inverted and written in the form

$$
\left\{
\begin{array}{c}
\alpha_1 \\ \alpha_2 \\ \alpha_3 \\ \alpha_4 \\ \alpha_5 \\ \alpha_6 \\ \alpha_7 \\ \alpha_8 \\ \alpha_9 \\ \alpha_{10} \\ \alpha_{11} \\ \alpha_{12}
\end{array}
\right\}
=
\begin{bmatrix}
1 & 0 & 0 & 0 & 0 & 0 & 0 & 0 & 0 & 0 & 0 & 0 \\[1mm]
\dfrac{-1}{a} & 0 & 0 & \dfrac{1}{a} & 0 & 0 & 0 & 0 & 0 & 0 & 0 & 0 \\[2mm]
\dfrac{-1}{b} & 0 & 0 & 0 & 0 & 0 & 0 & 0 & 0 & \dfrac{1}{b} & 0 & 0 \\[2mm]
\dfrac{1}{ab} & 0 & 0 & \dfrac{-1}{ab} & 0 & 0 & \dfrac{1}{ab} & 0 & 0 & \dfrac{-1}{ab} & 0 & 0 \\[2mm]
0 & 1 & 0 & 0 & 0 & 0 & 0 & 0 & 0 & 0 & 0 & 0 \\[1mm]
0 & 0 & 1 & 0 & 0 & 0 & 0 & 0 & 0 & 0 & 0 & 0 \\[1mm]
0 & \dfrac{-1}{b} & 0 & 0 & 0 & 0 & 0 & 0 & 0 & 0 & \dfrac{1}{b} & 0 \\[2mm]
0 & \dfrac{-3}{a^2} & \dfrac{-2}{a} & 0 & \dfrac{3}{a^2} & \dfrac{-1}{a} & 0 & 0 & 0 & 0 & 0 & 0 \\[2mm]
0 & 0 & \dfrac{-1}{b} & 0 & 0 & 0 & 0 & 0 & 0 & 0 & 0 & \dfrac{1}{b} \\[2mm]
0 & \dfrac{2}{a^3} & \dfrac{1}{a^2} & 0 & \dfrac{-2}{a^3} & \dfrac{1}{a^2} & 0 & 0 & 0 & 0 & 0 & 0 \\[2mm]
0 & \dfrac{3}{a^2b} & \dfrac{2}{ab} & 0 & \dfrac{-3}{a^2b} & \dfrac{1}{ab} & 0 & \dfrac{3}{a^2b} & \dfrac{-1}{ab} & 0 & \dfrac{-3}{a^2b} & \dfrac{-2}{ab} \\[2mm]
0 & \dfrac{-2}{a^3b} & \dfrac{-1}{a^2b} & 0 & \dfrac{2}{a^3b} & \dfrac{-1}{a^2b} & 0 & \dfrac{-2}{a^3b} & \dfrac{1}{a^2b} & 0 & \dfrac{2}{a^3b} & \dfrac{1}{a^2b}
\end{bmatrix}
\left\{
\begin{array}{c}
u_i \\[1mm] v_i \\[1mm] \left(\dfrac{\partial v}{\partial x}\right)_i \\[2mm] u_j \\[1mm] v_j \\[1mm] \left(\dfrac{\partial v}{\partial x}\right)_j \\[2mm] u_k \\[1mm] v_k \\[1mm] \left(\dfrac{\partial v}{\partial x}\right)_k \\[2mm] u_l \\[1mm] v_l \\[1mm] \left(\dfrac{\partial v}{\partial x}\right)_l
\end{array}
\right\}
\tag{viii}
$$

The non-zero terms of the symmetric matrix

$$[\bar{k}] = \iint [Q]^t[D][Q]\,dx\,dy \tag{ix}$$

are as follows:

$$\bar{k}_{2,2} = abC_x$$

$$\bar{k}_{4,2} = \tfrac{1}{2}ab^2C_x$$

$$\bar{k}_{7,2} = \dot{a}bC_1$$

$$\bar{k}_{9,2} = \tfrac{1}{2}a^2bC_1$$

$$\bar{k}_{11,2} = \tfrac{1}{3}a^3bC_1$$

$$\bar{k}_{12,2} = \tfrac{1}{4}a^4bC_1$$

$$\bar{k}_{3,3} = abC_{xy}$$

$$\bar{k}_{4,3} = \tfrac{1}{2}a^2bC_{xy}$$

$$\bar{k}_{6,3} = abC_{xy}$$

$$\bar{k}_{8,3} = a^2bC_{xy}$$

$$\bar{k}_{9,3} = \tfrac{1}{2}ab^2C_{xy}$$

$$\bar{k}_{10,3} = a^3bC_{xy}$$

$$\bar{k}_{11,3} = \tfrac{1}{2}a^2b^2C_{xy}$$

$$\bar{k}_{12,3} = \tfrac{1}{2}a^3b^2C_{xy}$$

$$\bar{k}_{4,4} = \tfrac{1}{3}(ab^3C_x + a^3bC_{xy})$$

$$\bar{k}_{6,4} = \tfrac{1}{2}a^2bC_{xy}$$

$$\bar{k}_{7,4} = \tfrac{1}{2}ab^2C_1$$

$$\bar{k}_{8,4} = \tfrac{2}{3}a^3bC_{xy}$$

$$\bar{k}_{9,4} = \tfrac{1}{4}a^2b^2(C_1 + C_{xy})$$

$$\bar{k}_{10,4} = \tfrac{3}{4}a^4bC_{xy}$$

$$\bar{k}_{11,4} = \tfrac{1}{6}a^3b^2(C_1 + 2C_{xy})$$

$$\bar{k}_{12,4} = \tfrac{1}{8}a^4b^2(C_1 + 3C_{xy})$$

$$\bar{k}_{6,6} = abC_{xy}$$

$$\bar{k}_{8,6} = a^2bC_{xy}$$

$$\bar{k}_{9,6} = \tfrac{1}{2}ab^2C_{xy}$$

$$\bar{k}_{10,6} = a^3bC_{xy}$$

$$\bar{k}_{11,6} = \tfrac{1}{2}a^2b^2C_{xy}$$

$$\bar{k}_{12,6} = \tfrac{1}{2}a^3b^2C_{xy}$$

$$\bar{k}_{7,7} = abC_y$$

$$\bar{k}_{9,7} = \tfrac{1}{2}a^2bC_y$$

$$\bar{k}_{11,7} = \tfrac{1}{3}a^3bC_y \tag{x}$$

$$\bar{k}_{12,7} = \tfrac{1}{4}a^4bC_y$$

$$\bar{k}_{8,8} = \tfrac{4}{3}a^3bC_{xy}$$

$$\bar{k}_{9,8} = \tfrac{1}{2}a^2b^2C_{xy}$$

$$\bar{k}_{10,8} = \tfrac{3}{2}a^4bC_{xy}$$

$\bar{k}_{11,8} = \frac{2}{3}a^3 b^2 C_{xy}$

$\bar{k}_{12,8} = \frac{3}{4}a^4 b^2 C_{xy}$

$\bar{k}_{9,9} = \frac{1}{3}(a^3 bC_y + ab^3 C_{xy})$

$\bar{k}_{10,9} = \frac{1}{2}a^3 b^2 C_{xy}$

$\bar{k}_{11,9} = \frac{1}{4}a^4 bC_y + \frac{1}{3}a^2 b^3 C_{xy}$

$\bar{k}_{12,9} = \frac{1}{5}a^5 bC_y + \frac{1}{3}a^3 b^3 C_{xy}$

$\bar{k}_{10,10} = \frac{2}{5}a^5 bC_{xy}$

$\bar{k}_{11,10} = \frac{3}{4}a^4 b^2 C_{xy}$

$\bar{k}_{12,10} = \frac{9}{10}a^5 b^2 C_{xy}$

$\bar{k}_{11,11} = \frac{1}{5}a^5 bC_y + \frac{4}{9}a^3 b^3 C_{xy}$

$\bar{k}_{12,11} = \frac{1}{6}a^6 bC_y + \frac{1}{2}a^4 b^3 C_{xy}$

$\bar{k}_{12,12} = \frac{1}{7}a^7 bC_y + \frac{9}{15}a^5 b^3 C_{xy}$

Finally the element stress matrix $[S]$ can be obtained from the expression

$$[S] = \frac{1}{h}[D][Q][C]^{-1} \tag{xi}$$

BEHAVIOUR AND DESIGN OF CURVED GIRDER BRIDGES

CONRAD P. HEINS, Jr.

University of Maryland

SYNOPSIS

A comprehensive analytical and experimental research program is currently being conducted at the Civil Engineering Department, University of Maryland, U.S.A. Details of the experimental tests of (1) curved girder system, (2) plate models, (3) Box beam model are given. These tests have provided data which when correlated with the theoretical data gave good agreement. Thus, the application of the analytical techniques in developing design information has some credibility.

Preliminary bridge design information is given for evaluation of: (1) Internal Girder System Forces and Distortions; (2) Maximum Normal Stresses; (3) Impact Factors; (4) Composite I-Girder Properties.

INTRODUCTION

The location of highway bridge structures was often selected at the most convenient site, with the alignment of the highway system predetermined by this selection. This approach to planning highway systems has now been greatly modified due to the need for complex interchanges and high traffic densities. Thus, the bridge engineer is now required to design a structure to fit the highway alignment.

In order to meet the highway alignment the structures should be curved, which would therefore require analysis and design information for the bridge engineer.

In order to establish such information, a research program has been established at the University of Maryland under the sponsorship and advisement of the Maryland State Roads Commission and the Federal Highway Administration. This program contains the following research areas:

1. Static Load Analysis
    (*a*) I-Girder Slab Bridge System with constant elevation, radial supports; Fourier Series — Slope Deflection Technique
    (*b*) I-Girder Slab Bridge System with varying elevation, skew supports; Space Frame Matrix Technique.
    (*c*) Box Girder Slab Bridge System; Space Frame Matrix Technique
    (*d*) Single Curved Girder
    (*e*) Local Loading Effects on Curved Slabs
2. Ultimate Load Analysis
    (*a*) I-Girder Slab Bridge System
3. Dynamic Load Analysis

4. Behaviour of Composite Beams due to Torsion
   (*a*) Torsional Properties
   (*b*) Shear Connector Spacing
5. Experimental Static Loading Studies
   (*a*) Stiffened and Unstiffened Curved Steel Plate Models
   (*b*) Single Curved Steel I-Girder
   (*c*) Single and Multi-Span Curved Steel I-Girder System, bare steel frame, steel frame with noncomposite and composite deck
   (*d*) Plexiglass Curved Box Beam Model
   (*e*) Torsion Tests of Straight Composite WF Members
6. Design Information — I-Girder Bridges
   (*a*) Preliminary Determination of Maximum Forces and Distortions of Curved Girders as Related to Straight Girders
   (*b*) Preliminary Determination of Maximum Norma Stresses
   (*c*) Impact Factors
   (*d*) Composite I-Girder Properties.

STATIC LOAD ANALYSIS

The determination of the resulting stresses and deformations in any girder bridge system requires consideration of the interaction of the deck, girders and diaphragms. Compounding this problem is the curvature of the girders, which therefore requires inclusion of the torsional mode.

A technique was developed, 'Slope Deflection — Fourier Series',[2] which does consider interaction of the various elements and includes bimoment effects, i.e., warping torsion. This technique, however, is limited in that the bridge system must have radial supports and constant support elevation. The application of this technique[3,10,20] has, therefore, permitted determination of all internal girder forces and distortions throughout a bridge system.

Due to the restriction of the 'Slope Deflection Technique', the stiffness matrix technique was adopted to investigate the effects of (1) skew supports, (2) varying support elevations, (3) box girders. The behaviour of box girder members was also studied by applying the 'Slope Deflection Technique'.[16] The major disadvantage in applying the 'Stiffness Technique' is the approximation of the warping phenomena and the idealisation of the deck slab. This research work is currently being undertaken.

In addition to studying curved girder bridge systems, the behaviour of a single curved girder was investigated.[4,19] This study provided an insight on the magnitude of the stresses and distortions of a curved girder. Also a closed formed solution of the differential equation[4] was obtained.

The behaviour of a curved bridge system could be determined by idealising the structural members and slab as an equivalent orthotropic plate.[5,6,11] A solution of the orthotropic plate equation would then yield distortions and finally forces. The solution of the orthotropic plate equation,[6,11] utilising the finite difference technique, was applied to a model study of a stiffened steel plate.[5]

The use of the orthotropic plate equations was extended to include membrane effects. This would, therefore, allow examination of the local loading influences on curved deck structures.[14] The solution of the coupled differential equations was obtained by finite difference method. A comprehensive model study was conducted in order to obtain data for comparison with the theory.[22]

ULTIMATE LOAD ANALYSIS

At present the prediction of the ultimate load carrying capacity of curved I-girder slab bridge systems is being performed. A relationship between the ultimate strength of the curved system, single curved girder, and a straight girder will be developed. The results of this investigation will be available in September 1971.

DYNAMIC LOAD ANALYSIS

The dynamic response of a single span, simply supported, curved highway bridge, traversed by simulated highway loadings, was obtained by two techniques: (1) Series Solution, (2) Lumped Mass. The bridge was idealised as a single curved girder of equivalent rigidity and mass of the system. The vehicle was considered as a single axle and two axle system with sprung and unsprung masses.[8,12] The bridge response was obtained in the form of spectrum curves as a function of various parameters, of which the following had the most influence on the dynamic response; frequency ratios, weight ratios, speed parameters and radius of curvature. The study showed that, in general, the constant-force solution was adequate for estimating the dynamic response.

The constant-force solution was applied to develop impact factors for both the flexural and torsional response of a bridge.

BEHAVIOUR OF COMPOSITE BEAMS DUE TO TORSION

The analytical determination of forces and distortion in curved girder systems require a knowledge of the system bending and torsional properties. If the individual girders are non-composite, the torsion properties of typical girder sections can be determined.[23] However, most I-girder slab bridges are composite, thus the evaluation of these girder sections require modification of the general thin walled theory.[24] The modifications to the general equations were performed and applied in the determination of torsional properties of typical bridge members.[17] A statistical study of these girder properties has resulted in a series of equations relating warping normal stresses to bending stresses.

The application of the thin walled theory to composite girders assumes complete interaction of the slab and girders. This will, therefore, require a shear transfer device or shear connectors. The determination of the magnitude of the forces, induced by both torsion and bending, and the required spacing of the shear connectors is presently being studied.

EXPERIMENTAL STATIC LOADING STUDIES

Throughout the development of the various analytical techniques, which have been completed, a series of model tests were performed in order to verify or indicate modifications of the proposed equations.

There were five categories of test models, each designed to provide data for the respective theory under examination. The categories were:

1. Curved Steel Plate Models
2. Single Curved I-Girder Model
3. Curved I-Girder System Models
4. Curved Box Girder Model
5. Straight Composite Girder Models

The steel plate models were designed, fabricated, and tested to provide data to correlate with the orthotropic plate theory as applied to (*a*) First Order Theory and (*b*) Second Order Theory. The First Order Theory would be applicable to curved bridge systems analysed as an equivalent orthotropic plate.[5,6,11] The Second Order Theory would be useful in studying local loading influence on curved bridge decks.[22] The models were subjected to measured vertical concentrated loads at various positions and the resulting model deflections and strains were then measured. Comparisons between theoretical and test data were good.[11,22] The model descriptions are as follows:

Primary Bending Model — Top deck was 16 gauge steel with five 1½ in curved steel bubb tee sections attached at 3 in intervals. The inside radius was $R_i = 33 \cdot 5$ in with a curvature of $\theta_T = 120°$. The supports for the model were radial and simple and positioned at $\theta_T = 83°$ and $101 \cdot 5°$ for the two tests which were conducted.

Local Loading Model — Top deck was ¼ in steel and the inside radius was $R_i = 135$ in with a curvature of $\theta_T = 35°$ and width of 45 in. The radial supports were simple and the angular supports were (1) fixed, (2) free, (3) elastic. The plate was tested as an isotropic plate and an orthotropic plate by attaching interior angular ribs ½ in x 1 in @ 9 in spacings. The exterior angular elastic supports were provided by attaching a ½ in x 1 in rib to the isotropic plate and a ½ in x 2 in rib to the stiffened plate.

A single curved 7I 15.3 steel girder was examined under vertical and torsional loadings. The girder radius, $R_i = 50 \cdot 0$ ft, and curvature $\theta_T = 31°$, the girder ends were embedded into a concrete mass to provide fixed supports. The resulting experimental data was compared to theory[4,19] providing good results. A 3 in x 36 in curved reinforced concrete slab was attached to the same I-girder by means of channel shear connectors spaced at 12 in. The girder was subjected to varying loads and the results compared to theory.[7]

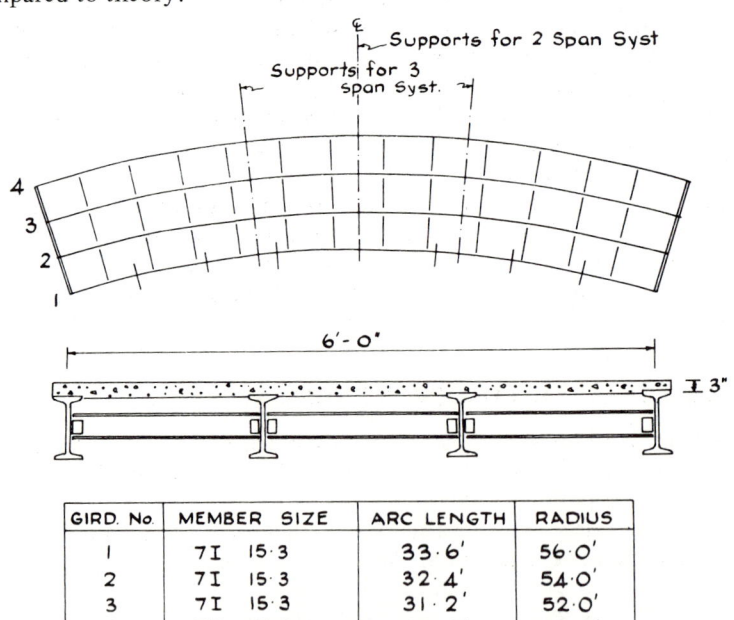

GIRD. No.	MEMBER SIZE	ARC LENGTH	RADIUS
1	7I 15·3	33·6'	56·0'
2	7I 15·3	32·4'	54·0'
3	7I 15·3	31·2'	52·0'
4	7I 15·3	30·0'	50·0'
DIAPHRAM	4I 7·7	—	—

FIG. I   CURVED I GIRDER MODEL

In order to substantiate the 'Slope Deflection Theory',[2,3,20] a series of curved I-girder bridge model systems were tested.[13] Three basic structural models were tested: (1) Steel Girder Model, (2) Steel Girder with Non-Composite Concrete Deck, (3) Steel Girder with Composite Concrete Deck, as shown in Fig. 1.

The steel girder model, which was also used for the other two structures consisted of 4 − 7I15.3 girders @ 2·0 ft in radial direction with R_i = 50·0 ft, L_i = 30·0 ft. Radial diaphragms 4I7.7 @ 2·3 ft angular spacing were employed. The model was tested in a single span (L = 50 ft), two span (L = 25 ft), and three span (L = 16·7 ft) configuration. The number of diaphragms used was also varied for each model and angular spacings consisted of (I) 13 @ 2·3 ft, (II) 6 @ 5·0 ft, (III) 4 @ 7·5 ft and (IV) 2 @ 15·0 ft.

The non-composite model used the same structural frame variations, except a 3 in reinforced concrete slab was placed on top of the girders.

The composite model also used the same structural frame and the various configurations, however, a 3 in reinforced concrete slab was placed on top of the girders and attached to the girders with 2 x ½ in round studs @ 10 in spacings.

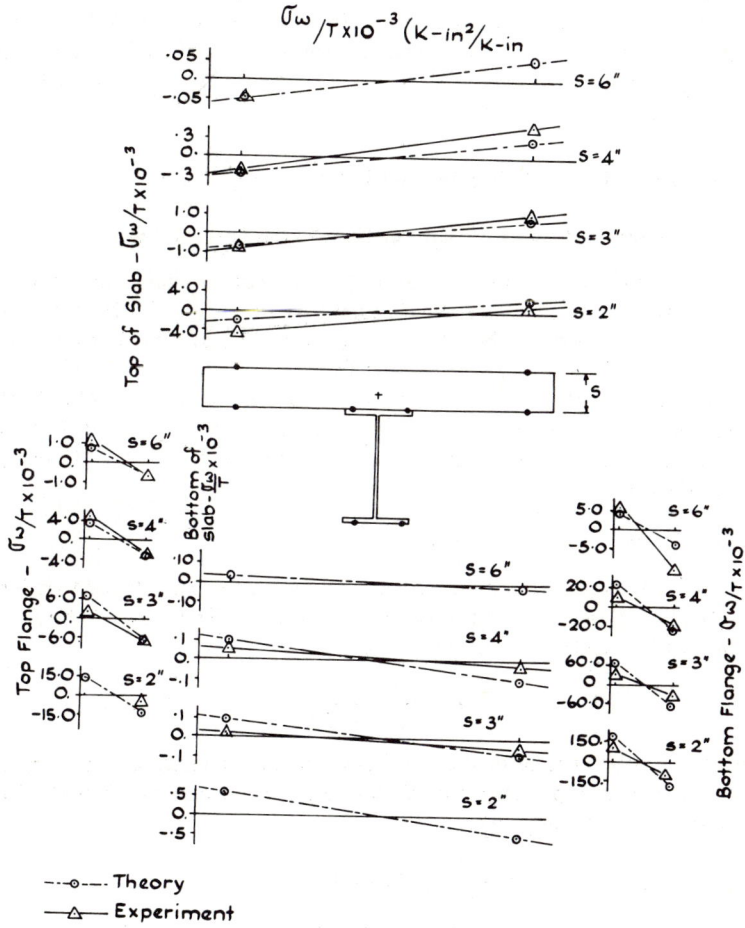

FIG. 2 THEORY VS EXPERIMENT COMPOSITE GIRDER TORSION TEST

The results of the testing of these model systems were compared to theory[13] resulting in excellent agreement. The results of the tested indicated negligible diaphragm action for the concrete-girder models. The maximum normal stresses relative to warping stresses were:

Steel Frame Model	$\sigma w = (20 \text{ to } 25\%)\sigma_{max}$
Non-Composite Model	$\sigma w = (10 \text{ to } 15\%)\sigma_{max}$
Composite Model	$\sigma w = (10 \text{ to } 12\%)\sigma_{max}$

The development of the 'Slope Deflection Technique' was considered most suitable for open cross-sections, however, the general theory was not limited to such sections and could incorporate closed cross-sectional members. In order to examine the theory, as applied to closed sections, a plexiglass curved box model was tested. The model had $R_i = 63$ in, $R_o = 81$ in and a curvature $\theta = 55°$ with three cells 3 in x 3 in x 1/8 in thick, spaced at 6 in on centre. The model was tested as a single and two span structure. The test data in comparison with theory show good correlation.[16] The results indicate that warping effects are negligible.

The use of thin walled theory[24] was substantiated[39] by testing a series of composite girders under torsional loadings. The specimens were 12 WF 27 beams with 2 in, 3 in, 4 in, and 6 in reinforced concrete slabs attached with stud shear connectors. The specimens were 19·0 ft long and were pinned or fixed at one support and free at the other support. Figure 2 describes the comparison between theory and analytical warping stresses for the various members.

DESIGN INFORMATION – I-GIRDER BRIDGES

1. Preliminary Determination of Maximum Forces and Distortions

In order to utilize various computer programs[10] which are available in analysing curved bridge systems, the cross-sectional properties are required. The determination of the required properties, therefore, requires an estimate of the maximum internal girder forces. The evaluation of such forces has been determined for typical bridge systems when subjected to specified AASHO truck loadings.[7] The result of this investigation has resulted in a series of equations[7,15] which permit determination of maximum girder forces and distortions. The following will list these equations and applications.

Amplification factor – K_1

All the internal forces and deformations for a single curved and straight girder have been evaluated, using the various computer programs.[4,7] The ratio of the reactions for these two girders gives the following:

$$K_1 = \frac{f(S.C.G.)}{f(S.S.G.)} \tag{1}$$

This factor describes the immediate effect of curvature relative to a straight member. The graphical representation of this data and its analysis gives the following general equations (2) through (8).

$$K_{moment} = \frac{0·15}{n}(L/R) + 1 \tag{2}$$

$$K_{St.Torsion} = (4000\,n)\,(L/R)^2 - 600\,(L/R) \tag{3}$$

$$K_{warping\,torsion} = (50\,n)\,(L/R)^2 - 3\,(L/R) \tag{4}$$

$$K_{bimoment} = [(35\,n)\,(L/R)^2 - 15\,(L/R)] \times 10^3 \tag{5}$$

$$K_{shear} = 1{\cdot}0 \tag{6}$$

$$K_{rotation} = \left[n^2\,(2L/R)^3 - (n-.4)\,(L/R)^2\right] \times 10^{-2} \tag{7}$$

$$K_{deflection} = S\,.\,e^{P}\,(L/R) \tag{8}$$

Where:

S and P are evaluated using Fig. 3
$n = R/100 \quad R \geqslant 100$ ft
R = Radius — ft
L = Span length — ft

Distribution factor — K_2

The evaluation of the true distribution of load to each girder, and thus realistic values of internal forces, can be considered by analysing the curved girders as a system. The number of trucks used in the analysis would be dependent on the number of girders as described previously. The ratio of these resulting maximum forces to those in a single curved girder gives:

$$K_2 = \frac{f(Sy.C.G.)}{f(S.C.G.)} \tag{9}$$

A plot of this ratio vs. R and L will yield the following general equations (10) through (30). In all instances the parameters R and L refer to the mid length and corresponding radius of the system. These equations have been evaluated for the four, six, and eight girder systems.

Four girder system

$$K_{moment} = (n + 2)\,\frac{0{\cdot}4L}{R} + 0{\cdot}48 \tag{10}$$

$$\left\{\begin{array}{l} K_{St.\,Torsion} = \dfrac{0{\cdot}085}{M}\,(R/L) \qquad\qquad R/L \geqslant 3{\cdot}3 \tag{11a} \\[3mm] K_{St.Torsion} = \dfrac{0{\cdot}64}{n+1} \qquad\qquad\qquad R/L \leqslant 3{\cdot}3 \tag{11b} \end{array}\right.$$

$$K_{warping\ torsion} = \frac{0{\cdot}35}{M^2}\,(R/L) + 0{\cdot}3 \tag{12}$$

$$\left\{\begin{array}{l} K_{bimoment} = \dfrac{0{\cdot}11}{M}\,(R/L) \qquad\qquad L \leqslant 70 \text{ ft} \tag{13a} \\[3mm] K_{bimoment} = \dfrac{(M-1)}{6}\,(R/L) \qquad\quad L \geqslant 70 \text{ ft} \tag{13b} \end{array}\right.$$

$$K_{shear} = (n + 2)\,\frac{0{\cdot}4L}{R} + 0{\cdot}55 \tag{14}$$

$$\left\{\begin{array}{l} K_{rotation} = \dfrac{0{\cdot}098}{M}\,(R/L) \qquad\qquad R/L \geqslant 3{\cdot}3 \tag{15a} \\[3mm] K_{rotation} = \dfrac{0{\cdot}74}{n+1} \qquad\qquad\qquad R/L \leqslant 3{\cdot}3 \tag{15b} \end{array}\right.$$

$$K_{deflection} = -Y\,Ln\,(L/R) + X \tag{16}$$

Six girder system

$$K_{\text{moment}} = (n + 3) \frac{0 \cdot 4L}{R} + 0 \cdot 6 \tag{17}$$

$$K_{\text{St.Torsion}} = \frac{0 \cdot 105}{M} (R/L) \qquad R/L \geqslant 3 \cdot 3 \tag{18a}$$

$$K_{\text{St.Torsion}} = \frac{0 \cdot 68}{n+1} \qquad R/L \leqslant 3 \cdot 3 \tag{18b}$$

$$K_{\text{warping torsion}} = \frac{0 \cdot 15}{M^2} (R/L) + 0 \cdot 1 \tag{19}$$

$$K_{\text{bimoment}} = \frac{0 \cdot 11}{M} (R/L) \qquad L \leqslant 70 \text{ ft} \tag{20a}$$

$$K_{\text{bimoment}} = (M - 1)/6 \ (R/L) \qquad L \geqslant 70 \text{ ft} \tag{20b}$$

$$K_{\text{shear}} = (n + 2 \cdot 5) \frac{0 \cdot 4L}{R} + 0 \cdot 65 \tag{21}$$

$$K_{\text{rotation}} = \frac{0 \cdot 11}{M} (R/L) \qquad R/L \geqslant 3 \cdot 4 \tag{22a}$$

$$K_{\text{rotation}} = \frac{0 \cdot 84}{n + 1} \qquad R/L \leqslant 3 \cdot 4 \tag{22b}$$

$$K_{\text{deflection}} = -Y \, Ln \, (L/R) + X \tag{23}$$

Eight girder system

$$K_{\text{moment}} = (\frac{n}{2} + 3) (\frac{0 \cdot 4L}{R}) + 0 \cdot 54 \tag{24}$$

$$K_{\text{St.Torsion}} = \frac{0 \cdot 115}{M} (R/L) \qquad R/L \geqslant 3 \cdot 3 \tag{25a}$$

$$K_{\text{St.Torsion}} = \frac{0 \cdot 72}{n+1} \qquad R/L \leqslant 3 \cdot 3 \tag{25b}$$

$$K_{\text{warping torsion}} = \frac{0 \cdot 2}{M^2} (R/L) + 0 \cdot 2 \tag{26}$$

$$K_{\text{bimoment}} = \frac{0 \cdot 11}{M} (R/L) \qquad L \leqslant 70 \text{ ft} \tag{27a}$$

$$K_{\text{bimoment}} = \frac{(M - 1)}{6} (R/L) \qquad L \geqslant 70 \text{ ft} \tag{27b}$$

$$K_{\text{shear}} = (n + 3) (0 \cdot 3 \, L/R) + 0 \cdot 6 \tag{28}$$

$$K_{\text{rotation}} = \frac{0 \cdot 116}{M} (R/L) \qquad R/L \geqslant 3 \cdot 2 \tag{29a}$$

$$K_{\text{rotation}} = \frac{0 \cdot 86}{n + 1} \qquad R/L \leqslant 3 \cdot 2 \tag{29b}$$

$$K_{\text{deflection}} = -Y \, Ln \, (L/R) + X \tag{30}$$

where Y and X are evaluated using Fig. 4.
$\quad\quad\quad\quad n = R/100 \quad R \geqslant 100'$
$\quad\quad\quad\quad M = L/50 \quad R \geqslant 50'$
$\quad\quad\quad\quad R$ = Radius — ft
$\quad\quad\quad\quad L$ = Span length — ft

Reduction factor K_3

Because many bridge structures are continuous, it is desirable to obtain some factors, which can be applied to the simple span data, to give preliminary forces in continuous spans. This factor can be written as:

$$K_3 = \frac{f(Sy.C.G.)N}{f(Sy.C.G.)} \tag{31}$$

N = Number of spans (2 or 3)

Utilising a computer program,[3] the maximum forces in a two or three span curved bridge system of four, six and eight girders, were evaluated under various critical loadings. A study of all the data and the resulting K_3 values give the following values

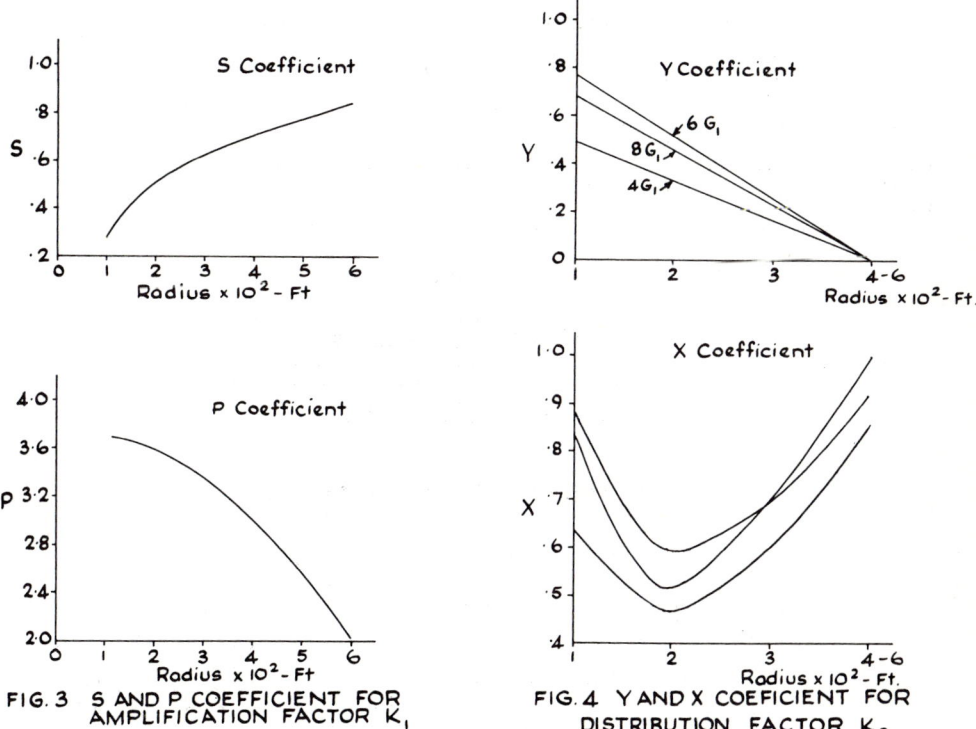

FIG. 3 S AND P COEFFICIENT FOR AMPLIFICATION FACTOR K_1

FIG. 4 Y AND X COEFICIENT FOR DISTRIBUTION FACTOR K_2

listed in Table 1. The data is described relative to number of spans and is independent of number of girders. It should be emphasized that the two and three span girder systems must all contain equal span lengths with a maximum given span length of 100 ft. For example, for a three span system the total maximum bridge length would then be 3 x L = 300·0 ft.

EVALUATION OF GIRDER FORCES AND DEFORMATION

With the various factor equations available, it is now possible to evaluate preliminary forces in a curved girder bridge, relative to the forces in a straight girder. The procedure is as follows:

1. Evaluate maximum function "*F*" for a single straight girder of length L and load *Peq*. (Fig. 3). These functions would be F bending, F shear, and F deflection. The remaining functions F rotation, F St. Venants, F warping and F bimoment are all assumed to be equal to one.
2. Evaluate Amplification Factors K_1 equations (2) through (8), for the mid-span length L and Radius R of the bridge system.
3. Evaluate Distribution Factors K_2, equations (10) through (30), for the given mid-span length L, number of girders in system and radius R.
4. Select Reduction Factor K_3, from Table 1, if system is a continuous span.
5. Determine maximum function F of curved girder system, i.e.

$$\text{Max. Moment} = M \text{ straight Beam} \times (K_1 \times K_2 \times K_3)$$

or in general

$$F \text{ curved Max.} = F \text{ straight Max.} \times (K_1 \times K_2 \times K_3) - \qquad (32)$$

TABLE 1

K_3 – Reduction Factor for Maximum Function in Two and Three Span Bridges

No. of Spans	$K_{Bending\ Moment}$	$K_{Deflection}$	$K_{Rotation}$	$K_{St.\ Venant}$	$K_{Warping\ Torsion}$	$K_{Bimoment}$	K_{Shear}
Two Span	·75	·70	·70	·75	·45	·35	1·00
Three Span	·65	·60	·70	·75	·40	·35	·90

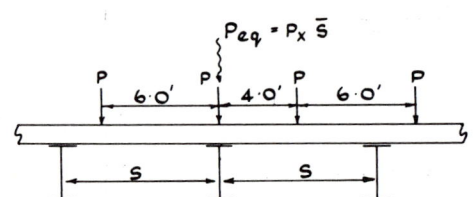

Transverse Loading Position

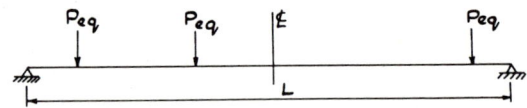

Longitudinal Loading Position

FIG. 5   LOADING   CONDITION

The original spacing of the girders was equal to 7·0 ft. If the girders are spaced other than 7·0 ft, a modification to the basic equations are necessary. Therefore, the following Spacing Factor, $\bar{S}$, has been evaluated, as shown in Fig. 5.

$$\bar{S} = 1·29 \text{ for } S = 7 \text{ ft or } 8 \text{ ft}$$
$$\bar{S} = 1·57 \text{ for } S = 9 \text{ ft or } 10 \text{ ft}$$

The spacing factor is then applied to the wheel load $P(4{\cdot}0$ kip or $16{\cdot}0$ kip) to evaluate Peq.

2. Preliminary Determination of Maximum Normal Stresses

A study of the torsional and bending properties of typical composite bridge members[17] provides a series of equations which can be used to estimate the maximum normal stress in a composite section. This stress would be induced by the girder bending moment M_b and warping moment Bi. The total stress would be of the form:

$$\sigma_T = \sigma_b + \sigma w \tag{33}$$

The warping and bending stresses are equal to:

$$\sigma_b = M_b/Z \text{ or}$$
$$\sigma_b/M_b = 1/Z \tag{34}$$

$$\sigma w = Bi\,\frac{Wn}{Iw} \text{ or}$$
$$\sigma w/Bi = Wn/Iw \tag{35}$$

The terms $1/Z$ and Wn/Iw are known quantities for various bridge members; a plot of these data will then give the relationship between (σ_b/M_b) and $(\sigma w/Bi)$. Performing a linear regression analysis, the following trends are indicated and are of the form:

$$\sigma w/Bi = K\sigma_b/M_b \tag{36}$$

WF Beam Section

$$\text{Steel Flange} \quad \left(\frac{\sigma w}{Bi}\right)_s = 1{\cdot}15\left(\frac{\sigma_b}{M_b}\right) \tag{37a}$$

$$\text{Concrete Slab} \left(\frac{\sigma w}{Bi}\right)_c = {\cdot}002\left(\frac{\sigma_b}{M_b}\right) \tag{37b}$$

Plate Girder Section

$$\text{Steel Flange} \quad \left(\frac{\sigma w}{Bi}\right)_s = {\cdot}652\left(\frac{\sigma_b}{M_b}\right) \tag{38a}$$

$$\text{Concrete Slab} \left(\frac{\sigma w}{Bi}\right)_c = {\cdot}062\left(\frac{\sigma_b}{M_b}\right) \tag{38b}$$

The subscripts s, c indicate the location of the stress, i.e., bottom of steel flange (s) and top of concrete slab (c).

The total normal stress can then be written as:

$$\sigma_T = \sigma_b\left(1 + K \cdot \frac{Bi}{M_b}\right) \tag{39}$$

The determination of the moments. Bi and M_b can be obtained by applying the approximate force equations.

3. Impact Factors

The solution of the dynamic response of typical curved bridge systems has been reported.[8,12] These solutions have permitted determination of impact factors relative to bending and torsional modes. The curves are a function of the radius of the bridge, central angle and span length.

Partial results from this study are shown in Figs. 6 through 9. Figures 6 and 7 represent impact factors for typical composite I-beam bridges. Figures 8 and 9 are

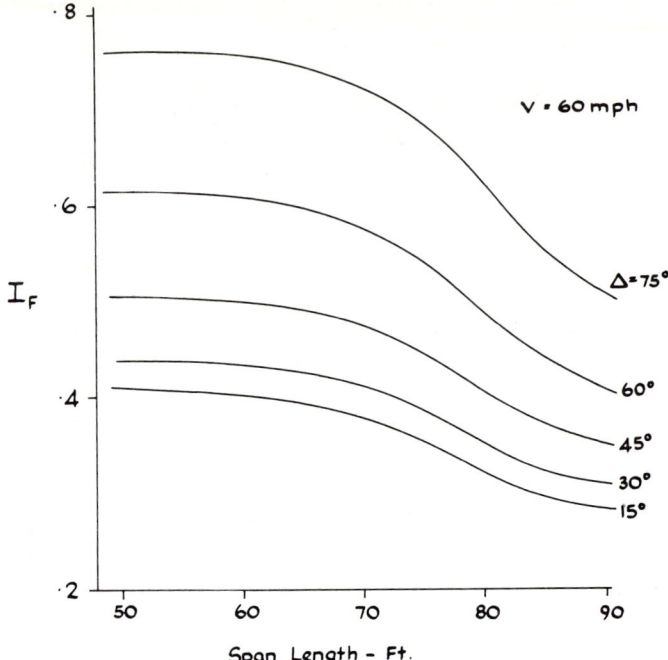

FIG 6 FLEXURAL IMPACT FACTOR — I Bm.
COMPOSITE BRIDGES.

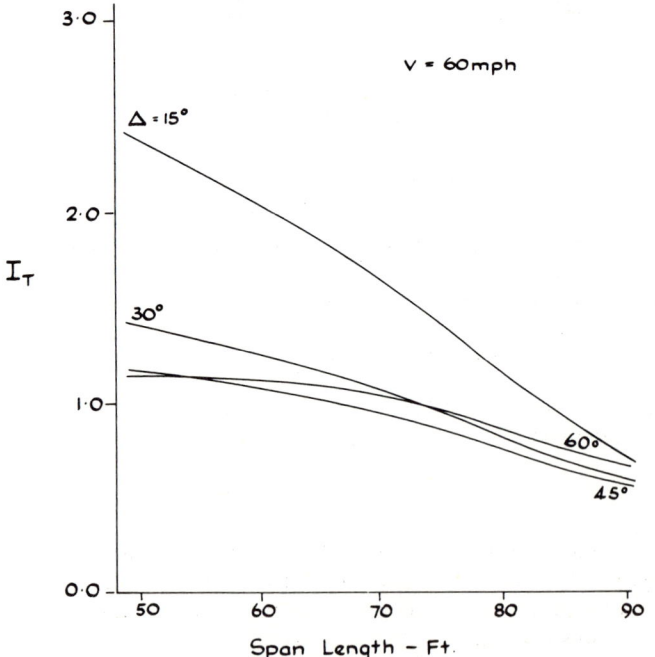

FIG. 7 TORSIONAL IMPACT FACTOR — I Bm.
COMPOSITE BRIDGES

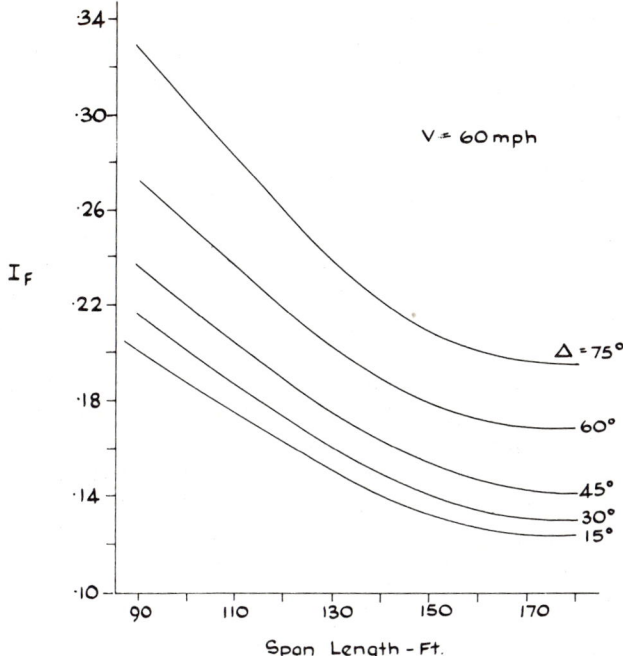

FIG. 8 FLEXURAL IMPACT FACTOR – WELDED GIRDER
COMPOSITE BRIDGES.

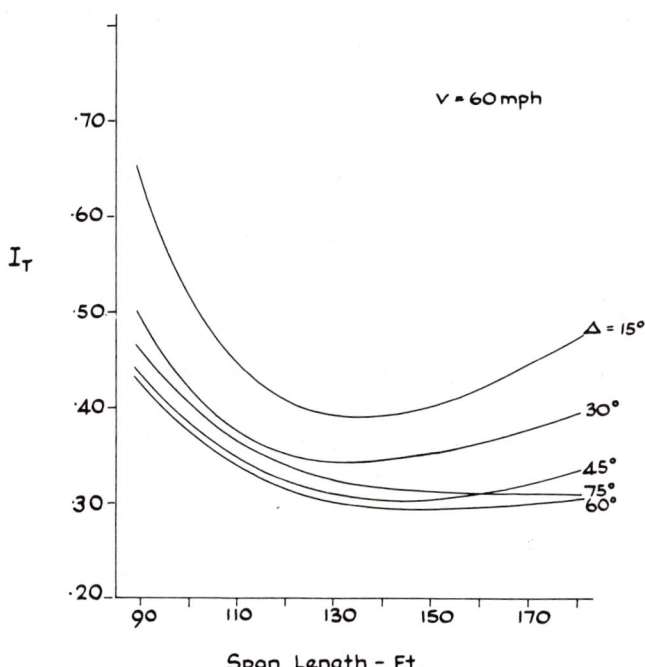

FIG. 9 TORSIONAL IMPACT FACTOR – WELDED GIRDER
COMPOSITE BRIDGES.

the impact factors for typical welded composite girder bridges. The application of these curves, for preliminary design, is as follows:

(*a*) Compute the maximum static values of the bending moment, St. Venant's torque, warping torque and bimoment using preliminary equations.

(*b*) For a given span length, L, enter appropriate figure to obtain the value of I_F, corresponding to the chosen value of the central included angle.

(*c*) For a given span length, L, enter appropriate figure to obtain the value of I_T, corresponding to the chosen value of the central included angle.

(*d*) To obtain the maximum total value of the bending moment, multiply the corresponding static value obtained in step (*a*) by $1 + I_F$.

(*e*) To obtain the maximum total values of the St. Venant's torque, warping torque and bimoment, multiply the corresponding static values obtained in step (*a*) by $1 + I_T$.

This method of approach towards establishing a working criteria for impact is promising and yields guidelines for preliminary design. It should be noted that this design criteria is based on a constant force solution and the actual highway bridge is idealised as a single beam with cross-sectional parameters in accordance with the BPR standard design plans. It should be remembered that rather large values of I_T, described in some of the figures, are for small values of the central included angle Δ, and are of no practical importance since these values correspond to small values of maximum static effects.

4. Composite I-Girder Properties

In order to determine design stresses or distortions, the girder section property is required. If the girder is subjected to torsion and bending, the torsional properties, in addition to bending properties, will be necessary. The exact solution of the torsional properties is complex[17] and not suitable for design. However, by idealising the composite section a series of simplified equations can be developed and used for design.[24]

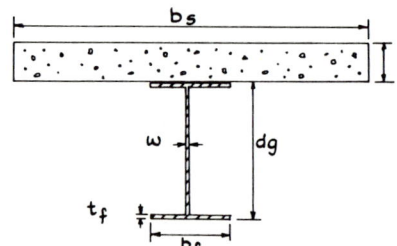

FIG. 10   TYPICAL COMPOSITE SECTION

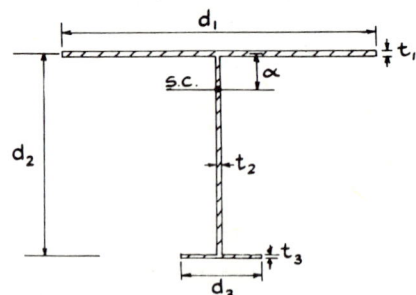

FIG. 11   IDEALIZED COMPOSITE SECTION

Figure 10 describes a typical composite girder and pertinent dimensions. Neglecting the top girder flange and modifying the concrete slab thickness, as shown in Fig. 11, the following dimensions are defined:

$$t_1 = nt_s$$
$$t_2 = w$$
$$d_2 = d_g + t_s/2 - t_f/2$$
$$d_3 = b_f$$
$$t_3 = t_f$$
$$n = Ec/Es$$
$$m = Gc/Gs$$

The resulting torsional properties are given as follows:

Shear centre:

$$\alpha = \frac{d_3^3 t_3}{(d_1^3 t_1 + d_3^3 t_3)} d_2 \tag{40}$$

Normalized warping functions:

$$\text{Slab:} \quad Wn_c = \frac{\alpha d_1}{2} \tag{41}$$

$$\text{Beam:} \quad Wn_s = \frac{(d_2 - \alpha)}{2} d_3 \tag{42}$$

Warping statical moments:

$$\text{Slab:} \quad Sw_c = \frac{\alpha d_1^2 t_1}{8} \tag{43}$$

$$\text{Beam:} \quad Sw_s = \frac{(d_2 - \alpha) d_3^2 t_3}{8} \tag{44}$$

Warping stiffness:

$$Iw = \frac{\alpha^2}{12} t_1 d_1^3 + (d_2 - \alpha)^2 \frac{t_3 d_3^3}{12} \tag{45}$$

Torsional constant:

$$K_T = \frac{1}{3} \left[d_3 t_3^3 + d_2 t_2^3 + m d_1 t_s^3 \right] \tag{46}$$

where $\quad m = G_c/G_s$

With the evaluation of these torsional parameters, the resulting stresses in the composite section can be evaluated. These stress equations are as follows:

Pure torsional shearing stress:

$$\text{Slab:} \quad \tau_{st_c} = G_c t \theta' \quad \text{or} \tag{47}$$

$$\tau_{st_c} = t \frac{Mz}{K_T} \times m \tag{48}$$

$$\text{Steel:} \quad \tau_{st_s} = G_s t \theta' \quad \text{or} \tag{49}$$

$$\tau_{st_s} = t \frac{Mz}{K_T} \tag{50}$$

Warping shearing stress:

$$\text{Slab:} \quad \tau w_c = -\frac{EcSw}{nt}\theta''' \tag{51}$$

$$\text{Steel:} \quad \tau w_s = -\frac{EsSw}{t}\theta''' \tag{52}$$

Warping nominal stress:

$$\text{Slab:} \quad \sigma w_c = E_c W_n \theta'' \tag{53}$$

$$\text{Steel:} \quad \sigma w_s = E_s W_n \theta'' \tag{54}$$

Constant a:

$$a = \left[\frac{E_s I_w}{G_s K_T}\right]^{\frac{1}{2}} \tag{55}$$

where $I_w K_T$ are according to Equations (45) and (46) and $\theta', \theta'', \theta'''$ are derivatives of the rotation along the girder.

REFERENCES

1. Heins, C. P. and Looney, C. T. G. The analysis of curved orthotropic highway bridges by the finite difference technique. *University of Maryland Civil Engineering Report*, No. 13, Jan. 1967.
2. Heins, C. P. The presentation of the slope-deflection method for the analysis of curved orthotropic highway bridges. *University of Maryland Civil Engineering Report*, No. 15, June 1967.
3. Bell, L. C. and Heins, C. P. The solution of curbed bridge systems using the slope-deflection Fourier series method. *University of Maryland Civil Engineering Report*, No. 19, June 1968.
4. Spates, K. R. and Heins, C. P. The analysis of single curved girders with various loadings and boundary conditions. *University of Maryland Civil Engineering Report*, No. 20, June 1968.
5. Hails, R. L. and Heins, C. P. The study of a stiffened curved plate model using the finite difference technique. *University of Maryland Civil Engineering Report*, No. 22, June 1968.
6. Heins, C. P. Behaviour of a stiffened curved plate model. ASCE Annual Meeting and Structural Engineering Conference (Pittsburgh, Pa., Oct. 1968). ASCE Preprint #700.
7. Siminou, J. and Heins, C. P. Design of curved girder bridges. *University of Maryland Civil Engineering Report*, No. 25, June 1969.
8. Vashi, K. M. and Heins, C. P. Impact factors for curved highway bridges. *University of Maryland Civil Engineering Report*, No. 32, June 1969.
9. Committee on Flexural Members. Design of curved girders. ASCE National Transportation Convention (Washington, D.C., July 1969). ASCE Preprint #946.
10. Bell, L. C. and Heins, C. P. Curved girder computer manual. *University of Maryland Civil Engineering Report*, No. 30, Sept. 1969.
11. Heins, C. P. and Hails, R. L. Behaviour of stiffened curved plate model. *J. struct. Div. Am. Soc. civ. Engrs*, Vol. 95, ST 11, Nov. 1969, pp. 2353–2370.
12. Vashi, K. M. Schelling, D. R. and Heins, C. P. *Impact factors for curved highway bridges.* (Highway Research Record No. 302). Washington, D.C., Highway Research Board, 1970.
13. Bonakdarpour, B., Bell, L. C. and Heins, C. P. Analytical and experimental study of a curved bridge model. *University of Maryland Civil Engineering Report*, No. 34, Feb. 1970.
14. Lee, H. W. and Heins, C. P. Local loading on curved reinforced concrete slabs. *University of Maryland Civil Engineering Report*, No. 35, April 1970.
15. Heins, C. P. and Siminou, J. Preliminary design of curved bridges. *Eng. J. Am. Inst. Steel Constr.*, Vol. 7, No. 4, April 1970.
16. Bonakdarpour, B., Bell, L. C. and Heins, C. P. The behaviour of a curved box beam model bridge. *University of Maryland Civil Engineering Report*, No. 36, June 1970.
17. Kuo, J. T. C. and Heins, C. P. Torsional properties of composite steel bridge members. *University of Maryland Civil Engineering Report*, No. 37, June 1970.
18. Heins, C. P. Analytical-experimental study of a curved bridge model. Proceedings of Speciality Conference on Steel Structures, ASCE, (Columbia, Mo., June 1970).

19. Heins, C. P. and Spates, K. R. Behaviour of a single horizontally curved girder. *J. struct. Div. Am. Soc. civ. Engrs*, Vol. 96, ST7, July 1970, pp. 1511–1524.
20. Bell, L. C. and Heins, C. P. Analysis of curved girder bridges. *J. struct. Div. Am. Soc. civ. Engrs*, Vol. 96, ST8, August 1970, pp. 1657–1673.
21. Kuo, J. T. C. and Heins, C. P. Experimental studies of composite members subjected to torsion. *University of Maryland Civil Engineering Report*, No. 39, Feb. 1971.
22. Lee, H. W. L. and Heins, C. P. Large deflections of curved plates. *J. struct. Div. Am. Soc. civ. Engrs*, 1971 (accepted for publication).
23. Heins, C. P. and Seaburg, P. A. *Torsion analysis of rolled steel sections.* Bethlehem, Pa., Bethlehem Steel Co., 1963.
24. Vlasov, V. Z. *Thin walled elastic beams*; 2nd ed., trans. from the Russian by Y. Schectman. Washington, D.C., National Science Foundation, 1961.

A HIGHER ORDER INPLANE PARALLELOGRAM ELEMENT AND ITS APPLICATION TO SKEWED CURVED BOX-GIRDER BRIDGES

R. G. SISODIYA and Y. K. CHEUNG

University of Calgary

SYNOPSIS

The inplane stiffness for a parallelogram element is derived for nodal parameters u_i, v_i and $\theta_{zi}\left(=\dfrac{\partial v}{\partial x}\right)$ by assuming linear displacement functions for u and linear and cubic displacements for v along the two sets of parallel sides of a parallelogram. This element gives good accuracy with even coarse mesh for the analysis of beams, when the cubic variation of v corresponds to the vertical deflection.

A parallelogram shell element is also derived simply by combining an in-plane and a plate bending[1] stiffnesses. A straight box-girder bridge on skew supports and a curved box-girder bridge on skew support are analysed and the results are compared with those obtained experimentally and by using lower order elements.

INTRODUCTION

It has been found that the standard rectangular in-plane element with 8 degrees of freedom (u and v at each node) is inadequate for the analysis of beams and bridge structures.[2] This element is very stiff and usually in order to produce accurate results the depth of the beam has to be divided into at least two elements.[2,3] Also the accuracy deteriorates rapidly when the aspect ratio of the sides of the rectangle deviates from unity.

Several higher order elements have been developed in the past. Among these are:
(a) a rectangular element[4] using u, v and θ_z (with θ_z at alternate nodes and

$$\theta_z = \frac{\partial u}{\partial y} \text{ or } \frac{\partial v}{\partial x}),$$

(b) a rectangular element[5] using u, v and $\theta_z\left[\theta_z = \frac{1}{2}\left(\frac{\partial v}{\partial x} - \frac{\partial u}{\partial y}\right)\right]$,

(c) a rectangular element[6] using $u, v, \dfrac{\partial u}{\partial y}$, and $\dfrac{\partial v}{\partial x}\left[\text{or } u, v,\left(\frac{\partial u}{\partial y} + \frac{\partial v}{\partial x}\right)\text{and}\left(\frac{\partial v}{\partial x} - \frac{\partial u}{\partial y}\right)\right]$,

and

(d) a triangular element[7] using $u, v, \dfrac{\partial u}{\partial x}, \dfrac{\partial v}{\partial y}, \left(\frac{\partial u}{\partial y} + \frac{\partial v}{\partial x}\right)$ and $\dfrac{1}{2}\left(\frac{\partial u}{\partial y} - \frac{\partial v}{\partial x}\right)$ as nodal degrees of freedom.

The primary objective of this paper is to produce a suitable element for the analysis of arbitrarily shaped box-girder bridges and to demonstrate that accurate solutions can be obtained with reasonably short computer running time. In order to achieve this end, the authors have observed that the following points should be taken into account:

(i)   Since a box girder bridge is essentially a spatial plate structure, the adequate number of nodal degrees of freedom should include $u, v, w, \theta_x, \theta_y$ and θ_z. A suitable bending element is available using w, θ_x and θ_y, therefore it is natural that the plane stress element should include u, v and θ_z as its nodal parameters.

(ii) Elements with higher number of nodal parameters are not to be recommended because such elements tend to increase the band width of the stiffness matrix, since a corresponding reduction in the total number of nodals is not always possible because of thickness variations.

(iii) The chosen plane stress element should be able to represent the beam behaviour accurately when only one element is used through the depth of the web, and also should yield good accuracy even for large aspect ratio of element sides (say $1:4$), so that the number of mesh divisions along the span can be kept to a minimum. In this way both the band width and the total number of unknowns in the overall stiffness matrix can be made as small as possible.

The finite element solution presented here uses a higher order plane stress parallelogram element PPLC3 which conforms with the above requirements.

BEAM TYPE FINITE PARALLELOGRAM ELEMENT (PPLC3)

(A) Displacement Functions

In this element the three degrees of freedom at each node are

$$\{\delta_i\} = \left\{ \begin{array}{c} u_i \\ v_i \\ \left(\dfrac{\partial v}{\partial x}\right)_i \end{array} \right\} \tag{1}$$

Such a choice of the nodal parameters will allow u to vary linearly in the ξ and η directions (Fig. 1) and v to vary linearly in the η direction and as a cubic in the ξ direction. The cubic variation in ξ, of course, corresponds to the variation of vertical

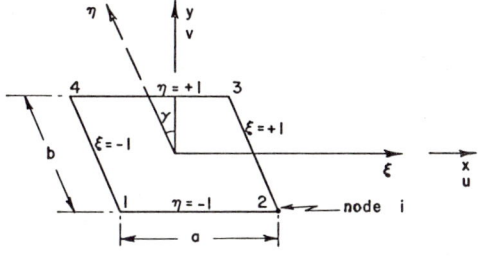

FIG. I   A PARALLELOGRAM ELEMENT

deflections in a simple beam element. The displacement functions chosen are

$$
\left\{ \begin{matrix} u \\ v \end{matrix} \right\} = \sum_{i=1}^{4} \begin{bmatrix} f_{i1} & 0 & 0 \\ 0 & f_{i2} & f_{i3} \end{bmatrix} \{\delta_i\} \tag{2}
$$

where

$$
f_i^1 = \frac{1}{4}(1 + \xi\xi_i)(1 + \eta\eta_i)
$$

$$
f_{i2} = \frac{1}{8}(1 + \xi\xi_i)(2 + \xi\xi_i - \xi^2)(1 + \eta\eta_i) \tag{3}
$$

$$
f_{i3} = -\frac{\alpha}{8}\xi_i(1 + \xi\xi_i)(1 - \xi^2)(1 + \eta\eta_i)
$$

and

$$
\alpha = \frac{1}{4}[(x_2 - x_1) + (x_3 - x_4)]
$$

These displacement functions satisfy the compatibility conditions for a general quadrilateral, however the constant strain criterion has been satisfied only for rectangular and parallelogram elements and not for a general quadrilateral element.

(B) Stiffness and Stress Matrices

For a parallelogram with skew angle γ and sides a, and b (Fig. 1), the x, y coordinates are related to ξ, η coordinates by

$$
\left\{ \begin{matrix} x \\ y \end{matrix} \right\} = \begin{bmatrix} \frac{a}{2} & -\frac{b}{2}\sin\gamma \\ 0 & \frac{b}{2}\cos\gamma \end{bmatrix} \left\{ \begin{matrix} \xi \\ \eta \end{matrix} \right\} \tag{4}
$$

and therefore the Jacobian matric $[J]$ is given by

$$
\left\{ \begin{matrix} \frac{\partial}{\partial\xi} \\ \frac{\partial}{\partial\eta} \end{matrix} \right\} = \begin{bmatrix} \frac{a}{2} & -\frac{b}{2}\sin\gamma \\ 0 & \frac{b}{2}\cos\gamma \end{bmatrix} \left\{ \begin{matrix} \frac{\partial}{\partial x} \\ \frac{\partial}{\partial y} \end{matrix} \right\} \tag{5}
$$

Using inverse of $[J]$ and (2) and (3), the strain matrix ξ can be written as

$$
\{\epsilon\} = \left\{ \begin{matrix} \frac{\partial u}{\partial x} \\ \frac{\partial v}{\partial y} \\ \frac{\partial v}{\partial x} + \frac{\partial u}{\partial y} \end{matrix} \right\} = \sum_{i=1}^{4} [B]_i \{\delta\}_i \tag{6}
$$

while the stress vector $\{\sigma\}$ is given by

$$
\{\sigma\} = [D]\{\epsilon\} = \sum_{i=1}^{4} [D][B]_i \{\delta_i\} \tag{7}
$$

and the stiffness matrix can then be written as

$$[K] = \begin{bmatrix} K_{11} & K_{12} & K_{13} & K_{14} \\ \cdot & & & \\ \cdot & & & \\ \cdot & & & \\ K_{41} & \cdots & \cdots & K_{44} \end{bmatrix} \tag{8}$$

where

$$[K]_{ij} = \int_A \{[B_j]^T [D] [B_i]\} \, dA \tag{9}$$

After the completion of this work, the authors came across a report by Scordelis and Williams[8] in which a rectangular element developed along similar lines was mentioned. The explicit displacement function used was, however, not given in the report.

PARALLELOGRAM SHELL ELEMENT

Box-section bridges are frequently thin cellular structures, and can be analysed by using thin flat shell elements. The stiffness of a flat parallelogram shell element can be written as

$$[K]_e = \begin{bmatrix} [K_p] & [0] \\ [0] & [K_b] \end{bmatrix} \tag{10}$$

in which $[K_p]$ is the in-plane stiffness of the element for nodal parameter u_i, v_i and θ_{zi} and $[K_b]$ is the plate bending stiffness for nodal parameters w_i, θ_{xi} and θ_{yi}.

Usually the nodal parameters are arranged in the order of u, v, w, θ_{xi}, θ_{yi} and θ_{zi}, therefore a rearrangement of Eq. (10) is then necessary.

EXAMPLES

The accuracy of the plane stress element PPLC3 is checked first by solving some simple examples and subsequently the parallelogram shell element derived by combining PPLC3 and a plate bending stiffness is used to analyse different types of box-girder bridges.

(A) Simply Supported and Cantilever Beams

Different cases of simply supported and cantilever beams are shown in Table 1 and 2 respectively. Two types of mesh, 1 x 2 and 1 x 4 are used and for each mesh type, the beams are analysed for different span lengths, which is equivalent to using elements with different aspect ratios. The same examples are solved by the standard rectangular element PPL2 with two degrees of freedom u_i and v_i at each node. It can be seen from the tables that the element PPLC3 gives much better results as compared to that by PPL2 with the same type of mesh and that the aspect ratio has negligible effect on its accuracy.

(B) A Skew Strip of a Long Wall

A skew strip cut out of a continuous wall under uniform pressure (Fig. 2), is selected to check whether the parallelogram element can truly represent a case of constant stress in one direction. Table 3 shows the results obtained by using PPLC3 element along with that by element PPL2 and one dimensional analysis (or column analysis)

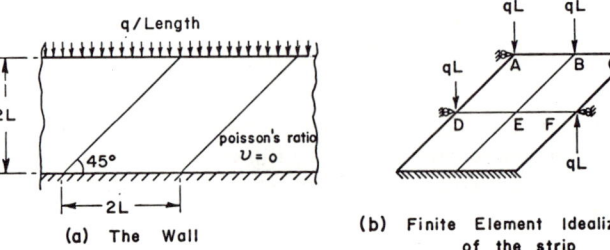

(a) The Wall

(b) Finite Element Idealization
of the strip

FIG. 2   SKEW   STRIP   INSIDE   A   CONTINUOUS   WALL

of the strip. The results are practically identical and it is therefore verified that this element can represent constant stress situation accurately.

The constant strain criterion is also checked analytically as follows. General polynomials are assumed for displacements such that the strains at any point are constant

$$u = A_1 + A_2 x + A_3 y$$
$$v = A_4 + A_5 x + A_6 y \tag{11}$$

The nodal parameters for node i of the parallelogram element for this displacement field are

$$u_i = A_1 + A_2 x_i + A_3 y_i$$
$$v_i = A_4 + A_5 x_i + A_6 y_i \tag{12}$$

$$\left(\frac{\partial v}{\partial x}\right)_i = A_5$$

Upon substituting these nodal parameters into Eq. (2) it is found that the displacement functions inside a parallelogram will be the same as given by Eqs. (11) which were for constant strains. This means that the parallelogram element will represent constant strain problems exactly and thus the constant strain criterion is satisfied.

(C) Beam with Inclined Faces at the End

The beam shown in Fig. 3 is analysed for different mesh divisions and the results are compared in Table 3. The same beam is analysed theoretically by assuming it simply supported at the two supports. After the superior performance of PPLC3 in Example A in the form of a rectangle, it is indeed disappointing to see that the accuracy has deteriorated significantly as a parallelogram. However the results do converge when a fine mesh is used and the convergence is much faster when compared with the standard parallelogram element PPL2 with linear displacement functions. Since such parallelograms are usually used for top and bottom slabs which do not really behave as beams, the present example represents a very severe test for the element.

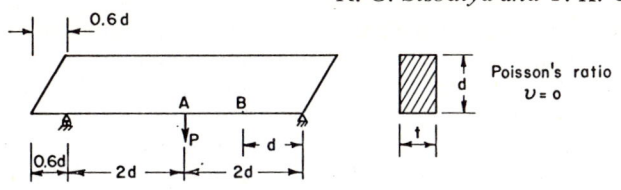

(a) Beam and the dimensions

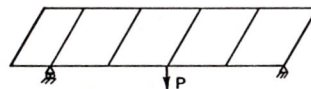

(b) Finite Element Idealization (Variable number of equal divisions along the depth)

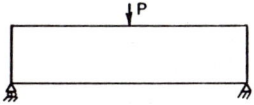

(c) Beam approximation

FIG. 3   BEAM WITH INCLINED END FACES

(D) Straight Skewed Box-Girder Bridge Model

A skewed box-girder bridge model (Fig. 4) tested and analysed using finite differences by Ghali[9] was analysed previously by using lower order elements.[3] The results of the bridge model of reference [3] are those obtained by using two lower order elements

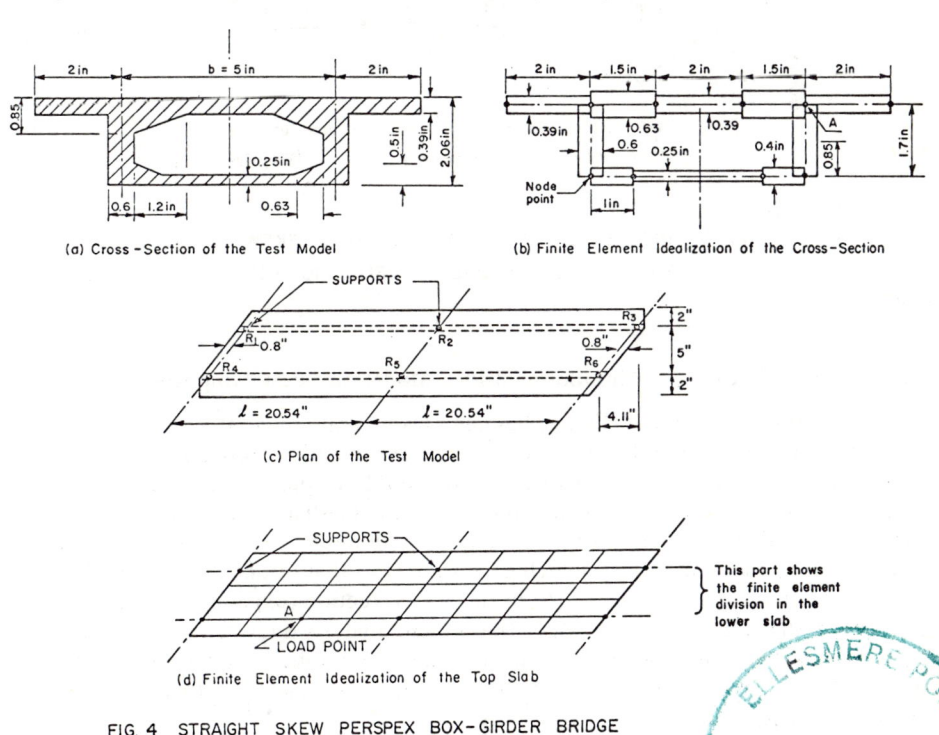

(a) Cross-Section of the Test Model

(b) Finite Element Idealization of the Cross-Section

(c) Plan of the Test Model

(d) Finite Element Idealization of the Top Slab

FIG. 4   STRAIGHT SKEW PERSPEX BOX-GIRDER BRIDGE

along the depth and 20 divisions along the span. In the present analysis only one element is used along the depth (Fig. 4(*b*)) and the number of divisions is 10 along the span (Fig. 4(*d*)). Thus the total number of equations is reduced to nearly half and the band width is reduced to five-sixth of that in reference [3].

The influence lines of the vertical deflections of the webs are plotted in Fig. 5(*a*) and the influence lines of the longitudinal strains in Fig. 5(*b*). The present results are almost identical to those obtained for the fine mesh analysis of reference [3].

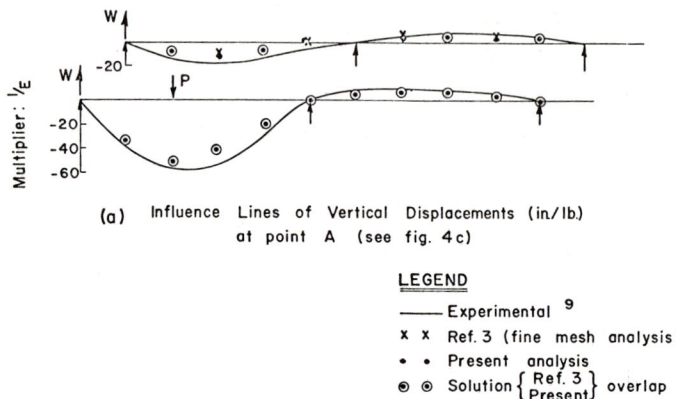

(a)   Influence Lines of Vertical Displacements (in./lb.)
at point A   (see fig. 4 c)

LEGEND

———— Experimental [9]

x x   Ref. 3 (fine mesh analysis)

• •   Present analysis

⊚ ⊚   Solution $\left\{\begin{matrix} \text{Ref. 3} \\ \text{Present} \end{matrix}\right\}$ overlap

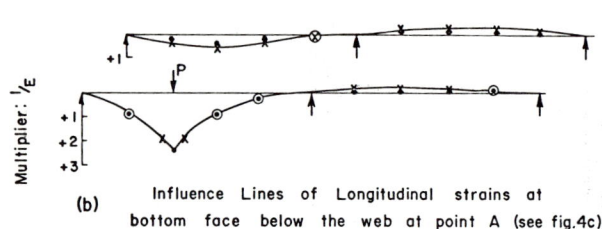

(b)   Influence Lines of Longitudinal strains at
bottom face below the web at point A (see fig.4c)

FIG. 5   INFLUENCE   LINES   OF   STRAIGHT   SKEW
BOX - GIRDER   BRIDGE   MODEL   (FIG. 4)

(E) A Curved Box-Girder Aluminium Bridge Model on Skew Supports

An aluminium bridge model (Fig. 6) has been tested experimentally and analysed[2, 10] by lower order element and it was shown that the top and bottom slabs of a curved bridge with relatively large radius of curvature could be approximately divided into an assemblage of parallelogram elements. In the fine mesh analysis of reference [2], the webs were divided into two lower order elements along the depth and the number of divisions along the span was 38 which corresponds to that obtained by subdividing into two all the elements in Fig. 7(*b*) in the span direction (except for the elements adjacent to the end cross-sections). In the present analysis the web is divided into only one element along the depth (Fig. 7(*a*)) and the number of divisions along the span is 18 for fine mesh analysis (Fig. 7(*b*)), while for coarse mesh analysis the number of divisions along the span is only 10.

Fig. 8 shows vertical deflections of the webs, tangential strains above the webs at upper surface of the top slab and the reactions due to load P at point A in Fig. 6(b). The co-relation between the fine mesh analysis of reference [2] and the present analysis is good. The discrepancy between the experimental and analytical displacements has been attributed to bad experimental results.[2,10]

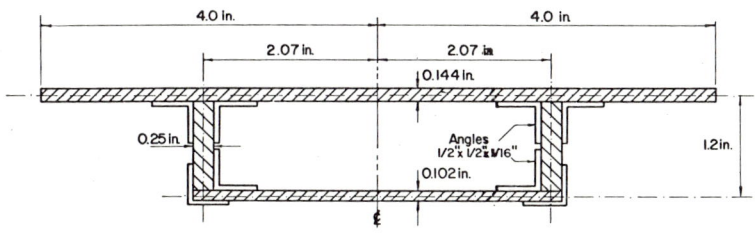

(a) CROSS-SECTION

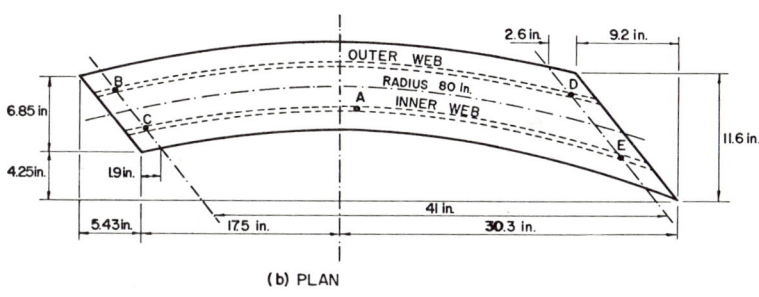

(b) PLAN

FIG. 6 CURVED BOX-SECTION ALUMINUM BRIDGE MODEL

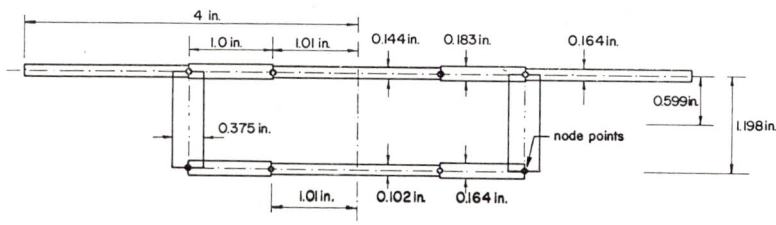

(a) CROSS-SECTION

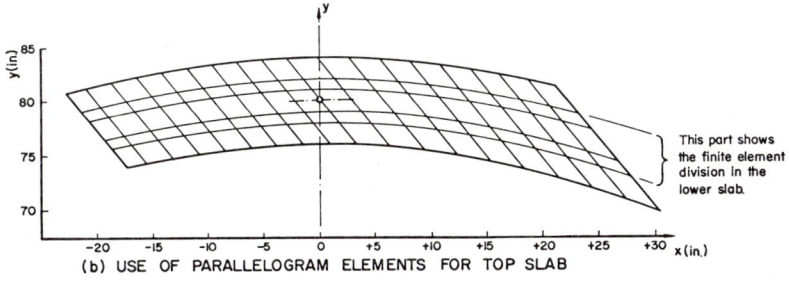

(b) USE OF PARALLELOGRAM ELEMENTS FOR TOP SLAB

FIG. 7 FINITE ELEMENT IDEALIZATION

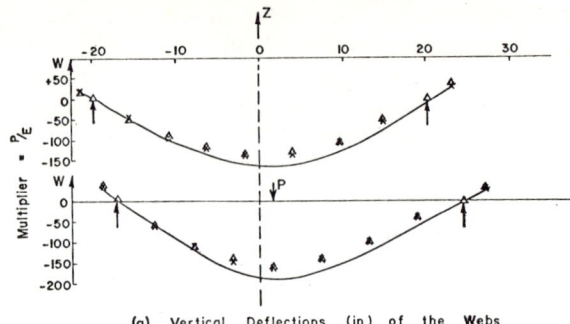

(a) Vertical Deflections (in) of the Webs

REACTIONS	(MULTIPLIER = P)			
POINT	EXPERIM-ENTAL[10]	ANALYTICAL		
		Ref 2	Fine Mesh	Coarse Mesh	Fine Mesh
B	-.131	-.169	-.211	-.191
C	+.606	+.669	+.705	+.632
D	+.657	+.647	+.662	+.647
E	-.127	-.147	-.168	-.157

LEGEND

——— Experimental [10]

x x Ref. 2 (fine mesh analysis)

• • Present Analysis
         (coarse mesh.)

⊚ ⊚ Present Analysis
         (fine mesh.)

△ △ Present Analysis
         (coarse & fine mesh overlapping)

(b) Tangential Strains at Upper Surface of Top Slab above Webs

FIG. 8   DEFLECTIONS, REACTIONS, AND STRAINS DUE TO
VERTICAL LOAD 'P' AT POINT A (FIG. 6b)
OF ONE SPAN BRIDGE MODEL ON A CURVE

CONCLUSIONS

The parallelogram element derived here gives good accuracy for the analysis of beam type structures. The accuracy is better for rectangular than for parallelogram elements.

The improved accuracy of the plane stress element provides us with the basis for an efficient shell element and therefore an economical analysis of complex box girder bridge structures. For bridges with arbitrary shaped decks, no real difficulty is imposed as the analysis can still be carried out using parallelogram elements and some triangular elements.

NOTATIONS

A — area of an element
$[B]$ — matrix relating strains and nodal parameters
$[D]$ — matrix relating stresses and strains
E — Young's modulus
$[J]$ — Jacobian matrix

$[K]$	– stiffness matrix
L	– span length
P	– load
PPL2	– abbreviation for parallelogram element in-plane stiffness derived by assuming linear displacement functions for two nodal parameters u_i and v_i at each node
PPLC3	– abbreviation for parallelogram element in-plane stiffness derived by assuming all linear displacement functions except v cubic in ξ direction (Fig. 1)
R	– reaction
a, b	– sides of a parallelogram element
d	– depth of a beam
f	– assumed shape functions
i	– integers
t	– thickness
u, v, w	– translations in the x, y and z directions, respectively
x, y, z	– axes
α	– a constant
ξ, η	– skew coordinates for a parallelogram element
$\{\delta\}$	– displacement vector
ϵ	– strains
σ	– stresses
λ	– a constant
γ	– skew angle
θ	– rotation which can be replaced by a vector along any of the axes x, y or z

Subscripts

b	– refers to bending stiffness of an element
e	– refers to an element
i	– refers to a node
p	– refers to in-plane stiffness of an element
x, y, z	– axes

REFERENCES

1. Dawe, D. J. Parallelogrammic elements in the solution of rhombic cantilever plate problems. *J. Strain Anal.,* Vol. 1, No. 3, 1966, pp. 223–230.
2. Sisodiya, R. G., Cheung, Y. K. and Ghali, A. Finite element analysis of skew curved box-girder bridges. *Publs int. Ass. Bridge struct. Engng,* Vol. 30, part 2, 1970.
3. Sisodiya, R. G., Ghali, A. and Cheung, Y. K. *Finite element analysis of skew box-girder bridges.* University of Calgary Research Report, July 1970.
4. MacLeod, I. A. New rectangular finite element for shear wall analysis. *J. struct. Div. Am. Soc. civ. Engrs,* Vol. 95, ST3, March 1969, pp. 309–409.
5. Scordelis, A. C. *Analysis of continuous box-girder bridges.* University of California, Berkeley, Structural Engineering and Structural Mechanics Report No. SESM 67-25, Nov. 1967.
6. Pole, G. M. and Felippa, C. A. Discussion on 'New rectangular finite element for shear wall analysis'. *J. struct. Div. Am. Soc. civ. Engrs,* Vol. 96, ST1, Jan. 1970, pp. 140–147.
7. Tocher, J. L. and Hartz, B. J. Higher-order finite element for plane stress. *J. Engng Mech. Div. Am. Soc. civ. Engrs,* Vol. 93. EM4, August 1967, pp. 149–172.
8. Willam, K. J. and Scordelis, A. C. *Computer program for cellular structures of arbitrary plan geometry.* University of California, Berkeley, Structural Engineering and Structural Mechanics Report No. SESM 70-10, Sept. 1970.
9. Ghali, A. Analysis of continuous skew concrete girder bridges. *First International Symposium on Concrete Bridge Design (Toronto, 1967).* American Concrete Institute Publication SP-23, 1969, pp. 136–169.
10. Sisodiya, R. G. *Analysis of box-girder bridges by the finite element method.* M.Sc. thesis, University of Calgary, 1969.

TABLE 1

Simply Supported Beam

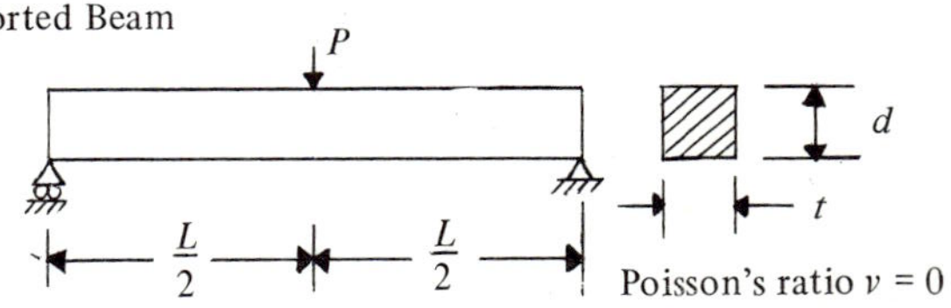

Poisson's ratio $\nu = 0$

Mesh (Half Span)	Span L (Multiplier d)	Average Vertical Displacement at Midspan (Multiplier -P/Et)					Stresses at 3/8 Span (Bottom Face) (Multiplier P/td)					
		Beam Theory*	FEM – PPL2		FEM – PPLC3		Beam Theory*	FEM – PPL2		FEM – PPLC3		
			Value	% of Beam Theory	Value	% of Beam Theory		Value	% of Beam Theory	Value	% of Beam Theory	
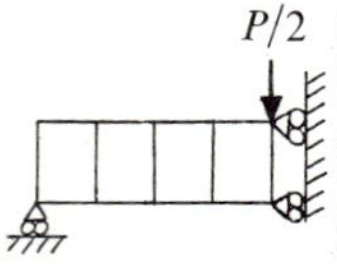 P/2	4	18·19	16·50	90·7	18·30	100·1	4·5	4·000	88·9	4·499	100·0	
	8	132·59	88·33	66·6	130·42	98·4	9·0	5·999	66·7	8·996	100·0	
	16	1033·39	344·11	33·3	1007·69	97·5	18·0	5·998	33·3	17·85	99·2	
	32	8210·99	910·11	11·1	8191·9	99·8	36·0	3·992	11·1	36·48	101·3	
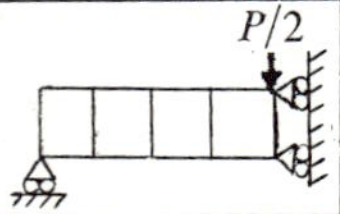 P/2	4	18·19	12·38	68·1	17·49	96·1	4·5	3·000	66·7	4·500	100·0	
	8	132·59	44·23	33·4	124·26	93·7	9·0	3·000	33·3	8·997	100·0	
	16	1033·39	114·677	11·1	968·99	93·8	18·0	1·998	11·1	18·02	100·1	

* Including Shear Corrections

TABLE 2

Cantilever Beam

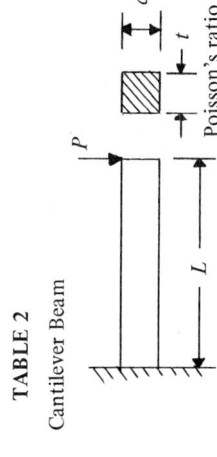

Poisson's ratio $v = 0$

Mesh	Span L (Multiplier d)	Average Vertical Displacement at the Free End (Multiplier -P/Et)					Stresses at L/4 from Fix End (Bottom Face) (Multiplier -P/td)					
		Beam Theory*	FEM – PPL2		FEM – PPLC3		Beam Theory*	FEM – PPL2		FEM – PPLC3		
			Value	% of Beam Theory	Value	% of Beam Theory		Value	% of Beam Theory	Value	% of Beam Theory	
	2	36·38	32·00	88·0	35·38	97·3	9	8·000	88·9	8·999	100·0
	4	265·18	175·98	66·4	259·67	97·9	18	11·998	66·7	17·994	100·0
	8	2006·8	687·75	33·3	2027·9	98·1	36	11·995	33·3	35·934	99·8
	16	16,422·0	1805·6	11·0	16·177	98·5	72	7·919	11·0	72·174	100·2
	2	36·38	24·00	66·0	33·76	92·8	9	6·000	66·7	9·000	100·0
	4	265·18	88·00	33·2	247·47	93·3	18	6·000	33·3	17·996	100·0
	8	2066·8	229·1	11·1	1941·59	93·9	36	3·996	11·1	36·122	100·3

* Including Shear Corrections

TABLE 3

Results of Strip in Fig. 2

Point	Vertical Displacements $\left(Multiplier\ \dfrac{-qL}{Et}\right)$		
	Column Analysis	*FEM − PPL2*	*FEM − PPLC3*
A	2	2·000003	2·000010
B	2	2·000005	2·000009
C	2	2·000005	2·000007
D	1	1·000001	1·000004
E	1	1·000003	1·000006
F	1	1·000004	1·000007

Stresses correct to five significant figures at all points.

TABLE 4

Results of Beam in Fig. 3

| Mesh | | Vertical Displacement at Point A $\left(\text{Multiplier} - \dfrac{P}{Et}\right)$ | | | | | Horizontal Stresses at Point B $\left(\text{Multiplier} \dfrac{P}{td}\right)$ | | | |
| | Beam Theory with Shear Correction | FEM – PPL2 | | FEM – PPLC3 | | Beam Theory | FEM – PPL2 | | FEM – PPLC3 | |
		Value	% of Beam Theory	Value	% of Beam Theory		Value	% of Beam Theory	Value	% of Beam Theory
5 × 1	18·19	7·72	42·4	10·74	59·2	3	1·140	38·0	1·556	51·9
5 × 2	18·19	9·44	51·8	15·21	83·6	3	1·547	51·6	2·541	84·7
5 × 4	18·19	10·09	55·4	17·27	95·0	3	1·675	55·8	2·959	98·6

STRUCTURAL BEHAVIOUR OF A REINFORCED CONCRETE BOX GIRDER BRIDGE

J. G. BOUWKAMP, A. C. SCORDELIS and S. T. WASTI

University of California, Berkeley

SUMMARY

The construction, instrumentation and testing of a large, reinforced box girder model are described. The system of data acquisition and methods of data reduction are treated, and pertinent data from some important loading cases dealing with deflections, longitudinal moments at the instrumented sections, and the distribution of the moments among the girders are tabulated. Comparisons are made with results obtained from an analysis based on the folded plate equations, and general conclusions are drawn from the results.

Introduction

The box girder bridge is a smooth, functional structure with a high resistance to torsional moments. Through extensive use and with the experience gained in construction, the box girder bridge has become economically competitive and is at the same time aesthetically attractive in appearance. It is estimated that well over sixty per cent of the total bridge deck area built every year in California is of the box girder type. In 1970 this area amounted to over 6 500 000 ft^2.

Current design methods for box girder bridges are based on considering either typical repeating units of the cross-section as independent continuous beams or the entire cross-section as a whole unit. For transverse and longitudinal slab moments, as well as for the study of wheel loads on bridge decks, empirical expressions are used.

Much analytical work has been done, especially at the University of California, to study the structural behaviour of box girders when subjected to loads of different kinds. General computer programs have been developed for performing the analyses of box girder bridges, using the folded plate, the finite element, the finite segment, and other approaches, and comprise the subject matter of a separate paper at this conference by Scordelis.[1]

The present method of box girder design in the United States employs for live loads the American Association for State Highway Officials (AASHO) standard HS 20—44 truck with a width of 10 ft. For the purpose of most studies, the spacing between axles is taken as 14 ft to produce maximum stress. In each span of the box girder bridge, one HS 20—44 truck is placed per traffic lane in a position that will produce maximum stress.

AASHO Specifications[2] propose a design method wherein a box girder bridge is considered to be made up of a number of identical I-shaped interior girders plus two exterior girders. Each girder is designed as a separate member by applying to it a certain fraction of a single longitudinal line of wheels from the standard truck. Until 1959 this fraction, known as the distribution factor N_{WL}, was given by the relations

$$N_{WL} = S/5 \text{ for interior girders}$$

and

$$N_{WL} = S_1/5 \text{ for exterior girders}$$

Here S is the flange width in feet of the interior girder, which is also equal to the average width of the cell, and S_1 is the top flange width in feet of the exterior girder, which is also equal to half the cell width plus the cantilever overhang.

In 1959 the AASHO Specifications changed the distribution factors to $S/7$ and $S_1/7$ thus separating the concrete box girder bridge from a concrete tee beam bridge or a slab on steel stringers.

In December 1967 the State of California adopted a design specification in which the distinction between S_1 and S was abolished and the total value of the distribution factor N_{WL} for the 'whole-width unit' was given by

$$N_{WL} \text{ (total)} = \frac{\text{Deck width in feet}}{7}$$

These changes were indications of the recognition of the structural efficiency of the box girder section, but they also called into question the whole process of bridge design based on distribution factors. Scordelis and Meyer[3] have pointed out that the most important variable not taken into account by the AASHO specifications is the number of traffic lanes on the bridge, and have demonstrated that a two lane box girder bridge of conservative design could be changed into an unconservatively designed three lane bridge by minor adjustments in the widths of the barrier curbs alone. Other factors such as span, total width, number of cells, and continuity or fixity at the supports also influence the load distribution and should be considered.

Analytical models for box girder bridges generally assume that the plates forming the box girder are elastic and isotropic, and that a linear relationship between forces and deformations exists. The true response for a concrete box girder bridge is highly complicated, and involves a non-isotropic structure made up of two materials, concrete and steel. Under increasing load the concrete cracks and stress redistributions occur. Time dependent effects such as shrinkage and creep in the concrete also affect displacements and internal stresses and are a function of the environment of the prototype.

Both analysis and design find their ultimate confirmation in the behaviour of a real structure. Only by experimental observations of the behaviour of actual reinforced concrete box girder bridges under controlled conditions can it be ascertained if the proposed analytical methods, based on the simplified analytical model described above, adequately predict the load distribution properties, the magnitudes of displacements, internal forces and moments, and the effects of interior diaphragms.

The box girder bridge model of the present study is the largest model of its kind instrumented, built, and tested from service load conditions to failure.

Description of Model

The box girder bridge prototype chosen was a typical two lane, two span structure, 203 ft long, 34 ft wide, and 4 ft 10 in in depth. It had four cells, a centre bent with a circular column support, and two end diaphragms. Many reinforced concrete bridges of this kind are in actual service in California as part of the highway network.

For the present investigation, a 'direct method' model with a linear scale relation

of 1:2·82 was chosen. The scale was determined by replacing the standard longitudinal reinforcement of an actual box girder bridge, i.e., a No. 11 deformed bar of
nominal cross-sectional area 1·56 in^2 by a No. 4 deformed bar of nominal cross-
sectional area 0·20 in^2 in the model. The large scale thus enabled the use of standard
high strength deformed steel bars as reinforcement, and concrete with small aggregate
rather than a mortar mix for the model material.

The only difficulty in satisfying the similitude relations lay in having to make the
unit weight of the model equal to 2·82 times that of the prototype while keeping the
elastic modulus and Poisson's ratio unchanged. The best solution was the artificial
addition of extra dead weight which, added to the weight of the model, would result
in 2·82 times the weight of the model itself. Various schemes of realising this considerable loading (equivalent for the model bridge in the present study to an extra
weight of about 1·57 kips/ft) were examined in detail.

It was finally decided that the most convenient and least expensive way of providing the extra dead load for proper simulation of prototype conditions lay in renting
steel billets of approximately 9 x 9 x 65 in dimensions from a local manufacturer and
placing these suitably within the cells of the box girder bridge model during its
construction.

The box girder bridge model, Figs 1 and 2, was identical to the prototype in shape
but not in size, being 72 ft long, 12 ft wide, and 1 ft 8$\frac{9}{16}$ in depth. For convenience
of access and observation, it was decided to fix the height of the model so that the
distance of the soffit from the level of the test floor was 5 ft. This was done to allow
unimpeded views of crack formations in the tensile zones, which could also be easily
marked, at any stage of the loading. The abutments at each end provided for the
simply supported end condition and, if desired, for restraint against longitudinal
movement. Both the abutments and the square footing of the central column were
anchored to the test floor by means of prestressed steel rods. Views of the end abutments and designations for significant longitudinal and transverse sections are also
shown in Figs 1 and 2. Thus, the girders of the bridge model are identified as girders

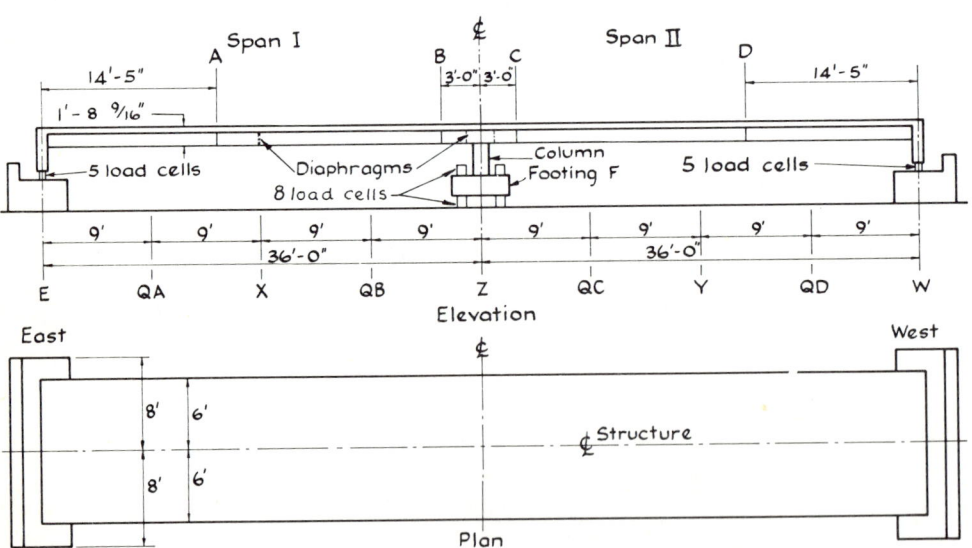

FIG. I   DIMENSIONS OF BOX GIRDER BRIDGE MODEL
WITH TRANSVERSE LOCATIONS

1 to 5. The longitudinal sections at the cantilever edges were identified as 0 and 6, respectively.

Transverse sections A and D are locations for maximum positive moment under the dead load, while sections B and C on either side of the centre bent diaphragm are in the region of maximum negative moment. Sections X and Y are the mid-span sections, while section Z is the bridge centreline. Quarter span sections are designated in accordance with proximity to another transverse section; thus QA is the quarter span section nearest section A. Sections E and W represent sections through the East and West ends of the box girder bridge model. F denotes the square footing of the interior support as a location.

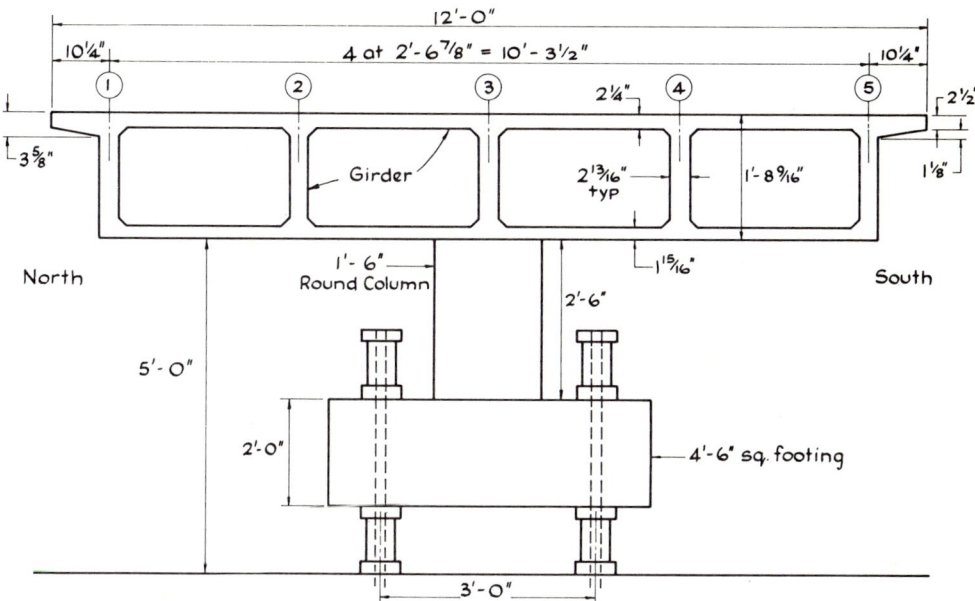

FIG. 2 TYPICAL SECTION OF BOX GIRDER BRIDGE MODEL

The two spans of the bridge model are designated Span I and Span II, respectively. Span I has a mid-span diaphragm at Section X.

All reinforcing bars for the box girder bridge model were standard quality deformed bars. The No. 4 bars had a nominal yield stress of 60 ksi and the No. 2 bars had a nominal yield stress of 48 ksi. The main reinforcement in the maximum positive moment regions consisted of 55 − No. 4 bars in each span, and the main reinforcement in the negative moment region consisted of 82 − No. 4 bars over the centre bent. There were a few No. 2 continuous bars running along the length of the bridge model in the top and bottom slabs. Transverse reinforcement in the top and bottom slabs consisted of closely spaced No. 2 bars. Web longitudinal reinforcement consisted of continuous No. 3 bars for the upper part, and continuous No. 2 bars for the middle and lower part. No. 2 bars were used to form stirrups for the webs.

Instrumentation

The fundamental measurements in any study of a structural model are reactions, deflections and strains. In view of the large number of reaction, deflection and strain readings envisaged during the testing of the box girder model from the start of the experimental program to its ultimate failure, it was decided at the very outset to record measurements electronically.

Reactions were measured at each of the five girders at the two abutments of the bridge model, and at four locations under the central footing. The end reactions were recorded by load cells centred under the girder webs. As the central footing was to be anchored to the floor by means of prestressed steel rods, hollow load cells were also fitted above this footing to indicate the change of tensile force in each rod. The difference between the two load cell readings at each location represented the net reaction at that point.

Vertical deflections of the box girder bridge model along the length were measured at several locations by means of potentiometers. These locations included transverse sections X, QB, Z, QZ and Y. At each of these sections five potentiometers were placed, one under each girder, except at Z, where the central column under girder 3 allowed only four potentiometers, resulting in a total of 24 potentiometers.

Longitudinal strain measurement was confined to sections A and B in the span with the diaphragm and to sections C and D in the other span. As has already been observed, section A is in the region of maximum positive moment in Span I and section D is in the region of maximum positive moment in Span II. Sections B and C are at distances of 3 ft, respectively, on each side of the bridge centre support, and provide information on the negative moments at those locations. At section D, i.e., in the undiaphragmed section, some transverse strains in the steel reinforcement were also measured.

It was decided to use strain meters for the measurement of concrete compressive strains, and weldable waterproofed strain gauges for the measurement of strains in the steel reinforcing bars. The strain meter and the weldable gauge were known for their reliability in strain measurement, and also eliminated the uncertain effects of laboratory waterproofing methods on the final results. The strain meter was a special recently developed miniature version of a type of strain meter extensively used for the measurement of compressive strains in full scale concrete structures. It consisted of a cylindrical barrel (gauge length 4·5 in, diameter 0·625 in), with a flange of 0·75 in diameter at each end. The relative movement of the flanges resulted in a strain output from the gauge bridge inside the barrel. The barrel was wrapped in cloth so as to prevent bonding to the concrete.

The commercially available weldable waterproofed strain gauge with a gauge length of 1·09 in consisted of a hermetically sealed and mechanically protected nickel-chrome strain filament housed in a small stainless steel cylindrical shell (diameter 0·03 in). The shell had two 0·06 in wide flanges allowing spot welding of the gauge to the ground reinforcing bar. The gauge was designed to perform under severe conditions of moisture and shock, the filament being inherently shielded in the cylindrical shell by highly compacted magnesium oxide insulating powder.

In addition, in order to measure a representative amount of strain in the steel reinforcement over a long gauge length of the order of 1 ft, a 'clip-gauge' consisting of a spring steel strip instrumented with a strain-gauge bridge was developed. The clip-gauge could be attached externally to plugs brazed to a reinforcing bar before placing in concrete, and provided a reliable check on values of strain measured by the weldable gauges.

Clip gauges were attached to the reinforcement at each girder centreline at sections A, B, C and D. All instrumentation was properly calibrated before installation in concrete. In particular, the weldable strain gauges were calibrated in a special rig after installation at diametrically opposite locations on a reinforcing bar.

Locations and types of measuring devices and gauges at the instrumented sections A, B, C and D of the box girder bridge model are given in Figs 3 and 4.

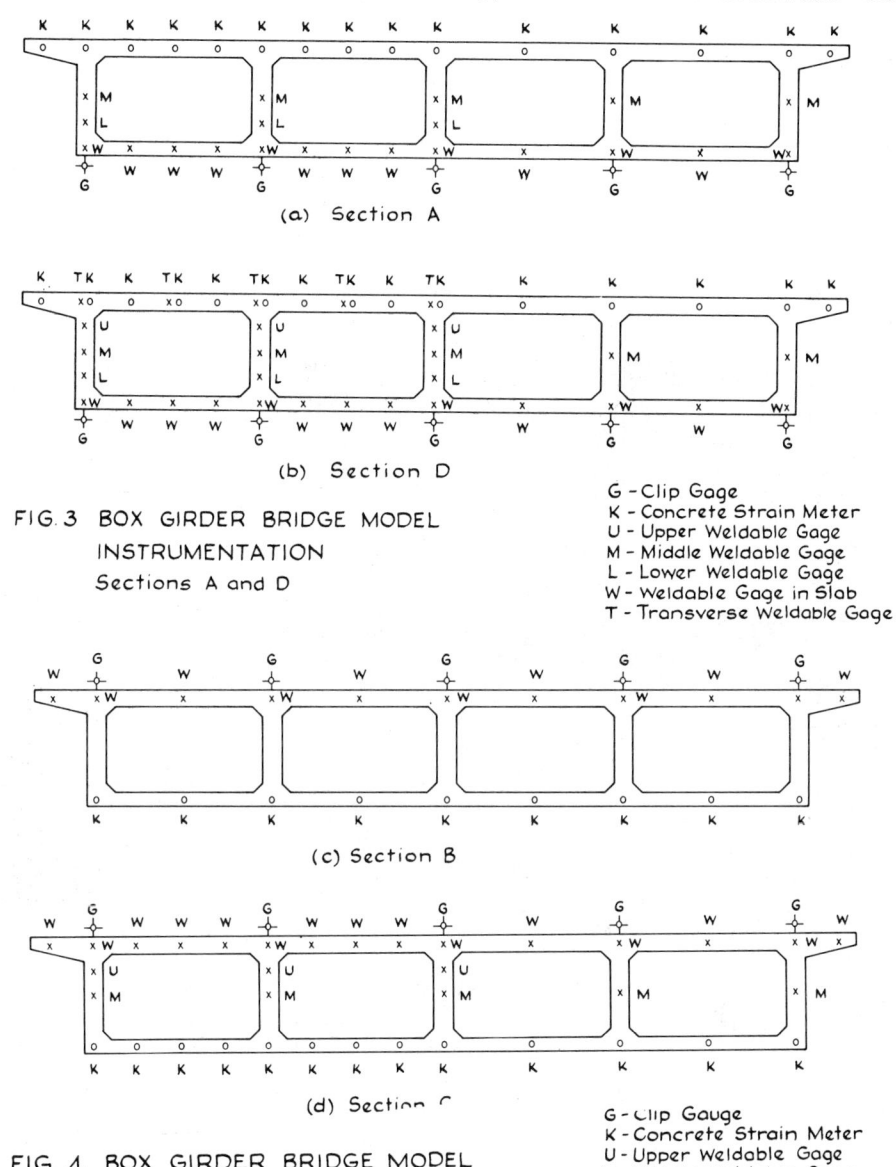

(a) Section A

(b) Section D

FIG. 3 BOX GIRDER BRIDGE MODEL
INSTRUMENTATION
Sections A and D

G – Clip Gage
K – Concrete Strain Meter
U – Upper Weldable Gage
M – Middle Weldable Gage
L – Lower Weldable Gage
W – Weldable Gage in Slab
T – Transverse Weldable Gage.

(c) Section B

(d) Section C

FIG. 4 BOX GIRDER BRIDGE MODEL
INSTRUMENTATION
Sections B and C

G – Clip Gauge
K – Concrete Strain Meter
U – Upper Weldable Gage
M – Middle Weldable Gage
L – Lower Weldable Gage
W – Weldable gage in slab

Construction of Box Girder Bridge Model

The size and complexity of the box girder bridge model, and the importance of completing the job in a minimum amount of time, dictated that the construction of the
box girder bridge model be done under contract. The structural design of the box
girder bridge model was identical to that of the prototype. The reinforcing steel was
specially purchased, and was cut, bent, and placed by the contractor under supervision of the project staff. All instrumentation was purchased, calibrated, and
installed in place by the project staff.

The mix for the concrete used in the box girder bridge model was designed by the project staff, and the ready-mix concrete was supplied by a local dealer. The size of aggregate used was $\frac{3}{8}$ in, and the slump specified for the concrete mix was 5 in $\pm \frac{1}{2}$ in.

The two end abutments and the centre column with footing were cast first. Subsequently, the forms for the bottom slab, diaphragms, and sides of the model were built. After placement of the gauged and ungauged reinforcement and concrete strain meters, the inner forms for the cells were installed, and the concrete for the bottom slab, girder webs, and diaphragms was placed. No vibration was permitted in the vicinity of the gauged sections. After curing for 10 days, the inner forms were removed. The box girder bridge model at this stage is shown in Plate 1.

PLATE 1. BOX GIRDER BRIDGE MODEL AFTER REMOVAL OF INNER FORMS

To allow for the placing of steel billets within the cells, the lead wires of the gauges and strain meters were braided in a manner that would also prevent their getting damaged when the framework for the top slab was subsequently pushed down and made to fall on top of the steel billets. After an inch of sand had been spread in the cells, six pairs of steel billets per cell per span were placed end to end to provide a uniformly distributed load. The total weight of the 96 billets each of approximate dimensions 9 x 9 x 65 in and the sand was 113·1 kips, which compared favourably with the extra load for simulation of prototype behaviour calculated as 116 kips.

Next, the forms for the top slab were fabricated, and after the two layers of rein-forcement for the $2\frac{1}{4}$ in thick top slab were placed in position and the gauges and strain meters checked, the top slab was cast in the same manner as the bottom slab and cured for 10 days. Load cells were placed in the appropriate reaction locations after the concrete strength of the top slab exceeded 2500 psi. At this stage, the forms holding the top slab in place were pushed down by means of rods with washers which had been provided with the forms. The forms for the top slab were thus detached from the bottom surface of the slab and made to fall into the cells.

On the bridge deck, centrelines of the girders and transverse sections were marked. Finally, after zero readings were taken, the shoring and bottom formwork were removed for the taking of dead load readings, and the bridge was whitewashed for the observation of crack propagation.

Experimental Program

The main objective of the test program was to obtain information on load distribution in reinforced concrete box girder bridges under conditions of working loads. Working loads would result in total design stresses of 24 ksi in the tensile steel at the sections of loading. Bearing in mind, however, that the tensile stresses in the tensile reinforcement at these sections due to the self-weight and extra dead load of the bridge model alone was about 12 ksi, it was decided to consider two levels of working loads — those producing total stresses in the steel of 24 ksi, and those resulting in total tensile steel stresses of 30 ksi at the sections of loading. The advantage of the latter stress level was that 50% higher values of live load stresses and strains could be registered for a total increase in the bridge model stresses of only 6 ksi.

In terms of the actual experimental data, it was convenient to divide the experimental program into seven phases, from the dead load condition [Phase 0] through the 24, 30, 40, 50 and 60 ksi stress levels [Phases I to V] to the failure condition [Phase VI].

The box girder bridge model had a loading frame at each mid-span enabling live loads to be applied at each of the girders by means of jacks singly and in various combinations. Each phase of the experimental program for live loads comprised firstly the application of equal loads on each girder at both mid-spans to produce the same order of nominal steel stress at sections of maximum positive and negative moment. These loads were termed 'conditioning loads'. Subsequently, after the removal of the conditioning loads, point loads were applied in several combinations. The conditioning loads were chosen to produce nominal total tensile steel stresses of 24, 30, 40, 50 and 60 ksi at the sections of loading, and to represent the successive deterioration of the box girder bridge model due to the effects of overload. The point loads, however, were chosen in all cases to produce maximum total stresses of the order of the working stresses, i.e., 24 and 30 ksi total tensile stress in the reinforcement.

The loading phase involving the application of the conditioning loads to produce the 30 ksi tensile steel stress was chosen as the most representative from the point of view of assessing actual box girder bridge behaviour for design purposes. Following the conditioning loads, single and combined point loads were applied to the box girder bridge model for three different types of support conditions: simply supported ends, centre bent restrained against torsional movement, and bottom of end diaphragms restrained against longitudinal movement. The two latter support conditions modelled a bridge with a three-pedestal support at the bent and a bridge with the end diaphragms cast directly on vertical piles.

In addition to the point loads, scaled down truck and heavy construction vehicle loads and a moving load were applied to the bridge in this loading phase.

For the application of live loads to the bridge in each span as envisaged by the experimental program, two identical loading frames were designed by the project staff and fabricated. Scaled down AASHO trucks, heavy construction vehicles and accessories for loading, changes in support conditions, etc., had also to be fabricated and assembled.

To apply the conditioning loads and to permit the application of any set of combined point loads, five hydraulic 20 ton jacks at each mid-span were connected by means of steel pipes fitted with valves to a common manifold. Each of the two manifolds was connected by means of high pressure hoses to an air pressure hydraulic pump system. Load cells were used to check that each jack delivered an equal load when all valves were open. Calibrations for jack load versus pump pressure were made earlier, and confirmation of the load value in each case was obtained from load cells. The use of ten identical jacks eliminated the need for moving the loading jacks into position, and all that was necessary was the opening or closing of the valves as the loading scheme required.

The arrangement of jacks and a general view of the bridge before testing can be seen in Plate 2. As some readings were taken in two steps (zero and maximum) and others in several increments, the total number of loading steps recorded was about 600 for 218 combinations of loads.

PLATE 2. BOX GIRDER BRIDGE MODEL LOADING ARRANGEMENT

Data Recording

All live load readings were obtained by means of a data acquisition and scanning system comprising a portable computer of 8 k storage, a digital voltmeter unit, a teletype unit, and four terminal boxes.

The computer was programmed to convert gauge readings to direct values of strain in micro inch/inch, deflection in inches, and reaction or load in kips. The calibration factors for each device were fed into the computer to make those conversions.

The basic routine for the calibration and output was as follows:

After the bridge was loaded at any stage, the computer scanned each of the 237 gauge or meter channels five times and averaged the readings in a period of four minutes. The output, in which each reading was the difference from a datum established before each session of loading, was obtained by means of the teletype on punched paper tape and was also simultaneously printed on paper. As a simple check on the stability of the equipment, the first four readings recorded the variation of the standard resistors from datum readings taken for each session of testing. The next two readings gave the values of the loads on the bridge as registered by the load cells of the loading jacks. All these readings were carefully scrutinized at each loading and unloading step.

Data Reduction

As the bridge model experimental program was finished only in September 1970, the data reduction is still in progress. Four separate computer programs are being used in an overlay system to reduce the large volume of experimental data.

The first program, called DATAMAN, classifies and prints the data according to category of measuring device, rounds off readings to the desired accuracy, and separates readings given by unbalanced gauges, or garbled readings. The latter are separately catalogued and edited, and refer only to strain data in steel or concrete. For any loading case, the number of strain readings edited has never exceeded 7 channels out of 175 channels recording strains. The DATAMAN program also produces the data in the form of punched cards.

The punched card decks are fed into the ADJSTRN program, the main purpose of which is to pass smooth parabolas to fit the strains registered in each bay of the top and bottom slabs, and least squares lines through the strains registered in the girder webs and the slabs. This program prints out the measured and adjusted strain ordinates at each gauged location, and their ratio. The program can discard any measured strain value if its ratio does not satisfy a given criterion, e.g., lie between 0·80 and 1·20, and return to pass a parabola to fit the remaining points.

The MOMENT program integrates the compressive strains at each instrumented section and sums up the tensile forces in the steel reinforcement. The box girder bridge model is considered to be divided into three identical I-shaped interior girders and two exterior girders that lack half a bottom flange each. For each loading case, all compressive and tensile quantities for each girder are calculated at each of the four instrumented sections, A, B, C and D, by the MOMENT program. The resisting longitudinal moment for each girder and the whole section is printed out in actual value and as a percentage of the total moment at the section for four different axes: the compression flange mid-depth, the line of main tensile reinforcement, the neutral axis of the uncracked box girder cross-section, and the neutral axis of each girder as obtained from experiment. The last axis is not a horizontal straight line.

The REACTION program sums up the readings of the reactions separately at the east and west abutments and at the centre footing. Several checks for equilibrium are made, and the moments obtained at various sections from the measured reactions are tabulated and compared with values obtained from continuous beam formulas.

Experimental and Analytical Results

Due to restrictions of space, only two types of loading are considered in the present paper.

First, the conditioning loads, in which equal loads act over each girder web at both mid-span sections, are treated. The conditioning loads, along with the self weight and extra dead load of the model cause nominal maximum stresses in the tensile reinforcement of 24, 30 and 40 ksi, respectively; however, the readings have been normalised for purposes of comparison to values of 20 kips per girder, or 100 kips per span.

Second, a point load acting on an exterior girder web at the diaphragmed mid-span, after conditioning loads causing nominal maximum stresses of 30 ksi in the tensile reinforcement have been applied and removed, is considered.

The point load has been normalised to a value of 100 kips, and results have been given for simple end supports, for the centre bent restrained against torsion, and for the end supports restrained against longitudinal movement.

Comparisons relating only to magnitude of the deflections and moments are made with the MUPDI program developed at the University of California, in which the direct stiffness method is used to analyse the bridge as an uncracked folded plate system. This theory assumes that the bridge model has simply supported ends.

Tables 1 and 2 show the normalised deflections for each girder of the diaphragmed mid-span (section X) for three different conditioning loads and for the same point load with three different support conditions, respectively.

Table 3 shows the distribution of moments in kip-ft and as percentages to each girder at each of the instrumented sections, A, B, C and D for three stress levels of the conditioning loads. The moments for both the analysis and the experiment were taken about the same axis (the neutral axis of the uncracked box girder cross-section) for comparison.

Table 4 gives the distribution of moments in kip-ft and as percentages at the same instrumented section and for the same value of point load but for three different support conditions. As for Table 3, all moments were taken about the neutral axis of the uncracked box girder cross-section.

Comparisons of Experimental and Analytical Results

As a substantial volume of the data has not yet been reduced, only preliminary observations can be made on the agreement between experiment and theory. All the comparisons made in the preceding tables incorporate high values of load and levels of stress which have especially been chosen to provide an insight into the experimental behaviour of box girder bridges. Under service loads, actual box girder bridges have large factors of safety, and deflections and moments produced in them by truck and vehicular loads are very small.

The values given in the tables suggest the following general conclusions:

1. Mid-span deflections increase with increasing levels of tensile stress, and decrease with restraints against torsion or longitudinal movement. The rate of increase in deflections is higher in the tensile stress range 24 ksi to 30 ksi than in the tensile stress range 30 ksi to 40 ksi. Under conditioning loads, the ratio of experimental values to theoretical values for the deflections at 24, 30, and 40 ksi stress levels is uniformly about 30%, 50%, and 70% higher than unity. For the load on an exterior girder, no such uniformity obtains, but the ratio of deflections is again greater than unity for the two cases of simple end supports where comparison can be made. The higher experimental

values are, of course, directly attributable to cracking of the concrete, ignored in the elastic folded plate theory which considers a homogeneous uncracked cross-section.

2. The agreement in the values of moments in kip-ft at the instrumented sections under conditioning loads is greatest at sections B and C for the conditioning loads causing 24 ksi total tensile stress. The moment at section D is greater than the analytical value, and the moment at section A is smaller than the analytical value, but overall agreement obtains with the analytical results. This agreement deteriorates somewhat for conditioning loads causing higher total tensile stresses, the experimental moment tending to increase. As with the deflections, the unit increase is smaller between the 30 and 40 ksi levels than between the 24 and 30 ksi stress levels. For the load on an exterior girder, general agreement obtains, between theory and experiment, for the two cases of simple end supports; the values measured at section A continue to be lower than those given by the theory.

3. As opposed to the kip-ft values of the moments, the percentage distribution of the longitudinal moment to the girders obtained from experiment agrees remarkably well with theory, except for the exterior girders at section D, irrespective of the loading or support conditions or maximum stress levels. In fact, the distribution improves a little at stress levels higher than working stresses. As the measured values of deflection in the region of section D agreed with values obtained for similar types of loading in the region of section A, it is possible that some of the gauges and meters in the exterior girders at section D gave faulty readings.

A detailed and complete set of comparisons between analytical and experimental results is in progress and will be presented along with other findings and conclusions in a set of research reports[4] to be published later this year (1971).

Loading of Bridge Model to Failure

For the final loading phase to failure, it was decided to use two loading rams, each of 200 kips capacity and a maximum stroke of 10 in at each mid-span. The rams were centred on steel plates with neoprene bearing pads at girders 2 and 4, respectively. The nominal mid-span load causing failure through the formation of a collapse mechanism with plastic hinges at the mid-span and at the centre bent support was calculated as 192 kips. The loads were successively increased and readings were taken at each stage. The maximum tensile stress level at a load of about 150 kips per span was 60 ksi. The load per span was increased to 160 kips per span and then to 170 kips per span, at which point yielding of the bottom slab steel started in the undiaphragmed mid-span, causing a local slab failure under girder 2, which was directly loaded. The girder pushed downward under the load causing a major crack in the bottom slab that extended longitudinally about 5 ft in the direction towards the centre bent. With further loading, the local failure progressed, the girder moved downward, until with an audible sound the bearing plate of the ram at girder 2 in the undiaphragmed mid-span punched through the top slab, and the load dropped to 125 kips.

No damage of this kind was observed at the diaphragmed mid-span where the behaviour of the gauges continued to be fairly linear. The deflection in the undiaphragmed span was maintained by the insertion of solid timber struts at the mid-span section there, the rams were unloaded, and a new set of zero readings taken. Subsequently, only the diaphragmed span was loaded. At a load between 170 and 180 kips yielding began to take place at the midspan and continued uniformly. Cracks began to widen, and longitudinal shear cracks appeared on the north and south faces of the bridge model at the junction of the top slab with the girders. Under

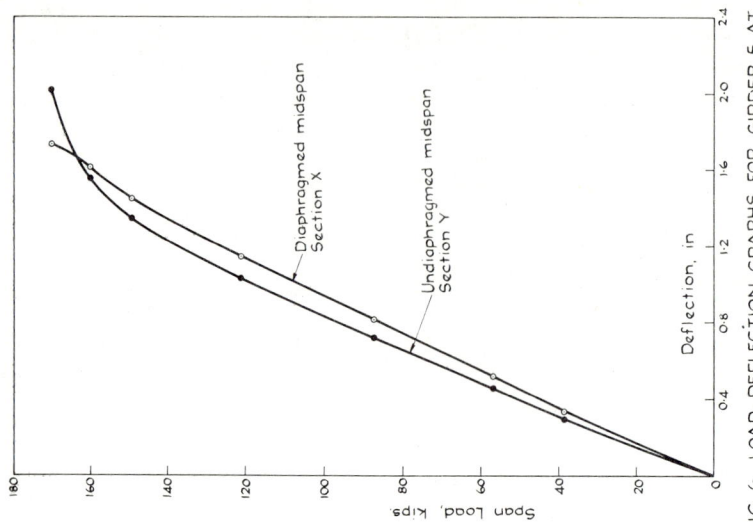

FIG 6 LOAD-DEFLECTION GRAPHS FOR GIRDER 5 AT MIDSPAN SECTIONS DURING LOADING TO FAILURE

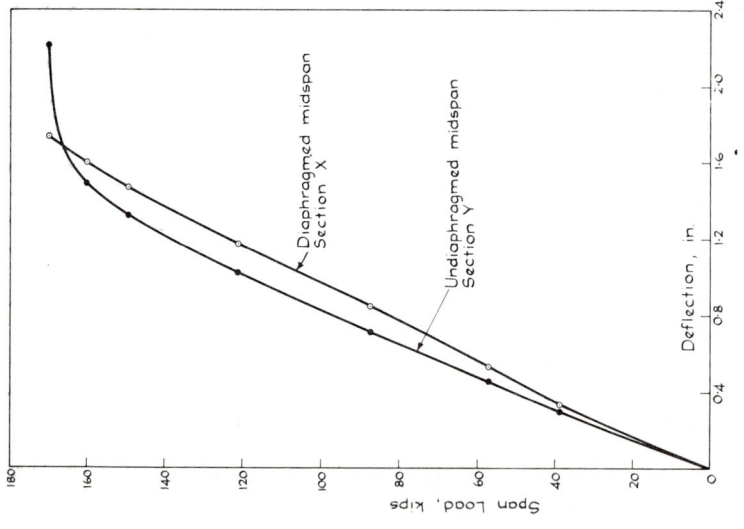

FIG 5 LOAD-DEFLECTION GRAPHS FOR GIRDER 1 AT MIDSPAN SECTIONS DURING LOADING TO FAILURE

decreasing load the second hinge formed at a section about 8 ft away from the bridge centreline, and the collapse mechanism was clearly visible. The reason for the formation of the second hinge at this distance lay in the fact that the top slab reinforcement was cut off in this region. The final data recorded on the data acquisition system was for a midspan deflection of 10 in, by which time the load had fallen to 80 kips.

Graphs for the first part of the loading to failure in which both spans were loaded comprise Figs 5 and 6, which give the load-deflection relations for girders 1 and 5 at the diaphragmed and undiaphragmed midspan sections X and Y.

ACKNOWLEDGEMENTS

Research work described in this paper was sponsored and financed jointly by the U.S. Department of Transportation, Federal Highway Administration, Bureau of Public Roads, and the California Division of Highways of the California Department of Public Works. The opinions, findings, and conclusions expressed in this paper are those of the authors and not necessarily those of the sponsors.

REFERENCES

1. Scordelis, A. C. Analytical solutions for box girder bridges. Paper submitted to Conference on Developments in Bridge Design and Construction, Cardiff, 1971.
2. American Association of State Highway Officials. *Standard specifications for highway bridges.* 10th ed. Washington, D.C., 1969.
3. Scordelis, A. C. and Meyer, C. *Wheel load distribution in concrete box girder bridges.* University of California, Berkeley, Structural Engineering and Structural Mechanics Report no. SESM 69-1, January 1969.
4. Bouwkamp, J. G., Scordelis, A. C. and Wasti, S. T. Structural behavior of a two span reinforced concrete box girder bridge model. Volumes 1 and 2. University of California, Berkeley, Structural Engineering and Structural Mechanics Reports, 1971 (to be published).

TABLE 1

Deflections in Inches at Diaphragmed Midspan Under Normalised
Conditioning Loads at Each Midspan

| Girder | Analytical Results | Normalised Experimental Results for Conditioning Loads Causing Total Nominal Tensile Stress in Reinforcement at Loaded Midspans of | | | | | |
| | | 24 ksi | | 30 ksi | | 40 ksi | |
	Theory	Expt.	E/T	Expt.	E/T	Expt.	E/T
1	0·478	0·627	1·312	0·722	1·510	0·815	1·705
2	0·478	0·619	1·295	0·716	1·498	0·820	1·715
3	0·478	0·610	1·276	0·709	1·483	0·809	1·692
4	0·478	0·608	1·272	0·716	1·498	0·812	1·699
5	0·478	0·609	1·274	0·705	1·475	0·807	1·688

TABLE 2

Deflections in Inches at Diaphragmed Midspan Under Normalised
Point Load on Exterior Girder 1

| Girder | Analytical Results | Normalised Experimental Results for Point Load on Exterior Girder 1 Causing Total Nominal Tensile Stress in Girder Reinforcement of 30 ksi for Following Support Condition | | | | | |
| | | Simply Supported Ends | | Centre Bent Restrained Against Torsion | | Ends Restrained Against Longitudinal Movement | |
	Theory	Expt.	E/T	Expt.	E/T	Expt.	E/T
1	1·059	1·421	1·342	1·363	1·287	1·112	1·050
2	0·883	1·181	1·337	1·151	1·305	0·877	0·993
3	0·743	0·965	1·299	0·967	1·301	0·658	0·886
4	0·633	0·785	1·240	0·802	1·267	0·475	0·750
5	0·543	0·603	1·110	0·645	1·188	0·303	0·558

TABLE 3

Distribution of Moments (kip-ft and %) to Each Girder at Each Instrumented
Section Due to Normalised Conditioning Loads

Section	Girder	Analytical Results Theory		Normalised Experimental Results for Conditioning Loads Causing Total Nominal Tensile Stress in Reinforcement at Loaded Midspans of					
				24 ksi		30 ksi		40 ksi	
		K-Ft	%	K-Ft	%	K-Ft	%	K-Ft	%
A	1	73	16·2	68	16·3	72	16·4	68	15·7
	2	102	22·6	91	21·7	95	21·9	96	22·1
	3	102	22·4	94	22·5	98	22·5	100	23·1
	4	102	22·6	100	23·8	101	23·1	102	23·6
	5	73	16·2	66	15·7	70	16·1	67	15·5
	Σ	452	100·0	419	100·0	436	100·0	433	100·0
B	1	−75	16·2	−72	14·8	−75	15·0	−77	15·1
	2	−105	22·6	−112	22·9	−112	22·4	−115	22·6
	3	−104	22·4	−110	22·5	−112	22·3	−116	22·6
	4	−105	22·6	−110	24·4	−126	25·0	−123	24·1
	5	−75	16·2	−75	15·4	−77	15·3	−80	15·6
	Σ	−464	100·0	−488	100·0	−502	100·0	−511	100·0
C	1	−75	16·2	−69	14·6	−74	14·7	−78	15·3
	2	−105	22·6	−114	24·2	−122	24·1	−125	24·4
	3	−104	22·4	−109	23·0	−114	22·5	−117	22·7
	4	−105	22·6	−108	22·9	−116	22·9	−112	21·8
	5	−75	16·2	−72	15·3	−80	15·8	−81	15·8
	Σ	−464	100·0	−472	100·0	−506	100·0	−513	100·0
D	1	74	16·3	108	22·0	106	20·6	109	20·6
	2	102	22·5	105	21·4	116	22·5	119	22·6
	3	101	22·4	108	22·0	116	22·5	120	22·6
	4	102	22·5	107	21·9	112	21·8	114	21·7
	5	74	16·3	62	12·7	65	12·6	66	12·5
	Σ	453	100·0	490	100·0	515	100·0	528	100·0

TABLE 4

Distribution of Moments (kip-ft and %) to Each Girder at Instrumented Section A Due to
Normalised Point Load on Exterior Girder 1 at Diaphragmed Midspan

Section	Girder	Analytical Results Theory		Normalised Experimental Results for Point Load on Exterior Girder 1 Causing Total Nominal Tensile Stress in Girder Reinforcement of 30 ksi for Following Support Condition					
				Simply Supported Ends		Centre Bent Restrained Against Torsion		Ends Restrained Against Longitudinal Movement	
		K-Ft	%	K-Ft	%	K-Ft	%	K-Ft	%
A	1	116	19·8	109	20·7	110	21·1	92	23·0
	2	146	24·9	125	24·0	126	24·2	98	24·6
	3	128	21·8	118	22·5	117	22·4	90	22·6
	4	115	19·6	109	20·7	107	20·5	78	19·5
	5	82	13·9	63	12·1	61	11·8	41	10·3
	Σ	587	100·0	524	100·0	521	100·0	399	100·0

FINITE ELEMENT ANALYSIS OF BRIDGE CURVED IN PLAN

Prof. F. SAWKO
University of Liverpool
P. A. MERRIMAN
W. S. Atkins and Partners

SYNOPSIS

The paper deals with the analysis of isotropic and orthotropic bridge decks curved in plan. Stiffness matrices for segmental and quadrilateral elements for plate bending are used in general finite element slab programs for the analysis of simply supported and continuous isotropic and orthotropic bridges. Comparison with published results demonstrates the high degree of accuracy obtainable with the authors' segmental plate bending element. Tests on continuous uniform and varying section perspex slabs on point strips are described and used as further evidence for the validity of the proposed element.

INTRODUCTION

Urban interchanges often necessitate the design and construction of bridges curved in plan to a small radius. The increased demand for these bridges has stimulated research into accurate methods of analysis which would in turn lead to economic design of these structures.

Three different types of deck are usually employed. These are true slab bridges, composite slab and beam bridges with curved ribs or true spine beam bridges. This paper deals with the analysis of the first two types, where structural action can be idealised to that of a curved isotropic or orthotropic slab.

Methods of Analysis for Bridges Curved in Plan

(a) The grillage approach

The grillage approach for curved bridges was developed by Sawko[1] who showed that the approximation of a curve by a series of straight members could lead to a serious misinterpretation of the computer output. He therefore developed the stiffness of a curved beam by the flexibility method, and included it in a general computer program for the analysis of grillages consisting of straight and curved members. The application of the program to bridge design was demonstrated successfully on Lofthouse interchange,[2] which consisted of a number of prefabricated steel box girders with an *in situ* composite concrete slab, and the accuracy of the approach for slab analysis was verified by comparison with published solution.[3]

(b) Series solution for simply supported bridge decks

Yoshimura[4,5] presented a series solution for the differential equation in polar co-ordinates for a radially supported curved orthotropic bridge deck under uniform load. Coull,[6] using a similar approach, produced a reduced function of isotropic decks under concentrated loading. The deflection in the spanwise direction was expressed as a Fourier series, the unknown coefficients being only functions of radial co-ordinates.

The applied point load was included as a corresponding series, producing shear discontinuities in the radial direction. Conditions of compatibility were satisfied along the line load and the coefficients calculated. Experimental tests on perspex and asbestos models[6] confirmed the accuracy of the series solution for decks with different Poisson's ratios.

(c) The finite element method

The most suitable plate bending element for bridges of arbitrary shape is the triangle. Results of investigations on different stiffness matrices for triangular elements have been published[7-10] and more sophisticated techniques are continuously being developed to ensure normal slope compatibility of adjoining elements and curvature compatibility at the nodes. The limitation of triangular elements is the inaccurate representation of the curved edges by a series of chords and of section properties for curved orthotropic bridges, where girder stiffening is in the form of continuous concentric circular areas. Both effects are a possible source of error.

The authors felt that the development of a successful annular segment would overcome the disadvantages encountered with triangular elements because of the following reasons.

   (i) Exact boundary conditions would be simulated.
  (ii) Orthotropic properties could be expressed without any stiffness discontinuities.
 (iii) Experience with rectangular elements has demonstrated their superior convergence characteristics compared with corresponding triangular elements.

A successful annular plate bending element has been recently developed by the authors and reported in technical literature.[11] A brief summary of the deformation

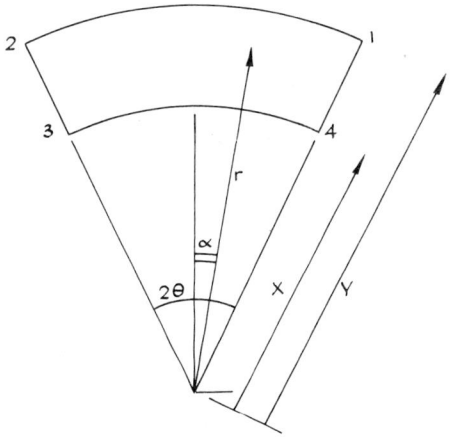

FIG. 1. NOTATION FOR ANNULAR SEGMENT

function used will thus be sufficient. Over the annular segment, Fig. 1, the deflection function was expressed in polar co-ordinates as a function of r and α as follows:

$$w = (a + b \cos \alpha + c \sin \alpha + d \sin 2\alpha)(e + fr + gr^2 + hr^3)$$

At each node $w, \dfrac{\partial w}{\partial r}, \dfrac{1}{r}\dfrac{\partial w}{\partial \theta}, \dfrac{1}{r}\dfrac{\partial^2 w}{\partial r \partial \theta}$, were specified as the unknown displacements.

The choice of these displacements ensured that deflections and normal slopes were compatible along the boundaries.

The strain energy of the element was calculated in polar co-ordinates and the principle of minimum potential energy applied to determine the familiar relationship.

$$F = ku$$

The stiffness element thus derived will be used in the examples that follow. Mention must also be made of a quadrilateral element outlined by Allwood[12] which was derived using curvilinear co-ordinates. This was also applied in subsequent examples to isotropic bridge decks to compare the solutions for the two types of element.

Examples of Curved Bridge Analysis

Search of technical literature revealed a complete absence of full scale or large scale tests on curved bridge decks which could form a basis for comparison of various methods of analysis. It has, therefore, been necessary to make use of published results for small scale tests and to conduct additional experiments for curved continuous decks. Four different types of structure have been investigated.

(a) Simply supported isotropic bridge decks

Sawko[3] compared the curved grillage analysis with the two model tests and series solution by Coull.[6] The results confirmed the accuracy of both methods for predicting the longitudinal moments ($M\theta$) at the centre line of a curved bridge under a point load.

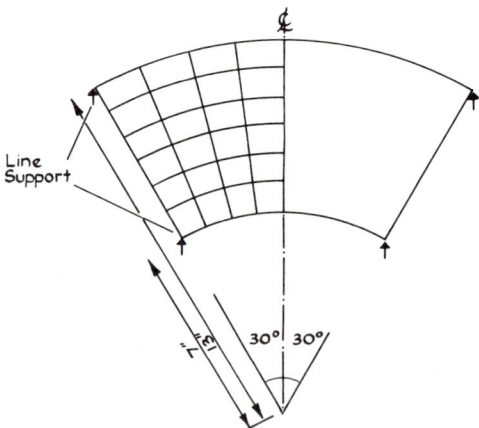

$E = 4 \cdot 6 \times 10^5 \text{ lbs/sq.ins.}$
$v = 0 \cdot 35$
$t = 0 \cdot 168 \text{ ins.}$

FIG. 2. PROPERTIES OF PERSPEX MODEL AND MESH USED
FOR ANALYSIS

In Table 1 the centre line deflections for the annular segment, quadrilateral element and series solution are tabulated for the perspex model shown in Fig. 2 under three concentrated loads. The small difference between the series solution and finite element results demonstrates the accuracy of both quadrilateral and annular segment elements for the analysis of curved isotropic slabs.

For the asbestos model good correlation was obtained[6] between the experimental results and the annular segment for longitudinal (*Mθ*) and radial moments (*Mr*). It was found generally that the peak moments approximately 10% higher are obtained with the annular segment compared with the series solution.

(b) Simply supported orthotropic bridge deck

The real test for the annular segment is the case of orthotropic slab where the main flexural members lie along radial lines. Bending moments for orthotropic properties are simply obtained as follows:

$$
\begin{aligned}
Mr &= \\
M\theta &= \\
Mr\theta &=
\end{aligned}
\begin{bmatrix}
DR & D1 & 0 \\
D1 & DT & 0 \\
0 & 0 & 2DRT
\end{bmatrix}
\begin{bmatrix}
(-\dfrac{\partial^2 w}{\partial r^2}) \\[2mm]
(-\dfrac{1}{r^2}\dfrac{\partial^2 w}{\partial \alpha^2} + \dfrac{1}{r}\dfrac{\partial w}{\partial r}) \\[2mm]
\dfrac{\partial}{\partial r}(\dfrac{1}{r}\dfrac{\partial w}{\partial \alpha})
\end{bmatrix}
$$

where *DR* — flexural rigidity/unit length in radial direction.
  DT — flexural rigidity/unit length in circumferential direction.
  D1 — flexural rigidity due to the coupling of the curvatures in the orthogonal directions due to Poisson's ratio/unit length.
  DRT — torsional rigidity/unit length.

The introduction of these flexural constants was tested by comparing the results with a series solution for a uniformly loaded orthotropic slab with an included angle of 30° analysed by Yoshimura.[4] The ratio of *DT/DR* varied from 0·25 to 4 and *v* was assumed zero. Good agreement is obtained with the annular segment for deflections and moments shown in Figs. 3—5.

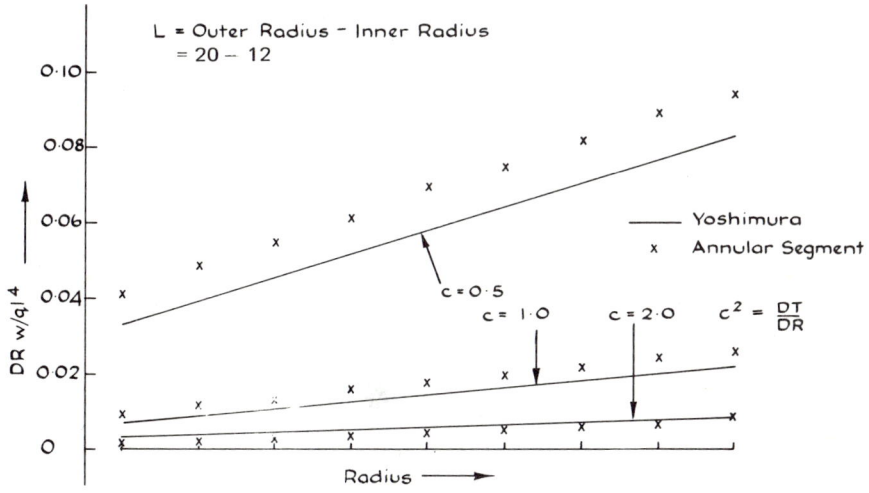

FIG. 3. DEFLECTIONS FOR ORTHOTROPIC SLABS UNIFORMLY LOADED

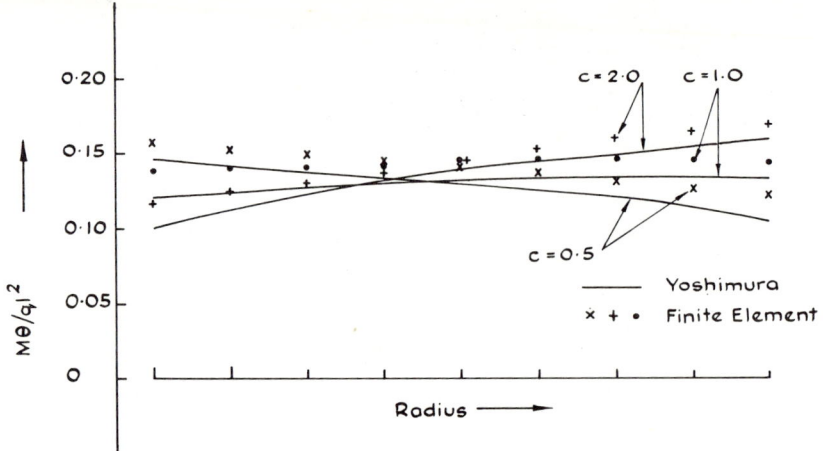

FIG. 4. LONGITUDINAL MOMENT $M\theta$ FOR ORTHOTROPIC
SLABS UNIFORMLY LOADED

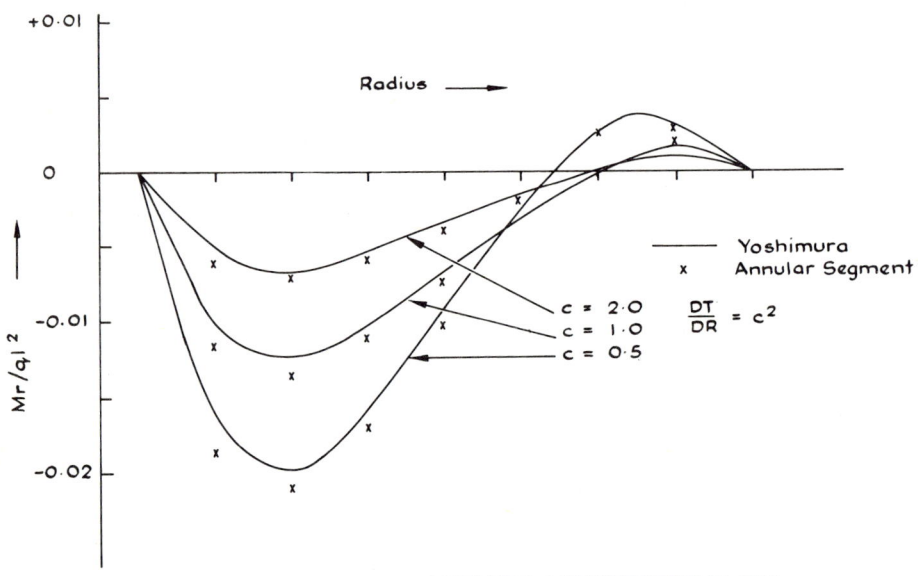

FIG. 5. RADIAL MOMENTS FOR ORTHOTROPIC SLABS
UNIFORMLY LOADED

(c) Continuous isotropic bridge deck

No suitable model test was available to investigate the accuracy of annular segment
elements for the analysis of continuous slabs on discrete supports. It has, therefore,
been necessary for the authors to design and test a model which could be used for
that purpose.

Details of the model are shown in Fig. 6. It will be observed that the model was
supported on eight discrete supports along four radial lines. Considerable cross-
bending between supports was thus introduced to produce a critical assessment of

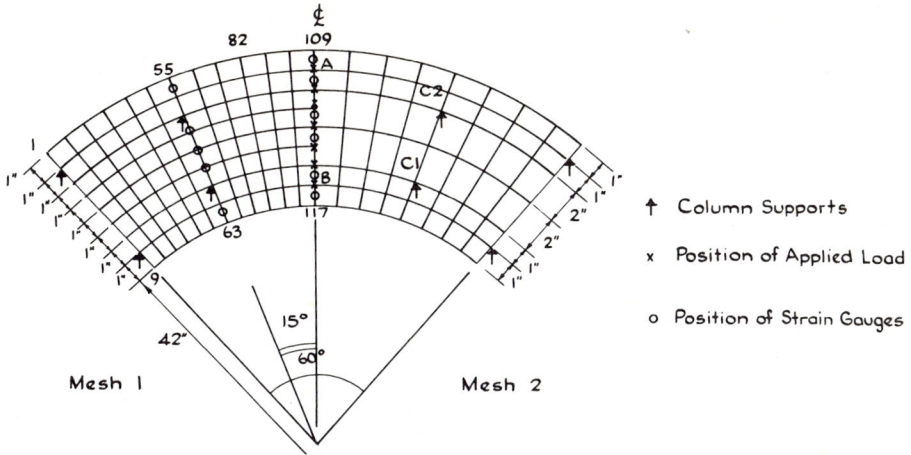

FIG. 6. LAYOUT OF CONTINUOUS CURVED SLAB MODEL WITH THE TWO MESH SIZES

the behaviour of the annular element. Circumferential and radial strains were measured, and radial and circumferential moments calculated from the formulae:

$$Mr = \frac{Eh^2}{6(1-v^2)}\,(\epsilon r + v\epsilon\theta), \quad M\theta = \frac{Eh^2}{6(1-v^2)}\,(\epsilon\theta + v\epsilon r)$$

where h = thickness of slab ¾in
$\quad\quad$ ϵr = radial strain
$\quad\quad$ $\epsilon\theta$ = circumferential strain.
Values of Young's Modulus and Poisson's ratio were obtained from standard tests, and the values used in the analysis were: F = 440 000 lb/in² v = 0·4.

Figure 6 also shows the two different mesh sizes employed for finite element analysis. The coarser division was only used for the quadrilateral element. The corresponding grillage idealisation is shown in Fig. 7.

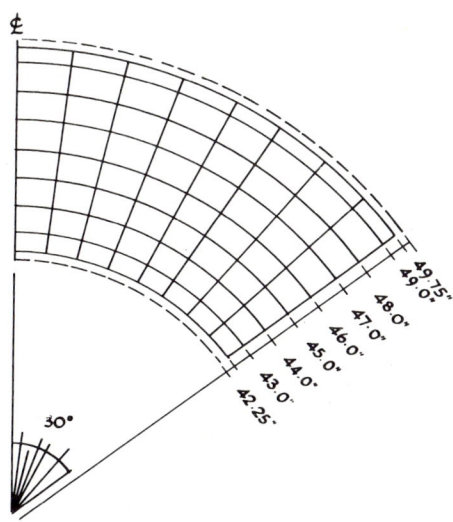

FIG. 7. MESH USED FOR GRILLAGE ANALYSIS OF CONTINUOUS CURVED BRIDGE DECK

Deflection results for the annular segment and quadrilateral are tabulated in Table 2 for the two central loading cases (*A* and *B*). All results are almost identical. In Fig. 8 the experimental deflections are in close agreement with the annular segment.

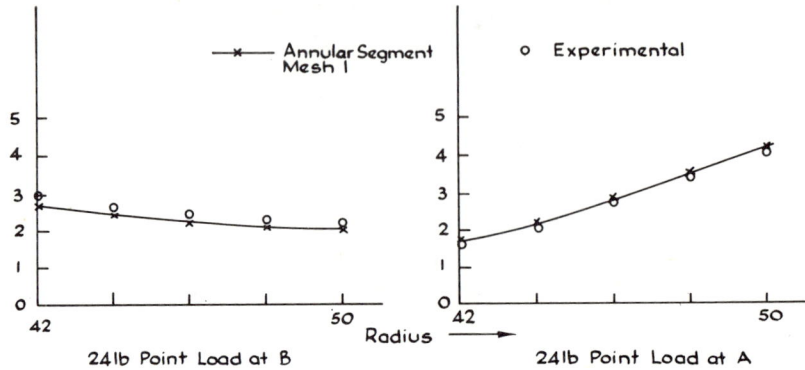

FIG. 8. TRANSVERSE DEFLECTION PROFILES ACROSS 109–117

Results for circumferential moments are compared in Figs. 9–10, for the three analyses. The peak moments appear to be underestimated by the grillage approach whereas the quadrilateral predicts lower moments for the outer load and higher moments for the inner. For the annular segment the small difference between the two meshes demonstrates the rapid convergence of the element. All three methods of analysis are in good agreement with the experimental moments and are accurate

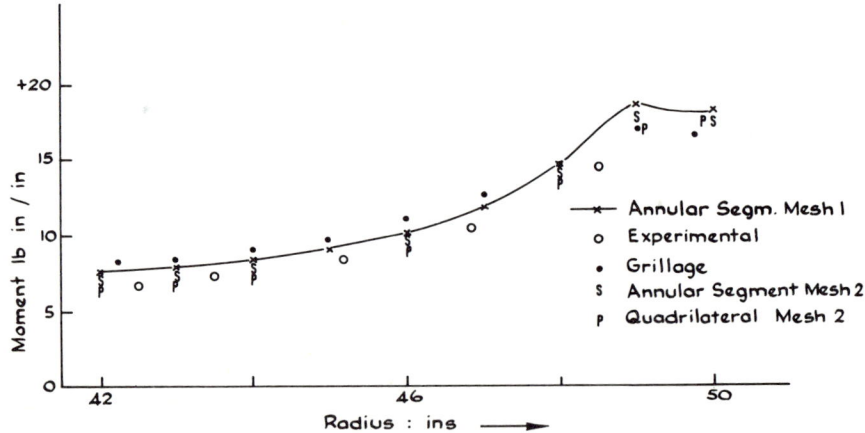

FIG. 9. LONGITUDINAL MOMENT (*Mθ*) ACROSS 109–117 (24 LB POINT LOAD AT *A*)

enough for design purposes. The radial moments and the centre line and circumferential moments along the inner supports are shown in Figs. 11–12 for the outer loading case. The large discrepancy between analytical and experimental results at the supports is probably explained by the concentration of moments at the columns due to the additional stiffness of the metal discs used for metal supports.

Over the annular segment the shear Qr and $Q\theta$ are calculated by standard plate formula. As for moments, shears were calculated at the four nodes of the annular

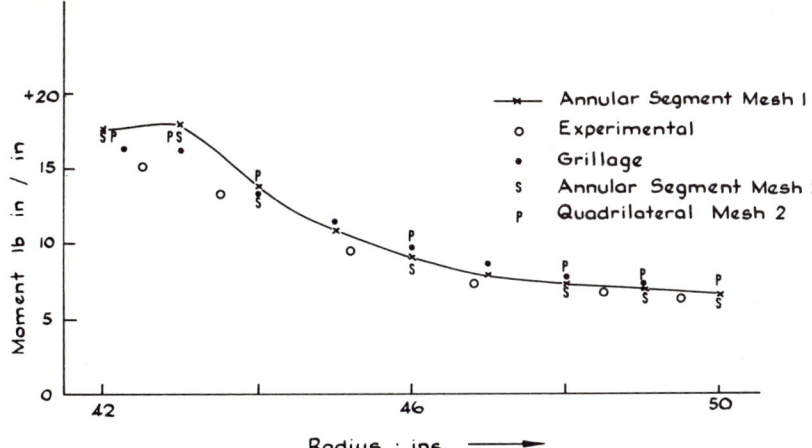

FIG' 10. LONGITUDINAL MOMENT ($M\theta$) ACROSS 109−117 (24 LB POINT LOAD AT *B*)

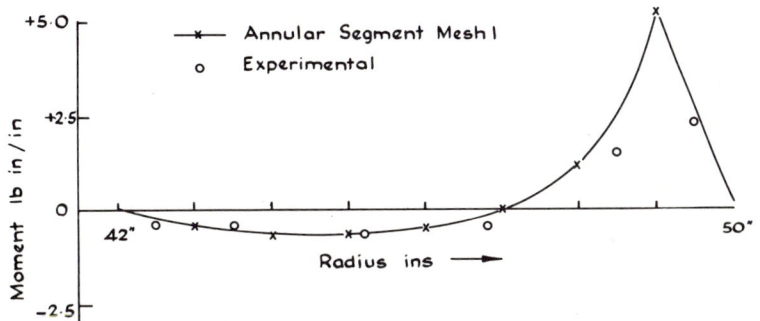

FIG. 11. TRANSVERS MOMENT (*Mr*) ACROSS 109−117 (24 LB POINT LOAD AT *A*)

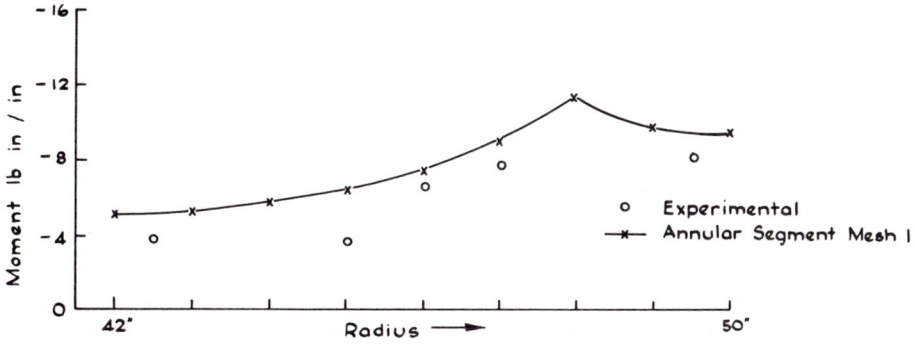

Fig. 12. LONGITUDINAL MOMENT ($M\theta$) ACROSS 55−63 (24 LB POINT LOAD AT *A*)

segment and averaged. Where shear discontinuities are encountered in the form of line supports or columns, the contributions are averaged on both sides of the discontinuity. The column reactions were obtained directly from the vertical nodal forces for the element.

In the curved bridge deck tested, the shears $Q\theta$ are large in comparison with Qr except in the vicinity of the columns and the concentrated loads. This is to be expected as the load has to be transmitted to the columns in the circumferential direction. At the centre line of the slab the shears are compared with those for the grillage analysis in Fig. 13. The two sets are consistent. It should be mentioned that in the finite element approach the sum of shears $Q\theta$ across radial section in the centre span was not equal to half the applied load. At the two edges, forces equal

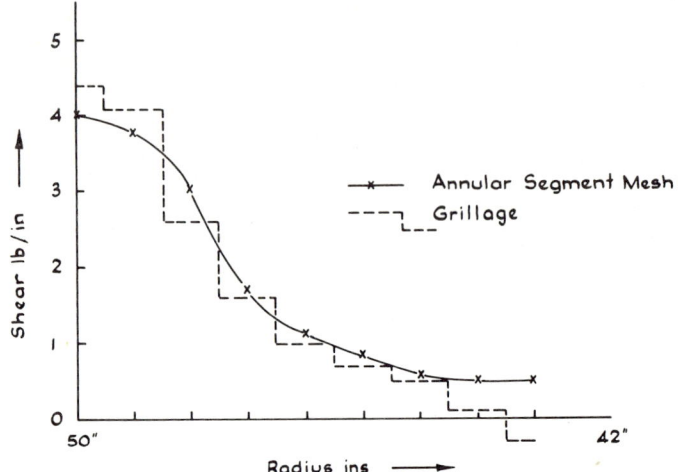

FIG. 13. DISTRIBUTION OF SHEARS ACROSS SECTION 109–117
(MESH 1) (24 LB POINT LOAD AT *A*)

in magnitude to the value of the torsional moments $Mr\theta$ had to be added for equilibrium. This apparent anomaly is explained in detail in reference (12).

Finally, the principal moments are plotted in Fig. 14 at a radial section where the torsional moments are greatest. Again the convergence of the annular segment is demonstrated and the accuracy of the quadrilateral element confirmed.

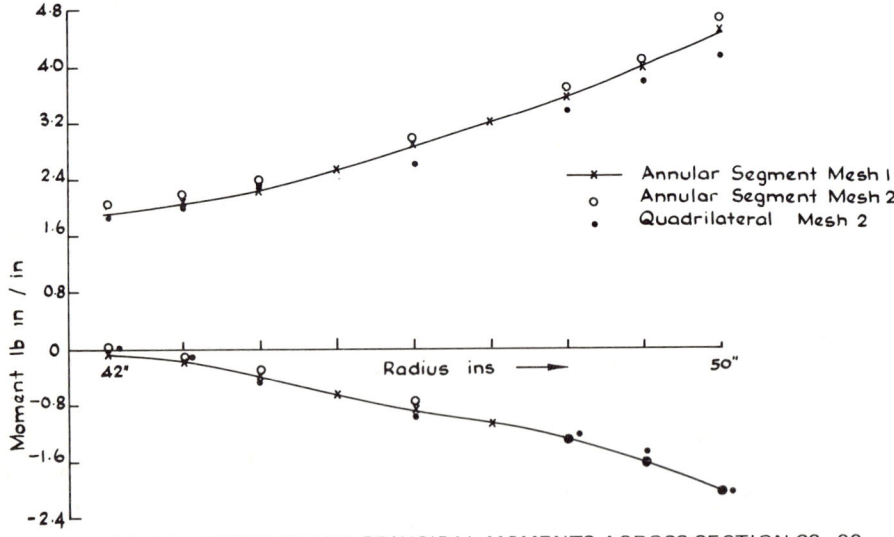

FIG. 14. MAXIMUM AND PRINCIPAL MOMENTS ACROSS SECTION 82–90
(MESH 1) (24 LB POINT LOAD AT *A*)

(d) Varying section bridge deck

The model was constructed using the plan of the one described in section (c), the depth of the slab varying in the circumferential direction from 0·4 in at the centre line to 1·25 in over the inner columns. The main purpose of this test was to assess the accuracy of the simulation of the structure by a number of stepped elements in the circumferential direction.

The strain gauges, mesh sizes, standard tests for material properties were the same.* For the analysis the flexural properties were averaged over each element.

In Table 3 the deflections for the two central loading cases are shown. There is little difference between the deflections for the two meshes of the annular segment. However, the quadrilateral appears stiffer than the annular segment for load on the inner radius. This is probably due to the effective stiffening of the inside edge by the representation of the curve to a series of chords. For several point loads on the centre line, the experimental deflections shown in Table 4 are in excellent agreement with the annular segment.

The longitudinal moments are shown in Fig. 15 for the load on outer radius. The authors believe that sufficient results have been given to confirm the use of stepped elements for this type of analysis.

Further tests are described for both models[13] with a concentrated load in the end span and off the centre line in the centre span. Correlation was of the same degree of accuracy as that demonstrated in this paper.

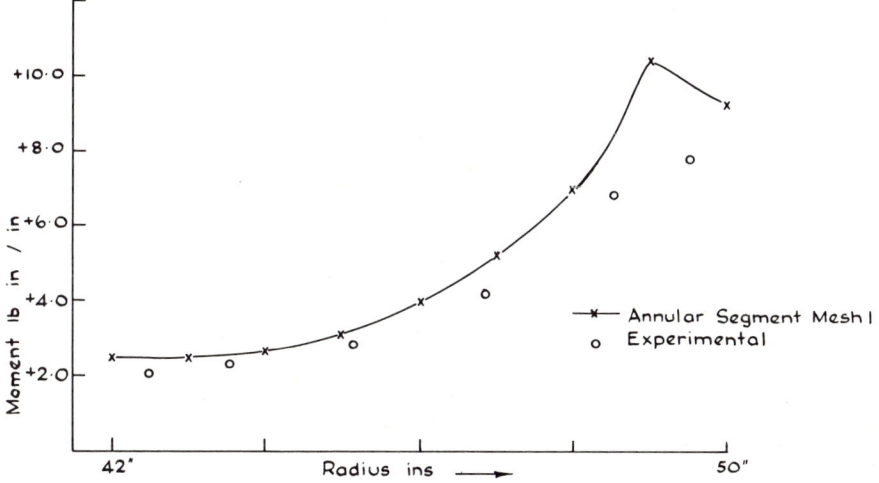

FIG. 15. LONGITUDINAL MOMENT $(M\theta)$ ACROSS 109–117 FOR VARYING SECTION MODEL (19·2 LB POINT LOAD AT A)

Column Reactions

Vertical reactions were not measured during test because of difficulties with the scale of model. In Table 5 the reactions of the centre supports are tabulated for the two models to illustrate the redistribution of reactions due to the varying-section in the circumferential direction. Results of the grillage and annular segment finite element analyses are in close agreement for all loading cases.

*E = 500 000 lb/in², $\nu = 0.4$

Conclusions

On the basis of rigorous testing of various methods of analysis the following general conclusions can be drawn.

1. The series solution produces accurate results for curved bridge decks, but is restricted to simply supported decks.

2. Grillage idealisation produces a remarkable degree of accuracy even for complex layouts. Arithmetic computation is, however, required for obtaining actual bending moments from computer output.

3. Finite element analysis using quadrilateral element produces results with a good degree of accuracy.

4. Finite element analysis using annular segment elements produces results with a high degree of accuracy for isotropic and orthotropic slabs.

Acknowledgement

The work described in this paper forms part of studies into modern methods of analysis and design of bridge structures at the Department of Civil Engineering, University of Liverpool.

REFERENCES

1. Sawko, F. Computer analysis of grillages curved in plan. *Publs. int. Ass. Bridge struct. Engng*, Vol. 27, 1967, pp. 151–170.
2. Deuce, T. L. G. The Lofthouse Interchange bridges. *Proceedings of the Conference on Steel Bridges (London, 1968)*, London, British Constructional Steelwork Association, [1969], pp. 117–122.
3. Sawko, F. Discussion on 'Analysis of curved bridge decks'. *Proc. Instn civ. Engrs*, Vol. 38, Dec. 1967, pp. 775–778.
4. Yoshimura, J. Bending of curvilinear orthotropic circular ring sector plate. *Civil Engineering in Japan*, 1962, pp. 72–74.
5. Yoshimura, J. Bending of curvilinear orthotropic circular ring sector plate. *Civil Engineering in Japan*, 1964, pp. 23–27.
6. Coull, A. and Das, P. C. Analysis of curved bridge decks. *Proc. Instn civ. Engrs*, Vol. 37, May 1967, pp. 75–85.
7. Bell, K. A refined triangular plate bending element. *Int. J. numer. Meth. Eng.*, Vol. 1, No. 1, Jan./March 1969, pp. 101–122.
8. Shieh, W. Y. J., Seng-Lip Lee and Parmalee, R. A. Analysis of plate bending by triangular elements. *J. Engng Mech. Div. Am. Soc. civ. Engrs*, Vol. 94, EM5, Oct. 1968, pp. 1089–1107.
9. Butlin, G. A. and Ford, R. *A compatible triangular plate bending finite element*. University of Leicester Engineering Department Report 68–15, Oct. 1968.
10. Clough, R. W. and Tocher, J. L. Finite element stiffness matrices for analysis of plate bending. *Proceedings of the Conference on Matrix Methods in Structural Mechanics (Wright-Patterson Air Force Base, Ohio, 1965)*, 1967, pp. 515–546.
11. Sawko, F. and Merriman, P. A. An annular segment finite element for plate bending. Int. J. numer. Meth. Eng., in press.
12. Sawko, F., Cope, R. J. and Merriman, P. A. Shears in finite elements. In preparation.
13. Merriman, P. A. *The analysis of varying section and curved bridge decks by finite element methods*. Ph.D. thesis, University of Liverpool, Oct. 1970.

TABLE 1

Deflections across mid-span of perspex model
due to 10 lb at (*a*) Outer load
(*b*) Central load
(*c*) Inner load

	Radius in	Quadr.	Ann. Segm.	Coull
(*a*)	13	0·888	0·851	0·876
	11	0·582	0·559	0·578
	9	0·355	0·344	0·353
	7	0·198	0·192	0·194

Outer Load

	Radius in	Quadr.	Ann. Segm.	Coull
(*b*)	13	0·459	0·445	0·457
	11	0·344	0·333	0·342
	9	0·241	0·236	0·241
	7	0·154	0·154	0·155

Central Load

	Radius in	Quadr.	Ann. Segm.	Coull
(*c*)	13	0·194	0·192	0·195
	11	0·163	0·162	0·163
	9	0·150	0·151	0·157
	7	0·168	0·170	0·169

Inner Load

TABLE 2

Comparison of deflections (in) for central
point loads on uniform slab

Radius at ₵ in	Quadr.	Annular Mesh 1	Segment Mesh 2
50	0·106	0·105	0·105
42	0·045	0·045	0·044

60 lb load at A

50	0·050	0·050	0·049
42	0·065	0·066	0·067

60 lb load at B

TABLE 3

Comparison of deflections (in) for central
point loads on varying section slab

Radius at ₵ in	Quadr.	Annular Mesh 1	Segment Mesh 2
50	0·152	0·155	0·152
42	0·065	0·067	0·067

60 lb load at A

50	0·071	0·075	0·072
42	0·101	0·107	0·105

60 lb load at B

TABLE 4

Comparison of deflections (in) for several point loads (28·8 lb)
on the centre line for varying section slab

Load at	Experimental			Annular Segment		
	Inner Edge	Outer Edge	Ratio O/I	Inner Edge	Outer Edge	Ratio O/I
A	0·0332	0·0730	2·20	0·0326	0·0745	2·28
111	0·0369	0·0700	1·95	0·0337	0·0650	1·93
112	0·0394	0·0610	1·55	0·0354	0·0560	1·58
113	0·0417	0·0547	1·31	0·0380	0·0495	1·30
114	0·0447	0·0476	1·06	0·0410	0·0435	1·06
115	0·0476	0·0425	0·89	0·0455	0·0390	0·86
B	0·0520	0·0347	0·67	0·0515	0·0350	0·68

TABLE 5

Column reactions for point loads on centre line

60 lb at	Inner Central Supp. C1		Outer Central Supp. C2	
	Annular Segment	Grillage	Annular Segment	Grillage
A	0·53	1·20	−43·92	−44·58
111	−6·40	−6·27	−36·00	−36·77
113	−21·60	−21·27	−20·00	−20·97
115	−37·20	−36·25	−3·88	−5·25
B	−45·00	−43·73	+4·24	+2·60

Uniform Model

A	−15·56	−13·27	−35·76	−38·24
111	−22·12	−20·33	−28·40	−30·78
113	−35·88	−34·45	−13·34	−15·55
115	−50·28	−48·56	1·76	−0·43
B	−57·80	−55·69	9·72	7·11

Varying Section Model

THE TORSIONAL STIFFNESS OF MULTICELLULAR RECTANGULAR BOX SECTIONS

J. K. GOODALL

G. Maunsell & Partners

SYNOPSIS

A simple and accurate solution is derived for the torsion constant of a multicellular box section of unrestricted wall thickness. Its characteristics are compared with those of the same solid section and the solution is considered in relation to its limiting values and to the approximate solutions that are commonly adopted. It also enables the maximum shear stresses to be evaluated.

Notation

a	mean cell depth	k_0	torsion coefficient = $(1-\phi/n)$
c	mean cell width	ϕ	shape function
b	overall width of section	$\phi_{\lim}$	limiting value of ϕ and = $(1-\sigma)/(1-\sigma\eta).\,2\eta/(1-\eta)$
b'	sum of mean cell widths = nc	η	function of section geometry = $(w-\sqrt{(w^2-1)})$
d	overall depth of section	ω	function of section geometry = $(1+c/a.\,t_{wi}/t_f)$
v	void height	σ	function of section geometry = $(1-t_{wi}/t_{wo})$
t_f	flange thickness	τ_{f_r}	shear stress in the flange of cell r
t_{wi}	inner web thickness	$\tau_{w_{r(r+1)}}$	shear stress in the web between cells r and $(r+1)$
t_{wo}	outer web thickness	τ_e	shear stress in the flanges of cells 1 and n
ρ	void ratio = v/d	p, q	variables relating to the section properties
n	number of cells	θ	torsional rotation per unit length
G	shear modulus	δ_r	membrane height in cell
J	torsion constant	M_t	torsional moment

Many of the medium span concrete bridges now being designed to carry motorways incorporate cellular decks.[1,2] The development of powerful methods of analysis has enabled the structural properties of these decks to be fully exploited and has consequently demanded an accurate assessment of their torsional stiffness.

At least one comprehensive method exists[3] for any thick or thin-walled section but it is not very suitable for a rapid solution of complicated sections with a large number of voids. The solution by Wittrick[4] of the particular section considered here is, like the simplified approximations frequently adopted, restricted to thin-walled members which may not necessarily be applicable to thick-walled concrete sections. Since this solution is extended and simplified for application to such sections its derivation by membrane analogy is outlined briefly.

347

THIN-WALLED SECTIONS

The method of applying the membrane analogy to determine the torsional stiffness of thin-walled tubular members with internal webs has been described for the general case by Timoshenko.[5] The particular case of a rectangular section with regularly spaced webs is considered as in Fig. 1.

Making the basic assumption that the wall thickness is small and that the surfaces of the membrane are plane then the shear stress across a wall at any point is constant and is defined as:

$$\tau_{f_r} = \frac{\delta_r}{t_f}, \; \tau_{w_{r(r+1)}} = \frac{\delta_r - \delta_{(r+1)}}{t_w} \tag{1}$$

These stresses are induced by the torque M_t acting on the section. This is equal to twice the contained volume and is expressed in terms of the shear stress in the flanges as:

$$M_t = 2 a c t_f \sum_{r=1}^{n} \tau_{f_r} \tag{2}$$

The torsional rotation of the section per unit length is proportional to the integration of the induced shear stress in the walls of the section. As the rotation of each cell is equal to the rotation of the whole, then:

$$2 a c G \theta = a \tau_{w_{r(r-1)}} + 2 c \tau_{f_r} + a \tau_{w_{r(r+1)}} \tag{3}$$

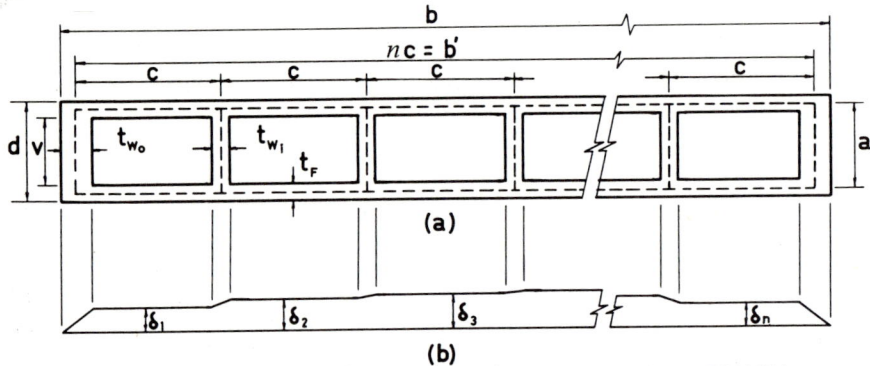

FIG. I MULTICELLULAR BOX SECTION (a) AND SECTION THROUGH
MEMBRANE (b)

which is re-expressed as:

$$c \frac{t_{wi}}{t_f} G \theta = -\frac{1}{2} \tau_{f(r-1)} + \left(1 + \frac{c}{a} \frac{t_{wi}}{t_f}\right) \tau_{f_r} - \frac{1}{2} \tau_{f(r+1)} \tag{4}$$

Summing this general expression for n cells gives:

$$n c \frac{t_{wi}}{t_f} G \theta = \frac{c}{a} \frac{t_{wi}}{t_f} \sum_{r=1}^{n} \tau_{f_r} + \frac{1}{2}(\tau_{f_1} + \tau_{f_n}) \tag{5}$$

Because the section is symmetrical the shear stresses associated with the end cells are identical, i.e.:

$$\tau_{f_1} = \tau_{f_n} = \tau_e$$

so that:

$$\sum_{r=1}^{n} \tau_{f_r} = n a G \theta \left(1 - \frac{1}{n} \frac{\tau_e t_f}{c t_{wi} G \theta}\right) \tag{6}$$

Since the torsion constant is related to the torque and the rotation in the form $M_t = JG\theta$ it can be expressed in terms of the shear stress in the end cell flanges by substituting the above in equation (2) so that:

$$J = 2na^2 ct_f \left(1 - \frac{1}{n} \frac{\tau_e t_f}{ct_{wi}G\theta}\right) \tag{7}$$

However, since the shear stresses τ_e in the outer cells are dependent upon the geometry of the section an evaluation of $(\tau_e t_f / ct_{wi}G\theta)$ is required for a complete solution of the torsion constant. The shear stress in the flange of any cell can be expressed in terms of the shear stress in the flange of the first cell. This is derived from equation (4) as:

$$\frac{1}{2}\tau_{f_r} = \frac{1}{2}\tau_{f_1} \cdot p_r - c\frac{t_{wi}}{t_f}G\theta \cdot q_r \tag{8}$$

The variables p_r and q_r are dependent upon the proportions of the section and the number of cells. They are generated in the form:

$$p_r = 2\omega p_{(r-1)} - p_{(r-2)}$$
$$q_r = 2\omega q_{(r-1)} - q_{(r-2)} + 1 \tag{9}$$

where

$$\omega = \left(1 + \frac{c}{a} \cdot \frac{t_{wi}}{t_f}\right)$$

Since the shear stresses in the flanges of the end cells are the same they can be equated for a finite number of n cells as:

$$\frac{1}{2}\tau_e = \frac{1}{2}\tau_e \cdot p_n - c\frac{t_{wi}}{t_f}G\theta \cdot q_n \tag{10}$$

This provides the solution of $(\tau_e t_f / ct_{wi}G\theta)$ as a function of the section geometry which, more conveniently termed the shape function ϕ, is given as:

$$\phi = \frac{\tau_e t_f}{ct_{wi}G\theta} = \frac{2 \cdot q_n}{p_n - 1} \tag{11}$$

The basic values of p and q are found from an inspection of equation (3) expressed in terms of the first two cells. They are:

$$p_1 = 1 \qquad q_1 = 0$$
$$p_2 = (2\omega - \sigma) \quad q_2 = 1 \tag{12}$$

where

$$\sigma = \left(1 - \frac{t_{wi}}{t_{wo}}\right)$$

The iterative expression for the shape function in equation (11) is solved for n cells to give:

$$\phi = \frac{(1 - \sigma)(1 - \eta^n)}{(1 + \eta^{n+1}) - \sigma\eta(1 + \eta^{n-1})} \cdot \frac{2\eta}{(1 - \eta)} \tag{13}$$

where

$$\eta = (\omega - \sqrt{(\omega^2 - 1)})$$

The substitution of the cell dimensions and of the relevant value of the shape function into equation (7) enables the torsion constant for any thin-walled section of the form in Fig. 1 to be evaluated. This is quite satisfactory for steel tubes that obviously qualify as thin-walled sections but a great many sections are of concrete where the wall thicknesses can be such that the application of the thin-wall theory results in a loss of accuracy.[3]

By taking a more general view of the torsional stiffness of rectangular box sections the overall effect of thicker walls can be evaluated by a further application of the membrane analogy and a refinement of the above solution obtained.

THICK-WALLED SECTIONS

The torsion constant for a multicellular rectangular box can be expressed as a coefficient of bd^3 in the familiar form:

$$J = kbd^3 \qquad (14)$$

The torsion coefficient, k, is evaluated for a typical multicellular box from the above theory and plotted in Fig. 2 against the overall breadth to depth ratio, b/d, of

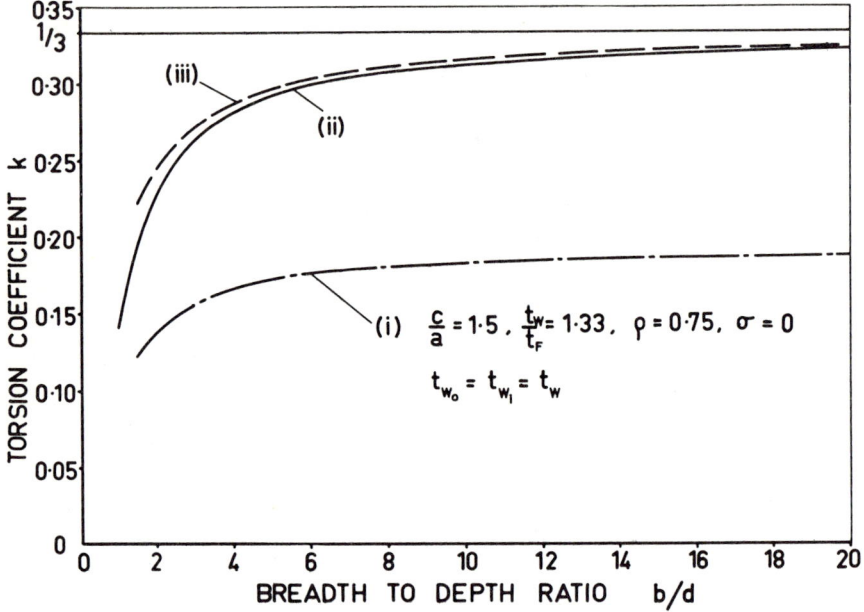

FIG 2 COMPARISON OF THE TORSIONAL STIFFNESS OF A MULTICELLULAR AND SOLID SECTION FOR $\sigma = 0$

the section as curve (i) and compared with curve (ii) which relates to a solid rectangular section of the same shape.[6] From these two curves, which are of a similar form, it is reasonable to suppose that the relationship between the torsion constant and the external shape of both the solid and cellular rectangular sections is similar and that as the flange thickness of the multicell box is increased the torsion coefficient approaches that of the solid section.

It is known that the contribution of the internal webs to the torsional stiffness is small and is confined basically to the ends of the section. This enables an assessment of thick-walled boxes to be made by considering a section that is wide compared to its depth; the end effects, and hence the webs, can then be ignored and only the flanges considered. In any wide, thin section the flange thickness is the controlling factor and by extending the treatment of an infinitely long, solid rectangular section[5] to a voided section with walls of equal thickness it is seen by inspection that the internal and external sections are additive and that:

$$J = (\frac{1}{3}bd^3 - \frac{1}{3}bv^3) = \frac{1}{3}(1 - \rho^3)bd^3 \tag{15}$$

where the void ratio, $\rho = v/d$.

The torsion coefficient plotted in curve (i) is then adjusted by the factor $(1-\rho^3)$ and re-plotted as a coefficient of $(1 - \rho^3)bd^3$ as curve (iii) where the close agreement between curves (ii) and (iii), which also exists for other sections, illustrates the similarity between the torsion constant of a solid section and a multicellular box when the latter is expressed in terms of the void ratio, ρ. The discrepancies apparent at the lower values of b/d arise from the end effects and from the contribution of the webs, their effect diminishing as the section becomes wider.

The expression derived for thin-walled sections can be modified and expressed in terms of the void ratio by substituting the section dimensions as follows:

$$2a = (1 + \rho)d, \quad 2t_f = (1 - \rho)d$$
$$a^2 \approx \rho d^2, \quad nc = b' \tag{16}$$

so that:

$$6na^2ct_f \approx (1 - \rho^3)b'd^3 \tag{17}$$

This conveniently corresponds to that portion of equation (7) which relates to the basic geometry and enables the expression for the torsion constant of a thick-walled section to be re-expressed in its final form as:

$$J = \frac{1}{3}k_0(1 - \rho^3)b'd^3 \tag{18}$$

where the torsion coefficient k_0, which is a function of the stress distribution around the section, is defined as:

$$k_0 = (1 - \frac{\phi}{n}) \tag{19}$$

and some typical values for the particular case of $\sigma = 0$ are given in Fig. 5.

This rearrangement overcomes the limitation of the original derivation in its application to thick-walled sections. The inequality is eliminated because the variation in shear stress across the flanges is taken into account in the evaluation of the idealised voided strip. Although this refinement has only been applied to the flanges the error resulting from ignoring the webs is negligible as the shear stresses in the webs are low and very nearly constant with the exception of the two outer webs which form only a small proportion of the section perimeter.

LIMITING VALUES

Those sections with only a few cells and which are deep in comparison with their overall width have a torsional stiffness which is sensitive to any variations in their geometry. This can be seen in Fig. 3 in which the torsion coefficient is plotted as a function of the number of cells and the parameter $c/a.t_{wi}/t_f$ for a range of sections which are restricted for purposes of comparison to the particular case of $t_{wi} = t_{wo}$.

The spectrum of curves is bounded by the limiting values in which the lower bound is provided by the single cell (where $\phi = 1/\omega$ with t_{wi} taken as t_{wo}) with the contribution from the inner webs being ignored and only the enclosing section considered. This is a reasonable approximation[1] when the webs are thick and the size of

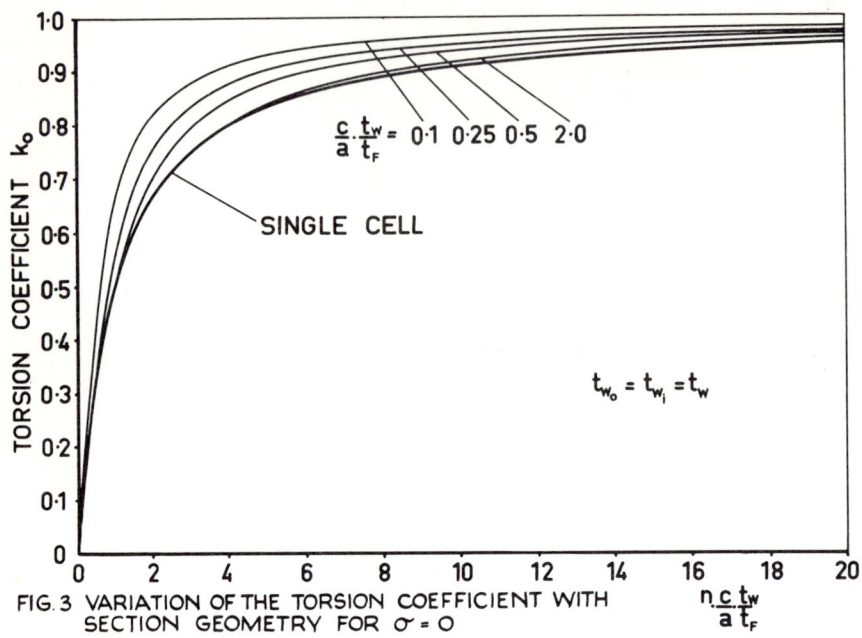

FIG. 3 VARIATION OF THE TORSION COEFFICIENT WITH
SECTION GEOMETRY FOR $\sigma = 0$

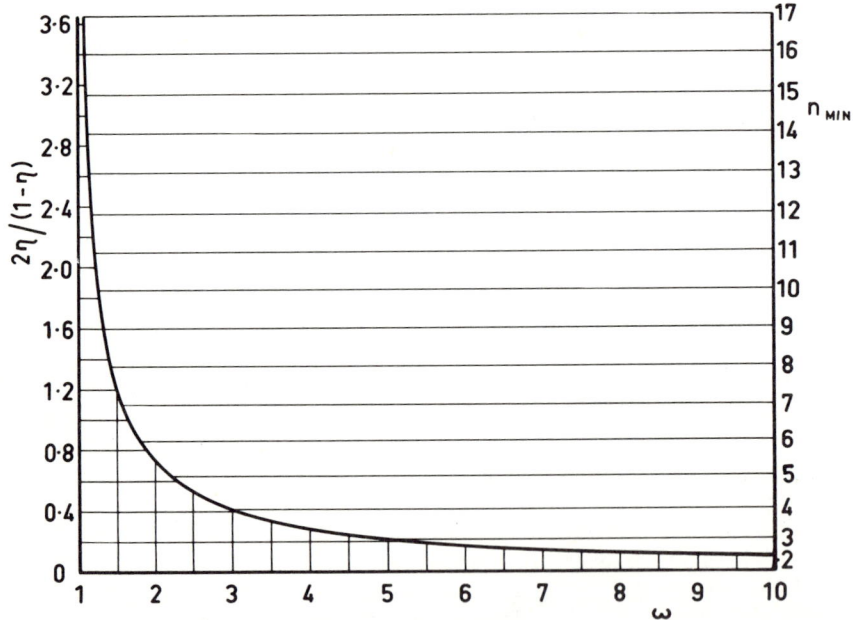

FIG. 4 LIMITING VALUE OF THE SHAPE FUNCTION IN RELATION
TO THE SECTION GEOMETRY FOR $\sigma = 0$

VALUES OF ω

η	1.1	1.2	1.4	1.6	1.8	2.0	2.25	2.5	3.0	4.0	5.0	6.0	8.0	10.0
1	0.091	0.167	0.286	0.375	0.444	0.500	0.556	0.600	0.667	0.750	0.800	0.833	0.875	0.900
2	0.167	0.286	0.444	0.546	0.615	0.667	0.714	0.750	0.800	0.857	0.889	0.909	0.933	0.947
3	0.249	0.397	0.566	0.660	0.720	0.762	0.799	0.826	0.863	0.903	0.925	0.939	0.955	0.965
4	0.329	0.492	0.654	0.735	0.785	0.818	0.848	0.868	0.897	0.927	0.944	0.954	0.966	0.974
5	0.403	0.568	0.716	0.785	0.826	0.854	0.878	0.894	0.917	0.942	0.955	0.963	0.973	0.979
6	0.468	0.628	0.760	0.820	0.855	0.878	0.898	0.912	0.931	0.952	0.962	0.970	0.978	0.982
7	0.525	0.676	0.794	0.846	0.876	0.895	0.912	0.925	0.941	0.958	0.968	0.974	0.981	0.985
8	0.573	0.714	0.819	0.865	0.891	0.908	0.923	0.934	0.948	0.964	0.972	0.977	0.983	0.987
9	0.614	0.744	0.839	0.880	0.903	0.919	0.932	0.941	0.954	0.968	0.975	0.980	0.985	0.988
10	0.649	0.769	0.855	0.892	0.913	0.927	0.939	0.947	0.959	0.971	0.978	0.982	0.987	0.989
12	0.704	0.807	0.879	0.910	0.927	0.939	0.949	0.956	0.966	0.976	0.981	0.985	0.989	0.991
14	0.745	0.835	0.896	0.923	0.938	0.948	0.956	0.962	0.970	0.979	0.984	0.987	0.990	0.992
16	0.776	0.855	0.909	0.932	0.946	0.954	0.962	0.967	0.974	0.982	0.986	0.988	0.992	0.993
18	0.801	0.871	0.920	0.940	0.952	0.959	0.966	0.971	0.977	0.984	0.988	0.990	0.993	0.994
20	0.821	0.884	0.928	0.946	0.956	0.963	0.969	0.974	0.979	0.986	0.989	0.991	0.993	0.995
25	0.857	0.907	0.942	0.957	0.965	0.971	0.976	0.979	0.983	0.988	0.991	0.993	0.995	0.996
30	0.881	0.923	0.952	0.964	0.971	0.976	0.980	0.982	0.986	0.990	0.992	0.994	0.996	0.996
35	0.898	0.934	0.959	0.969	0.975	0.979	0.982	0.985	0.988	0.992	0.994	0.995	0.996	0.996
40	0.910	0.942	0.964	0.973	0.978	0.982	0.985	0.987	0.990	0.993	0.994	0.995	0.997	0.997
45	0.920	0.948	0.968	0.976	0.981	0.984	0.986	0.988	0.991	0.994	0.995	0.996	0.997	0.998
50	0.928	0.954	0.971	0.978	0.983	0.985	0.988	0.989	0.992	0.994	0.996	0.996	0.997	0.998

FIG. 5. TABULATED VALUES OF THE TORSION COEFFICIENT k_0 for $\sigma = 0$

the cells large in terms of the overall width; for those sections in which $t_{wi} \ll t_{wo}$ the approximation is obviously very much better. The upper bound of $k_0 = 1$ is only valid for very thin, wide sections in which the webs are very thin and are spaced at very close centres. This is the ideal value sometimes adopted in equivalent plate analysis[2] but is only an approximation for the majority of practical cases in which the section depth is appreciable. Both of these limiting values are aggravated by applying thin-wall theory to sections with thick walls.[3]

The accurate solution of a particular section which lies between these extremes is simplified considerably for the majority of the practical cases encountered. For any specific values of ω and σ the value of the shape function ϕ reaches a limiting value which is obtained from equation (13) as:

$$\phi_{\lim} = \frac{(1 - \sigma)}{(1 - \sigma\eta)} \cdot \frac{2\eta}{(1 - \eta)} \tag{20}$$

and remains unchanged if the number of cells is greater than a minimum determined by:

$$n \geqslant - \frac{\log(1 + \eta) . 10^3}{\log \eta} \tag{21}$$

This is shown in Fig. 4 where the same curve also serves to demonstrate the relationship of the function $2\eta/(1 - \eta)$ with ω. This is modified to the limiting value of the shape function by the factor $(1 - \sigma)/(1 - \sigma\eta)$ and is equal to it for the case where $t_{wi} = t_{wo}$. Since this modifying factor is primarily a function of σ and is little affected by ω it can be deduced that the limiting value of ϕ usually applies because in the majority of practical sections the number of cells is inversely proportional to the value of ω.

SHEAR STRESSES

The maximum shear stresses induced by a torsional moment acting on the section are readily obtained from this analysis.

The maximum stress in the flanges, which will be at the centre of the section, can be taken as the maximum possible as this is achieved even for very few cells. In this case the end effects can be ignored and the average stress in the flange wall is given as:

$$\tau_{f\max} = aG\theta \tag{22}$$

By considering the constant shear flow around the walls of the end cell it is obvious that the shear stress in the webs is a maximum in the end wall and is given by:

$$\tau_{w\max} = \frac{t_f}{t_{wo}} . \tau_e \tag{23}$$

and so from equation (11):

$$\tau_{w\max} = c(1 - \sigma)\phi G\theta \tag{24}$$

CONCLUSIONS

The torsional stiffness of a voided rectangular section expressed in the same form as the identical solid section has enabled a direct comparison of the two to be made; the results obtained from this analysis have demonstrated a close correlation.

When the basic form of expression, in which the stiffness is proportional to $(1 - \rho^3) . d^3$, is applied also to the flexural stiffness the actual properties of an orthogonally multicellular bridge deck can be fully expressed in the form of coefficients of this function. These coefficients then define the properties of the deck and can be considered in terms of the degree to which the actual deck deviates from the ideal; as the actual deck approaches this ideal the coefficients of flexural and torsional stiffness tend towards unity. Consequently any bridge deck of this cross-section, with continuous surfaces and full shear stiffness in both directions, can be related to an idealised thin plate by a flexural and torsional parameter determined from these coefficients. This enables multicellular bridge decks to be satisfactorily transformed into an equivalent orthotropic plate which can then be analysed by established plate analysis.

ACKNOWLEDGEMENTS

This paper is published with the permission of G. Maunsell & Partners.

REFERENCES

1. Hook, D. M. A. and Richmond, B. Precast concrete box beams in cellular bridge decks. *Struct. Engr*, Vol. 48, No. 3, March 1970, pp. 120–128.
2. Basu, A. K. and Dawson, J. M. Orthotropic sandwich plates. *Proc. Instn civ. Engrs*, supplement (iv), Paper 7275S, 1970, pp. 87–115.
3. Acton, J. E. A computer programme for the analysis of the effects of torsion. *Concr. constr. Engng*, Vol. 60, Aug. 1965, pp. 285–294.
4. Wittrick, W. H. Torsion of a multi-webbed rectangular tube. *Aircr. Engng*, Vol. 25, No. 298, Dec. 1953, p. 372.
5. Timoshenko, S. P. *Strength of materials. Part 2: advanced theory and problems*; 3rd ed. Princeton, N.J., Van Nostrand, 1956, pp. 250–239.
6. Timoshenko, S. P. and Goodier, J. N. *Theory of elasticity;* 2nd ed. New York. McGraw-Hill, 1951, p. 277.

SOME CONSIDERATIONS IN THE DESIGN OF COMPOSITE BRIDGES

J. C. CHAPMAN

Imperial College of Science and Technology

SYNOPSIS

The background to the British Code of Practice on Composite Construction (Pt. I —
Building and Pt. II — Bridges) is outlined. Some unresolved questions are raised in
relation to revisions to the Code. Topics discussed include deck analysis, ultimate load
design, longitudinal shear strength, shear connectors and composite columns. Attention is drawn to the possibility of unforeseen side-effects when the basis of design is
changed.

INTRODUCTION

Until 1965 there was no British Code of Practice which considered composite construction in steel and concrete, except in as much as BS 449 made some allowance
for the stiffening effect of concrete encasement on steel columns in buildings.

In 1963 a committee was formed to draft design recommendations for composite construction and in 1965 CP 117, Part I,[1] was issued. This Code of Practice
covered only simply supported beams in buildings, but was novel in that it permitted ultimate load design, with a considerably increased working stress.

In 1967 CP 117, Part II,[2] was issued, and this code gave recommendations for
the elastic design of composite bridge girders. The main innovation in this Code was
the provision of fatigue tables for shear connectors.

In 1968 a new committee was formed to draft comprehensive recommendations
for the design of bridges. This document will be divided into four parts — Loading,
Steel Bridges, Concrete Bridges, and Composite Bridges. It is intended that this
Code will supersede other Codes dealing with bridges. It is proposed that the Code
should be written in terms of 'limit states', and some difficulties are being experienced
in defining the limit states, in arriving at appropriate partial load factors corresponding to each state, and in suggesting suitable methods of calculation for determining
when the limit states are reached. Not the least of these difficulties lies in trying to
apply the rationale of limit state design to the inconsequencies of formula loading,
there being as yet insufficient data to define loading on a probability basis. Nevertheless it is hoped that the new Code will provide a better foundation for subsequent
editions. In drafting the Codes of Practice on composite construction recourse has
been had as much to recent, and in some cases unpublished, research as to practice.
This paper outlines some of the background to these Codes, and makes mention of
some open questions.

Basic Design Assumptions

The special property of a composite girder is that interface slip (and separation) occurs, causing a redistribution of direct stresses and of longitudinal shear forces. According to the British Codes of practice however the designer is not required to consider this redistribution, although the consequences of redistribution (some favourable, some unfavourable) have been carefully considered in drafting the recommendations.

The distribution of vertical shear between the slab and steel section is a matter for some speculation. The designer is required to assume that the total vertical shear is carried by the steel web, and no consideration is given to the effect of vertical shear on the slab or on the shear connection. This assumption is less than satisfactory in that its consequences are not fully understood, but no adverse results have yet been observed.

Design of Deck

It is required that a bridge deck slab be designed to resist the stresses due to composite action plus the stresses due to bending of the slab about its own axis. The method of analysis is not specified, but recognising that the analysis used might lack rigour, it is specified that in the absence of precise calculations the slab must resist a local moment in the direction of the span; also, where sagging moments can develop across a beam, this moment must be assumed to be at least 50% of the midspan moment. With the rapid advance of computer methods of analysis it is questionable whether these simplifying dispensations will much longer be justified.

Many designers feel that the local longitudinal slab bending stresses are of small importance. This feeling may perhaps be substantiated by the following arguments:

(1) The 3-dimensional effect: For patch loads of diameter greater than the slab thickness the stresses differ little from the 2-dimensional stresses. Thus the importance of the 3-dimensional effect depends on the loaded area. For small patches the 3-dimensional solution gives a smaller lower surface stress than the 2-dimensional solution.

(2) Plastic redistribution: Assuming that a design is made on the basis of elastic stresses, then the load factor corresponding to a yield line collapse mechanism would be greater for a patch load than for a uniform load or a line load. This might justify the use of higher working stresses in the case of a patch load.

(3) Triaxial effect: The stress will reach a peak at the centre of the loaded patch and the surface of the concrete will then be subject to biaxial stresses in the plane of the surface, plus a normal stress from the wheel. Thus the strength of the concrete will be somewhat augmented locally and this might justify a higher working stress in the concrete. This may not be a large effect because the surface of the concrete immediately outside the patch will be subject to stresses nearly as great and would not have the triaxial component.

(4) Spalling strain: Possibly the spalling strain will be increased when the high strain is localised.

In performing the deck analysis, and in designing the composite section, it is anticipated that the designer will wish to use the effective width concept in estimating the effect of shear lag on the stiffness and strength of the section. The effective width varies along the span and depends on the load distribution, cross-sectional properties, and boundary conditions, as well as on the plan dimensions of the slab. It is hardly practicable to take all these factors into account, and in CP 117, Part II, a single expression for an effective breadth is given in terms of the length/breadth ratio of the slab. This expression was derived from the numerical analysis of simply supported and continuous composite T-beams under uniform and point loading. In the new bridge Code it may be that the position in the span will be introduced as an additional parameter.

Again it can be argued that the effective width concept (as applied to the representation of shear lag) will become obsolete as computer programs which take into account both the extensional and flexural properties of the beam and slab, become economically available.

Forces due to shrinkage and temperature are assumed to be distributed in a linearly varying manner over a length given by a formula which was derived from an elastic slip analysis. It is interesting that when shear lag is included in this analysis the calculated shear force is not greatly reduced.[3] The British Code differs here from the German, which requires an anchorage sufficient to resist the force existing if slip did not occur, and from the American, which does not consider the question.

Elastic Design of Composite Section

The moments on the section are determined assuming a modular ratio of $1000/U_w$ for permanent loads and $500/U_w$ for transient loads, the concrete being assumed for this purpose to be uncracked, except that if the nominal tensile stress exceeds $U_w/7$ or 850 lb/in^2 the stiffness of the concrete is ignored (alternatively the midspan moments may be increased by 15%, the support moment remaining as for the uncracked section).

This method of calculating moments is less than logical and is at present under review. Much discussion has taken place with a view to proposing an alternative method which is both practical and realistic.

The shear connection is distributed according to the elastic shear diagram and the load spectrum. The shear connection in a bridge girder is normally subject to live load only and is designed accordingly. It follows that flexural failure would not necessarily precede connector failure under static overload; nevertheless the live load factor against connector failure will be at least 2·5. Since the whole of the connector loading is cyclic it follows that connectors must be designed against fatigue failure, and this makes it probable that flexural failure would in fact precede connector failure.

Ultimate Load Design of Composite Section

CP 117, Part I, permits the calculation of the ultimate moment of resistance by assuming rectangular stress blocks in the steel beam and in the slab. The steel stress is taken to be the yield stress and the concrete stress to be $4/9\ U_w$.

These recommendations were supported by experimental[4] and numerical investigations.[5] The investigations were intended to determine whether the limited strain capacity of concrete, or interface slip, would significantly reduce the strength of the composite section.

A computer program was written in which the steel and concrete stress-strain relations were represented by bi-linear curves, and the load-slip relation for the shear connection was represented by a continuous function. It was first shown that for full interaction the reduction in fully plastic moment was small for a range of positions of the plastic neutral axis. The effect of interface slip was then examined for a range of loading conditions, connector distributions, and other parameters (Fig. 1).

The investigation was intended to support the proposals of CP 117, Part I (Buildings). Nevertheless it should be possible to utilise the results in confirming that

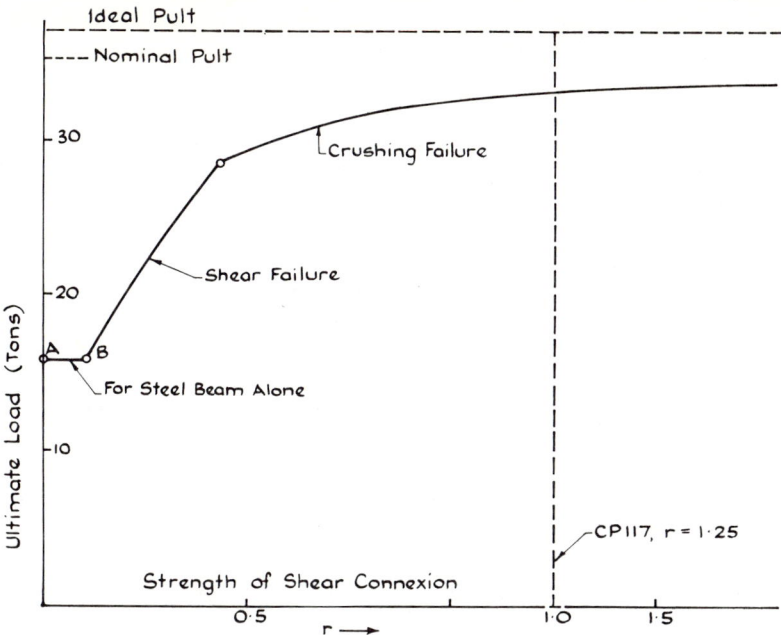

FIG. 1   ULTIMATE   LOAD - CONNEXION   STRENGTH (PT. LOAD AT MIDSPAN)

a shear connection designed against fatigue according to the live load shear envelope will provide the necessary factor against shear failure.

It was concluded for single spans that if the average force on the shear connectors is limited to 80% of their ultimate capacity when the ultimate moment is reached, then the design is satisfactory. The connectors may be distributed uniformly between discontinuities in the (elastic) shear diagram, the number of connectors in each section depending on the areas under the respective parts of the shear diagram. It was found

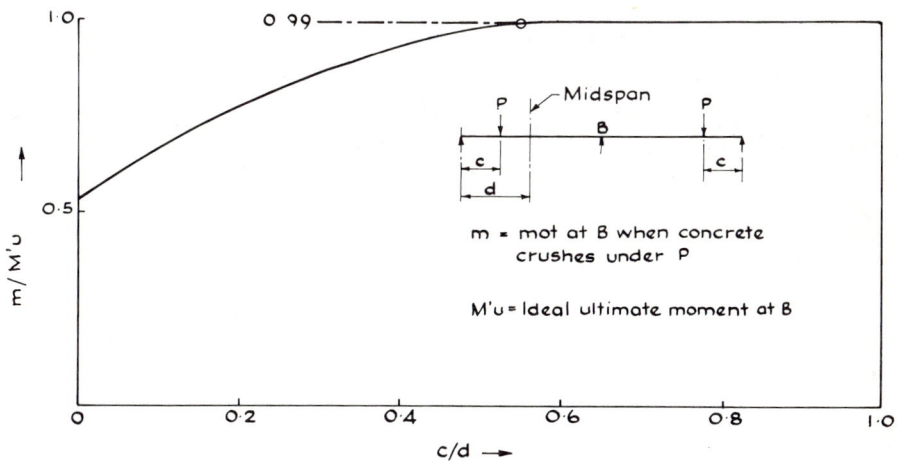

FIG. 2 (a)   REDUCTION IN SUPPORT HINGE MOMENT AT ONSET OF CRUSHING AT LOADED POINTS

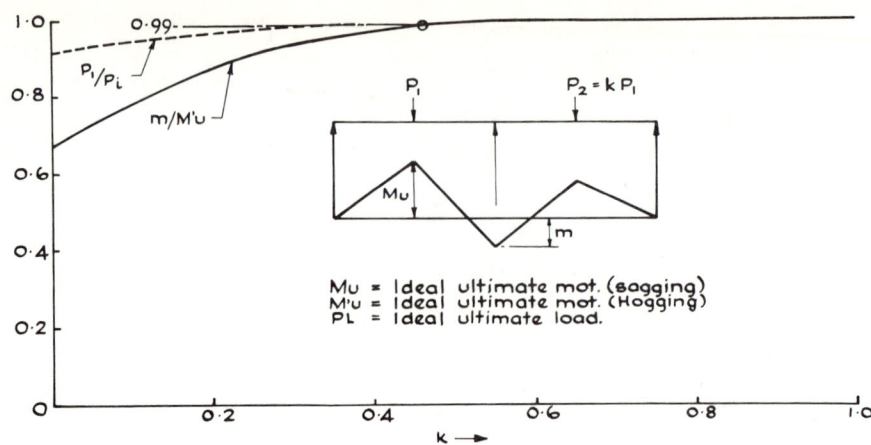

FIG. 2 (b) REDUCTION IN SUPPORT HINGE MOMENT AT ONSET OF
CRUSHING AT POSITION OF LOAD P_1

that flexural failure would then precede connector failure and that the reduction in ultimate moment due to slip would be small.

Similar calculations for continuous beams have shown that a similar criterion can be adopted for the distribution of connectors, the shear transfer in positive and negative moment regions being separately considered.

A more basic question in the ultimate load design of continuous girders is whether in some circumstances deterioration of a positive (sagging) hinge can begin before the negative hinge is fully developed. Calculations based on full interaction indicate that this situation is unlikely to occur in practice and that it should be possible to specify the limits of applicability of the method (Fig. 2).

Preliminary studies of the inelastic behaviour of composite beam and slab systems indicate that for this type of structure overall collapse is unlikely to govern design.

Design of Slab for Longitudinal Shear

The slab must be reinforced to transmit the shear flow from the connectors, and will receive the effect of any redistribution which takes place as a result of interface slip. The maximum possible shear flow is given by the capacity of the shear connectors. Transverse direct stresses also exist as a result of the slab attempting to bend in its own plane.

Shear failure can occur through the slab, through a haunch, or around the connectors. According to CP 117 the designer is required to consider number of possible failure surfaces and to satisfy various shear flow criteria.

The criteria were based on tests which were not intended to exhibit shear failure and might therefore be conservative. More recent tests (Fig. 3) in which the slab reinforcement has been systematically varied have indicated that this is probably so.

CP 117 does not permit that area of reinforcement which is required to resist bending to serve also as shear reinforcement. It can be argued however that tensile reinforcement does in fact contribute to shear resistance by virtue of the transverse compression induced in the concrete.[6] A conservative design approach is probably justified owing to the shortage of test data for girders in which slab moments are applied, particularly for the region of negative longitudinal moment. It must be

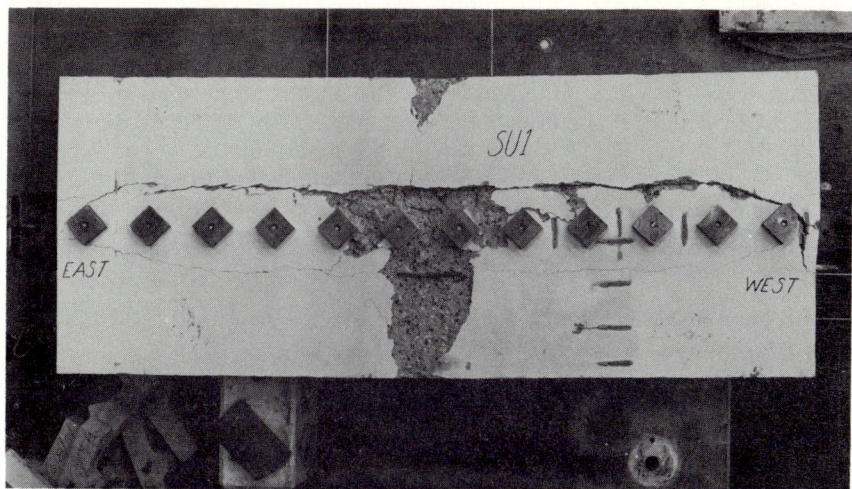

FIG. 3(a) HORIZONTAL SHEAR FAILURE IN SLAB WITHOUT HAUNCH UNDER UNIFORM DISTRIBUTED LOAD

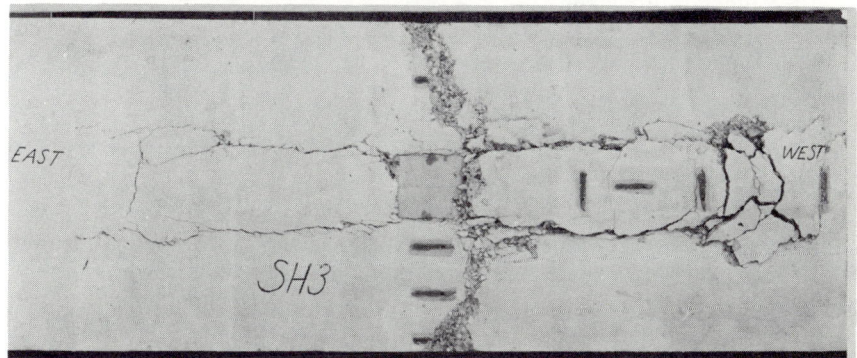

FIG. 3(b) HORIZONTAL SHEAR FAILURE IN BEAM WITH HAUNCH UNDER CENTRAL POINT LOAD

FIG. 3(c) HORIZONTAL SHEAR FAILURE IN HAUNCH

borne in mind that laboratory test results may benefit from concrete tensile strength which may not exist in practice.

According to CP 117, Part II, the shear flow is calculated assuming the slab to remain uncracked regardless of whether the slab is assumed to be cracked for the determination of moments on the composite section. This results in overestimating the shear flow and thus reducing the actual shear connector loading in the negative moment region. This serves to compensate (albeit not rationally) for the reduced static and fatigue strength of connectors which are embedded in cracked concrete and welded to the tension flange of the steel girder.

Static Strength of Shear Connectors

In drafting CP 117 it was decided that connector strengths should be determined directly from test results, rather than by calculation. The strength of a connector depends on many factors, including the following:

(1) The size, shape and attachment of the connector, including provision against interface separation.
(2) The concrete strength and reinforcement, and the slab dimensions, including the haunch if provided.
(3) The state of stress in the concrete surrounding the connector.
(4) The type of loading – for example, loading applied to the slab will restrict interface separation, whereas loading applied to the beam will increase separation. Uplift forces may also arise due to slab continuity, particularly where the girders are torsionally stiff. The applied shear may have two components.
(5) The thickness of the flange to which the connector is attached.

Thus a comprehensive and generalised theory of shear connector strength would be difficult to achieve. Most shear connector testing has been done with the connectors embedded in small slabs placed on both sides of a central steel member. The slabs have been variously reinforced, of various sizes, usually restrained by friction against separation, and nearly always unstressed except by the loading from the connectors. Notwithstanding the differences between push tests, and the differences between the conditions of any push test and the conditions pertaining in an actual structure, reasonably good correlation has been found between push tests and beam tests.

The connector static strengths specified in CP 117, Part I, are therefore taken from push tests, and are equal to 80% of the ultimate strength. The figure of 80% was chosen to limit the loss of interaction and to ensure that flexural failure would occur before connector failure. The latter criterion was accepted as axiomatic for CP 117, Part I, but was not so taken for CP 117, Part II. It will be remembered that the specified strength is to be used with a shear force corresponding to the ultimate strength of the beam. This means that the average connector force under working conditions (Part I) is about 40% of ultimate.

The strengths are tabulated in terms of three different concrete strengths, and the strength of the connector material is also specified. For connector sizes or types which are not tabulated, a standard form of push test is specified. This test will give conservative results for the larger connectors. For example, a ½ in stud fails by shearing, leaving the slab practically undamaged; a ¾ in stud fails by shearing, but causes extensive cracking of the slab; a heavily welded bar and hoop connector fails by crushing and splitting of the slab. It can be expected that a heavily reinforced bridge slab without haunch will provide much better containment for a strong connector than will the standard test slab. CP 117 therefore permits the test slabs to be in accordance with the slab with which the connectors are to be used in a particular case.

CP 117 also contains various provisions regarding embedment of connectors and spacing of reinforcement which are intended to prevent failure by pulling out of the slab (Fig. 4) or by bursting of a haunch (Fig. 5).

FIG. 4. PULL OUT FAILURE

FIG. 5. BURSTING OF HAUNCH

Connectors are sometimes described as 'stiff' or 'flexible' on the basis of slenderness. Thus a stud or channel connector would be described as flexible and a bar connector as stiff (or even rigid). These descriptions overlook the fact that interface slip depends on the deformation of the concrete as well as on the deformation of the connector. Thus a ½ in stud is stiffer than a ¾ in stud (assuming each to be loaded to the same fraction of its ultimate capacity) and a channel connector can be stiffer than a bar. Likewise if lightweight concrete is used the stiffness of all connectors diminishes and the loss of interaction for a given percentage of connector strength will be more pronounced. It may therefore be advisable in this case to design connectors at a reduced percentage of their ultimate strength. Not only will the stiffness at a given percentage of ultimate connector strength be reduced, but it can also be expected that the strength of some connectors will be less for lightweight concrete than for gravel concrete of the same strength. Failures for gravel and lightweight concrete can be compared in Fig. 6. The performance could be improved by placing gravel concrete locally around each connector.

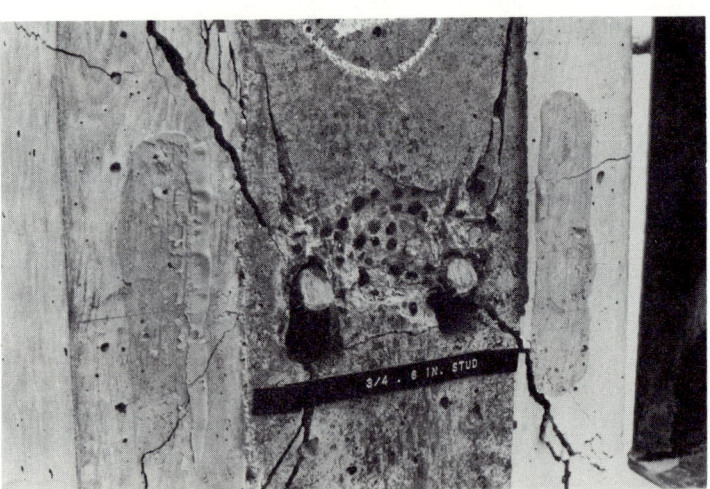

FIG. 6. PUSH TESTS ON GRAVEL AND LYTAG CONCRETE

Fatigue Strength of Shear Connectors

CP 117, Part II, tabulates connector strengths for a range of lives and ratios of minimum/maximum shear. For stud connectors the strengths are expressed as a percentage of the static ultimate strength and for channel, bar and T-connectors the strengths are expressed in terms of nominal shear stress in the connecting weld.

FIG. 7(a₁) FATIGUE FAILURE IN WEB OF CHANNEL CONNECTOR

FIG. 7(a₂) WELD FAILURE IN WEB OF CHANNEL

FIG. 7(b) BAR AND HOOP CONNECTOR — FLANGE FAILURE

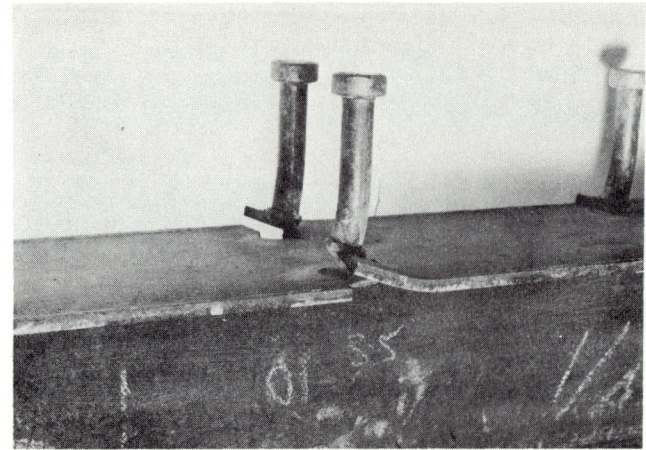

FIG. 7(c) BAR AND HOOP CONNECTOR — FLANGE AND WEB FAILURE

FIG. 8. PUSH TEST — ALTERNATING LOAD

The distinction between the two classes of connector was made because the fatigue tests on studs showed some correlation with concrete strength (and hence with connector strength), whereas little correlation could be found for the other types of connector.[7] It is probably possible to express the life with sufficient accuracy in terms of load range only and this might be done in future.

Four types of fatigue failure can be recognised — through the connector, through the weld, through the flange, and through the flange and web (Fig. 7). In an attempt to prevent flange failures, CP 117, Part II, limits the stud diameter to 1½ times the flange plate thickness and the leg length of a weld to ½ the flange thickness. Connector failures may remain undetected and lead to cumulative overloading and failure of other connectors. Where should such failures lie in the hierarchy of limit states? They do not initially amount to unserviceability; they are not readily repaired; they may lead to failure of the bridge. Presumably they should be regarded as major failures. What load (or frequency) factor should be placed on a predicted load spectrum?

The tabulated strengths are based on push tests (Fig. 8) and on simply supported beam tests. Thus the flange was either practically unstressed or stressed in compression. The effect of fatigue on connectors attached to a tension flange was not known (the flange would be designed for a welded attachment according to BS 153) and this is one reason why CP 117 requires that the shear be calculated assuming an uncracked slab. Since the actual shear will be reduced by cracking and by slip it was considered that the same strengths could be assumed for the tension region as for the compression region.

In view of the uncertainty regarding this question, tests were instigated on simply supported beams subjected to negative moment (Fig. 9). A preliminary analysis of the results indicates that the present design procedure, although not rational, appears

FIG. 9. FATIGUE TEST RIG FOR BEAMS UNDER NEGATIVE MOMENT

FIG. 10(a) FATIGUE FAILURE

FIG. 10(b) FATIGUE FAILURE

to be satisfactory. The types of failure previously noted can again be discerned, the web crack occurring where the shear is low and the flange stress high, and the weld failure where the shear is high and the flange stress low (Fig. 10).

Permanent Structural Formwork

A number of considerations arise if formwork is used structurally. If steel shutters are to be economic they must be thin, and the shear connection must be simple. A system of indented troughs which may be satisfactory under static loading may not be satisfactory under repeated loading. Although a waterproof membrane is obligatory, will this obviate corrosion of a thin continuous sheet? What should be the minimum permitted sheet thickness?

Concrete formwork which is to be used structurally must be adequately connected to the *in situ* concrete. Tests on a plank system in which a welded wire truss provides the bending strengths of the plank, and also connects the *in situ* concrete, have shown that although fatigue failures of the welded connections occur, the interface connection remains unimpaired. The longitudinal shear requirements around the shear connection must be maintained and reinforcement against longitudinal direct and bending stresses must be provided.

If full slab-depth pre-cast slabs are used pockets must be provided for shear connectors. These pockets could reasonably be filled with extra high strength concrete. An increase in slab size will reduce the number of transverse connections, some of which must have tensile strength. The design of these connections, having regard to tensile strength and crack width, seems to present some problems, as do the longitudinal joints, if these are required. An important question is whether such slabs require bedding; the local bearing provided at shear connectors is probably adequate structurally, but is corrosion then a problem?

Composite Columns

Tests on encased steel columns and on concrete filled tubes under equal end eccentricities correlate well with failure loads given by a numerical analysis which assumes full interaction between steel and concrete.[8, 9] A computer program has been used to generate design curves for columns with equal and unequal end eccentricities about one axis. An empirical interaction relation has been proposed for biaxial bending and an approximate procedure has been suggested to take account of frame action.[10] If sufficient corroboration can be found, these proposals for calculating the failure loads of composite columns would be suitable for design purposes. Recent tests show the biaxial bending proposals to be satisfactory. Computed failure loads for circular and rectangular filled tubes with equal end eccentricity are already available in tabular form.[11]

The triaxial behaviour of circular filled tubes has also been studied. It is easily shown for a short column that if at failure the tube develops its yield stress in circumferential tension (the longitudinal stress therefore being zero) the triaxial effect results in a strength considerably in excess of the sum of the separate uniaxial strengths of the steel and concrete. Tests show that for short columns a strength increase of the expected order does occur albeit at very high strain. A post-yield plastic flow analysis of strains measured in short columns indicates that at failure the circumferential stress in fact reaches about $\frac{3}{8}$ the yield stress, whereas the longitudinal stress remains at about $\frac{3}{4}$ the yield stress.[12] For more slender columns the triaxial effect diminishes.

The filled tube seems to be very suitable on a bridge column. It is self shuttering, compact and easily maintained. It is ductile and resistant to major and minor accidental loading. A range of strengths can be achieved with a standard diameter of varying the wall thickness.

CONCLUSION

Some of the thinking behind the existing and future British design Codes for composite construction has been discussed.

The results of research are being incorporated in Codes of Practice and this is commendable. We should take care however that in discarding well tried rules in favour of seemingly more rational approaches, we do not unwittingly lose safeguards whose existence may not have been apparent. A term such as 'design ecology' may be useful in this context.

The research worker is entitled to canvass has latest theories or test results, but may well be unaware of all the consequences of his proposals. It is right that he should be involved in the work of Code of Practice Committees, and he is likely to be the expert in his particular field. His colleagues on the Code Committee, recognising his expertise, may then feel they can accept his proposals without a very close examination, and this is a potentially dangerous situation. Committee members should scrutinise new proposals most critically, and the research worker will be grateful for this attention.

REFERENCES

1. British Standards, CP 117: *Composite construction in structural steel and concrete. Part 1: Simply-supported beams in building*, 1965.
2. British Standards, CP 117: *Composite construction in structural steel and concrete. Part 2: Beams for bridges*, 1967.
3. Chapman, J. C. and Teraszkiewicz, J. S. Research on composite construction at Imperial College. *Proceedings of the Conference on Steel Bridges* (*London, 1968*). London, British Constructional Steelwork Association, 1969, Paper 5.
4. Chapman, J. C. and Balakrishnan, S. Experiments on composite beams. *Struct. Engr*, Vol. 42, No. 11, Nov. 1964, pp. 369–383.
5. Yam, L. C. P. and Chapman, J. C. The inelastic behaviour of simply supported composite beams of steel and concrete. *Proc. Instn civ. Engrs*, Vol. 41, Dec. 1968, pp. 651–683.
6. Johnson, R. P. Longitudinal shear strength of composite beams. *J. Am. Concr. Inst.*, Vol. 67, No. 6, June 1970, pp. 464–466.
7. Mainstone, R. J. Shear connectors in steel-concrete composite beams for bridges and the new CP 117 Part 2. *Proc. Instn civ. Engrs*, Vol. 38, Sept. 1968, pp. 83–106.
8. Basu, A. K. Computation of failure loads of composite columns. *Proc. Instn civ. Engrs*, Vol. 36, March 1967, pp. 557–578.
9. Neogi, P. K., Sen, H. K. and Chapman, J. C. Concrete-filled tubular steel columns under eccentric loading. *Struct. Engr*, Vol. 47, No. 5, May 1969, pp. 187–195.
10. Basu, A. K. and Sommerville, W. Derivation of formulae for the design of rectangular composite columns. *Proc. Instn civ. Engrs*, suppl. Vol., Paper 7206 S, May 1969, pp. 233–280.
11. Sen, H. K. and Chapman, J. C. Ultimate load tables for concrete-filled tubular steel columns. CIRIA Technical Note 13, 1970.
12. Sen, H. K. *Triaxial effects in concrete-filled tubular columns.* Ph.D. thesis, University of London, 1969.

DEVELOPMENT OF DESIGN CRITERIA FOR COMPOSITE BOX GIRDER BRIDGES

ALAN H. MATTOCK

University of Washington, Seattle

INTRODUCTION

The 1969 edition of the AASHO Bridge Specifications[1] contains criteria for the design of single cell composite box girder bridges. The criteria were initially drafted by an *ad hoc* committee chaired by the author. They were subsequently reviewed by the Bridge Committee of the American Association of State Highway Officials and were adopted for inclusion in the AASHO *Standard Specifications for Highway Bridges.*[1] The drafting committee included engineers from state bridge departments, the U.S. Bureau of Public Roads, consulting practice, the steel industry and university faculties.

The type of bridge with which the criteria are concerned is shown in Fig. 1. It consists of rectangular or trapezoidal section steel girders made composite with a

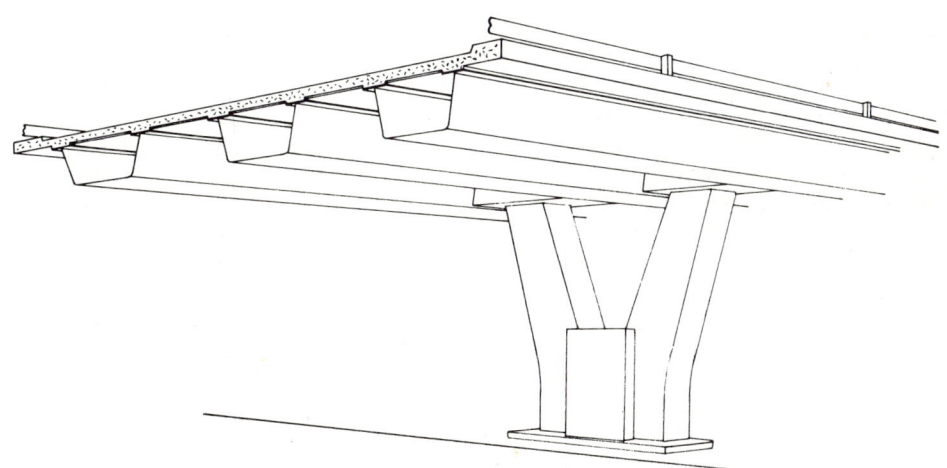

FIG. 1. TYPICAL COMPOSITE BOX GIRDER BRIDGE

reinforced concrete deck slab, to form single cell box girders. Diaphragms are provided only at the supports and composite action is ensured by the provision of stud shear connectors on the top flanges of the girders.

371

Due to the greater torsional stiffness of a closed box girder as compared to an I-section girder of similar flexural strength, a better lateral distribution of loads is achieved with this form of construction than is the case with a concrete deck slab on steel I-section girders. Consequently the design bending moment for each girder is less in the box girder bridge and economies in materials follow. Further economies in fabrication and erection are made possible by the elimination of diaphragms other than at the supports. The clean external appearance of this type of bridge is aesthetically pleasing.

The criteria are concerned with straight bridges of up to about 150 ft span. This category includes the great majority of highway bridges constructed in the United States. The objective of the drafting committee was to provide the design engineer with relatively simple design criteria, which, while being economic in design cost, would still enable the designer to realise the economies inherent in this form of construction.

LOAD DISTRIBUTION

In the AASHO Bridge Specifications, expressions are provided for the load for which each girder must be designed. The expressions vary with the type of bridge and primarily involve the spacing of girders. The first stage in the development of the design criteria was a study of the distribution of loads in single cell box girder bridges. This was carried out at the University of Washington. The objective was to develop simple expressions for the maximum loads carried by the girders, when this type of bridge is subject to the most unfavourable arrangements of AASHO Standard Trucks within its design lanes.

Load Distribution Study

The procedure adopted was as follows. Firstly, a computer program was written for the analysis of this type of bridge, treating it as an elastic folded plate structure. Secondly, the reliability of this method was confirmed by a comparison of the predictions of the computer program with the actual behaviour of a quarter scale model of an eighty foot span, two-lane highway bridge of the type under consideration. Finally, the computer program was used to calculate the maximum load per girder produced by various critical combinations of loading on thirty-one bridges having various spans, numbers of box girders and numbers of traffic lanes. The results of this computer study were used to develop a simple expression for the maximum load per girder suitable for use in design.

Computer Program – The computer program was written for the analysis of simply supported folded plate structures of general form. The structure was subdivided into a convenient number of plane plate elements, rigidly joined at their longitudinal edges. The stiffness method of analysis was used, allowing four degrees of freedom at each joint (1 rotational, 3 translational). The stiffness coefficients used were obtained from the exact solution of the folded plate problem produced by Goldberg and Leve.[2]

Model Test – Details of the two-lane, three girder, 80 ft span prototype bridge are shown in Fig. 2. It was designed for HS20-44 loading. The design was in accordance with the 1965 AASHO Bridge Specifications, except that a load distribution factor of s/6·5 was used.[†] It was assumed that the girders would act alone

---

[†] As compared to s/5·5 used for a concrete deck slab on steel I-section girders. s was taken as the spacing between the webs, measured in feet. The load was the load per web.

when carrying their own weight and that of the concrete deck, and that the composite section would resist the live load moments and those due to dead loads added after the deck had hardened.

Bridge Span – 80 ft.

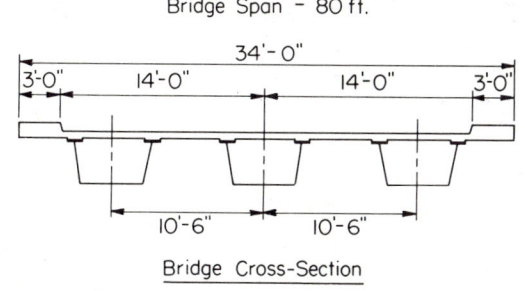

Bridge Cross-Section

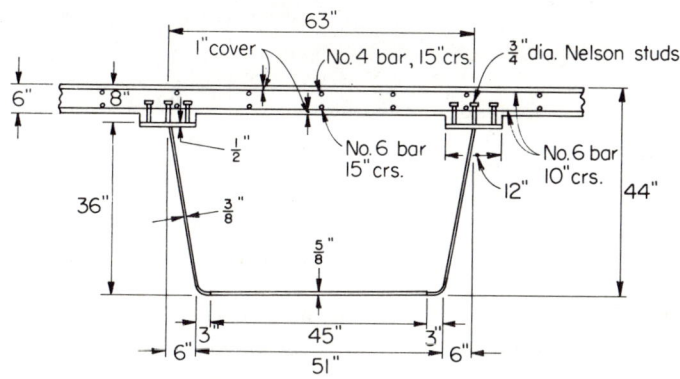

Typical Girder Cross-Section

FIG. 2. DIMENSIONS OF PROTOTYPE BRIDGE

The box girders were designed for a 36,000 lb/in^2 yield point steel, and an allowable stress of 20,000 lb/in^2 was used for flexure. The webs were made the minimum thickness permitted without the use of stiffeners. The concrete deck slab was made the thinnest possible to resist the design bending moments specified in the AASHO Bridge Specifications. The minimum thickness plates and deck slab were used so that the lateral distribution of loads obtained in the study should be the minimum likely to occur.

The model reproduced the prototype bridge to one quarter scale, using reinforced mortar to simulate the reinforced concrete deck slab. The bar sizes and spacing were reproduced exactly to scale. Complete details of the model study have been reported elsewhere.[3]

Both influence line tests and simulated truck loading tests were carried out on the model. In the influence line tests, a concentrated load was placed at nine successive locations across the width of the bridge at midspan. In the truck loading tests a group of six concentrated loads were applied to the bridge deck simultaneously, as seen in Fig. 3. The distribution and relative magnitude of these loads simulated to one quarter scale the wheel loads of the HS20-44 AASHO standard design truck. The loads were applied in both lanes of the bridge, in the extreme lateral positions which

FIG. 3. TRUCK LOADING TEST

it is considered in design that the truck can occupy. In all tests, measurements were made at midspan of girder deflection and of strain in the girder bottom plates. The average strain in the bottom plate of each girder may be regarded as a measure of the bending moment carried by that girder.

Close agreement was obtained between the calculated distributions of load and deflection in both the influence line and truck loading tests. In Fig. 4(a) the

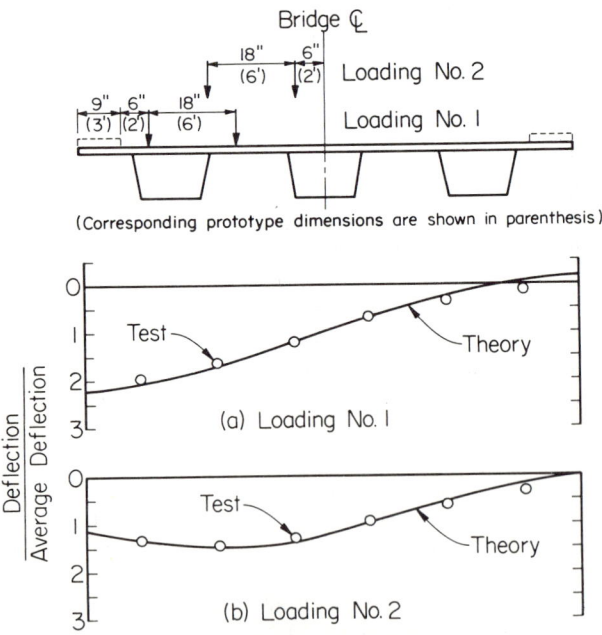

FIG. 4. MIDSPAN DEFLECTIONS OF MODEL BRIDGE IN TRUCK LOADING TESTS

measured deflections are compared with the calculated deflections for the truck loading placed as close to the curb as required by the AASHO Bridge Specifications. Similarly, in Fig. 4(b) the deflections are plotted for the case of the truck loading placed as close to the centre-line of the bridge as is required by the specifications.

By super-posing the results obtained in these tests it was possible to obtain the distributions of bottom plate strains when two standard trucks are placed on the bridge, (i) so as to produce maximum moment in an exterior girder, and (ii) so as to produce maximum moment in the interior girder. These are compared with the calculated distribution of strains in Fig. 5. Note that when either an exterior or an interior girder is subject to maximum moment, the strain distribution is uniform in the bottom plate of that girder. This was also found to be the case in other bridges studied analytically.

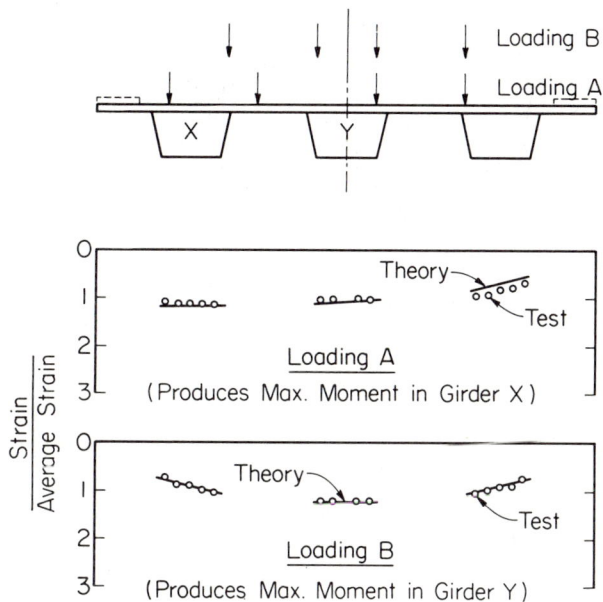

FIG. 5. DISTRIBUTION OF BOTTOM PLATE STRAINS FOR LOADINGS PRODUCING MAXIMUM GIRDER MOMENT

In Fig. 6, the experimentally determined distributions of load are compared with the distributions of load obtained from the calculated deflection influence lines, it being assumed that the total load is distributed between the three girders in proportion to their midspan deflections. It can be seen that the calculated transverse distribution of loads is in close agreement with the measured distribution, and in particular that the calculated maximum loads carried both by an exterior girder and by an interior girder are in very close agreement with the measured maximum loads.

The measured maximum loads carried by an exterior girder and by the interior girder are respectively equivalent to AASHO load distribution factors of s/6·91 and s/6·48. This result justifies the assumption of a load distribution factor of s/6·5 in the design of the prototype bridge.

After service load level tests on the model were completed, the deck slab was cut longitudinally between two of the girders, using a diamond saw. The two girder structure remaining can be regarded as a one-fifth scale model of a 100 ft span, two lane bridge. This bridge was also subjected to influence line and truck loading tests. Once again it was found that the calculated behaviour was in good agreement with the experimental behaviour.

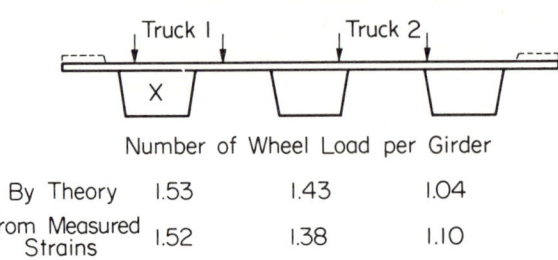

Loading A – For Max. Load on Exterior Girder X.

Number of Wheel Load per Girder

By Theory	1.53	1.43	1.04
From Measured Strains	1.52	1.38	1.10

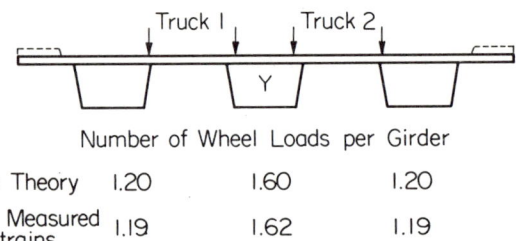

Loading B – For Max. Load on Interior Girder Y.

Number of Wheel Loads per Girder

By Theory	1.20	1.60	1.20
From Measured Strains	1.19	1.62	1.19

FIG 6. DISTRIBUTION OF LOADS– TRUCKS PLACED TO PRODUCE MAXIMUM GIRDER MOMENT

The close agreement between the observed behaviour of the model bridges and the behaviour predicted by the folded plate computer program indicated that the program is reliable and can reasonably be used to predict the behaviour of other bridges of this type.

Analytical Study – The folded plate computer program was used to calculate the behaviour of a series of thirty-one composite box girder bridges, covering a fairly wide range of spans, numbers of lanes and numbers of girders. The characteristics of the bridges considered are set out in Table 1. The design procedure used for these bridges was the same as that used for the prototype bridge on which the test model was based.

For each girder of each bridge, influence lines were calculated showing the load carried by the girder under consideration, as a unit load moves transversely across the bridge at midspan. These influence lines were based on the calculated deflection behaviour of the bridges, it being assumed that the load is divided between the girders in proportion to their centre-line deflections. This procedure is considered justified by the degree of agreement between the measured distribution of loads in the model bridge and the distribution of loads calculated using this assumption. It is also following the precedent set by lateral load distribution calculations made using the Guyon-Massonnet[4] theory. This assumption has also been discussed elsewhere.[5]

The maximum load carried by each girder of each bridge was calculated using the influence lines, the worst possible combinations of truck locations on the bridge deck being considered in each case. The AASHO Bridge Specifications allow a reduction of the maximum stress produced in any member as a result of simultaneous loading of several traffic lanes. This is equivalent to using in design whichever is the most critical of the following loadings, 100 per cent of the standard track loading on one or two lanes, 90 per cent on three lanes, or 75 per cent on four or more lanes. The maximum load per girder caused by each of these loadings was calculated using the influence lines. It was found that the difference between the maximum load per girder for interior girders and for the exterior girders was insignificant. The absolute maximum load per girder, W_M (which would be used in design), is listed in Table 1 for each bridge. The variation of W_M with various parameters reflecting the geometry of the bridges was studied. The outcome of this study was the following equation for W_L, the maximum load per girder to be used in design.

$$W_L = 0.1 + 1.7R + 0.85/N_W \qquad (1)$$

where

$$R = \frac{N_W}{\text{Number of Box Girders}}, \text{ but not less than } 0.50 \text{ nor more than } 1.50$$

$N_W = W_c/12$, reduced to nearest whole number

W_c = Roadway width between curbs (in feet)

Equation (1) predicts very closely the maximum load per girder which should be used in design. The average value of W_L/W_M for all thirty-one bridges investigated is 1.01.

It can be seen in Table 1 that the maximum load per girder calculated using folded plate theory increases slightly as the span decreases. However, in the interest of simplicity it was decided not to include the bridge span as an additional parameter in equation (1), even though the accuracy of the equation could have been improved by so doing.

The study was carried out on simple span bridges, but it is considered that equation (1) is also applicable to continuous bridges. Massonnet[4] showed analytically that when a continuous span is loaded to produce maximum midspan moments, then the lateral distribution of load at midspan is the same as for a simple span. It was found experimentally[5] that the lateral distribution of load at a support section in a continuous bridge is the same as that occurring at midspan when one span only is loaded, i.e. the same as in a simple span.

CROSS-SECTION DISTORTION STRESSES

When box girder bridges of the type under consideration are subjected to eccentric loads the cross-sections of the individual box girders become distorted as shown in Fig. 7. This distortion gives rise to secondary bending stresses, which are a maximum at the corners of the section. It was considered necessary to check the magnitude of these stresses, particularly with reference to possible fatigue problems. A study was made of the distortion stresses in the thirteen bridges indicated in Table 1, by A. C. Scordelis of the University of California. He, also, used a computer program based on folded plate theory to evaluate the behaviour of the bridges. A total of 80 harmonic terms of the Fourier series were used to represent each wheel load on the bridges. All these bridges were analysed for the loading conditions which produce the greatest distortions in the various girders. It was found that the maximum distortional stresses in the webs, the thinnest and most highly stressed parts of the section, ranged from 3,000 lb/in² to 6,000 lb/in² for the various bridges. Loading the opposite side of the bridge produces some reversal of stress. The range of stress is of interest from the point of view of possible fatigue effects. The total range of stress, considering the worst possible sequences of loading, was found to vary from 3,000 lb/in² to 11,000 lb/in². The maximum stresses and maximum range of stresses occur in the centre girder of those bridges with an odd number of girders. These

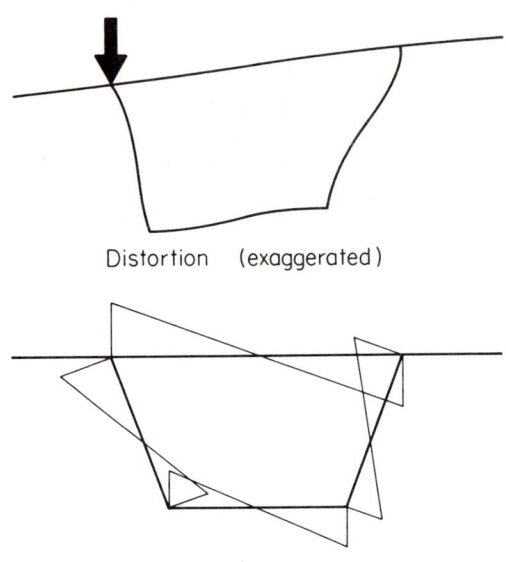

Distortion   (exaggerated)

Bending moments due to distortion

FIG. 7. DISTORTION OF BOX GIRDER DUE TO ECCENTRIC LOADS

stresses are within acceptable limits and should not cause any difficulty, providing the transverse bending stresses due to supplementary loadings such as utilities, are restricted to about 5,000 lb/in² as specified in the criteria. On the basis of the results of this study it is considered unnecessary to investigate distortion stresses in box girder bridges conforming to the limits on geometry laid down in the criteria.

VIBRATION STRESSES

Box girder bridges experience vibration and impact, with resulting dynamic stresses due to the passage of moving vehicles, in much the same way as other bridges of comparable span. However, if wide horizontal plate elements are used in the bridge section, local plate vibrations may be excited by the overall motion of the bridge. An analytical study was therefore made of the stresses in the flange plates caused by these vibrations, in bridges whose proportions conform to the limitations of the criteria. This study was carried out by W. H. Munse and W. H. Walker of the University of Illinois. The bridges chosen for study were those in which vibration would be most severe, i.e. those with wide thin bottom flange plates. The maximum stresses due to vibration were found to be moderate, being of the order of 3,000 lb/in² in tension and compression. These stresses occur at the centre-line of the bottom flange and at the web-bottom flange connection in exterior girders. The maximum stresses due to vibration and to distortion do not therefore occur at the same location. Furthermore the maximum stresses due to vibration and to distortion occur under different loading conditions. It was concluded that for bridges conforming to the criteria, secondary stresses due to plate vibration need not be taken into account explicitly in design.

BOTTOM FLANGE IN TENSION

In a wide tension flange, shear lag results in a non-uniform stress distribution across the width of the flange. Maximum stresses occur at the flange-web junctions with lower stresses at intermediate points. Data on the stress distribution in the tension flanges of a group of bridges was obtained incidentally by Scordelis when investigating the cross-sectional distorsion stresses. In Fig. 8 the ratio of the average flange stress to the maximum flange stress, is plotted against the ratio of the bridge span length to the tension flange width. On the basis of these results, the committee

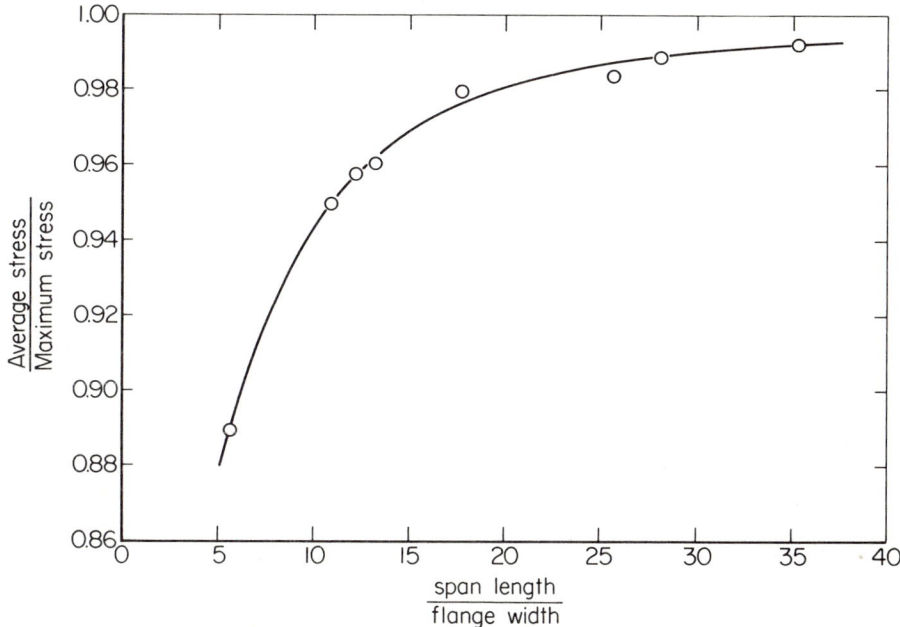

FIG. 8. STRESS VARIATION IN TENSION FLANGE

decided to limit tension flange plate widths to one fifth the span in a simple span bridge or one fifth the distance between points of contraflexure in a continuous span. This ensures that the actual maximum flange stress will not be more than about 13 per cent greater than the average flange stress in design calculations. For most bridges the difference between the maximum and average stresses will be less than 5 per cent.

BOTTOM FLANGE IN COMPRESSION

When a box girder is continuous over intermediate supports, compression will occur in the bottom flange. The provisions of the criteria concerning the design of the bottom flange to resist flexural compression are based on the non-dimensional plate buckling curve shown in Fig. 9. Extra strength available as a result of either strain hardening at low values of λ or the development of post-buckling strength at values of λ above 1·3, is not taken into account.

It is considered that the yield strength of the steel will be developed if λ is less than 0·6. This value of λ is reported by Beedle *et al.*[6], and is also in agreement with the treatment of plates in compression members in the AASHO Bridge Specifications.[1] The allowable stress is therefore $0·55F_y$ if λ is less than or equal to 0·6.

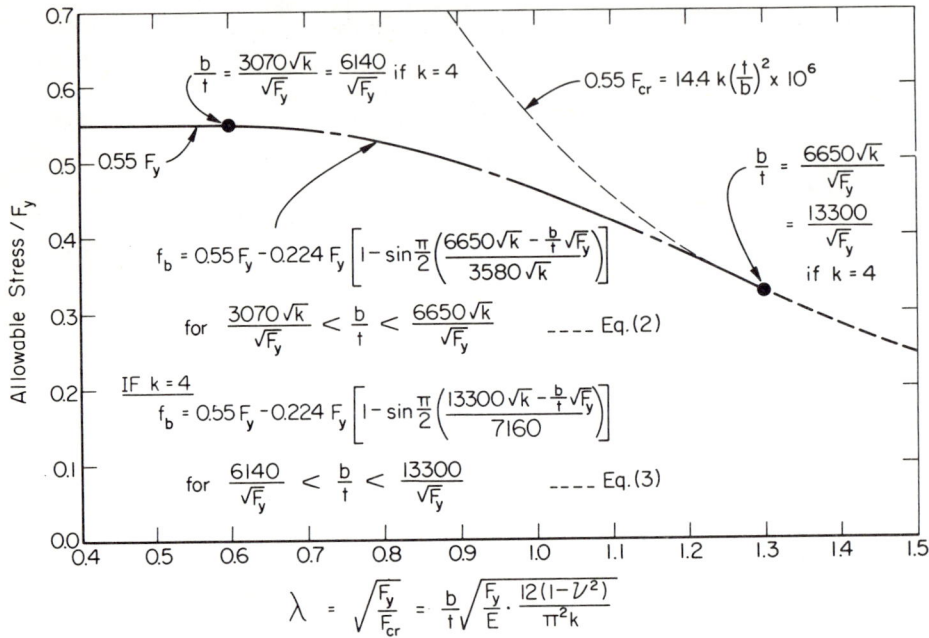

FIG. 9. NON-DIMENSIONAL PLATE BUCKLING CURVE FOR DESIGN

When λ is greater than 1·3 elastic buckling will occur and the allowable stress is taken as $0·55F_{cr}$, where F_{cr} is the elastic buckling stress. When λ is between 0·6 and 1·3 failure will be by buckling at a stress less than both the elastic buckling stress and the yield point of the steel, probably due to the effect of residual stresses and other imperfections.[6] The allowable stress in this region corresponds to a smooth transition curve tangential to $f_b = 0·55F_y$ at $\lambda = 0·6$, and to $f_b = 0·55F_{cr}$ at $\lambda = 1·3$, defined by Eq. (2) in Fig. 9.

Unstiffened Compression Flanges

In the case of a bottom flange plate without stiffeners, the plate is conservatively considered to act as if simply supported at the web-flange joints and the buckling coefficient k is taken as 4·0. The allowable stresses and limiting values of the flange width to thickness ratio, b/t, are then shown in Fig. 9. An arbitrary upper limit of 60 is set for b/t.

Longitudinally Stiffened Compression Flanges

The provisions for compression flanges with longitudinal stiffeners only are based on a treatment of the problem by Timoshenko and Gere,[7] considering that elastic instability constitutes failure. The moment of inertia I_s of each of the n equally spaced stiffeners, about its base, is required to be not less than,

$$I_s = \phi t^3 w \tag{4}$$

when

$$\phi = 0 \cdot 125\, k^3 \text{ when } n = 1$$

and

$$\phi = 0 \cdot 07\, k^3 n^4 \text{ when } n \text{ is greater than 1}$$
$$w = \text{width of flange between longitudinal stiffeners}$$

or distance from web to nearest longitudinal stiffener.

The buckling coefficient k for the stiffened flange may be assumed to have any convenient value, not exceeding 4. This value of k corresponds to the buckling of the plate panels between stiffeners. The allowable stress in the flange is then calculated using Eq. (2), but replacing b/t by w/t, and using the assumed value of k. It is thus possible to calculate directly the stiffener moment of inertia which will result in behaviour corresponding to the selected value of the buckling coefficient k.

The equation for I_s is an approximate expression based on an analytical study of the required longitudinal stiffener stiffness for various combinations of the value of k, number of stiffeners, and plate width to thickness ratio. In Table 2, the values of the buckling coefficient k, which are yielded by the equations of elastic stability when stiffeners having a moment of inertia of $I_s = \phi t^3 w$ are used, are compared with the initially assumed values of k used to calculate the coefficient ϕ. It can be seen that the actual values of k are very close to the initially assumed values. These values of k are the minimum values of k which would occur in a long panel and are therefore conservative. The variation in the allowable stress f_b, resulting from the variation in the actual value of k as compared to the assumed value of k, is generally less than the difference between the assumed and actual values of k, and may reasonably be neglected.

Compression Flange with Longitudinal and Lateral Stiffeners

The provisions for a compression flange with both longitudinal and transverse stiffeners are also based on considerations of elastic stability. The longitudinal stiffeners are required to have a minimum moment of inertia about their base of

$$I_s = 8t^3 w \tag{5}$$

The transverse stiffeners are required to have a minimum moment of inertia about the centroid of

$$I_t = 0 \cdot 10\, (n + 1)^3 w^3 \frac{f_s}{E} \cdot \frac{A_f}{a} \tag{6}$$

where A_f = area of bottom flange, including longitudinal stiffeners
   a = spacing of transverse stiffeners
   f_s = bending stress in the flange
   E = modulus of elasticity of steel, $(29 \times 10^6\ \text{lb/in}^2)$

Equation (6) was derived by considering the longitudinally stiffened plate to be replaced by a group of column elements of unit width. The sum of the buckling loads of the column elements is assumed equal to the total buckling load for the stiffened plate. The rigidity of transverse support required for columns, in order to make them buckle into half waves between transverse supports, without displacement of the supports, is known from the theory of elastic stability.[7] Transverse stiffeners having a moment of inertia equal to that given by Eq. (6), will supply support of the required rigidity to the equivalent column elements, and hence also to the plate. When applied to the case of a plate without longitudinal stiffeners, this method of proportioning transverse stiffeners was found to yield conservative results, as compared with the exact elastic theory solution available for certain cases.[7]

The allowable stress in a plate stiffened both longitudinally and transversely is given by Eq. (2) of Fig. 9, but replacing b/t by w/t, and k by k_1, where k_1 is given by

$$k_1 = \frac{[1 + (a/b)^2]^2 + 87\cdot 3}{(n + 1)^2 \, (a/b)^2 \, [1 + 0\cdot 1 \, (n + 1)]} \tag{7}$$

$$\text{for } a/b < 3.$$

Eq. (7) is the equation for the buckling coefficient for a longitudinally stiffened plate, buckling in half waves between transverse stiffness, when the longitudinal stiffeners have a moment of inertia equal to that given by Eq. (5). When two or more longitudinal stiffeners of the size specified are used, it is found that if the transverse stiffener spacing 'a' is made equal to $4w$, then k_1 will have its maximum value of 4. If a/b is made greater than 3, the panel will not buckle in a single half wave and Eq. (7) is invalid. When a/b exceeds 3 the value of k_1 will be approximately equal to that corresponding to $a/b = 3$. This value of k_1 is equal to the minimum value of k corresponding to the use of longitudinal stiffeners only which conform to Eq. (5). This indicates that to serve a useful purpose, the transverse stiffeners must be spaced significantly less than 3 times the flange width apart. When one longitudinal stiffener only of the size specified is used, no advantage is gained by the use of transverse stiffeners. This is because the minimum value of k, when one longitudinal stiffener only of this size is used, is 4.

General

The equations for allowable stress appear complicated, but they are easy to use in practice if design charts are prepared, similar to that shown in Fig. 10. In this figure, the allowable bottom flange compression stress is plotted against the ratio of the spacing of the longitudinal stiffeners to the plate thickness, w/t, or the ratio of the width of an unstiffened flange to its thickness, b/t. With a known bottom flange thickness, the longitudinal stiffener spacing and the corresponding 'k' value necessary to make the allowable flange stress equal to or greater than the actual flange stress, can be determined by inspection. The required longitudinal stiffener size can then be calculated directly using Eq. (4), if longitudinal stiffeners only are used. Alternatively, the required transverse stiffener size and spacing can be calculated using Eqs. (6) and (7), if longitudinal and transverse stiffeners are used.

DESIGN STUDIES

Design studies were made by R. S. Fountain of the U.S. Steel Corporation, to determine the economic feasibility of this type of construction. He found that box girder bridges of this type should prove to be economical for spans greater than about 80 or 90 ft.

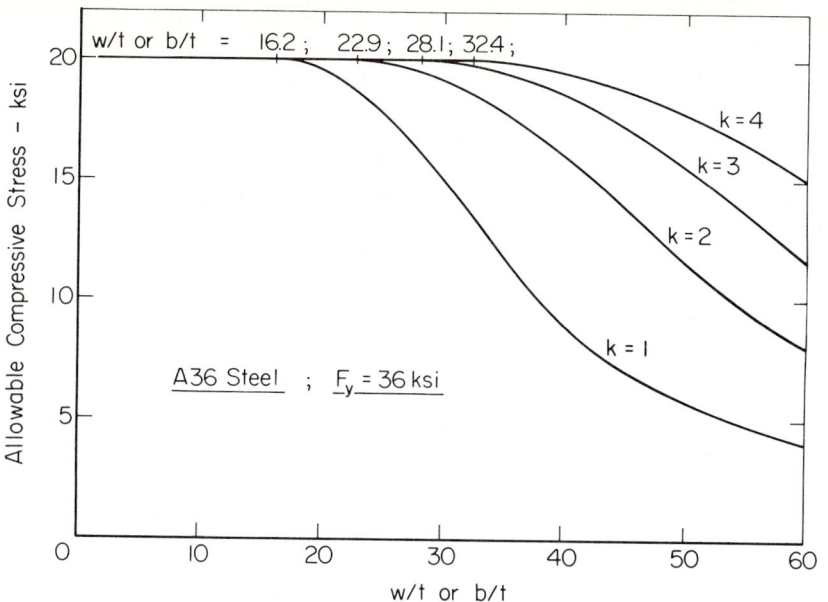

FIG. 10. TYPICAL DESIGN CHART FOR ALLOWABLE COMPRESSIVE STRESS

For bridges of this type it is found that the specified design load per lane of bridge width increases as the number of box girders used increases. This is shown in Fig. 11. Hence, by using the least number of box girders practicable to support a given width of bridge, it should be possible to obtain designs requiring a minimum

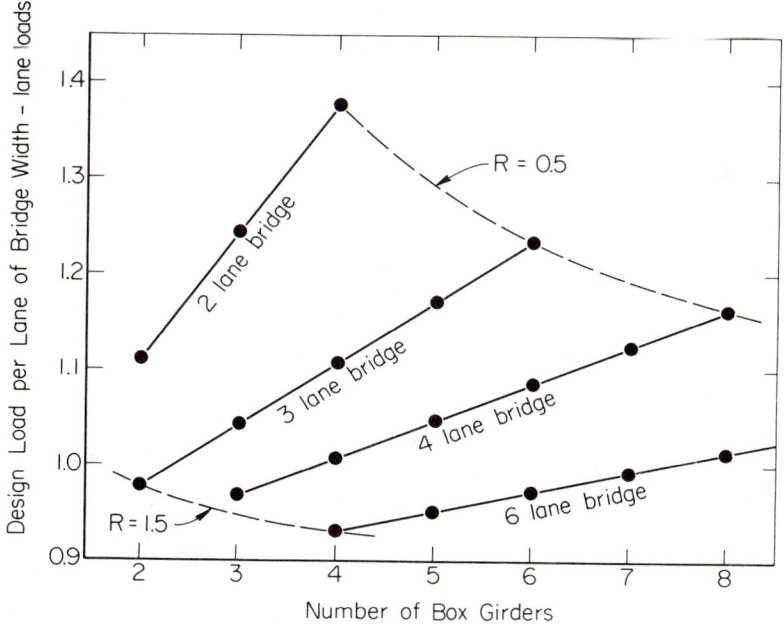

FIG. 11. EFFECT OF NUMBER OF BOX GIRDERS USED ON THE DESIGN LOAD PER LANE OF BRIDGE WIDTH

amount of steel. Such designs also will require the least number of pieces to be fabricated and erected. When a small number of girders are used, the individual girders will be wider than when a greater number of girders are used. The minimum number of girders practicable in a given case may therefore result from considerations of fabrication and of transportation to the construction site.

Fountain also made a study of the total weight of steel used per ft^2 of bridge deck, in this type of bridge. He found that the weight of steel is minimised if a span to depth ratio of 25 is used and that compared to plate girder construction the box girder bridge requires about 15 per cent less steel. The minimum total weight of steel per ft^2 of bridge deck was found to vary almost linearly from about 18 lb/ft^2 at 90-ft span, to 30 lb/ft^2 at 150-ft span. An increase in span to depth ratio to 40 was found to require about a 35 per cent increase in weight of steel. Use of shallower sections may still of course be justified where there are problems of restricted clearance or where the bridge approaches would be unduly expensive if a deeper bridge cross-section were used. The bridges with a span to depth ratio of 40 were also found able to meet the AASHO Bridge Specification limitation on live-load deflections of span/800.

CONCLUDING REMARKS

This paper has presented the background to the criteria for the design of steel-concrete composite multi-box girder bridges of moderate span, contained in the Tenth Edition of the AASHO *Standard Specifications for Highway Bridges*.[1] The provisions are simplified and conservative, to encourage their use in the design of a frequently encountered category of bridges, for which more complicated and expensive design procedures are not economically justifiable.

REFERENCES

1. American Association of State Highway Officials, *Standard Specifications for Highway Bridges*, 10th ed. Washington, D.C., 1969.
2. Goldberg, J. E. and Leve, H. L. Theory of prismatic folded plate structures, *Publs. int. Ass. Bridge struct. Engng,* vol. 16 1957, pp. 59–86.
3. Mattock, A. H. and Johnston, S. B. Behavior under load of composite box-girder bridges, *J. struct. Div. Am. Soc. civ. Engrs,* vol. 94, ST10, October 1968, pp. 2351 – 2370.
4. Massonnet, C. Contribution an calcul des ponts à poutres multiples. *Annls. Trav. Publ. Belg.,* vol. 103, nos. 3, 5, 6, June, October, December 1950, pp. 377–422, 749–796, 927–964.
5. Mattock, A. H. and Kaar, P. H. Precast-prestressed concrete bridges – 6. Test of half-scale highway bridge continuous over two spans. *J. Res. Dev. Labs Portld Cem. Ass.,* vol. 3, no. 3, September 1961, pp. 30–70.
6. Beedle, L. S., *et al. Structural steel design*. New York, Ronald Press, 1964. Art. 17.3.
7. Timoshenko, S. P. *Theory of elastic stability*; 2nd ed. in collaboration with J. M. Gere, New York, McGraw-Hill, 1961.

TABLE 1 – MAXIMUM LIVE LOAD PER BOX GIRDER

Bridge No.	N_w Number of Lanes	G Number of Box Girders	Span (ft)	W_M[1] (Wheel Loads)	W_L[2] (Wheel Loads)	$\dfrac{W_L}{W_M}$
1*	6	4	50	2·87	2·79	0·97
2	6	4	75	2·85	2·79	0·98
3*	6	4	100	2·79	2·79	1·00
4*	6	4	150	2·79	2·79	1·00
5	5	4	75	2·41	2·40	1·00
6	5	4	150	2·39	2·40	1·00
7	4	4	50	2·08	2·01	0·97
8	4	4	75	2·02	2·01	1·00
9	4	4	100	1·96	2·01	1·03
10	4	4	150	2·07	2·01	0·97
11*	4	5	50	1·69	1·67	0·99
12	4	5	75	1·64	1·67	1·02
13*	4	5	100	1·62	1·67	1·03
14*	4	5	150	1·53	1·67	1·09
15	3	3	50	2·18	2·08	0·96
16*	3	3	75	2·12	2·08	0·98
17	3	3	100	2·12	2·08	0·98
18*	3	3	150	2·03	2·08	1·03
19	3	4	50	1·69	1·66	0·98
20*	3	4	75	1·64	1·66	1·01
21	3	4	100	1·60	1·66	1·04
22*	3	4	150	1·57	1·66	1·06
23*	3	2	75	3·04	2·93	0·96
24	2	2	50	2·18	2·23	1·02
25*	2	2	75	2·16	2·23	1·03
26	2	2	100	2·12	2·23	1·05
27*	2	2	150	2·11	2·23	1·06
28	2	3	50	1·70	1·66	0·98
29	2	3	75	1·63	1·66	1·02
30	2	3	100	1·58	1·66	1·05
31	2	3	150	1·55	1·66	1·07

[1] W_M = Max. load per girder calculated using folded plate theory.
[2] $W_L = 0·1 + 1·7R + 0·85/N_w$; $R = N_w/G$; Average $W_L/W_M = 1·01$.
* Bridges included in study of distortion stresses.

TABLE 2

Number of Stiffeners 'n'	Assumed Value of 'k'	ϕ^1	Value of 'k'[2] calculated by Elastic Theory for $I_S = \phi t^3 w$
1	1	0·125	1·19
	2	1·00	1·99
	3	3·50	3·02
	4	8·00	4·15
2	1	1·1	0·91
	2	8·9	1·91
	3	30·3	3·03
	4	71·8	4·15
3	1	5·7	0·90
	2	45·5	1·98
	3	153	3·05
	'4	362	4·12
4	1	17·9	0·89
	2	143	1·92
	3	482	2·89
	4	1140	3·75
5	1	43·8	0·87
	2	350	1·82
	3	1180	2·65
	4	2800	3·63

[1] $\phi = 0\cdot125\,k^3$ for $n = 1$
  $= 0\cdot07\,k^3 n^4$ for $2 \leqslant n \leqslant 5$

[2] Tabulated values correspond to $w/t = 36$,
 k will increase slightly for larger values of w/t.

DESIGN OF COMPOSITE BRIDGE BEAMS FOR LONGITUDINAL SHEAR

R. PAUL JOHNSON

University of Cambridge

SYNOPSIS

The origin of the design methods for longitudinal shear in the British Code of Practice for Composite Beams for Bridges is explained. Proposals are made for the revision of these methods in accordance with recent research, and for their presentation in terms of Limit State Design Philosophy. The scope of the paper is limited to uncased steel girders with in-situ concrete slabs.

INTRODUCTION

Composite bridge beams in Great Britain are usually designed in accordance with CP 117 : Part 2 : 1967[1], here referred to as 'CP 117'. Almost half of this Code is concerned with the design of the junction between the concrete slab and the steel girder, a region of concentrated and complex stress. The present paper gives an account of the origin of the clauses on longitudinal shear, and discusses possible revisions and additions to them, presented in a form suitable for incorporation in a future code of practice based on Limit State Design Philosophy, and so comparable with the draft Unified Code for the structural use of concrete.[2]

It is difficult to predict the full consequences in design of the many inter-related changes that are proposed, so it will be necessary to compare the new and the old methods in trial designs.

Topics are considered here in a sequence thought to be appropriate for a new Code, determined from a study of flow charts for typical design operations. It is assumed that the reader is familiar with CP 117, and all clause numbers quoted thus, §10.2, refer to it.

LIMIT STATE DESIGN PHILOSOPHY FOR COMPOSITE BRIDGE BEAMS

Design of bridge decks is normally governed by the Unserviceability Limit States, for which methods of analysis based on elastic theory are appropriate. For longitudinal shear, the most important of these limit states is fatigue failure within the design life of the bridge. The other limit state of primary interest is 'inelastic behaviour' of steel or concrete under static overload, which may cause the subsequent distribution of stress-resultants under service load to differ from that assumed in the fatigue design calculation, and so lead to a premature fatigue failure, or may cause local damage. These are here referred to as the 'Fatigue' and 'Yield' limit states, respectively. The unserviceability limit states of excessive Deflection, Crack Width, and Vibration are unlikely to influence design for longitudinal shear.

In order to develop limit state design procedures and to compare them with existing methods, it was necessary to assume characteristic loadings and material strengths and values of the partial safety factors γ_m (materials) and γ_s (loads). Attention was concentrated on highway-bridge loadings, and on design for dead and live loads (Combination 1 of CP 117, §5). The values assumed are given below and in Table 1.

Characteristic values:

Dead load: as in current design to CP 117.
Static live load: HA and HB as in B.S.153[3]
Repeated live load: Ministry of Transport Live-load Spectrum.[4]
Compressive strength of concrete: specified works cube strength, u_w.
Static strength of steel: specified yield or proof stress, f_y.
Static and fatigue strengths of shear connectors: tabulated values, based on push-out and fatigue tests (similar to those given in CP 117).

The static overloads at which it was assumed that 'inelastic behaviour' might occur are the characteristic loads multiplied by the γ_s values for the Unserviceability Limit State (Table 1). The corresponding limiting or design stresses were taken as the design yield strength ($f_y/1.1$) for steel and 0.5 times the design cube strength for concrete (for which $\gamma_m = 1.0$). It was assumed that the Live-load Spectrum and the tabulated fatigue strengths of connectors incorporated the necessary margin of safety against fatigue failure; that is, they were taken as design values.

For the Design Loads, longitudinal shears and loads per connector can be calculated by elastic theory for the Yield and Fatigue limit states, but no method is known for checking whether the local stresses around and within shear connectors then exceed the design stresses given above. A similar problem exists in the design of bolted and welded joints, and a similar solution was used. Empirical design procedures were developed, based on three main criteria: the avoidance of significant inelastic behaviour in service of the connection as a whole and the achievement of adequate margins of safety against failure, as demonstrated by tests; and the need for continuity with earlier methods that have been shown by experience to give serviceable structures.

At the Unserviceability Limit State, it was assumed that the design of shear connectors and their welds is normally governed by fatigue, but that provision must be made for a check that the Yield Limit State is not reached under static overload. This makes it possible to base the fatigue requirements on range of stress only, for the static check ensures that the maximum load per connector is not excessive. There is no evidence of failure of slab reinforcement during fatigue tests on connectors, so that its design can be based on static strength only.

It is implied in CP 117 that structures designed by the working-load design methods of that Code always have sufficient ultimate strength. It is not obvious that this is so with the new design philosophy and methods, so a tentative method of checking the shear connection at the Collapse Limit State will be given. It is hoped that trial designs will show that this check is not necessary.

DESIGN OF THE DECK SLAB

The basic problem is to design the deck slab and its reinforcement to resist the forces imposed on it by the shear connectors without excessive slip or separation and without longitudinal splitting, local crushing or bursting. Particular care is necessary where there is a free concrete surface adjacent to a connector, as at an end or at a side of a slab or in a haunch.

Haunches

The only haunches covered by CP 117 are those (§10.6) having a width at the slab soffit not less than one and a half times their depth. This rule would allow the cross-section shown in Fig. 1. It is suggested that a more logical requirement would be for the sides of a haunch to lie outside a line drawn at 45° from the outside edge of the connectors (Fig. 2).

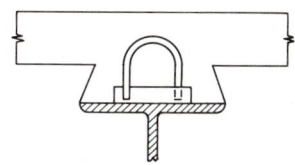

FIG. 1. CROSS-SECTION OF HAUNCH

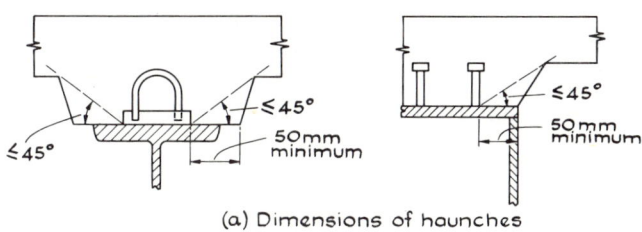

(a) Dimensions of haunches

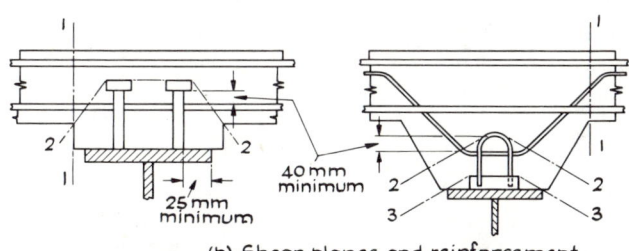

(b) Shear planes and reinforcement

FIG. 2. HAUNCHES AND SHEAR PLANES

Calculation of Longitudinal Shear

This should be based on elastic theory and an uncracked slab, as in §10.1, but the loading now considered is higher than in CP 117 (e.g. 1·5 times HA rather than HA), so that the value of shear per unit length (Q) is not directly comparable.

Transverse Reinforcement

It is suggested that the clause corresponding to §10.2 in CP 117 could be as follows: 'For Combination (1) loading the longitudinal shear force (Q) per unit length of composite beam on any shear plane through the concrete shall satisfy:

$$Q \not> k_1 L_s (s u_w)^{\frac{1}{2}}, \text{ and} \tag{1}$$

$$Q \not> 0 \cdot 7 \, A_t f_{yd} + k_2 L_s \, (s u_w)^{\frac{1}{2}} \quad \text{where} \tag{2}$$

k_1 is a constant equal to 0·8 for normal-density concrete and 0·6 for lightweight concrete,

k_2 is a constant equal to 0·14 for normal-density concrete and 0·10 for light-weight concrete,

L_s is the length of the shear plane under consideration. Typical shear planes are shown in Fig. 1 (around perimeter of connectors) and Fig. 2 (for haunches),

s is a constant stress of 1 N/mm², expressed in·units consistent with those used for the other quantities,

u_w is the specified design cube strength of the concrete, but not more than 45 N/mm²,

A_t is the sum of the cross-sectional areas per unit length of beam of all of the reinforcing bars cut by the shear plane being considered. Top transverse reinforcement may be included only when the shear plane crosses the whole thickness of the slab, and

f_{yd} is the design yield strength of the reinforcement.'

The 'Fig. 1' referred to would be a metricated version of Fig. 1 of CP 117. The proposed Fig. 2 is given as Fig. 2 of this Paper.

There are substantial changes here. Before discussing them, some explanation is necessary of the corresponding formulae in CP 117. Those in Part 1[5] are, in lb/in² units:

$$Q_u \not> 7 \cdot 5 \, L_s (u_w)^{\frac{1}{2}} \tag{3}$$

$$Q_u \not> n A_{bt} f_y + 2 \cdot 8 \, L_s (u_w)^{\frac{1}{2}} \tag{4}$$

The symbols L_s and u_w are as defined above;

Q_u is the ultimate longitudinal shear per unit length of beam,

f_y is the yield strength of the reinforcing bar steel,

A_{bt} is the cross-sectional area per unit length of beam of transverse reinforcing bars in the bottom of the slab, and

n is the number of times that each of these bars is intersected by the shear plane,

so that Q_u, f_y, and A_{bt} differ from Q, f_{yd}, and A_t, defined above.

Due to the term $(u_w)^{\frac{1}{2}}$, the coefficients in Eqs. (3) and (4) depend on the units used. The problem is avoided in the new Eqs. (1) and (2) by the use of the symbol s, but it is convenient to retain lb/in² units for the discussion of Eqs. (3) and (4). Part 1 of CP 117 was drafted before much research on longitudinal shear was available, so part of the ultimate-strength design method of the American Concrete Institute for vertical shear was adopted, as follows. For beams with a high proportion of longitudinal reinforcement and a high shear/moment ratio (i.e., a low shear-span/depth ratio), the nominal ultimate shear stress v_u for shear-compression failure was originally given[6] as

$$v_u \not> 9 \, (u_c)^{\frac{1}{2}} \tag{5}$$

and $$v_u = p_w f_y + 3 \cdot 5 \, (u_c)^{\frac{1}{2}}, \tag{6}$$

where p_w is the proportion of transverse reinforcement in the form of stirrups at 90° to the beam axis, and u_c is the cylinder strength of the concrete

In practice, Eq. (6) gives the web reinforcement required, and Eq. (5) prevents the use of amounts of web steel so large that failure of the concrete compression zone occurs before yield of the web steel. Putting $Q_u \not> v_u L_s$, $p_w = nA_{bt}/L_s$, and $u_c = 2u_w/3$ in (5) and (6) gives Eqs. (3) and (4).

Local stresses in the concrete are very high near the base of a shear connector, where most of the shear is transferred, and there is risk of longitudinal splitting. Adekola's study of the problem[7] led to the empirical rule, $A_{bt} \not< 0 \cdot 004\, t$, where t is the depth of the slab. The rule given in CP 117:Part 1 is

$$A_{bt} \not< Q/4f_y, \tag{7}$$

which can be related to Adekola's result as follows.

The effective breadth of the slab is limited in CP 117:Part 1 to $L/3$, where L is the span, and at flexural failure the mean compressive stress in concrete is taken as $4u_w/9$. Thus the maximum possible force in the slab (for a neutral axis at the steel-concrete interface) is $(4u_w/9)t\,(L/3)$. If this occurs at midspan, the mean value of Q_u is $4u_w tL/27 \div L/2$, or $8u_w t/27$. Then from Eq. (7),

$$A_{bt}/t \not< 0 \cdot 074\, u_w/f_y.$$

The lowest value of u_w/f_y used in practice is about $0 \cdot 05$, which gives $A_{bt} \not< 0 \cdot 0037\, t$. This is for an extreme combination of neutral-axis depth, u_w, and f_y, and it can be seen that Eq. (7) usually requires more bottom transverse reinforcement than the $0 \cdot 4$ per cent found by Adekola.

By the time CP 117:Part 2 was written, more data from tests on composite beams was available[8,9,10]. These showed that the rules in Part 1 for bottom transverse reinforcement were conservative. It was deduced that Eqs. (3) and (4) could be replaced (in lb/in² units) by:

$$Q_u \not> 15L_s(u_w)^{\frac{1}{2}} \tag{8}$$

$$Q_u \not> 2nA_{bt}f_y + 3L_s(u_w)^{\frac{1}{2}} \tag{9}$$

In Part 2, Q is a working-load value, say Q_w, found by the elastic theory. Assuming a load factor of three, and putting $a_t = nA_{bt}$ (as in CP 117), then from (8) and (9),

$$Q_w \not> 5L_s(u_w)^{\frac{1}{2}} \tag{10}$$

$$Q_w \not> 2a_t f_y/3 + L_s(u_w)^{\frac{1}{2}}, \tag{11}$$

as given in §10.2 of CP 117.

Equation (11) is for regions of longitudinal sagging moment. Allowance is made for a possible reduction of longitudinal shear strength of concrete due to transverse cracking in regions of hogging (negative) moment by multiplying the term $L_s(u_w)^{\frac{1}{2}}$ by $(1 - 10f_{tc}/u_w)$, where f_{tc} is the calculated tensile stress at the top surface of the slab. It is assumed that concrete makes no contribution to shear strength if $f_{tc} > u_w/10$.

Bottom transverse reinforcement to resist positive transverse moments is usually provided in bridge decks, but not in buildings, so that in CP 117:Part 2, A_{bt} became a_t, now defined as the bottom reinforcement additional to that required for flexure. This assumption, which tests show to be conservative, was partly offset by the use of a load factor for longitudinal splitting of two, rather than three, so Eq. (7) appears in §10.3 in the form

$$a_t \not< Q_w/2f_y \tag{12}$$

New Rules for Transverse Reinforcement

The new proposals, Eqs. (1) and (2), are now explained and related to Eqs. (10) and (11), which are re-stated in a comparable form:

$$Q_w \not> 0 \cdot 415 L_s (s u_w)^{\frac{1}{2}} \tag{13}$$

$$Q_w \not> 0 \cdot 67 a_t f_y + 0 \cdot 083 L_s (s u_w)^{\frac{1}{2}}, \tag{14}$$

where $s = 145$ when lb/in^2 units are used.

The new design method is basically as in CP 117, as is evident from the similar form of the equations, but there are changes:

1. No reduction is made in the assumed shear strength of the concrete in negative moment regions, so that Eq. (2) is applicable throughout the length of a continuous beam. This should simplify design. The reasons are:
(a) The design longitudinal shear, Q, is calculated for an uncracked section. Cracking in a negative moment region is likely to reduce the shear flow Q more than it reduces the total shear strength, for in practice the 'steel' term $0 \cdot 7 A_t f_{yd}$ in Eq. (2) predominates.
(b) There is no evidence from tests on continuous composite beams[10,11,12,13] that longitudinal shear strength of a slab is significantly reduced by longitudinal negative moment.

2. Comparison of the definitions of A_t in Eq. (2) and a_t in Eq. (14) shows that reinforcement provided to resist transverse moment is now assumed to contribute also to longitudinal shear strength, if it crosses the relevant shear plane. This should save steel. The reason is that a study of tests on about eighty shear spans in unhaunched beams, with and without negative transverse bending[13] showed that the critical shear plane usually crosses the whole depth of the slab, and that shear strength is related to the total amount of transverse reinforcement rather than to that part of the bottom reinforcement that is provided to resist shear. The reason is probably that tensile force in any reinforcement is associated with an equal transverse compression, which increases the longitudinal shear strength of the concrete.

The new definition of A_t is assumed to be applicable also in regions of positive transverse moment, for which little test data is available. Reasons are that significant positive transverse bending is likely to occur only in regions of positive longitudinal moment, in which shear connectors are required to extend into the compression zone of the slab (as in §11.1.5 of CP 117), and that there is a separate requirement for minimum bottom transverse reinforcement, to prevent splitting. For shear planes of type 2−2 in Fig. 2(b), the definition of A_t excludes the top transverse reinforcement, to avoid ambiguity in situations where the top steel is level with the tops of the connectors. Shear plane 2−2 is thus likely to be more critical than 1−1 in narrow-flanged girders, but not in wide-flanged or box girders, where the new method should lead to a reduction in the total requirement for transverse reinforcement.

3. New coefficients of the terms in Eqs. (3), (4), and (7) were required, to take account of the change to Limit State design. The region concerned is one where states of stress are too complex for precise calculation, as mentioned earlier, and in fatigue tests it is the connectors that fail, not the transverse reinforcement, so the only clear evidence available is that from static tests to failure.

An ultimate-strength design method has recently been deduced[13] from all available tests on uncased simply-supported and continuous beams in which longitudinal shears were high. The coefficients were initially chosen to be in sympathy with that method, and were then modified after comparison with those in CP 117 to ensure that the departure from current practice was not too great. Comparisons at both the Collapse and the Unserviceability Limit States are made in Table 2. The values there given are now derived, and will then be discussed.

The design equations can be summarised as:

$$Q \triangleright k_3 A_t f_{yd} + k_2 L_s (su_w)^{\frac{1}{2}} \triangleright k_1 L_s (su_w)^{\frac{1}{2}} \qquad (15)$$

and, to prevent splitting, $\qquad Q \triangleright k_4 A_{bt} f_{yd}.$ $\qquad (16)$

For normal-density concrete and the Unserviceability Limit State, Eqs. (1) and (2) give the proposed coefficients k_1, k_2, and k_3 as 0·8, 0·14, and 0·7, and the value proposed for k_4 is 4·0 (Table 2, line 4).

The first step is to deduce from Eqs. (1) and (2) the corresponding expressions for the Collapse Limit State, for which all the γ_s values (Table 1) exceed those for the Unserviceability Limit States by 15 per cent. Thus if elastic behaviour continued until collapse, the ultimate longitudinal shear Q_u for propped construction would be 1·15Q. For unpropped construction, the ratio Q_u/Q depends on how much of any increase in the dead load at the Collapse Limit State is resisted compositely — an interesting problem in design philosophy. At present it is assumed that the same ratio 1·15 can be used.

The longitudinal shear per unit vertical shear is increased by the inelastic action that will occur at loads exceeding those at the Yield Limit State. The shape factor for a composite section is usually around 1·3, so 15 per cent increase of load is not sufficient to cause full plasticity. The increase of longitudinal shear in excess of that given by elastic theory was assumed to be 20 per cent, so that

$$Q_u = 1·2 \times 1·15Q = 1·38Q \qquad (17)$$

Putting $Q = Q_u/1·38$, $f_{yd} = f_y/1·1$, and $k_2 = 0·14$ in Eq. (2) we have

$$Q_u \triangleright 0·87 A_t f_y + 0·19 L_s (su_w)^{\frac{1}{2}} \qquad (18)$$

Similarly, from Eq. (1) with $k_1 = 0·8$,

$$Q_u \triangleright 1·1 L_s (su_w)^{\frac{1}{2}}, \qquad (19)$$

and from (16) with $k_4 = 4$, $\qquad Q_u \triangleright 5·0 A_{bt} f_y.$ $\qquad (20)$
(Table 2, line 1).

The Author's ultimate-strength design equations[13] for total transverse reinforcement A_t are, in the present notation,

$$Q_u \triangleright 0·8 A_t f_y + 0·22 L_s (su_w)^{\frac{1}{2}} \qquad (21)$$

and $\qquad A_t f_y \triangleleft 0·55 \text{ N/mm}^2.$ $\qquad (22)$

It was recommended that half of A_t should be placed near the bottom of the slab, and that the top steel should be the other half or that required for flexure, whichever was greater. There was no evidence from tests for any upper limit on Q_u, such as that given in Eq. (19), but subsequently the Author analysed[14] a series of tests by Davies[15] on composite beams, one of which, having only 75 per cent of A_t as given by Eq. (21), reached a value $Q = 1·04 L_s (su_w)^{\frac{1}{2}}$ without sign of shear failure (Table 2, line 2).

The second comparison is with the modified CP 117 : Part 1 equations on which the Part 2 equations were based. Replacing nA_{bt} in Eq. (9) by A_t, these are:

$$Q_u \not> 1\cdot 25 L_s (su_w)^{\frac{1}{2}}, \tag{23}$$

$$Q_u \not> 2 A_t f_y + 0\cdot 25 L_s (su_w)^{\frac{1}{2}}, \tag{24}$$

and

$$Q_u \not> 4 A_{bt} f_y. \tag{7) bis}$$

These coefficients are given in line 3 of Table 2.

Finally, Eqs. (1) and (2) will be compared with the CP 117 equations, (10), (11), and (12). The ratio Q/Q_w must be estimated. Elastic theory is applicable, so for unpropped construction and HA loading, $Q/Q_w = 1\cdot 5$ (Table 1). For propped construction, a dead/live load ratio of $1\cdot 0$ (say) and a mean γ_s for dead load of $1\cdot 25$, $Q \simeq 1\cdot 37 Q_w$. Thus $Q/Q_w = 1\cdot 45$ is a reasonable estimate. From Eq. (10) with $f_y = 1\cdot 1 f_{yd}$ and $s = 145$ lb/in^2 (i.e., 1 N/mm^2),

$$Q \not> 0\cdot 6 L_s (su_w)^{\frac{1}{2}}, \tag{25}$$

from (11),

$$Q \not> 1\cdot 07 a_t f_{yd} + 0\cdot 12 L_s (su_w)^{\frac{1}{2}}, \tag{26}$$

and from (12),

$$Q \not> 3\cdot 2 a_t f_{yd}. \tag{27}$$

These results are summarized in Table 2, and are now discussed in relation to the values proposed for the four constants. The overall limit on Q, k_1, is higher than in CP 117 : Part 2 and in Part 1 $(0\cdot 63)$, but lower than the Part 1 value $(1\cdot 25)$ that was used in deriving the formulae in Part 2. It is at the edge of the region explored in tests on beams of normal-density concrete.

The 'concrete' coefficient, k_2, is close to that in Part 2, and lower than that deduced from research on ultimate-load behaviour $(0\cdot 22)$. The 'steel' coefficient, k_3, reflects changes in the method of calculating A_t. It is slightly above that proposed in the ultimate-strength design method, which includes all transverse reinforcement, and well below the CP 117 : Part 1 value. The new definition of A_t includes more reinforcement than that used in Part 2, so the new coefficient is also set below the Part 2 value, to avoid too great a discontinuity with existing practice.

In lightweight concrete, the ratio of shear strength to compressive strength is lower than in normal-density concrete, so that the coefficients k_1 and k_2 given for deck slabs of this material are 0·6 and 0·10, about 25 per cent lower than the values (Table 2, line 4) for normal-density concrete. This reduction is as given in Clause 6.12.2.4(b) of the Draft Unified Code.[2] Coefficients k_3 and k_4, for reinforcement, are unaltered.

The coefficient for the minimum reinforcement to prevent splitting is $1/k_4$. The new value is seen from Table 2 to be 25 per cent lower than in CP 117 Parts 1 and 2, and there is a further reduction because in Part 2, A_{bt} excluded steel required for flexure, whereas now it is included. This reduction is due mainly to the change in design philosophy, and must be reconsidered in the light of trial designs.

Detailing of the shear connection

One alteration to the empirical rules given in §10.5, §11.1.5, and §11.1.6 of CP 117 is suggested. The rule that connectors shall project not less than 2 in. above bottom transverse reinforcement was intended to provide resistance to uplift forces. A more rational requirement would be for the surface of a connector that resists uplift forces (i.e., the underside of a hoop, top flange of a channel, or head of a stud) to extend not less than 40 mm (say) clear above the bottom transverse reinforcement, nor less than 40 mm into the compression zone of the concrete flange in regions of sagging longitudinal moments (see Fig. 2). A similar rule is required for haunches.

STRENGTH OF SHEAR CONNECTORS

It is assumed that the characteristic strengths of shear connectors given in a new Code will be essentially the 'ultimate capacities' given in Tables 2 to 4 of CP 117, modified to incorporate recent research, and including values for concrete strengths up to 52·5 N/mm^2 (7600 lb/in^2).

The presentation of fatigue data can be greatly simplified, as may be seen from Table 3, which could replace Tables 3 and 4 of CP 117. It has been found that in

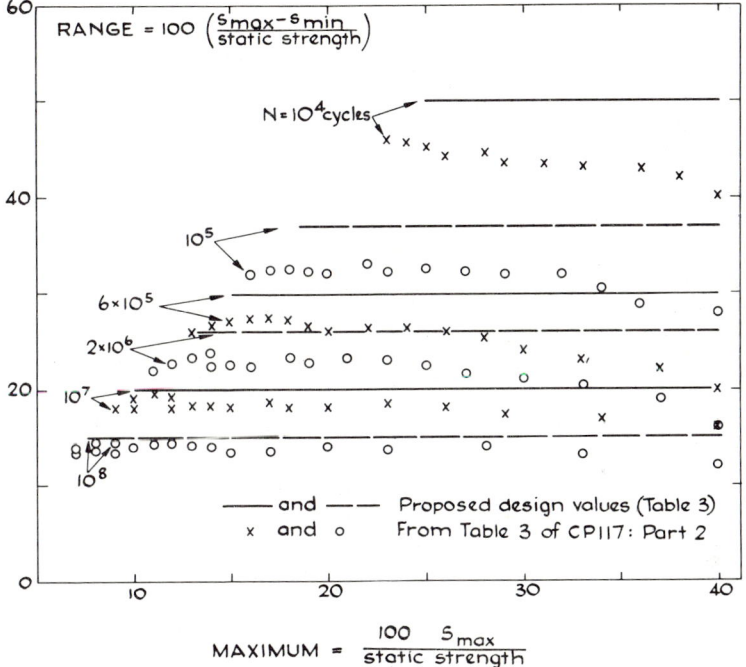

FIG. 3. FATIGUE STRENGTH OF STUD CONNECTORS

design for the Fatigue Limit State, one can assume that connector life depends only on the range of stress, and not on the maximum stress. For studs, the designer needs to know only the range of shear load ΔS, ($= S_{max} - S_{min}$), expressed as a percentage of the static strength. Similarly, ranges of nominal shear stress on the weld throat, Δf, are given for other types of connector. The author is grateful to Dr. J. B. Menzies for these values, which are based on his re-examination of all available information from tests. Design ranges at a given number of cycles are slightly higher than before, as may be seen from Fig. 3, in which the reduced capacities for studs given in CP 117 are shown in a plot of range against maximum load. The proposed design values of ΔS appear as horizontal lines. A similar design method was adopted by AASHO in 1968.[16]

DESIGN OF THE SHEAR CONNECTION

It is envisaged that connectors will be designed initially to satisfy the Fatigue Limit State, and that checks for static strength will be made at both the Unserviceability and the Collapse Limit States. Where fatigue is unlikely to govern, as in footbridges, the initial design could be for static strength at the Unserviceability Limit State, with a check at Collapse.

General Requirements

Three possible changes in §11.1 of CP 117 are now discussed. To limit the stress concentration in a plate due to a shear connector, and to improve its protection against corrosion, it is suggested that the distance between the edge of a shear connector and the edge of the plate to which it is welded should be not less than 25 mm. This new rule is distinct from the existing requirement for the concrete cover to be not less than 50 mm, as shown in Fig. 2.

The present §11.1.2 allows, in effect, that the spacing of connectors may be kept constant over any length where the design spacing does not vary by more than ± 5 per cent from its mean value. Theoretical studies by Mr. D. Osborne-Moss are in progress at Imperial College, London, to find out if this figure can be increased. It seems likely that an increase will be possible. This would simplify the detailing of connectors.

The placing of connectors in regions of peak negative moment involves the use of welds that reduce the fatigue strength of the girder flange. Tests to find out if this design penalty could be avoided by bunching the connectors near the points of contraflexure have recently been completed at Lehigh University. Their results may make it possible to relax the rule in §11 of CP 117 that the maximum longitudinal spacing of connectors may not exceed 0·61 m (24 in.).

Fatigue Limit State

As mentioned earlier, it is likely that the influence of maximum stress on fatigue behaviour can be neglected in beams having adequate static strength; but it will be necessary to check this in trial designs. Otherwise, the design method can be similar to the present §11.3.3, if loading spectra and numbers of cycles are specified for highway bridges, as has been done for railway bridges in the past.

Unserviceability Limit State

The design procedures could be as in §11.3.5, with the 'allowable working load per connector' of $0·4 P_u$ replaced by a 'design load' of about $0·67 P_u$ (i.e., $\gamma_m = 1·5$ for this limit state). A rough comparison between these values can be made by considering unpropped construction and HA loading, for which $\gamma_s = 1·5$. Existing design would give a static load per connector of $0·4 P_u \times 1·5 = 0·6 P_u$ at the Unserviceability Limit State. Thus an increase of about 10 per cent is suggested, on the grounds that connectors show little inelastic action at this load, and that such action tends to reduce the load per connector.

Collapse Limit State

This subject is not covered in CP 117. The following discussion illustrates the problems involved in specifying a procedure for checking the shear connection at the Collapse Limit State.

A simple solution would be to require that the number of shear connectors, N_f say, should be sufficient in each beam to develop its ultimate moment of resistance calculated by full-interaction plastic theory, M_f. But when design is governed by serviceability or fatigue criteria, as is usual in bridge beams, M_f may be far in excess of the design ultimate strength, M_d say, corresponding to the γ-values specified for the Collapse Limit State. For economical design, it is suggested that the number of connectors, N_d say, need be sufficient only to develop the moment M_d.

Studies by J. W. Baldwin and others in North America of the relationship between M and N have shown that it is on the safe side to assume that the increase of flexural strength of a beam above the non-composite value M_p is proportional to the number of connectors provided; that is,

$$(M - M_p)/(M_f - M_p) = N/N_f. \tag{28}$$

The use of partial-interaction design in accordance with Eq. (28) is allowed in the current A.I.S.C. Code.[17] Adopting this result, the design number of shear connectors is given by:

$$N_d \geqslant N_f(M_d - M_p)/(M_f - M_p) \tag{29}$$

The next problem is to decide the length of beam over which these connectors may be spread. Design envelopes of longitudinal moments for the Unserviceability Limit State will be available. A possible procedure, involving little additional calculation, would be to scale up these envelopes by the ratio of the γ_s values at the Collapse and

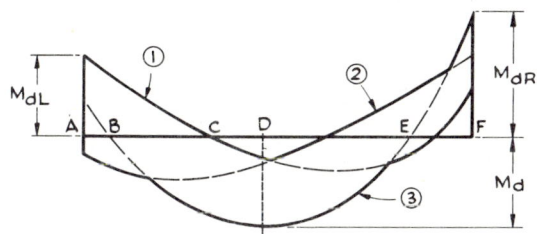

FIG. 4. ENVELOPE OF LONGITUDINAL MOMENTS FOR
ONE SPAN OF A COMPOSITE BEAM

Unserviceability Limit States, which is likely to be 1·15. This gives moments in equilibrium with the collapse loads, but takes no account of inelastic redistribution of moments.

The design moment envelope must now be broken down into its main constituent curves. An idealised example, for a beam of uniform section, is given in Fig. 4. An envelope for one span is shown, built up from curves 1, 2, and 3, due to three different design ultimate loadings. A possible design procedure is to check that sufficient connectors are provided between any cross-section of peak bending moment (a 'critical cross-section') and the adjacent point or points of contraflexure for the loading that causes the peak moment considered. These lengths are termed 'shear spans'. Thus in Fig. 4, the shear span for moment M_{dL} is AC, that for M_{dR} is EF, and those for M_d are BD and DE.

Thus a possible check at the Collapse Limit State involves these steps:

1. Identify critical cross-sections and the bending-moment distributions due to the loadings that cause the peak moments M_d at these points, and hence find the associated shear spans.

2. At each critical section, calculate M_p, M_f, and the corresponding number of connectors (of the size used in the design) for full interaction, N_f, assuming $\gamma_m = 1{\cdot}25$ (i.e., 80 per cent design as in CP 117:Part 1).

3. Find N_d as given by the existing design for each of the shear spans.

4. The shear connection is satisfactory at the Collapse Limit State if $M_d \leqslant M_f$ at each critical cross-section and inequality (29) is satisfied in each shear span.

This does not solve all the problems, but it provides a basis for discussion, and can usefully be tried out in trial designs.

CONCLUSIONS

The preceding discussion of a single aspect of the design of composite bridge beams illustrates the problems of re-writing design methods that are necessarily empirical in terms of a new design philosophy. The work is presented because the author believes that the comments of experienced designers can be of great value at this stage in the development of new design methods, and because the next stage must be to use the proposed methods in trial design. These will inevitably show up many problems not adequately covered, but should also indicate classes of structure for which one or other of the Limit States never governs, and so lead to 'deemed to satisfy' clauses that can greatly reduce the work of the designer. It will also be found that there are significant differences between new and existing designs in certain respects, and it will be necessary to check that these are due to deficiencies in the old method, rather than in the new one.

The help of practising engineers will be essential for this work. It is hoped that they, at least, will find useful the background information given here.

The Author acknowledges with thanks the valuable contribution to the development of these proposals made by the members of the Sub-Committee 'Composite Bridges', under the Chairmanship of Dr. J. C. Chapman.

TABLE 1

Limit State	*Collapse*	*Unserviceability*
Partial safety factors for materials, γ_m		
Structural steel and reinforcement	1·1	1·1
Concrete	1·4	1·0
		(see text)
Shear connectors – static	1·25	1·5
Shear connectors – fatigue	–	1:0
Partial safety factors for loads, γ_s		
Dead (steel)	1·20	1·05
Dead (concrete)	1·32	1·15
Dead (surfacing, etc.)	2·0	1·75
Live, HA	1·75	1·5
Live, HB	1·44	1·25
Live-load spectrum	–	1·0

TABLE 2

Coefficient	k_1	k_2	k_3	k_4
Collapse Limit State				
(1) Proposed	1·1	0·19	0·87	5·0
(2) Ultimate-strength method[13]	>1·04	0·22	0·8	–
(3) CP 117:Part 1 (modified)	1·25	0·25	2·0	4·0
Unserviceability Limit State				
(4) Proposed	0·8	0·14	0·7	4·0
(5) CP 117:Part 2	0·6	0·12	1:07	3·2

TABLE 3

No. of cycles, N	10^4	10^5	6×10^5	2×10^6	10^7	10^8
Stud connectors						
$100 \, \Delta S/$ (static strength)	50	37	30	26	20	25
Channel, bar, tee, and horseshoe connectors						
Δf N/mm^2	195	145	115	100	80	60

REFERENCES

1. British Standards, CP 117. *Composite construction in structural steel and concrete, Part 2: Beams for bridges.* 1967.
2. British Standards Institution. *Draft code of practice for the structural use of concrete.* London, Sept. 1969.
3. British Standards, BS 153. *Steel girder bridges.* 1958.
4. Ministry of Transport, Great Britain. *Provisional fatigue requirements for steel bridges.* Technical memorandum (Bridges) No. BE 16, 1969.
5. British Standards, CP 117. *Composite construction in structural steel and concrete. Part 1: Simply-supported beams in building.* 1965.
6. Report of ACI-ASCE Committee 326. Shear and diagonal tension. *Proc. Am. Concr. Inst.,* Vol. 59, Nos. 1–3, Jan.–March 1962, pp. 1–30, 277–333, 353–395.
7. Adekola, A. O. *Interaction between steel beams and a concrete floor slab.* Ph.D. thesis, University of London, 1959.
8. Teraszkiewicz, J. S. *Static and fatigue behaviour of simply supported and continuous composite beams of steel and concrete.* Ph.D. thesis, University of London, 1967.
9. Sen, H. K. Technical Note on the influence of shear on the behaviour of composite beams. (Under preparation, for Construction Industries Research and Information Association, London.)
10. Johnson, R. P., van Dalen, K. and Kemp, A. R. Ultimate strength of continuous composite beams. Proc. Conf. "Structural Steelwork" (1966), pp. 27–35, British Constructional Steelwork Association, London, 1967.
11. Daniels, J. H. and Fisher, J. W. *Static behavior of continuous composite beams.* Lehigh University, Department of Civil Engineering, Report 324.2, 1967.
12. Van Dalen, K. *Composite action at the supports of continuous beams.* Ph.D. thesis, University of Cambridge, 1967.
13. Johnson, R. P. Longitudinal shear strength of composite beams. *Proc. Am. Concr. Inst.,* Vol. 67, No. 6, June 1970, pp. 464–466.
14. Johnson, R. P. Transverse reinforcement for composite beams in buildings. (Synopsis). *Struct. Engr.,* Vol. 49, No. 5, May 1971.
15. Davies, C. Tests on half-scale steel-concrete composite beams with welded stud connectors. *Struct. Engr.,* Vol. 47, No. 1, Jan. 1969, pp. 29–40.
16. American Association of State Highway Officials. *Interim Specifications, Bridges and Structures.* Washington D.C., 1968.
17. American Institute of Steel Construction. *Specification for the design, fabrication, and erection of structural steel for buildings.* New York. 1969.

THE SELECTION OF BOX BEAM ARRANGEMENTS IN BRIDGE DESIGN

D. J. LEE

G. Maunsell & Partners

SYNOPSIS

Many different arrangements of box beams are possible in bridge construction and a logical classification system is proposed.

Factors affecting the proportioning of box beam bridges are reviewed in the light of practical requirements of prestressed concrete and composite construction. The application of these ideas is illustrated by some recent designs by the author's firm.

Multicellular precast concrete decks are discussed critically in an attempt to further design ideas in this field. Some proposed composite highway bridges in South Wales incorporate advanced features and these are described.

The paper concludes that spectacular box beam designs are feasible.

1. Introduction

The use of box beam construction in bridges has increased markedly in recent years for various reasons. This paper sets out to examine the merits or otherwise of this form of construction and considers the various arrangements of single and multiple box girders.

2. A Classification System for Box Beam Arrangements

The use of box beams, of single or multiple cells together with arrangements whereby several box beams are combined to form a complete bridge deck, opens up a very wide range of possible forms. It is convenient to suggest a simple classification system for box beam arrangements and such a system is shown in Fig. 1.

The system conveniently establishes the type and order of any given box structure. For instance, that part of Mancunian Way illustrated in Fig. 2 is a continuous double box classified under [2−2]. The part of Westway shown (Fig. 3) comprises single boxes for the slip roads [1−2] and a pair of double boxes for the elevated structure [2−3]. When the deck slabs merge these would become a compound system [1−2/2−3/1−2].

3. Behaviour of a Box Beam

A box beam carries load which can be considered by the engineer in a somewhat different way to other forms of bridge deck such as slabs, orthotropic plates or

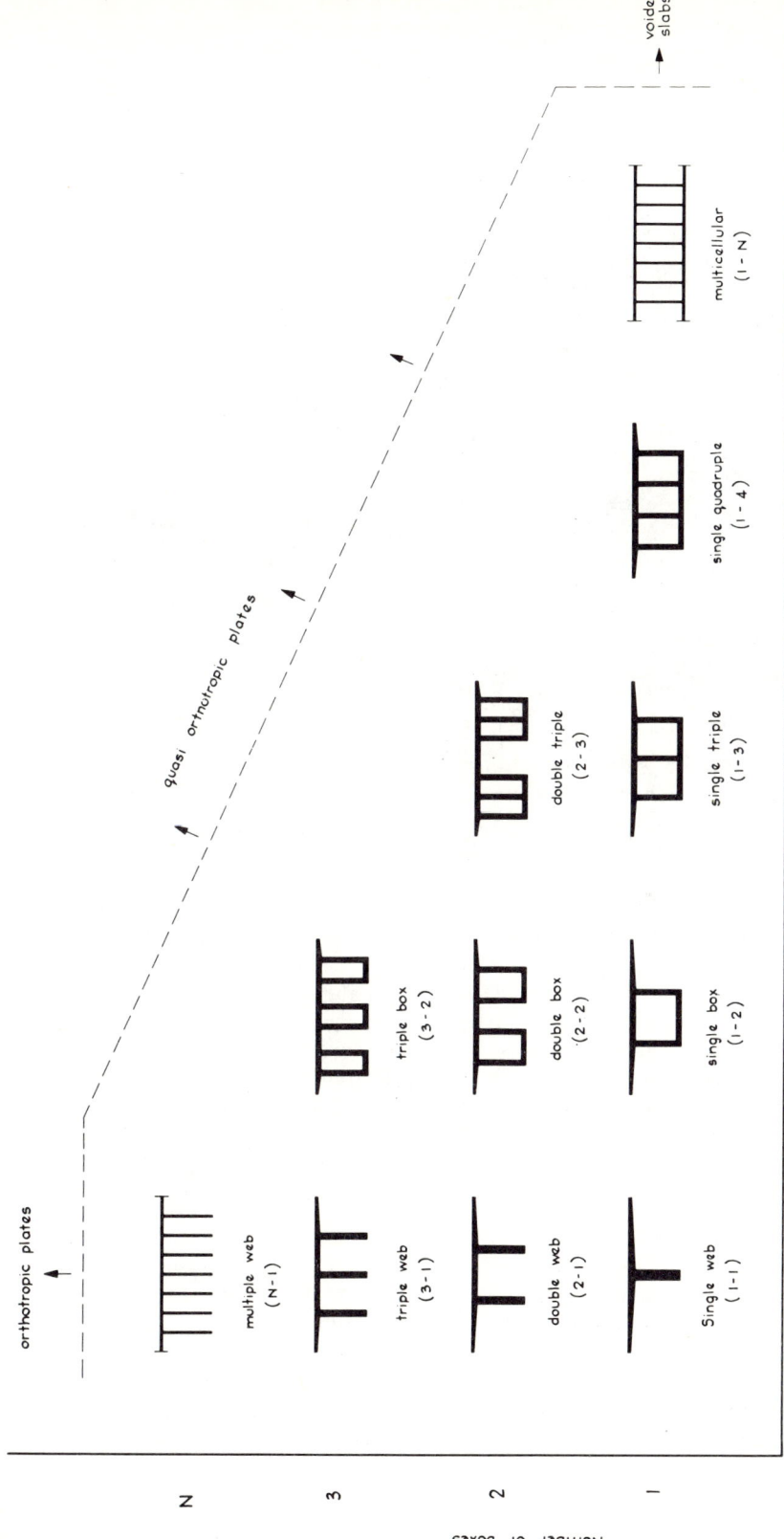

FIG. 1

FIG. 2

grillages. In the latter forms of construction the bending moments and shear stresses can be taken together in analysing the cross-section properties. In the traditional I-beam, the shape is efficient in bending but when heavy shear and torsional stresses are to be resisted great care is required in dealing with principal stresses, shear flow and stability.

In a box structure, direct shear is shared between the webs of the box and these may be vertical, inclined or curved, provided the lateral components of inclined shear forces are taken into consideration.

In addition a box beam, being a closed section, has a very much greater torsional rigidity than other types of beam. Torsional rigidity in a box is of the same order of importance as bending rigidity since resistance can be developed by the flanges and webs working together in pure torsion. The torsional strength of an I-beam is very low and torsional stresses are principally resisted by the lateral bending strength of the flanges.

4. Shear Lag and Shear Flow

Because of the form of a box beam the distribution of direct shear stress cannot be considered uniform across the top and bottom flanges but is higher at the web and flange junctions than it is mid way between webs. The proportioning of the shape and thickness of box beam elements is important. Carefully shaped haunches in concrete box beams may be necessary to adequately smooth the flow of shear stress from one element to another.

The torsional resistance of a box beam is porportional to the area enclosed by the median line of the box and this is affected by width as well as depth. Therefore the torsional rigidity of a box beam can be increased or decreased not only by

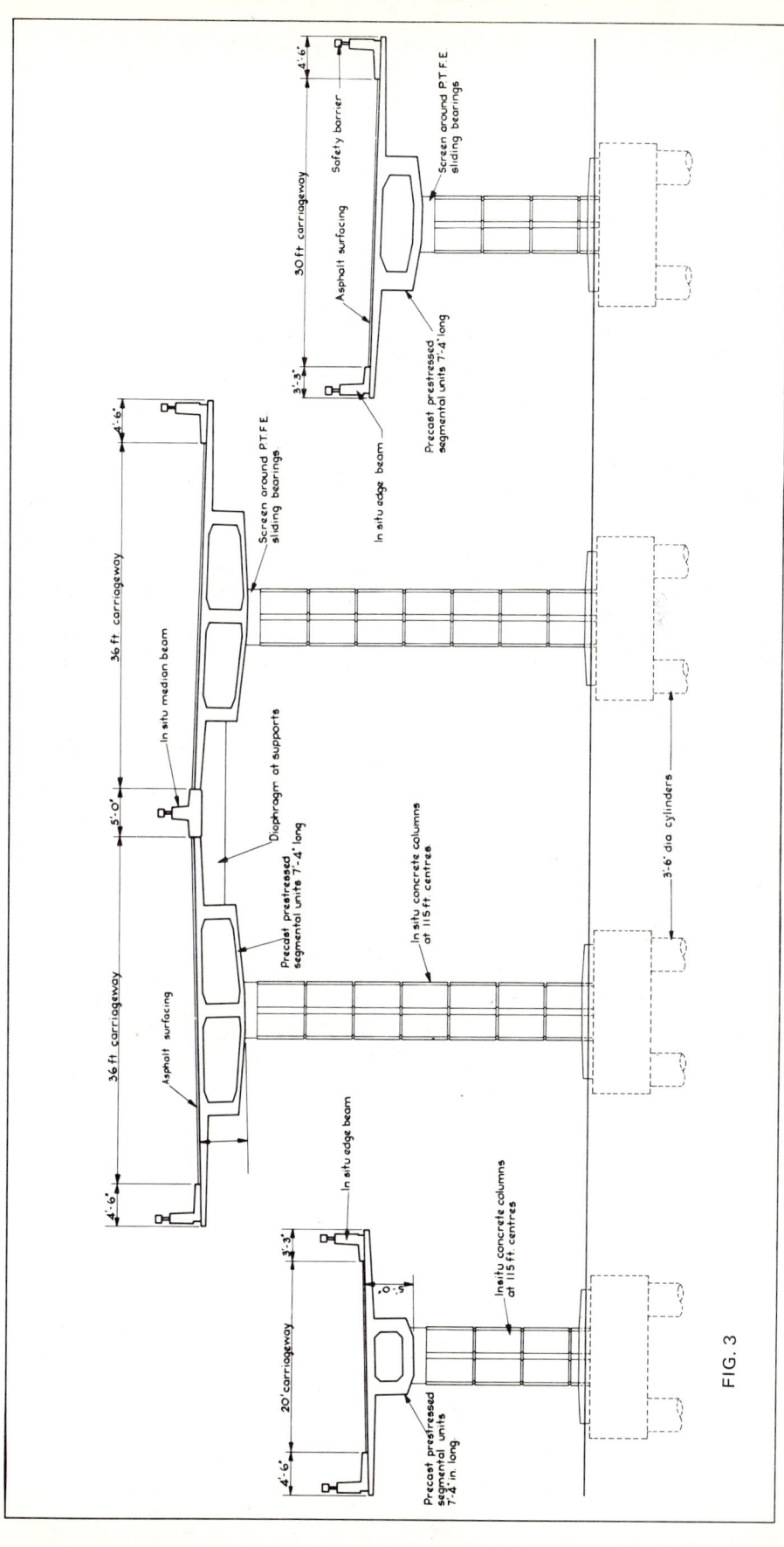

FIG. 3

altering the span/depth ratio but by altering the width/span ratio. From this it also follows that as varying depth girders are used to resist varying bending moments there may be cases where changes in geometrical dimensions of the box are related to changes in torsional loading on the beam.

5. Cantilever Flanges

Because the torsional rigidity of the box beam is so great it is not generally necessary to provide a box which is as wide as the deck; hence the outer part of the deck is very frequently cantilevered beyond the outer web of the box. For analysis purposes the bending moments developed from concentrated loads such as imposed by wheels of vehicles are theoretically independent of the distance of the point load from the root of the cantilever flange. This is reflected in the usually adopted Westergaard formula where the maximum root moment developed by a point load p is p/π per unit length. This somewhat striking result is of course accounted for by the fact that the distance of the point load from the encastre end determines the length of dispersion in the longitudinal direction. A similar result is achieved by a number of investigations when the cantilever is of varying depth and whether or not a stiffened outer edge is incorporated. For carrying heavy vehicles, therefore, and for isolated point loads it means the designer is free to choose a large cantilever without affecting the moments arising therefrom. For the self-load of the deck itself the determinate bending moments of the pure cantilever are instrumental in arriving at the optimum length of cantilever adopted. Nevertheless, for both reinforced concrete and prestressed concrete a fairly large cantilever can be economical. Consideration of long-term creep deflection suggests limiting the length of cantilever adopted for a reinforced concrete slab. For a prestressed concrete slab where creep deflections can be controlled it may be economical to adopt a fairly substantial cantilever, provided that the slab is checked for shear.

6. Spacing of Webs

Local loading on flange plates is similarly derived independently of the span and therefore the spacing of web supports to the flange plates is independent of local loading and is determined by an optimisation of the dead load requirements. The spacing of webs is more strongly determined by the vertical shear forces being carried by the total web area and in addition by the practical requirement to accommodate prestressing cables which for purposes of continuity in multi-span beams can be located in the webs, their thickness being determined by a convenient arrangement of cables and the concreting around the cable profiles and shear reinforcement. If prestressing cable splices are required then the anchorage areas necessary may require an extra thickness of webs in the area where the splices are located, which in turn may be simplified by adopting a somewhat thicker uniform web throughout the length of the box. There is also relevance in the fact that the webs may be thicker than the flanges because they have to accommodate the torsional shear stress which is additive to the vertical shear force.

7. Diaphragms

In concrete construction it is practically necessary to adopt reasonable thicknesses of elements which have an influence on the distortional behaviour of the box beam when subject to incidental symmetrical and asymmetrical loading. For this reason concrete box girders may not require diaphragms between pier positions. It is only in very large box girders or in box girders having thin concrete webs where problems of box distortion may require intermediate stiffeners or intermediate diaphragms. A somewhat different situation exists in steel box beams where the thin plate may not only undergo substantial distortion under eccentric loading but intermediate stiffening is required for fabrication and erection of the boxes.

8. Distortion and Warping of Box Beams

As has been discussed above, the spacing of diaphragms affects the distortion of a cross-section to a box beam. If the diaphragms are widely spaced then it is necessary to evaluate the secondary bending and shear stresses which are most apparent near the corners of the section. This is influenced by the width/depth proportions of the box. It can be shown for boxes with flanges and webs of similar thickness that the distortional stresses are a minimum when the box is square, but in many cases it is inconvenient to adopt square shaped boxes.

For practical reasons flanges are nominally horizontal and the webs vertical. However, the most efficient torsion resisting closed section is a circular tube and the nearer the shape of box approaches a circular shape, the more efficient it will be in respect of distortional stresses. For other cases adjustments can be made to the relative thickness of webs and flanges.

Warping, which describes the distortion of a cross-section in the longitudinal direction generated by torsional loading, is a problem which is not usually severe in closed boxes compared with open sections such as I or T beams. If, however, restraints to warping are built in (such as at transverse diaphragms), then the stiffness of the restraint can be instrumental in developing quite high direct stresses in the flanges and webs.

9. The Shape of the Boxes

It is usual for box girders to have vertical webs for simplicity of fabrication and construction. However, it is convenient in dealing with a relatively wide deck supported on a single box for the webs which provide line supports to the deck to be located in positions which give a satisfactory distribution of transverse bending moments. In other words the top of the webs can be relatively widely spaced under the top flange and if carried down vertically would lead not only to a very wide and hence thick and expensive bottom flange but provide excessive torsional rigidity in the completed box section. For this reason the outer webs are often inclined to the vertical to reduce the span of the bottom flange. This may have considerable planning advantages. For instance, in elevated highway construction the sloping webs may improve dispersion of daylight and give better natural lighting intensity at ground level. If webs are very heavily inclined a streamlined result is achieved which may be appropriate where aerodynamic considerations have to be taken into account. It can be readily appreciated, however, that the shape which provides considerable area of section at the top tends to throw the centroid of the section towards the top of the box. This is satisfactory for a simply supported box beam in prestressed concrete where the lever arm of the prestressing cables in the bottom of the box can be made as large as possible, thus increasing the efficiency of prestress. It may not, however, be so appropriate for continuous beams.

For prestressed concrete in continuous beams another factor arises from the parasitic or secondary stresses. The prestressing profile may well be substantially non-concordant owing to the high location of the neutral axis. The parasitic stresses generally induce a sagging moment in continuous beams which offsets the primary prestressing at mid span but provides a corresponding benefit near the support sections. Thus for convenience in prismatic continuous beams it is possible to adopt a uniform section of box and further to adopt the same number of prestressing cables at mid span and support which is apparently contrary to the requirements of the basic primary bending moment diagram.

Shear lag also has an influence in respect of outstanding cantilever flange area since the flange is not fully operative in the support region. Whereas at mid span the whole of the cross-section can be taken to be resisting bending moments, at supports an effective section has to be assumed. This is the same as saying that the neutral axis is effectively located at a lower position at support than it is at mid span and thus a readjustment of the prestressing moment and also of the parasitic moment will take place. The number of independent variables involves an unwieldy number of expressions to embrace all the various configurations of box beams. In the practical prestressed concrete examples which have been designed by the author's firm it is possible to adopt a uniform section which gives nominal stresses. These can then be refined later in the design process when all the secondary adjustments can be taken into account. For composite steel girders, minimising the costs of fabrication also suggests nominally prismatic sections.

10. Parasitic Moments

Parasitic moments generated by the cable profile not being concordant are one aspect of the problem but the situation is not quite so simple on those grounds alone. The prestressing forces inevitably have friction loss components which affect the values and more importantly the method of erection can influence the generation of parasitic moments.

11. Serial Construction

It is frequently appropriate to employ progressive span-by-span methods of construction where a continuous beam it built up one construction stage at a time, a joint between stages being located at some point in every span.[1] Some examples have located the joint at the support but usually they are some distance away either at the fifth, quarter or third points. Depending on the way the supporting falsework is connected to preceding stages and whether or not intermediate props are inserted, there is nevertheless a considerable difference in the self-load moments generated by such span-by-span methods of construction compared to those dead load moments which would arise if the complete continuous beam was theoretically allowed to materialise instantaneously. These matters have been reviewed in a recent paper.[2] For concrete construction it might also be mentioned that creep and shrinkage of the concrete has an influence on the moments developed. It has been found, however, that by judicious location of the construction stage and particularly for prestressed concrete which can be load balanced under permanent load to minimise changes in deflection of the structure during service, this aspect may be reduced to a very minor consideration.

12. Additional Support to Cantilever Flanges

During discussion of cantilever flange design it was noted that it seemed appropriate that the cantilevers should be relatively large and this would not incur a great penalty except that of generation of dead load. Hence it follows that a cantilever flange may be designed with a secondary system of supports. This may take the form of cantilever beams such as were used at Hammersmith Flyover[3] or even the provision of transverse struts from some position in the soffit of the cantilever slab to a suitable footing in the wall of the outer web or bottom flange.

It will be realised that such a sequence of strutting arrangements allows thrust to be communicated into the bottom flange and may well be arranged to provide a beneficial effect. In the longitudinal direction such struts or cantilevers may have small torsional rigidity. However, if the strutting arrangements are designed as an inclined lattice then not only may additional stability be available in the transverse direction but the lattice becomes the medium whereby longitudinal torsional effects may be transmitted to supports. The lattice arrangement becomes in effect a pierced or trussed outer web of a box beam. It has not been used very much, possibly owing to difficulty and cost of construction. One example is the Transverse Illtal Bridge in the Saar. Transverse strutting arrangements, however, are used both in steel and concrete bridges.

13. Segmental Construction

For concrete work the segmental technique has relevance in box beam applications. Box beam shapes are stable in both the vertical and lateral directions and provide resistance to lateral forces which may be prompted by slight inaccuracies in the location of prestressing forces at the segmental joints.

For bridges designed in accordance with the British Department of the Environment rules, tensile bending stress is not allowed at working loads. Stemming from the original work of Freyssinet and Guyon, this rule applies to both *in situ* and precast concrete. In Germany, however, partial prestressing is allowed and in the USA tensile stresses are permitted to certain limits. There can be a saving of up to 10–13 per cent of the quantity of prestress in achieving the required load factor at the expense of the bending stress condition at working load.

Thus it can be seen that in Britain the no tension rule indirectly tends to encourage the use of segmental work. If tensile bending stresses are not permitted at working load and sufficient prestress must be provided to achieve this, then it can be seen that precast segments are directly comparable with monolithic concrete. It may be added that the author is not aware of any code of practice in the world which allows tensile bending stresses at normal working load in segmental structures.

For box beams where torsional loading is high, the segments may require to be mechanically bonded together rather than reliance being placed entirely on the friction of the concrete surfaces.

14. Curved Beams

Modern bridge designs are frequently tailor-made to suite the alignment. This means that curved decks, eccentric supports and skew abutments may be required. Box beams are very well suited to resist the torsional stresses which are induced by the self-load of the structure as well as eccentricities of live load brought about by the application of superimposed loading well away from the line of centroid of the cross-section. The development of torsional parasitic moments is an important consideration in the design of curved beams of prestressed concrete.

Another aspect of the problem of building up curved bridge decks arises when precast girders are being used. It is convenient if longitudinal girders are made in straight lengths and a curved deck formed of these beams requires polygonal jointing of the beams to negotiate the curve. A varying width of overhanging flange at the outer edges can be added *in situ*.

For concrete structures the usual way of dealing with curved box beams is to construct them *in situ* or of precast segmental units. The latter is of particular use when the width of box is of the same nominal section throughout its length and it is merely the curvature in plan and profile which is introducing the geometrical problems.

15. Prestressed Concrete Box Beam Decks

The foregoing discussion has suggested that for box beams the concept of breadth is as important as depth and span. This three dimensional appreciation has been applied to several examples of prestressed concrete box beam construction.

For a shallow box with cantilever deck the proportions have to take account of local loading, particularly in the vicinity of piers. In the case of Westway Section One the breadth/depth ratio is approaching the practical maximum for laterally unprestressed concrete designed for up to 45 units of HB loading (Figs 4 & 5).

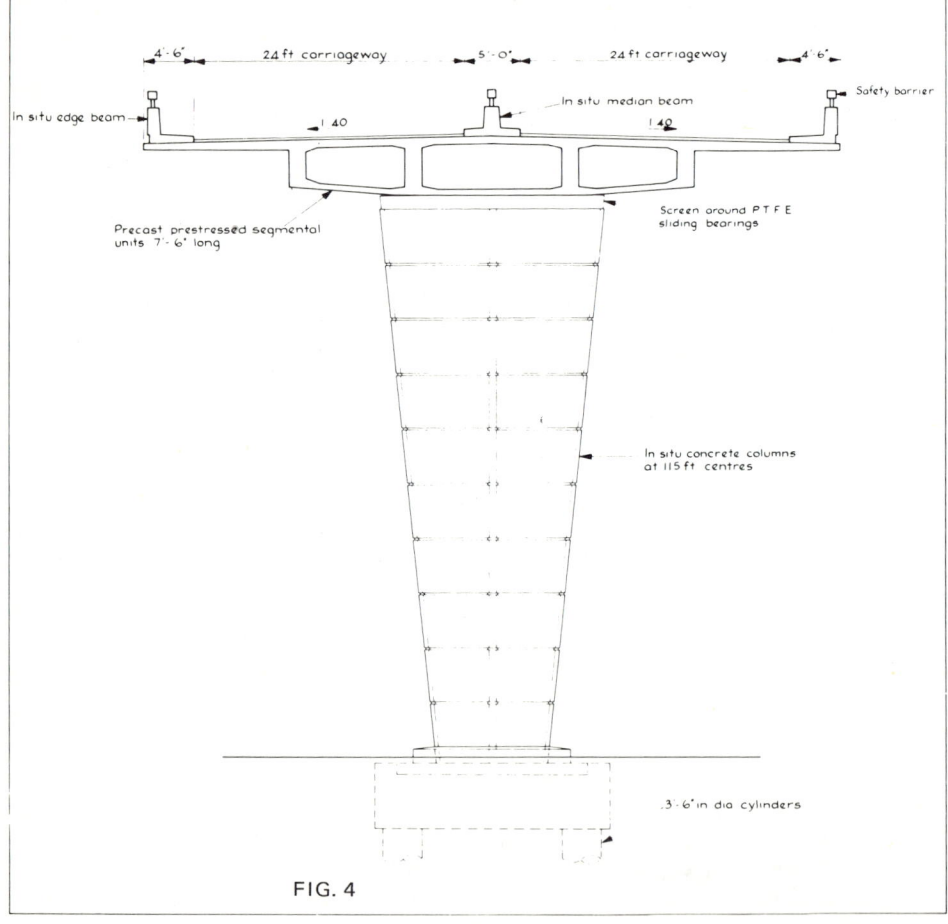

FIG. 4

For this case the highway is of dual 24 ft carriageways, the total width 64 ft and a structural depth of 5 ft supported by a central column system in continuous spans of 115 ft. The sliding bearings are located under the inner webs. The outer webs can shed eccentric load into the support region either, when the load is at the pier, by pure cantilever action or, when it is sufficient distance along, by bending and torsion dispersion. When the eccentric load is near the support then care is required to ensure local stresses are not exceeded.

FIG. 5

Where the width and depth can be greater, the span can also be increased. Such is the case for Section Five where the span is 204 ft for a width of 94 ft and a depth of 10 ft 3 in (Fig. 6).[4] The diaphragm is omitted in the outer cells at supports which clarifies the load dispersion problem for the slow lanes. The arrangement with sloping webs greatly reduces the apparent bulk of a large structure (Fig. 7) which is very useful aesthetically apart from theoretical structural reasons.

When a highway deck is wider still, as in the case of the West Gate Approaches a slightly different structural arrangement has to be resorted to to keep step with structural expediency.

16. West Gate Bridge Approach Viaducts

The approach viaducts of this project comprise nearly 5,000 ft of prestressed concrete box girder arranged in typical spans of 220 ft (Fig. 8). Ten spans comprise the west approaches and thirteen spans the east approaches and both are horizontally curved; 4,500 ft radius on the west side and 2,900 ft on the east side. The prestressed concrete design developed to overcome the problems has some special features and a novel sequence of construction. Figure 9 shows a view of the various elements making up the concrete bridge. The deck is relatively wide allowing for eight traffic lanes plus two breakdown lanes and hard shoulders, and has a total width of 118 ft 6 in. A single box supported on single central columns has been adopted. Comparative design estimates indicated that the design adopted would be economical despite the

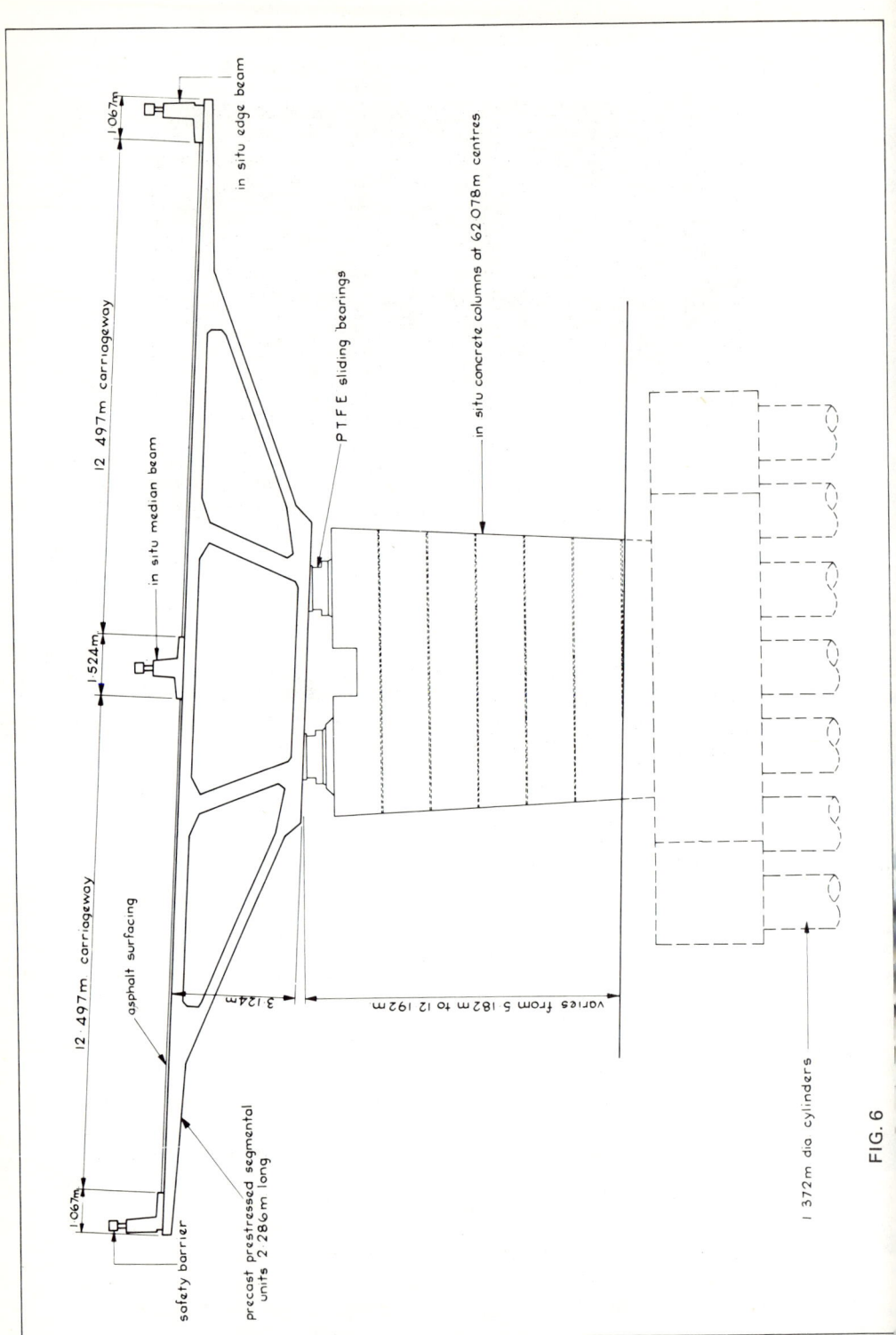

FIG. 6

great width of deck. It offers the aesthetic advantage of a single spine and central columns on the horizontal curves. It was found that the arrangement of elements in the deck slab and the cantilever flanges would be economic if

FIG. 7

they were dissociated from the first stage erection of the spine beam. In addition the main spine beam has been shaped in a somewhat unusual manner. The shape adopted gives a parasitic moment of the order of 3% of the maximum design moment owing to the low position of the neutral axis.

The spine is self-supporting under the first stage of prestress which keeps to a reasonable minimum the weight on the falsework during construction and permits rapid erection. The dead load bending moments are designed on the basis of span by span construction with each span cantilevering 55 ft beyond each column. The bending moments are progressively altered as the cantilevers and deck slabs are added, after which the second stage longitudinal prestress is applied, which causes a change in the parasitic moments. On the completion of the structure the loading from surfacing, balustrades, etc., together with live load effects and differential settlement allowances are added to produce the final bending moment envelope.

The maximum torsion in the spine is produced by a combination of dead weight eccentricity owing to the horizontal curvature of the structure, the live load eccentricity and wind loading.

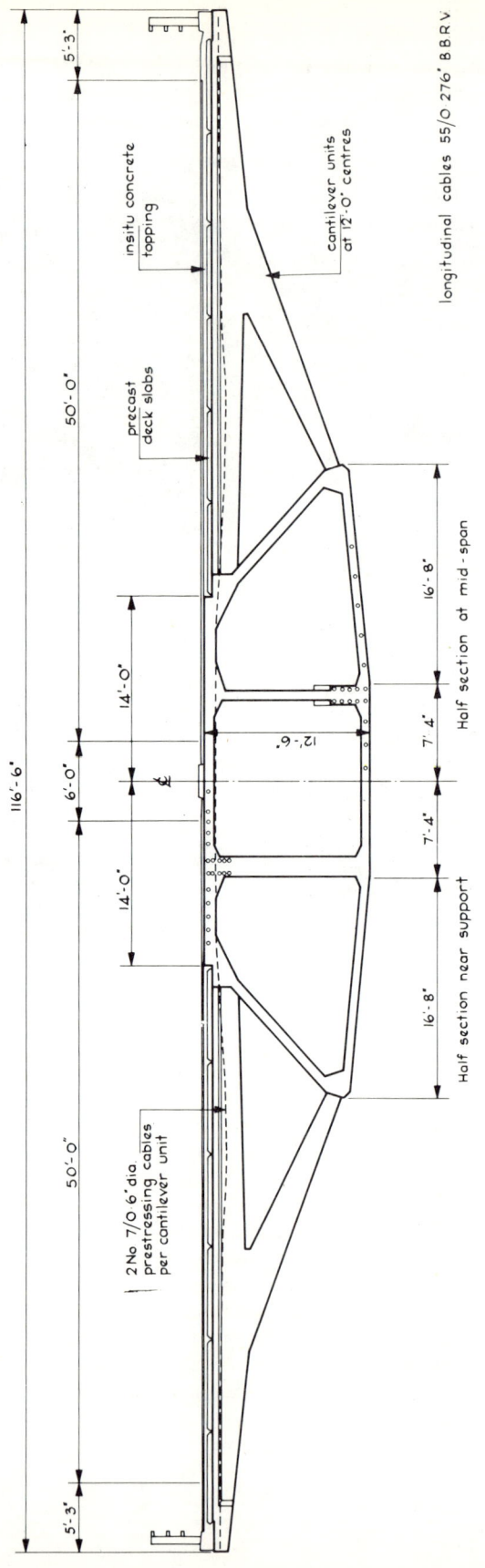

FIG. 8

Labels within figure:

116'-6"

5'-3"

insitu concrete topping

50'-0"

precast deck slabs

cantilever units at 12'-0" centres

longitudinal cables 55/0 276' BBRV

14'-0"

6'-0"

₵

14'-0"

12'-6"

16'-8"

7'-4"

Half section at mid-span

7'-4"

16'-8"

Half section near support

50'-0"

2No 7/0 6'dia. prestressing cables per cantilever unit

5'-3"

The thurst from the lower leg of the cantilever brackets under live load induces a residual compression in the bottom flange but there is some torque transmitted to the web.

The distribution of loading in the spine beam has been very carefully analysed and checked against a perspex model test at Imperial College.

FIG. 9

A 1/9th scale micro concrete model of a cantilever unit and three half-spine units was made and tested at the Cement and Concrete Association. From this the working load strains were checked and the failure mechanism was obtained.

The author would like to congratulate the Association's staff on this particular testing programme .

In the field, one completed spine beam unit with cantilevers was loaded to simulate the action under transverse moments.

17. Development of Precast Multicellular Bridge Decks

It is important that the analysis of structures should not be allowed to take precedence over the design of structures. The practical problems of bridges in service, such as drainage, access and cost of maintenance should all receive full consideration and be treated as just as important factors in the evolution of design as esoteric structural theory.

The author's firm has developed a type of box beam which is intended to overcome certain problems inherent in multicellular deck construction of the precast type. This has been nicknamed the 'top hat' beam and was developed primarily for Westway. The analysis work and the thinking behind the design is contained in a recent paper.[5]

The top hat beam has been developed as a means of providing a multicellular slab deck with excellent load distribution properties, and yet reducing site operations to a minimum. Experience has justified these hopes (Fig. 10). However, like all box beam designs, care in manufacture is required. It is economic because of the shallow structural depth. It can be classified as a $[1 - N]$ system.

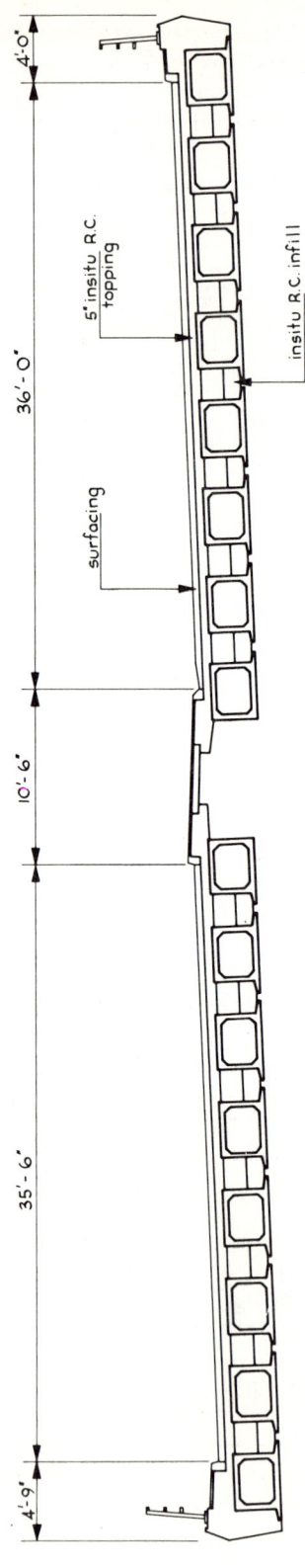

FIG. 10

4'-0"

36'-0"

5" insitu R.C. topping

insitu R.C. infill

surfacing

10'-6"

35'-6"

4'-9"

The following points emerge from experience with top hat beam cellular decks:

(i) The precast beam manufacturer is asked to produce something relatively complex to save time and trouble at site. In the author's opinion this is putting much of the responsibility of total production where it should be — in the factory.

(ii) The webs are on the slender side so that reinforcement is required to resist shear forces. The design requires modification to accommodate deflected tendons and tendons debonded at the ends have been the usual method of detailing.

(iii) The edge detail requires a special edge beam or the provision of *in situ* edge concrete. The stopping off of the bottom flange has proved very easy in practice but nevertheless the detailing of different reinforcement and off-setting the prestress to avoid lateral bowing puts up design costs per bridge.

(iv) The design behaves like a monolithic multicellular deck despite the lack of transverse prestress. It is very suitable for precast beam decks requiring to resist heavy live loading and for skew crossings.

Top hat beams are being used on the M53 Motorway. Typical beams 114 ft long are shown being erected for an overbridge on that project (Fig. 11).

The M-range inverted T-beam has been developed by the M.o.T. & C. & C.A.[6] for spans greater than the standard range of Concrete Society inverted T-beams and with the ability to form a cellular deck.

FIG. 11

When used to form a cellular deck, however, with *in situ* concrete placed top and bottom the construction does not act (and is not expected to act) as a monolithic section. The results of the model test carried out by the Cement and Concrete Association were compared with analysis by a grillage computer program, a method of analysis based on rigorous theory and confirmed by appropriate model tests by Professor Sawko at Liverpool and by Professor Cusens at Dundee. The C. & C.A. suggests that in order to correlate with the test results the torsion parameters should be halved. The author is not happy about the validity of this procedure since it confuses design rules with mathematical analysis. The deck cannot be fitted into the classification system as the change under increasing load from a [1 − N] system to quasi-orthotropic or shear connected beams is not predictable.

The inverted T-beam deck in multicellular form is not as efficient as it could be. In fact the inclusion of the bottom *in situ* concrete only reduces the live load moment carried by one beam by approximately 17% whilst increasing the dead load moment by approximately 12%. The model test also confirms that the bottom slab might crack between the beams.

Other comments on this type of cellular construction are:

(i) Because of the narrowness of the top flange and the heaviness of the deck formed, the construction depth is fairly large.

(ii) The holes placed in the webs for transverse steel are so large (good for easy threading) that the amount of steel that can be placed in the webs to resist shear forces is severely curtailed, e.g. the web holes with their taper extend for a distance of approximately 300 mm leaving a solid length of only 300 mm between holes for the steel. The webs are slender so that for the longer spans a fair amount of shear steel is required for which there is little room in the webs. Modifying the holes can add up to 10% to the cost of the beams.

(iii) The shape of the beam is such that it is not well suited for use at the edges without additional concrete. The edge detail can become complex, expensive and impose unnecessary load on the edge beam.

(iv) The shuttering and detailing of end diaphragms is complex due to the shape of the beams.

Because of the various points discussed above, the author has not specified multicellular deck construction using 'M' beams. The author's firm is responsible for a bridge employing 'M' beams discretely with a concrete deck spanning over the top and without diaphragms (Fig. 12). This bridge is currently under construction. It will be noted that this type of deck can be conveniently classified as a [N−1] system.

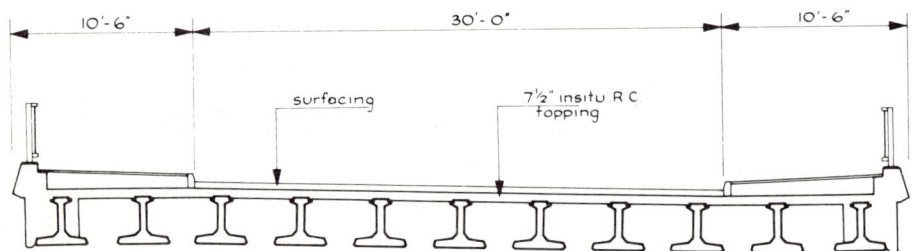

FIG. 12

It can be seen that there is scope for the development of a new type of standard precast beam which could be used to form a deck with a minimum of site work. Such ideas have prompted development of the U-beam (Fig. 13). In the completed deck the beams become torsionally stiff and have good load distributing properties, although not as good as the top hat deck. A critical examination of the U-Beam deck suggests that:

(i) Although the simplest of the three beams to shutter and cast there is a slight complexity required at the beam ends to enable a satisfactory diaphragm to be built.

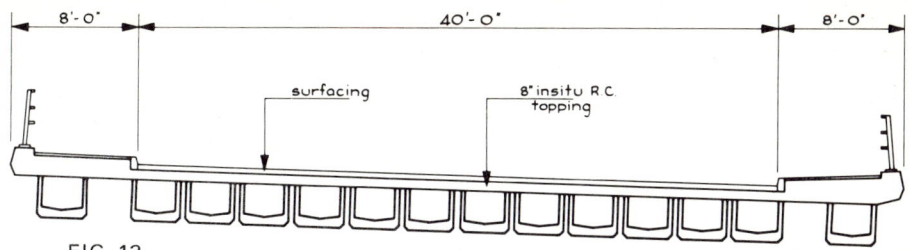

FIG. 13

(ii) The proposed web thickness, whilst being greater than the other two types per metre width of deck, requires modification to permit the use of deflected tendons.

(iii) Care is still required to ensure that the edge beam is not overloaded. The design, however, allows for varying the spacing or size of beam at the edges and there is no wasted *in situ* concrete.

The design attempts to achieve and combine:

(*a*) low cost of construction
(*b*) economy of structural depth
(*c*) good load distribution properties
(*d*) simplicity of manufacture and erection
(*e*) facilitation of drainage and service requirements
(*f*) elimination of intermediate diaphragms or transverse stressing.

In the author's opinion the most economic solution for medium span bridge decks lies in a balance between the desiderata of structural analysis, the precast beam manufacturer and the contractor.

It will be noted that a deck formed of such U-beams can be logically classified as a [*N*−2] system.

The U-beam design is currently being intensively developed and modified.

18. Composite Box Beam Bridges

Steel bridges in the small to medium span range pose the problem of economical fabrication and erection. The manufacture off site in the fabrication yard suggests that some form of standardisation will be in the interests of economy. The use of repetitive elements with simple high strength friction grip bolted connections is a logical application of standardisation.

The author's firm has prepared a series of designs for the Neath — Abergavenny Trunk Road which at the time of writing have been sent out to tender. It is not therefore possible to give any information about the relative costs or the reception the designs are given by the contracting industry. However, the designs have been formulated with the aim of giving as much standardisation as possible in the single contract. Several are based on simple composite box girder configurations and a brief description is given of two bridges of each of two types. One type is a single box system [1−2] and the other is a quadruple system [4−2].

19. Two Bridges for the Llandarcy Interchange

One bridge is a two-span composite structure, continuous over two spans carrying a two-lane slip road (Fig. 14). The roadway is carried by a 300 mm thick reinforced concrete slab which cantilevers 3 m beyond the main steel box girder forming a [1−2] system (Fig. 15). Stud shear connectors promote composite action of the girder and deck slab.

The ends are simply supported, one end being fixed, the other being free to move longitudinally. The centre support consists of a steel column pinned at the base and free to rotate about a transverse axis at the top.

The steel girder is constructed of welded high yield plate to form a trapezoidal cross-section. The girder is divided into three lengths, the longest being 32 m, for

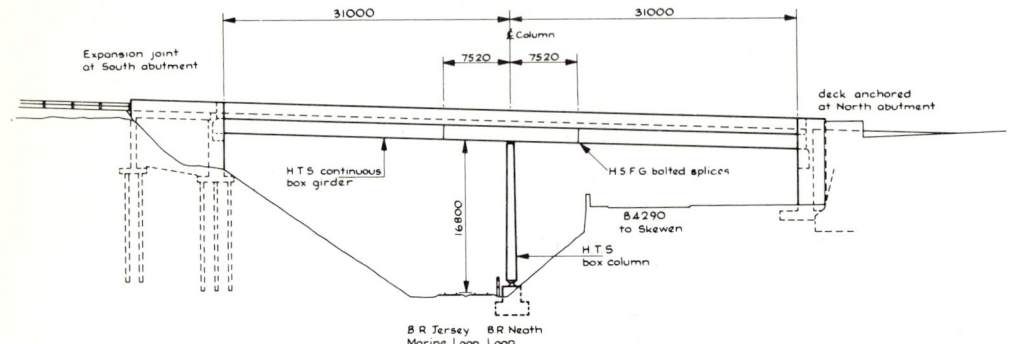

FIG. 14

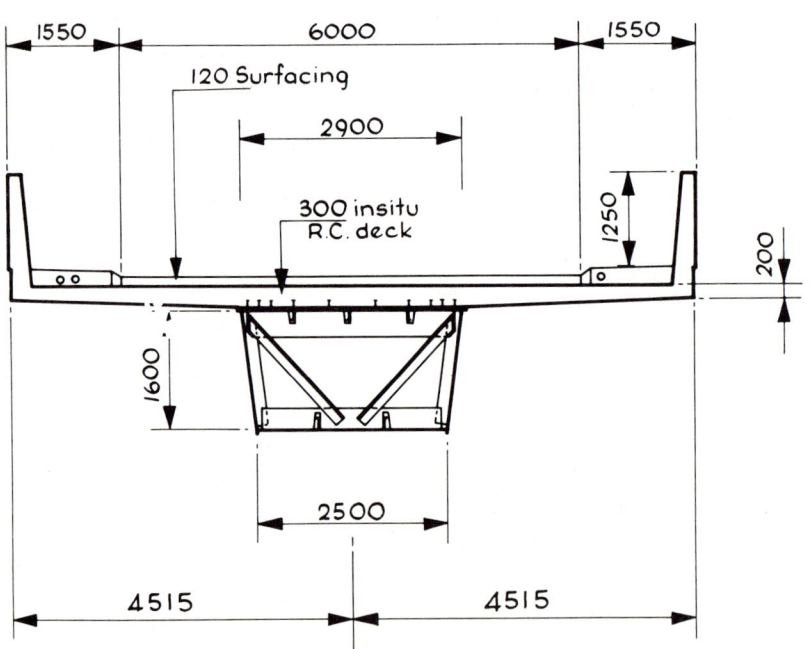

FIG. 15

transportation purposes. They will be jointed on site by high strength friction grip bolts. The girder is supported on six bearings. The two at each end are mounted outside the webs on a short transverse cross beam whilst the centre pair are located just inside the line of the webs.

The bridge has been designed to carry either two lanes of HA loading or 1 HB truck in one lane and $\frac{1}{3}$ HA in the other. The stresses have been limited to those permitted by CP 117. The main bending moments under various stages of composite and non-composite action were calculated using a plane frame computer program. Most critical stresses occurred when both lanes were loaded with HA. The HB loading caused the greatest twist which is transferred to the girder via the internal bracing. The twist is taken out by the bearings both at the central column and at the abutments. The spacing of the bearings is such that under no condition of loading is it possible to produce uplift on any of them. The slab is also most critically loaded under the action of the HB truck.

The other bridge is similar in construction except that it is continuous over four spans (Fig. 16). The main difference is that the columns are made of reinforced concrete and cantilever from the ground. The two longest columns are fixed to the superstructure by rocker bearings whilst the short column has sliding bearings beneath the girder.

20. Saltings Viaduct

This is a seventeen span viaduct of continuous composite construction built on a cross-fall and with a horizontal and vertical curve (Fig. 17). The reinforced concrete deck slab which is 250 mm thick is supported on four mild steel girders (Fig. 18). The system is [4—2] and provides for a dual carriageway with cantilevered footpaths. The cantilever on the inside of the curve is longer than the outer to ensure that the parapet does not obscure the sight lines.

The bridge is fixed at one abutment and free at the other. The girders are supported on individual columns to be built in groups of four from a common pile cap. The bearings are anticlastic and allow longitudinal sliding and rotation about both axes.

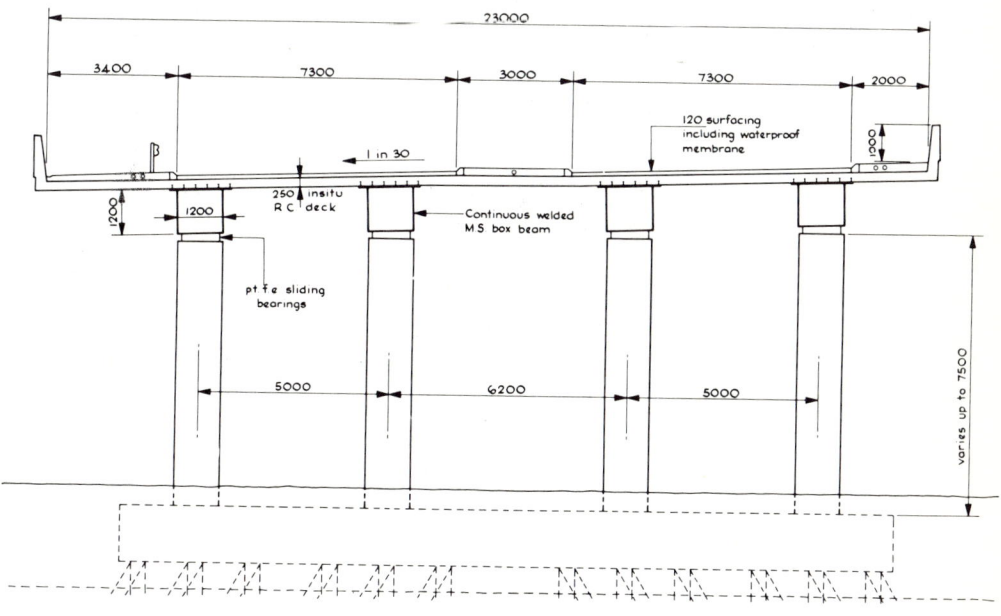

FIG. 18 TYPICAL SECTION OF SALTINGS VIADUCT

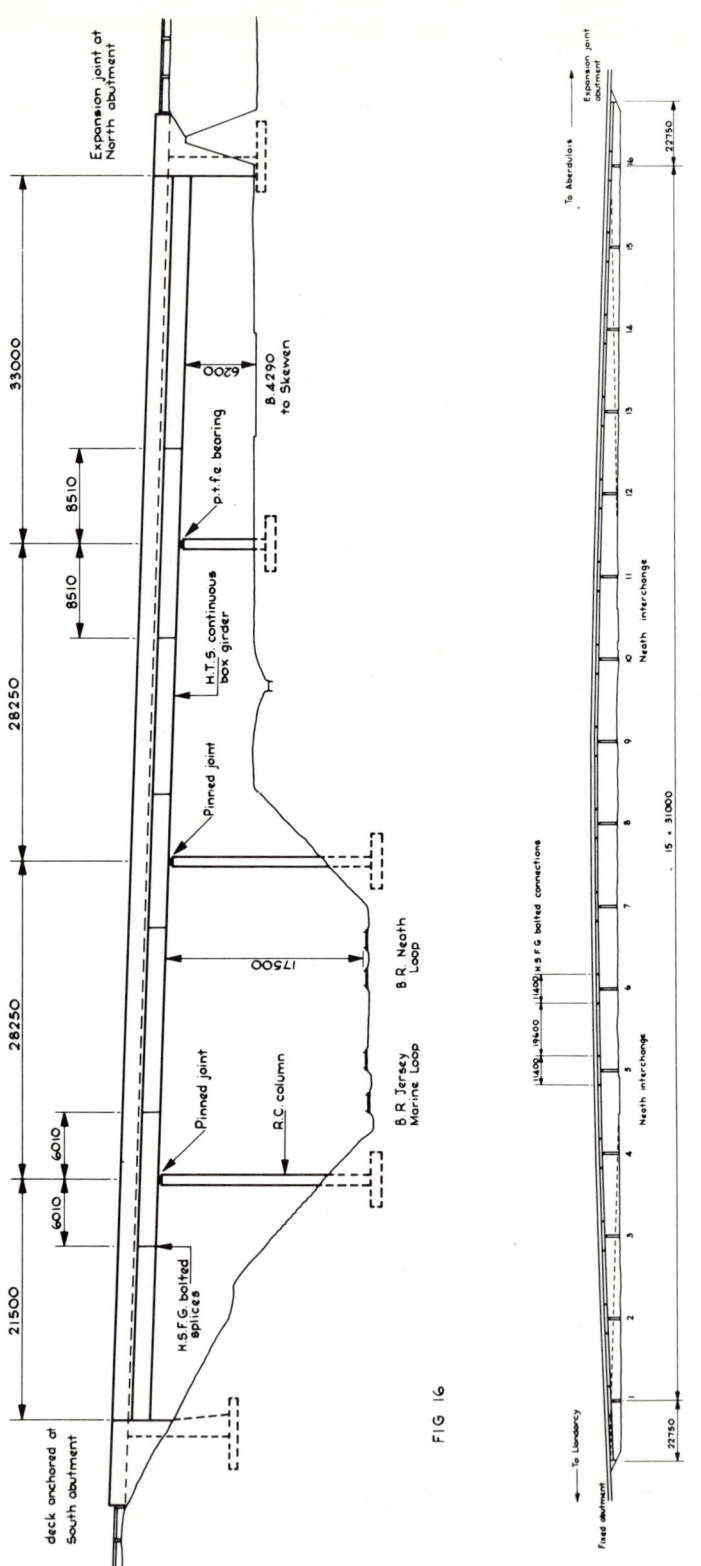

FIG 16

FIG. 17

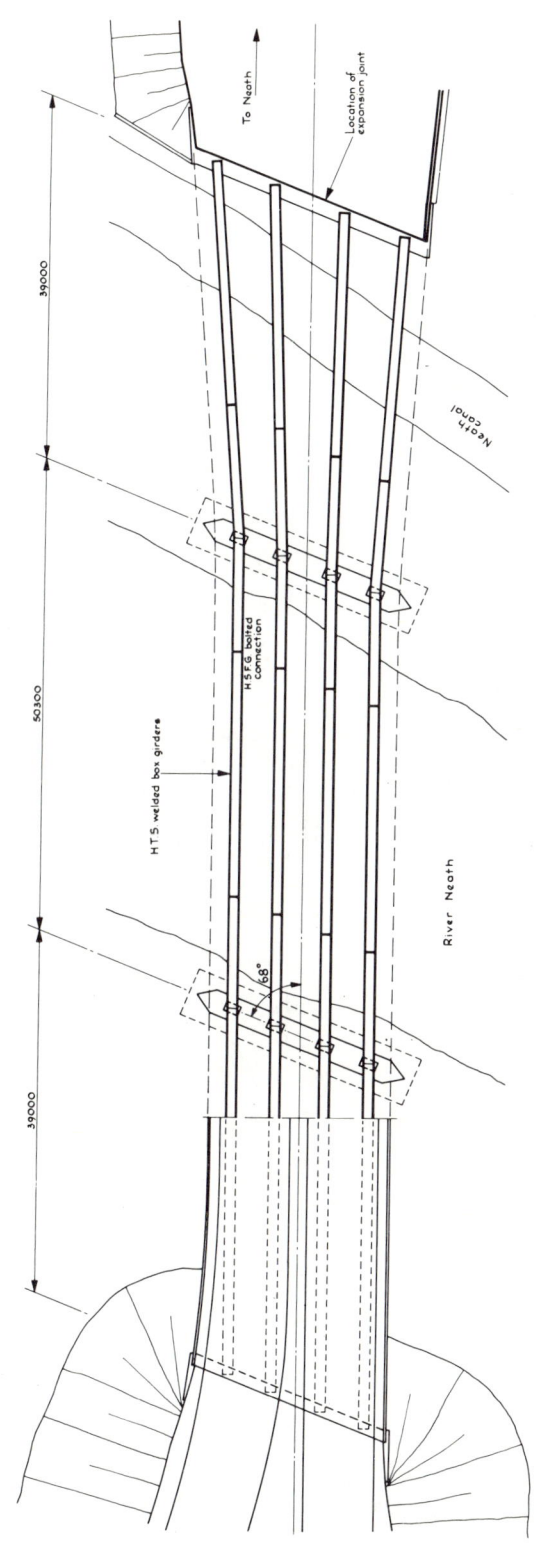

FIG. 19

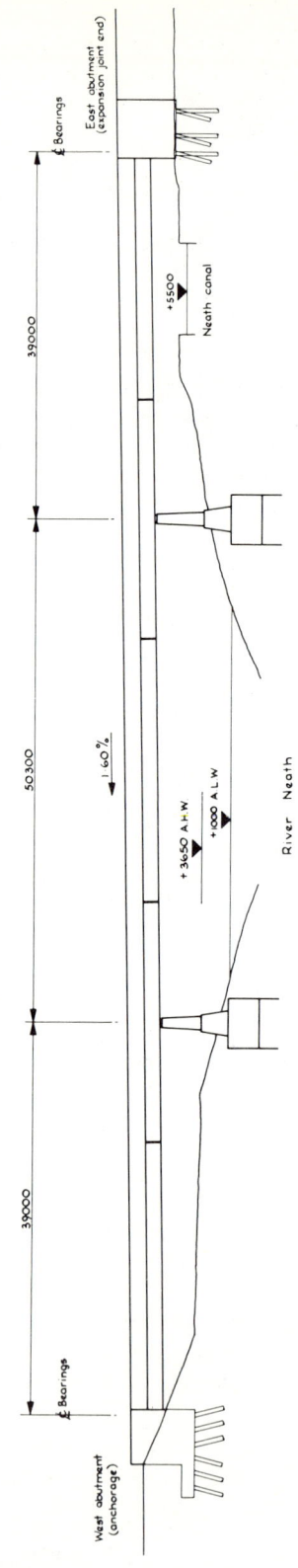

FIG. 20

There are no transverse beams between the girders so that the only connection between them is via the reinforced concrete deck. Critical loading occurs in the deck with HB loading whilst for the girders it is under full HA. The most critical slab loading occurs with an HB vehicle at mid span half way between the outer and inner girder.

The girders are fabricated from plate to form a 1·2 m square cross-section. Each span comprises one 19·7 m section at midspan and an 11·4 m section over the supports. These sections are spliced together on site by means of high strength friction grip bolts. 22 mm stud connectors spaced uniformly over the top flange of the box in a 300 mm square grid are sufficient to provide composite action.

An Armco barrier is mounted on the kerb on the side of the longer cantilever to prevent heavy traffic parking on the footway.

Calculation of the bending moments in the main deck and girder system was done with the aid of a grid analysis program which was arranged so that the girder was represented by a single longitudinal grid member with rigid transverse arms. The deck slab was split up into transverse strips represented by single transverse members which were supported at the extremities of the rigid arms. Thus the slab was effectively supported above the webs in the grid analysis instead of the centre line of the girder.

Unlike the Llandarcy Interchange bridges where asymmetrical loading is resisted by torsion of the spine box, in this bridge induced torsion is a secondary effect. Rotation of the girders at the column position is not transmitted down the columns. However, horizontal forces are transmitted to each column by the bearings.

21. Bridge over the River Neath

This bridge carries a four-lane dual carriageway across the River Neath (Figs 19, 20, 21). The superstructure consists of a composite deck and steel girder arrangement similar to Saltings Viaduct. The three skew spans are 39 m, 50 m and 39 m long respectively. The main difference between the two bridges is that the river bridge has high tensile

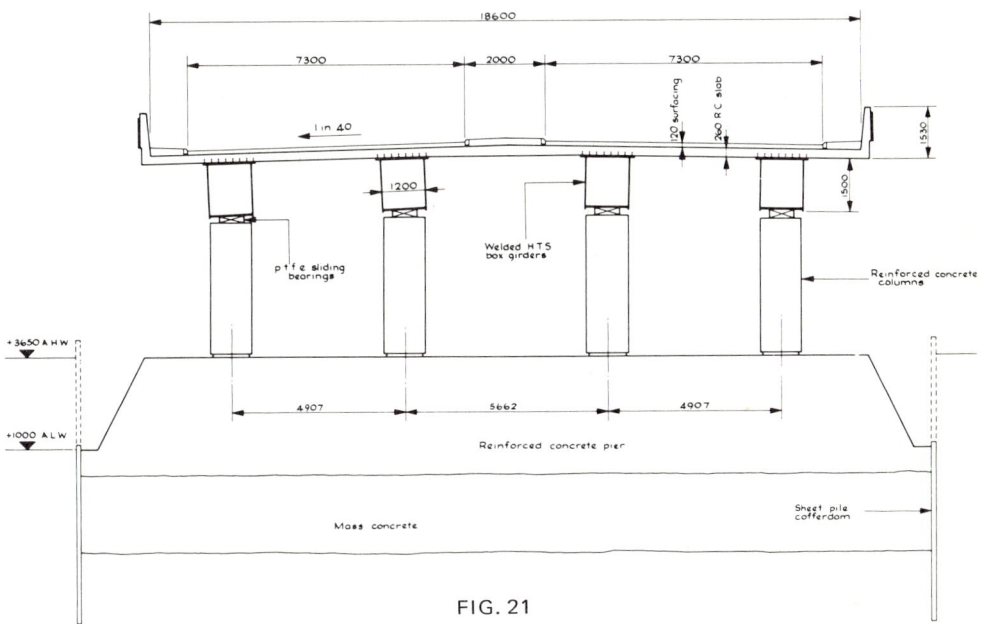

FIG. 21

steel girders which are slightly deeper in cross-section. Horizontal forces are taken out at the abutments only. In other respects the structural behaviour and methods of calculation were as for the viaduct and indicate that standardisation of detail and principle can be applied to a formal river crossing.

22. Conclusion

Although much of this paper has concerned itself with bridge decks, there is every likelihood that box beams will find application in a broader environmental sense, particularly as solutions to urban transportation problems. Most existing elevated roads rely on a deck type structure with accessories such as lighting, crash barriers and signs added to the structure carrying the highway. Noise and fumes of traffic are held to be debasing the quality of the environment and criticism (which verges on the emotional rather than rational observation) is levelled at urban elevated motorways as a whole. It is even proposed that urban highways be constructed underground to solve the environmental problem. Of course tunnel solutions will be required in towns but they should be judged on commonsense economic comparisons. It is much cheaper to construct a box in the air than tunnel through the ground. The design of intersections and slip connections is a problem with urban tunnels just as much as bridges. It may be that a box beam enclosing the traffic might provide a rational solution to urban transportation (Fig. 22). The noise and

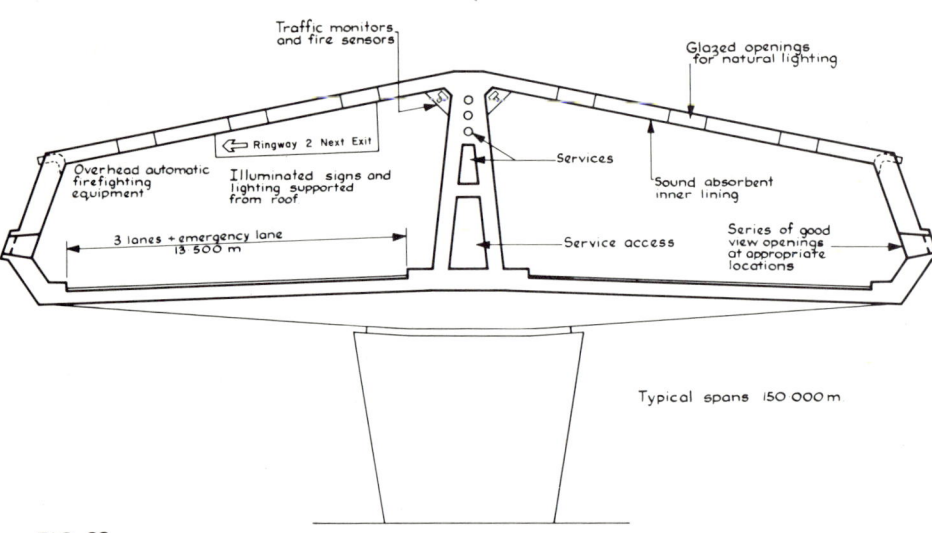

FIG. 22

safety barrier would be part of the structure as would the layout of lighting, traffic control and signing. Natural ventilation vents would be located in the upper part of the webs. The climate would have no effect on road surface conditions. The highway would be dry, fog and ice-free at all times, leading to a reduction in accidents. The roof would be available for engineering services. A towing machine could be provided which draws broken down vehicles into and along the reserved shoulder to the nearest exit. Similar remotely controlled apparatus could be devised for policing, fire-fighting and rescue.

The concept of an integrated motor duct could be realised without solving any new technical problems. The use of a structural concrete box beam would allow curves, ramps and intersections to be dealt with.

Finally, the steel torsion box has a vital role to play in large span suspension bridges. The author's firm prepared an entry for the recent Messina Straits competition comprising a double span suspension bridge of unique design. The two spans were each 1,325 m (5,200 ft), 20% larger than any clear span existing (Fig. 23) and

FIG. 23

accommodate railway loading and four lanes of highway by a steel torsion tube (Fig. 24).

The merits of box structures apply equally to steel and concrete. There is every reason to believe that such structural forms will be of advantage to future developments in structural materials.

REFERENCES

1. Bingham, T. G. and Lee, D. J. The Mancunian Way elevated road structure. *Proc. Instn civ. Engrs*, Vol. 42, April 1969, pp. 459–492.
2. Lee, D. J. Bending moments in beams of serial construction. *Proc. Instn civ. Engrs*, Vol. 38, Dec. 1967, pp. 621–637.
3. Rawlinson, Sir Joseph and Stott, P. F. The Hammersmith Flyover. *Proc. Instn civ. Engrs*, Vol. 23, Dec. 1962, pp. 565–600.
4. Lee, D. J. The design of Section Five. *Struct. Engr*, Vol. 48, No. 3, March 1970, pp. 109–120.
5. Hook, D. M. A. and Richmond, B. Precast concrete box beams in cellular bridge decks. *Struct. Engr*, Vol. 48, No. 3, March 1970, pp. 120–128.
6. Manton, B. H. Standard beam sections for concrete bridges: prestressed inverted T-beams for spans 15 m to 29 m. Ministry of Transport/Cement & Concrete Association, 1970.

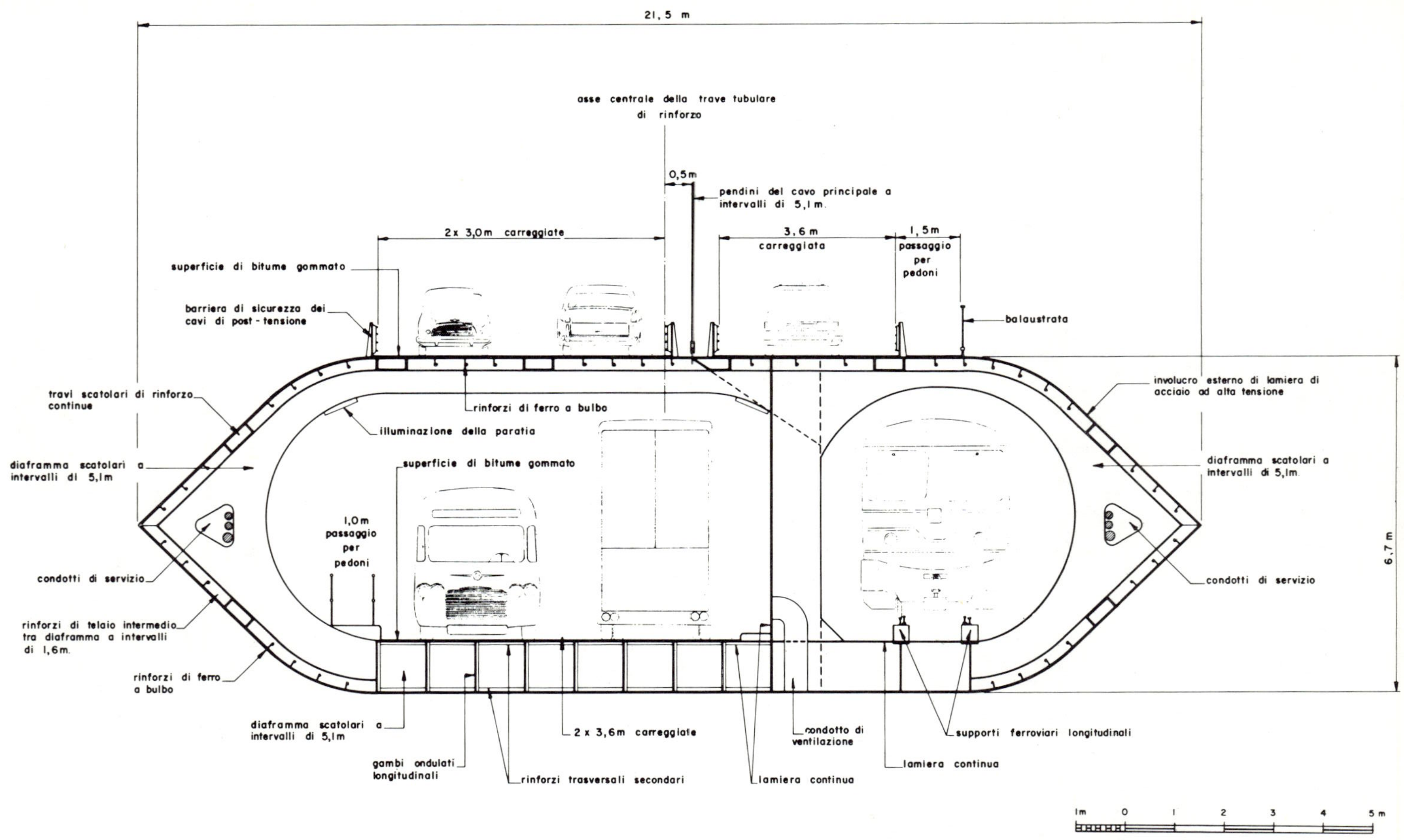

FIG. 24

Labels (as they appear around the figure):

6,7 m

involucro esterno di lamiera di acciaio ad alta tensione

diaframma scatolari a intervalli di 5,1m

condotti di servizio

balaustrata

supporti ferroviari longitudinali

lamiera continua

1,5m passaggio per pedoni

condotto di ventilazione

lamiera continua

pendini del cavo principale a intervalli di 5,1m

asse centrale della trave tubulare di rinforzo

3,6 m carreggiata

2 x 3,6 m carreggiate

0,5m

rinforzi trasversali secondari

2 x 3,0 m carreggiate

21,5 m

rinforzi di ferro a bulbo

illuminazione della paratia

superficie di bitume gommato

superficie di bitume gommato

1,0 m passaggio per pedoni

barriera di sicurezza dei cavi di post-tensione

gombi ondulati longitudinali

travi scatolari di rinforzo continue

diaframma scatolari a intervalli di 5,1m

condotti di servizio

rinforzi di telaio intermedio tra diaframma a intervalli di 1,6m

rinforzi di ferro a bulbo

diaframma scatolari a intervalli di 5,1m

1m 0 1 2 3 4 5m

SPECIAL FEATURES IN THE DESIGN OF BOX GIRDERS

B. P. WEX

Freeman, Fox & Partners

1. INTRODUCTION

In recent years owing to their economic and aesthetic appeal the use of steel box girders in bridge construction, largely pioneered in Germany, has become more popular outside that country.

Spans have increased, hence the depths of webs and the widths of flanges have also grown whilst the thickness of plating used has not increased commensurately, the necessary stability for parts in shear and compression being supplied by suitable stiffening. This paper is concerned with a discussion of problems in bridges whose flanges have a total width to thickness ratio of perhaps 200 to 500 and webs possessing total depth to thickness ratios of 700 or more.

Practice in advanced engineering always is ahead of bodies who draw up national codes and specifications. Illustrative of this is the fact that German orthotropic deck plate construction developed before any codes were written for it. Indeed, if the codes had been written first orthotropic decks might never have emerged. It is not surprising, therefore, that no UK code of practice is devoted to the problems of modern box girder bridges although certain portions of BS 153 are applicable. Other national codes, e.g. German, have more relevance but they are difficult to apply in many respects.

Because a government appointed committee is at present, as part of its brief, examining the details of the design and construction of box girder bridges, this paper does not specifically lay down methods of design. Instead a number of the major problems facing designers of such bridges are discussed and methods by which they may be solved indicated in general terms.

The aims of the paper are:

For designers not familiar with box girder problems —

Briefly, to survey some existing steel design codes and comment on their suitability on some important points.

To describe the main problems involved in multi-stiffened thin walled steel structures and generally to describe a philosophy of design method.

To indicate how these methods might be simplified for design office use.

For researchers —

To indicate regions in which extensive test programmes would help.

A further general aspect of box girder design receiving brief mention is the subject of erection, since this frequently may cause stresses well in excess of those occurring in service.

2. SUITABILITY OF BS 153 (1958)

2.1   General

Many aspects of the total BS 153 specification (i.e. parts 1, 2, 3 and 4 are appropriate to box girders. Designers must, however, remember that it is written for simply supported bridges up to 300 ft span, hence no guidance is given for the design of continuous structures, whereas most box girder bridges are continuous over several spans. BS 153 does not specifically refer to requirements for box girders as such and simple plate girder theory is not applicable in many cases.

2.2   Compression flanges

Thin stiffened plates form the compression flanges of box girders but BS 153 only deals with unstiffened plates. An equivalent strut method based on the Perry Robertson formula in BS 153 may frequently be used to deal with stiffened compression plating, as described in section 9, but this is not invariably the case and certain safeguards must be applied. In many cases this equivalent strut approach can yield uneconomic structures.

2.3   Webs

For reasons of economy depth/thickness ratios considerably in excess of the value permitted by 153, are often required in medium and long span bridges. Such webs have to be provided with multiple longitudinal stiffening in addition to the more usual transverse components, but the rules in BS 153 are limited to two longitudinal stiffeners. The complex problems of stress distribution, interaction and stability in deep thin webs close to supports where in continuous structures maximum moments and shears co-exist cannot be covered in a specification concerned with simply supported design. The BS 153 formula for equivalent stress deals only with one aspect of stress concentration in points of concentrated load or support.

2.4   Fabrication

Many fabrication details, again not peculiar to box girders but of special relevance to them, are not covered by BS 153. In particular, fabrication distortion and flatness tolerance are important considerations closely allied to the use of welding.

3. SUITABILITY OF GERMAN CODES

3.1   General

Two German Codes of Practice, DIN 1073 and DIN 4114, contain many clauses which are relevant to box girder design, particularly with reference to such subjects as shear lag, orthotropic deck plates, and the behaviour of stiffened plates in compressions and shear.

3.2   DIN 1073

3.2.1   This specification contains simple rules for calculation of effective width of flange plates in plate girders and boxes. Only one parameter is used, the ratio of plate width to span. By reducing the width of flanges in the prescribed manner an allowance can be made for the shear lag effects leading to reasonably accurate design.

3.2.3 However, the shear lag problem in actual fact is more complicated. Besides being a function of geometry it depends also on the degree of orthotropy of the plate and on the type and position of loading. The DIN rules were formulated for an isotropic plate only and it is felt that some allowance may also be necessary for increased shear lag effects in orthotropic plates.

3.2.4 This specification also deals with requirements for design of orthotropic steel plate decks.

3.3 DIN 4114

3.3.1 DIN 4114 contains sections devoted to the analysis of the stability of unstiffened and stiffened steel plates, and provides useful rules for the design of such plates and their stiffeners.

3.3.2 The method covers numerous loading patterns and geometrical considerations, it is based on calculating the Euler critical stress of a strip of plate and multiplying this by a factor given in the code derived from the elastic buckling theory of orthotropic plate. Where this factored stress exceeds the limit of proportionality of the material it is reduced to allow for plastic behaviour, and the resultant reduced stress reaches a maximum of the yield stress when the critical stress is infinite. Further reduction factors are then applied to the reduced stress to obtain a working stress for design purposes. This reduction factor depends on the type of loading, being high where direct compression is dominant, and lower where shear predominates.

3.3.3 Whilst DIN 4114 is much more comprehensive than other specifications, it is somewhat complicated to apply, it ignores initial deformations in the plates (presumably on the basis that this sort of effect is catered for in the reduction factors used to obtain working stress) and does not provide rules for the design of transverse stiffeners.

3.4 Other German work

It may be mentioned that much work in analysis of stiffened plates has been carried out in the last 10 years by Klöppel and his collaborators. The results, which contain solutions for buckling of orthotropic plates with longitudinal stiffeners of uniform and varying spacing under direct and shear stress, have been published in graphical form.

4. PHILOSOPHY OF APPROACH TO BOX GIRDER DESIGN

4.1 Particular box girder problems

Box girder design involves consideration of many effects which are of second order importance or non-existent in conventional single web plate girder design. In addition to bending and shear an eccentric loading on the box produces torsional, warping and distortional stresses. These secondary stresses must be fully considered but can be minimised by the provision of intermediate diaphragms or cross frames. Furthermore, load bearing diaphragms which have to be provided at bearings, although they are often in fact single web plate girders, require special consideration due to their geometry and the high stress levels, (involving bi-axial compression and shear) to which they are subject. Thin stiffened plates forming the flanges and webs of box girders require special analyses, taking into account initial lack of flatness and residual fabrication stresses, either explicitly or implicitly.

4.2   Shear lag

Shear lag, whilst not peculiar to box girders alone, has particular relevance to them. Basically, this phenomenon arises due to the shear distortion of girder flanges preventing uniform distribution across the flanges of the longitudinal forces transferred from the webs. In particular, near concentrated loads or reactions the flange stress at the web to flange intersection is substantially higher than away from it. The increase is frequently as much as 10% or 20% of the average stress and may be more in flanges which have very high breadth to length ratios or form part of the continuous span system, both features being common in box girder bridges.

4.3   General remarks on philosophy

It is apparent from the earlier sections of this paper that the designer faced with the problems of a box girder must look beyond the simple clauses of BS 153 to obtain guidance. Some guidance, mainly in the form of empirical rules, may be found in the German DIN standards. In the rest of this section the general behaviour of box girders is discussed as are the criteria that may be used in their design. Much of this work requires complex mathematical analysis and is not suitable for either preliminary or routine design for which simplified methods are highly desirable. Section 9 will indicate a possible line of approach to such simplification.

4.4   Criteria of collapse

4.4.1   In the past design tended to be based on factors of safety, whereas it is now considered more logical to use a load factor approach. If this method is used considerable care will be required in the choice of load factors, otherwise very un-economical structures will result. Furthermore, only realistic and not academic criteria of collapse must be used. Four main possibilities occur in a bridge super-structure:

(a)   Yielding of a tension flange.
(b)   Buckling* or yielding of a compression flange (either plate and/or stiffeners).
(c)   Buckling* or yielding of a web (either plate and/or stiffeners).
(d)   Buckling* of a diaphragm over a support.

The above criteria, whilst by no means solely applicable to 'box girder' problems, nonetheless are particularly relevant to this form of structure. Other phenomena such as fatigue also demand consideration but, although incomplete, the existing BS 153 does give much data which is applicable to box girder bridges. Appropriate loading spectra are, however, required; these are given in a DoE technical memorandum, and will be included in the revised BS 153.

4.4.2   Local yielding of a tension flange is most unlikely to cause collapse; provided the material possesses adequate ductility local high stress concentrations across the flange will be redistributed; until the whole flange reaches the yield stress no serious effect will occur. Even when the whole flange reaches yield stress,

---

*   Here the term buckling implies inability to carry further load, this is not the same as distortion at critical stress.

provided no buckling occurs elsewhere, an overall redistribution of moments may take place in a continuous span to enable it to carry further load. Nevertheless overall yielding of the tension flange is undesirable and may be considered to be a criterion of failure.

4.4.3   Local buckling of a compression flange may promote progressive buckling of the whole flange; hence the first local buckle might be assumed to be a criterion of failure. In fact this is conservative, if, due to causes such as shear lag there is a considerable portion of plate in the flange still capable of carrying axial load, when the outer sections reach the buckled condition. If a compression flange is in general yield (i.e. the buckling stress is greater than the yield stress) the same considerations apply as for tension flanges.

4.4.4   Web buckling is a different phenomenon; a web is required to carry a combination of shear and direct stress, but a buckled web is not necessarily a failed web, since it may carry shear by means of a tension field. Design of the web is then frequently governed by deflections rather than stress. Shear yielding of a complete web may also be considered to be a criterion of failure, since there is no alternative stress path. This, however, would be a most unusual condition throughout the web of a large box girder.

4.4.5   Whilst failure of a diaphragm over a support will almost inevitably lead to unserviceability at least, local buckling of a diaphragm need not necessarily constitute failure, since alternative stress paths frequently exist. Care must be taken with this assumption since the alternative paths are likely to throw very heavy stresses on, for example, the web/diaphragm connection.

4.4.6   Based on the above arguments a logical approach appears to be to design each element independently to be on the point of collapse under the factored applied load, ideally this should all result from plastic overall analysis but at present elastically obtained moments and shear form the applied load on each element. Subsequent sections indicate the approach for particular cases.

4.5   Calculation of stresses in webs and flanges

4.5.1   The basic stresses in the webs and flanges may be calculated using the Engineer's Theory of bending, assuming the gross section to be effective, and subjecting the bridge to its calculated loads multiplied by the relevant load factors.

4.5.2   The direct stresses calculated as in 4.5.1 will be increased by cross sectional warping due to eccentric load, by shear lag, and in the case of the compression flange, by load shed from the web if the latter has buckled.

4.5.3   The shear stresses calculated as in 4.5.1 may be increased by torsion due to eccentric load. In addition the shear stress distribution in webs close to supports bears little relation to that predicted by simple theory.

4.5.4   Various methods exist for calculating the effects mentioned in 4.5.2 and 4.5.3. and are fully described elsewhere. Probably the most powerful and general method of analysis is using a finite element approach, although cheaper and simpler methods are available in some cases.

5.   COMPRESSION FLANGES

5.1   General behaviour of a stiffened plate flange in compression

5.1.1   In box girder construction a compression flange will normally consist of a comparatively thin plate with a number of fairly closely spaced longitudinal stiffeners. Transverse stiffeners will be at much wider centres, and span between webs (Fig. 1).

5.1.2   If the plate is initially perfectly flat with no residual stresses, and a gradually increasing longitudinal compression is applied to it, it will remain flat until

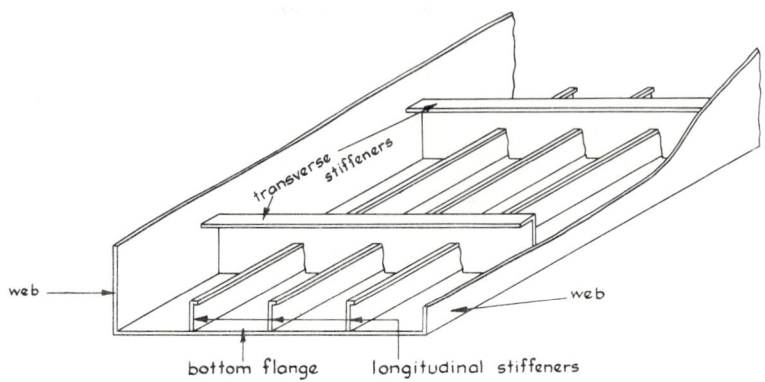

FIG. I   TYPICAL ARRANGEMENT OF COMPRESSION PLATE

at a particular value of the stress it will suddenly buckle. Depending on the relative sizes of plate and stiffeners this buckling may be of any of the four forms shown in Fig. 2, i.e.

(a)   Buckling of an individual plate panel.
(b)   Buckling between transverse stiffeners.
(c)   Buckling between longitudinal stiffeners.
(d)   Overall buckling.

In some of these cases the stiffened plate is capable of carrying further load; for example if (a) occurs, the stiffeners may well carry the load without help from the plate until a further buckling mode occurs. The value of the stress to cause buckling of a perfectly flat plate is called the elastic critical stress, and may be likened to the Euler stress for the simple strut.

5.1.3   The behaviour of a compression flange is complicated by the presence of initial out of flatness and residual stresses, always present to some degree due to steel making and fabricating processes. In this case any longitudinal compression at all will cause deflections of the plate out of plane, and it is probable that, due to this, yield of the extreme fibres of plate or stiffeners will occur before the theoretical elastic critical stress is attained. This is analogous to the behaviour of a real strut as compared to an Euler strut.

5.2   Design criteria for a compression flange

5.2.1   A compression flange of a girder supporting an externally applied load must not collapse under the action of this load multiplied by the relevant load factor. This collapse could be any of the following:

(a)   Overall yielding.
(b)   Buckling due to elastic critical stress.
(c)   Buckling as a result of yielding of extreme fibres of plate plus stiffener due to initial imperfections and residual stresses.
(d)   Torsional instability of stiffeners.

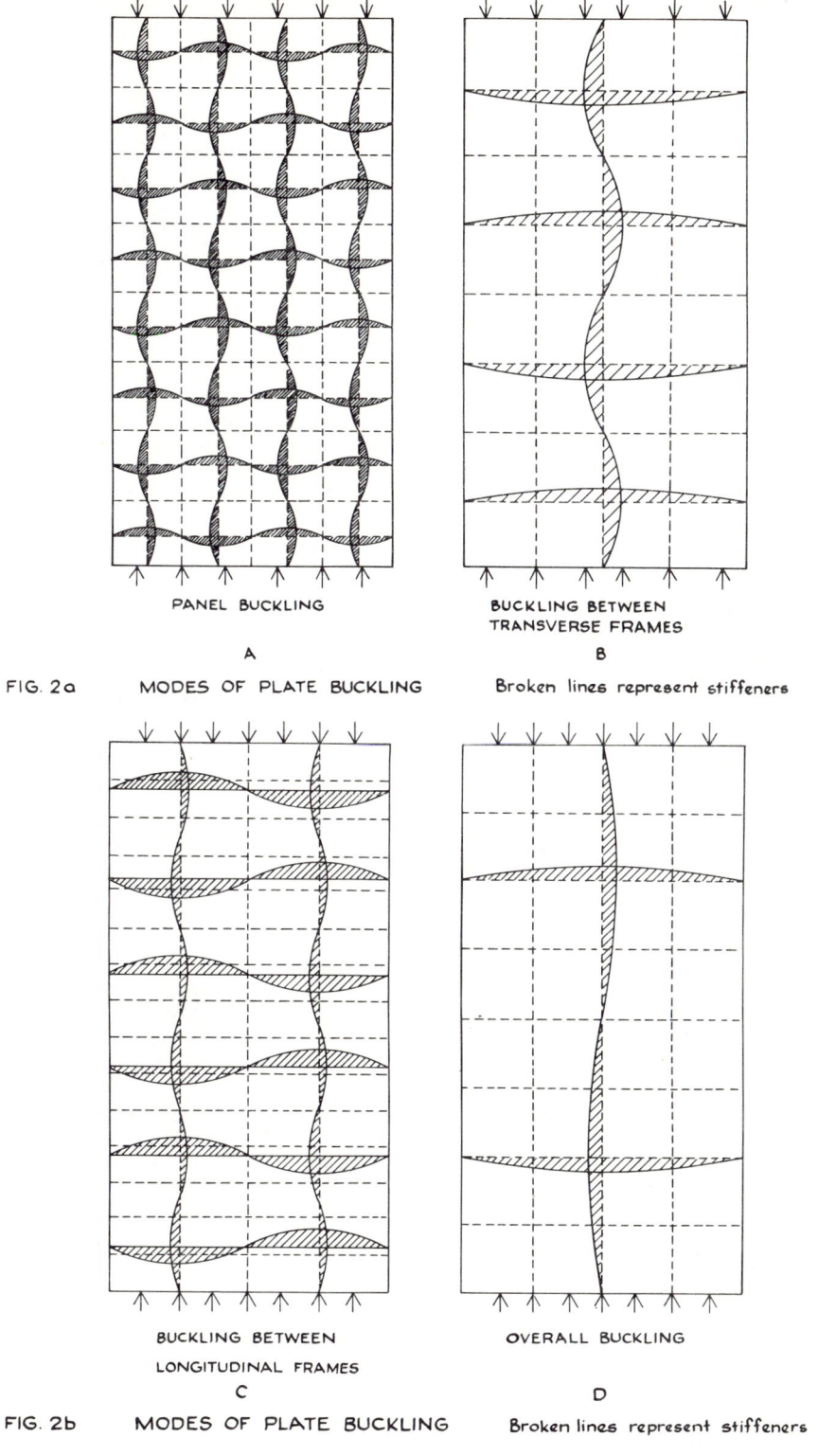

PANEL BUCKLING

A

BUCKLING BETWEEN
TRANSVERSE FRAMES

B

FIG. 2a   MODES OF PLATE BUCKLING   Broken lines represent stiffeners

BUCKLING BETWEEN

LONGITUDINAL FRAMES

C

OVERALL BUCKLING

D

FIG. 2b   MODES OF PLATE BUCKLING   Broken lines represent stiffeners

Items (b) and (c) are really inter-related and with complete information on the behaviour of stiffened plate panels it would only be necessary to consider one criterion of buckling which might either be of the plastic or elastic type depending on panel and stiffener properties.

5.2.2 Simple methods of calculating the elastic critical stresses are available, but the theoretical assessment of the effects of initial imperfections and residual stresses is in general a complex mathematical procedure. Computer programs are available which solve the basic large deflection equations of the equivalent orthotropic plate, taking into account the interaction between in-plane forces and out-of-plane effects — this is, however, costly and, subject to certain safeguards, a simplified approach may sometimes be justified and is described later.

5.3   Stresses to be used when applying the design criteria

5.3.1   The actual flange stresses which should be considered in applying the criteria of section 5.2 are the basic factored direct stresses (increased if necessary by load shed from the web), together with coexistent shear. The increased axial stresses due to warping and shear lag tend to be concentrated near the webs and thus occur in a comparatively well stiffened zone. Hence they are unlikely to affect the overall buckling behaviour, but could have a significant effect on the factor against buckling of the individual plate panels close to the web.

5.3.2   Under certain circumstances additional stresses can occur in the compression flange. For example, at a pier there may be a transverse compression due to a horizontal compression in the diaphragm. Shear stresses will arise due to the bending shear as well as from torsional load. Transverse bending will occur due to distortion of the box from normal loading. Many of these stresses can be of considerable magnitude.

6. WEBS

6.1   General behaviour of a thin stiffened web

6.1.1   The general configuration of a web of a box girder is similar to that of a compression flange inasmuch as it is stiffened in both directions. It frequently differs in that the longitudinal stiffeners may be at significantly wider centres than in the flange and the plate is usually thinner.

6.1.2   A web in general will be required to carry some combination of shear stress plus direct stress due to bending of the box. The direct stress will vary from a maximum compression at one edge to a maximum tension at the other. It is worth considering separately the behaviour of an initially flat web under the effects of direct stress and shear stress.

6.1.3   Under the effects of direct stress the plates will behave in much the same way as the compression flange; if a buckle includes the longitudinal stiffeners, however, the cross section of the buckled plate will no longer be a symmetrical half wave but will tend to bulge in the zone of compressive stress and remain substantially flat in the tension zone. A web, buckled by direct stress, is still capable of carrying shear, assuming of course that the adjacent compression flange is unbuckled thus, in effect, controlling the compression strain in the web.

6.1.4   The behaviour of the plate under a pure shear stress is quite different; as is well known there is a critical stress in shear, but this stress seldom determines the ultimate capacity of the web. Once it is exceeded, rather than carrying the load as a shear stress, the web develops a series of diagonal waves which, provided adequate vertical stiffeners are used, form in effect an N type truss with tension being taken diagonally across the panels framed by the stiffeners. To rely on this action careful detailing of the web to flange intersection is required.

6.1.5   As a result of the foregoing, there is some justification in working to a higher proportion of critical stress in webs than in flanges. In order, however, to maintain an adequate overall safety margin if the web should buckle, the compression flange must be sufficiently strong to take any direct load shed from the web due to incipient buckling.

6.1.6   It is, of course, possible that the web might yield in shear before it buckles; whilst this is not frequently so in the typical proportions of deep box girders it must be checked since in this case the reserve of strength is not available.

6.2   Design criteria for a web

6.2.1   A web of a girder subjected to the factored load system must not suffer overall yielding, although limited yielding due to high local combined stress (e.g. Mises Hencky criterion) may sometimes be tolerated. Many people find the idea of the waves which develop with a tension field unacceptable and one way of limiting transverse web deflection is to limit the direct stresses to some way below critical. Additionally, torsional instability of stiffeners and yielding of fibres of plates or stiffeners due to the effects of in plane loads on an initially deformed plate must not occur. In this respect it may be noted that the very thin plates used in webs may contain considerable fabrication distortions and residual stresses.

6.2.2   Methods are available for calculating critical stresses of stiffened webs due to varying direct stress, but these are laborious (although easily evaluated by computer). Shear critical stresses may also be calculated but may have little relevance, particularly if a large deflection analysis can be carried out. As with flanges, a simplified approach may be justified in some cases, and is described later.

6.3   Stresses to be used when applying the design criteria

6.3.1   Stresses in the web which should be considered in applying the criteria of section 6.2 are:

(a)   The direct longitudinal stress.
(b)   Shear stress due to vertical loads.
(c)   Shear, warping and distortional stresses due to eccentric loads.
(d)   Transverse compressive stresses at points of support.

6.3.2   In calculating the direct longitudinal stress, the effects of shear lag in the flanges should be taken into account, since these may substantially increase the stress in the web/flange intersection. Warping stresses also will be directly additive to the direct stresses.

6.3.3   The total value of the shear force on the cross section may be calculated using the Engineer's Theory of bending, and the average value of shear stress calculated from this. It is not, however, possible to calculate the maximum value of shear stress close to a bearing from conventional theory and special analyses (probably using finite element methods or deep beam theory) will have to be carried out to estimate this.

6.3.4   Eccentric loads produce three effects; warping stresses which have been mentioned in 6.3.2, shear stresses which may be calculated assuming a uniform shear flow round the box, and distortional stresses producing transverse bending. The last mentioned are minimised if adequate diaphragms or cross bracings are provided.

6.3.5   At the bearing, a significant transverse compression occurs in the webs. Since a bearing diaphragm carries a considerable component of compression in a direction transverse to the web longitudinal centre line, compatibility considerations require that the web also must carry this compression. In practice, the web is usually too thin to sustain it, and unless closely spaced transverse stiffeners are provided will buckle. This is, however, a 'controlled' buckle since the diaphragm plate restrains it and hence is perhaps more realistically thought of as an initial imperfection when applying the criterion of paragraph 6.2.1. If, with this assumption, the extreme fibres of the web are overstressed, adequate transverse stiffeners must be added to stop the controlled buckling. It is possible to calculate (possibly using finite element methods) how far the zone of controlled buckling spreads and hence how far the transverse stiffeners are required.

6.3.6   When examining possible yielding of the web the Mises Hencky criterion may be adopted taking all possible effects into account; these combined stresses can reach very high magnitudes in the bottom of the web at bearings over which the box is continuous.

7.   BEARING DIAPHRAGMS

7.1   General introduction

7.1.1   In a box girder, particularly a large one, the stress pattern over the area of a bearing diaphragm is extremely complex. Consider, for instance, the cross section shown in Fig. 3. The webs carry the vertical shear force from the applied loading system, and this is transferred, via the diaphragm, to the bearings. A horizontal

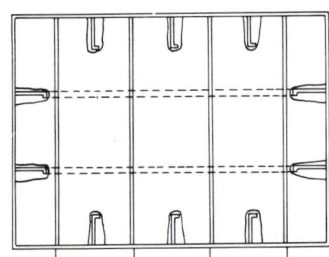

FIG. 3   TYPICAL CROSS-SECTION AT DIAPHRAGM

compression occurs at the bottom of the diaphragm due to the overhang of the bearings by the webs with the consequent 'beam effect' of the diaphragm. The finite length of the bearing raises questions of the distribution of stress along the length, and this is profoundly affected by fabrication inaccuracies. Unless the bearings are very accurately fabricated and offered up to the box there is a considerable risk that one will be loaded significantly more heavily than the other. It should also be noted that if the webs slope, the shear will have a horizontal component, thus further modifying the transverse stress distribution in the diaphragm.

7.1.2   It is possible within certain limitations to analyse a bearing diaphragm using finite element techniques; one limitation is the accuracy with which the shear stress distribution in the webs can be estimated. (In some cases of bearing diaphragms it appears that the resulting stresses within the diaphragms are not too sensitive to the applied shear distribution patterns). However, it is not easy to decide the significance of the stresses so computed since almost certainly local areas of stress in excess of yield will be indicated (for example just over the outer edge of the bearing.) In general, provided certain safeguards are taken, this will merely cause a minor redistribution of stress with no adverse affect, although the high stresses resulting from inaccuracies in setting the bearing could be more serious.

7.2   Design criteria

7.2.1   A diaphragm must be designed so that under the applied factored loads, it does not buckle, either in part or overall, and also so that no significant areas of yield occur. Whilst it has been shown that a partially buckled diaphragm sometimes possesses a considerable reserve of strength it is virtually impossible to estimate this reserve with any accuracy.

7.2.2   In almost all practical cases it will be necessary to frame the diaphragm with stiffeners to achieve the criteria of 7.2.1. Individual panels thus formed are, in general, subject to a complex stress field with biaxial direct stresses and shear stresses acting together. Provided the stiffeners divide the diaphragms into rectangular panels simple methods exist for calculating the critical stresses of such a panel subject to a biaxial stress system parallel to the stiffeners. Large deflection analysis has indicated that unless the shear stress is a substantial portion of the critical shear stress it generally has little effect on the biaxial direct critical stress, although in areas of high stress its effect on the combined stress should be examined.

7.2.3   The individual stiffeners must, from compatibility considerations, carry similar stresses to the plate panels and hence must be safe from buckling. It is frequently possible to design them as simple struts, as far as normal buckling is concerned, but the possibility of torsional instability must also be examined.

7.2.4   Local stiffeners will be required over the bearings to reduce the intensity of stress in these zones and to provide adequate buckling resistance in such highly stressed areas. These local stiffeners should preferably be symmetrically disposed about the diaphragm, whereas the overall stiffeners in paragraph 7.2.3 in general need not be.

7.2.5   Some of the stress concentration phenomena in diaphragms can be reduced by choice of geometry; for instance minimising the slope of the webs and the overhang of the webs beyond the bearings can reduce substantially the horizontal component of compression in the diaphragm, although such action may cause certain other undesirable effects.

7.2.6   When calculating loads on the diaphragm, it must of course be noted that, even in the case of perfect fabrication and erection, bearings may be loaded unevenly due, for instance, to wind load, eccentric live load, etc.

7.2.7   In certain cases reaction loads may be applied out of the plane of the diaphragm, for example due to temperature movement of the girder if the diaphragm is situated over a roller bearing.

7.2.8   By their nature, box girders lend themselves to carrying services, and so openings are necessary in bearing diaphragms to accommodate them. It goes without saying that such openings should wherever possible be located in zones of low stress.

7.2.9   As in the case of flanges and webs the effects of initial deformations and residual stress on the behaviour of the diaphragms must be investigated.

8.   SIMPLIFIED APPROACH

8.1.   Introduction

8.1.1   In a practical design office it may be possible to carry out the detailed analysis outlined in previous sections once, but to use it for a step by step process of approaching a design is likely to be uneconomical. It is well worth while, therefore, to seek a simplified approach to producing a preliminary design.

8.2   Compression flanges

8.2.1   The stiffeners may be treated as individual struts spanning between adjacent transverse stiffeners. In calculating the strut properties, an effective width of plate is taken into account.
8.2.2   The capacity of the flanges may be estimated by applying "Perry Robertson" type theory to these struts. In this application, collapse may be deemed to occur at the attainment of failure stress in either the extreme fibres of the stiffeners or the associated plate; this failure stress may be yield or some lower stress resulting from instability. The estimate of flange collapse thus obtained will sometimes be a conservative value, since the stabilising influence of the transverse stiffness of the plate is ignored. Use of this method presupposes that the transverse stiffeners are adequately stiff to restrain the flange.
8.2.3   Comparison with methods that are more theoretically exact indicates that this approximate method is accurate for wide flanges, but, as would be expected, becomes increasingly conservative as the flange width is reduced.
8.2.4   With regard to buckling of the plate between longitudinal stiffeners a possible method is to fix a limit on the breadth/thickness ratio of elements between longitudinal stiffeners.

8.3   Webs

8.3.1   A similar approach may be applied to webs, but will in general yield even more conservative values. It ignores the fact that web failure does not necessarily constitute total failure, and also ignores the stabilising effect that regions of the plate in the tensile zone have on those in the compressive zone. Some allowance may possibly be made by carrying out the analysis on a stiffener not closest to the compression flange, but at present there is insufficient evidence to quote any empirical rules.
8.3.2   As with flanges, use of this method presupposes adequate transverse stiffeners, and longitudinal stiffeners spaced sufficiently closely to preclude plate buckling between them.
8.3.3   The simplified method completely ignores the effect of the shear stress in the web on the capacity to carry direct stress. As has been implied previously, provided this shear stress is some way below its critical value, there is little interaction between it and the direct stress in producing buckling. It must, however, be noted again that the actual shear stress distribution is complex and the peak stress at the bottom of the web may be as much as 30% or 40% higher than the average. Hence local buckling or local yield due to high combined stress must be considered.

8.4   Diaphragms

8.4.1   Due to the large range of possible sizes and shapes it is virtually impossible to develop any simple general rules for diaphragms. If the webs slope the consequent horizontal stress may easily be found, and if the webs overhang the bearings the diaphragm may be treated as a deep beam to estimate the compression at the base. The actual design of the stiffened plate must, however, follow the methods of section 7.

8.4.2 It should also be noted that a simple plane stress finite element analysis of a diaphragm is a simple and straightforward exercise and should be carried out whenever possible.

9. ERECTION

9.1 It is not intended to discuss erection in any detail in this section. However, certain erection problems affect the design of box girders and so cannot be separated from this subject.

9.2 Box girders are ideally suited to cantilever erection; problems of lateral or torsional stability of the girder are unlikely to arise due to the torsionally stiff configuration. However, cantilever erection does give rise to special problems which should be anticipated at the design stage.

9.3 If the roadway deck is to be orthotropic plate type the bending stress in flanges and webs induced during cantilever erection are likely to be much higher than those sustained during service.

9.4 Diaphragms in particular are subjected to unusual loads during erection. Whilst the support reaction is generally lower than that occurring in service, bearings are sometimes temporarily set under different points, or eccentrically, and the large slope at the support during cantilever erection imposes indeterminate extra stresses.

9.5 A further practical point, particularly relevant to cantilever erection, concerns reserves of strength. A continuous girder in service possesses a significant reserve of strength since local failure will frequently cause a redistribution of stress without complete failure. For example if a diaphragm suffers a small local buckle in service the load on it will probably be reduced by transferring some to the adjacent piers. This is clearly not so in a cantilever erection case where no alternative stress path exists.

10. FUTURE DEVELOPMENTS

10.1 As will be appreciated from the preceding sections, theoretical analysis tends to be complex and is probably unduly conservative. Since little test data exist it is imperative that a comprehensive programme of tests is drawn up to attempt to validate various approaches to the problem. In particular tests are required for:
  (a)   Wide multi-stiffened plates under compressive loading (i.e. flange).
  (b)   As (a) under shear loading and with varying combinations of biaxial compression, and compression varying across the width (i.e. webs).

Where possible, tests should be on full sized specimens with realistic initial distortions and reasonable welds. Additionally, some tests are required on complete cross sections to investigate the intereaction of (a) and (b) above and the limits within which tension fields in the webs may be assumed to act.

It is essential that relatively straightforward design criteria should be developed from such tests to avoid wasteful use of valuable engineering manpower in performing abstruse calculations.

It is only by a factual approach to the ultimate strength of box girders, as opposed to solely theoretical predictions based on complex theories of unproven validity that it will be possible to exploit the intrinsic economy of this form of structure.

11. CONCLUSIONS

11.1 The design of a box girder bridge is not a simple task to be undertaken lightly by anybody working literally to rules in codes of practice. Present day codes simply do not contain adequate clauses, and whilst revisions are in hand it will be some time before they are available. Furthermore, even when they are, it will be essential to have a 'feel' for the type of structure involved.

11.2 Whilst it is straightforward to calculate the average stresses in flanges and webs by conventional theory, the actual distribution of these stresses is significantly affected by shear lag, torsion, warping and deep beam effects. An accurate assessment of these effects requires complex mathematical analysis.

11.3 This full analysis presupposes that a basic design is available which may then be subjected to the mathematical treatment. Hence, simple rules for a first design are required.

11.4 As with any structural design, good detailing is a *sine qua non* and particularly so with large thin walled boxes which are sensitive to initial deformations and residual stresses.

11.5 Erection should be considered at the design stage.

11.6 Any new standard codes should be reasonably easy to apply but should not preclude the use of sophisticated mathematical treatment in special cases.

11.7 Tests are urgently required to establish factually the ultimate strength of web and compression flanges. Only by this means can the intrinsic economy of the box girder form be fully exploited.

GATESHEAD VIADUCT: THE ANALYSIS OF A COMPLEX MULTICELLULAR BOX STRUCTURE

W. J. R. SMYTH and SRINIVASAN

Ove Arup & Partners

SYNOPSIS

This paper deals with the structural analysis of the deck of a viaduct which is a multicellular box structure of prestressed concrete. Two complementary methods of analysis, one of which is three dimensional and the other two dimensional are described and compared with the results from model tests.

INTRODUCTION

Urban motorways are typically complex and so are the structures which are associated with them. If the designer takes the view that he must design the right structure to take account of the physical, economic and aesthetic constraints he may find himself faced with some formidable problems of analysis which may tax to the utmost his resources of time, staff and computer facilities.

This paper deals with a problem of this kind from the limited point of view of the structural analysis.

The scheme is described in a booklet published by the Cement and Concrete Association in December 1969, and is under construction.

VIADUCT STRUCTURE

The A.1 Viaduct at Gateshead is an elevated motorway about 1000 m long, generally 19·5 m wide but up to 38 m wide at the interchanges. Structurally it consists of a multicellular prestressed concrete box. Figure 1 shows a part plan at one of the two junctions of slip roads with the viaduct; section C-C is typical of the main structure and section A-A of the ramps. The broken circles indicate the bearing positions and the typical span is 27 m. It can be seen that the deck structure is supported on central columns and that the structure is wide in relation to its span with a particularly large transverse cantilever at the second crosshead to the left of section C-C.

The structure to be analysed had the following characteristics:—

1. The proportions of a typical span are such that it falls into an 'intermediate' category in the same sense of the word as is used for shells — it cannot be treated as a beam with subsequent corrections, nor can it be treated as a slab.
2. There is a large number of untypical spans, and the ramp junctions are very complicated.

441

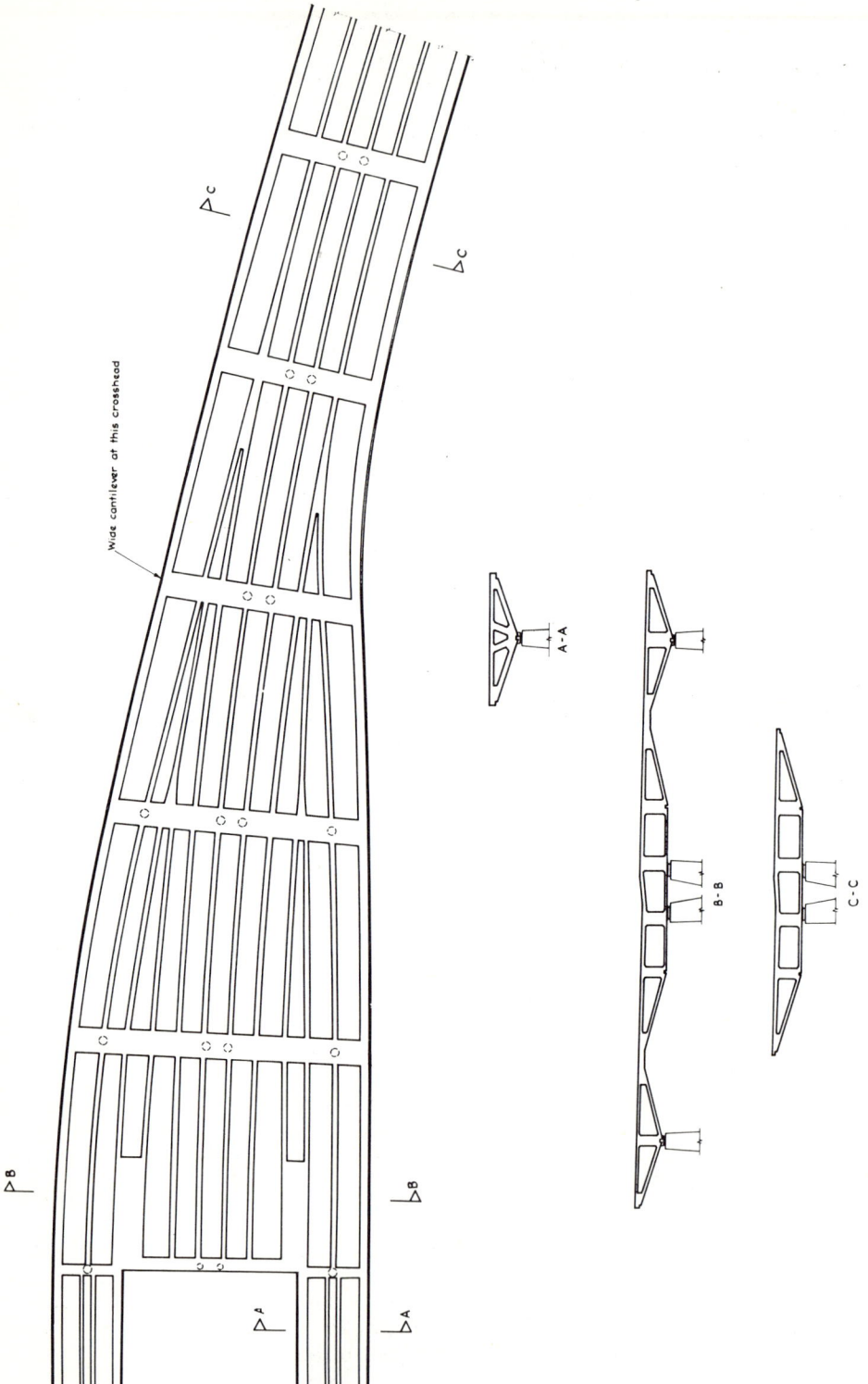

FIG. 1 PLAN & SECTIONS OF SOUTHERN RAMP INTERSECTION PLAN DRAWN WITH TOP DECK REMOVED TO SHOW RIBS

3. A fair number of different loading cases had to be examined.

4. Span by span construction had to be taken into account.

These characteristics imply a difficult analysis with many different cases to be examined.

The strategy which evolved for the final analysis of the structure was as follows:—

1. A typical span would be analysed by a method which took the three dimensional action of the structure into account.

2. A typical span would be analysed using an equivalent grillage and the results from (1) used to check, modify or calibrate the grillage.

3. The equivalent grillage would then be used to analyse the more complex parts of the structure at the slip road junctions.

4. A model test would be used to confirm the analytical results.

THREE-DIMENSIONAL ANALYSIS

A number of methods was investigated and found to be unsuitable. The method which was adopted could make use of existing computer programs and the deficiencies of the conceptual model could readily be seen because they arose from physical assumptions, rather than mathematical approximations, and so could be taken more easily into account when translating the results back into designs for the real structure.

The conceptual model was based on two main assumptions. The longitudinal action of each individual plate between plate junction lines is represented by simple beam theory, in other words straight lines between junctions remain straight. This seems to be a satisfactory assumption for plates with the proportions of those in the structure. A more radical assumption is that the plates are not connected continuously along their edges but at a number of discrete points, at which all forces and moments are transmitted. As the number of these points increases, the model approaches nearer to the physical reality and to establish how many points were

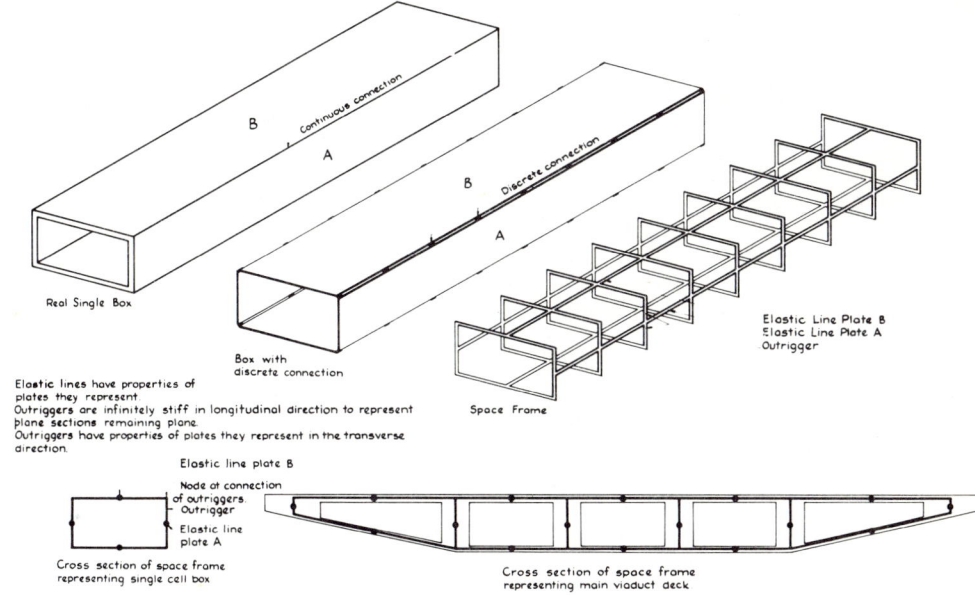

FIG. 2 DEVELOPMENT OF SPACE FRAME CONCEPTUAL MODEL

required for a satisfactory solution some very simple cases such as I-beams or single cell boxes were analysed with various numbers of connections along the span.

Once this discrete point connected model is assumed, it can be represented by a space frame and Fig. 2 shows the way in which the space frame is developed. Each plate can be represented in a longitudinal direction by its properties as a beam, concentrated on its centre line. A series of rigid outriggers at right angles to the central line connect it to the edge of the beam at the node points where the plates are supposed to be connected. These outriggers approximately represent the property plane sections remain plane, so that the ends of the outriggers will have the correct deformation in the longitudinal direction at the node points on the outer edges on the corresponding beam. This produces a bending moment diagram for the plate which has steps in it because the shears between the edges of adjacent plates are applied at the node points instead of continuously along the outer edge, but it is not difficult to draw a reasonably accurate bending moment diagram through the steps, provided that they are not too large in relation to the maximum value of the bending moment. In the transverse direction the outrigger has to be given the bending stiffness of a width of plate equal to the distance between outriggers. Again this produces transverse moments which, if plotted longitudinally, will give a stepped diagram.

We now have a model which can be analysed using programs which were available in 1966, and whose physical implications are quite clear, so that the results can be interpreted by engineers.

A space frame with a large number of nodes is needed to give a reasonable representation of the structure and inherently requires much more computer capacity than a grillage. It was therefore only used to check a limited number of cases for comparison with the corresponding grillage analysis.

EQUIVALENT GRILLAGE ANALYSIS

The equivalent grillage approach is effectively two dimensional, whereas the real structure has a complex action in three dimensions. For example, the real structure has transverse forces acting in it which cannot be taken account of in the grillage, nor can warping although it may not be significant in this particular case. The section acts transversely as a kind of Vierendeel girder, and to simulate this the

Bending deformation
(extension, contraction
of flanges)

Pseudo-shear deformation
(bending of flanges accompanied
by some bending of webs)

FIG. 3   GRILLAGE ANALYSIS.

TRANSVERSE SECTIONS THROUGH ONE CELL OF A MULTICELL
BOX SHOWING DEFORMATIONS REPRESENTED BY BENDING AND SHEAR
OF TRANSVERSE MEMBERS OF GRILLAGE.

grillage analysis had to take into account not only the bending deformation of the transverse members of the grillage, but a pseudo shear deformation.

In a solid beam the shear deformation is normally not significant, but a single beam representing a Vierendeel girder has to have shear and bending deformations represented. The equivalent beam appears to have a shear flexibility which is out of all proportion to its bending flexibility. This pseudo shear flexibility really represents the bending of the top and bottom flanges of the Vierendeel girder as illustrated in Fig. 3. An existing computer program for grillage analysis had to be modified to allow for the shear deformation.

The plan of one of the equivalent grillages is shown in Fig. 4 as drawn directly by a drum plotter using the computer output in order to give a visual check on the geometrical input.

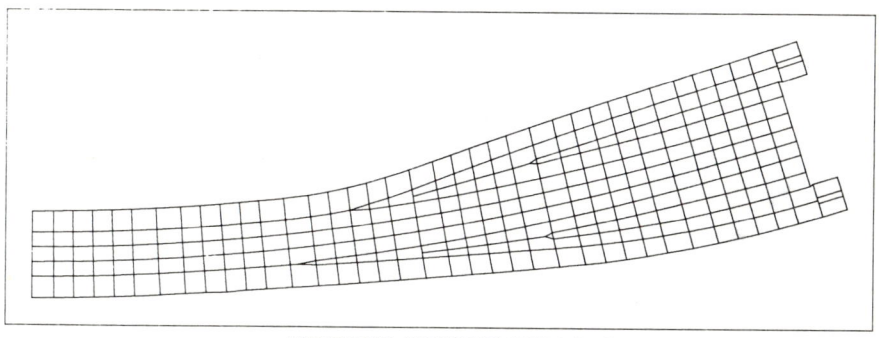

NORTHERN RAMP INTERSECTION SPANS 4-9   2

FIG. 4 GRILLAGE ANALYSIS — COMPUTER PLOT TO CHECK GEOMETRICAL INPUT

PHYSICAL MODELS

1. *Elastic Model*

The conceptual model of the equivalent grillage was checked for typical straight spans by comparing results with those from the space frame model. The intersections of the slip roads with the main viaduct are too complex to be checked by this means and a physical model was made of one of them. Most of the working drawings had been completed before the model results were available but they gave a final confirmation of the results of the analyses before the structure was built.

The model was made and tested by the Cement and Concrete Association at Wexham Springs. Because of space limitations, there was a difficult choice between having a sufficient length of structure represented and having a large enough scale. A 1:48 scale model of the structure was made in perspex to represent five spans of the actual structure. 96 electrical resistance strain gauges were installed on the model and the strains were recorded automatically by electronic logging equipment, which unfortunately set an upper limit to the number of gauges.

2. *Ultimate Load Model*

Other tests were made at Loughborough University under Professor Brock to examine the ultimate load capacity of the structure, as part of a research programme.

Models to a scale of 1:24 were made of one span of the viaduct with a cantilever at each end, to be able to simulate the effect of continuity. The models were made from a plaster with similar characteristics to those of concrete, reinforced with wires and mesh conforming to the reinforcement details of the structure. The prestressing cables were represented by unstressed steel. The self weight and uniform live loads were applied by springs and the abnormal HB load by a special rig on top of the model. Strain gauges were fixed on the top and the bottom along the ribs at the mid and third points of the span and the reactions were measured by load cells at the supports. The load was applied in increments of a quarter of the design load and the models behaved elastically for load cycles up to three-quarters of the design load. The model collapsed at just over twice the design load by the ribs failing in shear, due to effects which would not occur in the real structure. The results from the tests showed fair agreement with the analysis and as the tests were carried out quickly the information was available at an early stage.

3. *Actual Structure*
The structure as built has acoustic gauges cast in.

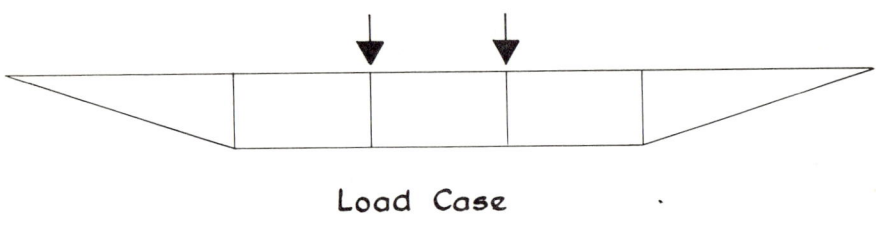

Load Case

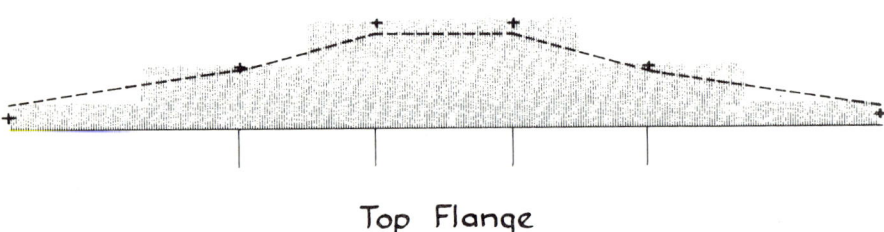

Top Flange

Results from space frame shown by broken line.
Results from grillage shown shaded.
Results from model test shown by crosses.

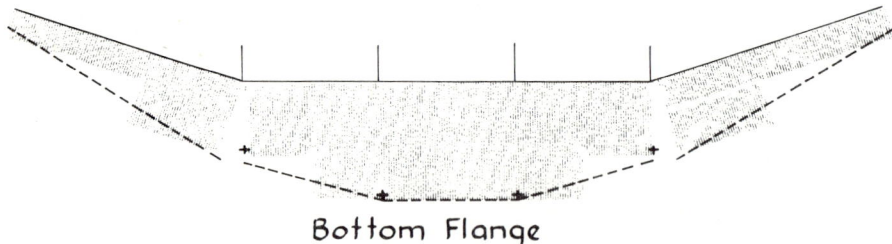

Bottom Flange

FIG. 5   LONGITUDINAL STRESSES AT MIDSPAN

Note on figures 5 to 10

Results from space frame shown by broken line
Results from grillage shown shaded
Results from model test shown by crosses.

CONCLUSIONS

Typical comparative results from the various analyses and tests are shown in Figs. 5 to 10. In each case the results from the space frame analysis are shown by a broken line, the grillage analysis by shading and the model test results by crosses. These results are for line loads on one rib, or on two for the symmetrical cases, and a smoothing process has been applied to discontinuities. The grillage seems to give quite good results for longitudinal actions, but needs supplementary information for transverse actions. The results need to be interpreted by an engineer who understands the way in which closed hollow sections work — the relationship between longitudinal warping and the distortion of the cross section for example.

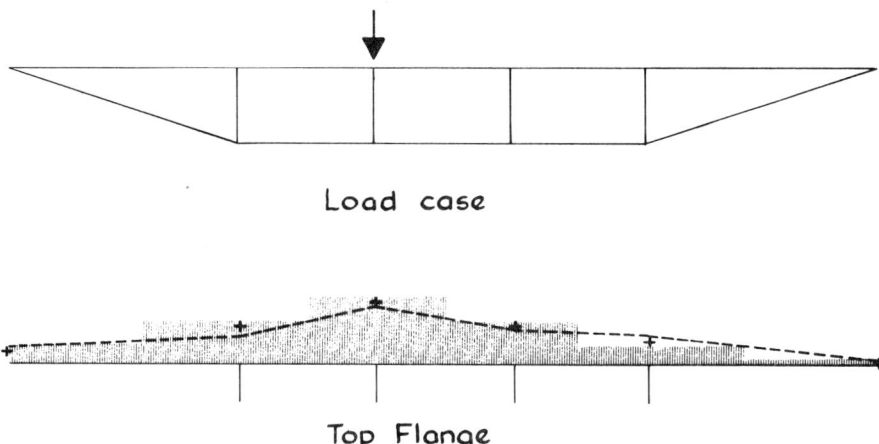

Load case

Top Flange

Results from space frame shown by broken line
Results from grillage shown shaded
Results from model test shown by crosses.

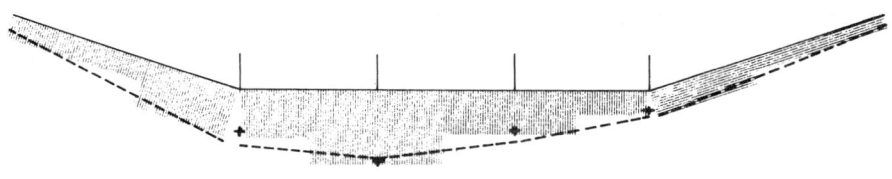

Bottom Flange

FIG. 6 LONGITUDINAL STRESSES AT MIDSPAN

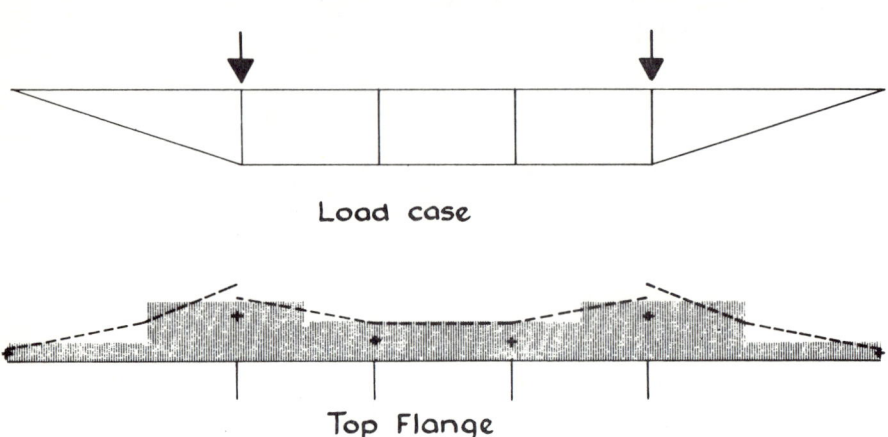

Load case

Top Flange

Results from space frame shown by broken line
Results from grillage shown shaded
Results from model test shown by crosses.

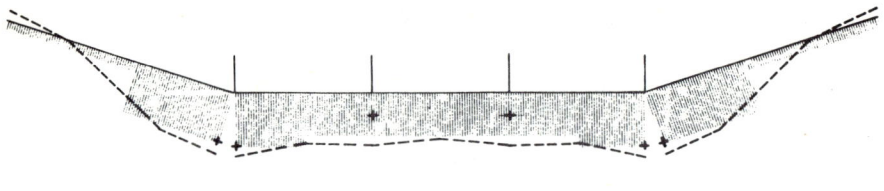

Bottom Flange

FIG. 7 LONGITUDINAL STRESSES AT THIRD POINT

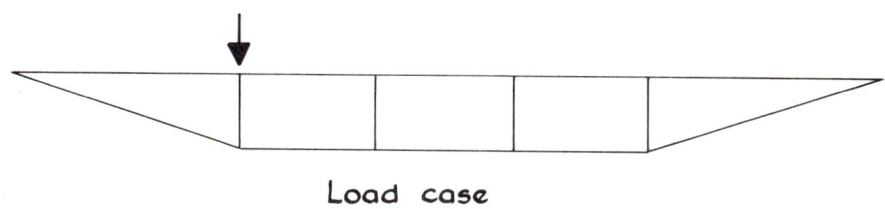

Load case

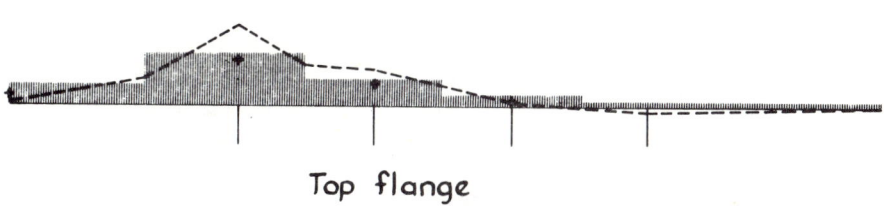

Top flange

Results from space frame shown by broken line
Results from grillage shown shaded
Results from model test shown by crosses

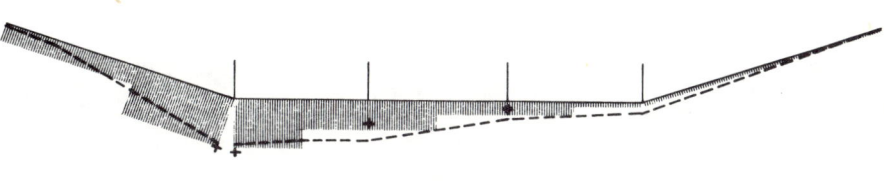

Bottom flange

FIG. 8 LONGITUDINAL STRESSES AT THIRD POINT

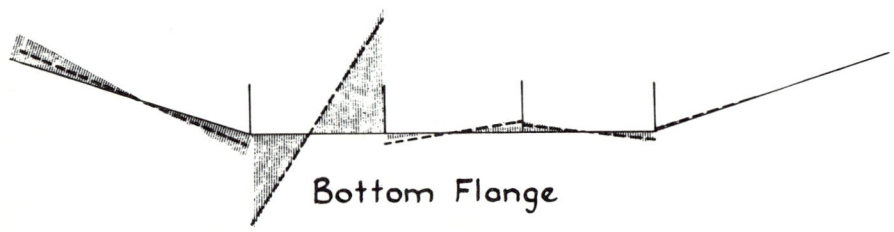

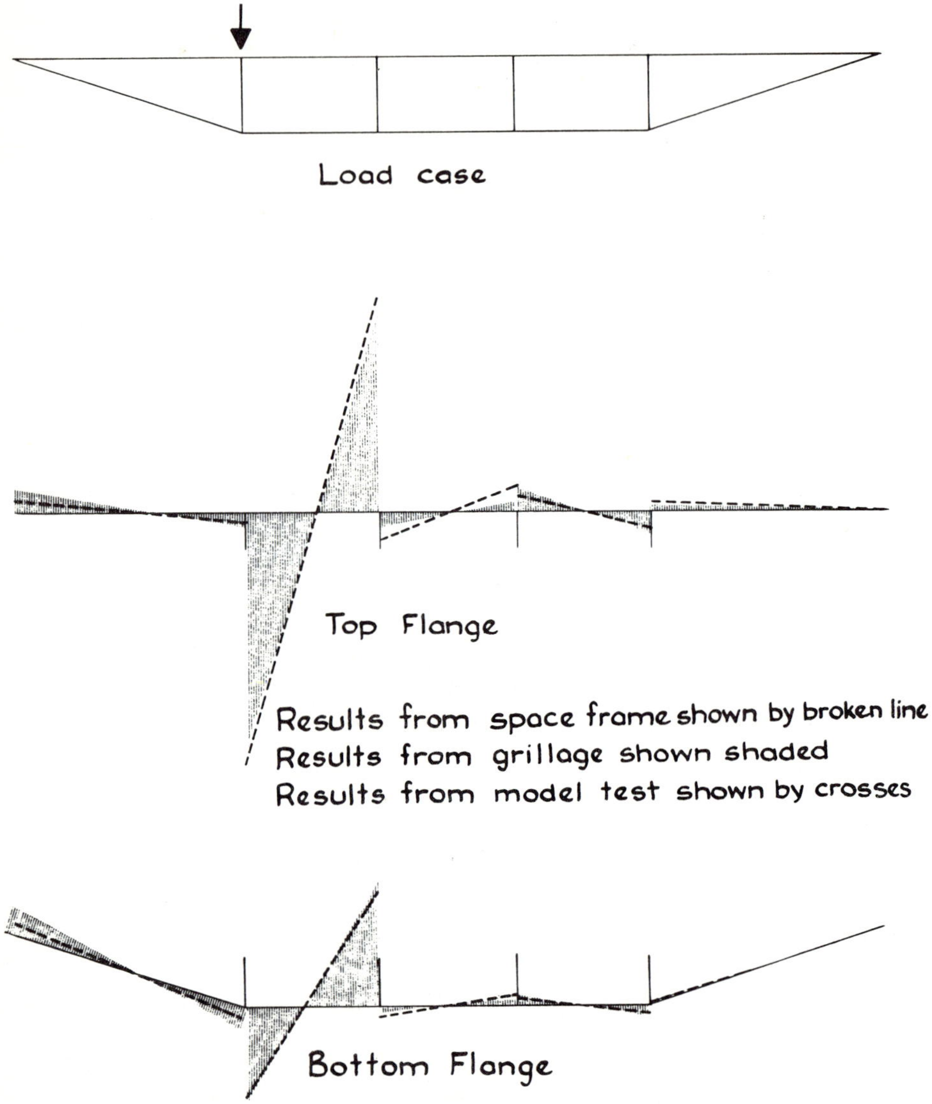

Load case

Top Flange

Results from space frame shown by broken line
Results from grillage shown shaded
Results from model test shown by crosses

Bottom Flange

FIG. 9 TRANSVERSE MOMENTS AT THIRD POINT

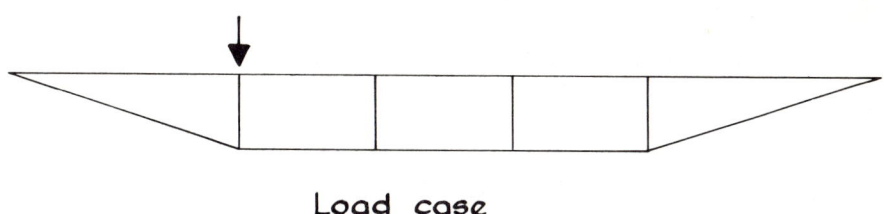

Load case

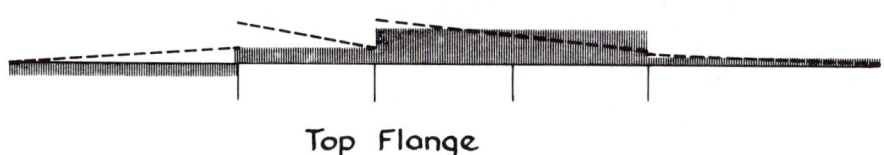

Top Flange

Results from space frame shown by broken line.
Results from grillage shown shaded.
Results from model test shown by crosses.

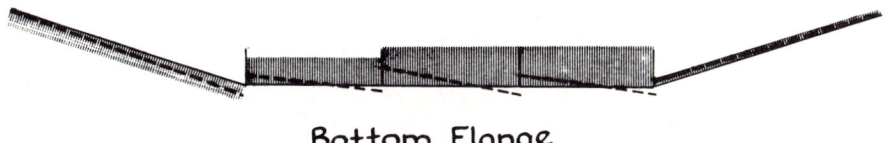

Bottom Flange

FIG. 10 TRANSVERSE FORCES AT THIRD POINT

FIG. 11 VIEW OF STRAIGHT SPANS DURING ERECTION

FIG. 12 VIEW OF COMPLETED STRAIGHT SPANS

FIG. 13 VIEW OF SOFFIT OF STRAIGHT SPANS, SHOWING THE RELATION OF PIER TO THE DECK

FIG. 14 VIEW OF THE UNDERSIDE AT THE INTERSECTION OF RAMP WITH THE MAIN STRUCTURE

CABLE STAYED CONCRETE BRIDGES

ANTHONY F. GEE

Mott, Hay & Anderson

The paper contains what the author believes to be a comprehensive review of the cable stayed concrete bridges completed to date or currently under construction. Two other such bridges are described which are actively being designed, with one of which the author himself is concerned, and in the second part of the paper he discusses the factors which influence the choice of solution to some of the specific problems arising in the design of such bridges.

Only the design and construction of the superstructures of the cable stayed spans of these bridges are described, no reference being made to the often very interesting foundations or approach spans. Likewise, apart from the historical introduction, the author has attempted to avoid duplication of the contents of the very many excellent published works on cable stayed steel bridges, although several of the problems encountered are common to both materials.

INTRODUCTION

It is becoming clear that the cable stayed girder is extending the competitive spans for concrete construction, given adequate foundation conditions, into the range between 200 m and 300 m where previously only an arch form was feasible.

There are now at least a dozen bridges of this type completed during the last decade or in various stages of design and construction, half of which are road bridges having spans in this range, and it would appear to be an opportune moment to carry out a detailed examination and comparison of these structures.

HISTORY

There are reliable indications that primitive forms of suspension bridge figure prominently amongst the very earliest forms of civil engineering construction. It is likely that the cable stayed bridge was evolved in prehistoric times since it is an equally logical form of structure, but in modern times it does appear to have evolved from the suspension bridge rather than to have been developed in parallel.

The first modern iron suspension bridge was probably James Findlay's 21·3 m span Jacob's Creek bridge in Pennsylvania, USA, in 1796, although there are claims for the Winch bridge, a footbridge of unknown span over the Tees at Middleton which was reputedly built in 1741 and collapsed in 1802. Certainly Sir John Rennie and Alan Smith's 39·6 m span bridge over the Humber at Hookstow in 1807 and Findlay's famous Merrimac bridge, built in 1810 in Massachusetts, USA and still existing, preceded any cable stayed bridges, but there was an unauthenticated cable stayed footbridge with a span of 33·5 m reportedly built in England in 1817, probably following the theoretical work of C. J. Löscher of Fribourg, who published a proposal for such a design in 1784.

Samuel Brown's suspension bridge over the Tweed in 1820 represented a tremendous step forward with a span of 137 m but unfortunately it collapsed 6 months later in a high wind. In 1821 another Frenchman, Poyet, again published a theoretical treatment of the cable stayed bridge and in 1824 a 78 m span bridge was built over the river Saale near Nienburg. It, too, had a short life, collapsing under a crowd of people the following year, but we were now approaching the end of the trial and error stage.

In 1836, James Dredge used inclined hangers in his 45·7 m span suspension bridge over the Avon at Bath, which still stands, and 2 years later Thomas Motley's 36·6 m span Twerton bridge also over the Avon a few miles away became the first authenticated successful cable stayed bridge.

Although the second half of the nineteenth century was a golden era of British bridge building, the cable stayed design does not appear to have been favoured. There are several possible reasons for this, but two predominate: many of the really great bridges carried railways and suspended bridges were not considered suitable for the heavy loads and vibrations encountered. Also the low tensile strength of the irons and steels available meant that the size and weight of the cables, or chains as they were more likely to be, was very considerable.

This was not significant in a suspension bridge, in fact it was a positive advantage in stiffening the structure, but the resulting sag had a very serious effect on the behaviour of a cable stayed bridge. This led to the combined system adopted in Barlow's Lambeth bridge in 1861 and Ordish's Albert bridge in 1873, wherein the action of the catenary chain was almost solely to take the sag out of the "straight" suspension cables.

During the first half of this century the large suspension bridges stole the limelight and it was not until the post war years of reconstruction in Europe that the cable stayed bridge appeared to make its impact felt. Even then, for technical or political reasons, steel held sway as the constructional material and it was a considerable sensation when Morandi's winning design for a bridge over Lake Maracaibo appeared in 1957 with a 400 m cable stayed centre span in prestressed concrete.

That this design as originally conceived was never built was a matter of great regret but, in a way, it is probably just as well because anything which followed would have been a considerable anticlimax.

It is probably a little known fact, however, that even when the severely modified scheme was built it was not the first cable stayed concrete bridge; not by nearly forty years! (See Tables on page 472 and page 473).

TEMPUL AQUEDUCT, JEREZ DE LA FRONTERA, SPAIN (Fig. 1)

This remarkable structure has, as befits such a landmark in bridge construction, a classic configuration of two towers each with a single backstay and forestay, towers and stays being equally spaced to give a ratio of main span to side spans of three to one. The designer had to be one of the masters and of course it was, the great Professor Torroja, who was responsible for many original conceptions particularly in prestressed concrete.

It is probable that he conceived this as a prestressed beam with the tendons transformed well outside the section. By thus using a relatively small prestressing force at a large eccentricity he would alleviate the problem of high relaxation losses associated with the low tensile strength of the cables. Since it is known, however, that he introduced the ties to replace two piers which were originally

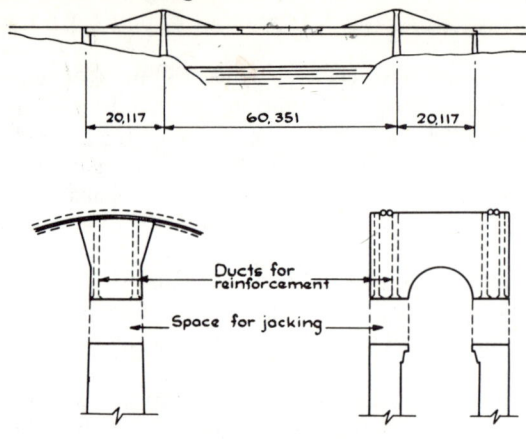

FIG. I   TEMPUL AQUEDUCT

intended but which were found too difficult to construct in the deep water, it is possible that his first intention was to provide intermediate supports in the main span and thus truly a cable stayed bridge.

The towers are very low to provide an economic solution but he may have been worried by stability problems since he stressed his ties by jacking up the tops of the towers. After the side spans and cantilevers had been cast *in situ* the saddles carrying the cables over the towers were raised a distance of 250 mm which was apparently just sufficient to lift the ends of the side spans off their supports. He must then have provided some form of anchorage at these piers before placing the suspended span which was probably precast.

LAKE MARACAIBO BRIDGE, VENEZUELA (Fig. 2)

Even though the original concept became somewhat diminished in realization, this bridge still represented a truly formidable undertaking. Nearly five and a half miles long across water which was for most of its width over 30 m and in places up to 45 m deep to the bottom of the silt, it offered a tremendous challenge. However, its very scale did afford the opportunity for investment in extremely heavy and expensive plant which was the key to its construction. Books could be and have been written about this bridge, but here we are concerned only with the super-structure of the seven spans which form the central navigation opening.

From each of the six towers a single pair of backstay and forestay cables support three cell box girder cantilevers 79·5 m from the tower, beyond which they extend another 15 m to support the 46 m long suspended spans each comprising four T-beams.

The X-shaped piers and the A-frames forming the towers were cast *in situ* using climbing shuttering, together with approximately 50 m of deck over the piers.

Steel lattice girders were then supported on the outer legs of the X's and at their front ends on tubular steel trestles on temporary piled foundations. This falsework extended for the full width and length of the cantilevers which were then also cast *in situ*.

Huge prefabricated sections of shuttering, together with reinforcement, pre-stressing ducts and, where possible, tendons were prepared on shore and lifted into position.

LAKE MARACAIBO, VENEZUELA (PHOTO BY COURTESY OF THE CEMENT AND
CONCRETE ASSOCIATION)

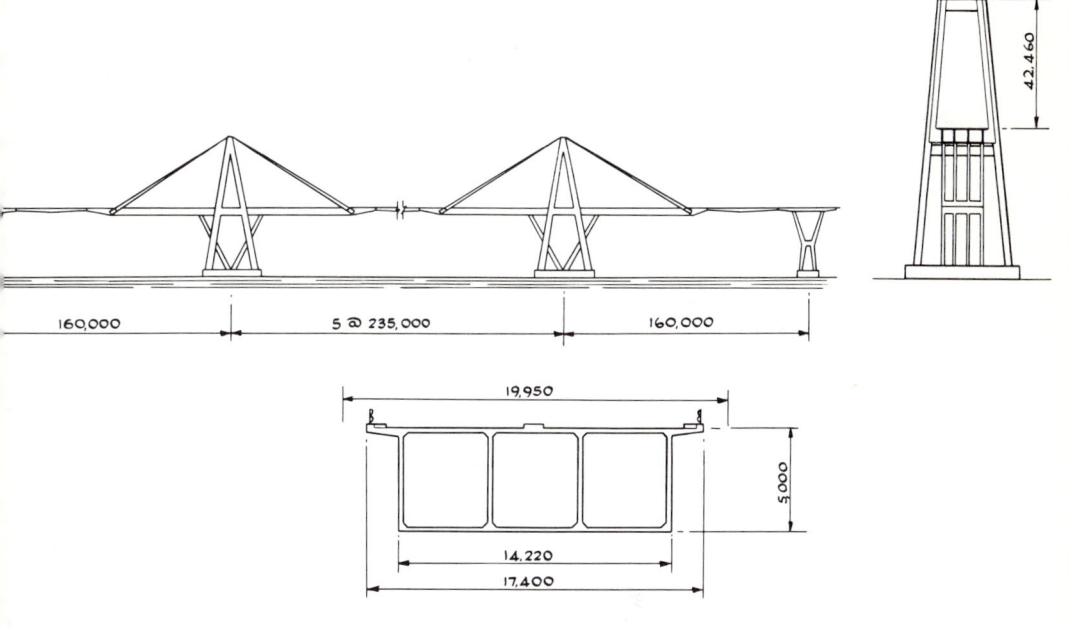

FIG. 2　　MARACAIBO BRIDGE

Temporary bridges were erected between the ends of the cantilevers together with suspended walkways along which the cables were winched through the bottom of one anchorage over the tower to the other anchorage.

Large amounts of kentledge were placed on the deck against which the stays were tensioned before the precast suspended span beams were erected by massive floating cranes.

Stressing was from below the anchorages using threaded pull-rods tapped into the sockets and anchored in a crosshead which was displaced by two 250-ton hydraulic jacks.

Although the original conception included a proposal to sheath the cables in a concrete casing, this was not in fact carried out.

Despite the comparatively modest width of the deck, the transverse beams into which the stays were anchored were formidable members: they were 6 m wide by 2 m deep, each contained 70 prestressing cables of eighteen 7 mm wires and the prefabricated reinforcement cage alone weighed 60 tons.

CANAL DU CENTRE FOOTBRIDGE, OBOURG, BELGIUM (Fig. 3)

This remarkable structure provides access to a cement works and was very properly built by the owners to replace the existing steel bridge and demonstrate the versatility of concrete.

It was erected complete on prepared foundations in 24 days: it is only fair to point out, however, that the existing bridge was left in place and used to support the prefabricated deck units of the new bridge during erection.

It took eight days to erect the eight precast reinforced concrete double-T units, each 16·75 m long, two days to erect the two legs of the tower, ten days to place the cables, anchor them and stress the outer pair of stays which merely serve to

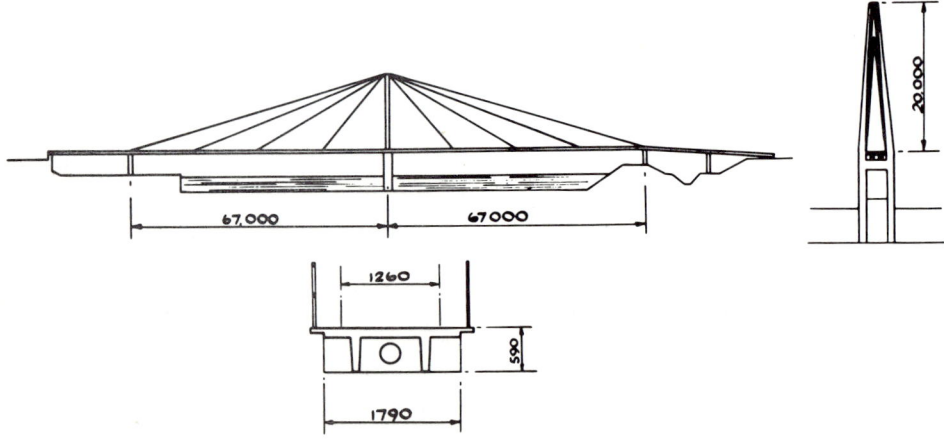

FIG. 3   CANAL DU CENTRE FOOTBRIDGE

hold the hinged tower in position, three more days to adjust the tensions in the remaining stays to obtain the correct profile to the completely articulated deck and finally one day to complete the joints between the deck units.

The cables were socketed into crossheads jacked down threaded U-bolt shackles which were in turn anchored by the pins which formed the hinged joints between the deck units.

POLCEVERA VIADUCT, GENOA, ITALY (Fig. 4)

Morandi was quick to follow up the success of the Maracaibo bridge with an almost identical design for a high level viaduct carrying the Genoa-Savona motorway.

Once again the piers and towers were cast *in situ* together with the sections of deck above the piers. The five cell box girder cantilevers were also cast *in situ* but this time in cantilever construction using temporary prestressing cables outside the section of the deck running over stools at the piers. This enabled the construction of the cantilevers to proceed concurrently with the towers.

When the massive transverse beams, which this time were of box section, the stays bifurcating above deck level, had been cast and stressed, the single forestay

POLCEVERA VIADUCT, GENOA (PHOTO BY COURTESY OF THE CEMENT AND CONCRETE ASSOCIATION)

and backstay cables were erected from suspended catwalks. As they were stressed the temporary prestressing was progressively reduced.

The stays were then encased in ducts surrounded by *in situ* concrete which was in turn prestressed by secondary cables. The space between the cables and the concrete was grouted up and the stress in the main cables increased as the secondary cables were released. Finally, the secondary cables were restressed. In this way the casing remains in compression under applied loading even allowing for creep and shrinkage.

The 36 m long precast suspended span beams were brought in over the completed deck and launched into position.

ANSA DE LA MAGLIANA VIADUCT, ROME, ITALY (Fig. 5)

At about the same time Morandi was engaged in the design of a viaduct to carry the motorway from Rome to Fiumicino airport over an area of swamp land formed by an elbow of the river Tiber which could not be avoided because of the adjacent railway line from Rome to Pisa.

Although of much smaller actual span than the Maracaibo bridge and the Polcevera viaduct, Morandi by using only one tower created an effective span of the same order, the main cantilever being supported by the forestays at a point 69 m from the tower and extending a further 13 m to support a 63 m long suspended span.

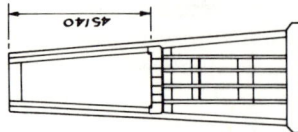

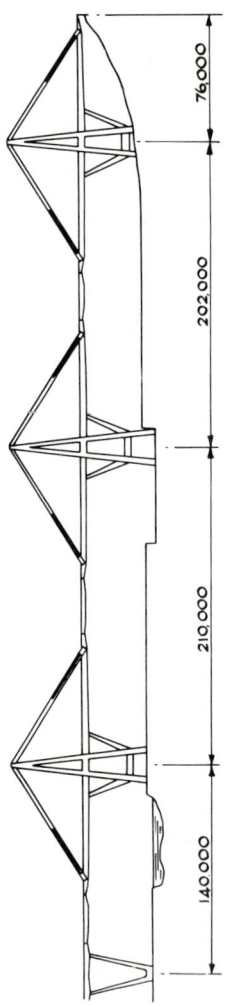

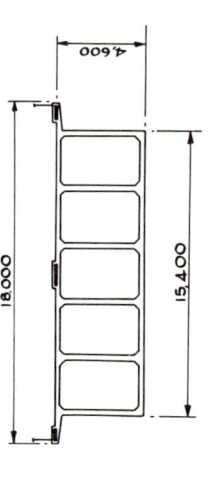

FIG. 4   POLCEVERA VIADUCT

There were additional problems involved in this design: it was much wider and at the site of the main span the road lay on a curve of 475 m radius.

In many respects the solution is similar to the earlier designs with single pairs of forestays and backstays from a portal tower supporting a seven cell box girder cantilever which in turn supports at its extremity a suspended span comprising seven T-beams but because of the poor foundation conditions which were the raison d'etre of the bridge, Morandi abandoned the fixity he achieved from the X-shaped piers and A-frame towers and adopted a fully articulated structure. Nevertheless, despite the poor ground, the whole bridge was cast *in situ.*

The tower portal is hinged at its feet and, because of the different inclinations of the forestays and backstays, leans backwards nearly 2·5 m from the vertical.

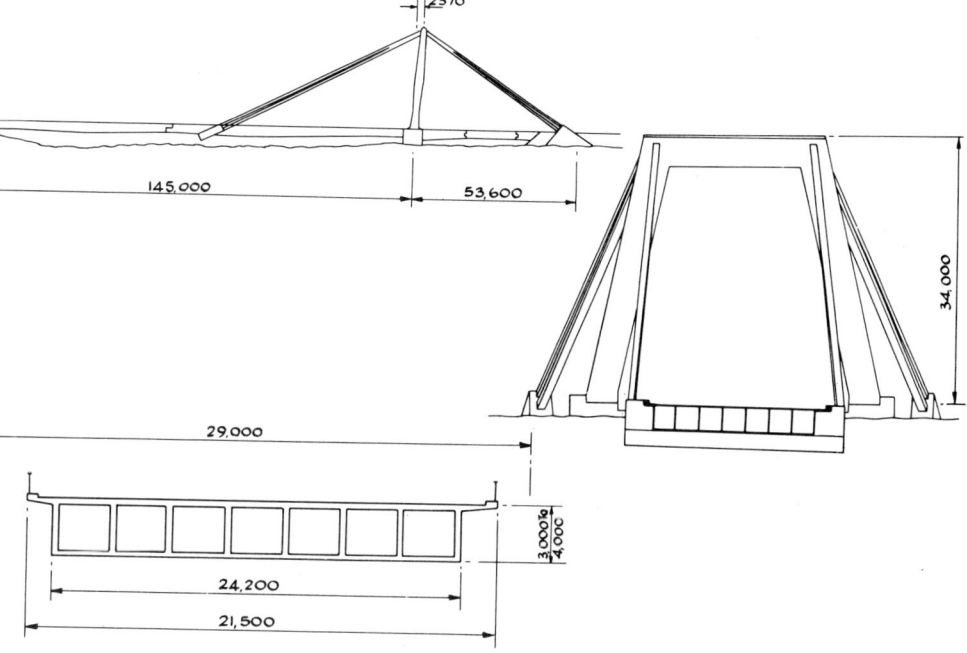

FIG. 5   ANSA DE LA MAGLIANA VIADUCT

Similarly, the cantilever span is hinged at the tower and the anchor span, which rests on the ground, at two positions.

All these hinges carry the full horizontal thrust of the deck by means of large radius steel lined concrete surfaces extending the full width of the deck and stability is restored by a wide gravity abutment.

The backstays, which are widely splayed to accommodate the curvature of the deck, are anchored independently in large gravity blocks which nevertheless rely on extensions of the abutment for their resistance to horizontal movement.

The transverse beam into which the forestays are anchored is again of box section, the cables being bifurcated as in the Polcevera viaduct. Because the deck is wider, this beam is a little matter of 8 m deep and 2·7 m wide and is prestressed with seventy-six cables each of sixteen 5 mm wires.

This marked Morandi's first use of parallel wire prestressing cables in the main stays. Two different types of tendon were used and it seems probable that the smaller ones were used to apply the secondary prestress to the *in situ* casing to the stays.

PIPE BRIDGE, PRETORIA, SOUTH AFRICA (Fig. 6)

This prosaically named bridge No. 2348 of the Transvaal Roads Department carries two 685 mm diameter water mains with provision for the future installation of a further 865 mm diameter pipe.

It was cast entirely *in situ,* the most interesting feature being that the hinge between the two deck sections, which again must transmit the horizontal thrust, was achieved by contiguous casting of two mating surfaces. Presumably advantage

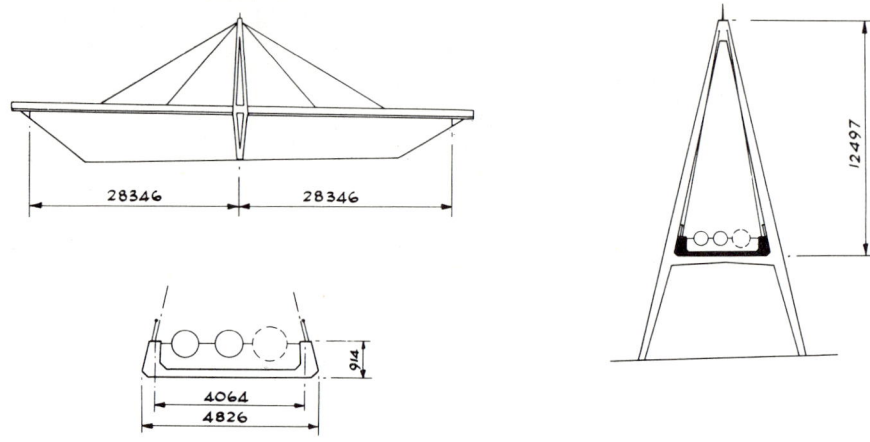

FIG. 6   PRETORIA PIPE BRIDGE

was taken of the fact that the type of bridge precludes the incidence of transient loading but one would have expected significant rotation at this joint between the loaded and unloaded conditions.

The stays were formed from parallel wires anchored by a prestressing system not familiar in the United Kingdom in which the wires are looped around a capstan type boss into which is tapped a stressing rod. The other ends of the wires are usually crimped and embedded as a dead end anchorage but in this case the cables were prefabricated from a continuous length of wire, presumably completed by welding or brazing, and stressed simultaneously from both ends. The wires were made up into a circular cable with spacer bars, the interstices packed with grease and wrapped with protective tape and the whole drawn into a steel tube.

The stressing rods pass through the deck to fabricated steel bearing plates mounted on the soffit and prevented from sliding by a large number of inclined bolts anchored in the deck concrete.

RIVER BARWON FOOTBRIDGE, GEELONG, AUSTRALIA (Fig. 7)

Although designed for only pedestrian traffic, this bridge also carries a 1·1 m diameter sewer pipe, totally enclosed by casting a PVC pipe into the trapezoidal beam which forms the deck.

The side spans and the lower parts of the towers were cast *in situ* in conventional manner. The main span was cast *in situ* by cantilevering in 3 m sections and the whole deck was made continuous by a joint at mid-span, being anchored at one of the towers. The upper legs of the towers were precast in two halves and bolted together at the cross member after erection. Freyssinet flat jacks were inserted in the joints at the base of the towers for future correction of distortions in the bridge profile caused by creep and shrinkage. Each of the precast tower legs contains a concrete hinge just above the base.

The stay cables, which are anchored at both deck and tower, were formed from galvanized seven-wire strand. They are not cased and do not even have any protective

BARWON RIVER FOOTBRIDGE (PHOTO BY COURTESY OF THE CEMENT AND CONCRETE ASSOCIATION)

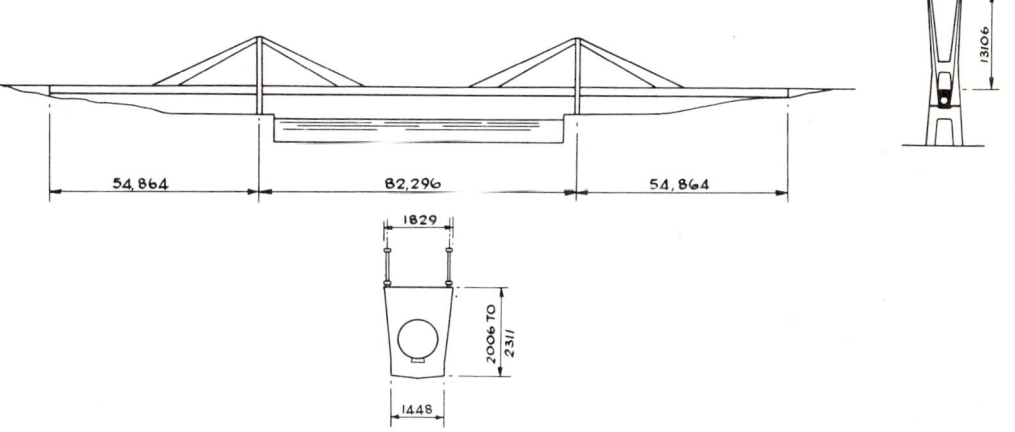

FIG. 7   BARWON RIVER FOOTBRIDGE

wrapping, although at the lower end each cable is threaded through a steel tube as a safeguard against vandalism.

With a main span length to breadth ratio of 45, this might be one of the most slender concrete bridges ever constructed and wind stresses in the deck must be significant.

RIVER PARANA BRIDGE, CORRIENTES, ARGENTINA (Fig. 8)

Although Morandi's name is not officially associated with this bridge, his influence appears to have been felt in its design. The familiar A-frame legs to the portal tower together with the splayed columns of the pier providing fixity to the deck appear again but in this case there are two backstays and two forestays from each tower.

The big change from, and almost certainly improvement over, previous Morandi designs is in the elimination of the heavy transverse beams at cable anchorage points. This has been achieved by adopting a deck section comprising twin box girders supporting the roadway slab and joined by 300 mm thick intermediate diaphragms at between 12 m and 16 m centres, the cable planes lying on the centrelines of the box girders.

The towers and splayed column legs are cast *in situ* together with a 53 m long section of deck symmetrical about each tower, complete with the roadway slab, which is formed from precast channel units spanning between the box girders, and the cantilevered footways.

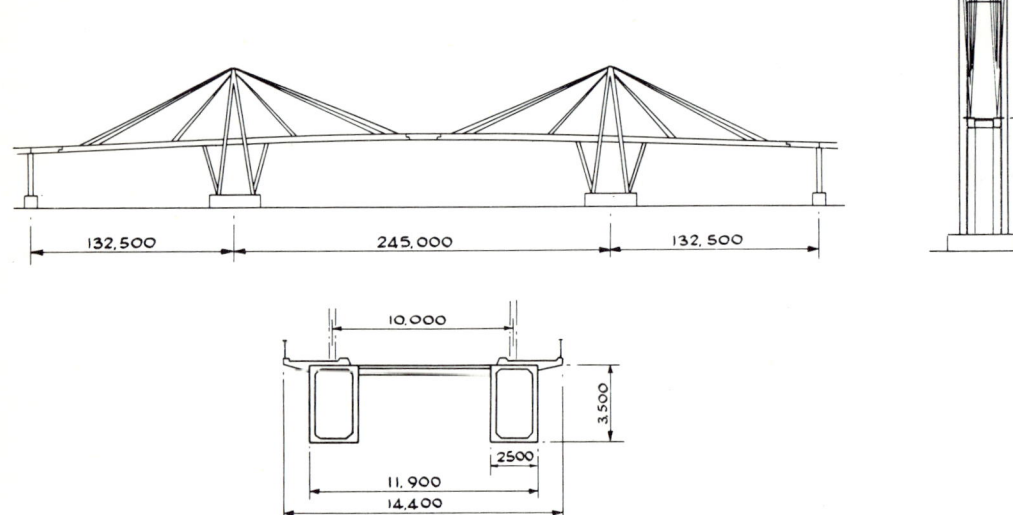

FIG. 8   CORRIENTES BRIDGE

The remaining 86 m of each cantilevered box girder is erected from twenty-five precast units, sixteen of which are 14 m long whilst those in the region of the cable anchorages vary from 1·5 m to 2·75 m. The units are erected in cantilever with resin bonded joints and a system of temporary support cables is employed to assist the cantilevering.

The main cables are spread and anchored in pairs in adjacent units so as to distribute the large concentrated forces as much as possible. The inner stays, comprising four cables, are anchored at points 61·25 m and 63·75 m from the towers and the outer stays, comprising six cables at points 100 m, 102·5 m and 105 m from the towers.

The first temporary cable is anchored 39·7 m from the tower and progressively stressed while construction proceeds a further 12 m, the second cable being anchored 51·7 m from the tower and progressively stressed while construction proceeds a further 10·5 m, at which time the first cable will have become unloaded and can be removed.

The first pair of permanent cables are now anchored and fully stressed and a further 13·5 m of deck erected, which will enable the second temporary cable to be removed. The second pair of permanent cables are then anchored, the load equalized between all four cables and the precast roadway slab and the cast *in situ* footway cantilevers are added up to a point 60 m from the tower.

A similar procedure is now repeated: the third temporary cable is anchored 76·5 m from the tower and stressed while construction proceeds a further 12 m, and the fourth temporary cable is anchored 88·5 m from the tower and stressed while construction proceeds a further 13·25 m allowing the third cable to be removed.

The first two pairs of the second permanent cables are then anchored and fully stressed and the remaining 8·75 m of cantilevered box girder erected, by which time the last temporary cable will have become unloaded and can be removed. The last pair of permanent cables are then anchored and the loads equalized and the precast deck channels and *in situ* footway cantilevers added to the end.

Finally, three 20 m long suspended spans are cast *in situ* simultaneously at each end of the two cantilever structures and the surfacing and railing applied throughout.

By following this somewhat complex sequence, it is apparently possible to avoid any temporary prestressing of the box girders and to obtain reuse of the temporary support cables.

The permanent cables are socketed and anchored by means of horseshoe washers at the tower. At the deck anchorages, the sockets are each retained by a crosshead which is jacked down a pair of high tensile threaded rods anchored to the beams.

Uncomfortably large openings, up to 940 mm x 1700 mm in plan, have had to be left in the top flange of the box girders to permit the passage of the sockets, crossheads and jacks at each anchorage position.

Immediately above the anchorage, the cables pass through a zinc-lined former to prevent slight changes of direction of the cables causing unacceptable bending in the rods.

WADI KUF BRIDGE, BEIDA, LIBYA (Fig. 9)

This is yet another long span bridge designed by Morandi. It is the largest concrete cable stayed span so far and is probably the third longest concrete span of all time.

The familiar A-frame portal and inclined pier legs support a single box girder of varying depth with a single pair of forestay and backstay cables.

These are attached to the deck at points 97·5 m from the tower by the equally familiar inclined transverse box beam, 6·5 m deep by 2·5 m wide and stressed with twenty-one cables of sixteen 5 mm wires even for this narrow bridge. The main girders extend a further 16 m beyond the cables to support a 55 m suspended span.

The piers, towers and main cantilever box girders are cast *in situ*, the latter in free cantilever assisted by temporary support cables attached at points approximately 57 m, 86 m and 93 m from the tower.

The main sidespan cantilevers extend to the abutments where they are anchored down through very large doubly-hinged rockers. The single central suspended span is formed from two precast I-beams lifted directly into position.

Nineteen wire strand and conventional proprietary prestressing anchorages are used for both temporary and permanent supporting cables, the former being taken directly through ducts in the webs of the box girder and systematically reused in the permanent stays. There are also some supplementary multiple wire tendons in the permanent stays, probably for the adjustment of stress in the casing, although they are rather surprisingly located at the top of the cables.

The main cables are cased using a most ingenious method comprising precast segmental comb units 1·740 m wide by 1·040 m deep which are offered up from

BRIDGE AT WADI-KUF, LIBYA (REPRODUCED BY COURTESY OF PROF. MORANDI)

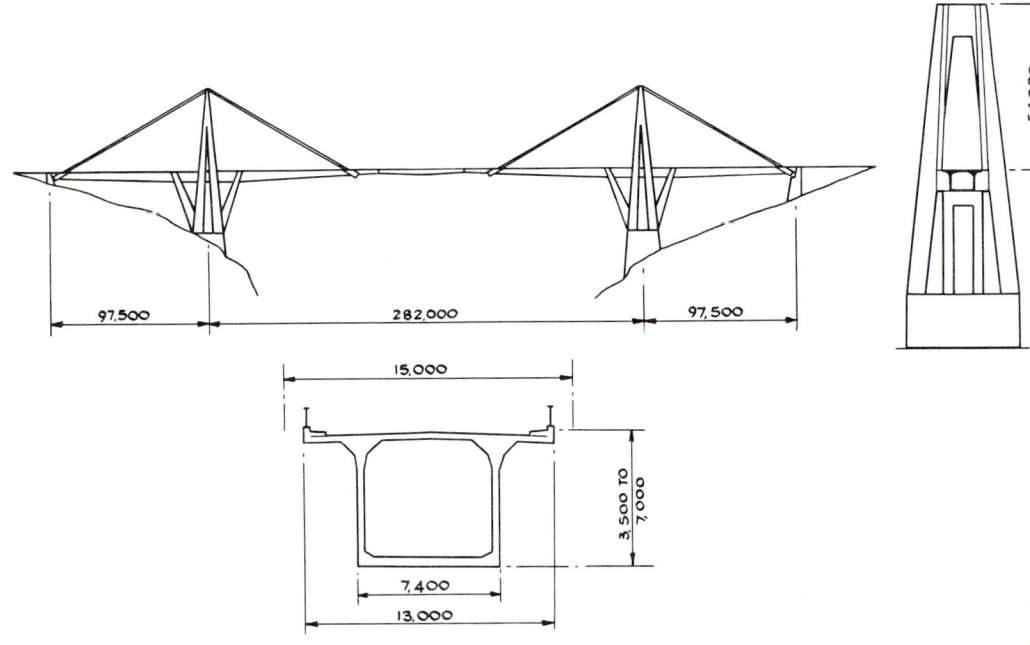

FIG. 9   WADI KUF BRIDGE

below. They are sealed by lids cast *in situ* and a number of such segments each 2m long are bonded together using epoxy resin. A 200 mm wide *in situ* joint every 10 m accommodates any casting tolerances.

The casings are then prestressed sufficiently to prevent tension developing in them under live load and the deep slotted ducts formed by the combs are grouted up.

RIVER WAAL BRIDGE, TIEL, HOLLAND (Fig. 10)

This is a road toll-bridge being constructed by a concessionaire over the river Waal which forms the main shipping route from Rotterdam and Europort to Germany, Switzerland and France.

A large clear span with no temporary works was therefore required. Furthermore during the winter the whole area is usually flooded for several weeks at a time and speed of erection was a prime consideration.

RIVER WAAL BRIDGE, TIEL, HOLLAND

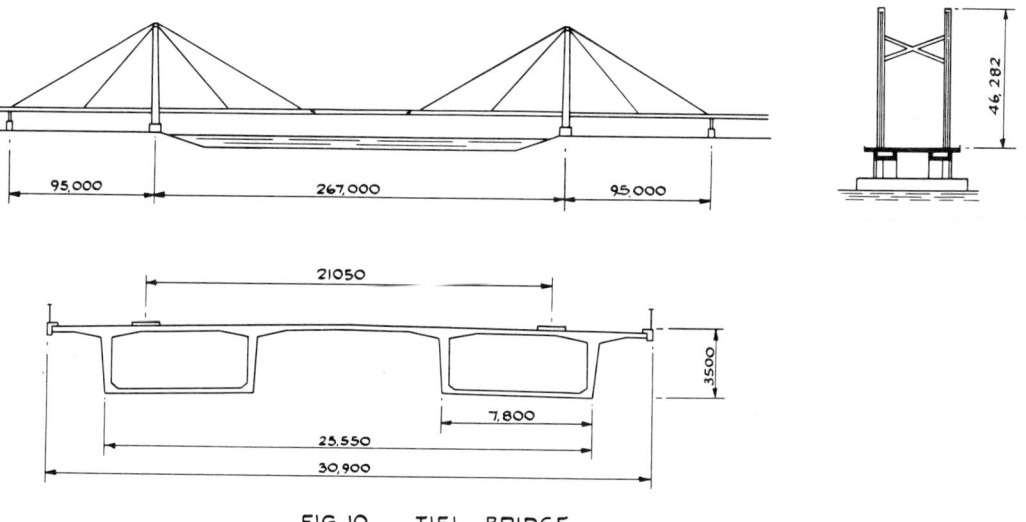

FIG. 10   TIEL   BRIDGE

The deck is formed from twin box girders connected by the deck slab and transverse diaphragms at the piers and cable anchorages.

The main cantilevers, which are supported by two forestay cables at 47·5 m and 95 m from the towers and extend a further 6 m to pick up the suspended span, are continuous not only with the 95 m side span but also with the first 77·5 m approach span on each side.

The lower backstay cables support the sidespans at their mid-points and the upper backstays are anchored over the first shore pier, so that their function is to stabilize the tops of the towers from which the other points are effectively suspended.

Because of this, the forestays and backstays, although of the same total size, are split into different proportions and it is necessary to divert some tendons from upper to lower stays in the tower saddles.

Dutch practice allows the slow lane to be physically separated from the two fast lanes of a three-lane carriageway and advantage has been taken of this fact to accommodate both the tower legs and the stays within the separating strip and thus enable the cables to be anchored, although eccentrically, within the box girders themselves.

On the north side the side span and first approach span are constructed from precast segments assembled on centering with *in situ* joints while on the south side they are wholly cast *in situ.*

The main span is then constructed in cantilever in 4 m sections, each supported by temporary stays from temporary pillars erected on the main piers. The 65 m long suspended span beams are precast in lightweight concrete.

The permanent stays are cased by a variation of the technique being used on the Wadi Kuf bridge. In this case precast hexagonal elements are assembled on scaffolding with *in situ* joints and the cables threaded through them, stressed and grouted.

Multiple strand cables are used for the permanent stays and longitudinal stressing of the box girders, multiple wire for transverse stressing of the deck and diaphragms and high tensile bars for the temporary stays and vertical stressing of the box girders.

MOUNT STREET FOOTBRIDGE, PERTH, WESTERN AUSTRALIA (Fig. 11)

This footbridge is included in a large road complex forming the northern approaches to the famous Narrows bridge. The interchange includes a number of prominent structures and this design was chosen to counter possible insignificance in this comparatively small one.

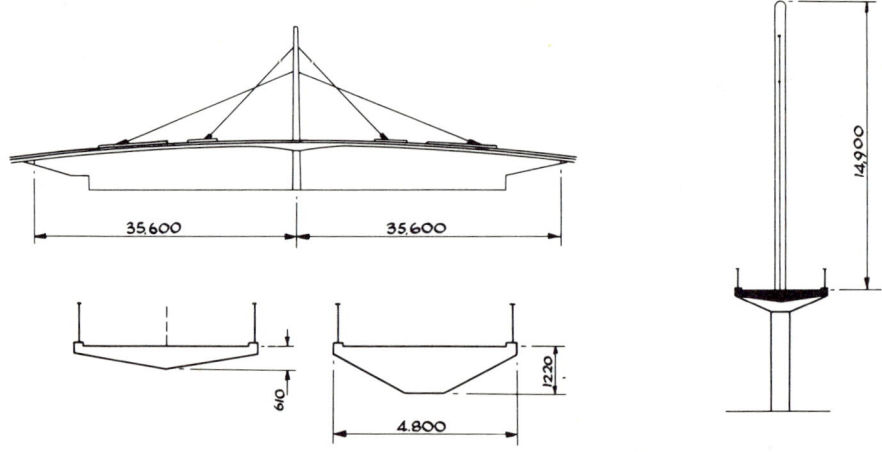

FIG. II   MOUNT STREET FOOTBRIDGE

Certainly the unusual cable arrangement can have been chosen only for aesthetic reasons since there is no structural advantage: equalizing the length of the cables is of no benefit when the inclinations are so different.

The solid prestressed deck section is cast *in situ* together with all substructure and the single central mast assembled from two precast units stressed down to the deck girder.

The parallel wire stay cables are anchored in conventional proprietary prestressing anchorages using dead end anchorages in the deck and jacking at the tower to which all cables are rigidly anchored.

The cable forces necessary to achieve a satisfactory distribution of dead load bending moment are very different from the static reactions, casting doubt on the long-term behaviour of the structure but its short-term design was confirmed by tests on a large one-fifth scale model constructed principally to verify its dynamic behaviour.

The tests indicated extreme sensitivity to the assumed stiffness of the tower and better correlation with the experimental results was obtained by assuming a hinge at the base of the mast in the analysis.

The results of the dynamic tests have not been published.

M25 OVERBRIDGE, LYNE, NR. CHERTSEY, ENGLAND (Fig. 12)

This, the only railway bridge in the list, presented the designers with a very difficult bridging problem. The railway is to cross the proposed motorway at an acute skew of about 62° from the normal whilst on a curve, headroom is extremely restricted if expensive approach works are to be avoided on either road or railway and the location is considered to be of such exceptional scenic beauty as to require a bridge of particular aesthetic merit.

M25 OVERBRIDGE AT LYNE, ENGLAND

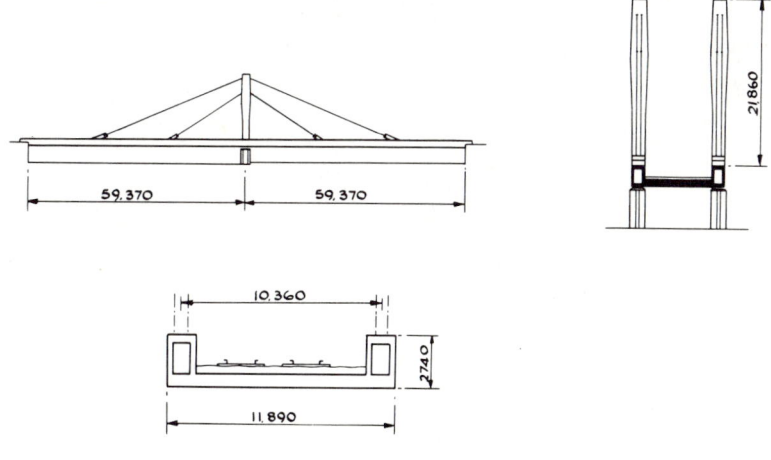

FIG. 12   M 25 OVERBRIDGE LYNE.

The solution chosen is a bold one although its construction will be simplified by the green field nature of the site and the ability to divert the line during construction.

Consequently *in situ* construction will be used throughout for the bridge which consists basically of two plane frames each comprising a continuous box girder supported from a single central mast by two forestay and backstay cables, the frames being connected by a slab at the level of the bottom flange of the girders which acts as the rail carrier.

The bridge is in many respects similar to the Pretoria pipe bridge on a larger scale, except that the deck is also prestressed.

All stay cables, which are parallel wires using conventional prestressing anchorages and are uncased, and deck tendons will be stressed simultaneously to control stresses in the tower and deck as the bridge is raised off its centering.

The stay cables will be jacked at the towers, the lower cables being anchored to the mast itself while the upper ones terminate in an anchor block mounted on rockers.

At the lower end, all cables have dead end anchorages in separate anchor blocks which are themselves stressed down independently to the deck to facilitate replacement in the event of damage to the cables due to a derailment or other cause.

RIVER FOYLE BRIDGE, LONDONDERRY, NORTHERN IRELAND (Fig. 13)

This bridge bears little similarity to any of the other designs although naturally it shares common features with several of them.

Like Ansa de la Magliana viaduct, the main span is supported asymmetrically from a single tower.

MODEL OF THE RIVER FOYLE BRIDGE

The deck is a torsion beam supported essentially on a single central cable plane as in the Mount Street footbridge.

The mast combines the features of an A-frame such as is used in the Obourg footbridge or the Pretoria pipe bridge with the single mast which appears in the Mount Street footbridge and, in a sense, in the Lyne railway bridge.

With these last two it also shares the common feature that the deck is fully continuous whilst it is similar to the Obourg footbridge in that the backstays anchor the tower to rigid points.

It will be seen, therefore, that the design employs many of the features of the smaller bridges at the scale of the larger ones.

The proposed method of construction, however, is apparently appropriate to its size since it contains much in common with the bridges at Corrientes, Wadi Kuf and Tiel.

The tower will be cast *in situ* up to deck level and the approach spans cantilevered 35 m into the main span at each end.

These approach spans are to be erected in balanced cantilevers using precast segments bonded with epoxy resin: a mid-span joint will be cast *in situ* and continuity cables placed in the soffit to cater for stresses due to superimposed loads. All deck prestressing cables run in internal ducts in either the top or bottom flanges of the box section.

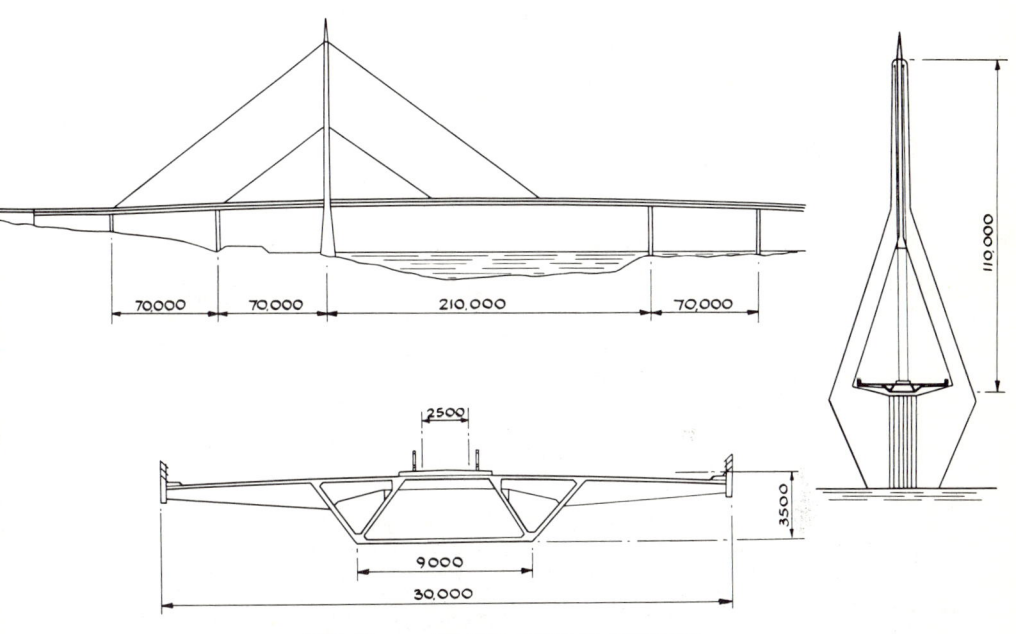

FIG. 13   RIVER FOYLE BRIDGE

The A-frame part of the tower will then be cast on falsework supported off the deck and the lower backstay cable placed in position.

The first temporary support cable will be fixed to the deck 35 m from the tower which will enable erection of the deck to proceed a further 35 m using temporary prestressing in the top of the section.

Meanwhile the mast part of the tower will proceed using slipform construction.

The lower permanent forestay cable is anchored 70 m from the tower and cantilevering can then proceed a further 35 m.

During this stage the temporary top prestressing will be replaced by the permanent bottom "continuity" cables and the first temporary support cable removed.

When the tower is complete, the upper backstay will be placed in position and the first temporary support cable, suitably lengthened, reused for the second temporary stay fixed to the deck 105 m from the tower.

Bridge	Completed	Type	Over	Place	Designer	Contractor	Spans (m)	Towers	Type	Height (m)
TEMPUL	1925	aqueduct	River Tempul	Jerez de la Frontera, Spain	Torroja		20·1–60·3–20·1	2	Portal	4·3
MARACAIBO	1962	road	Lake Maracaibo	Maracaibo, Venezuela	Morandi	Precomprimido & Julius Berger	160–5 @ 235–160	2	Portal (A-frame in elevation)	42·5
OBOURG	1966	foot	Canal du Centre	Obourg, Belgium	Vandepitte	C.B.R.	2 @ 67	2	A-frame (hinged)	20·0
POLCEVERA	1967	road	River Polcevera	Genoa, Italy	Morandi & Cherubini	Societa Italiana per Condotte d'Aqua	86–202–210–140	2	Portal (A-frame in elevation)	45·1
MAGLIANA	1967	road	River Tiber	Rome, Italy	Morandi	Sagi	53·6–145·0	1	Portal (hinged)	34·0
PRETORIA	1968	pipe	Johannesburg – Pretoria motorway	Pretoria, S. Africa	Van Niekerk, Kleyn and Edwards	Labour Constructions	2 @ 28·3	1	A-frame (hinged)	12·5
BARWON RIVER	1969	foot	River Barwon	Geelong, Australia	J. L. van der Molen	D. A. Constructions	54·9–82·3–54·9	2	H-frame (hinged)	13·1
CORRIENTES	under construction	road	River Parana	Corrientes, Argentina	Amman and Whitney & Porter and Ripa	Ferrocemento, Empresso Umberto Girola, Impressit & Sideco S.A.C.I.C.	132·5–245·0–132·5	2	Portal (A-frame in elevation)	47·4
WADI KUF	under construction	road	Wadi Kuf valley	Beida, Libya	Morandi	Construzioni Stradali	97·5–282·0–97·5	2	Portal (A-frame in elevation)	54·1
TIEL	under construction	road	River Waal	Tiel, Holland	Van Hattum en Blankevoort	Hollandsch Aannemingsbedrijf Zanen Verstoep & Van Hattum en Blankevoort	95–267–95	2	H-frame	64·3
MOUNT STREET	under construction	foot	Mitchell Freeway	Perth, Australia	Main Roads Department, W. Australia	Citra	2 @ 35·6	1	Mast	14·9
LYNE	–	rail.	M25 motorway	Chertsey, England	British Rail (SR)	–	59·4–59·4	1	Twin mast	23·3
FOYLE	–	road	River Foyle	Londonderry, N. Ireland	Mott, Hay & Anderson		70–210	1	A-frame and mast	110·0

	Stays	Cable Planes	Arrangement	Type	Size (mm)	Sheathed	Anchored at Tower	Stressed	Deck	Width (m)	Roadway (m)	Depth (m)	Articulation
(Tempul)	1	2	–	Steel wire rope	2 No. 63	No	No	Tower (Jacked)	–	–	–	2·1	Suspended Span (20·1)
(Maracaibo)	1	2	–	Locked coil wire rope	16 No. 74	No	No	Deck	3-cell rectangular box	17·4	2 @ 7·2	5·0	Suspended Spans (46·0)
(Obourg)	4	2	Radiating	Alternate lay wire rope	1 No. 49 / 1 No. 49 / 1 No. 35 / 1 No. 35	No	No	Deck	Double-tee	2·1	–	0·6	Continuous live load
(Polcevera)	1	2	–	Locked coil wire rope		Yes	No	Deck	5-cell rectangular box	18·0	2 @ 7·3	4·6	Suspended Spans (36·0)
(Magliana)	1	2	–	Parallel wire	76 No. 12/7 & 36 No. 4/5	Yes	No	Deck	7-cell rectangular box	24·2	22·6	varies 3·0–4·0	Hinged at Tower and Suspended Span (63·0)
(Pretoria)	2	2	Radiating	Parallel wire (Loeba)	1 No. 16/8 / 1 No. 12/8	No	No	Deck	U-section	4·8	–	0·9	Hinged at Tower
(Barwon River)	2	2	Fan	Strand	1/2″	No	Yes	Deck	Trapezoidal with circular duct	1·8	–	varies 2·0–2·3	Continuous
(Corrientes)	2	2	Radiating	Locked coil wire rope	6 No. 100 / 4 No. 100	No	Yes	Deck	Twin rectangular box	14·4	8·3	3·5	Suspended Spans (20·0)
(Wadi Kuf)	1	2	–	Strand & Parallel wire (CCL)	90 No. 1⅛″ & 18 No. 4/5	Yes	No	Deck	Rectangular box	13·0	11·0	varies 4·0–7·0	Suspended Span (55·0)
(Tiel)	2	2	Radiating	Strand (Freyssinet)	f'stays—40 No. & 12 No. 12/0·6″ b'stays—16 No. & 36 No.12/0·6″	Yes	No	Deck	Twin rectangular box	30·9	2 @ 12·4	3·5	Suspended Span (65·0)
(Mount Street)	2	1	Crossed	Locked coil wire rope		No	Yes	Tower	Trapezoidal	4·8	–	varies 0·6–1·2	Continuous
(Lyne)	2	2	Fan	Parallel wire (BBRV)	2 No. 139/7 / 2 No. 139/7	No	Yes	Tower	Twin rectangular box	11·9	–	2·7	Continuous
(Foyle)	2	2	Harp	Parallel wire (BBRV)	7 No. 195/7 / 7 No. 217/7	No	Yes	Tower	3-cell trapezoidal box	29·0	2 @ 11·3	3·5	Continuous

The cycle is then repeated until the cantilever extends 175 m from the tower with both permanent forestay cables in position and abuts the 35 m cantilever from the approach spans.

Finally, the set of temporary prestressing cables which have been used twice in the top of the section over the temporary support cable positions will be permanently installed as the bottom continuity cables across the *in situ* joint at this point.

Only the spine beam section, which provides the necessary high torsional strength to the completed superstructure, is erected in cantilever, the transverse ribs and deck slab being added later.

This enables an efficient section with a low neutral axis to be used for canti-levering, the slots for the ribs still further lowering the neutral axis locally to enable a reasonable compressive stress to be maintained at the joints whilst a moderate tension is permitted within the units themselves.

The permanent stays are parallel wire cables anchored in conventional pro-prietary prestressing anchorages: dead end anchors will be used at the deck with jacking at the tower through stressing rods.

Each stay is in fact two cables to give room for the anchorages and this arrangement conveniently allows for a central footway on the bridge.

Torsional deflection of the deck is not sufficient to create any significant difference in cable forces.

Jacking at the tower will allow stressing equipment to be concentrated at two jacking points rather than being dispersed to four positions, an important consideration when such large jacks are required and when a good deal of cable adjustment will be needed. It will also clearly serve to improve communications to maintain compatible forestay and backstay extensions.

The deck is anchored at the mainpier so that the tower is not stressed by expansion movement of the deck. The two adjacent piers to which the backstays are anchored each functions as a flexible strut and tie.

It is proposed to use high strength concrete having a characteristic strength of about 60 N/sq mm throughout the deck to reduce self weight to a minimum.

This is the most effective means of economy since self weight accounts for about 95% of the total load which the cables, tower and deck itself are designed to carry.

The quality of aggregates available in the area is very high and preliminary tests indicate that there will be no difficulty in achieving these strengths.

CABLE ARRANGEMENTS

Clearly the arrangement of the cables is only relevant where there is more than one forestay or backstay.

The radiating pattern of stays enables the minimum total cable force to be achieved for a given height of tower. On the other hand, the tower is an unstabilized strut carrying the maximum load for its full height. Because of the concentration of lead at the tower top, it is almost impossible to anchor the cables to the tower using a radiating pattern.

The harp arrangement enables each cable to be optimized, separates the tower anchorages and not only stabilises the tower but reduces the loads carried in each part except the lowest. It results in a larger total cable force but can lead to standardization of anchorage and other details.

The fan is a compromise between these two extremes or more often is likely to develop from a radiating pattern through the space required to accommodate tower anchorages.

The optimum cable volume for a given vertical reaction is obtained with an inclination of 45° but this does not take account of the cost of the tower.

This will tend to reduce the optimum angle but anchorages form a high proportion of the total cost of cables so that there is an incentive to keep the cable forces down by using as high as inclination as possible.

Since a concrete tower is a highly efficient compression member, it is unlikely that the totally optimized inclination can ever be as low as the 25° to 30° which appears to be favoured in steel cable stayed girders.

It is sometimes suggested that the horizontal component of the cable forces represents a gratuitous prestress in the deck but even at an angle of 30° this amounts to only about 0·1 N/sq mm per metre of span supported.

It is also, of course, arguable whether a uniform prestress is necessarily advantageous in a beam subjected along its length to complete reversal of moment.

CABLE PLANES

The total cable forces required to support the dead load of the deck are independent of the transverse location of the cables. However, two widely spaced planes of cables are a highly inefficient means of carrying non-uniform superimposed loads.

Provided that sufficient torsional strength can be incorporated in the deck to transfer asymmetric loads to a single central plane of cables, this will represent a significant saving in total cable forces.

If a central reserve is not an inherent part of the deck, however, the necessary increased width of the bridge may completely offset this.

The biggest saving of all will undoubtedly result from the elimination of the very heavy transverse beams which are necessary when the cables are anchored at the extremities of the bridge section.

It should be appreciated that the maximum bending moment in such a transverse beam is as great as that in the main longitudinal member if the width between cables approaches 40% of their longitudinal spacing, while the maximum shear is always as great whatever the proportions.

TOWER TYPE

The appropriate type of tower depends very much on the configuration adopted for the cables.

A tower having significant longitudinal stiffness is required for stability with a radiating pattern whilst maximum flexibility is needed with a harp arrangement, particularly if the cables are anchored to the tower.

Transverse stability must be ensured in either case although the transverse forces transmitted by the cables are small.

If the tower is anchored back to a fixed point it may be necessary for it to be effectively hinged at the base to prevent undesirable stresses arising from temperature movements, but generally the difficulty of construction will make it worthwhile to design out this feature.

In order to maintain tower members acting basically in axial compression, the transverse configuration of the tower should follow the cable planes: two legs, perhaps of a portal frame, for two cable planes, a central mast for a single central cable plane.

Unfortunately, a paradox arises in that a torsion beam type of superstructure is necessary with a single cable plane and a central mast will almost certainly prevent the continuity essential to this form of construction.

CABLES AND FATIGUE

The basic cost of parallel wire cables is very much lower than that of locked coil wire rope and the facility for direct anchorage without socketing greatly simplifies the detailing and operation of stressing anchorages.

Stranded wire rope is a compromise for which standard anchorages are readily available but the unit sizes are small, making the total anchorage assemblies very cumbersome.

The reluctance to employ parallel wire cables in steel cable stayed girders has arisen from the doubtful fatigue properties of the anchorages.

In a large concrete bridge, however, the stress range is likely to be only 5% to 10% of the total or even less, compared to typically 25% for a large steel bridge, and the problem is not significant.

CABLE CASING

For the same reason, sheathing of the cables in concrete casing is of limited value since, by reducing deflections, it only improves the behaviour of the deck under a very small proportion of its total load.

Casing actually increases cable forces and hence anchorage and tower loads.

It is undoubtedly one method of protecting the cables against corrosion but modern synthetic treatments are available which should prove equally effective.

AERODYNAMICS

A concrete cable stayed girder will have approximately the same natural frequency of vibration as a similar bridge in steel.

Consequently many of these bridges will be sensitive to resonant vibration at a critical wind speed within the normal range.

However, compared with an equivalent steel bridge, a concrete superstructure has two great advantages: its damping properties are usually much better and its mass is much greater.

Consequently, although vibrations may occur, their amplitude is unlikely to be significant: a low critical wind speed will not impart sufficient energy and a high critical wind speed will not remain steady for a sufficient length of time.

The cables, however, may suffer from aerodynamically induced resonant vibrations.

Aeolian vibrations of a single cable may arise under a critical combination of cable mass, length and tension and wind speed. A smooth cylindrical section is the most sensitive.

If two separate cables are placed close together as in the Corrientes, Foyle and Lyne bridges, "buffeting" of the leeward cable may be caused by vortices shed alternately from the upper and lower surfaces of the adjacent cable.

In this case, however, vibrations can be eliminated by using rigid intermediate spacers between the cables to increase the natural frequency and raise the critical wind speed beyond the range of steady flow conditions.

CONSTRUCTION

The method of cantilevering is obviously appropriate to the erection of the deck of a large cable stayed bridge over water or otherwise where *in situ* construction is not practical.

There is, however, a problem in accommodating the reversals of bending moment which must occur between erection and service conditions at sections between the cable support positions, a notable distinction between steel and prestressed concrete stayed girders.

This can be overcome by using temporary stay cables or temporary prestress or a combination of the two.

TEMPORARY CABLES

In construction of this kind, necessary temporary support cables fall into the no-man's-land of responsibility between the designer's permanent works and the contractor's normal temporary works.

On the whole it is probably advisable to leave the method of support to the discretion of the contractor and make no special provision in the permanent structure at the design stage.

At both deck and tower, suitable external steel fabrications can be devised in which to anchor the temporary stays and it will usually be possible to leave temporary holes in the concrete if necessary for the passage of the cables.

THE FUTURE

There can be little doubt that a span of over 300 m will very shortly be built in this form of construction.

It is significant that at least three of the prize winning designs in the Great Belt bridge competition were for concrete cable stayed spans up to 484 m in length.

This included a scheme by E. W. H. Gifford and Partners for a span of 450 m carrying both road and rail and it must be remembered that in his paper to the Institution of Structural Engineers, as long ago as October 1962, Mr. Gifford referred to a proposal for a 458 m span cable stayed concrete bridge across the Solent.

Morandi continues prolifically to produce novel designs for major crossings, particularly in South America, and has undoubtedly gained more experience than anyone else in this form of construction.

There would appear to be absolutely no reason why, given the right site, two river Foyle bridges could not be placed back to back to produce a span of 350 m.

Dead load is so predominant that the design would be virtually unaffected and the method of construction would remain quite unaltered.

ACKNOWLEDGEMENTS

The author is grateful to Mott, Hay & Anderson for permission to publish this paper and is indebted to the following organizations for assistance in compiling the descriptions of the various bridges: Van Niekerk, Kleyn and Edwards; J. L. Van der Molen; British Ropes Ltd; Cable Covers Ltd; Van Hattum en Blankevoort N.V.; Maunsell and Partners (Pty.) Ltd; Simon Carves (BBRV) Ltd; British Rail (Southern Region).

SOME ASPECTS OF MODERN POST-TENSIONING

J. E. LONG

PSC Equipment Ltd

As with most fields of Structural Engineering, post-tensioning techniques and equipment are being constantly developed. The object of this paper is to review some of these developments, to highlight certain site difficulties, to comment upon common Specification clauses, and to present a few examples of interesting recent applications.

Wire and Strand

Low relaxation wire and strand, of normal strength, is now commercially available, in sizes up to 18 mm diameter, in which the relaxation loss is 2% or less, after 1000 hours at 20°C, from an initial load of 70% of the material breaking strength. Strand of similar relaxation, but with increased breaking load, known in North America as 270K grade, is also available in sizes up to 15·2 mm diameter, and can be used in standard anchorages.

For all these low relaxation materials, temporary overstressing appears to have little or no value, at least from the point of view of relaxation, although many Specifications still call for it. Indeed, temporary overstressing is of questionable merit when applied to stress-relieved materials of normal relaxation, since locking off at a slightly higher load is generally preferable. Temporary overstressing can extend the actual stressing time by one-third, so there ought to be a definite case for it, and it is impossible with some modern systems having automatic 'lock-off'.

With the exception of 'Dyform' strand, none of these high strength or low relaxation materials is visually distinguishable on site from the standard product, which means that mixing them on a job may introduce complications.

The use of large 19 wire non-stress-relieved strand, with diameters over 22 mm, appears to be diminishing, and only one British manufacturer now produces systems for it. Whereas stress-relieved strand cuts readily, and can be rewound quite satisfactorily if any of the individual wires become displaced, this material has to be bound before cutting and cannot be rewound, so that Specifications calling for the rejection of unravelled strand ought to distinguish between the two.

This material plays off the reel with a pronounced curvature, and may be more susceptible to corrosion than stress relieved strand.

A fairly considerable torque is induced in these large strands during stressing, which necessitates a special detail in the jack to prevent rotation of the piston. The limit of proportionality is low on first loading, although the load/deflection relationship differs on subsequent loadings, making it difficult to correlate load and extension on site.

Wire and strand are now available coated with PVC or polypropylene, either as a tight protective layer or as a loose sleeve. Neither arrangement can be stressed

478

directly, and the coating has to be removed to attach the jack and permanent anchorage, thus introducing a corrosion protection problem. Loose sheathed strand is widely used in North America for flat slab construction and there is at least one application as stressed stirrups in a bridge, but in this country, the major application to date appears to have been in external cables, passing around deflectors. Test work carried out by P.S.C. Equipment Limited on p.t.f.e. coated deflectors of 1·4 m radius, deviating the strand through 0·1 radians, indicates that there is a danger of the plastic coating cracking in the vicinity of the deflector. Although the strand may at first move through the sheath, eventually the sheath itself moves on the deflector. Both of these effects have been seen on site, the latter resulting in puckering of the plastic between deflectors, but the friction loss during stressing is much reduced from that occurring in conventional ducting.

Obviously, the coating should not be relied upon to provide corrosion protection, especially in the vicinity of the deflectors and the anchorages. The long term durability and fire resistance of these plastics coating remains to be established, in particular there appears to be a risk of the release of chlorine from PVC when heated, which would cause rapid corrosion of the prestressing steel.

Systems

The trend is towards larger systems, and cables of 4000 kN are used fairly regularly in bridges (Fig. 1). Even larger cables, up to about 10,000 kN are available, although at present their use has been restricted largely to nuclear pressure vessels. Jacks for these cables are themselves of considerable size, and normally have to be handled with cranes.

FIG. 1 STRESSING 19/15·2 mm DYFORM TENDONS TO ABOUT 4400 kN. (JACK CAPACITY 7000 kN, MASS 800 kg)

The economics of these larger systems is far from clear cut. Cost comparisons of those components and operations, peculiar to post-tensioning, indicates that on this limited basis the advantage may rest with quite small systems. But these costs may themselves represent perhaps only 5% of the total cost of the bridge, so that the wider implications, such as speed of operation and rapid release of formwork must be considered, and it is on this basis that the larger systems show benefits.

Additionally, there may be sound technical reasons for choosing a large system, for example, it may be easier to achieve eccentricity with one large cable, than with a number of smaller ones, and elastic losses may be eliminated in some cases.

It is virtually impossible to check the load measuring devices on large systems completely on site. If the job is sufficiently important and extensive to warrant the expenditure, it is possible to construct a rig for the purpose, and, with some systems, two jacks can be operated nose-to-nose, one actively and one passively, if the characteristics of the passive jack are known. However, major changes in jack performance usually result from mechanical damage, which is obvious, and the necessity for daily checks, as required by some Specifications, is questionable. Steady minor changes in a jack can be detected satisfactorily on the basis of no more than two or three calibrations in a year. Generally, the jack performance is likely to be known far more accurately than the modulus of elasticity of the concrete, for example.

However, pressure gauges can be damaged fairly easily and a periodic check on these is desirable, using perhaps a master gauge or even a dead weight tester. Many pumps have two simultaneously reading gauges, as an insurance against the malfunction of one gauge.

FIG. 2 COUPLERS FOR 12/15 mm TENDONS

Cable coupling in continuous structures is increasingly popular (Fig. 2) and even the largest systems in bridges now can be coupled. In the greater majority of cases, it seems most unlikely that the first stage anchorage will lift, when the second stage cable is coupled to it and stressed, so that the first stage cable is frequently grouted shortly after stressing. Certainly this is desirable from the safety aspect.

A Gaussian distribution of anchorage static rupture strengths is found. The method of test is outlined in BS.4447:1969, and, although no figures are given, it appears that the characteristic strength of any system should be taken as 90% of the rupture strength of the tendon.

When it is considered that tendon failures can be anticipated to occur at lower loads in 1 in 20 of the anchorage, it can be seen that the practice of stressing above 80% becomes inadvisable.

It should also be remembered that a fretting or fatigue failure can occur, if the load in an unbonded tendon fluctuates at the anchorage. BS.4447:1969 specifies that an anchorage should be capable of resisting 1 million cycles of loading $60 \pm 2\frac{1}{2}\%$ of the characteristic strength of the tendon. Testing has so far not been extensive, but raising the mean or the range shortens the fatigue life, as would be anticipated. Obviously, this is of great importance in an unbonded application in a bridge.

The design of the reinforcement for end blocks still seems to give rise to problems, and so much steel is provided, on occasions that it is difficult to place the concrete. P.S.C. Equipment Limited, have used the design method due to Zielinski and Rowe, since its publication, for all sizes of tendons with complete success. Because their research work was performed with anchor loads of about 800 kN, there may be natural reservations about applying the method to loads perhaps ten times greater. However, end block reinforcement details have been designed and successfully tested,[1] for a single cable of about 11,000 kN rupture load, symmetrically placed in the prism, although the method appears slightly less conservative for larger systems. Except where major errors of design or placing have occurred, failure in an end block normally consists of cracking in the cover concrete, which can be broken away safely and made good.

Cable Making and Placing

In some cases it seems possible to distinguish between cutting wires and strands for cable making and after stressing, although few Specifications do this. For some large systems, the last 1·5 m of cable is simply for attaching the jack, and the localised change of elastic properties caused by flame cutting may be acceptable, provided that the correct precautions are observed during cutting. Certainly, flame cutting in the vicinity of the anchorages is hazardous, and anchorages have been damaged doing this.

Made up cables have a natural lay, and in the case of strand this means that the entire cable rotates through 180° in a distance of about 10 m − 12 m, so that the strand which began on top is then underneath. There is no doubt that this also occurs in circular ducts, and is probably accentuated by undulations in the cable profile, regardless of the presence or absence of spacers. However, it is not detrimental, particularly when the cable is stressed as a single unit, and the effort expended on some sites in attempting to avoid twisting is really wasted and largely ineffective. Also, the practice of identifying individual wires or strands, in a cable stressed as a unit, as required by some Specifications, seems unnecessary.

Whenever possible, it seems preferable to thread the cable in the duct before concreting. Cable threading can damage the duct, and, if found before concreting, such damage can be rectified. Also, the additional weight and rigidity of the cable gives the duct greater stability during concreting, and the cable can be freed by pulling it to and fro, if the duct should leak slightly.

Large cables are heavy, perhaps 35 kg/m, so that careful consideration is needed in their fabrication and handling. Various techniques have been evolved for drawing the requisite number of wires or strands, directly from reels, through a former and into the duct, using a winch. In cables composed of several rings of strands it is generally advisable to make the inner strands slightly longer, to keep them in their correct positions and to facilitate identification and anchorage threading. With some jacks, these central strands have to be longer for temporary anchoring.

Cable making and threading is fairly commonly carried out on the Continent by the prestressing specialist sub-contractor, but this practice has only just started in this country.

Ducting

It is easy to show that many ducts are in a state of near hydraulic balance during concreting, that if the weight of duct and enclosed strand only just exceed the weight of concrete displaced, but, in spite of this, most reported duct displacements are downwards. Depending upon the flexibility of the duct, adequate support and positive location is required, at spacings probably not exceeding one metre.

The duct alignment ought to be checked after cable threading, if this is done before concreting.

At present, there is no Standard for the performance of ducting. One of P.S.C.'s associated Companies carried out some interesting preliminary tests, in which several types of commercial prestressing ducting were subjected to repeated reverse bending, internal hydraulic pressure when externally unsupported, loading to produce identation under a reinforcing bar, and loading when spanning empty as a beam. For example, some spirally wound duct leaked at an internal pressure of about 0.15 N/mm^2 and seams opened at 0.40 N/mm^2. The weight of a man standing on an 8 mm diameter reinforcing bar would completely bury it sideways into this duct, and, when empty and spanning one metre, the duct would collapse under a central force of 45 N

A recent publication by the Western Concrete Reinforcing Steel Institute,[2] on post-tensioned box girder bridges construction in California, shows that the friction coefficients achieved on site with various types of duct, are $u = 0.25$ and $K = 0.0007$ per metre, the latter value in particular being about one-fifth of typical British values. If the recommendations on duct placing from the same source,[3] were followed, this may in some measure explain the improvement, since great emphasis is placed upon the close examination and testing of the duct at all stages of construction. For example, an air test is recommended on the duct both before and after first lift concreting. Of the holes detected before concreting, those less than about 5 mm diameter can be sealed by moderate wrapping with waterproof tape, but larger holes should be repaired with a split metal sleeve taped on to the duct. Indentations should be cut out and treated as large holes. Obviously, it is possible to carry such duct examination to extremes, but it is money well spent in avoiding subsequent difficulties in stressing and grouting.

Excessive wrapping with tape should be avoided, particularly in the vicinity of the anchorage, because failures have occurred in cases where the anchorage has been supported largely on tape, with insufficient bearing area on concrete. With most systems, the duct and anchorage should be co-axial for a distance of about 10 times the duct diameter behind the anchorage, to avoid anchoring problems.

Rusting of the duct, and also of the strand, can cause almost 50% increase in friction in extreme cases, and this has been substantiated by laboratory experiments,[4] and site trials.[5] Careful storage of materials is the prime requirements, but significant rusting can continue after concreting if stressing is delayed, and the use of a vapour phase inhibitor in the duct then should prove beneficial.

Water-soluble oils, detergents, graphite powders and dispersions, greases and plastic linings have all been used successfully in ducts to reduce friction. Some of these are pumped into the duct and applied fairly indiscriminately to the tendon as well, so that they have to be washed out after stressing in order not to impair the grout bond. There is no bond between grout and plastic lining to ducts, but simple tests with corrugated ducting appears to indicate that the mechanical key of the grout, due to the shape, and profile of the duct, provides a more than adequate substitute.

Continental practice in segmental bridge construction seems generally to be to match cast each segment against its immediate neighbours, and then glue them together on site, with only a thin joint between.

In most segmental bridges constructed in this country, a fairly wide gap is left between segments, which is later packed with dry mortar or similar filler. One major problem is ensuring continuity and alignment of the duct across this gap. Inflatable rubber ducting has been tried, but this has tended to bulge in the gap and so interconnect ducts within the filler. Another technique is to leave the steel ducting projecting from the precast segments and then bridge the gap with a standard duct coupler, but the projecting ends are damaged in handling, and, in the event of any misalignment between segments, the coupler cannot be fitted. A recent development has been a variable length plastic unit, with sufficient articulation to accommodate minor misalignment.

It is normal to provide an air vent at all high points in a length of duct. Less common is the practice of providing a drain point in the first valley, to prevent the duct filling with rain, when the anchorage is located obliquely in a deck pocket, although this is equally important. Development is in hand of a site technique for providing a vent at any point in a completed length of duct.

Stressing

Few site operations can give rise to as much discussion as the attempt to reconcile calculated load with measured extension, in post-tensioning. Part of the trouble seems to stem from the fact that there is no predetermined acceptable degree of correlation. By inference, A.C.I. Standard $318 - 63$ suggests that the measured extension can be accepted if it is within 5% of the calculated value, which seems reasonable. By far the commonest cause of discrepancies is excessive duct friction, but the influence of other factors should not be ignored, for example steel is sometimes received in which the modulus of elasticity varies by 10% or more from the average value commonly used in the calculations.

Jack and anchorage friction can account for between about 3% and 8% loss of load. In a number of systems, it does not appear to vary directly with load, as suggested by CP 115:Part 2:1969, and, when expressed as a percentage loss, can even vary inversely with load. An average figure is best determined for each system, although it can only be an average, due to variations in the surface condition of the tendon and anchorage, and in the jack performance. Even load cell readings on site cannot eliminate the anchorage effect.

Obviously, extension measurements have to start at a small initial load, to eliminate slackness in the tendon, and some correction has to be made for the initial extension achieved but not measured. One method of doing this, is to measure the extension obtained at low load in several small increments, and then assume that the extension achieved in the early increments of load is the same. In some systems, extension measurements include the contribution due to a short length of tendon actually outside the anchorage, and due allowance may have to be made for this, and also for the 'pull-in' at the dead end where appropriate. In this latter connection it should be remembered that the pull-in increases roughly linearly with load and, where duct friction losses are high, the dead end load may be small, resulting in low pull-in. Whenever possible, the actual value should be measured.

Grip 'pull-in' at the stressing anchorage seems to cause some difficulty, apparently from the mistaken belief that it is an absolute quantity. In fact, if a large number of stressing operations are monitored, a Gaussian distribution of grip pull-in results; for example, the average value for over 1230 anchorages on one site was 8 mm, with a standard deviation of 2 mm, but values as low as 1½ mm and as high as 19 mm were recorded. Often restressing is called for, in specifications, when the pull-in exceeds a certain figure. Due to reverse friction effects, it appears that the loss of load will be confined to the vicinity of the anchorage, and in the majority of cases will not affect the critical section. Restressing to the same load, or even a higher load, will produce much the same pull-in, so that the real value of the exercise should be carefully considered.

In multi-wire and multi-strand tendons, in which the elements are stressed singly, it is normal for those stressed last in each anchorage to have lower extensions than those stressed first. Presumably, this is due to a progressive build up of lateral pressure between the elements and the side of the duct, which can cause up to one-third increase in curvature friction, and has been reproduced in laboratory experiments.[4] This effect is less marked, but not necessarily eliminated, when spacers are used.

To avoid slips, wedges should always be clean, free from rust, well lubricated, and of the correct size. Nevertheless, slips and breakages may well occur and there should be some acceptable level of loss predetermined. A.C.I. Standard 318 − 63 requires that the total loss of prestress due to unreplaced broken tendons must not exceed 2% of the total prestressing force.

Grouting

Many specifications give some grout characteristics, such as bleeding, reabsorption, and compressive strength, in great detail, and much effort is expended on site in producing a mix to comply with these requirements, which may well be important. However, characteristics which seem at least as important for example fluidity and volume change, are seldom specified.

Interest can tend to wane during the actual injection process, which is often less rigidly specified, but is of paramount importance. For example, it has been shown[6] that the influence of the strength of the grout, on the ultimate strength of a post-tensioned beam is small, and it has been suggested that adequate corrosion protection is provided once the tendon is encased in grout, and the actual grout properties are less important. But, in both cases, the grout has to be dense and in the correct place. Within limits, it appears that rigorous and restrictive specification of the grout properties is not necessary, and perhaps more attention should be paid to the techniques involved.

Much useful information can be obtained monitoring the pressure rise during grouting. This should be at a reasonably steady rate, which is not calculable unfortunately, but the general trend should be clear after two or three ducts have been completed. A general, but abnormally rapid rise, can be due to stiffening grout, a sudden rapid rise indicates a blockage, and a drop to an abnormally slow rise can result from a duct leakage.

The time to discover a blockage is obviously before grouting and not during the actual operation, so that some preliminary effort ought to be made to establish that the duct is clear. Opinion is divided on the merit of flushing through steel ducts with water. Certainly it tends to collect in the lower corrugations and cannot be removed completely by subsequently blowing through with air, although it is not clear whether it is then forced out by the grout and merely mixes with it over a short distance, or whether the mixing is more general and likely to impair the bond. However, the presence of a general film of water on the walls of the duct may reduce drag during grouting, and prevent loss of water from the grout.

If grouting is carried out at high pressures, there is a tendency for the grout to segregate, leakage becomes excessive and the grout may rush past, and fail to fill voids. The normal maximum is 1 N/mm^2.

It is difficult to ensure that a duct is fully filled, and air will always be trapped in small pockets in corrugations. Some Engineers calculate the theoretical volume of grout required to fill the duct and then check this against pump delivery rate and time of pumping, but wastage is high and this method cannot be reliable. Another technique sometimes specified is to seal all outlets, when grouting appears to be complete, and raise the pump pressure to the maximum permissible value. Experience shows that the amount of additional grout introduced is small, indeed in a long duct the maximum pressure may be reached during the actual pumping operation, and water can be expelled from the grout. Trapped air may simply compress, and then expand, when the pressure is removed.

An expanding additive may be of value in producing a controlled volumetric expansion of the grout in the duct to fill any small voids. The commonest material used is aluminium powder, which reacts with the alkali present in the grout to produce hydrogen, and this still raises the bogey of hydrogen embrittlement of the tendon. However, as long ago as 1964, the Prestressed Concrete Development Group issued a notice to its members, stating in effect that the molecular hydrogen produced by this reaction would not attack the steel.

It seems highly desirable to have separate flushing equipment standing by, in case any difficulty is experienced during grouting, but rather surprisingly this is rarely specified.

Cement grout is still by far the commonest material, probably due to the high cost of chemical resins. Some work undertaken by P.S.C. Equipment, to secure post-tensioning cables at the foot of columns, indicates the value of polyester resin as a grouting material. Both in works trials and on the site, it was demonstrated that this material could be deposited at about 0°C, in wet vertical ducts of 105 mm diameter, either by tremie tube or by conventional grout access points at the foot of the column, and the full prestressing load, in some cases over 3000 kN, could be carried on a bonded length of less than 2 metres.

Tie Backs

The application described above is closely analogous to the construction of tie backs, and other similar anchoring devices, using post-tensioned cables. A hole is drilled into the ground, often under-reamed at the lower end, into which a tendon is dropped. This is then grouted over a sufficient length to produce the requisite bond, normally with cement and water, although chemical resins can be used in particularly wet or aggressive surroundings. The top end of the tendon is anchored into the main structure, using fairly conventional post-tensioning techniques, and then stressed, to induce the necessary holding down force. Finally, the hole is grouted to provide corrosion protection.

Stressing is normally carried out only to about 50% of the characteristic strength of the tendon, which means that an overload test can be safely performed to 75% of the characteristic strength, to prove the security of the system.

Future Trends

It is difficult to see post-tensioning systems growing much beyond their present size, since the capital cost of developing a system of say 15,000 kN would be considerable. Most of the stimulus for the development of large systems results from the construction of pressure vessels in nuclear power stations, where the need for heavy force concentrations arises. The consumption of post-tensioning equipment in one modern station is roughly equivalent to one year's entire consumption in British bridges.

Doubtless, even the very largest systems will be used eventually in bridges, in fact designs have been prepared incorporating 10,000 kN tendons. If the present pattern continues some of the smaller existing systems will die a natural death.

The impact of metrication upon post-tensioning is likely to be slight. The larger wires have always been in metric sizes, whereas strand sizes have been in Imperial dimensions. The normal strand sizes will remain unchanged, although described in metric units, but the first steps are being taken towards the rationalisation of sizes and breaking loads throughout Europe.

Dyform strand will be metricated this year, with generally increased breaking loads. Although the full implications are not clear, because no tests have been performed, it seems certain that the existing anchorages will continue to be satisfactory, because the increases in strand diameters and breaking loads are relatively slight.

REFERENCES

1. Harvey, J. T. C. New developments in the Freyssinet System. 6th International Congress, Fédération Internationale de la Précontrainte (Prague, 1970).
2. Western Concrete Reinforcing Steel Institute. *Post-tensioned box girder bridges: design and construction.* Burlingame, California, 1969.
3. Prestressed Concrete Manufacturers Associaton of California Inc. & Western Concrete Reinforcing Steel Institute. *Recommended practice for grouting post-tensioning tendons.* 1967.
4. Wyatt, K. J. *Measurement of friction on corrugated curved prestressing ducts.* Commonwealth Experimental Building Station, Australia, Technical Record No. 332.
5. Nielson, H. K. and Nilsson, B. A. John F. Kennedy tunnel, Antwerp. Coefficient of friction for frictional loss of prestressing force, measured on about 2000 cables. 6th International Congress, Fédération Internationale de la Précontrainte (Prague, 1970).
6. Soroka, I. and Geddes, J. D. Cement grouts and the grouting of post-tensioned prestressed concrete. *Bull. Dep. civ. Engng Univ. Newcastle,* No. 26, 1965.

AN ULTIMATE LOAD METHOD OF DESIGN FOR PLATE GIRDERS

K. C. ROCKEY

University College, Cardiff

1. SYNOPSIS

The paper presents an ultimate load method of design for plate girders having webs reinforced by both transverse and longitudinal stiffeners and subjected to both shear and bending. In addition to dealing with the design of conventional girders in which both web and flanges have the same properties, it can deal with the design of hybrid and composite girders. One important feature of the design method is that it allows for the influence which the rigidity of the flanges has upon the post buckled behaviour of the webs. The method has been checked against available test data and found to accurately predict the ultimate load of the girders.

Symbols

t	thickness of web plate
t_f	thickness of flange plate
d	clear depth of webplate between flanges
b	clear width of webplate between stiffeners
$\alpha = \dfrac{b}{d}$	aspect ratio of panel
I	flexural rigidity of flange members about an axis passing through their centroid and perpendicular to the web plate
c	position of plastic hinge, see Figure 2.
V_B	ultimate shear load provided by Basler collapse mechanism
V_{exp}	experimental ultimate shear load
V_{ult}	theoretical ultimate shear load
V_u	ultimate shear load for pure shear case only
W	applied load
W_{ult}	theoretical ultimate shear load
W_B	collapse load according to Basler mechanism
W_{exp}	experimental ultimate load
M	applied bending moment
M_u	applied bending moment to cause collapse when acting alone
τ	applied shear stress
τ_{cr}	critical shear stress $= K \dfrac{\pi^2 E}{12(1 - \mu^2)} \left(\dfrac{t}{d}\right)^2$ where K is a non-dimensional parameter

τ_{cre}	reduced critical shear stress — see equation (8)
τ_{yw}	shear yield stress of web material
τ_{ult}	ultimate shear stress
σ_{yw}	tensile yield stress of web material
σ_{mc}	maximum comparison stress in Huber Von Mises plasticity condition
σ_{yf}	tensile yield stress of flange material
z_f	plastic modulus of flange
E	Young's Modulus of Elasticity
μ	Poisson's ratio
θ	inclination of diagonal of panel with respect to flanges
σ_{crb}	compressive critical bending stress for a panel loaded in pure bending
σ_{crc}	critical direct stress for a panel loaded in pure compression
$\sigma_{mb},$ $\sigma_{mc},$ τ_m	compressive bending stress, direct compressive stress and shear stress which when acting as a combined system cause buckling

All other terms are defined as they appear in the text.

2.1   Introduction

The economic design of plate girders frequently involves the use of thin web plates reinforced by longitudinal and vertical stiffeners. Since more often than not these stiffened webs will buckle before the girders collapse, the use of ultimate load methods of design are essential if full advantage is to be taken of the post buckling load carrying capacity of the web plates. The present paper presents an ultimate load design method for plate girders which has been developed from an extensive study into the ultimate load behaviour of plate girders.[1-5]

An extensive study of the behaviour of shear webs carried out by Skaloud and the writer[1-3] has shown that a shear web may fail in a number of different ways, see Fig. 1. From Fig. 1 it will be noted that if the webplate has no initial imperfections then prior to buckling it does not impose any lateral loading upon the boundary members. However, once the web has buckled it is no longer capable of carrying any further compressive loading across the diagonal *ab* and as a result the web has to carry all additional shear loads by a diagonal tensile membrane action, this action being referred to as a 'truss type action'. This membrane action imposes a lateral load upon the flanges and it is possible for this membrane action to cause the flanges, and therefore the girder, to fail due to the development of plastic hinges in the flanges. Whether or not this type of failure will occur depends upon the stiffness of the flanges and upon the magnitude of the membrane loading which varies with the elastic buckling stress of the web; with an increase in the buckling stress there is a decrease in the membrane action. If the flanges are sufficiently stiff then the boundary members comprising of the flanges and the stiffeners permits the web to develop a full membrane action until it yields. Once this stage has been developed, the web cannot carry any further shear load and any further loads imposed upon the girder have to be carried by the flanges and stiffeners acting as a Vierendeel frame.

The first significant step in the development of plastic methods of design for plate girder webs, loaded in shear, occurred when Basler and his associates at Lehigh University[6-9] presented their design method some ten years or so ago. They

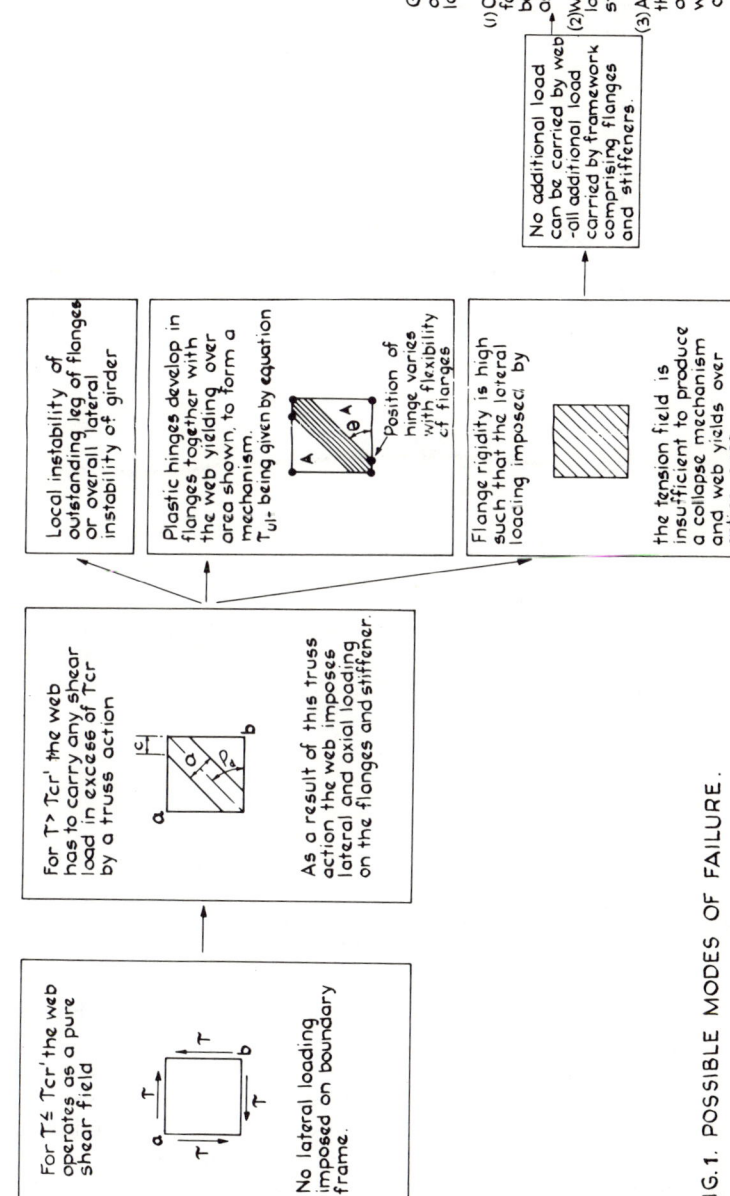

FIRST STAGE

SECOND STAGE

THIRD STAGE

PURE SHEAR BEHAVIOUR

POST BUCKLED BEHAVIOUR

COLLAPSE MODES.

For T ≤ Tcr the web operates as a pure shear field

No lateral loading imposed on boundary frame.

For T > Tcr the web has to carry any shear load in excess of Tcr by a truss action

As a result of this truss action the web imposes lateral and axial loading on the flanges and stiffener.

Local instability of outstanding leg of flanges or overall lateral instability of girder

Plastic hinges develop in flanges together with the web yielding over area shown, to form a mechanism.
Tult being given by equation

Position of hinge varies with flexibility of flanges

Flange rigidity is high such that the lateral loading imposed by

the tension field is insufficient to produce a collapse mechanism and web yields over entire area.

No additional load can be carried by web - all additional load carried by framework comprising flanges and stiffeners.

Girder then fails on the application of further load when either:-

(1) Creation of an overall frame failure mechanism of the boundary frame operating as a Vierendeel frame.

(2) Web tears under tensile loading initiated at a stress concentration.

(3) Attachments fail in the case of a bolted or riveted girder or weld fractures in the case of welded girders.

FIG. 1. POSSIBLE MODES OF FAILURE.

assumed that the webplate would fail due to the development of an inclined plastic band anchored against the vertical stiffeners, see Fig. 3. In developing this mechanism, Basler *et al* assumed that the flanges of most girders were too flexible to withstand the membrane action imposed by the buckled web. However, it has been shown in references[1-3] that their 'collapse model' can both significantly underestimate and overestimate the strength of plate girders.

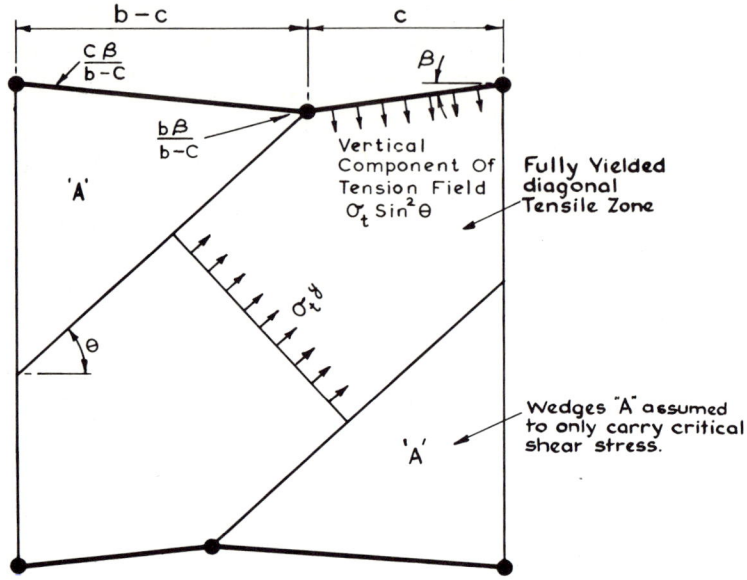

FIG. 2   COLLAPSE MECHANISM PROPOSED BY
ROCKEY & SKALOUD (3)

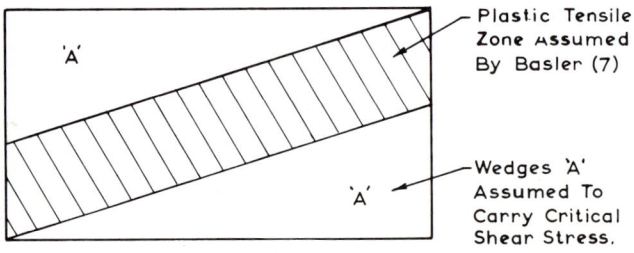

FIG. 3. COLLAPSE MODE ASSUMED
IN BASLER THEORY.

More recent research at Lehigh by Ostapenko *et al*[10-12] has developed the Basler model and shown how it can be applied to the design of plate girder webs reinforced by both transverse and longitudinal stiffeners; one very important contribution being the development of formulae for flange and stiffener design. Unfortunately, the model assumed still does not allow for the influence of transverse flange rigidity upon the behaviour of buckled webs.

The present paper indicates how the ultimate load method of design for transversely stiffened shear webs as proposed by Skaloud and the writer[3,4] can be adapted to deal with the loading cases of combined shear and bending and also to deal with the design of plate girder webs reinforced by both transverse and longitudinal stiffeners.

3.1 Ultimate load design of webs loaded in shear

In reference,[1-3] Skaloud and the writer have shown that when a shear web plate buckles prior to yielding, it fails with the development of a diagonal tension band which is fully yielded together with the development of plastic hinges in the flange members to form a mechanism, see Fig. 1 and 2. It was established experimentally in Reference 1 that the width of the diagonal band and therefore the position of the plastic hinges varies with the I/b^3t ratio where I is the flexural rigidity of the compression flange about an axis through its centroid and perpendicular to the web plate, b is the spacing of the transverse stiffeners and t is the thickness of the web.

Subsequently in Reference 4 a theoretical solution based on the collapse model shown in Fig. 2 was developed and it was established that this method of analysis was capable of predicting with very good accuracy the failure load of transversely stiffened plate girders loaded in shear. In section 3.2, this solution will be briefly presented, whilst in section 3.3 and 3.4 it will be extended to deal with the design of plate girder webs reinforced by both longitudinal and transverse stiffeners and also with the design of hybrid girders.

3.2 Design of shear webs

The behaviour of a plate girder web loaded in shear can be divided into three stages.

Stages I and II

In Stage I, which only applies to a perfectly flat plate, the applied shear stress is less than the critical shear buckling stress and therefore the web panel carries the applied load by a pure shear action.

The second mode of action results from the fact that in a buckled web the compressive stresses cannot increase and any additional load has to be carried by a tensile truss action.

With normal welded plate girders which have webs with significant permanent deformations, no buckling phenomena will be observed and the loadings which are associated with Stage II occur as soon as load is applied to the girder.

Failure occurs when the diagonal tension band, see Figs 1 and 2, yields and the boundary members develop sufficient plastic hinges to result in a failure characterised by one of the three possible forms of failure listed under Stage III.

Stage III

(a) If the lateral membrane loading on the flange is sufficient to develop plastic hinges in the flanges, then failure will be due to the development of a mechanism consisting of a yielded diagonal strip together with plastic hinges in the tension and compression flanges; see Figs 1 and 2.

(b) If, however, the membrane loading corresponding to a yielded web is not sufficient to develop plastic hinges in the flanges then failure will occur when either

(1) the web material fractures, such as occurs in an aluminium web
(2) the framework comprising the flanges and the stiffeners, acting as a
    Vierendeel frame develops a 'frame' mechanism
(3) the compression flange buckles laterally or torsionally.

3.2.a Theoretical basis

Stage I

For an initially plane web, for loading below the buckling stress τ_{cr}, the stress state is assumed to be one of pure shear. Obviously the value of τ_{cr} will vary with the flexural and torsional rigidity of both the flanges and the stiffeners. However, since most conventional welded steel girders have flanges of low torsional rigidity it is reasonable to assume that the shear web is simply supported on all edges. If, however, a tubular flange is employed it would be necessary to use the corresponding buckling stress.[13]

As stated earlier, following buckling, the web is unable to withstand any further compression loading and any additional loading has to be carried by a tension field action. The present solution does not attempt to deal with the very complicated stress field which occurs in the elastic post buckled range, it is solely concerned with the final collapse mode. This is essential, if a comparatively simply design procedure is to be developed since observation of the collapse behaviour of girders indicates that the stress and deflection distributions vary quite rapidly at loads close to the ultimate.

The experimental evidence resulting from the earlier study by Skaloud and the writer[1-3] is that at collapse the web develops a tension band as shown in Fig. 2 in which the angle of the tension band is equal to the inclination of the geometrical diagonal and that the tension band is symmetric with respect to the geometric diagonal. The width of this diagonal tension load is assumed to be such that the intercept of its boundary with the flange coincides with the position of the plastic hinges in the flanges. In the case of girders with very stiff flanges, the above assumptions will clearly result in slightly lower bound solutions for values of $\alpha > 1$, since in such cases the inclination of the tension field will be closer to $45°$ than the inclination of the diagonal. This point will be discussed again later in the paper.

Thus we see that there are two stress regions, see Fig. 4.

(1) two triangular wedges in which the critical shear stress is assumed to act.
(2) a yielded diagonal strip.

The tension stress σ_t is assumed to act uniformly over the diagonal band, yielding occurring when σ_t reaches a value $\sigma_t{}^y$.

The stress condition, see Fig. 5, in the diagonal web strip is given by

$$\sigma_\xi = \tau_{cr} \operatorname{Sin} 2\theta + \sigma_t{}^y$$

$$\sigma_\eta = \tau_{cr} \operatorname{Sin} 2\theta \tag{1}$$

$$\tau = \tau_{cr} \operatorname{Cos} 2\theta$$

Using the Huber Von Mises plasticity condition, the material yields when $\sigma_{mc} = \sigma_{yw}$ where

$$\sigma_{mc} = \sqrt{\sigma_\xi{}^2 + \sigma_\eta{}^2 - \sigma_\xi \sigma_\eta + 3\tau^2} \tag{2}$$

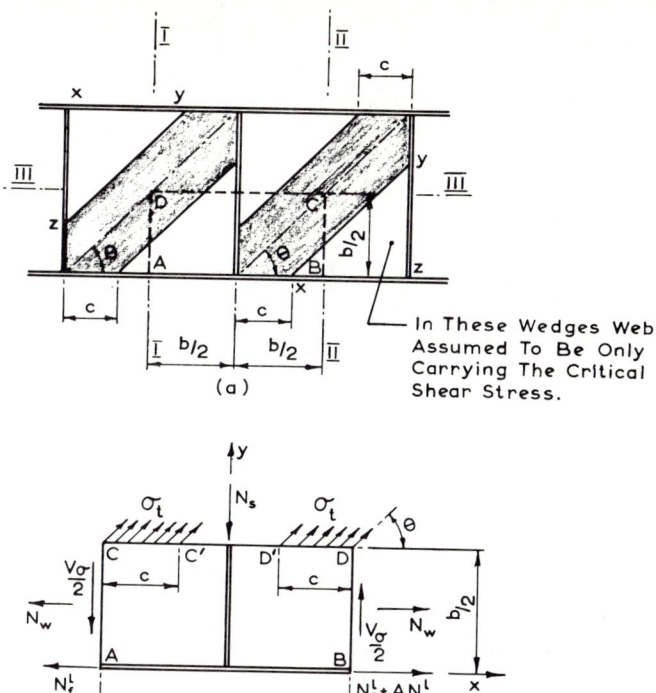

In These Wedges Web
Assumed To Be Only
Carrying The Critical
Shear Stress.

(a)

(b)

FIG. 4 .

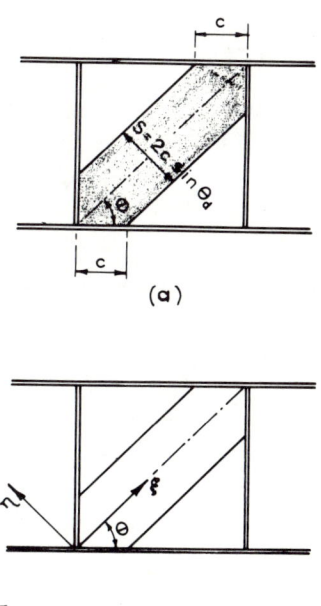

(a)

(b)

FIG 5

Substituting equations (1) into (2) and rearranging yields

$$\sigma_t{}^y = \frac{-3}{2}\tau_{cr}\,\text{Sin}\,2\theta + \sqrt{\sigma_{yw}{}^2 + \tau_{cr}{}^2\left[\left(\frac{3}{2}\,\text{Sin}\,2\theta\right)^2 - 3\right]} \tag{3}$$

The vertical component V_σ of the diagonal stress $\sigma_t{}^y$ is given by equation (4)

$$V_\sigma = 2ct\,\text{Sin}^2\theta\left(-\frac{3}{2}\tau_{cr}\,\text{Sin}\,2\theta + \sqrt{\sigma_{yw}{}^2 + \tau_{cr}{}^2\left(\left(\frac{3}{2}\,\text{Sin}\,2\theta\right)^2 - 3\right)}\right) \tag{4}$$

The total shear force V_{ult} is equal to the sum of V_σ and the shear force V_{cr} necessary to cause the plate to buckle.

$$V_{ult} = V_{cr} + V_\sigma = \tau_{cr}\,dt + 2ct\,\text{Sin}^2\theta\left(-\frac{3}{2}\tau_{cr}\,\text{Sin}\,2\theta + \right.$$

$$\left.\sqrt{\sigma_{yw}{}^2 + \tau_{cr}{}^2\left(\left(\frac{3}{2}\,\text{Sin}\,2\theta\right)^2 - 3\right)}\right) \tag{5}$$

Since $\tau_{yw} = \sigma_{yw}/\sqrt{3}$, equation (5) may be rewritten as equation (6)

$$\frac{\tau_{ult}}{\tau_{yw}} = \frac{\tau_{cr}}{\tau_{yw}} + 2\sqrt{3}\frac{c\alpha}{b}\,\text{Sin}^2\theta\left(-\frac{\sqrt{3}}{2}\,\text{Sin}\,2\theta\left(\frac{\tau_{cr}}{\tau_{yw}}\right) + \right.$$

$$\left.\sqrt{1 + \left(\frac{\tau_{cr}}{\tau_{yw}}\right)^2\left(\frac{3}{4}\,\text{Sin}^2(2\theta) - 1\right)}\right) \tag{6}$$

The position of the plastic hinge in the flanges may be theoretically determined using the collapse mechanism shown in Fig. 2. This mechanism assumes that the hinge coincides with the edge of the diagonal strip and that the loading consists of the vertical component of the diagonal tensile membrane stress $\sigma_t{}^y$. The solution of this simple mechanism reduces to the solution of the cubic equation given in equation (7).

$$\left(\frac{c}{b}\right)^3 - \left(\frac{c}{b}\right)^2 + \frac{4z_f{}^\sigma yf}{b^2 t\,\text{Sin}^2\theta\,(\sigma_t{}^y)} = 0 \tag{7}$$

where z_f denotes the plastic modulus for flange assembly. It is proposed that when the web buckling stress is less than half the shear yield stress, a depth of web plate

$$z = 30t\left(1 - \frac{2\tau_{cr}}{\tau_{yw}}\right)$$

be assumed to act with the flange assembly.

It is of interest to consider how equation (6) satisfies a number of the limiting conditions.

(1) *Very thin webs and rigid flanges*

For very thin webs $\tau_{cr} \to 0$, in which case

$$\frac{\tau_{ult}}{\tau_{yw}} = 2\sqrt{3}\,\alpha\frac{c}{b}\,\text{Sin}^2\theta$$

Since for rigid flanges, $\frac{c}{b} = 0.5$, then for square web panels in which $\theta = \frac{\pi}{4}$ one obtains the value for τ_{ult} of

$$\frac{\sqrt{3}}{2}\tau_{yw} \text{ or } \frac{\sigma_{yw}}{2}$$

which agrees with the value obtained from Wagner's Theory[14] for a complete tension field.

(2) *Very thick webs*

When webs are very thick, then $\tau_{cr} \to \tau_{yw}$ and the terms inside the main brackets reduces to zero so that equation (6) reduces to $\tau_{ult} = \tau_{yw}$; which again is as to be expected.

(3) *Very flexible flanges*

Finally, if $\frac{c}{b} \to 0$, as would be the case if the flanges had zero stiffness, and could not withstand any lateral loading, then

$$\tau_{ult} = \tau_{cr}$$

Thus we have seen that equation (6) satisfies the extreme boundary conditions exactly.

When $\sqrt{3\tau}_{cr}$ exceeds the limit of proportional stress of the material then the effective modulus E_r is less than the modulus of Elasticity E. This reduces the critical shear stress; and to allow for this Basler and his colleagues have recommended that τ_{cr} be replaced by τ_{cre} when

$$\tau_{cr} > \frac{0\cdot 8\sigma_y}{\sqrt{3}},$$

τ_{cre} being obtained from equation 8.

$$\frac{\tau_{cre}}{\tau_{yw}} = 1 - \frac{0\cdot 16\tau_{yw}}{\tau_{cr}} \qquad (8)$$

Using equation (8) in conjunction with equation (6) the relationship between the ratio τ_{ult}/τ_{yw} and the depth to thickness ratio for different values of c/b have been plotted in Fig. 6 for the case of $\propto$ equal to 1. The values of τ_{ult}/τ_{yw} as derived from Basler's ultimate load expression, see equation 9, have also been plotted

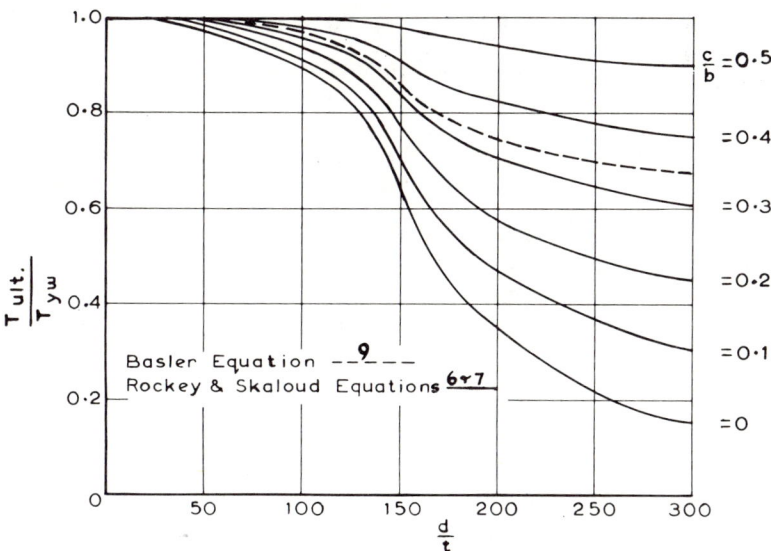

FIG. 6

using the same relationship between τ_{cre}, τ_{cr} and τ_{yw}, and it is clearly seen that for very flexible flanges Basler's equation overestimates the strength of the girder and for relatively stiff flanges it underestimates the strength, this being particularly true for larger values of α.

$$W_{ult} = dt \left[\tau_{cr} + \frac{\sqrt{3}\,\tau_{yw}}{2\sqrt{1+\alpha^2}} \left[1 - \frac{\tau_{cr}}{\tau_{yw}} \right] \right] \qquad (9)$$

The present design procedure, see equations (7) and (8), has been checked[4,5] against existing experimental data and as will be noted from Fig. 7 very good correlation has been obtained.

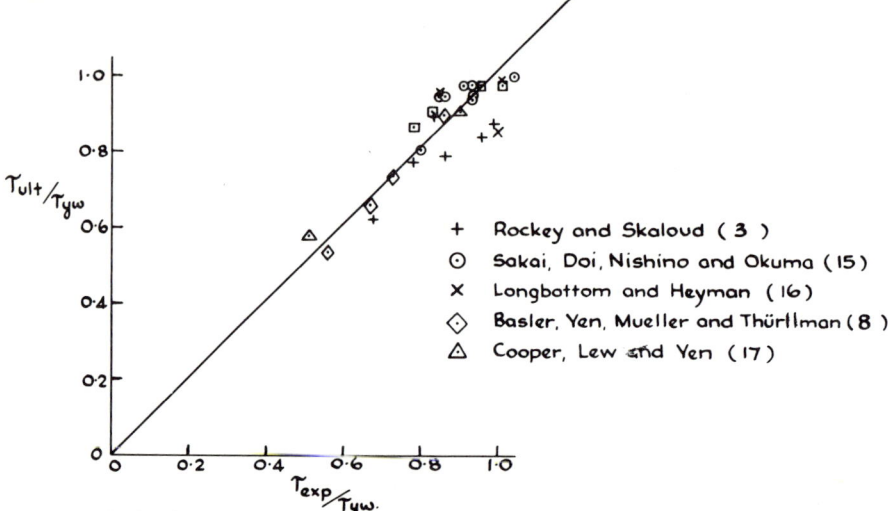

FIG: 7   COMPARISON OF PREDICTED ULTIMATE SHEAR STRESS (τ_{ult}) AND EXPERIMENTAL ULTIMATE SHEAR STRESS (τ_{exp}) FOR WEBS LOADED IN SHEAR.

3.2   Transversely stiffened web plates loaded in shear and bending.

Web plates are normally subjected to a combination of shear and bending and in the present section an ultimate load method of design is proposed for this case of loading.

Three important additional factors have to be considered when determining the failure load of a web plate loaded in shear and bending, these are

(1) The reduction in the buckling stress of the web due to the presence of a bending stress or a direct stress.

(2) The influence of the inplane bending stresses upon the value of the diagonal tensile membrane stress $\sigma_t{}^y$ which is developed in the diagonal strip.

(3) The reduction in the magnitude of the plastic modulus z_f of the flanges due to the presence of the axial compressive and tensile stresses.

For the case of a web plate subjected to combined shear and bending the reduction in the buckling stress τ_{cr} due to the presence of a bending stress σ, can be calculated with reasonable accuracy from equation (10).

$$\left(\frac{\sigma_{mb}}{\sigma_{crb}}\right)^2 + \left(\frac{\tau_m}{\tau_{cr}}\right)^2 = 1 \qquad (10)$$

where σ_{crb} = critical bending stress when the plate is subjected to pure bending

τ_{cr} = critical bending stress when the plate is subjected to pure shear

σ_m , τ_m = critical bending and shear stresses when acting together

For the case of all edges being simply supported σ_{cr} and τ_{cr} can be determined from equations (11) and (12)

$$\sigma_{crb} = 23 \cdot 9 \left(\frac{\pi^2 E}{12(1 - \mu^2)} \right) \left(\frac{t}{d} \right)^2 \tag{11}$$

$$\tau_{cr} = \left(5 \cdot 35 + \frac{4d^2}{b^2} \right) \left(\frac{\pi^2 E}{12(1 - \mu^2)} \right) \left(\frac{t}{d} \right)^2 \text{ when } b \geqslant d \tag{12a}$$

$$\tau_{cr} = \left(5 \cdot 35 \frac{d^2}{b^2} + 4 \right) \frac{\pi^2 E}{12(1 - \mu^2)} \left(\frac{t}{d} \right)^2 \text{ when } b \leqslant d \tag{12b}$$

The plastic modulus z_f will be reduced by the presence of the axial force and for flanges having a simple rectangular cross section the following relationship may be employed to determine the reduced modulus z_{fr}

$$z_{fr} = z_f \left[1 - \left(\frac{\sigma}{\sigma_{yf}} \right)^2 \right] \tag{13}$$

where σ is the axial stress in the flange and σ_{yf} is the yield stress for the flange material.

As stated earlier when a plate girder web buckles in shear it loses its capacity to carry any additional compressive load, likewise when a panel loaded in direct compression buckles the central area of the panel is unable to carry any further direct stress, and any additional direct load has to be carried by the web material adjacent to the flanges and stiffeners.

Since the stress distribution in a yielded panel subjected to shear and bending is very complex it is assumed in the proposed design procedure that after the plate buckles, the flanges alone carry the additional bending moments. Furthermore, it is assumed that the web carries the additional shear loads by the development of a diagonal membrane stress $\sigma_t{}^y$.

When a webplate is loaded by direct bending stresses as well as by shear stresses, the value of the diagonal stress $\sigma_t{}^y$ at which yielding occurs is changed from that given in equation (3) to the value obtained from equation (14).

$$\sigma_t{}^y = \frac{1}{2} \left[(3\tau_m \text{ Sin } 2\theta + \sigma_m \text{ Sin}^2 \theta - 2\sigma_m \text{ Cos}^2 \theta) \right.$$

$$\left. + \sqrt{(+3\tau_m \text{ Sin } 2\theta + \sigma_m \text{ Sin}^2 \theta - 2\sigma_m \text{ Cos}^2 \theta)^2 - 4 \left[\sigma_m{}^2 + 3\tau_m{}^2 - \sigma_{yw}^2 \right]} \right] \tag{14}$$

Figs 8 and 9 show how the value of $\sigma_t{}^y$ varies with the presence of a direct stress. It will be noted that the presence of a tensile bending stress reduces the capacity of $\sigma_t{}^y$ more significantly than does the compressive bending stress. Because of these factors one would expect to observe a spreading of the diagonal band either side of the neutral axis and that the diagonal band in the tension area, because of its reduced $\sigma_t{}^y$ value, would be wider than the band in the compression zone.

It should be appreciated that the bending stresses, (σ) in the tension zone continue to grow after the plate buckles and therefore the value of $\sigma_t{}^y$ in the tension zone will continue to be affected, and that this area will yield first.

The experimental tests conducted by Rockey and Skaloud have shown that the distance '*c*' giving the position of the 'central' plastic hinges in the tension flanges is larger than the distance '*c*' which occurs in the compression flange. This is to be expected because of the reduced σ_t^y value and the presence of the tension forces in the tension flange which will have the tendency to keep the flanges straight.

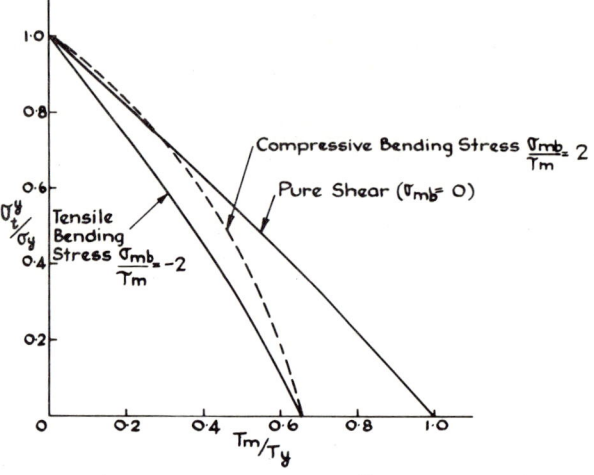

FIG. 8   VARIATION OF σ_t^y/σ_y RATIO WITH T_m/T_y RATIO

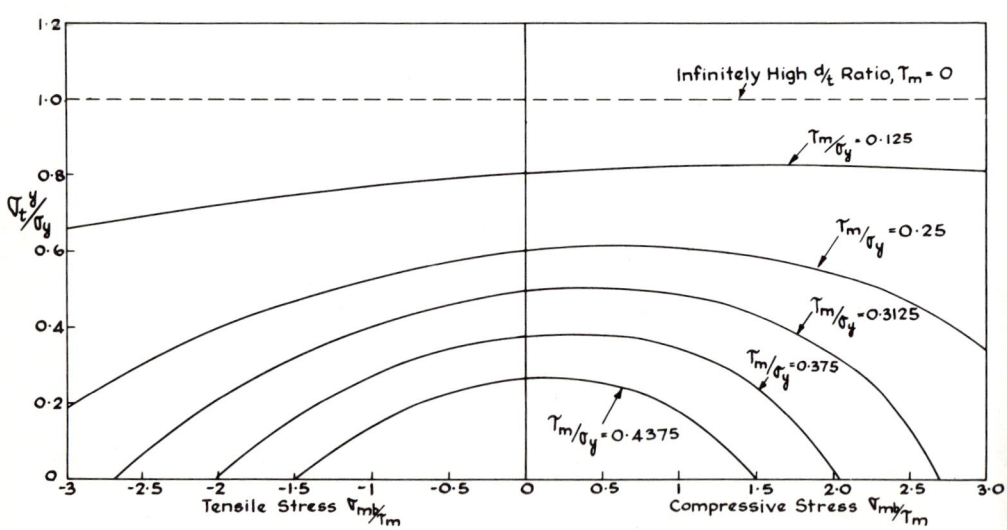

FIG: 9   VARIATION   OF   THE   RATIO σ_t^y/σ_y WITH THE RATIO σ_{mb}/T_m FOR DIFFERENT VALUES OF THE BUCKLING STRESS T_m.

In the present design procedure the values of τ_m and σ_m which are to be used in equation (14) are the stresses at which the plate buckles under the combined loading, τ_m being the average shear stress across the web and σ_m is the compressive bending stress at the web/flange junction.

A general expression for the ultimate load can be obtained by combining equations (7), (13) and (14). In section 3.2 the position of the central hinge was obtained by solving equation (7)

$$\left(\frac{c}{b}\right)^3 - \left(\frac{c}{b}\right)^2 + \frac{4z_f\sigma_{yf}}{b^2 t \, Sin^2\theta \, (\sigma_t{}^y)} = 0 \tag{7}$$

When a bending stress acts with the shear stress, the value of z_f which has to be used in equation (7) is the reduced value z_{fr} as given by equation (13) and $\sigma_t{}^y$ is the modified value of $\sigma_t{}^y$ given by equation (14). Equation (7) thus becomes

$$\left(\frac{c}{b}\right)^3 - \left(\frac{c}{b}\right)^2 + \frac{4\sigma_{yf}}{b^2 t \, Sin^2\theta} \, \frac{z_f\left[1 - \left(\frac{\sigma}{\sigma_{yf}}\right)\right]^2}{\sigma_y{}^t} = 0 \tag{15}$$

It will be noted from Figs 9 and 10 that $\sigma_y{}^t$ varies with the value AM which is the ratio of the applied bending stress to the applied shear stress before buckling occurs.

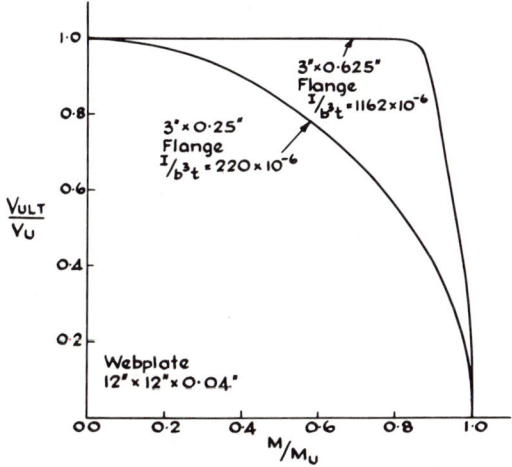

FIG.10 INTERACTION DIAGRAMS - DEMONSTRATING INFLUENCE OF FLANGE RIGIDITY UPON SHAPE OF DIAGRAM

The value of the flange bending stress σ in equation (15) can, for values of σ up to σ_{yf}, be obtained from equation (16).

$$\sigma = \frac{M_y}{I_z} = \frac{(WF)(d + 2t_f)}{2I_z} = W_{ult} \, (q) \tag{16}$$

Where I_z = Moment of inertia of a section comprising the flanges only, F is a factor depending upon the type of loading, which for the case of a centrally loaded simply supported girder $= \frac{l}{2}$ where l is the distance of the section to the nearest support. Now for a centrally loaded, simply supported girder,

$$W_{ult} = 2dt \left[\tau_{cr} + \frac{2C}{d} \, Sin^2\theta \, \sigma_t{}^y \right] \tag{17}$$

Substituting (15) and (16) into (14) yields

$$\left(\frac{c}{b}\right)^3 - \left(\frac{c}{b}\right)^2 + \frac{4}{b^2 t \, Sin^2\theta} \, \frac{z_f\left[\sigma^2{}_{yf} - q^2 d^2 \, t^2 \left[\tau_{cr} + 2\frac{c}{d} \, Sin^2\theta\sigma_t{}^y\right]^2\right]}{\sigma_y{}^t\sigma_{yf}} \tag{18}$$

which reduces to an equation of the form

$$\left(\frac{c}{b}\right)^3 - (A)\left(\frac{c}{b}\right)^2 + B\left(\frac{c}{d}\right) + D = 0 \tag{19}$$

The solution of this equation leads to the solution of $\frac{c}{b}$ and hence the position of the hinge c which when substituted into equation (17) will give the ultimate load W_{ult}.

The above solution does not require the use of any assumed interaction relationship between the shear load ratio (V_{ult}/V_u) and the moment ratio (M/M_u), since this relationship is incorporated in the solution.

Figure 10 gives typical interaction curves between the ratio (V_{ult}/V_u) and M/M_u which have been obtained using equation (18) for two girders from which it will be noted that the shape of the interaction curves is greatly influenced by the flange rigidity parameter and the slenderness of the web (d/t). For girders with very stiff flanges the loss in shear strength with applied bending stresses is not significant, but with the case of girders with relatively flexible flanges, the influence of the bending stresses upon the z_f value becomes critical and there is a steady loss in shear load carrying capacity. The manner in which the position of the central hinges varies with the M/M_u ratio is clearly shown in Fig. 11. It will be noted that with 'flexible'

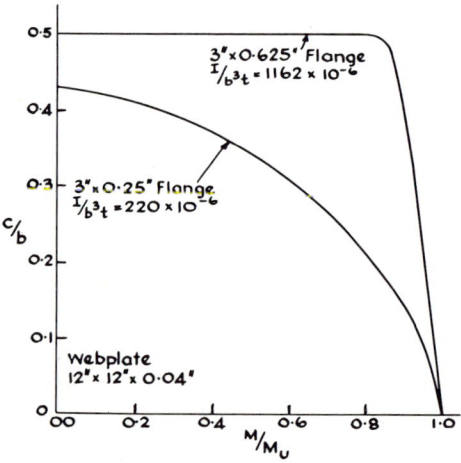

FIG: 11   VARIATION OF THE c/b RATIO WITH THE MOMENT RATIO M/M_u

flanges, as the applied bending stress increases the effective shear stress at buckling decreases and the position of the central hinge moves towards the stiffeners. The reduction of the width of the diagonal tension band together with the reduction in the critical shear stress, which is the stress acting in the triangular areas either side of the diagonal band, means that the shear load capacity decreases steadily with the (M/M_u) ratio.

3.3   Web plates reinforced by both transverse and longitudinal stiffeners and subjected to shear.

3.3.1   Pure shear

Figure 12 shows the typical collapse pattern which can be assumed to occur in a longitudinally reinforced web plate subjected to shear.

Consider Panel 1, this panel will impose lateral loading on the flange and it can be assumed that a yield zone will develop as indicated with hinges forming in the flanges as shown, the position of the internal hinge (C_1) in the flange varying with the rigidity of the flange and the buckling stress in the panel. However, since the

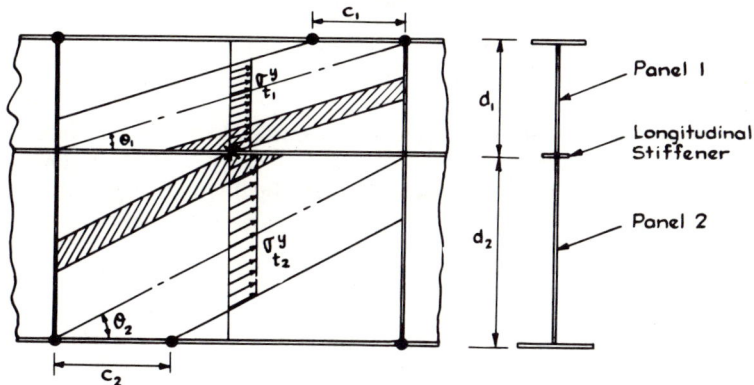

FIG. 12   PROPOSED COLLAPSE MECHANISM FOR A LONGITUDINALLY REINFORCED WEB SUBJECTED TO SHEAR

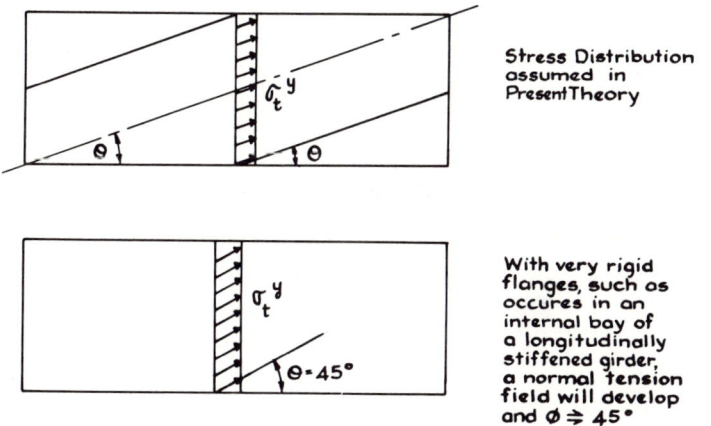

FIG. 13 PROPOSED COLLAPSE MECHANISM FOR A LONGITUDINALLY REINFORCED WEB SUBJECTED TO SHEAR.

adjacent panel will act as a very stiff flange, at the position of the longitudinal stiffener the position of the hinge can be assumed at $0.5b$. Thus for a panel such as 1, the shear load V_1 will be given by equation (20). In equation (20), the subscripts 1 signify Panel 1.

$$V_1 = \left[\tau_{cr_1} d_1 t + t \, \text{Sin}^2 \theta_1 \sigma_{t_1}{}^y (C + 0.5b) \right] \tag{20}$$

For panel '2', a similar procedure can be followed, the load V_2 for this panel can be calculated from equation (21).

$$V_2 = \left[\tau_{cr_2} d_2 t + (0.5b + C_2) \, \text{Sin}^2 \theta_2 \sigma_t{}^y \right] \tag{21}$$

where C_2 is the position of the hinge in the tension flange, calculated using equations (6) and (7).

The total shear load $V = V_1 + V_2$

$$V = \left[\tau_{cr_1} d_1 t + \tau_{cr_2} d_2 t + \left[C_1 + 0.5b\right] \sigma_{t}{}^y{}_1 \operatorname{Sin}^2 \theta_1 + (C_2 + 0.5b) \sigma_{t_2}{}^y \operatorname{Sin}^2 \theta_2\right] (21)$$

When two or more longitudinal stiffeners are employed the shear load carried by the internal panels can be calculated from equation (6) and (7) assuming $c/b = 0.5$ in these cases.

In such cases if the inclination of the diagonal θ is used then a lower bound solution would be expected, since for such internal panels it would be reasonable to expect that the inclination of the tensile membrane field would approach $45°$, as indicated in Fig. 13. However, further research studies are required before this further refinement could be accepted.

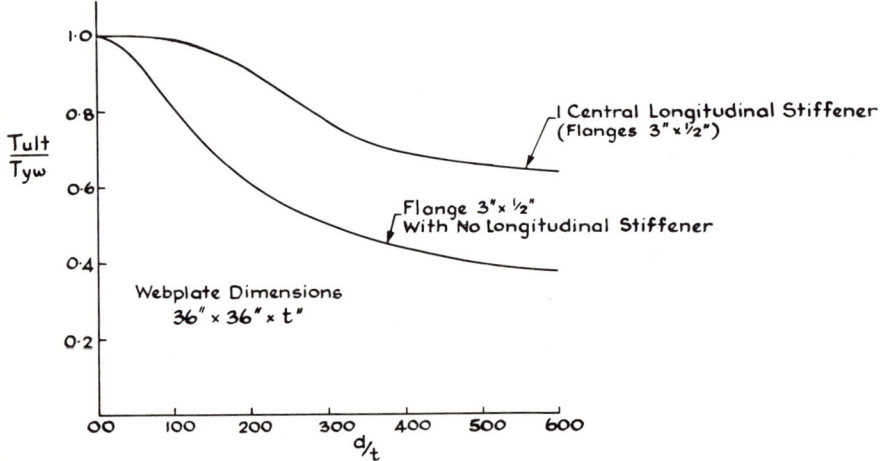

FIG.14   INFLUENCE OF A LONGITUDINAL STIFFENER ON THE SHEAR ULTIMATE STRENGTH OF A WEB.

Figure 14 shows the significant gain in ultimate shear load which can be achieved by employing a longitudinal stiffener at mid depth.

3.4   Webplates reinforced by both transverse and longitudinal stiffeners and subjected to shear and bending.

Figure 15 shows a typical panel subjected to a combined linearly varying axial direct stress and a shear stress τ. The linearly varying axial stress distribution can be replaced by an axial stress together with a pure bending stress as shown. For example, in Panel 'I' there will be a direct compressive stress of σ_{mc1} and a pure bending stress at the flange/web junction of τ_{mb1}.

The critical stresses σ_{mc}, σ_{mb} and τ_m which will cause buckling under their combined action can be predicted with reasonable accuracy by equation (23).

$$\left(\frac{\sigma_{mc}}{\sigma_{crc}}\right) + \left(\frac{\sigma_{mb}}{\sigma_{crb}}\right)^2 + \left(\frac{\tau_m}{\tau_{cr}}\right)^2 = 1 \qquad (23)$$

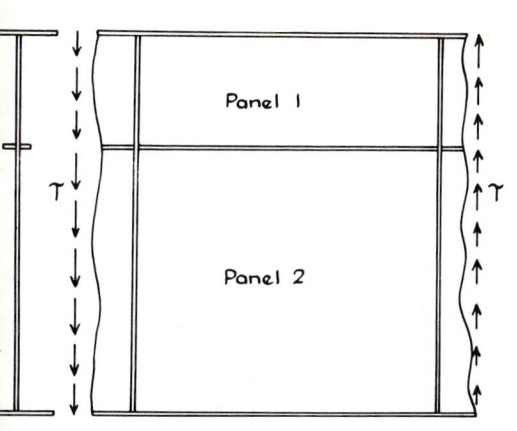

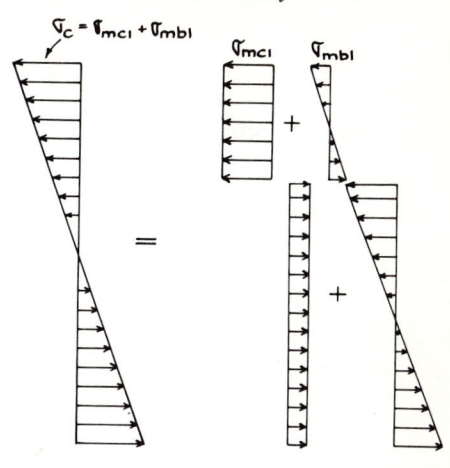

Actual Direct
Stress Distribution

Equivalent Direct
Stress Distribution

**FIG. 15   STRESS DISTRIBUTION IN PANELS OF A PLATE GIRDER SUBJECTED
TO SHEAR AND BENDING**

where:

σ_{crc} = The critical uniform direct axial stress to cause buckling see equations (24) and (25)

σ_{crb} = The compressive edge stress causing buckling in the panel when loaded in pure bending, see equations (11) and (26)

τ_{cr} = The uniform shear stress to cause buckling, see equations (12a), (12b) and (27).

$$\sigma_{crc} = 4\left[\frac{\pi^2 E}{12(1-\mu^2)}\right]\left[\frac{t}{d}\right]^2 \qquad \text{when all edges are simply supported} \qquad (24)$$

$$\sigma_{crc} = 5.41\left[\frac{\pi^2 E}{12(1-\mu^2)}\right]\left[\frac{t}{d}\right]^2 \qquad \text{when one longitudinal edge is clamped, the others simply supported} \qquad (25)$$

$$\sigma_{crb} = 41.7\left[\frac{\pi^2 E}{12(1-\mu^2)}\right]\left[\frac{t}{d}\right]^2 \qquad \text{when the compressive longitudinal edge is clamped, the other simply supported} \qquad (26)$$

$$\tau_{cr} = \left[7.07 + \frac{3.91}{(b/d)^{\frac{3}{2}}}\right]\left[\frac{\pi^2 E}{12(1-\mu^2)}\right]\left[\frac{t}{d}\right]^2 \qquad \text{when one longitudinal edge is clamped, the others simply supported} \qquad (27)$$

Once the critical stresses, σ_{mc}, σ_{mb} and τ_m for the individual panels have been determined, the stress distribution at buckling will be known and the collapse load for each of the panels determined using the basic equations (7) and (14). In a longitudinally stiffened web plate this will involve an iterative procedure since the axial stresses in the flanges will vary with the shear load. Thus the use of either a desk calculating machine or a small computer is highly desirable.

CONCLUSION

The paper establishes an ultimate load method of design for transversely and longitudinally reinforced web plates. In particular, it is shown that the flexural stiffness of the flange members has a significant influence upon the post buckled behaviour of web plates and the design method allows for the interaction which occurs between the buckled webplate and the flanges.

ACKNOWLEDGEMENT

This research project has been sponsored in part by the Construction Industry Research and Information Association, in part by the British Constructional Steelwork Association and by the University College, Cardiff.

REFERENCES

1. Rockey, K. C. Factors influencing ultimate behaviour of plate girders. *Proceedings of the Conference on Steel Bridges (London, 1968)*. London, British Constructional Steelwork Association, 1969.
2. Rockey, K. C. and Skaloud, M. Influence of flange stiffness upon the load carrying capacity of webs in shear. Final Report of the 8th Congress of the International Association for Bridge and Structural Engineering (New York, Sept. 1968).
3. Rockey, K. C. and Skaloud, M. Influence of the flexural rigidity of flanges upon the load carrying capacity and failure mechanism of web in shear. *Acta tech. C.S.A.V.,* vol. 14, no. 3, 1969, pp. 295–313.
4. Rockey, K. C. and Skaloud, M. The ultimate load behaviour of plate girders loaded in shear. University College, Cardiff, Department of Civil and Structural Engineering Report, 1969.
5. Rockey, K. C. and Skaloud, M. A survey of the post buckled behaviour of web loaded in shear. University College, Cardiff, Department of Civil and Structural Engineering Report, 1969.
6. Basler, K. *Strength of plate girders.* Ph.D. thesis, Lehigh University, 1959.
7. Basler, K. Strength of plate girders in shear. *J. struct. Div. Am. Soc. civ. Engrs,* vol. 87, ST7, Oct. 1961, pp. 151–180.
8. Thürlimann, B. Static strength of plate girders. *Mém. Soc. r. Sci. Liège,* vol. 8, 1963, pp. 137–175.
9. Basler, K., Yen, B. T., Mueller, J. A. ana Thürlimann, B. Web buckling tests on welded plate girders. *Bull. Weld. Res. Coun.,* no. 64, Sept. 1960.
10. Chern, C. and Ostapenko, A. Ultimate strength of plate girders under shear. *Fritz Engineering Laboratory Report* no. 328.7, Lehigh University, August 1969.
11. Dimitri, J. R. and Ostapenko, A. Pilot tests on the ultimate static strength of unsymmetrical plate girders. *Fritz Engineering Laboratory Report* no. 328.5, Lehigh University, 1968.
12. Schueller, W. and Ostapenko, A. Static tests on unsymmetrical plate girders main test series. *Fritz Engineering Laboratory Report* no. 328.6, Lehigh University, 1968.
13. Cook, I. T. and Rockey, K. C. Shear buckling of rectangular plates with mixed boundary conditions. *Aeronaut. Q.,* vol. 14, Nov. 1963, pp. 349–356.
14. Wagner, H. Flat sheet metal girders with very thin metal webs. Parts I, II and III. *Tech. Memo. natn. advis. Comm. Aeronaut., Wash.,* nos. 604, 605 and 606, Feb. 1931.
15. Sakai, F., Doi, K., Nishino, F. and Okumwa, T. Failure tests of plate girders using large sized models. (In Japanese) University of Tokyo, Department of Civil Engineering, Structural Engineering Laboratory Report, 1966.
16. Longbottom, E. and Heyman, J. Experimental verification of the strengths of plate girders designed in accordance with the revised British Standard 153: tests on full-size and on model plate girders. *Proc. Instn civ. Engrs,* vol. 5, part 3, August 1956, pp. 462–521.
17. Cooper, P. B., Lew, H. S. and Yen, B. T. Welded constructional alloy steel plate girders. *J. struct. Div. Am. Soc. civ. Engrs,* vol. 90, STI, Feb. 1964, pp. 1–36.

ULTIMATE STRENGTH DESIGN OF PLATE GIRDERS

ALEXIS OSTAPENKO, CHINGMIIN CHERN
and SIAMAK PARSANEJAD

Lehigh University

SYNOPSIS

A method of ultimate strength analysis of plate girders, based on the assumption of beam action, tension-field action and frame action contributions, is briefly described. It is applicable to symmetrical, unsymmetrical, homogeneous or hybrid girders subjected to shear, bending or combined loads. Since due to the iterative processes used the method is not amenable to manual computation, presented here are formulas developed from the numerical output of a computer program based on the method. A tentative recommendation is made for precluding the development of fatigue cracks due to the back-and-fourth deflection of the web plate.

1. INTRODUCTION

Recognition of the fact that the web possesses considerable post-buckling strength led to a search for a method of determining the ultimate strength of plate girders. Basler offered a plausible theory[1,2,3] which was well substantiated by tests.[4] This theory was then accepted by AISC as the method for designing plate girders in buildings.[5] In a slightly streamlined version this theory was also incorporated in the load factor method proposed by Vincent for designing steel highway bridges.[6] Further developments of the theory were made, among others, by Fujii[7] and Rockey and Skaloud[8] who included the effect of the flange strength on the strength of the web plate.

However, all these theories were concerned with symmetrical plate girders, that is, with girders having flanges of equal area and therefore with the centroidal axis at the mid-depth of the web. Also, none of these theories could give a continuous description of the girder failure mode for a variable combination of shear and moment. To eliminate these deficiencies, a new approach was developed by Chern and Ostapenko.[9,10,11,12] This approach was also successfully extended to longitudinally stiffened plate girders.[13,14]

Due to the iterative procedures involved, this theory is hardly suitable for manual operations, and a computer must be employed. The numerical output from the computer was then utilised to develop simplified formulas for practical use. So far, this effort has been successful only for transversely stiffened plate girders.[15]

The purpose of the present paper is to briefly describe the general concepts of this theory and to present the simplified design formulas. A numerical example is given at the end to illustrate the procedure. A detailed description of the theory and a comparison of it with the available test results are given for various cases of loading in References[10,13].

2. ANALYTICAL MODEL

A typical plate girder panel, that is, a portion of the plate girder between two transverse stiffeners, is shown in Fig. 1. The forces acting on it are defined at the midlength of the panel as moment M and shear V. As the moment diagram illustrates, actually, a greater moment M_{max} exists at one end of the panel, and it also should be taken into consideration in design.

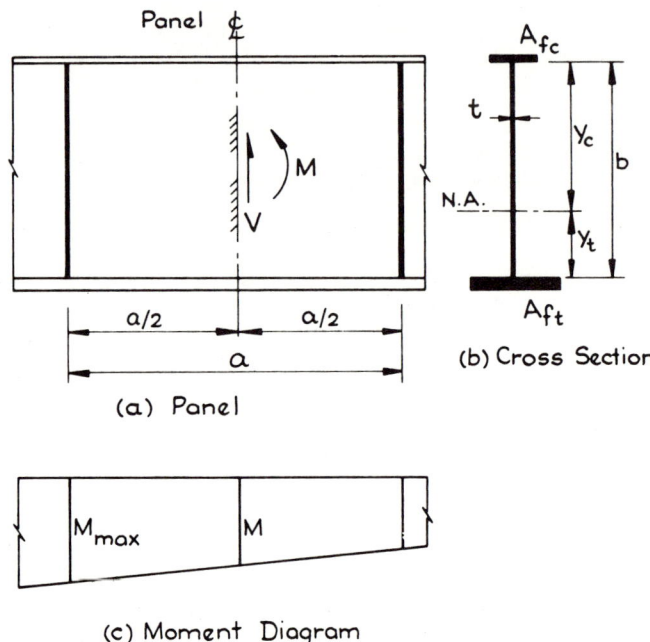

(a) Panel

(b) Cross Section

(c) Moment Diagram

FIG. I   PLATE GIRDER PANEL WITH DESIGN LOADS

Since for a particular arrangement of loads on a plate girder the moment in a panel is directly proportional to the shear in it, it is convenient to define the moment in terms of the shear span ratio μ.

$$\mu = \frac{M}{bV} \tag{1}$$

Thus,

$$M = \mu bV \tag{2}$$

Then, V is the only loading parameter needed in analysis.

The behaviour of a girder panel can be subdivided into two loading ranges — pre-buckling and post-buckling. The distribution of stresses of the two ranges is different, and therefore different analytical models should be employed in arriving at the ultimate load condition.

Beam Action

In the pre-buckling range it is assumed that the web plate is perfectly flat and does not deflect laterally. Then the ordinary beam theory is applicable to describe the distribution of stresses in the web plate and flanges as shown in Fig. 2. (For the sake of simplicity the shear stress is assumed to be constant rather than parabolic across the depth.) This state of stress exists up to the point of web buckling. As indicated in Fig. 2, it is assumed that the web plate is fixed at the flanges and simply supported at the transverse stiffeners. The buckling stress intensity is computed from an inter-action equation relating bending and shear stresses. The panel shear at the point of buckling is the 'beam action' shear $V_{\tau c}$.

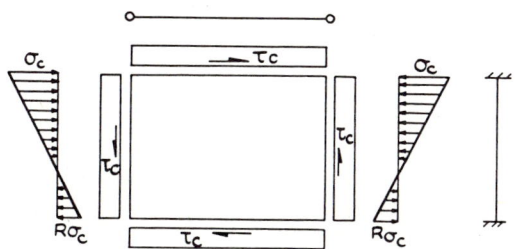

FIG. 2   BEAM ACTION (WEB BUCKLING STRENGTH)

Figure 5 shows an interaction diagram between moment and shear for a typical unsymmetrical plate girder panel. The shear and moment values are non-dimensional-ised by the ultimate shear and moment capacities for the cases of pure shear and pure bending, respectively. The right and left parts are for the larger portion of the web plate being under compression or tension as indicated by the small sketches under the diagram. The beam action shear contribution is given for various combinations of moment and shear by the cross-hatched area. It is seen to constitute only a portion of the total strength.

Tension Field Action

Behaviour of the web after buckling has been customarily described by the formation of a tension diagonal. It has been assumed that the stresses in the tension diagonal may be directly added to the stresses at buckling, and the ultimate condition is reached when the web yields under this combination.[2,7] These assumptions are also accepted as valid in the method proposed here. However, a pattern of stress distribu-tion in the tension field diagonal differing from those assumed by others is used.

A web plate panel under stresses assumed to exist in the post-buckling range is shown in Fig. 3. For simplification and with little loss of accuracy, the linearly vary-ing bending stress of Fig. 2 is replaced by the rectangular patterns shown at the sides. Yielding of the web in the cross-hatched triangle under the combination of the

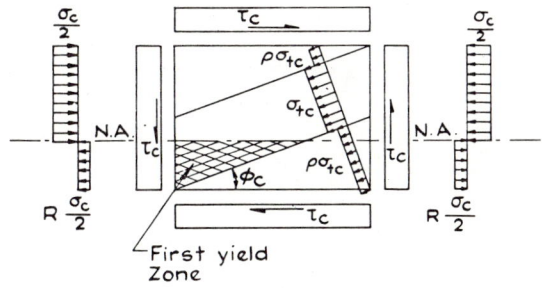

FIG 3   TENSION FIELD ACTION
(POSTBUCKLING WEB STRENGTH)

tensile bending stress $R\,(\sigma_c/2)$, the shear buckling stress τ_c and the tension field diagonal stress σ_{tc} limits the strength of the web.

The additional shear force needed to produce this condition, $V_{\sigma c}$, is given as the vertical component of the total tension field force. Its maximum value is obtained by optimising it with respect to the inclination of the tension field ϕ_c. The horizontal component of the tension field force and the increase in the panel moment due to it are assumed to be carried by the flanges. The contribution of the tension field action is substantial as shown in Fig. 5.

Frame Action

Since the flanges must deform to accommodate distortions of the web plate, they are assumed to form a plastic frame mechanism as shown in Fig. 4. Plastic moments at the ends of the flanges are computed by including a small portion of the web plate

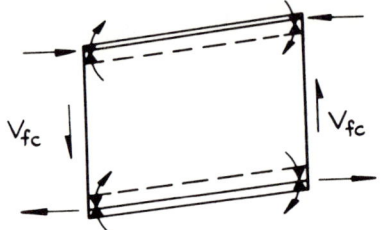

FIG 4   FRAME ACTION (FLANGE STRENGTH)

and considering the axial forces which develop due to the panel moment and the tension field action. The resultant frame action contribution to the shear carrying capacity of the girder panel, V_{fc}, is shown in Fig. 5 by the narrow vertically-hatched band.

The total shear strength of the panel is then found as the sum of the individual contributions as shown in the insert above the interaction diagram in Fig. 5.

$$V_{th} = V_{\tau c} + V_{\sigma c} + V_{fc} \tag{3}$$

Curve $Q_4 - Q_1 - Q_2$ gives this strength.

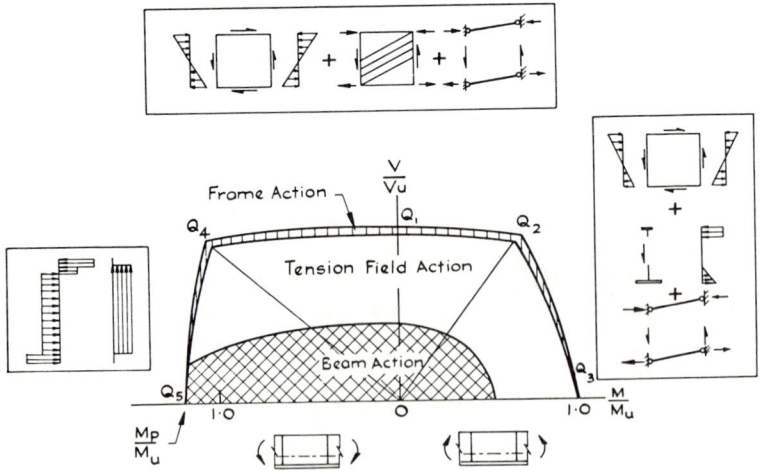

FIG. 5 SCHEMATIC INTERACTION DIAGRAM AND FAILURE MODEL

Compression Flange Failure

Since yielding or lateral buckling or local buckling limits the capacity of the compression flange to carry an axial force, the forces introduced into the flange by moment, tension field and frame action may not be able to develop to the level needed for failing the web as described above. Thus, the ultimate strength of the panel will be controlled by the failure of the compression flange. This case is depicted in Fig. 5 by curve $Q_2 - Q_3$.

The frame and tension field actions develop in this range only partially. Also, with an increasing moment-shear ratio, the tension field model degenerates into a reduction of the effective web area available to carry the moment. This case is depicted by the middle sketch of the right insert.

Tension Flange Failure

When the larger portion of the web is in tension, the panel will tend to plastify and in the limit reach the plastic moment capacity. This mode of failure is shown in Fig. 5 by curve $Q_4 - Q_5$. The corresponding panel stress distribution is in the left insert.

3. BENDING STRENGTH

Compression Flange Failure

As discussed above, the ultimate strength of a panel under pure bending and having the larger portion of the web in compression may be assumed to be controlled by the failure of the compression flange column.[11] A comparison of the available design methods[5,6] showed a good correlation with this assumption if the methods are somewhat modified to account for unsymmetrical and hybrid sections. With these modifications the ultimate moment is given by

$$Mu = \frac{I}{y_c} \sigma_{cf} \left\{ \frac{\sigma_{yw}}{\sigma_{cf}} \left[\frac{I_w}{I} - 0.002 \frac{y_{ct}}{A_{fc}} \left(\frac{y_c}{t} - 2.85 \sqrt{\frac{E}{\sigma_{yw}}} \right) \right] + \left(1 - \frac{I_w}{I} \right) \right\} \tag{4a}$$

where A_{fc} = area of the compression flange
I = moment of inertia of the section
I_w = moment of inertia of the web plate about the centroidal axis of the whole section
t = web thickness
y_c = distance from the centroidal axis to the compression edge of the web (see Fig. 1)
E = modulus of elasticity
σ_{yw} = yield stress of the web
σ_{cf} = buckling stress of the compression flange

The limitations on Eq. 4a are the following:

$$\text{if } \sigma_{yw} > \sigma_{cf}, \text{ use } \sigma_{yw} = \sigma_{cf} \tag{4b}$$

$$\text{and if } \left(\frac{y_c}{t} - 2.85 \sqrt{\frac{E}{\sigma_{yw}}} \right) < 0, \text{ use } \left(\frac{y_c}{t} - 2.85 \sqrt{\frac{E}{\sigma_{yw}}} \right) = 0 \tag{4c}$$

The buckling stress of the compression flange is computed as follows.

(1) *Lateral buckling* $\left(\dfrac{2c_c}{d_c} \leqslant 12 + \dfrac{L}{2c_c}\right)$:

$$\sigma_{cf} = \left(1 - \frac{\lambda_L^2}{4}\right)\sigma_{yc} \tag{5a}$$

$$\text{for } 0 < \lambda_L < \sqrt{2}$$

or

$$\sigma_{cf} = \frac{1}{\lambda_L^2}\sigma_{yc} \tag{5b}$$

$$\text{for } \quad \lambda_L \geqslant \sqrt{2}$$

and

$$\lambda_L = L\sqrt{\frac{\sigma_{yc}}{E\pi^2} \cdot \frac{A_{fc} + (1/3)y_ct}{I_f}} \tag{5c}$$

where the additional notation is

c_c = half width of the compression flange
d_c = thickness of the compression flange
L = unbraced length of the compression flange
I_f = moment of inertia of the compression flange about the vertical axis
σ_{yc} = yield stress of the compression flange.

(2) *Local (Torsional) buckling* $\left(\dfrac{2c_c}{d_c} > 12 + \dfrac{L}{2c_c}\right)$:

$$\sigma_{cf} = \left[1 - 0 \cdot 53 (\lambda_t - 0 \cdot 45)^{1 \cdot 36}\right]\sigma_{yc} \tag{6a}$$

$$\text{for } \quad 0 \cdot 45 \leqslant \lambda_t < \sqrt{2}$$

or

$$\sigma_{cf} = \frac{1}{\lambda_t^2}\sigma_{yc} \tag{6b}$$

$$\text{for } \quad \lambda_t \geqslant \sqrt{2}$$

where

$$\lambda_t = \frac{c_c}{d_c}\sqrt{\frac{12(1 - v^2)\sigma_{yc}}{0 \cdot 425\pi^2 E}} \tag{6c}$$

Tension Flange Failure

When the tension portion of the web is sufficiently larger than the compression portion ($R < -1$, see Fig. 2), the bending capacity of the panel may go up to the plastic

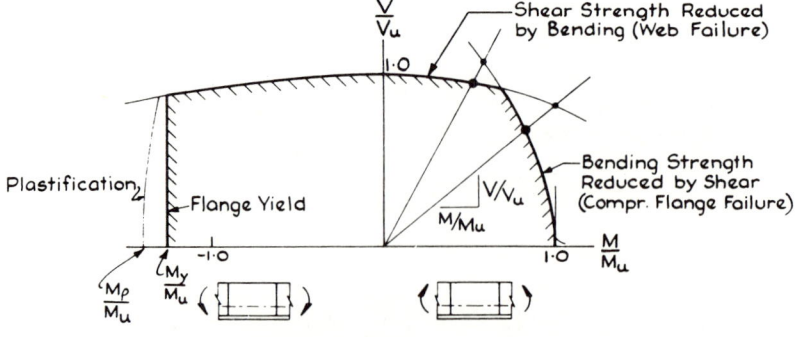

FIG. 6   DESIGN   INTERACTION   DIAGRAM

moment M_p. Due to some uncertainties in the behaviour of a very slender web, it is more conservatively assumed that the capacity is limited by the yielding of the tension flange, M_y. Thus, the left limit of the interaction diagram in Fig. 5 is changed as shown at the left end in Fig. 6.

$$Mu = \frac{I}{y_t}\sigma_{yt}\left[1 - \frac{I_w}{I}\left(1 - \frac{\sigma_{yw}}{\sigma_{yt}}\right)\right] \tag{7}$$

where y_t = distance from the centroidal axis to the tension edge of the web
(see Fig. 1)
σ_{yt} = yield stress of the tension flange

4. SHEAR STRENGTH

A parametrical study of the numerical output from a computer program based on the theory described earlier showed that the ultimate shear strength of a panel can be efficiently computed using simple formulas. Performing a statistical approximation for each individual contribution, a set of such formulas was evolved.

For the case of pure shear these formulas, after adding the individual contributions according to the form of Eq. (3), are as follows:

$$V_u = V_p\left\{\left[1 + 4\cdot3(0\cdot58 - \lambda)^{1\cdot56}\right] + \frac{\sqrt{3}}{2}\frac{A_{fc}d_c + A_{ft}d_t}{a\,A_w}\left(\frac{\sigma_{yf}}{\sigma_{yw}}\right)\right\} \tag{8a}$$

for $\lambda \leqslant 0\cdot58$ (strain-hardening range)

$$\text{or } V_u = V_p\left\{\left[1 - 0\cdot615(\lambda - 0\cdot58)^{1\cdot18}\right] + \left[\frac{9\cdot6\lambda - 0\cdot348}{\sqrt{\alpha^2 + 1\cdot6}}\right] + \frac{\sqrt{3}}{2}\frac{A_{fc}d_c + A_{ft}d_t}{a\,A_w}\left(\frac{\sigma_{yf}}{\sigma_{yw}}\right)\right\}$$

for $0\cdot58 < \lambda \leqslant \sqrt{2}$ (elastic-plastic range) (8b)

$$\text{or } V_u = V_p\left\{\left[\frac{1}{\lambda^2}\right] + \left[\frac{0\cdot9 - 0\cdot787/\lambda^2}{\sqrt{\alpha^2 + 1\cdot6}}\right] + \frac{\sqrt{3}}{2}\frac{A_{fc}d_c + A_{ft}d_t}{a\,A_w}\left(\frac{\sigma_{yf}}{\sigma_{yw}}\right)\right\} \tag{8c}$$

for $\lambda > \sqrt{2}$ (elastic range)

where A_{ft} = area of the tension flange
d_t = thickness of the tension flange
a = panel length (see Fig. 1)
α = a/b, aspect ratio of the panel
$V_p = \frac{1}{\sqrt{3}}\sigma_{yw}A_w$, plastic shear of the web
σ_{yf} = yield stress of the flanges, the same for both

and

$$\lambda = \frac{b}{t}\sqrt{\frac{12(1 - v^2)}{\sqrt{3}\pi^2 E}\frac{\sigma_{yw}}{k_v}} \tag{9}$$

where v = Poisson's ratio

The plate buckling coefficient k_v is computed from the following equations based on the assumption that the web is fixed at the flanges and simply supported at the transverse stiffeners as shown in Fig. 2.[10, 15]

$$k_v = \frac{5\cdot34}{\alpha^2} + \frac{6\cdot55}{\alpha} - 13\cdot71 + 14\cdot10\alpha \tag{10a}$$

for $\alpha < 1\cdot0$

or
$$k_v = 8.98 + \frac{6.18}{\alpha^2} - \frac{2.88}{\alpha^3} \qquad (10b)$$

for $\alpha \geqslant 1.0$

The first bracketted term in Eq. 8 a, b, c is the beam action contribution divided by the plastic shear, V_τ/V_p. The tension field action is the second bracketted term in Eq. 8b and c. There is no tension field action contribution in the strain-hardening range, Eq. 8a. The frame action is the same in all ranges; it is the last term starting with $\sqrt{3}/2$ in each equation.

In the frame action contribution, the flanges are assumed to be rectangular plates with the same yield stress σ_{yf}. If the yield stresses are different, the formula can be readily modified by multiplying each flange area, A_{fc} and A_{ft}, by its appropriate yield stress and removing σ_{yf}. For non-rectangular flanges, the plastic flange moments (Fig. 4) should be computed from proper equations.

5. STRENGTH UNDER BENDING AND SHEAR

As shown in Fig. 6, the ultimate strength of a panel under a combination of bending and shear is controlled by one of the two limiting conditions, the shear strength reduced by bending (web failure) or the bending strength reduced by shear (compression flange failure). These regions are indicated in Fig. 5 by curves $Q_4 - Q_1 - Q_2$ and $Q_2 - Q_3$, respectively. In the proposed design approach it is necessary to compute both strengths for the given moment-shear combination and use the lower value as the controlling one. The two rays in Fig. 6 emanating from the coordinate origin illustrate two alternate cases. The design values are indicated by the heavy dots.

Web Failure

The shear capacity for this case is again computed as the sum of beam, tension field, and frame actions. However, these are modified for the presence of bending stresses.

The beam action shear for the combined loads is then given by (subscript 'c' stands for 'combined')

$$V_{\tau c} = \tau_c A_w \qquad (11)$$

where τ_c is the shear buckling stress of the web subjected to shear and bending stresses as shown in Fig. 2. It can be computed with adequate accuracy from the following interaction equation.[12]

$$\left(\frac{\tau_c}{\tau_{cr}}\right)^2 + \frac{1+R}{2}\left(\frac{\sigma_c}{\sigma_{cp}}\right) + \frac{1-R}{2}\left(\frac{\sigma_c}{\sigma_{cp}}\right)^2 = 1.0 \qquad (12a)$$

with
$$\sigma_c = (\mu b A_w y_c/I)\tau_c \qquad (12b)$$

where σ_c = bending buckling stress at the extreme compression fibre of the web

  τ_{cr} = shear buckling stress for pure shear; it is the product of $(\sigma_{yw}/\sqrt{3})$ and the expression enclosed by the first pair of brackets in Eq. 8a, 8b, or 8c, whichever is applicable.

  R = ratio of the maximum tensile stress (or minimum compressive stress) to the maximum compressive stress (see Fig. 2). R is negative when the stress is tensile.

$$\sigma_{cp} = k_b \frac{\pi^2 E}{12(1-v^2)}\left(\frac{t}{b}\right)^2 \leqslant \sigma_{yw}, \text{ buckling stress under pure bending with} \qquad (13a)$$
the buckling coefficient conservatively taken for $\alpha = \infty$

$$k_b = 13.54 - 15.64R + 13.32R^2 + 3.38R^3 \qquad (13b)$$

Since σ_c is directly related to τ_c, Eq. 12 can be solved for τ_c

$$\tau_c = \tau_{cr} \frac{\sqrt{F(3-R)^2 + 16} - (1+R)\sqrt{F}}{2[2 + (1-R)F]} \tag{14a}$$

where

$$F = \left(\frac{\mu b y_c A_w}{I} \cdot \frac{\tau_{cr}}{\sigma_{cp}}\right)^2 \tag{14b}$$

The tension field action contribution was found to vary very little due to the application of bending stresses. Thus, it can be assumed with an accuracy of 2% that

$$V_{\sigma c} = V_\sigma \tag{15}$$

where V_σ is the tension field portion of Eq. 8.

The frame action contribution is usually quite small in ordinary plate girders (see Fig. 5). Thus, an approximate reduction factor for considering the effect of axial flange force should affect the total strength only insignificantly. The proposed formula is

$$V_{fc} = \left(1 - \frac{\sigma_c}{\sigma_{yc}}\right) V_f \tag{16}$$

where V_f is the frame action portion of Eq. 8.

The total ultimate shear is then found by adding the results of Eqs 11, 15 and 16.

Compression Flange Failure

When the strength of the panel is controlled by the failure of the compression flange and the moment is dominant, it is still convenient to give the strength as the panel shear consisting of the beam, tension field and frame action contributions.

The beam action shear is obtained from Eq. 11 as before.

Although frame action does not fully develop, the frame action due to its rather small contribution may be assumed — without a significant error, to be given by the same equation as for the web failure mechanism, that is, by Eq. 16.

The incomplete tension field action which constitutes the remainder of the panel strength required a special study. The following formula was developed from the numerical computer output and the associated parametric study:[15]

$$V'_{\sigma c} = \frac{(A_{fc} + 30t^2)(\sigma_{cf} - \sigma_c) - \mu V_{fc}}{B\left(\frac{V_p}{V_\sigma}\right)\left(\frac{180}{b/t}\right)\sqrt{\frac{33b}{\sigma_{yw}y_c}} + \mu} \geqslant 0 \tag{17a}$$

where

$$B = 0.338\lambda - 0.196 \tag{17b}$$
$$\text{for } 0.58 \leqslant \lambda < \sqrt{2}$$

or

$$B = 0.235\lambda - 0.05 \tag{17c}$$
$$\text{for } \lambda \geqslant \sqrt{2}$$

λ is given by Eq. 9 and σ_{yw} is in ksi.

The ultimate shear is then

$$V_{th} = V_{\tau c} + V'_{\sigma c} + V_{fc} \tag{18}$$

and the moment

$$M_{th} = \mu b V_{th} \tag{19}$$

Equation 18 becomes inaccurate as the shear on the panel approaches zero and may lead to an overestimate of the moment capacity. Thus, the moment should be checked not to exceed M_u. This situation is indicated in the right corner of the interaction diagram of Fig. 6.

Maximum Panel Moment

Since in a panel under bending and shear the moment at one end of the panel is greater than the design moment at mid-panel (Fig. 1), this maximum moment may control the panel strength. It seems reasonable and sufficiently accurate to keep M_{max} less than the moment which would produce yielding according to the ordinary beam theory. Then

$$V_{th} = \frac{I\sigma_{yf}}{yb(\mu + \frac{1}{2}\alpha)} \tag{20}$$

where σ_{yf} is the flange yield stress and y the distance from the centroid to the flange for the compression or tension side, whichever gives the smaller V_{th} and thus controls.

6. CONSIDERATION OF FATIGUE

When the load application is repeated many times as is the case for bridge and crane girders, design of plate girders for ultimate strength may be inadequate due to the danger of fatigue. A source of fatigue cracks peculiar to plate girders is the lateral flexing (back-and-fourth lateral deflection) of the web plate at each load application. Experimental studies have been conducted on symmetrical plate girders[16, 17] and recommendations were made for limiting the web slenderness ratio b/t to prevent the development of fatigue cracks due to this effect.[6, 18]

The rule proposed in Reference [6] for unsymmetrical girders was critically reviewed in Reference [19] in the light of some additional test results on unsymmetrical girders. It was found quite conservative but is being endorsed here till more research is conducted. It is recommended that the web satisfy the following requirements:

$$\frac{2y_c}{t} \leqslant \frac{36{,}500}{\sqrt{\sigma_{yw}}} \geqslant \frac{b}{t} \tag{21}$$

where σ_{yw} is in ksi.

7. NUMERICAL EXAMPLE

A numerical example is given here to illustrate the application of the proposed formulas to the determination of the bending, shear and combined strengths of an unsymmetrical plate girder panel.

Given are the following dimensions (in inches) and material properties:

Panel length　　　　: $a = 85{\cdot}0$
Panel depth　　　　: $b = 48{\cdot}07$
Compression flange: $2c_c \times d_c = 10{\cdot}0 \times 0{\cdot}75$
Tension flange　　: $2c_t \times d_t = 13{\cdot}0 \times 1{\cdot}384$
Web　　　　　　: $b \times t = 48{\cdot}07 \times 0{\cdot}183$
Unbraced length of the compression flange: $L = 85{\cdot}0$
Yield stress of the compression flange　　: $\sigma_{yc} = 34{\cdot}1$ ksi
　　　　　　　　　tension flange　　: $\sigma_{yt} = 34{\cdot}1$ ksi
　　　　　　　　　web　　　　　: $\sigma_{yw} = 35{\cdot}3$ ksi

Cross-Sectional Properties: $I = 15{,}166{\cdot}0$ in^4, $I_w = 2210$ in^4, $A_{fc} = 7{\cdot}5$ in^2, $A_{ft} = 18{\cdot}0$ in^2, $A_w = 8{\cdot}8$ in^2, $y_c = 31{\cdot}67$ in, $y_t = 16{\cdot}40$ in.

Non-dimensional parameters: $\alpha = a/b = 1{\cdot}77$, $b/t = 263$, $R = -(y_t/y_c) = -0{\cdot}52$.

Loading for the combination of bending and shear: $\mu = \dfrac{M}{Vb} = 2\cdot 65$.

Bending Strength

Compression Flange Failure

Check: $2c_c/d_c = 10/0\cdot75 = 13\cdot3$

$12 + L/(2c_c) = 12 + 85/10 = \underline{20\cdot5 > 13\cdot3}$

Thus, lateral buckling of the compression flange,

Eq. 5c: $\lambda_L = 85 \sqrt{\dfrac{34\cdot1}{29,600\pi^2} \left(\dfrac{7\cdot5 + (1/3)\,(31\cdot67)\,(0\cdot183)}{(1/12)\,(0\cdot75)\,(10\cdot0)^3} \right)} = \underline{0\cdot356 < \sqrt{2}}$

Eq. 5a: $\sigma_{fc} = 34\cdot1\,(1 - 0\cdot356^2/4) = 33\cdot0$ ksi.

Check (Eq. 4b): $\sigma_{yw} = \underline{35\cdot3\text{ ksi}} > \sigma_{cf} = \underline{33\cdot0\text{ ksi}}$. Thus, use $\sigma_{yw} = \sigma_{cf} = 33\cdot0$ ksi.

Check (Eq. 4c): $\dfrac{y_c}{t} - 2\cdot85 \sqrt{\dfrac{E}{\sigma_{yw}}} = \dfrac{31\cdot67}{0\cdot183} - 2\cdot85 \sqrt{\dfrac{29,600}{33}} = 83 > 0$ o.k.

Eq. 4a: $(Mu)_c = \dfrac{15,166}{31\cdot67}(33)\left\{ \dfrac{33}{33}\left[\dfrac{2,220}{15,166} - 0\cdot002\left(\dfrac{31\cdot67(0\cdot183)}{7\cdot5} \right)(88) \right] \right.$

$\left. + \left(1 - \dfrac{2,220}{15,166}\right) \right\} = 13,600$ kip-in.

$(Mu)_c = 13,600$ kip-in.

Yielding of Tension Flange

Eq. 7: $(Mu)_t = \dfrac{15,166}{16\cdot4}(34\cdot1)\left[1 - \dfrac{2,220}{15,166}\left(1 - \dfrac{35\cdot3}{34\cdot1}\right) \right] = 32,000$ kip-in.

$(Mu)_t = 32,000$ kip-in.

Since $(Mu)_c < (Mu)_t$, the compression flange failure mode governs.

$Mu = \underline{13,600\text{ kip-in.}}$

Shear Strength

For $\alpha = 1\cdot77 > 1\cdot0$,
Eq.10b: $k_v = 10\cdot44$

$V_p = 35\cdot3\,(8\cdot8)/\sqrt{3} = 179\cdot0$ kips

Eq. 9: $\lambda = 263 \sqrt{\dfrac{12(1 - 0\cdot3^2)}{29,600\pi^2\sqrt{3}} \dfrac{(35\cdot3)}{(10\cdot44)}} = 2\cdot24 > \sqrt{2}$

Eq. 8c: $V_u = 179\left\{ \left[\dfrac{1}{(2\cdot24)^2} \right] + \left[\dfrac{0\cdot9 - 0\cdot787/(2\cdot24)^2}{\sqrt{(1\cdot77)^2 + 1\cdot6}} \right] + \right.$

$\left. \dfrac{\sqrt{3}}{2} \cdot \dfrac{7\cdot5\,(0\cdot75) + 18\cdot0(1\cdot384)}{85\cdot0(8\cdot8)} \left(\dfrac{34\cdot1}{35\cdot3} \right) \right\}$

$= 179\,(0\cdot199 + 0\cdot343 + 0\cdot034)$

$= 35\cdot6 + 61\cdot5 + 6\cdot1 = 103\cdot2$ kips

$\quad (V_\tau) \quad (V_o) \quad (V_f)$

$V_u = 103\cdot2$ kips

Combined Shear and Bending Strength

Web Failure

For $R = -0.52$

Eq. 13b: $k_b = 24.80$

Eq. 13a: $\sigma_{cp} = 24.8 \left(\dfrac{\pi^2 (29{,}600)}{12(1 - 0.3^2)} \right) \left(\dfrac{0.183}{48.07} \right)^2 = 9.6 \text{ ksi} < \sigma_{yw} = 35.3 \text{ ksi}$

Use $\sigma_{cp} = 9.6$ ksi

For $\lambda = 2.24 > \sqrt{2}$

$\tau_{cr} = \left(\dfrac{35.3}{\sqrt{3}} \right) \left(\dfrac{1}{(2.24)^2} \right) = 4.05 \text{ ksi}$

Eq. 14b: $F = \left(\dfrac{2.65(48.07)(31.67)(8.8)}{15{,}166} \cdot \dfrac{4.05}{9.6} \right)^2 = 0.975$

Eq. 14a: $\tau_c = 4.05 \dfrac{\sqrt{0.975(3 + 0.52)^2 + 16} - (1 - 0.52)\sqrt{0.975}}{2[2 + (1 + 0.52)(0.975)]} = 2.83 \text{ ksi}$

$\tau_c = 2.83$ ksi

Eq. 11: $V_{\tau c} = 2.83(8.8) = 25.0 \text{ kips}$

$V_{\tau c} = 25.0 \text{ kips}$

According to Eq. 15, the tension field action shear is from the shear strength computations above

$V_{\sigma c} = V_\sigma = 61.5 \text{ kips}$

Eq. 12b: $\sigma_c = \dfrac{2.65(48.07)(8.8)(31.67)(2.83)}{15{,}166.0} = 6.72 \text{ ksi}$

Eq. 16: $V_{fc} = 6.1 \left(1 - \dfrac{6.72}{34.1} \right) = 4.9 \text{ kips}$

$V_{fc} = 4.9 \text{ kips}$

Eq. 3: $V_{th} = 25.0 + 61.5 + 4.9 = 91.4 \text{ kips}$

$V_{th} = 91.4 \text{ kips}$

Compression Flange Failure

For $\lambda = 2.24 > \sqrt{2}$

Eq. 17c: $B = 0.477$

Eq. 17a: $V'_{\sigma c} = \dfrac{[7.5 + 30(0.183)^2](33 - 6.67) - 2.65(4.9)}{0.477 \left(\dfrac{179}{61.5} \right) \left(\dfrac{180}{263} \right) \sqrt{\dfrac{33(48.07)}{35.3(31.67)}} + 2.65} = 56.0 \text{ kips}$

$V'_{\sigma c} = 56.0 \text{ kips}$

Eq. 18: $V_{th} = 25.0 + 56.0 + 4.9 = 85.9 \text{ kips}$

$V_{th} = 85.9 \text{ kips}$

Check Maximum Panel Moment

Eq. 20: $V_{th} = \dfrac{15{,}166(34.1)}{31.67(48.07)[2.65 + \frac{1}{2}(1.77)]} = 95.9 \text{ kips}$

Ultimate Strength

From web failure $\qquad$ $V_{th} = 91 \cdot 4$ kips

From compression flange failure $\qquad$ $V_{th} = 85 \cdot 9$ kips

From maximum panel moment $\qquad$ $V_{th} = 95 \cdot 9$ kips

therefore the compression flange failure is the mode of failure and the strength is

$$V_{th} = 85 \cdot 9 \text{ kips}$$

Eq. 19: $\quad M_{th} = 2 \cdot 65 \, (48 \cdot 07)(85 \cdot 9) = 10{,}920 \text{ kip-in}$

$$\underline{M_{th} = 10{,}920 \text{ kip-in}}$$

9. NOTATION

A_{fc}	=	Area of the compression flange.
A_{ft}	=	Area of the tension flange.
A_w	=	Area of the web.
E	=	Modulus of elasticity (Young's modulus).
F	=	Factor defined by Eq. 14b.
I	=	Moment of inertia of the girder cross section.
I_f	=	Moment of inertia of the compression flange about the vertical axis.
I_w	=	Moment of inertia of the web about the centroidal axis of the whole cross section.
L	=	Unbraced length of the compression flange.
M	=	Design moment at mid-panel.
M_{max}	=	Maximum moment in panel.
M_p	=	Plastic moment of the panel.
M_u	=	Ultimate moment of the panel under pure bending.
M_y	=	Moment causing yielding of the tension flange.
R	=	Ratio of the maximum tensile stress (or minimum compressive stress) to the maximum compressive stress of the web (negative when stress is tensile).
V	=	Design shear force at mid-panel.
V_f	=	Frame action shear under pure shear.
V_{fc}	=	Frame action shear under combined loads.
V_p	=	Plastic shear of the web.
V_σ	=	Tension field action shear under pure shear.
$V_{\sigma c}$	=	Tension field action shear under combined loads.
$V'_{\sigma c}$	=	Incomplete tension field action shear under combined loads.
V_τ	=	Beam action shear under pure shear.
$V_{\tau c}$	=	Beam action shear under combined loads.
V_{th}	=	Ultimate shear strength of the panel under combined loads.
V_u	=	Ultimate shear strength of the panel under pure shear.
a	=	Panel length.
b	=	Panel depth.
c_c	=	Half width of the compression flange.
d_c	=	Thickness of the compression flange.
d_t	=	Thickness of the tension flange.
k_v	=	Plate buckling coefficient for pure shear.
k_b	=	Plate buckling coefficient for pure bending.
t	=	Web thickness.
y_c	=	Distance from the centroidal axis to the compression edge of the web.
$\alpha = a/b$	=	Aspect ratio.
ϕ_c	=	Angle of inclination of the tension field.
λ	=	Web buckling parameter, Eq. 9.
λ_L	=	Lateral buckling parameter, Eq. 5c.
λ_t	=	Torsional (local) buckling parameter, Eq. 6c.
$\mu = \dfrac{M}{Vb}$	=	Shear span ratio.
υ	=	Poisson's ratio
σ_c	=	Bending buckling stress at the extreme compression fibre of the web.
σ_{cf}	=	Buckling stress of the compression flange column.

σ_{cp} = Web buckling stress under pure bending.
σ_{yc} = Yield stress of the compression flange.
σ_{yf} = Yield stress of the flange (same for tension and compression flange).
σ_{yw} = Yield stress of the web.
τ_c = Shear buckling stress under combined loads.
τ_{cr} = Theoretical shear buckling stress under pure shear.

REFERENCES

1. Basler, K. and Thürlimann, B. Strength of plate girders in bending. *J. struct. Div. Am. Soc. civ. Engrs,* Vol. 87, ST6, August 1961, pp. 153–181.
2. Basler, K. Strength of plate girders in shear. *J. struct. Div. Am. Soc. civ. Engrs,* Vol. 87, ST7, October 1961, pp. 151–180.
3. Basler, K. Strength of plate girders under combined bending and shear. *J. struct. Div. Am. Soc. civ. Engrs,* Vol. 87, ST7, October 1961, pp. 181–197.
4. Basler, K., Yen, B. T., Mueller, J. A. and Thürlimann, B. Web buckling tests on welded plate girders. *Bull. Weld. Res. Coun.,* No. 64, September 1960.
5. American Institute of Steel Construction. *Specification for the design, fabrication and erection of structural steel for buildings.* New York, 1963.
6. Vincent, G. S. Tentative criteria for load factor design of steel highway bridges. *American Iron and Steel Institute Bulletin,* No. 15, March 1969.
7. Fujii, T. On an improved theory for Dr. Basler's theory. Final Report of the Eighth Congress of the International Association for Bridge and Structural Engineering (New York, September 1968).
8. Rockey, K. C. and Skaloud, M. Influence of flange stiffness upon the load capacity of webs in shear. Final Report of the Eighth Congress of the International Association for Bridge and Structural Engineering (New York, September 1968).
9. Ostapenko, A., Yen, B. T. and Beedle, L. S. Research on plate girders at Lehigh University. Discussion on Theme IIc. Final Report of the Eighth Congress of the International Association for Bridge and Structural Engineering (New York, September 1968).
10. Chern, C. and Ostapenko, A. Ultimate strength of plate girders under shear. *Fritz Engineering Laboratory Report* No. 328.7, Lehigh University, August 1969.
11. Chern, C. and Ostapenko, A. Bending strength of unsymmetrical plate girders. *Fritz Engineering Laboratory Report* No. 328.8, Lehigh University, September 1970.
12. Chern, C. and Ostapenko, A. Unsymmetrical plate girders under shear and moment. *Fritz Engineering Laboratory Report* No. 328.9, Lehigh University. October 1970.
13. Ostapenko, A. and Chern C. Strength of longitudinally stiffened plate girders under combined loads. *Fritz Engineering Laboratory Report* No 328.10, Lehigh University, December 1970.
14. Ostapenko, A. and Chern, C. Strength of longitudinally stiffened plate girders under combined loads. To be published in the proceedings of the International Association for Bridge and Structural Engineering Colloquium 'Design of Plate and Box Girders for Ultimate Strength', London, March 1971.
15. Ostapenko, A., Chern, C. and Parsanejad, S. Strength formulas for design of steel plate girders. *Fritz Engineering Laboratory Report* No. 382.12, Lehigh University, January 1971.
16. Yen, B. T. and Mueller, J. A. Fatigue tests of large-size welded plate girders. *Bull. Weld. Res. Coun.,* No. 118. November 1966.
17. Mueller, J. A. and Yen, B. T. Girder web boundary stresses and fatigue. *Bull. Weld. Res. Coun.,* No. 127, January 1968.
18. Patterson, P. J., Corrado, J. A., Huang, J. S. and Yen, B. T. Fatigue and static tests of two welded plate girders. *Bull. Weld. Res. Coun.,* No. 155, October 1970.
19. Parsanejad, S. and Ostapenko, A. On the fatigue strength of unsymmetrical steel plate girders. *Bull Weld. Res. Coun.,* No. 156, November 1970.

COLLAPSE OF STEEL COMPRESSION PANELS

J. B. DWIGHT

Cambridge University

SYNOPSIS

The paper summarises recent research work at Cambridge, theoretical and experimental, on the behaviour of simple compression panels in the range $b/t < 80$. Formulae for panel strength are given. The adverse effects of distortion and residual stress caused by welding are considered. It is shown that in the approximate range $b/t = 45$ to 65 panels can shed their load rather violently after reaching maximum stress, particularly if long. This invalidates the widely used 'effective width' approach. It is proposed that thin-walled beams and columns should be designed on a simple elastic basis, using a fictitious yield stress.

Notation

t	= panel thickness
b	= panel width
b_e	= effective width
a	= panel length
c	= width of residual tension zone (Fig. 3)
A	= added section area in a weld
d	= initial panel distortion (Fig. 2)
L	= column length
f	= average applied compressive stress on panel
f_c	= elastic critical stress
f_r	= residual compressive stress due to welding
f_m	= average applied stress at maximum load
f_y	= yield stress
E	= modulus of elasticity
E_t	= tangent modulus
e	= apparent longitudinal strain on whole panel
$(b/t)_y$	= b/t at which $f_c = f_y$

Stresses are given in N/mm^2:
$100 \ N/mm^2 = 6\cdot47 \ ton/in^2 = 14\cdot5 \ kip/in^2 = 10\cdot2 \ kgf/cm^2$

1. INTRODUCTION

In the field of plate buckling a vast theoretical literature has been built up since the days of Bryan, but much of this has been to little avail in practical terms. Mathematical models used have not represented real cases. Much attention has been focussed on the elastic critical stress, a quantity which only has significance for very slender plates.

The prediction of true collapse loads has until now been virtually impossible because of the computational difficulties of dealing with elasto-plastic behaviour, complicated by the presence of imperfections. Designers have had to rely heavily on empirical test data. In aircraft everything gets tested. But in the field of heavy structures, such as bridges and ships, where prototype testing is impossible and the imperfections are worse, the lack of a sound theoretical background is a disadvantage.

Bigger computers are at last making realistic collapse analysis possible. A recent study has succeeded in producing practical answers for the simple case of an unstiffened compression panel. It is hoped that this will lead the way to the solution of other, more difficult cases. This paper summarises some results obtained for simple panels and goes on to discuss how they might be applied to design of members.

2. EFFECTS OF WELDING

The imperfections in fabricated panels arise largely from the shrinkage of the welds as they cool. Broadly speaking, transverse shrinkage causes distortion of the cross-section, while longitudinal shrinkage produces locked-in stress. Both these affect buckling. The smaller the weld, the less serious the effect. An intermittent weld produces less residual stress than a continuous one. With a multi-pass weld the distortion is worse than for a single-pass weld of the same final size, but the residual stress is less. Flame-cutting introduces stress too.[6]

2.1. Distortion

In a full penetration single-pass butt-weld between two plates the effect of transverse shrinkage is simply to draw them together, with very little angular movement. If, however, the weldment is offset relative to the plate thickness, a significant angular movement will occur as the weld cools, and this causes distortion.

Thus for a corner connection between two plates the relative movement will be less with a butt-weld (Fig. 1(a)) than with a fillet weld (Fig. 1(b)). In the case of a fillet-welded T-joint (Fig. 1(c)), transverse shrinkage puts a set in the plate. The severity of these angular movements depends on the size of weld and the amount of restraint provided by the rest of the member. Methods for predicting them have been developed by workers in Moscow[2] and Osaka.[1]

In a long rectangular panel with continuous edge welds the shape of the resulting distortion will be primarily cylindrical, dying away at the ends (Fig. 2). For typical longitudinally stiffened plating the distortion d in each panel, between stiffeners, may be of the order of $b/500$. With intermittent welds on alternate sides, as is commonly practised, the effect of the transverse shrinkage may be to produce up and down ripples, which are out of phase in adjacent panels.

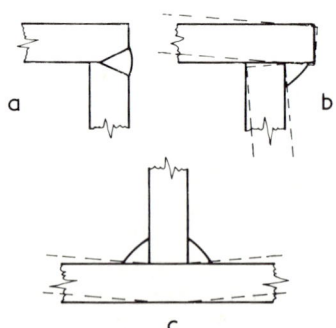

FIG.I DISTORTION CAUSED BY TRANSVERSE
SHRINKAGE OF A WELD.

Longitudinal shrinkage tends to produce overall bow in stiffened plating, but this is counteracted by the fabricator and can be ignored.

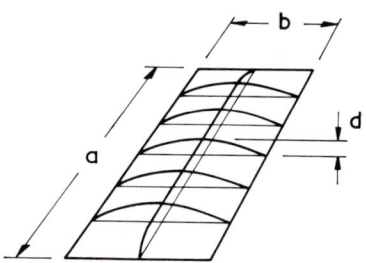

FIG. 2 TYPICAL DISTORTED SHAPE OF A PANEL
WITH CONTINUOUS EDGE-WELDS.

2.2. Locked-in Stress

The typical pattern[3] of longitudinal residual stress in a long welded plate (Fig. 3), resulting from longitudinal shrinkage, has been discussed elsewhere.[7,8,33] The extent of the (yielded) tension zones, extending typically $2t$ to $4t$ out from the weld, can be estimated from a knowledge of the welding parameters.[5,27] The stress f_r in the compression zone, which is what matters for buckling, then follows from statics.

Using an assumed rectangular stress block (curve 2) to replace the true stresses in the tension zone, the width c thereof is given by:[7]

$$c = \frac{CA}{f_y \Sigma t} \tag{1}$$

where A is the *added* cross-section area of weld and the summation Σt is extended to all the 'plates' which join at the weld (there are three at a T). The constant C is taken as 6500 N/mm² for manual stick welding with a.c.; its value for submerged arc welding is higher. Alternatively, instead of Eq. (1), it is possible to use an expression based on known heat input to the weld.[7,8]

The width thus obtained for the assumed rectangular stress block in each plate gives the same tensile force, and hence the same f_r, as would the true pattern of tensile stress. The error in assuming a rectangular pattern does not matter, as long as f_r is right. Other workers have assumed triangular[9] and trapezoidal[10] tension blocks.

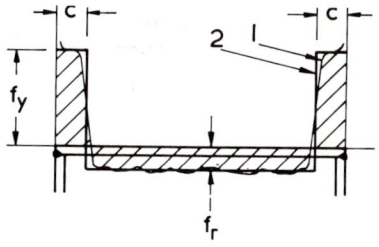

FIG. 3 PATTERN OF RESIDUAL STRESS IN A
PLATE WITH WELDED EDGES.
CURVE 1: ACTUAL .   CURVE 2: IDEALISED

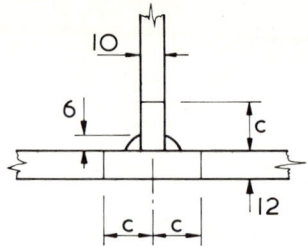

FIG. 4   ZONE OF LOCKED-IN TENSILE YIELD
AT A WELDED JOINT. TYPICAL VALUES
OF C WITH MANUAL WELDING ARE

Continuous, simultaneous   c = 35 mm
Continuous, one at a time   c = 20 mm
Intermittent staggered   c = 12 mm

The width of the tension zone decreases as the yield of the steel goes up, in such a way as to keep the total tensile force constant. For a given size of weld f_r is therefore independent of the steel used.

In a two-pass weld the second heat largely cancels out the effect of the first.[5,7,8] For the joint shown in Fig. 4 simultaneous welding of the two fillets leads to a bigger tension zone and hence a higher f_r than if they were welded separately. Using grade 43 steel with a.c. stick welding the estimated values of c are 35 mm and 20 mm respectively, so that the resulting value of f_r is 75% higher in one case than in the other. If the welds were intermittent, staggered with 80 mm weld and 240 mm miss on each side, the effective c would be about 12 mm.

With a doubly symmetrical cross-section, such as a box-member, f_r will be the same for each plate and can simply be taken as the total tension divided by the area of the compression material. In an unsymmetrical section the compressive stress may be non-uniform due to bending effects, although this will depend on the degree of restraint in fabrication. In typical stiffened plating one would expect the residual compression in the plate to be increased by this effect.

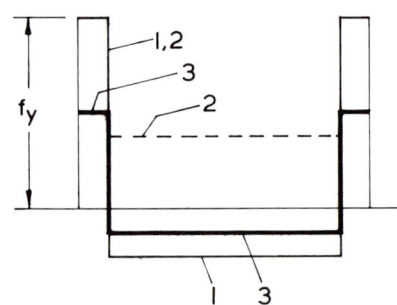

FIG. 5   SHAKE-OUT OF RESIDUAL STRESS BY
TENSILE STRAINING.

1: As welded pattern of stress.
2: With applied tension.
3: Final pattern after load is removed.

2.3. Shake-out

In Fig. 5 curve 1 shows the typical pattern of residual stress in an as-welded panel. If the whole panel is now pulled with a force T and then released, the pattern of residual stress will become altered to curve 3. This is because the applied tension is all taken by the unyielding portion (curve 2), whereas the whole plate unloads elastically when T is removed. The effect is to reduce the stresses, both tensile and compressive, in the ratio of approximately $(1 - T/btf_y)$ to one. The subsequent ability of the panel to resist buckling is improved.

This beneficial effect can easily occur in practical compression panels. In ships the top and bottom plating is continually subjected to alternating stress, thus inevitably causing some of the locked-in stress to get shaken out. In bridges the same may occur when box-members are subjected to reversed moments, during either service or erection.

It is also possible for residual stress to get shaken out during handling and transport. If stiffened plating is fabricated in long lengths and then lifted by its ends, the effect of the bending will be to reduce the level of locked-in compression in the plate when it is put down again. At the same time the length will be increased and a downward camber introduced. Messrs. Cleveland Bridge have reported that some fabricated panels for the Severn Bridge, having left Darlington dead to length and true, were found on arrival at site to have changed in length and become bowed, no doubt due to flexing in transit. This effect might be put to good use.

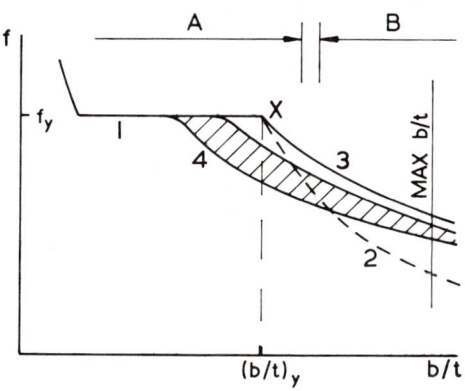

FIG 6 VARIATION OF PLATE STRENGTH WITH b/t.
1: Yield.   2: Elastic critical stress.
3: Collapse (ideal)   4: Collapse of real plates.

3. BASIC BEHAVIOUR

In Fig. 6 curves 1, 2 and 3 summarise the behaviour that would be obtained with perfect simply supported compression panels, that were initially flat and free of stress. Curve 1 denotes failure by yielding at f_y, curve 2 initial buckling at f_c and curve 3 collapse at f_m.

The elastic critical stress f_c (curve 2) is readily calculated,[36] but is of limited practical significance in the normal range of usage ($b/t < 80$). For a perfect panel the change-over from yield failure to buckling failure ($f_m = f_c = f_y$) would occur at a b/t value of $(b/t)_y = 1.9 \sqrt{E/f_y}$, which works out as follows for BS.4360 steels:

	$f_y(\text{N/mm}^2)$	$(b/t)_y$
Grade 43	250	54
Grade 50	350	45
Grade 55	450	40

For a flat panel containing welding stresses curve 2 becomes lowered by approximately f_r.

The accurate prediction of f_m is difficult by rigorous theory, even for a mathematically perfect panel. Minor imperfections, such as welding stresses and distortions, tend to reduce f_m, as is shown by the scatter band 4 in Fig. 6. Their effect is particularly marked in the practical range of b/t, and at $b/t = (b/t)_y$ the reduction in strength can be as much as 30%.

If the panel has rotational restraints at its edges, curve 2 will be lifted, but the effect on the collapse strength will be smaller.

4. DESIGN FORMULAE

T. Box[12] proposed the following, not unreasonable, formula for mild steel panels in 1883:

$$\sigma_m = \frac{80}{\sqrt{b/t}} \text{ ton/in}^2 \quad \left(= \frac{1236}{\sqrt{b/t}} \text{ N/mm}^2 \right)$$

Since then a variety of design methods have evolved.

4.1. Effective Width Concept

In the effective width method, which is widely used in structural design, above a certain b/t the true width b of a plate is replaced by an effective width b_e, the plate strength being then taken as $b_e t f_y$ ($= b t f_m$). In the method as originally proposed[13] by von Karman b_e was taken to be the width of plate for which $f_c = f_y$ (point X in Fig. 6). In fact this value, given by $b_e/t = (b/t)_y$, is too high. BS.153 employs a more realistic figure of about $0.85 (b/t)_y$ for plates that are free of residual stress (Fig. 2). A lower (varying) effective width is used for welded plates, a distinction not made in other codes.

A constant ('cut-off') effective width is not necessarily correct, and the A.I.S.I. code for light-gauge construction[14] adopts a formula which gives a b_e that increases steadily with b/t. The formula, however, is largely based on results for very wide plates up to $b/t = 400$, and it not necessarily correct for structural plating in the range $b/t < 80$; also, it refers to sheet, a different material.

The special significance of the effective width approach lies in its application to built-up members, whereby the actual section is replaced by the 'effective' parts of its component plates. The basic philosophy of this approach is questioned later.

4.2. Tangent Modulus Approach

Another approach, much used in airframe stressing and suitable for panels in region A in Fig. 6, is to treat the failing stress f_m as an 'inelastic buckling stress', which is obtained by using a modified modulus E' instead of E in the formula for the elastic critical stress. This approach lends itself to aluminium, where E' can be related to the shape of the stress-strain curve. Bleich[15] suggests putting $E' = \sqrt{EE_t}$, where E_t is the tangent modulus. Putting $E' = E_t$ is safe, and is convenient in that it leads to the 'equivalent slenderness ratio' concept, as is used in the new code for structural aluminium (CP.118); with this the failing stress is simply obtained by entering the (tangent modulus) column curve at $L/r = 1.65 \ b/t$.

The concept of a tangent modulus breaks down for structural steel with its elastic/perfectly plastic stress-strain curve. Richmond[16] would overcome this by employing a fictitious value of E_t, obtained by working back from the column curve. This approach lines up with German thinking (DIN 4114).

4.3. Maximum Stress Formulae

The tangent modulus kind of approach, when applied to panels in region B of Fig. 6, fails to take advantage of the post-buckled reserve of strength. This can be obtained from empirical formulae of the type used for aluminium (see CP.118):

$$f_m = C f_c^n f_y^{1-n} \tag{2a}$$

in which C and n are based on appropriate test data. In steel they would vary with the level of residual stress; rough values might be $C = 0.75$, $n = 0.3$.

In ship design use has been made of the following empirical formula:[11]

$$\frac{f_m}{f_y} = 1.18 \frac{(b/t)_y}{b/t} - 0.35 \left\{ \frac{(b/t)_y}{b/t} \right\}^2 \tag{2b}$$

5. ANALYTICAL TECHNIQUES

5.1. The Problem

The difficult analytical problem is to predict the collapse load of panels subjected to endlong in-plane compression. For panels lying in region A of Fig. 1 a successful theory must allow for the interaction between buckling and yield, which may reduce the collapse load to a value below f_c or f_y. In region B it must take account of the post-buckled reserve of strength. In either case the distortions and stresses produced in fabrication must be catered for.

Edge restraints are of two kinds: resistance to (a) rotation and (b) pull-in. Practical panels tend to be simply supported or nearly so, as far as (a) goes. Two important cases of (b) are: complete freedom to pull in (unstiffened box-column), and ability to pull in straight (wide stiffened plating).

5.2. Approximate Methods

Various writers have studied the elastic post-buckling behaviour of simple compression panels. The work has been based on the von Karman large deflection equations, using approximations to the true deflected shape, and has generally referred to simply supported panels. Much of it is well summarised in Reference[18]. Various expressions have resulted for the distribution of stress across the width of an elastic post-buckled panel.[37]

Estimates of panel strength can be made[19,20,21] by taking the elastic stress distribution, thus obtained, and assuming that when the edge stress reaches yield the panel is then carrying its maximum load (Fig. 7(a)). Ostapenko[21] has extended this idea to welded panels by assuming that for them collapse occurs when compressive yield is reached just outside the residual tension zone, this zone itself being then at low stress (Fig. 7(b)). This concept is supported by some Cambridge test observations[32] in which slip-lines were seen at the appropriate distance in from the weld in a welded plate, whereas a stress-relieved one had them right at the edge.

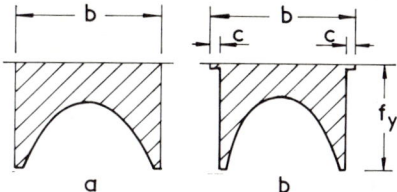

FIG. 7 STRESS DISTRIBUTION ACROSS WIDTH OF PANEL AT MAXIMUM LOAD.
(a) Unwelded (b) Welded

Another technique,[22] similar to the Merchant method for frames, is to draw an elastic loading line and a rigid-plastic unloading line on the same load-deflection plot. f_m is then estimated by using an empirical interaction to round off the corner where the curves cut. The plastic unloading line is based on a 'pitched roof' mechanism.

Both these approaches are unable to produce reliable predictions in the important region $b/t \simeq (b/t)_y$, where imperfections have a marked effect.

5.3. Elastoplastic Analysis

Ractliffe[23,24] developed a relatively simple elasto-plastic analysis for finding f_m which is less intuitive than the methods described above. He considers a square panel and assumes a sinusoidal deflected form. The unloaded edges are allowed to pull in while staying straight. The equilibrium of one quadrant of the panel is satisfied, taking into account the actions along the four sides thereof and including the effects of plasticity; no attempt is made to satisfy equilibrium at points within the quadrant. The method can allow for an initial out-of-flatness of the same form as the initial elastic buckle, and also for welding stresses. This has yielded good results which agree well with the more thorough analyses described below.

These are the energy methods developed by Graves Smith[25,26] and by Moxham,[27,28] both of which involved major efforts of computation. Both consider a simply supported panel, one buckle long, with the unloaded edges free to pull in. The plate is assumed to be strained between parallel platens. At each increment of strain the coefficients in the description of the deflected shape are adjusted so as to minimise the strain energy, and the load on the plate thus obtained. In this way the complete load-end shortening curve is determined and hence f_m.

Moxham divided the plate into 1620 volumes (18 across x 18 along x 5 through the thickness), and described the deflected form with a limited number of Fourier terms. Strain reversal was correctly treated. After some initial trials he adopted a standard length to width ratio a/b (Fig. 2) of 0·875, which gave a slightly lower f_m than that for a square panel. The program was able to allow for residual stresses, using the idealised pattern in Fig. 3 and assuming it to persist for the whole length of the panel. Three kinds of initial out-of-flatness were considered, all sinusoidal, as shown in Fig. 8.

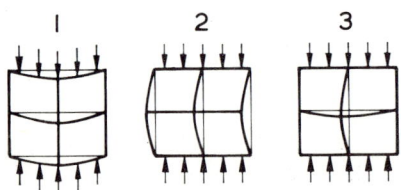

FIG. 8   TYPES OF INITIAL DISTORTION.

1: Longitudinal   2: Transverse.

3: Spherical

6. RESULTS OF COLLAPSE ANALYSIS

Some of the theoretical results obtained by Moxham[27] for simply supported panels are presented in this section. A full report is available.[28]

6.1. Effect of Imperfections

Fig. 9 compares load-end shortening $(f-e)$ curves for unwelded panels having 'longitudinal' and 'transverse' imperfections (see Fig. 8) of varying severity. The longitudinal imperfections are seen to have negligible effect on the maximum load, whereas the transverse ones can reduce it seriously. The greatest sensitivity is in the region where $f_c \simeq f_y$; at $b/t = 55$ a transverse imperfection of only $b/2000$ $(= t/36)$

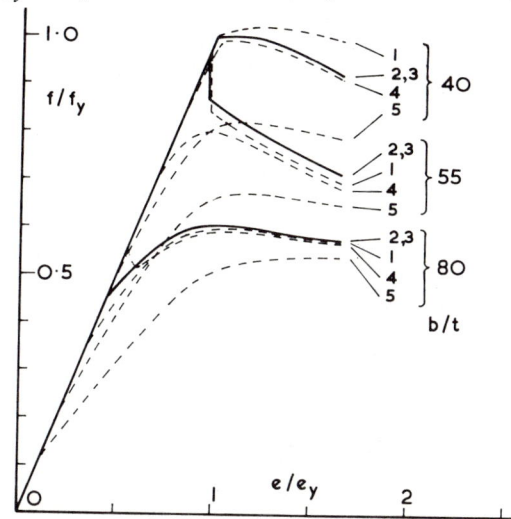

FIG. 9   LOAD-END SHORTENING CURVES FOR DISTORTED, UNWELDED PANELS (c = o).

f_y = 250 N/mm². a/b = 0.875
  1. d = b/200 (longitudinal)
  2. d = b/2000 (longitudinal)
  3. d = o
  4. d = b/2000 (transverse)
  5. d = b/200 (transverse)

causes a drop in strength of 17% as compared with a perfect panel. However, the extra strength with a longitudinal imperfection is of limited use, because of the violent nature of the failure.

Further computations have shown that the effect of a 'spherical' imperfection is very close to that of a transverse one of the same depth.

The typical distortion in an edge-welded plate is as shown in Fig. 2. For most of the way the imperfection is longitudinal, but it becomes 'hemispherical' at the ends; its depth is perhaps $b/500$. With intermittent welding the distortion will probably be in ripples. It is suggested that for practical strength prediction a spherical imperfection of $d = b/1000$ should be adopted. This smaller value is selected as an allowance for the fact that the shape of the initial distortion will differ from the preferred buckling mode.

6.2. Panel Stress-strain Curves

Fig. 10 shows the theoretical behaviour of short mild steel panels having an aspect ratio (a/b) of 0.875. The curves give average applied stress f plotted against proportionate change in length e, for various b/t. They are based on an assumed 'spherical'

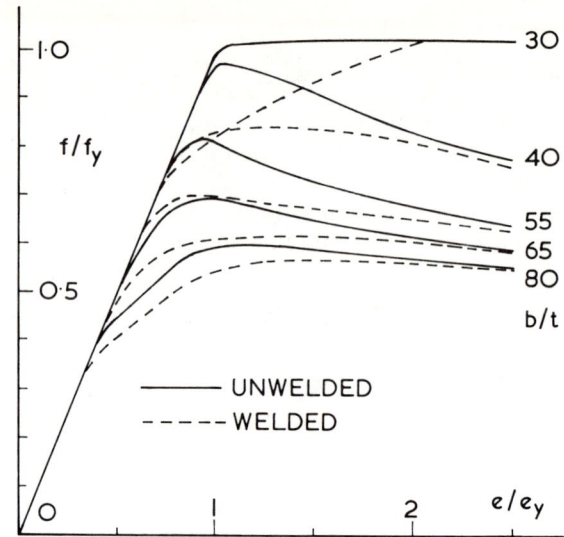

FIG. 10   LOAD-END SHORTENING CURVES FOR
UNWELDED (c = o) AND WELDED (c = 3t) PANELS.
Spherical imperfection, d = b/1000.
f_y = 250 N/mm². a/b = 0·875.

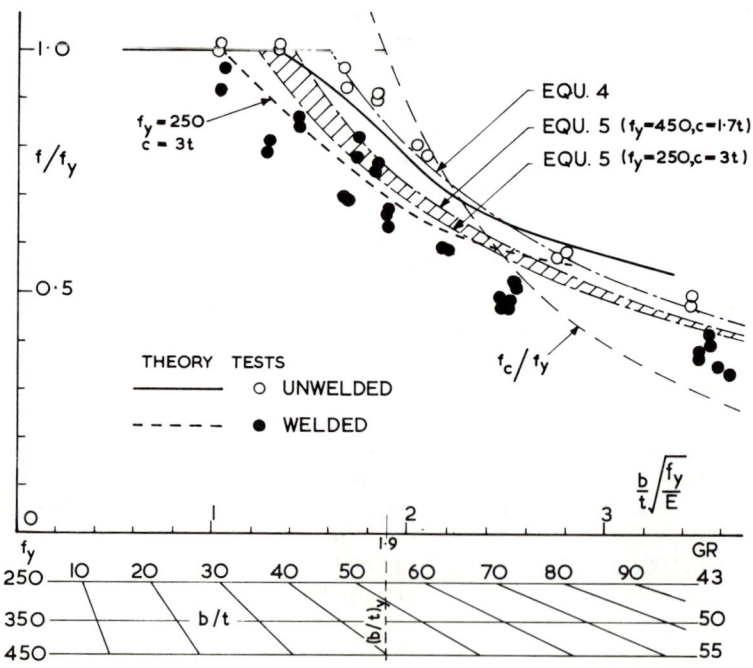

FIG. 11   COMPARISON OF THEORY WITH RESULTS OF
BOX-COLUMN TESTS.

out-of-flatness (Fig. 8) of $b/1000$. The full curves refer to initially stress-free panels; the broken ones refer to as-welded panels having $c = 3t$ (Fig. 3), a typical figure.

The stress-free panels in the range $b/t = 40$ to 65 show definite load shedding after reaching maximum load. Orthodox post-buckling behaviour is only seen at $b/t = 80$.

The welded panels have a reduced maximum load (except for $b/t = 30$), but unload more gently. At $b/t = 40$ to 55 the reduction in strength, compared to a stress-free panel, is 14%.

6.3. Panel Strength

The theoretical variation of strength (f_m) with b/t is plotted in Fig. 11. Curves are given for the stress-free case and the welded case $c = 3t$, based on a spherical out-of-flatness of $b/1000$.

The curves were computed for grade 43 steel ($f_y = 250 \text{ N/mm}^2$). With the non-dimensional presentation used, the curve for unwelded panels is also applicable to other grades.

7. EFFECT OF PANEL LENGTH

The above results were all computed for a short panel containing a single buckle. The values obtained for f_m apply equally well to a long panel, which can be considered as a number of short panels in series. But the panel stress-strain curves will be different for a long panel, as explained by Ractliffe.[23]

Fig. 12 considers two panels of the same b/t. The short one contains a single buckle. The other is four times as long, and might be thought to contain four identical buckles. Curve 1 is the panel stress-strain curve for the short panel.

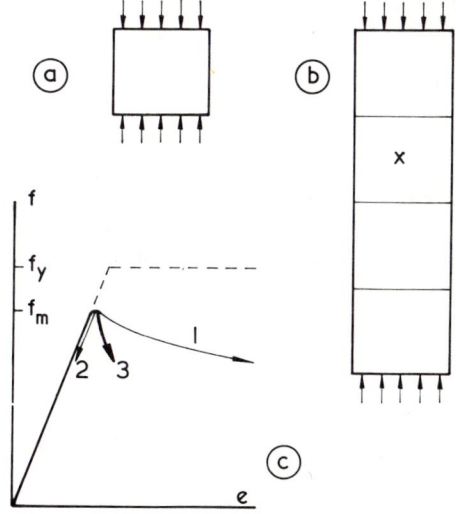

FIG. 12 EFFECT OF LENGTH ON PANEL BEHAVIOUR.
(a) Short panel containing single buckle.
(b) Long panel containing four buckles.
(c) Panel stress-strain curves.
1 : Curve for short panel.
2 : Unloading of non-critical buckles.
3 : Overall curve for long panel.

When the longer panel is subjected to increasing strain, the four buckles in it grow equally until maximum load is reached. Suppose buckle X has an infinitesimally lower f_m than the other three. After X reaches f_m further straining causes a fall-off in load; buckle X carries on down curve 1 with its length decreasing, but the others

FIG. 13 BUCKLED PANEL IN STIFF TEST RIG. SHOWING
LOCALISED FAILURE. FINGERS AT EDGES OF
SPECIMEN CAN PROVIDE S.S. OR CLAMPED EDGE
CONDITION

unload elastically and come back down curve 2 with their length *increasing*. The net result is that the falling branch of the stress-strain curve for the whole panel is as curve 3, and is steeper than it would be for a short one of the same b/t.

Only buckle X will be visible after the load has been removed. In tests to collapse on long panels it is indeed found that only one of the incipient buckles ever develops and that the rest of the plate stays virtually flat (Fig. 13). The phenomenon is analogous to necking in a tensile test. Ractliffe showed a photograph of a plate containing two developed buckles; this is rare, as are tensile specimens with two necks.

The possible violence of the load-shedding in longer panels (curve 3), which the above argument reveals, should be of concern to designers. It throws in question the whole concept of the effective width approach.

8. TEST RESULTS

8.1. Test on Box-columns

Tests[29,30,31,32] on some fifty square welded box-columns, of low L/r, have been summarised before.[33] They comprised as-welded and also stress-relieved specimens, with yields in the grade 43 to 50 range. The largest specimens carried over 800 tons (8 MN).

The collapse loads obtained are plotted in Fig. 11, for comparison with theory. In comparing the as-welded results it must be remembered that the width of the residual tension zone for the test columns was somewhat variable, with c ranging from $1.9t$ to $5.4t$.

The theoretical curves show good agreement.

8.2. Tests on Individual Plates

Tests on individual plates have been done by Ractliffe[23,24] and by Moxham.[27,34] In both cases the unloaded edges were supported by an array of specially designed fingers (Fig. 13) which were able to give either simple support or clamped edge condition, with freedom to pull in. The fingers took none of the endlong load, and were

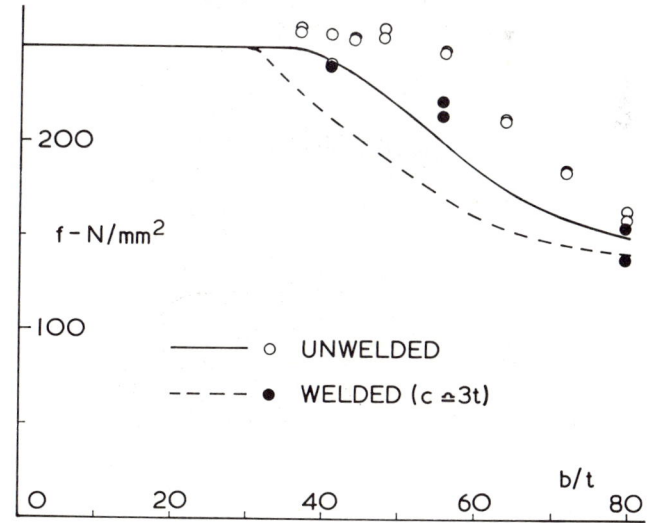

FIG. 14   COMPARISON OF THEORY WITH RESULTS OF
TESTS ON INDIVIDUAL SIMPLY SUPPORTED
PLATES. $f_y \simeq$ 250^c

better than a *V*-groove arrangement. Controlled residual stresses in the plates were introduced by means of edge welds. The aspect ratio *a/b* was 4·0.

Ractliffe tested 15 steel plates, 6·5 mm thick. As previously reported,[33] the simply supported results agreed well with the box-column findings.

Moxham's programme was on a model scale, and comprised some 150 tests on 3·2 mm mild steel sheet material. The full results are given elsewhere.[34] Their particular value lies in the many load-end shortening curves obtained. The failing stresses for the simply supported specimens, after allowing for the strength of the steel, were consistently higher than those for the box-columns and lay well above the theoretical predictions (Fig. 14). This is thought to be due to the nature of the sheet material used, which exhibited some strain-hardening and no definite yield in compression, and also to the extreme flatness of the specimens.

8.3. Shape of Buckle

Fig. 13 shows the typical 'pitched roof' form of the developed buckle in an individual plate test. The buckle is seen to be shorter than square, which supports Moxham's use of *a/b* < 1 in his theoretical work (see section 5.3).

In the box-column tests the buckles were of the same symmetrical form provided *b/t* was less than about 65. At higher *b/t* an entirely different asymmetric shape appeared[30], as has also been observed by Watson and Babb.[35] The asymmetric form was not seen in the individual plate tests.

8.4. Stress-strain Curves

A few of the experimental load-end shortening ($f-e$) curves obtained by Moxham for individual simply supported plates are given in Fig. 15. They refer to unwelded

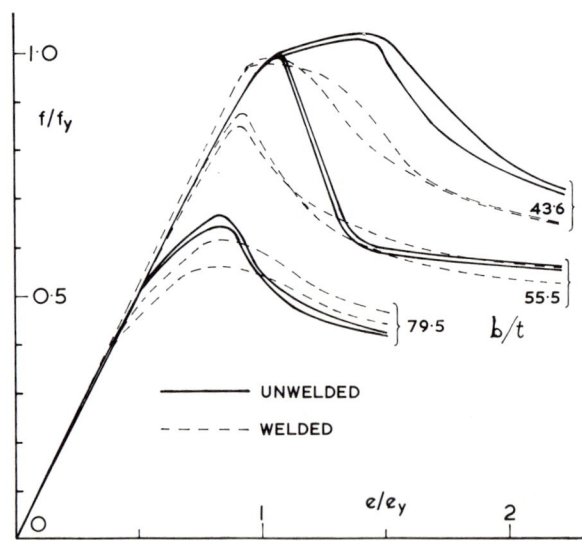

FIG. 15 EXPERIMENTAL LOAD-END SHORTENING CURVES FOR INDIVIDUAL SIMPLY SUPPORTED PLATES, UNWELDED (c = 0) AND WELDED (c ≈ 3t). $F_y \approx 250 N/mm^2$ a/b = 4.

and welded plates ($c \simeq 3t$) at three values of b/t. The point of most interest is their marked peakiness, when compared with the theoretical curves of Fig. 10. This is explained by the length effect discussed in section 7 above, the test panels having been $4b$ long, whereas the theoretical curves refer to short panels that are one buckle long. Allowing for this, the curves agree well in general shape with the theoretical ones. Ractliffe's tests gave similar results.

8.5. Effect of Edge Restraint

Fig. 16 compares $f-e$ curves for welded plates ($c \simeq 3t$) having simply supported and clamped edges respectively. The clamping is only seen to have a significant effect at b/t beyond about 70, where both f_c and f_m are lifted.

Typical panels tend to have only a small degree of edge restraint if any — far less than full clamping — and the above results suggest that the effect on the panel strength will be not worth considering.

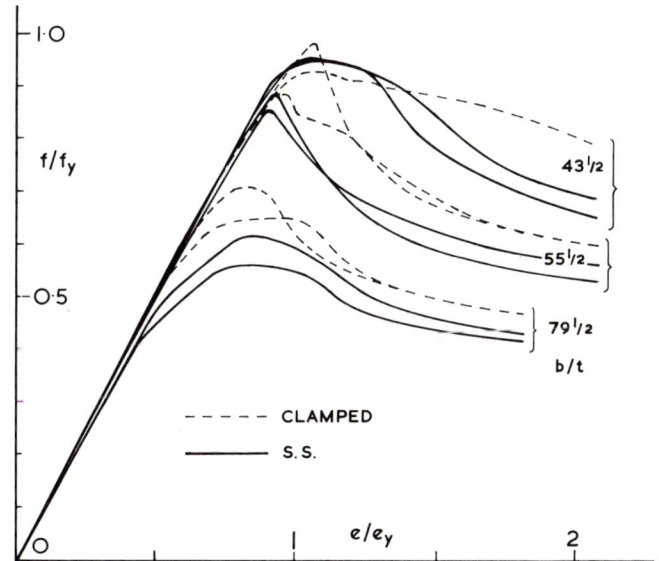

FIG.16 EXPERIMENTAL LOAD-END SHORTENING CURVES
FOR SIMPLY SUPPORTED AND CLAMPED PLATES.
WELDED ($c \simeq 3t$). $F_y \simeq 250 \text{N/mm}^2$ $a/b = 4$

9. PROPOSED RULES FOR SIMPLE COMPRESSION PANELS

The following design procedure is proposed for obtaining the strength of simple compression panels, based on the theoretical and experimental results discussed above. It is summarised in Fig. 17.

(a) Panel strength to be determined on a stress basis (f_m) and not by effective width.
(b) Panels to be taken as simply supported, and any effect of edge restraint ignored.

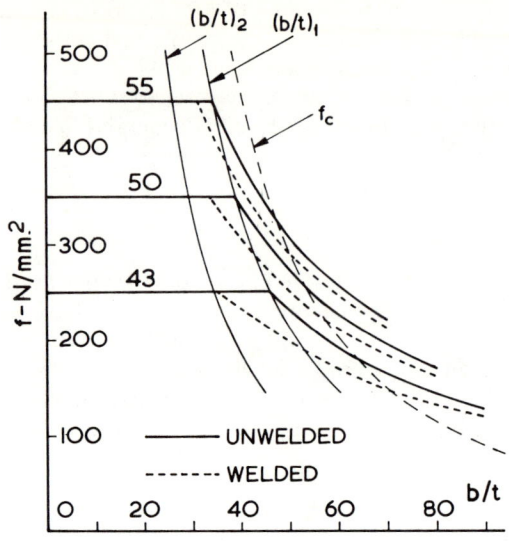

FIG. 17   PROPOSED   DESIGN RULE FOR PANEL
STRENGTH. $(b/t)_2$ = plastic design limit
$(b/t)_1$ = limit of yield attainment
for unwelded plates.
Full lines........unwelded
Broken lines.....welded.

(c) Stress-free (unwelded) panels: These may be stressed up to yield provided
that $b/t < (b/t)_1$, where:

$$(b/t)_1 = 0.85\,(b/t)_y = 1.6\,\sqrt{E/f_y} \qquad (3)$$

When b/t exceeds this value, the strength is given by:

$$f_m = \frac{(b/t)_1}{b/t}\,f_y \qquad (4)$$

For BS.4360 steels $(b/t)_1$ is as follows:

	$f_y(\text{N/mm}^2)$	$(b/t)_1$
Grade 43	250	46
Grade 50	350	39
Grade 55	450	34

(d) Welded panels: f_m to be obtained as follows, but not to exceed yield:

$$f_m = \frac{(b/t)_1}{b/t}\,f_y - \frac{(b/t)_y}{b/t}\,f_r \qquad (5a)$$

$$= \frac{(b/t)_y}{b/t}\left\{\frac{0.85b - 3.7c}{b - 2c}\right\}f_y \qquad (5b)$$

(e) Plastic design: In plastic design, when a compression panel is required to
exhibit a plastic plateau, b/t not to exceed $(b/t)_2$ where

$$(b/t)_2 = 0.75\,(b/t)_1 = 0.64\,(b/t)_y = 1.2\,\sqrt{E/f_y} \qquad (6)$$

Equations (4) and (5) are compared with theory and with the box-column tests in Fig. 11. The curve for Eq. (5) refers to a weld size that would give $c = 3t$ for mild steel.

These rules are intended for structural panels in the range $b/t < 80$. If applied to wider panels (say $b/t = 150$), they would be too safe, and it would be desirable to take edge restraints into account, both rotational and anti-pull-in.

In the case of unwelded panels the strengths given are close to those obtained from the BS. 153 effective widths. For welded panels the proposals now allow a graded reduction in strength depending on severity of welding, as against the fixed BS. 153 reduction for any welded panel. The proposals are slightly modified from those given[33] in 1969, as a result of more data; it is found that the simple reduction of f_m by an amount f_r is unsafe at low b/t and too safe at high b/t.

10. MEMBERS

An attempt is made to apply the above findings to the design of members.

10.1. Box-beams

Consider a square box-beam (Fig. 18(a)) in steel to grade 43, with $b/t = 60$. Suppose that the corner welding is light, so that residual stresses may be ignored. The greatest compressive stress that the top plate can carry is $f_m = 0.76 f_y$, from Eq. (4). Reference to the $f-e$ curves in Fig. 15 suggests that the plate will shed its load rapidly after reaching this. Therefore the maximum moment M that the beam can carry should be calculated on a simple elastic basis, taking an extreme fibre stress of f_m. This gives $M = 3600 \, t^3 f_y$.

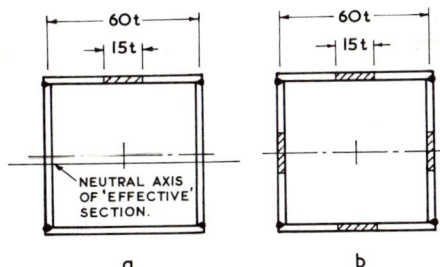

FIG. 18. BOX-SECTIONS: (a) beam (b) column.
"Ineffective" areas shown hatched.

If, instead, a conventional effective width calculation were used, in which one disregarded $15t$ of the width on the compression side and took the elastic section modulus of the rest of the section in conjunction with an extreme fibre stress of f_y, one would then get $M = 4050 \, t^3 f_y$. This is 12·5% too high.

The error of the effective width calculation is that it assumes the compression material to go on shortening at constant stress after reaching f_m, contrary to the evidence of Fig. 15. Admittedly the example chosen is severe. At high b/t, where there is an extended post-buckled range, f_m is indeed reached at a strain of f_y/E, or more if the panel is heavily welded; the effective width type of calculation may then be reasonable. But this only applies at the upper end of the practical b/t range of heavy construction. Panels in the region $b/t \doteq (b/t)_y$ collapse in a 'brittle' manner, and for them the effective width approach seems unsound.

If the full plastic moment of the member is to be developed, it is necessary to ensure that the compression material not only reaches yield, but also has a plastic plateau. In this case b/t should not exceed $(b/t)_2$ as given in section 9(e) above.

10.2. Box-column

For pin-ended struts the interaction between yield and overall instability can be conveniently allowed for by the Perry-Robertson type of formula. An initial bow is assumed, and as the strut flexes with increasing load failure is taken as occurring when yield at the extreme fibres occurs at the centre. If the assumed initial bow is realistic (say $L/1000$), the formula gives reasonable, but slightly conservative results for a stress-free member.

When the member is thin-walled, as well as being slender overall, the interaction between local and flexural buckling can be allowed for by again using the Perry-Robertson formula. Such a strut will fail when the extreme fibre stress at the middle reaches f_m, and it is therefore merely necessary to insert f_m instead of f_y in the formula. In other words, failure can be predicted by employing a strut-curve based on a fictitious yield stress of f_m.

Consider the square box-section that was discussed in the previous section (grade 43, $b/t = 60$, no residual stress). Suppose this is now used as a pin-ended strut with $L/r = 100$. Taking $L/1000$ initial crookedness and proceeding in the way proposed, we get a collapse load P of 0·55 times the squash load P_y. If, instead, the conventional effective width procedure is used, whereby we discard $15t$ from each side (Fig. 18(b)) and work with a full strength strut-curve, we get $P = 0·50 P_y$. This is an under-estimate of 10%.

In the practical use of the Perry-Robertson formula, as in the current BS. 153, an excessive initial bow is assumed ($L/410$ for a square box) in order to allow empirically for the effect of locked-in stress. The curve obtained is comparable to the many other strut-curves that are used. It is suggested that for dealing with thin-walled struts a curve of the usual form be employed, be it Perry-Robertson or of some other derivation, but drawn as if the yield stress were f_m and not f_y.

The use of a modified strut-curve based on a fictitious yield stress of f_m, for predicting local/flexural interaction, has been practised for some years in the light gauge field on both sides of the Atlantic (see Addendum No. 1 to BS.449, and the A.I.S.I. code). Heavy gauge designers would do well to follow suit. The idea is well supported by Graves Smith's results.[31]

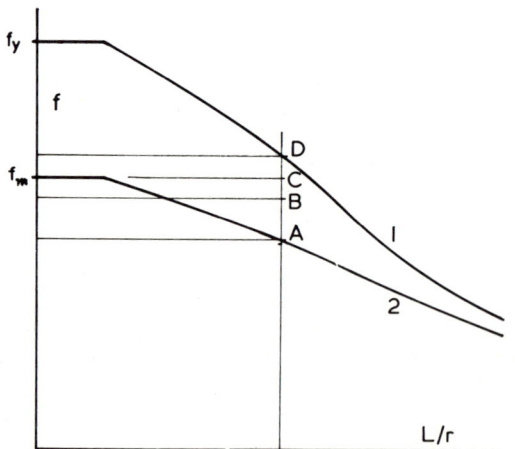

FIG. 19 BUCKLING OF THIN-WALLED COLUMNS.
1: BASIC STRUT-CURVE. 2: CURVE FOR
THIN-WALLED SECTION, BASED ON A
FICTITIOUS YIELD OF f_m.

10.2. Stiffened Plating

For wide stiffened plating carrying longitudinal thrust (Fig. 20(*a*)) the interaction between overall flexural buckling and local buckling of the plate itself can be treated in the general manner just described. But a refinement seems desirable.

In a box-column the thin material at the extreme fibres is subjected to the full augmentation of stress as the strut flexes. In stiffened plating the thin material, being closer to the centroidal axis than is the extreme fibre material at the toes of the

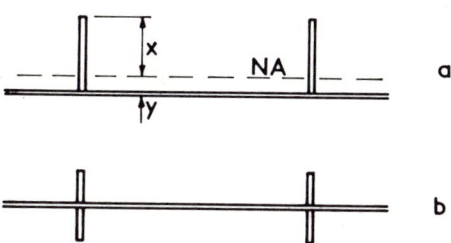

FIG. 20   STIFFENED PLATING.

stiffeners, enjoys a less severe stress augmentation. It therefore seems reasonable, for stiffened plating, to read off the critical stress at point B in Fig. 19, rather than at point A, where $BC/AC = y/x$. In the extreme case of thin material actually lying on the centroidal axis (Fig. 20(*b*)) it would be permissible to work right up to f_m (point C).

It is of course possible for B to lie above curve 1, in which case yield at the toes is the critical factor and the collapse loading must be read at point D. Torsional buckling of the stiffeners must be checked.

11. CONCLUSIONS

1. Designers should be aware of the 'brittle' behaviour of compression panels in the approximate range $b/t = 0.8$ to $1.2 (b/t)_y$, that is 45 to 65 for mild steel.
2. The maximum stress f_m that thin-walled panels can carry may be obtained from the design formulae in section 9. It is necessary to know the level of residual compression due to welding (section 2.2).
3. The 'effective width' approach should be abandoned.
4. The strength of thin-walled beams and columns should be predicted on a simple elastic basis, with the true yield stress replaced by a fictitious value equal to f_m. In the case of beams this will tend to give lower loads than the effective width method. But for columns the loads will be higher. The column treatment can be extended to cover wide stiffened plating.

ACKNOWLEDGEMENT

The author wishes to thank CIRIA for sponsoring much of the experimental work referred to in this paper. He also wishes to pay tribute to Dr. K. E. Moxham on whose results he has so freely drawn.

REFERENCES

1. Watanabe, M. and Satoh, K. Effect of welding conditions on the shrinkage distortion in welded structures. *Weld. Res.*, Vol. 26, No. 8, August 1961, pp. 377s–384s.
2. Vinokurov, V. A. Svarochiye deformatssi i napryazhenya. (Welding distortions and stresses). *Mashinostroenie*, Moscow, 1968. (H.M.S.O. translation in hand.)
3. Nishino, F., Ueda, Y. and Tall, L. Experimental investigation of the buckling of plates with residual stresses. *Symposium on Test Methods for Compression Members* (*Atlantic City, 1966*). American Society for Testing and Materials Special Technical Publication No. 419, 1967.
4. Rao, N. R. N., Estuar, F. R. and Tall, L. Residual stresses in welded shapes. *Weld. J.*, Vol. 43, No. 7, July 1964, pp. 295s–306s.
5. Tall, L. Residual stresses in welded plates – a theoretical study. *Weld. J.*, Vol. 43, No. 1, Jan. 1964, pp. 10s–23s.
6. Beedle, L. S. Ductility as a basis for steel design. *Conference on Engineering Plasticity* (*Cambridge, 1968*). Cambridge University Press, 1968, pp. 41–75.
7. Young, B. W. and Dwight, J. B. Residual stresses due to longitudinal welds and flame-cutting. C.I.R.I.A. Research Report, 1971.
8. Dwight, J. B. and Young, B. W. Residual stresses due to welding. Conference on Joints in Structures, Sheffield, 1971.
9. Fukumoto, Y. A study and investigation of the buckling of steel girders (in Japanese). Kansai Society for Study of Steel Girder Bridge Welding, Feb. 1969.
10. Ueda, Y. Elastic, elastic-plastic and plastic buckling of plates with residual stresses. *Fritz Engineering Laboratory Report*, No. 290.1, Lehigh University, 1962.
11. Frankland, J. M. The strength of ship plating under edge compression. E.M.B. Report No. 469, 1940.
12. Box, T. *A practical treatise on the strength of materials . . .* London, Spon, 1883.
13. Von Karman, T., Sechler, F. E. and Donnell, L. H. The strength of thin plates in compression. *Trans. Am. Soc. mech. Eng.* (Applied Mechanics), Vol. 54, Jan. 1932, pp. 53–57.
14. Winter, G. *Commentary on the 1968 edition of the specification for the design of cold-formed steel structural members.* New York, American Iron and Steel Institute, 1970.
15. Bleich, F. *Buckling strength of metal structures.* New York, McGraw-Hill, 1952.
16. Richmond, B. Discussion on 'Forth Road Bridge'. *Proc. Instn civ. Engrs*, Vol. 36, Jan. 1967, pp. 156–158.
17. Gerard, G. *Introduction to structural stability theory.* New York, McGraw-Hill, 1962.
18. Davidson, H. L. Post-buckling behavior of long rectangular plates. *Fritz Engineering Laboratory Report*, No. 248.15, Lehigh University, 1965.
19. Raslan, R.A.A.S. *The structural behaviour of corrugated plates.* Ph.D. thesis, Manchester University, 1969.
20. Abdel-Sayed, G. Effective width of thin plates in compression. University of Windsor, Ontario, 1969.
21. Vojta, J. F. and Ostapenko, A. Ultimate strength design of longitudinally stiffened plate panels with large b/t. *Fritz Engineering Laboratory Report*, No. 248.18, Lehigh University, August 1967.
22. Sherbourne, A. N. *The elastic-plastic behaviour of mild steel plates in compression.* Ph.D. thesis, Cambridge University, 1960.
23. Ractliffe, A. T. *The strength of plates in compression.* Ph.D. thesis, Cambridge University, 1966.
24. Dwight, J. B. and Ractliffe, A. T. The strength of thin plates in compression. *International Symposium on Thin Walled Steel Structures* (*Swansea, 1967*). London, Crosby Lockwood, 1969, pp. 3–34.
25. Graves Smith, T. R. *The ultimate strength of locally buckled columns of arbitrary length.* Ph.D. thesis, Cambridge University, 1966.
26. Graves Smith, T. R. The ultimate strength of locally buckled columns of arbitrary length. *International Symposium on Thin Walled Steel Structures* (*Swansea, 1967*). London, Crosby Lockwood, 1969, pp. 35–60.
27. Moxham, K. E. *Compression in welded web plates.* Ph.D. thesis, Cambridge University, 1970.
28. Moxham, K. E. Theoretical prediction of the strength of welded steel plates in compression. Cambridge University Report No. CUED/C-Struct/TR.2, 1971.

29. Anderson, J. K., Hamilton, J. A. K., Henderson, W., Roberts, Sir Gilbert and Shirley-Smith, H. Forth Road Bridge. 2 – Design. *Proc. Instn civ. Engrs*, Vol. 32, Nov. 1965, pp. 333–405.
30. Dwight, J. B. and Harrison, J. D. Local buckling problems in steel columns. *British Welding Research Association Report*, No. M9/63, 1963.
31. Dwight, J. B., Chin, T. K., Ractliffe, A. T. and Smith, T. R. G. Local buckling of thin-walled columns. *C.I.R.I.A. Research Report* No. 12, 1968.
32. Dwight, J. B. and Moxham, K. E. Further tests on welded box-columns. Cambridge University Report No. CUED/C-Struct/TR. 15. 1971.
33. Dwight, J. B. and Moxham, K. E. Welded steel plates in compression. *Struct. Engr*, Vol. 47, No. 2, Feb. 1969, pp. 49–66.
34. Moxham, K. E. Buckling tests on individual welded steel plates in compression. Cambridge University Report No. CUED/C-Struct/TR.3, 1971.
35. Watson, C. and Babb, A. S. Box columns; progress report on the determination of stiffening requirements. British Steel Corporation (Swinden Laboratory) Report No. D.TS.68/718/A.
36. Timoshenko, S. P. *Theory of elastic stability*; 2nd ed. in collaboration with J. M. Gere. New York, McGraw-Hill, 1961.
37. Falconer, B. H. and Chapman, J. C. Compressive buckling of stiffened plates. The Engineer, Vol. 195, No. 5080, June 5, 1953, pp. 789–791.

WHEEL LOAD DISTRIBUTION IN HIGHWAY AND RAILWAY BRIDGES

W. W. SANDERS, Jr.
Iowa State University, Ames

SYNOPSIS

A summary of two comprehensive studies of distribution of wheel loads to the floor systems of short and medium span concrete and steel highway and steel railway bridges is presented. The effects of many critical factors on the behaviour of the systems are indicated. Many of these factors affecting distribution are not considered in the present bridge specifications. Using results of these studies, proposals for new criteria to give more realistic designs are outlined.

INTRODUCTION*

Steel railway bridges have been built since the middle of the nineteenth century, and highway bridges have been constructed since about the turn of the century. Bridge engineers have attempted, from that time, to improve their methods of design, analysis, and construction. Analytical and experimental studies have been completed over the years that have improved the methods of design; however, only in recent years have studies been made in several areas that will lead to realistic and simple procedures for design. One of these latter areas is the distribution of live load to bridge floor systems.

This paper will outline two recent comprehensive studies of wheel load distribution in short and medium span bridges: one concerning steel railway bridges[1,2] and one concerning steel and concrete highway bridges.[3] The purpose of these studies was to develop more realistic design criteria for wheel load distribution which could be directly adopted in appropriate United States specifications.

It had been suggested that the specifications[4,5] in effect at the time of the studies, although giving satisfactory designs for service, were in general too conservative and limited in considering variables affecting behaviour. They provided no satisfactory consideration of such important variables as flexural and torsional stiffnesses of the floor slab and beams, the bridge span and the bridge width for determining distribution of wheel live loads to the beams. Neither did they provide consistent design criteria for all types of highway bridges nor, in the case of railway bridges, provide

---

* Although this paper is based largely on railway bridge research conducted at the University of Illinois and sponsored by the Association of American Railroads and on highway bridge research conducted at Iowa State University and sponsored by the National Cooperative Highway Research Program, the opinions and conclusions are those of the author and not necessarily those of the organisations that sponsored or conducted the original research.

any design benefits to a ballasted track or concrete deck for determining beam live loads. Thus, where warranted, changes have been recommended in the specifications for distribution of wheel loads in the design of floor systems for highway and steel railway bridges. Some of these recommendations were recently incorporated in the specifications for railway bridges.[6]

The investigations outlined herein relied significantly on theoretical methods and field test results of other investigators. These studies were used as the basis for the investigations. Modifications and extensions of the general theories were made so that these theories would apply to all bridge types considered. After correlation with field test results, extensive analytical studies were conducted relating all significant variables. From these results proposals for appropriate specification changes were developed and are presented.

This paper briefly outlines the types of bridges considered and the loadings used. In addition, analytical methods applicable to the load distribution study are indicated, and the results of the studies are discussed.

It was felt that a single study of a broad spectrum of highway or railway bridges would lead to a more uniform approach to any required changes in the load distribution criteria. It was with this in mind that these two studies were undertaken.

State of the Art

For more than 25 years, numerous researchers have studied the behaviour of bridge floor systems. Although most of these studies have been limited to theoretical behaviour, a significant number of field tests have been reported in the literature. Because of lack of space in this paper it will not be possible to summarise these studies; however, extensive bibliographies of available references in both areas are given in References 1 and 3. In addition, two significant reports detailing studies related to load distribution have been presented by Reese[7] and Aktas and Van Horn.[8]

BRIDGE TYPES AND LOADINGS

Types of Bridges

In determining load distribution in the floor system of bridges, it is necessary to classify the various types of bridges and floor systems into groups or categories for purposes of analysis. These categories are limited to those which fall within the geometrical restriction of the studies.

There are, in general, two types of floor systems in steel railway bridges. The first type consists of a number of transverse floor beams supported by heavy longitudinal edge girders. The second category is composed of a number of longitudinal beams which support timber ties or some other type of floor. These two categories of bridges are shown in Fig. 1 and will be classified generally as bridges with transverse floorbeams and bridges with longitudinal beams. The floor of these bridges may be open deck, a steel or wrought iron deck plate, or a concrete deck slab. The concrete slabs are assumed to be noncomposite with the beams since there are no requirements for composite action in the specifications.[6] Additional distribution of load is frequently obtained through the use of ballasted track and diaphragms.

The designs of the floor systems of highway bridges are quite varied. However, they may be classified, based on their assumed behaviour, into a few major categories. The various types of highway bridges have been classified, as shown in Fig. 2, into three categories: beam and slab, multi-beam, and concrete box girder bridges.

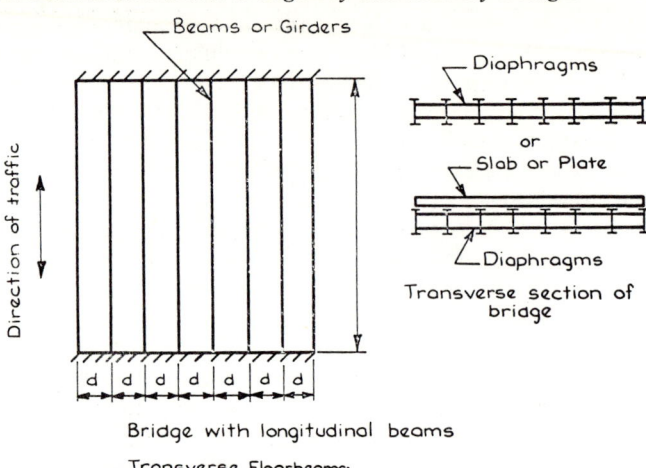

Bridge with longitudinal beams

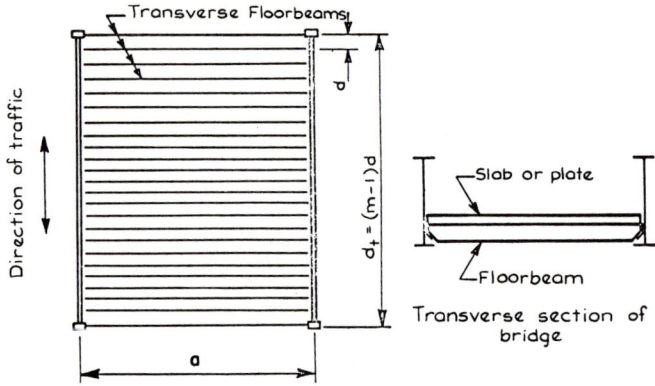

Bridge with transverse floorbeams

FIG. 1. TRANSVERSE AND LONGITUDINAL BEAM
STEEL RAILWAY BRIDGES.

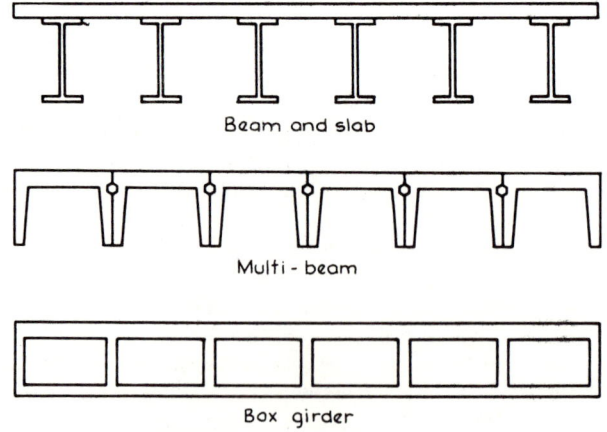

FIG. 2 INTERMEDIATE LENGTH HIGHWAY
BRIDGE TYPES.

Beam and slab bridge construction is characterised by separated longitudinal beams which support a deck slab. These beams are usually rolled steel beams, steel girders or prestressed concrete beams. A recent method of construction combines light gauge steel box sections compositely with the concrete deck. Multi-beam bridges are composed of several longitudinal beams placed side by side. The beams, which vary in shape, are usually constructed of precast prestressed concrete, connected by longitudinal shear keys, and tied together by transverse post-tensioned steel cables. Concrete box girder bridges are usually made of monolithically cast concrete. They are constructed of two continuous flanges with monolithic vertical webs.

The studies outlined herein were limited to the bridge types just mentioned. Although it is realised that there are many other types used (such as timber beams and floors), it is felt that these selected will include most of those bridges in the short and medium span range (up to 120 ft). Furthermore, the studies were limited to right simple span structures, although, based on other studies, approximations have been indicated for continuous structures.

Bridge Loadings

One of the major differences in the study of railway and highway bridges is the type of loading. Not only is the loading itself different, but the transverse positions of the loads are fixed in a railway bridge, but can vary significantly in a highway bridge. It should be noted that the effect of impact must be added to the standard live loads.

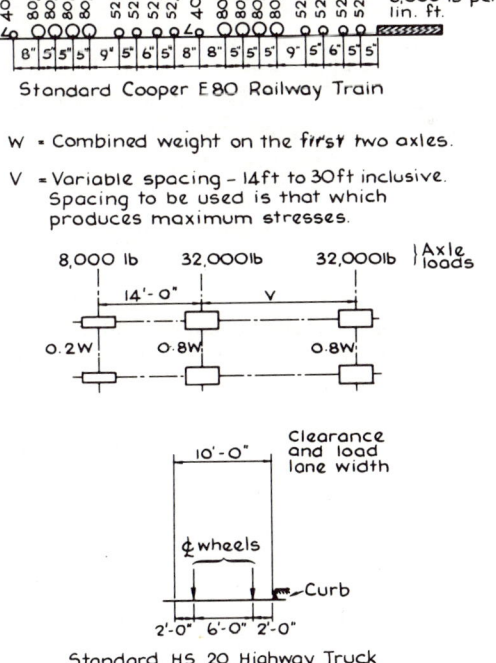

Standard Cooper E80 Railway Train

W = Combined weight on the first two axles.

V = Variable spacing – 14ft to 30ft inclusive. Spacing to be used is that which produces maximum stresses.

Standard HS 20 Highway Truck

FIG. 3. STANDARD BRIDGE LOADINGS.

The design for a steel railway bridge in the United States is based upon a Cooper E80 loading.[6] This loading, shown in Fig. 3, is composed of a number of heavy driving axles with intermediate lighter axles and a following uniform load. Although not indicative of present loadings, it has served as a basis for relating the loadings of diesel locomotive and standard and special heavy-duty cars. Since the standard clearance for railway cars is 16 ft, the positioning of track is usually consistent and can easily be determined.

The standard loading for highway bridges[5] depends upon the class of the highway. However, for all major highways, the primary loading is an HS 20-44. This loading (Fig. 3) is a simulated semi-trailer truck with a 14 ft axle spacing on the cab and a variable spacing on the trailer axle. The trailer axle spacing is varied to give the maximum loading condition in each design instance. The trucks are considered to occupy a width of 10 ft; however, this truck may be placed at any position in its lane to develop maximum stress. The number of lanes or trucks is determined by the width of roadway (between curbs).

LOAD DISTRIBUTION CRITERIA

In this section a brief summary of significant sections of the present United States load distribution criteria will be presented. In addition, a summary of the theoretical studies of load distribution and the resulting recommended changes are given.

Present Specifications

The present load distribution criteria[5,6] are significantly different for railway and highway bridges.

As noted earlier, several modifications in the criteria for steel railway bridges have been made as a result of the studies outlined herein. However, prior to this recent change, the beam design load[4] was determined by the following procedure:

(a) for bridges with longitudinal beams the load was divided equally between all beams within a 14 ft width centered under the track, and

(b) for bridges with transverse floorbeams, the load, P, was equal to $K\dfrac{Ad}{S}$,

where A was the axle load, d, the beam spacing, S, the axle spacing and $K = 1\frac{1}{8}$ for single track and $1\frac{1}{4}$ for double track.

At present, the criteria[6] have been changed for transverse floor beams to incorporate some of the factors not previously considered and several changes in those for longitudinal beams have been made to make them more consistent with those for concrete beam bridges. The present criteria are:

(a) for bridges with longitudinal beams the load is divided equally between all beams within a width equal to the width of tie plus twice the distance from bottom of tie to top of beam.

(b) for bridges with transverse floor beams the load, P, is equal to $1\cdot15\dfrac{AD}{S}$,

where D is $d\left(\dfrac{1}{1 + d/aH}\right)\left(0\cdot4 + \dfrac{1}{d} + \dfrac{\sqrt{H}}{12}\right)$. In this case, a is the beam span and H is a bridge stiffness factor defined later.

For highway bridges, the current specifications[5] are primarily based on the general type of supporting beam for the concrete deck and the beam spacing. A summary of some of the design load factors for beam moment in typical multi-lane bridges is shown in Table 1. The design load per beam is the fraction of the wheel load determined. For slab bridges, the design is based upon an equation for resisting moment. A special design criterion has just been added for the composite steel-concrete box girder bridges and is quite different from the above criteria. This is due to the fact that a specific study[9] was conducted on this bridge type.

It can be seen that, in most instances, many of the significant variables affecting distribution are not presently being considered.

THEORETICAL STUDIES

Field Test Correlations

The validity of the use of any theoretical procedure for predicting load distribution characteristics can be determined by comparing the results obtained from field tests of bridges to results as predicted by the theory. The results[1,3] show the validity of the theories selected. In each case the theories proposed for the type of bridge being studied were used to determine the moments in each beam element for the particular loading on the bridge. The results of these analyses were then compared with the results of the field test to determine the validity of the procedure in predicting actual behaviour.

After careful consideration of the applicability of many theoretical methods, the following were chosen for each bridge type. It should be noted that one primary consideration was their applicability across a broad spectrum of bridge geometries so that a single uniform approach could be given to load distribution. There is no doubt that for a specific bridge geometry or configuration different analytical methods might be more accurate for each case.

A detailed summary of each basic procedure or theory is given in the references listed. Modifications, except for concrete box girder highway bridges, are given in reports of the studies summarised herein.[1-3] In the case of box girder bridges, the original development was used directly.

Steel railway bridges:

(a) beams with transverse or longitudinal beams with slab — moment distribution procedure[10]

(b) beams with transverse floor beams with no slab — beam on elastic foundation theory[11]

(c) beams with longitudinal beams with no slab — moment distribution procedure with modification for diaphragms[10,12]

Highway bridges:

(a) beam and slab bridges — orthotropic plate theory[13-15]

(b) multi-beam bridges — articulated plate theory[16]

(c) concrete box girder bridges — theory of prismatic folded plate structures[17]

Studies of Steel Railway Bridges

Using the theories mentioned, numerous analytical studies were made of the behaviour of the two types of steel railway bridges. It was found that the five principal factors which determine their behaviour are: aspect ratio d/a (ratio of beam spacing to beam span), bridge stiffness factor $H(nI_b/ah^3)$, diaphragm stiffness ratio $r(I_b/I_d)$, depth of ballast b, and location of axle loads. Each of these factors affects the behaviour of the two types of bridges in a different manner.

The behaviour of a bridge with transverse floorbeams with a reinforced concrete floor is affected most by the aspect ratio, d/a, and the bridge stiffness factor, H. A low value of H corresponds to a relatively stiff slab in comparison to the floorbeams. As H becomes larger, the beams become stiffer in relation to the slab and the slab loses some of its effectiveness in distributing the load.

Several investigators have shown that for noncomposite structures the total midspan moment in the beams is related to the total midspan static moment just as the total stiffness of the beams is related to the total stiffness of the structure; This relationship, referred to as the stiffness approximation, can be expressed in a simplified form as $\dfrac{1}{1 + d/aH}$

The stiffness of the slab relative to the beams (indicated by H) not only affects the percentage of total static moment in the beams as indicated, but also affects the distribution of the load to the floorbeams. As the stiffness of the beams increases in proportion to the slab (a higher bridge stiffness ratio), the slab is not as effective in distributing the load. Figures 4(a) and 4(b) show this effect for various values of H and various values of the aspect ratio, d/a, for single track bridges. Although a double track bridge is considerably wider, the effect of this additional width is not too significant.

Because the aspect ratio and the bridge stiffness factor affect both the distribution of moment to the beams and the percentage of total moment in the beams, it is impossible to separate their effects. However, for a constant aspect ratio an increase in bridge stiffness factor not only increases the moment in the loaded beam but increases the total beam moment itself. Correspondingly, if only the beam spacing is increased (i.e., changing the aspect ratio), the percentage of beam moment in the loaded beam increases, but the total moment in the beams will decrease slightly.

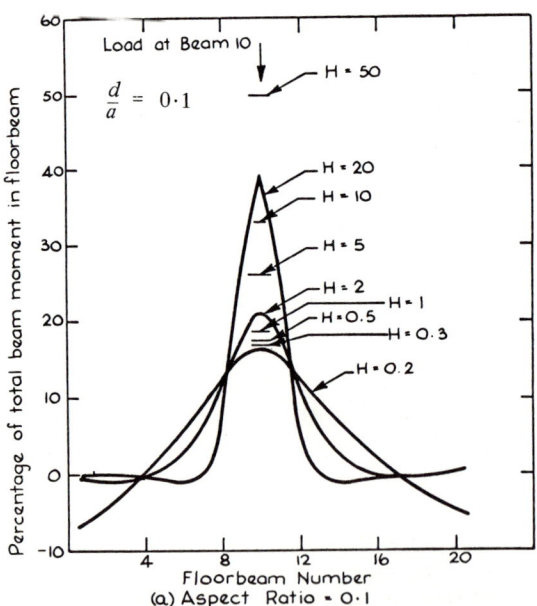

(a) Aspect Ratio = 0·1

FIG. 4a. PERCENTAGE OF TOTAL BEAM MOMENT IN EACH TRANSVERSE FLOORBEAM FOR VARIOUS VALUES OF BRIDGE STIFFNESS FACTOR (SINGLE TRACK - BALLAST DEPTH = 0.)

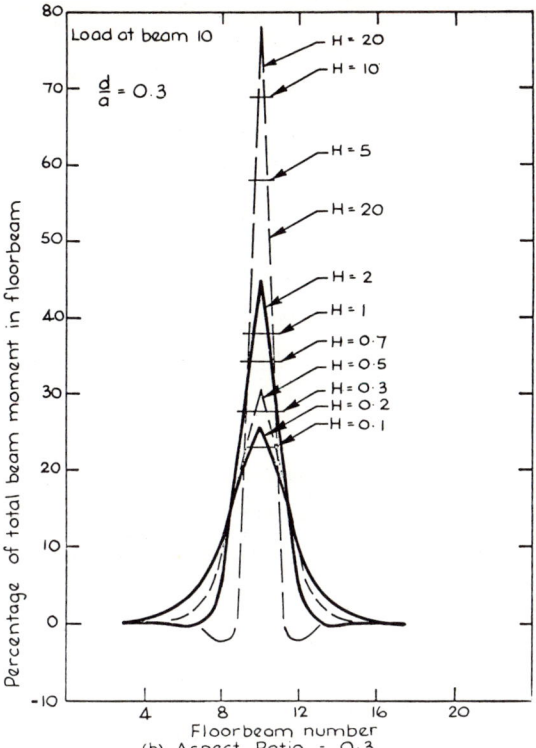

FIG. 4b. PERCENTAGE OF TOTAL BEAM MOMENT IN
EACH TRANSVERSE FLOORBEAM FOR VARIOUS VALUES
OF BRIDGE STIFFNESS FACTOR (SINGLE TRACK-BALLAST DEPTH=O).

The previous discussion was for bridges with no ballast. However, the effect of
ballast is very pronounced in bridges with transverse floorbeams as a result of the
ballast and the rail acting as a beam on a flexible base with the major distribution in
the direction of the rail, or longitudinally. The rail and ballast spread the load in
approximately a sine wave pattern. When the slab is very stiff in relation to the
beams (a low H), the main load distribution is obtained through the slab. However,
as the stiffness of the beams increases, the slab loses some of its distributing effect.
Figure 5 shows the effect of ballast depth, b, in reducing the moment in the loaded
beam. It can be seen that for the same bridge stiffness factor the beneficial effect of
the ballast is greater for smaller d/a ratios.

It was shown in the discussion of load distribution in transverse floor beam bridges
with slabs that the aspect ratio d/a and the bridge stiffness factor H were important
factors in determining the live load distribution to the beams. In bridges with longitudinal beams with slab the transverse position of the track with respect to the beams
is also very important. In most longitudinal beam bridges the track or tracks are
centered on the bridge. Thus, as the beam spacing is changed, the rail moves into a
different position with respect to the beams. This variation in relative position must
be kept in mind when comparing the behaviour of bridges with different beam
spacings.

Typical distribution curves of the total midspan beam moment for seven beam
single track bridges ($m = 7$) are shown in Fig. 6 for $d/a = 0.05$ and $d/a = 0.10$. The

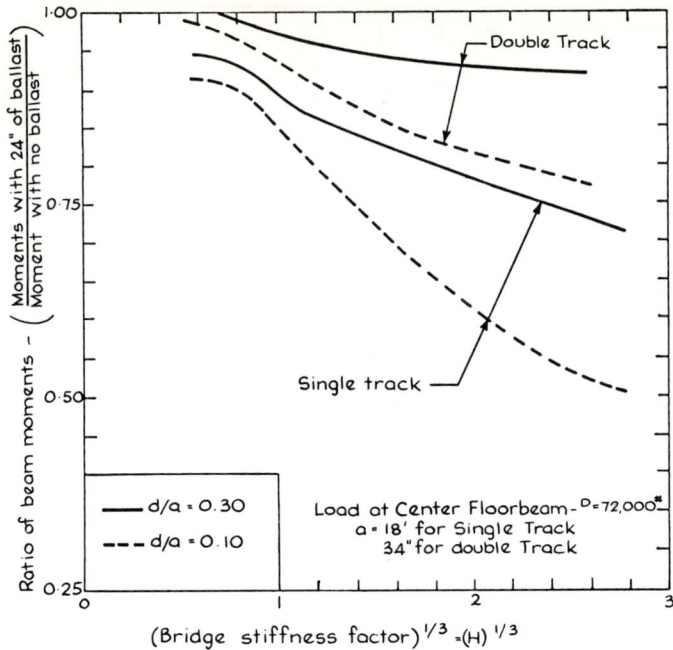

FIG. 5 EFFECT OF BALLAST ON MOMENT IN CENTRE
TRANSVERSE FLOORBEAM OF SINGLE AND DOUBLE
TRACK BRIDGES

effect of the slab on the transverse distribution of the moments can be seen to
diminish as the relative stiffness of the beams increases. Not only does more of the
total beam moment go to the beams under the track, but the total beam moment
itself increases. It should also be noted that the maximum percentage of beam
moment is generally in the beam under the centerline of the track. However, if the
beams are very stiff (high H) and the beam spacing is such that the rails are near
the beams adjacent to the centerline beam, the maximum moment may occur at
the adjacent beams (Fig. 6(a)).

The distribution of loads in bridges without slabs is discussed extensively in
References 1 and 2. In the case of transverse floorbeam bridges, the distribution of
load under the ties is nearly the same as the distribution of load to the floor and to
the beams. Because of the negligible effect of the plate and diaphragms, the behaviour
can be based completely on the beam-on-elastic foundation analysis.

For longitudinal beam bridges without slabs, the results[1] indicated that for any
variation in the diaphragm stiffness ratio (r) which would be allowed under the
specifications, there is little change in maximum individual beam moment.

Using results similar to those previously mentioned that were obtained for a
broad cross section of the critical variables, simplifications of the analytical procedure were developed so that the analysis would be usable in design offices. In
bridges with transverse floorbeams with slabs straight line approximations were
made of the variation in the load (or moment) distribution to the floorbeams at
various values of aspect ratio and the bridge stiffness factor from curves similar to
Fig. 4. The basis for the design recommendations for the bridges with transverse
floorbeams without slabs was a series of approximations of the beam-on-elastic

foundation equations. In the case of longitudinal beam bridges similar approximations were made in the analytical procedures.

Using the standard Cooper loading to sum for individual loads, specifications have been suggested[1] as a replacement for the section on load distribution in the

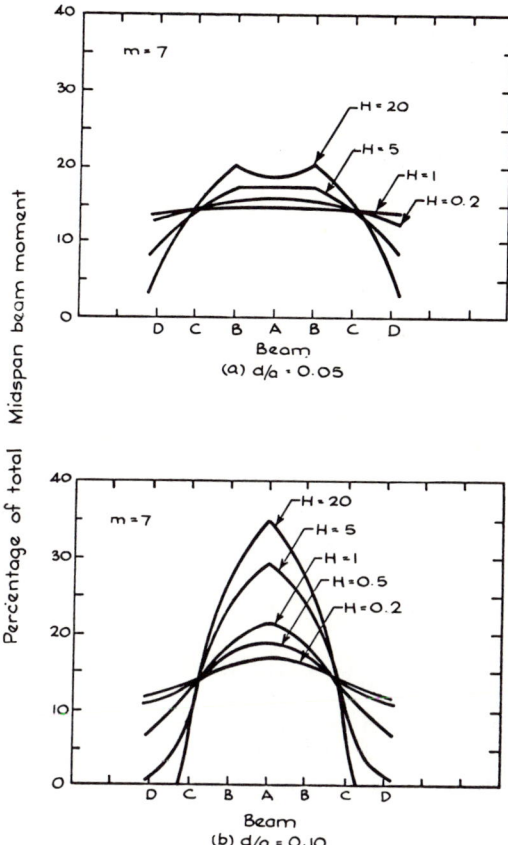

FIG. 6 MOMENT IN LONGITUDINAL BEAMS FOR VARIOUS VALUES OF BRIDGE STIFFNESS FACTOR (a = 40-LOAD AT MIDSPAN – b = 0)

AREA 'Specifications for Steel Railway Bridges.'[5] The major modifications recommended are shown in the following paragraphs. As noted earlier, nearly all parts of the recommendations in section (a) have recently been adopted as part of the specifications.[6]

Proposals for Specifications for Distribution of Live Load – Steel Railway Bridges

(a) Bridges with transverse steel beam:

Where an open or ballasted track is carried on transverse steel beams without stringers, the portion of any axle load on a single beam shall be as follows:

$$P = \frac{1 \cdot 15AD}{S} \ (B) \tag{1}$$

where

B = ballast factor

$$= 1 - \frac{b}{25}$$

D = effective beam spacing

$$= d\left(\frac{1}{1 + d/aH}\right)\left[0{\cdot}4 + \frac{1}{d} + \frac{\sqrt{H}}{12}\right] \text{ but not greater than } d \text{ or } S.$$

($D = d$ for bridges with no concrete deck slab or slabs less than 6 in thick or where the concrete slab extends over less than the centre 75% of the floorbeam.)

If d exceeds S, P shall be the reaction of the axle loads assuming that the flooring between beams acts as a simple span.

(b) Bridges with longitudinal steel beams:

Where the track is carried on longitudinal beams or girders with a ballast floor, the live load on each beam from one axle shall be

$$P = \frac{A}{m}(C) \tag{2}$$

where

m = the number of beams within a width of 14 ft for single track, but not to
exceed the distance between the track centres for multiple track

C, the distribution coefficient, shall be computed from Eq. (3).

$$C = \left(\frac{m}{1 + d/aH}\right)\left[\frac{1}{l} + \frac{\sqrt{H}}{a}\right] \tag{3}$$

but not greater than $\left[1 + \frac{a}{500}(d - 3)\right]$

with $l = 1{\cdot}15$ m for bridges at least 14 ft in width
$= m$ for bridges less than 14 ft in width.
(Note: For the calculation of end shear, $D = d$ and B and $C = 1$.)

Studies of Highway Bridges

Extensive analytical studies were conducted to determine the theoretical load distribution characteristics of each type of highway bridge considered in the program.[3] The initial analytical results provided the transverse variation of the longitudinal beam moment for numerous transverse positions of a single wheel, i.e. influence lines were generated. Thus, any combination of specific wheel positions could be considered for the determination of maximum beam moments. The use of these influence lines in combination with all of the loading conditions possible under the loading criteria yielded the maximum design moments.

The direct use of moments as a specification criteria would require significant changes in the design procedures. However, there is a direct relationship between the beam moment and the width over which a wheel load is distributed. This width is, in fact, used in the current specification in the distribution load factor equation, S/D. Thus, results of the load distribution studies were expressed in terms of D, the width of bridge over which one longitudinal line of wheels is distributed. If a satisfactory relationship between all of the variables and this width can be obtained, a more accurate and realistic distribution could be obtained without significantly altering the general distribution procedure.

For beam and slab bridges as indicated earlier, the orthotropic plate theory[13-15] was used to generate the analytical results. The method was selected because it can be used to express the load distribution properties of a bridge as a function of only a few generalised dimensionless variables so that investigation of a large variety of bridge properties becomes feasible. The theory assumes the beams and slab to be replaced by a continuous medium, which eliminates the requirement for knowing the specific bridge beam geometry in the theory formulation. The dimensionless parameters are α, a relative torsional stiffness parameter, and θ, a relative flexural stiffness parameter, where

$$\alpha = \frac{D_{xy} + D_{yx}}{2\sqrt{D_x D_y}} \quad \text{and} \quad \theta = \frac{W}{2L} \sqrt[4]{\frac{D_x}{D_y}}$$

(D_{xy} and D_{yx} are the torsional rigidities per unit width in the x and y directions, respectively; D_x and D_y are the flexural rigidities per unit width; W is the bridge width and L is the bridge span.)

In the above equations defining θ and α, it can be seen that θ, the relative flexural stiffness parameter, primarily depends on the aspect ratio of the bridge (W/L) for its sensitivity rather than the ratio of the flexural stiffnesses. It can also be seen that the aspect ratio of the bridge has no effect on the relative torsional stiffness parameter, α. Thus, if the cross-sectional geometries of the bridge remain the same, the parameter α is unchanged and, hence, is only a measure of distribution due to local torsional conditions in the bridge.

After using the orthotropic plate theory to study a number of standard and typical specially designed bridge plans provided by numerous highway agencies and referring to studies by others, the following ranges on the two critical parameters, θ and α, were selected. It was felt these would include almost all possible bridge configurations.

$$\theta: \ 0\cdot25 \text{ to } 1\cdot25$$
$$\alpha: \ 0\cdot20 \text{ to } 1\cdot00$$

Since finite loads must be used, it is necessary to fix the bridge width, W, so the maximum number of wheel loads, N_w, can be determined for the transverse section. Positioning combinations of the wheel loads for maximum eccentric and maximum central loadings, values of D were determined for many combinations of θ and α for bridge widths from 28 to 75 ft. Typical values of D for widths up to 51 ft are shown in Table 2.

The values of D could actually be used in the design of highway bridges. However, to use them the user must employ a three-way interpolation between the three parameters involved, i.e., the bridge width, the flexural stiffness parameter θ, and the torsional stiffness parameter α. Of course, this is highly impractical and the reduction of this table to a more usable form is outlined later in this section.

For multi-beam bridges the articulated plate theory was used. Since the transverse flexural rigidity is zero if the prestressing force is neglected, the only principal variable, other than N_L and W, is a combined flexural torsional stiffness parameter ϕ. The parameter is equal to

$$\phi = \frac{W}{2L} \sqrt{\frac{D_x}{D_{xy} + D_{xy}}},$$

and, for all multi-beam bridges, will fall within the range of $0\cdot1$ to $2\cdot0$. Using an approach similar to that for the beam and slab bridges, a table relating D to ϕ and the bridge width was prepared. For design office use this table could be used, but a single variable simplification is more desirable.

For beam and slab and multi-beam bridges, the analysis was carried out using a simple variation of only one or two parameters. However, due to the nature of the method of analysis[17] used for the solution of the box girder problem and the complexity of the cross section, each of seven critical variables had to be specified independently. For over 100 bridges, using typical combinations of the variables, values of D were determined using a computer program developed by Scordelis.[17]

In order to make all of the results suitable for incorporation in specifications, it was felt that a single parameter to determine D for each bridge width (or N_W) would be necessary. It was found that a single stiffness parameter, C, could be developed where

$$\text{for beam and slab bridges, } C = \theta/\sqrt{\alpha}$$
$$\text{for multi-beam bridges: } C = \sqrt{2}\phi$$

$$\text{or, in both instances, } \frac{W}{\sqrt{2L}} \sqrt{\frac{D_x}{D_{xy} + D_{yx}}}$$

for box girder bridges, it was found that the stiffness parameter was

$$C = 0 \cdot 55 \frac{W}{L} \left[1 + N_g \sqrt{\frac{d}{W}} \right] \cdot \left[\frac{1}{\sqrt{1 + N_d}} \right]$$

where

N_g = number of girders (vertical webs), and
N_d = number of diaphragms.

Variations of D with the bridge stiffness parameters for a three-lane bridge are shown in Figs. 7 to 9 for the three bridge types. It can be seen that the empirical equation is within 10% of all theoretical results.

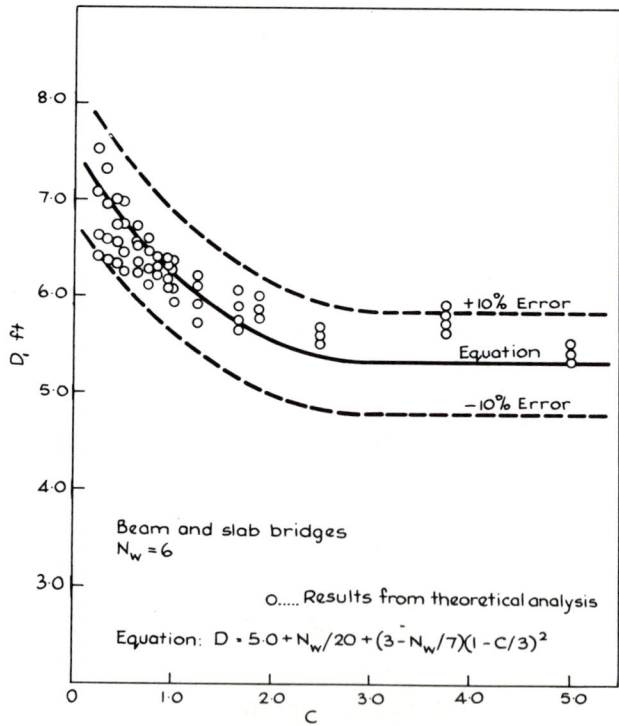

FIG. 7 VARIATION OF D WITH BRIDGE STIFFNESS PARAMETER C FOR BEAM AND SLAB BRIDGES, N_W = 6

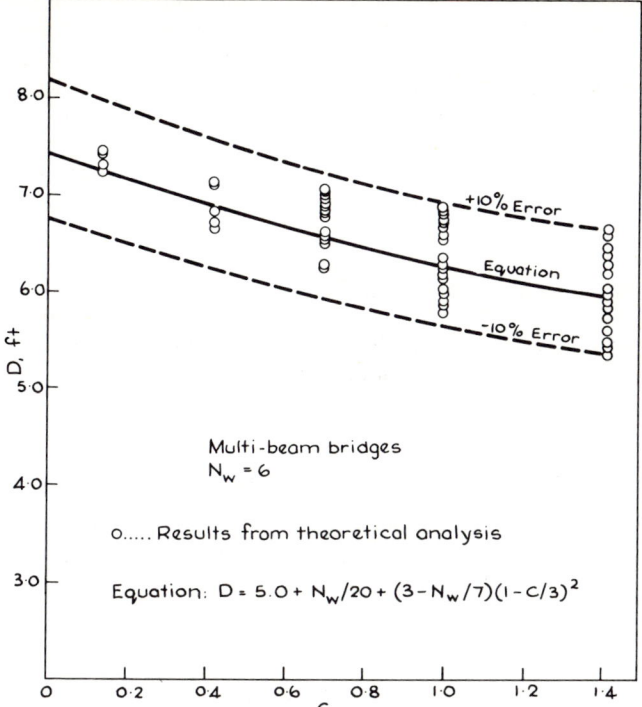

FIG. 8 VARIATION OF D WITH BRIDGE STIFFNESS
PARAMETER C FOR MULTI-BEAM BRIDGES, $N_w = 6$

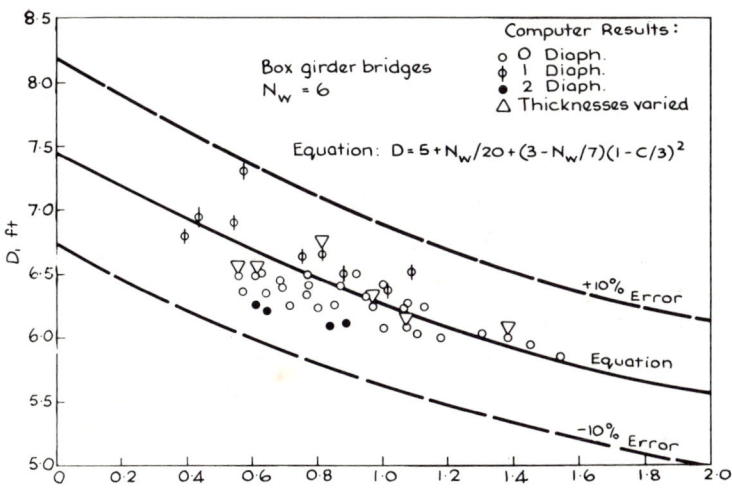

FIG. 9 VARIATION OF D WITH   BRIDGE STIFFNESS
PARAMETER C FOR CONCRETE BOX GIRDER BRIDGES, $N_w = 6$

Using this empirical equation as a basis for determining D, the following section has been recommended for inclusion in the AASHO Bridge Specifications.[5] Additional sections have been proposed,[3] but are not included here. It can be seen that these new criteria consider many more variables than the present criteria and would lead to more realistic designs.

Because of the extreme simplicity of the current requirements, it is obvious that any change to make them more realistic must entail some increase in complexity. The proposals presented are a balance between the need for an accurate distribution criterion and for a usable design office criterion. It should be noted that as the complexity of the bridge system increases, the simplification of the theoretical procedures requires more approximations. Thus, considerations of unusual conditions are required. The use of any theory outlined herein for a total computerised analysis of the basic behaviour will lead to the most accurate design and would be the optimum consideration. It is felt, though, that the changes proposed will lead to sufficiently accurate designs.

Proposals for Specifications for Distribution of Live Load — Highway Bridges

The live load bending moment for each beam shall be determined by applying to the beam the fraction of a wheel load (both front and rear) determined by the following relations:

$$\text{load fraction} = \frac{S}{D},$$

where S is

S_a for beam and slab bridges (for slab bridges, $S = 1$)

$\dfrac{12N_L + 9}{N_g}$ for multi-beam bridges, and the maximum of the two values for concrete box girder bridges

and the value of D determined by the following relationship:

$$D = 5 + \frac{N_L}{10} + \left(3 - \frac{2N_L}{7}\right)\left(1 - \frac{C}{3}\right)^2, \quad C \leqslant 3$$

$$= 5 + \frac{N_L}{10}, \qquad\qquad\qquad C > 3$$

where

S_a = average beam spacing, ft,
N_L = total number of design traffic lanes
N_g = number of longitudinal beams,
C = a stiffness parameter which depends upon the type of bridge, bridge and beam geometry, and material properties.

The value of C is to be calculated using the relationships given previously. However, for preliminary designs, C can be approximated using the values given in Table 3.

CONCLUSIONS

The purpose of the research summarised in this paper was to study the distribution of wheel loads in highway and steel railway bridges and to recommend, where warranted, changes in the AASHO 'Standard Specifications for Highway Bridges'[5] and the AREA 'Specifications for Steel Bridges.[6]

The results of the analytical studies and the development of empirical load distribution equations have been used to prepare specific recommendations for changes in the load distribution criteria. It is felt that with these new criteria, prediction of wheel load distribution will be more accurate and will more truly indicate the behaviour of the bridge types studied.

ACKNOWLEDGEMENTS

The research outlined herein was conducted by the Department of Civil Engineering, University of Illinois, and the Engineering Research Institute, Iowa State University. It was sponsored by the Association of American Railroads and the National Cooperative Highway Research Program.

The author would like particularly to acknowledge the guidance and assistance of W. H. Munse, Professor of Civil Engineering, University of Illinois, in the railway bridge research and the collaboration of H. A. Elleby, Associate Professor of Civil Engineering, Iowa State University, in the highway bridge research. Special thanks are given to the many researchers whose studies were used as the basis or stimulation for much of the work reported herein.

REFERENCES

1. Sanders, W. W., Jr. and Munse, W. H. Load distribution in steel railway bridges. *J. struct. Div. Am. Soc. civ. Engrs*, Vol. 95, ST12, Dec. 1969, pp. 2763–2781.
2. Sanders, W. W., Jr. *The lateral and longitudinal distribution of loading in steel railway bridges.* Ph.D. thesis, University of Illinois, 1960 (also published as Report ER-5, Association of American Railroads Research Center, Chicago, 1961).
3. Sanders, W. W., Jr. and Elleby, H. A. *Distribution of wheel loads on highway bridges.* National Cooperative Highway Research Program Report No. 83, Highway Research Board, Washington, D.C., 1970.
4. American Railway Engineering Association. *Specifications for steel railway bridges for fixed spans not exceeding 400 feet in length.* Chicago, 1964.
5. American Association of State Highway Officials. *Standard specifications for highway bridges.* 9th ed. (including 1966–1967 interim specifications). Washington, D.C. 1965.
6. American Railway Engineering Association. *Manual for railway engineering.* Chicago, 1970. Chapter 15 – Specifications for steel bridges.
7. Reese, R. T. *Load distribution in highway bridge floors – a summary and examination of existing methods of analysis and design and corresponding test results.* M.S. thesis, Brigham Young University, Provo, Utah, 1966.
8. Aktas, Z. and Van Horn, D. A. Bibliography on load distribution in beam – slab bridges. *Fritz Engineering Laboratory Report* No. 349.1, Lehigh University, Sept. 1968.
9. Mattock, A. H. and Johnston, S. B. Lateral distribution of load in composite box girder bridges. *Highway Research Records* No. 167, Highway Research Board, Washington, D.C., 1967, pp. 25–33.
10. Newmark, N. M. A distribution procedure for the analysis of slabs continuous over flexible beams. *Bull. Ill. Univ. Engng Exp. Stn*, No. 304, June 1938.
11. Hetényi, M. I. *Beams on elastic foundation . . .* Ann Arbor, University of Michigan Press, 1946.
12. Wei, B. C. F. Load distribution of diaphragms in I-beam bridges. *J. struct. Div. Am. Soc. civ. Engrs*, Vol. 85, ST5, May 1959, pp. 17–55.
13. Guyon, Y. Calcul des ponts larges à poutres multiples solidarisées par des entretoises. *Annls Ponts Chauss.*, Vol. 116, No. 24, Sept.-Oct. 1946, pp. 553–612.
14. Massonnet, C. Méthode de calcul des ponts à poutres multiples. *Publs int. Ass. Bridge struct. Engng*, Vol. 10, 1950, pp. 147–182.
15. Rowe, R. E. A load distribution theory for bridge slabs allowing for the effect of Poisson's ratio. *Mag. Concr. Res.*, Vol. 7, No. 20, July 1955, pp. 69–78.
16. Arya, A. S., Khachaturian, N. and Siess, C. P. Lateral distribution of concentrated loads on multibeam highway bridges. *Civ. Engng Stud. Univ. Ill. struct. Res. Ser.*, Report No. 213, 1960.
17. Scordelis, A. C. *Analysis of simply supported box girder bridges.* University of California, Berkeley, Structures and Materials Research Report SESM-66-17, 1966.

TABLE 1

Load distribution factors for highway bridges with
concrete floors[a].

Type of supporting beams	Fraction of wheel load[b]
Steel I-beam stringers and prestressed concrete girders	$S/5.5$[c]
Concrete T-beams	$S/6.0$
Timber stringers	$S/5.0$
Concrete box girders	$S/7.0$

[a] From Reference 5.
[b] If S exceeds from 10 to 16 ft, depending on beam type
simple beam reaction is used.
[c] S = average stringer spacing in feet.

TABLE 2

Typical theoretical results for beam and slab bridges value of D in
equation: load factor = S/D

		W, width of bridge in ft (N_w, No. of wheel loads)							
		28	33	37	39	41	45	49	51
θ	α	(4)	(4)	(4)	(6)	(6)	(6)	(6)	(8)
0.50	0.00	5.81	5.53	5.44	5.42	5.43	5.47	5.41	5.36
	0.04	5.89	5.70	5.63	5.63	5.64	5.67	5.57	5.53
	0.16	6.06[a]	6.07	6.10	6.09[a]	6.07	5.97	5.95	5.95
	0.36	6.17[a]	6.47	6.62	6.15[a]	6.24	6.29	6.36	6.03[a]
	0.64	6.28[a]	6.79	7.08	6.20[a]	6.36	6.54	6.71	6.08[a]
	1.00	6.38[a]	7.05	7.45	6.25[a]	6.46	6.72	6.98	6.12[a]

[a] Controlled by central loading; other values controlled by eccentric
loading.

TABLE 3

Values of K to be used in the relation: $C = K\dfrac{W}{L}$

Bridge type	Beam type and deck material	K
Beam and slab (includes concrete slab bridge)	Concrete deck:	
	Noncomposite steel I-beams	3.0
	Composite steel I-beams	4.8
	Nonvoided concrete beams	
	(prestressed or reinforced)	3.5
	Separated concrete box-beams	1.8
	Concrete slab bridge	0.6
	Separated steel box-beams	
	(composite box girders)	2.6
Multi-beam	Nonvoided rectangular beams	0.7
	Rectangular beams with circular voids	0.8
	Box section beams	1.0
	Channel beams	2.2
Concrete box girder	Without interior diaphragms	1.8
	With interior diaphragms	1.3

THE BEHAVIOUR OF ORTHOTROPIC STEEL DECK BRIDGES

PATRICK J. DOWLING

Imperial College of Science and Technology

INTRODUCTION

Steel decks have been used in bridge construction since the beginning of the century. Early forms included the buckled or dished plate and steel troughing. Although relatively sophisticated both in their conception and action as components (in the former membrane as well as flexural rigidity participated in the load carrying action, and in the latter the corrugated shape increased the stiffness of steel plate by a large amount) they were designed to act as independent load carrying members within the bridge deck systems and usually their sole function was to distribute the traffic loading to the main load carrying members. The battledeck which was developed shortly before the Second World War can be considered as the direct forerunner of the modern orthotropic deck, as it was the first welded stiffened steel plate deck in which the integral action between stiffeners and deck plate was taken into account in design. After the war the orthotropic deck bridge was developed and this form of bridge has now become a common form of construction in medium and long span bridges — or indeed in any bridge form where saving in weight or depth of construction are important parameters. The two main types of deck are the open stringer or torsionally weak deck and the closed stringer or torsionally strong deck. Seven major bridges using closed stringer steel decks have been built to date or are currently under construction in Britain. One example is the Erskine Bridge shown in Fig. 1.

Research carried out initially in Germany, and later in the United States, Great Britain, Holland and other European countries has provided a great deal of information on the behaviour of such decks. The recently circulated draft German code DIN 1073[1] is the first code to include design clauses which relate specifically to orthotropic decks and the new British bridge code currently being drafted will also deal with this topic. It is considered that the time is therefore opportune to summarize briefly what is known about the behaviour of orthotropic decks and to identify and, hopefully, provoke discussion on those areas which are less well documented.

ELASTIC BEHAVIOUR

Decks under in-plane loading

The deck acting as top flange of the main girders is subjected to in-plane compressive forces near midspan or tensile forces over the supports in a continuous

FIG. 1.   ERSKINE BRIDGE DURING CONSTRUCTION
SCOTTISH DEVELOPMENT DEPARTMENT

CONSULTANTS: FREEMAN FOX AND PARTNERS
CONTRACTORS: FAIRFIELD-MABEY LIMITED

bridge. These forces are developed through the shear connexion between deck and main girder webs and their distribution across the width of the deck is dependent upon the deck's shear rigidity, as well as the plan and sectional geometry of the bridge, and the applied loading. The ratio of extensional rigidity to shear rigidity in an orthotropic deck (an indication of which is given by the non-dimensional constant α expressing the ratio of stiffener to plate area) is greater than that in an isotropic plate (for which $\alpha = 0$) as the contribution of the stiffeners to the deck shear rigidity is small. As there is therefore relatively less shear stiffness to mobilize the available extensional stiffness the effects of shear lag are more marked in stiffened flanges than in unstiffened ones. It is also because stiffeners do not contribute significantly to the shear stiffness of the deck that the presence of cross girders has little effect on the shear lag behaviour of the stringer-deck plate combination. In many cases the bridges in which steel decks may be used will be relatively large span ones for which the critical parameter B/L will be small, and the deck is therefore likely to be fully effective in acting with the main girders except perhaps in the region of the supports. However, in short span bridges, such as moveable bridges or temporary flyovers, the distribution of in-plane forces across the deck may be much less uniform. Shear lag in orthotropic decks has been studied by Ferahian[2], Kondo *et al*[3] and Abdel Sayed[4]. Work carried out recently at Imperial College[5] has made use of programs developed there[6] to extend the study to include among other parameters effects of variation in cross section, support conditions and loading. The results shown in Fig. 2 may be used to estimate the effective breadth of stiffened or unstiffened flange for the two limiting cases of a simply supported and encastre main girder. Uniformly-distributed loading is treated only as it is considered to be more representative of the type of loading to which the main girders are subjected. It may be seen that the effective breadth at midspan of a simply supported girder is more sensitive to the parameter α than either the midspan or support region in an encastre girder of equal length. For any particular value of B/L it is permissible to interpolate linearly over the practical range of α (from 0 to 1) to estimate the effective breadth. It should be noted that in common with most effective breadth design curves or tables in current codes, the values are based on the behaviour of an isolated tee beam. Table 1 shows, however, that the effective breadth values derived for a bridge of typical cross sectional shape, Fig. 3b, do not deviate significantly from those of Fig. 2 when the width B is chosen as indicated in Fig. 3. Some guidance for the designer who must choose the support condition which most closely approximates his own case is provided by the results given in Fig. 4. This shows the variation in effective breadth along the length of a three span continuous bridge, uniformly loaded along its entire length. Two values of flange aspect ratio are considered and it can be seen the effective breadths for the centre span fall between those given for the simply supported and encastre cases in Fig. 2.

Using the effective breadth of orthotropic flange the main girders may be proportioned in the normal way. The deck itself, of course, must be designed to resist the in-plane forces from the Primary System together with the forces from the other systems which may occur independently or simultaneously.

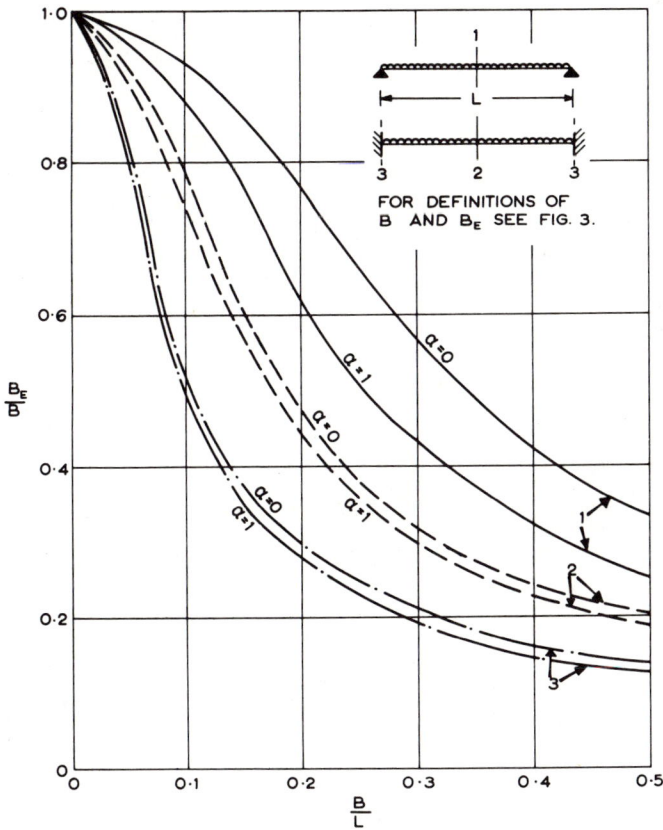

FIG. 2. SHEAR EFFECTIVE BREADTH OF STIFFENED FLANGES

B/L	SUPPORT		MID SPAN		
	T Fig. 3a	∏ Fig. 3b	T Fig. 3a	∏ Fig. 3b	
0·1	0·51	0·53	0·78	0·81	
0·25	0·25	0·26	0·38	0·38	α = 0
0·5	0·14	0·15	0·20	0·20	
0·1	0·49	0·49	0·74	0·76	
0·25	0·23	0·24	0·36	0·36	α = 1
0·5	0·13	0·14	0·19	0·19	

TABLE 1. COMPARISON BETWEEN EFFECTIVE BREADTH
VALUES, $\frac{B_E}{B}$, FOR A TWIN GIRDER BRIDGE AND
SINGLE GIRDER BRIDGE WITH A UNIFORMLY
LOADED ENCASTRE SPAN

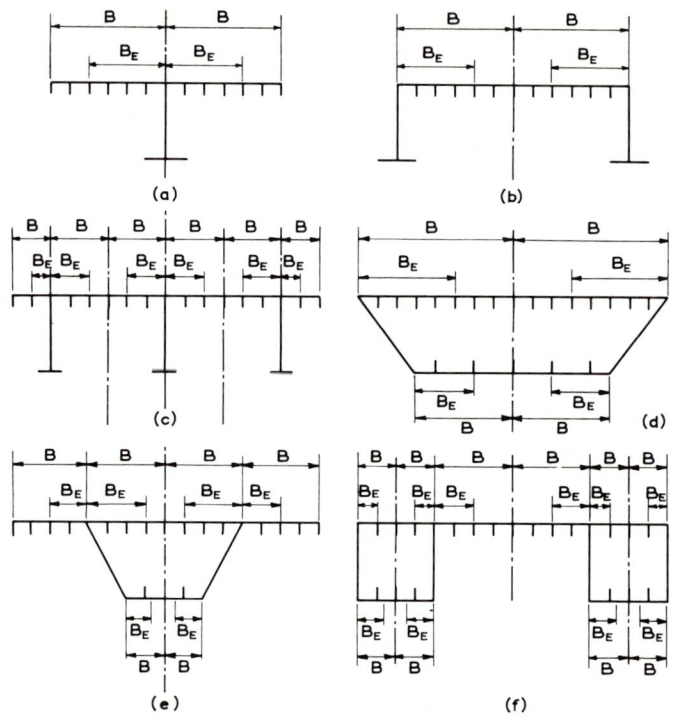

FIG. 3. APPLICATION OF EFFECTIVE BREADTH CURVES TO TYPICAL
BRIDGE CROSS SECTIONS

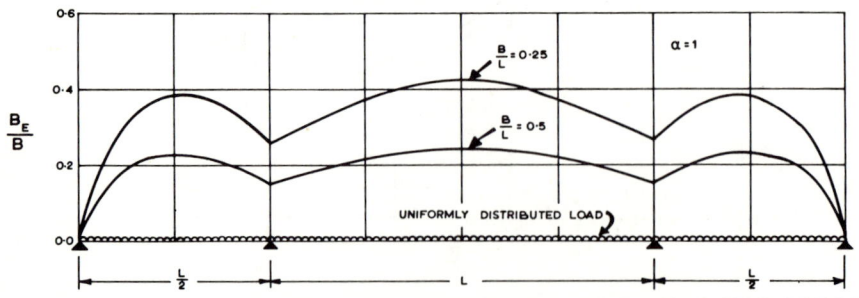

FIG. 4. VARIATION OF EFFECTIVE BREADTH IN A CONTINUOUS SPAN BRIDGE

Decks under lateral loading — overall behaviour

As might be deduced from the name orthotropic deck, the most commonly used approach for analysing the deck under lateral loading is the substitute plate method. Of the available variations in approach, that proposed and developed by Pelikan and Esslinger[7] is the most popular. In this method the stringer-deck plate combination is replaced by an orthotropic plate which is supported on flexible cross girders represented by beam line supports. A Levy Series solution of the homogeneous equation is used to derive expressions for influence lines for the bending moment in the stringers for the case when cross girders do not deflect. A second stage of the analysis uses influence coefficients for beams on elastic supports to calculate modified stringer and cross girder moments caused by flexibility of the cross girders. This method, although amenable to hand calculation, is time-consuming and fairly limited in scope. Numerical solutions using finite difference approximations to describe the orthotropic plate equation have been used by the author to analyse the steel deck of Kut Bridge[8] and similar techniques have also been described by Heins[9]. The advantage of this method is that when computerized a wide range of deck layouts and support conditions can be analysed. In particular the facility to generate influence surfaces for any stringer or cross girder location in the deck enables the designer to rapidly select the critical loading combinations.

The behaviour of steel decks under lateral load and the accuracy of the ortho-tropic plate method in describing such behaviour can be illustrated by a study of the following test results[10]. The panel (Fig. 5) tested was designed by the firm of Flint and Neill in conjunction with the Cleveland Bridge Construction Co. and represented a portion of the deck of a proposed Swedish suspension bridge. The dimensions are shown in Fig. 6. It can be seen that the deck plate and trough wall thicknesses are of the minimum sizes used in American and European practice, but are less than those currently in use in this country[11]. (See Fig. 13.) The load used in design was a 10-ton wheel load with an area of contact of 24 in x 6 in but the panel was also tested using the 11¼ ton load (including impact) acting over 15 in x 3 in as specified in the present British specification[11].

When the British wheel load was placed directly over one stringer at midspan of the panel the deflected profile of the deck plate across the transverse centre line was as shown in Fig. 7. (The results have been scaled down to correspond to a load level of 10 tons so that a direct comparison may be made with the Swedish wheel load.) It can be seen that the orthotropic plate theory, while tending to overestimate the distribution achieved, provides reasonable agreement with the experimental results. The distribution as obtained by the substitute plate method is critically dependent on the values of rigidities used in the analysis, and more particularly in the case of a closed stringer deck on the choice of torsional rigidity as well as flexural rigidity in the direction of span. It has been shown[7] that the value of torsional rigidity used should not be that calculated on the assumption of the closed trough retaining its cross sectional shape but should be a reduced value which allows for the effect of distortion of the trough and the flexibility of the plate between troughs across which the vertical shear is transmitted. The non-dimensional reduction factor is derived by equating the work done in twisting an idealized non-distorting isolated trough unit subjected to asymmetrical sinusoidally distributed shears along the edges of its deck plate flange with the sum of the work done in torsion of an actual unit whose cross section deforms and that done in bending each of the unit's components. No account is taken of the width of the wheel in this calculation yet this must be an important factor, as the transverse distribution is less dependent on the plate flexibility for a wide wheel which spans from the loaded to the

(a) UNDERSIDE OF TEST PANEL

(b) DEFLEXION DIAL GAUGES UNDERNEATH MIDSPAN SECTION

FIG.5 FULL SCALE CLOSED STRINGER STEEL DECK TEST PANEL

adjacent troughs than for a narrow wheel which does not. In the example illustrated
the suggested reduction factor amounts to 0·33. It is therefore somewhat surprising
that with the resultant fairly crude estimate of the torsional rigidity the
comparison between theory and experiment is as close as obtained.

As the trough was constructed from thin plate a significant variation in
longitudinal stress across the bottom occurred due to warping when the trough was
loaded eccentrically. Although this variation in stress is acceptable it should be
noted that it may in certain cases produce a situation where the maximum longitudinal stringer stress is obtained when a wheel load straddles one wall of a trough.

A more exact analysis can be achieved by using the method of finite elements to
study the three dimensional behaviour of the stiffened plate panel. A comparison
between experimental results and those derived using Lim's program[6] is provided
in Fig. 7. The elements used are rectangular elements with both extensional and
flexural stiffness. The agreement between theory and experiment is seen to be

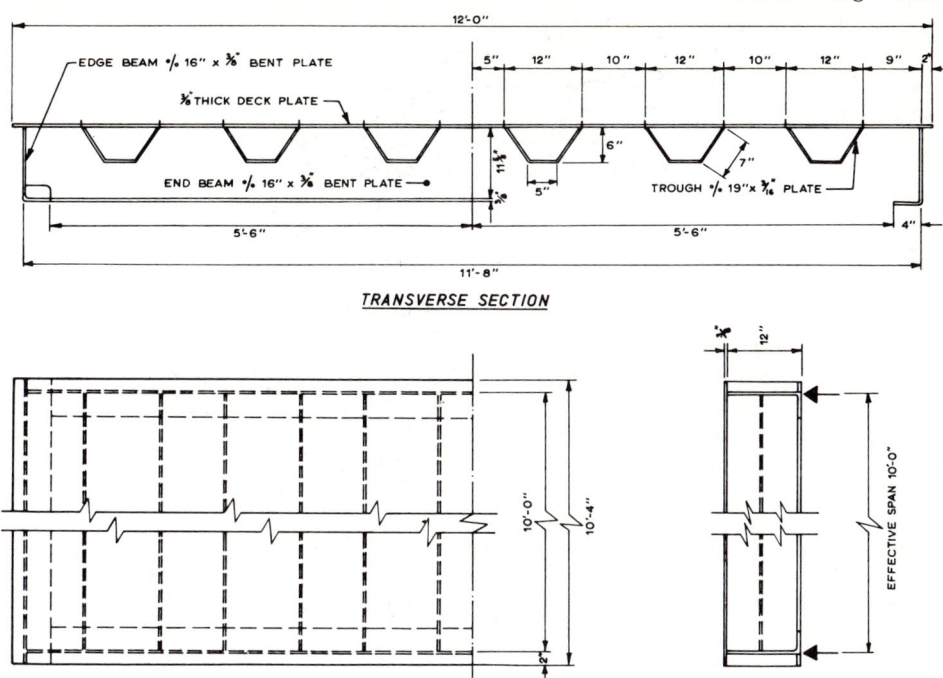

TRANSVERSE SECTION

PART PLAN VIEW · PART END VIEW

FIG. 6. DIMENSIONS AND DETAILS OF TEST PANEL

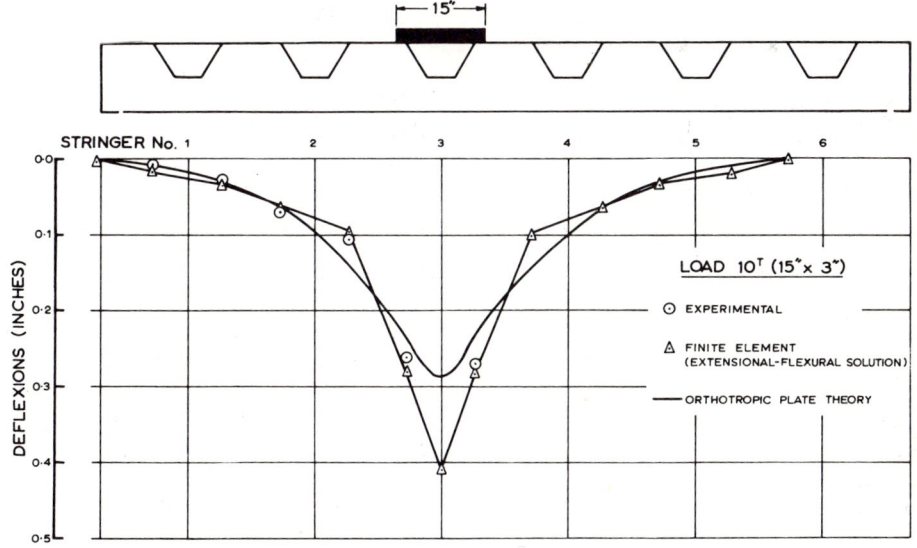

FIG. 7. DEFLECTED PROFILE OF PANEL AT MIDSPAN SECTION

excellent. It should also be noted that the deflexion of the plate underneath
the load can be accurately estimated by this three dimensional finite element
approach if sufficient elements are provided locally whereas such behaviour cannot
be predicted directly by the orthotropic plate approach.

Stresses at the bottom of the stringer and in the plate of the deck caused by
overall bending of the panel are plotted in Fig. 8 for both loads. In the case of the
orthotropic plate approach these stresses are obtained by applying the moments
obtained in the idealized plate to the actual discontinuous, eccentrically stiffened
plate panel. Two difficulties are encountered in this calculation. In the first instance
a considerable variation of longitudinal moment in the transverse, or weak, direction
occurs over a distance equal to the spacing of the stringers. Thus if the maximum
moment obtained from the substitute plate is used to calculate stresses a very
conservative value of stress may be obtained as can be seen from the point labelled
'maximum moment' in the figure. If the moment is averaged over the width of the
stringers, however, the experimental and theoretical results show closer agreement.
The second difficulty which arises is the choice of section moduli which must be
used in conjunction with the average moment to calculate stresses. A semi-empirical
approach proposed by Pelikan[7] may be used for the purpose of calculating an
effective width of deck plate, which in the case of closed stringer decks gives
a width of plate approximately equal to the spacing of the stringers. Averaging
the moments and using the section properties of a trough deck plate unit of
width equal to stringer spacing leads to the results shown by the horizontal
lines in Fig. 8. It can be seen that the agreement with experiment for the stresses
at the bottom of the stringers is good, but is less satisfactory for the deck plate
stresses. This latter result is not very surprising as the top section modulus is much
more sensitive to choice of effective width than the bottom one. However, the poor

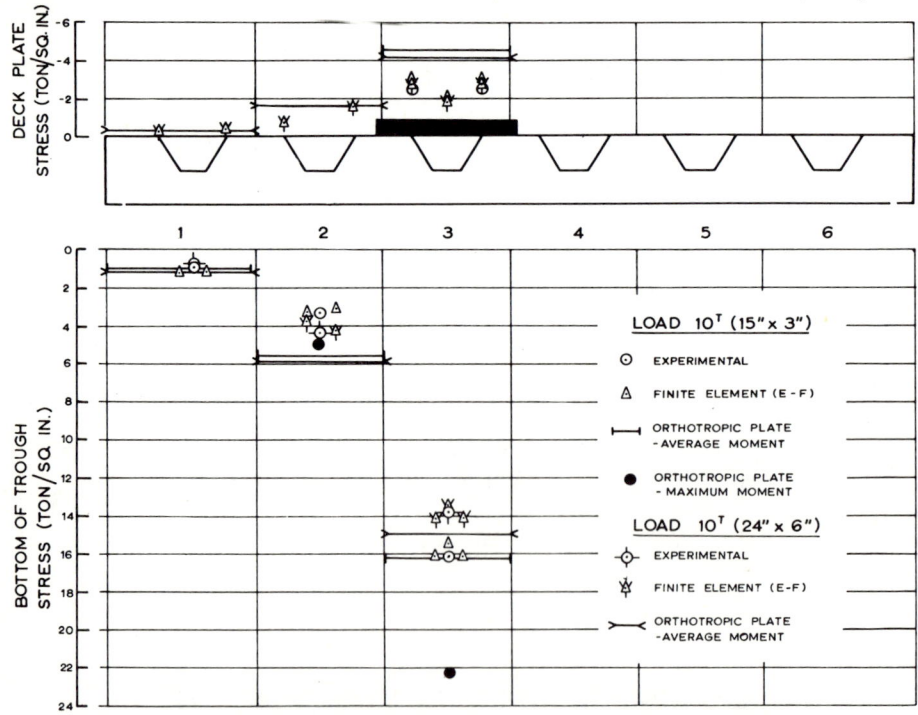

FIG. 8. <u>OVERALL BENDING STRESSES IN PANEL AT MIDSPAN SECTION</u>

agreement of deck stresses is not quite so important in design as the bottom stresses are generally more critical. The agreement between test and finite element results on the other hand is excellent for both deck plate and bottom of trough stresses.

The finite element approach can now be used to study in more depth the actual load distributing behaviour of this panel. The deck plate and adjacent troughs can be considered to contribute to the load distribution in two ways. In the first instance they contribute by virtue of their extensional rigidity and receive their load by shear forces in the plane of the deck plate; that is, the troughs behave as eccentrically loaded columns. Such action would occur even if the deck plate had no flexural rigidity. In the second instance the adjacent troughs also resist, by bending as beams, the shear normal to the plane of the deck which is distributed to them by virtue of the flexural rigidity of the deck plate. It is only this latter type of distribution which is described by the orthotropic plate approach while the existence of the former distribution can only be recognized by a judicious choice of effective breadth of deck plate acting with the trough. The finite element method using elements which only have extensional rigidity is used to study the former behaviour.

Consider the panel to be loaded over one stringer by the narrow 10-ton load. Using static beam distribution to transfer the load to the web line supports almost the total load would be carried by the webs of the loaded stringer. The extensional solutions for stresses in the deck plate and stringer are shown in Fig. 9. Also

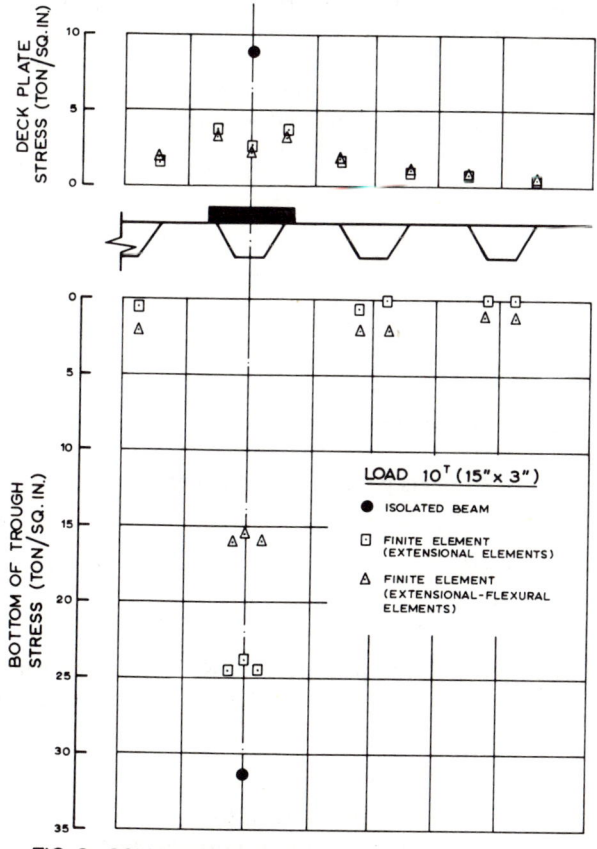

FIG. 9. COMPARISON BETWEEN THEORETICAL MIDSPAN BENDING STRESSES.

plotted is the simple beam solution which assumes the deck plate to be cut midway
between stringers so that the adjacent stringers and deck plate do not contribute
in any way to the distribution. It can be seen that the wide flange action gives
stresses which are less than the isolated beam stress by 25 per cent. If an extensional-
flexural solution is used then the trough stresses are reduced by a further 25
per cent while the deck plate stresses are only slightly affected. Thus it
can be seen that in this particular example the contributions to distribution of the
wide flange effect and the plate effect are of almost equal significance. It seems almost
fortuitous therefore that the orthotropic plate approach gives good comparison with
experiment in this case. However, with steel decks made from thicker plate material
the relief of stress in the bottom of a directly loaded trough would become more
dependent on the transfer of vertical shear than on the wide flange effect and the
use of the orthotropic plate approach would become more easily justifiable.

   Despite these criticisms the orthotropic plate approach is a more suitable
approach for design purposes as a three-dimensional finite element analysis of a
section of a steel deck consisting of deck plate, stringers and cross girders would
require an exorbitant amount of computer time and storage space. Besides as has
been shown if the rigidities are calculated as proposed by Pelikan reasonable
agreement with test results is obtained. In fact there is a need to simplify and
not complicate the approach for designers, particularly for use in the initial stages
of design. An attempt in this direction has been made by the author with the
shear key analogy method[10]. Based on the observation that a point of contraflexure
occurs in the deck plate approximately midway between stringers a set of formulae
which gives the proportion (k) of an applied wheel load (P) which is carried by a
stringer loaded symmetrically midway between cross girders may be derived. For
the case where only three stringers are considered to participate in the transverse
distribution the following formula applies.

$$k = \frac{\frac{1}{3} + \psi}{1 + \psi}$$

where,

$$\psi = \frac{D_x}{H} \left(\frac{a}{\ell}\right)^2$$

$D_x = \dfrac{EI}{a}$ flexural rigidity of stringers per unit width

H = torsional rigidity of stringers per unit width as defined by Pelikan (7, 13)
a = spacing of stringers
s = spacing of cross girders
ℓ = effective span of stringers
   = s for simply supported stringers
   = 0·7s for stringers continuous over equally spaced rigid cross girders

The method has yet to be developed to provide, if possible, a comprehensive
approach for all load cases and deck layouts likely to be encountered in practice.

Decks under lateral loading – local behaviour

Deflexions and bending stresses which are confined to the vicinity of the individual
wheel loads must be considered in the design of orthotropic decks. In particular local
deflexions of the deck plate between stringers must be restricted (generally to a value

of span/300 to ensure a satisfactory performance of the deck surfacing. Measurements which have been made on full scale bridges (7, 8, 12) as well as in laboratory specimens [7, 10, 12] show that local stresses may under traffic loading reach very high values. In open stringer decks the significant deflexions and stresses occur mainly in the deck plate and may be calculated using small deflexion plate theory by treating the deck as an isotropic plate (the deck plate) supported on elastic line supports (the stringers). Stresses evaluated on this basis show excellent agreement with those measured in the laboratory[10]. In closed stringer decks large stresses may also occur in the stringer walls. These may be estimated, as suggested by Esslinger[7], by calculating the vertical shear midway between stringers using orthotropic plate theory, and doing a secondary frame analysis on a unit strip of stringer to get transverse 'load transfer' bending stresses. To these must be added the local 'load carrying' stresses caused by the bending of the deck plate directly beneath the load.

A more accurate theoretical analysis is obtained using the finite element method. The maximum theoretical transverse stress in the deck plate of the test panel without surfacing was 25 ton/in² under the Swedish loading. In the design a 2-inch layer of surfacing was used on the deck and the transverse stress was less than 14 ton/in² (using a 45° distribution of load through the depth of surfacing). The maximum longitudinal bending stress in the deck plate reached a value approximately 50 per cent of the maximum transverse bending stress. The experimental stresses in the deck plate without surfacing reached only 12 ton/in² as the rigidity of the test pad was such that the load was carried to a large extent by the hard points in the deck; that is, the line of intersection of trough walls and deck plate. (Tests in later models in which the pad was designed to reproduce a uniform pressure showed excellent agreement with theory.) This result illustrates the sensitivity of local stresses to pressure distribution and area of contact. Transverse bending stresses of 10 ton/in² were measured in the walls of the loaded trough.

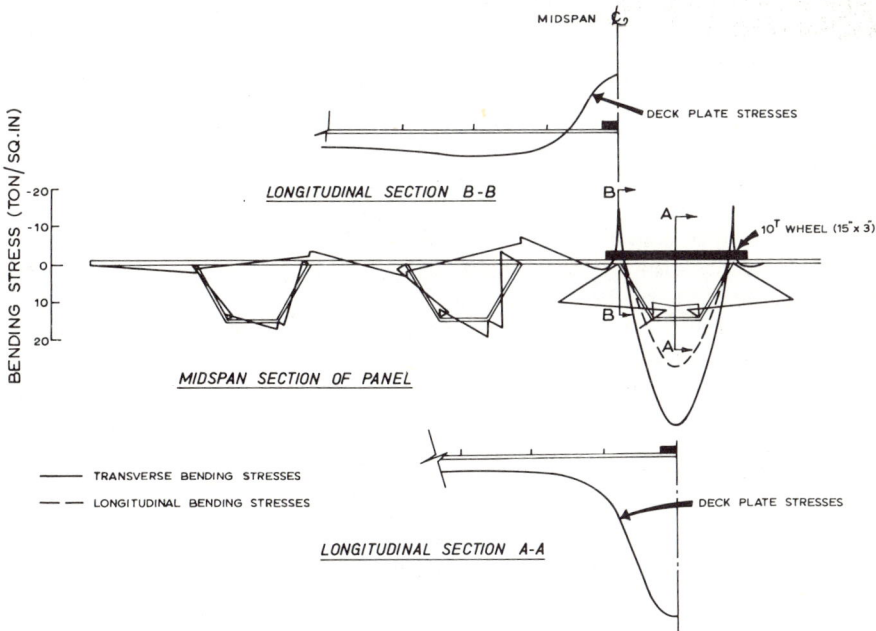

FIG. 10. THEORETICAL LOCAL BENDING STRESSES IN DECK PANEL UNDER WHEEL LOAD

The theoretical local stresses in the panel when the 10-ton wheel with a contact area of 15 in x 3 in was placed over one stringer are shown in Fig. 10. Because of the very high intensity of loading the stresses can be seen to reach quite unacceptable values. The loading currently being proposed for the new British code will have a reduced pressure and will be more in line with the maximum wheel loads encountered in practice. It should also be noted from the figure that the stresses at the welded intersection of trough and deck plate change sign a short distance away from the position of the wheel — a point of some importance in the evaluation of the fatigue life of such details as it increases the stress range.

While the stresses measured in unsurfaced test panels subjected to controlled wheel loads of known pressure distribution and contact areas are amenable to calculation and compare well with analysis, those stresses which occur in actual decks under real traffic loading are less amenable to such direct analysis. It is well known[12, 13] that the surfacing may relieve stresses in the steel deck by distributing the wheel load through its depth, thus increasing the effective area of contact at surface of the deck plate. In addition, the surfacing by acting compositely with the deck plate can increase its rigidity and so reduce the stresses further[12]. However, these actions are dependent on such parameters as temperature, duration of loading, rigidity of the tyre walls, etc., and so the situation is even more complicated. Efforts are currently being made in Britain, notably at the Road Research Laboratory, to clarify this situation and it is to be hoped that the results of this work may be incorporated in the new code. It should be mentioned here that the real significance of these local stresses is in relation to the fatigue life of the deck and this aspect will be discussed later.

Decks under combined in-plane and lateral loading

The procedure normally adopted in practice is to calculate the stresses in the deck under in-plane and lateral loading separately and to superimpose the stresses so obtained to give the total stresses in the deck. This procedure is not strictly correct as it neglects the interaction between the two forms of loading and a program developed by Aalami[14] may be used to check the errors involved.

Consider a panel of the type shown in Fig. 6 to be loaded not only by a central 10-ton wheel load but also subjected to an in-plane uniform displacement in the direction of stringer span corresponding to an average stress of 8 ton/in^2 along the end perpendicular to the direction of stringer span as shown in Fig. 11. The order of loading assumed is that the in-plane load is applied first and then the lateral loading. If the deck is initially flat then the in-plane loading which is below the critical buckling load of the deck produces no lateral deflexion. When the 10-ton wheel is applied, however, the in-plane compressive force tends to increase the deflexion, Curve 2, beyond that which would occur if just the lateral load were considered, Curve 1. This is due, of course, to the additional moment introduced by virtue of the eccentricity of the in-plane load with respect to the deflected panel. A corresponding increase in stringer bending moment would also occur. In the example illustrated the increase in deflexions and moment is approximately 14 per cent for the load $P_\varrho = 10$ ton and $P_i = 4\cdot3$ ton/in.

The effect will be exaggerated if initial deformations are considered. For example an assumed maximum initial deformation of span/500 in the stringers between supports, with the deformation acting in a direction sympathetic to those caused by the lateral load, produces the load deflexion Curve 4 for a point at mid-span. This may be compared to Curve 3 given by the simple theory which neglects the effects of in-plane load on lateral deflexions completely. The increase in

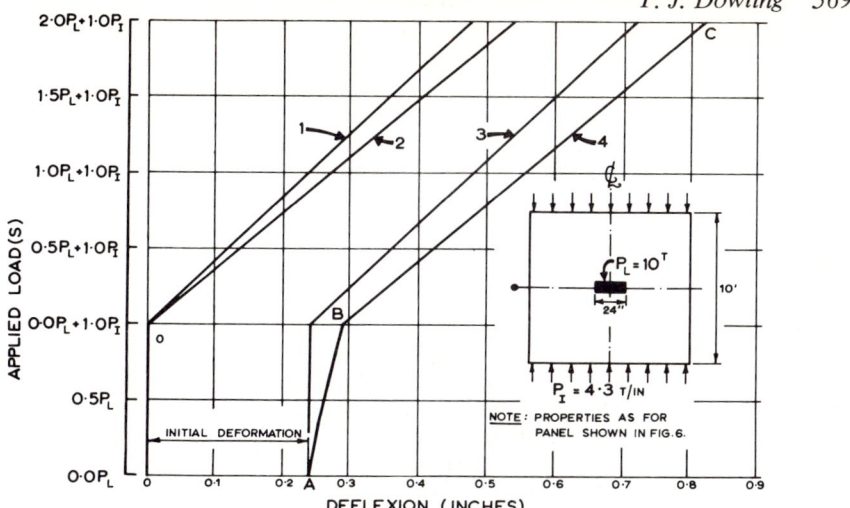

FIG. 11. <u>INTERACTION OF TRANSVERSE AND INPLANE LOAD IN A STIFFENED DECK PANEL.</u>

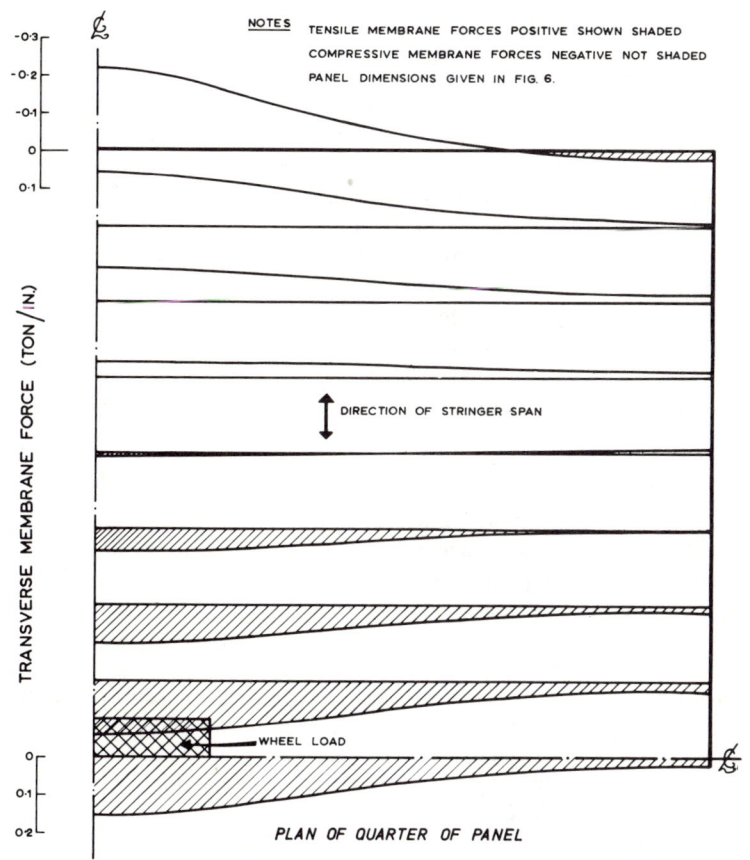

PLAN OF QUARTER OF PANEL

FIG. 12. <u>THEORETICAL TRANSVERSE INPLANE FORCES IN DECK PLATE OF STIFFENED DECK PANEL LOADED BY CENTRAL WHEEL LOAD.</u>

deflexions caused by the interaction of the applied loadings is 33 per cent when the loads $P_\varrho = 10$ tons and $P_i = 4\cdot3$ ton/in. It should be noted in the case of Curve 4 that the rate of increase in deflexion is positive within the region AB of the curve due to the column effect, but that it is slightly negative in the region BC due to the restraining large deflexion, or membrane, effect.

While the case considered may be an extreme one the exercise serves to illustrate the behaviour of a deck panel under combined loading and suggests that in certain cases it may be prudent to investigate such interaction. A simple initially deformed column calculation may suffice to estimate the magnification of deflexions and moments due to interaction in many cases.

It is worth noting that the magnitude of the membrane stresses in the deck due to the overall large deflexion effect is small under the normal design loads. Fig. 12 shows the distribution of membrane forces in the transverse direction calculated by the program referred to above. The tensile forces near midspan are a maximum under the load while the forces at each end are compressive reaching a maximum at the centre of each end where a transverse stress of approximately 1 ton/in^2 occurs in the 3/8 in thick plate. As in the case considered the edges parallel to the span were clamped against movement in their own plane the forces along those edges are non-zero. When these edges are free to move in their own plane, a self-equilibrating set of membrane forces occur, similar in distribution to those shown.

FATIGUE BEHAVIOUR

Because the local stresses produced in a steel deck are mainly caused by live loading and are therefore fluctuating stresses, the fatigue behaviour of the welded orthotropic deck needs special consideration. Two details which warrant particular attention are the transverse welds which connect stringers and cross girders and the longitudinal welds which connect the stringers with the underside of the deck plate. The choice of detail can have a major influence on the economics of this form of construction on account of the large amount of welding involved. Efforts have been made, therefore, to devise economical details which have a satisfactory fatigue performance. Fillet welded connections between lipped troughs and deck plate which require no edge preparation of the stringers have been shown by Toprac[15] to be unsatisfactory in fatigue. Preliminary tests at Imperial College suggest that the different in fatigue life between fillet welded troughs with walls bevelled to fit flush with the underside of the deck plate and troughs similarly prepared but connected to the deck plate by penetration welds is not very significant. Elliott[17] has estimated the cost of the former connexion to be only 0·6 that of the latter. It is this more economical connexion which has been used almost exclusively in the British bridges built to date (Fig. 13).

Whereas, it is normal to thread open stringers through cross girders to make up prefabricated panels of about 60 ft in length, it was the practice in some of the earlier British bridges to fabricate closed stringer panels by welding short lengths of troughs to the cross girders. This was because of the limitation on the length of troughs which could be pressed on the brake presses available to steel fabricators. More recently V shaped stringers have been used as these can be readily manufactured in longer lengths. Thus, recent British bridges have used a stringer-cross girder inter-section detail of the type shown in Fig. 13b in the prefabricated deck panels. Little information is available on the fatigue performance of these alternate details but it would appear that it is the latter detail which is to be preferred.

There is, in general, a dearth of published work on the fatigue performance of orthotropic steel decks and clarification is needed on the relevant performances

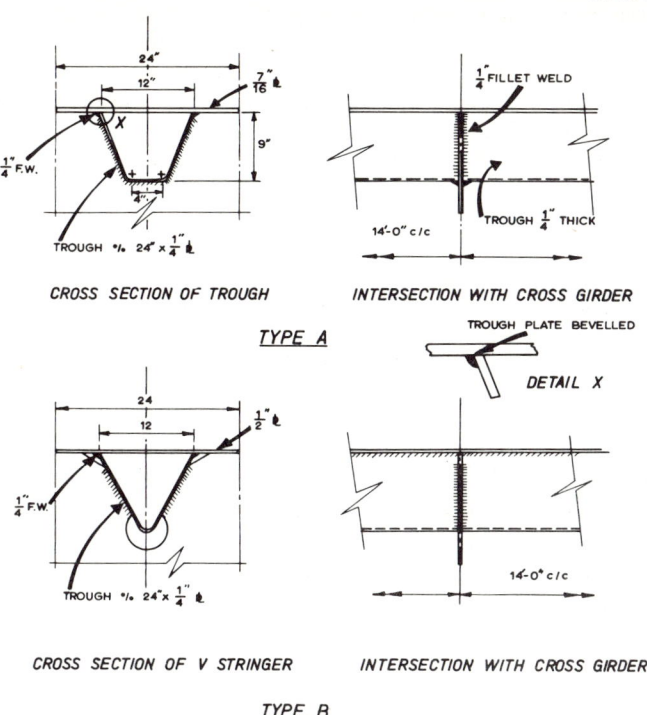

CROSS SECTION OF TROUGH

INTERSECTION WITH CROSS GIRDER

TYPE A

DETAIL X

CROSS SECTION OF V STRINGER

INTERSECTION WITH CROSS GIRDER

TYPE B

FIG. 13. TYPICAL DECK DETAILS USED IN BRITISH ORTHOTROPIC STEEL BRIDGES

of the various available details in fatigue, and more generally on the loading spectrum and stress range to which steel decks are subjected together with information on the influence of such parameters as residual stresses and the effects of the interaction between the shear and bending stresses to which these welded details are subjected.

ULTIMATE LOAD BEHAVIOUR

Decks under in-plane loading

The capacity of a steel deck in uniaxial tension is reached when the stress throughout the deck reaches the yield stress. Its capacity in compression is a more complex problem as it involves the non-linear buckling behaviour of a stiffened steel plate. No research dealing specifically with the capacity in compression of stiffened plates of bridge deck proportions has been reported, although related experimental work on stiffened ship's plating may be found in the literature[17, 18]. As the deck plate span between stringers to thickness ratio is normally ≤ 25 the problem is simplified as plate buckling between stringers does not occur in advance of the stringer buckling. Henry[19] has suggested that in the absence of more accurate methods the transverse rigidity of the deck plate might be ignored and the capacity of the deck based on the column capacity of a stringer deck plate unit spanning between cross girders. It is suggested that the onset of first yield might be taken

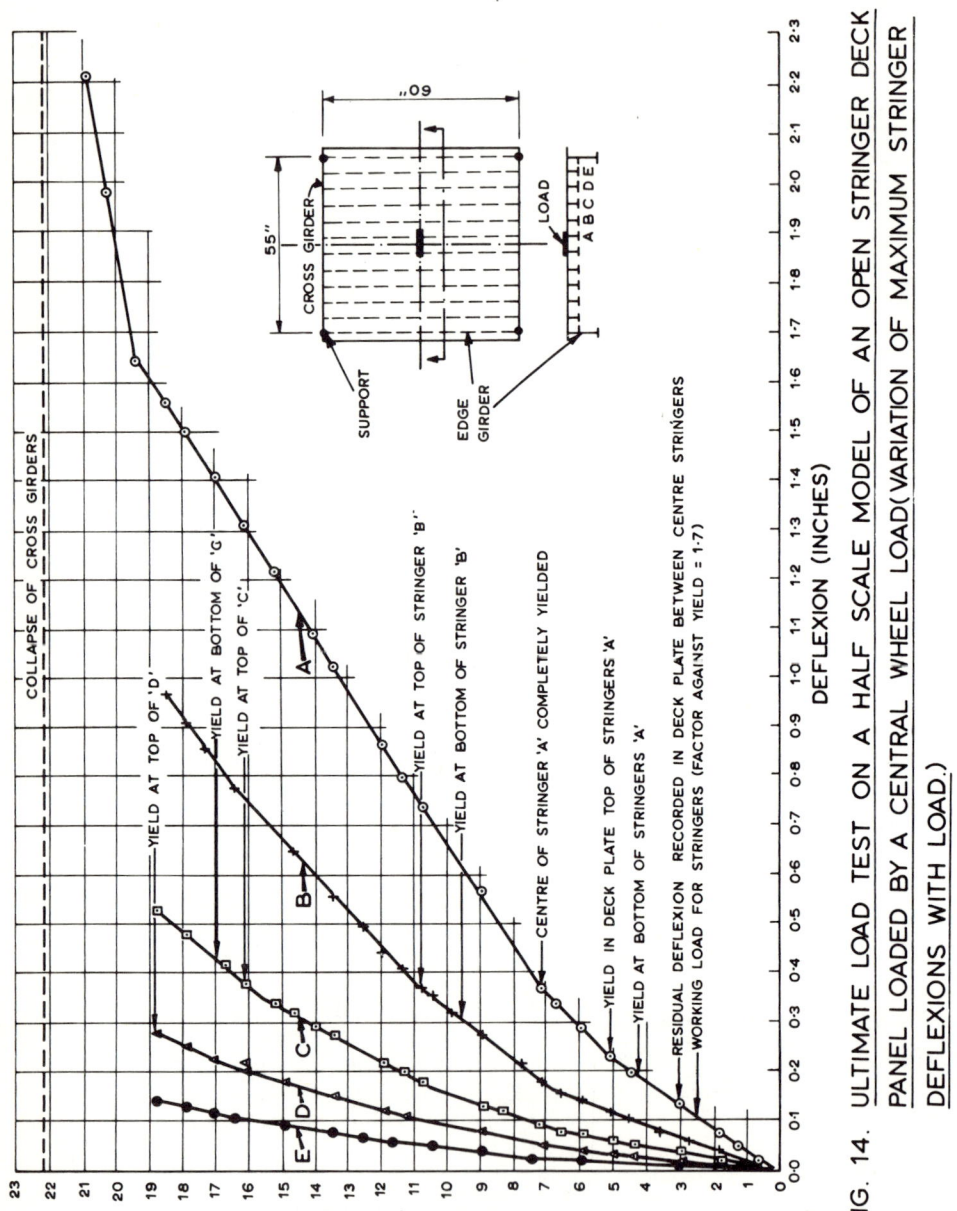

FIG. 14.   ULTIMATE LOAD TEST ON A HALF SCALE MODEL OF AN OPEN STRINGER DECK PANEL LOADED BY A CENTRAL WHEEL LOAD(VARIATION OF MAXIMUM STRINGER DEFLEXIONS WITH LOAD.)

as a suitable criterion to assess the decks' capacity without much loss of accuracy. Account must be taken of initial deformation and the possible effects of residual stresses on the capacity of the column. In the case of open stringers it is possible for torsional buckling of the stiffener about the line of contact with the deck to precipitate collapse while local buckling of the stringer walls in closed stringer construction must also be considered. An inelastic beam column program developed at Imperial College is currently being used to investigate this behaviour.

Decks under lateral loading

Tests by Kloppel[20], Pelikan[7], the author[10] and others have shown that the stiffened steel deck has a large reserve under wheel loading. This capacity is provided by the membrane strength of the deck plate — a reserve which, as was shown in Figs 11 and 12, is not utilised under normal loading. The results of a test on a 1/2 scale steel deck panel stiffened by inverted tee stiffeners are plotted in Fig. 14. This figure shows the growth of deflexions at midspan of the stringers with increase in load. The wheel load which was applied at the centre of the panel covered the two central stringers. These central stringers reached yield in their bottom flange at a load of 4·3 tons and had developed plasticity across the entire depth at midspan when the load reached 7·2 tons. This represented 3·0 times the working load for which the stringers were designed. At this stage the load deflexion curve for the central stringers showed a slight increase in the rate of deflexions but continued to behave linearly with increasing load. Simultaneously the load deflexion curve for stringer B increased its deflexion rate and began to carry the additional load. At this stage

FIG. 15   VIEW OF CROSS GIRDERS AFTER COLLAPSE OF HALF
SCALE STEEL DECK TEST PANEL

load was being transferred to stringers B by the plate partially in flexural and partially in membrane action, that is, the tensile membrane stresses indicated by Fig. 11, were beginning to play a significant part in the load distribution. As plasticity was reached at the centre of each stiffener in turn the additional load was transferred to the adjacent stiffeners as shown in the diagram. Finally, at a load of approximately 22·2 tons the plate in the vicinity of midspan of the cross girder at each end buckled and precipitated collapse of the cross girders, Fig. 15. This load represented a load factor of approximately 10 when compared to the design wheel load. As may be seen from Fig. 16 the deflexions were very large at this stage and would of course be clearly unacceptable in design. Nevertheless, due recognition of this capacity against overload from wheel loads should be made in selecting the load factors in any rational design procedure.

FIG. 16.   PERMANENT   DEFORMATIONS   IN   STEEL   DECK   PLATE
AFTER   ULTIMATE   LOAD   TEST

Decks under combined in-plane and lateral loading

As with the decks under in-plane loading little experimental information is available on the capacity of steel bridge decks under combined loading. However, some relevant work has been done at Lehigh University on stiffened plate panels of the type used in ships[21]. The analytical work is based on the capacity of an isolated beam-column consisting of a stiffener and an associated width of deck plate, and includes the effects of residual stresses and the effects of the uniform lateral pressure in reducing the effectiveness of the plate between stiffeners. Williams[18] has drawn attention to the biaxial stresses in the plate of a stiffened ship's bottom and has suggested that these may have a significant influence on the collapse behaviour of such stiffened panels. Tests are currently being planned at Imperial College to study all aspects of this complex problem.

CONCLUSIONS

Elastic behaviour

The finite element method employing rectangular shell elements can be used to analyse accurately the three dimensional overall and local elastic behaviour of orthotropic steel decks subjected to both lateral and inplane loading. The substitute orthotropic plate method may also be used to predict the approximate overall behaviour if the values of rigidities and the procedure for calculating stresses suggested by Pelikan[13] are employed. In general the increasing application of numerical solutions and availability of high speed digital computers have made all aspects of the elastic analysis of such decks a relatively straightforward procedure. It would appear that the need now is to provide the designer with simple design curves or formulae based either on the result of numerical analyses or on concepts such as the shear key analogy for use in the initial stages of design.

The inplane loading in the deck causes by its composite action with the main girders may have a significant effect on the maximum stringer deflexions and stresses under design lateral load particularly when the stringers are initially deformed.

Fatigue behaviour

Very little has been published on this important topic, although some work has been done in Britain in recent years. It is to be hoped that the results of this work will be made known in the near future so that designers can be given some guidance in the design of the critical welded connexions between stringers and deck plate and stringers and cross girders.

Ultimate behaviour

The whole field of ultimate load behaviour of bridge structures has been given an impetus as a result of the adoption of limit state philosophy in the new bridge code. Work currently in progress should provide much of the information needed to select a rational set of load factors to be used in the design of orthotropic decks against the limit state of collapse.

REFERENCES

1. Draft German Standard, DIN 1073. *Steel road bridges,* April 1969.
2. Ferahian, R. H. *Shear lag and buckling in plated grids.* M.Sc. thesis, University of London, 1963.
3. Kondo, K., Komatsu, S. and Nakai, H. Theoretical and experimental researches on the effective width of girder bridges with steel deck plate. *Trans. Japan Soc. civ. Engrs,* no. 86, Oct. 1962, pp. 1–19.
4. Abdel-Sayed, G. Effective width of steel deck-plates in bridges. *J. struct. Div. Am. Soc. civ. Engrs,* vol. 95, ST7, July 1969, pp. 1459–1474.
5. Lim, E. C. *Shear effective breadth in stiffened steel bridge decks.* M.Sc. thesis, University of London, Oct. 1970.
6. Lim, P. T. K. and Moffatt, K. R. General purpose finite element program. Proceedings of the Bridge Program Review Symposium (London, Jan. 1971).
7. Pelikan, W. and Esslinger, M. Die Stahlfahrbahn Berechnung und Konstruktion. *M.A.N. Forsch-Hft,* no. 7, 1957.
8. Dowling, P. J. Research on orthotropic steel plate bridge decks. *British Constructional Steelwork Association Conference on Structural Steelwork (London, Sept. 1966).*
9. Heins, C. P. and Looney, C. T. G. *The solution of continuous orthotropic plates on flexible supports as applied to bridge structures.* 2 vols. University of Maryland, March 1966.
10. Dowling, P. J. *The behaviour of stiffened plate bridge decks under wheel loading.* Ph.D. thesis, University of London, October 1968.
11. British Standards, B. S. 153. *Steel girder bridges.* 1954.
12. Douwen, A. A. van. Dutch report on orthotropic decks. *Proceedings of the Conference on Steel Bridges (London, 1968).* London, British Constructional Steelwork Association, 1969.
13. American Institute of Steel Corporation. *Design manual for orthotropic steel plate deck bridges.* New York, 1963.
14. Aalami, B. and Chapman, J. C. Large deflexion behaviour of rectangular orthotropic plates under transverse and in-plane loads. *Proc. Instn civ. Engrs,* vol. 42, March 1969, pp. 347–382.
15. Davis, H. L. and Toprac, A. A. *Fatigue testing of ribbed orthotropic plate bridge elements.* University of Texas, Centre for Highway Research, Research Report 77–1, May 1965.

16. Elliott, P. Can steel bridges become more competitive? *Proceedings of the conference on Steel Bridges (London, 1968)*. London, British Constructional Steelwork Association, 1969.
17. Wah, T. and Sherman, R. *Study and behaviour of grillages in elastic and plastic ranges.* Naval Ships Systems Command Research Report. Southwest Research Institute, Dec. 1969.
18. Williams, D. G. *Analysis of a doubly plate grillage under in-plane and normal loading.* Ph.D. thesis, University of London, July 1969.
19. Henry, W. J. *Limit state analysis of Kut Bridge superstructure.* M.Sc. thesis, University of London, Nov. 1967.
20. Klöppel, K. Zur orthotropen Platte aus Stahl. *Die neue Köln-Mülheimer Brücke: Zusammenstellung der wissenschaftlichen Beiträge.* Köln, F. W. Waltking, 1951.
21. Ostapenko, A. Longitudinally stiffened plate panels under lateral and axial loads (ship bottom plating). Final report on research project. *Fritz Engineering Laboratory Report* no. 248.28, Lehigh University, August 1969.

THE EXTENSION OF A FATIGUE CRACK IN A MODEL STEEL DECK PLATE UNDER CYCLIC LOADING

M. S. G. CULLIMORE and R. W. HORSINGTON
University of Bristol

SYNOPSIS

The model deck was subjected to local cyclic variations of loading, producing bi-axial bending. The growth of the fatigue crack caused by this loading was measured by an indirect method of combining continuous observation of strain with a theoretical prediction of the relation between that strain and the crack length. It was found that a linear relation existed between the length of crack and the number of cycles of load after its initiation, and the rate of growth of the crack for various applied loads was predicted.

NOTATION

x, y, z	Cartesian Coordinates
u, v, w	displacement components
$\epsilon_x, \epsilon_y, \epsilon_z$	direct strain components
$\gamma_{xy}, \gamma_{yz}, \gamma_{zx}$	shear strain components
N	number of load cycles after crack initiation
P	amplitude of load cycle
$\epsilon_R(N)$	amplitude of strain cycle at the reference point
$\alpha(N)$	$= \epsilon_R(N)/\epsilon_{R(0)}$ Relative Amplitude of strain cycle
$2b$	mean crack length
$\overline{w}(x, y)$	deflection function for Rectangular plate element
$\overline{w}(x)$	deflection function for beam element
$H_{ij}^{(n)}(x)$	Hermite polynomial
a, c	plate element dimensions
t	plate thickness

INTRODUCTION

The deck plates of a box girder bridge are subjected not only to overall compression from dead loading, but also to fluctuating local bending by wheel loading, which will normally produce a state of bi-axial bending. Although there are numerous references in the literature to the propagation of fatigue cracks in plates under plane stress,[1,2] very little work appears to have been published on plates subjected to pulsating flexure. In a laboratory study of the fatigue strength of plates under this type of loading it is desirable to simulate the deck panel by a specimen which is cheap and simple to produce, thus enabling a large number of specimens to be tested. A standard steel Rectangular Hollow Section was chosen as being suitable propor- tions and of a shape to which an axial force, simulating the overall compression, could conveniently be applied.

Although the initiation of a fatigue crack in a bridge deck would not normally constitute failure, it is a significant event in the life of the structure. The growth of the crack under continued loading will, however, eventually cause the panel to become unserviceable on the grounds of excessive local deflections. Knowledge of the rate of growth of the crack is therefore essential in determining the remaining useful life of the panel subsequent to the first appearance of a crack.

The start of a crack in a specimen subjected to cyclic stress is commonly difficult to detect. In the case of the hollow section specimens the problem is aggravated by the inaccessibility of the point of origin of the crack, which lies on the inside surface of the tube. As this prevents the attachment of any crack gauges an indirect means of detection is required.

The method employed is based on the general principle that if a structure is sub- jected to a cyclic pattern of load of constant amplitude then, provided that the yield stress of the material is nowhere exceeded, the strain response of the structure at some arbitrary reference point will also be a cyclic pattern of constant amplitude. An alteration in the geometry of the structure, such as the introduction of a crack, will be detectable as a discontinuity in the amplitude of the strain cycle at the reference point. Continuous monitoring of the strain amplitude (ϵ_R) will therefore give the number of load cycles necessary to cause a crack to start and subsequent observa- tions will provide a relation between ϵ_R and the number of cycles of load (N). Thus, if a relation between the crack length and the strain at the reference point is known, the rate of crack growth in terms of the number of load cycles can be obtained.

Calculation of Strains in the Cracked Plate

A solution for the strains on the upper surface of the tube was obtained by the method of finite elements. The plate forming the upper flange of the tube may be considered to be an arrangement of rectangular elements subjected to a concentrated load at the centre and elastically supported by the side walls. The presence of two axes of symmetry make it necessary to consider only one quarter of the plate (Fig. 1).

The method used to generate stiffness matrices for the elements was that developed by Bogner, Fox and Schmit[3] in which the displacement state of the element is defined by interpolation between the nodal values of the degrees of freedom.

$$\overline{w}(x,y) = \sum_{i=1,2} \sum_{j=1,2} \left[H_{0i}^{(1)}(x).H_{0j}^{(1)}(y).(w)_{ij} + H_{1i}^{(1)}(x).H_{0j}^{(1)}(y)(w_x)_{ij} \right.$$
$$\left. + H_{0i}^{(1)}(x).H_{1j}^{(1)}(y)(w_y)_{ij} + H_{1i}^{(1)}(x).H_{1j}^{(1)}(y)(w_{xy})_{ij} \right] \tag{1}$$

in which the Hermite polynomials are:

$$H_{01}^{(1)}(x) = \frac{1}{a^3}(2x^3 - 3ax^2 + a^3), \qquad H_{02}^{(1)}(x) = -\frac{1}{a^3}(2x^3 - 3ax^2),$$

$$H_{11}^{(1)}(x) = \frac{1}{a^2}(x^3 - 2ax^2 + a^2x), \qquad H_{12}^{(1)}(x) = \frac{1}{a^2}(x^3 - ax^2),$$

and the expressions for $H_{ij}^{(1)}(y)$ are obtained by replacing x by y and a by c. The degrees of freedom w, w_x, w_y and w_{xy} represent the values of w, $\dfrac{\partial w}{\partial x}$, $\dfrac{\partial w}{\partial y}$ and $\dfrac{\partial^2 w}{\partial x . \partial y}$ at the corners of the element.

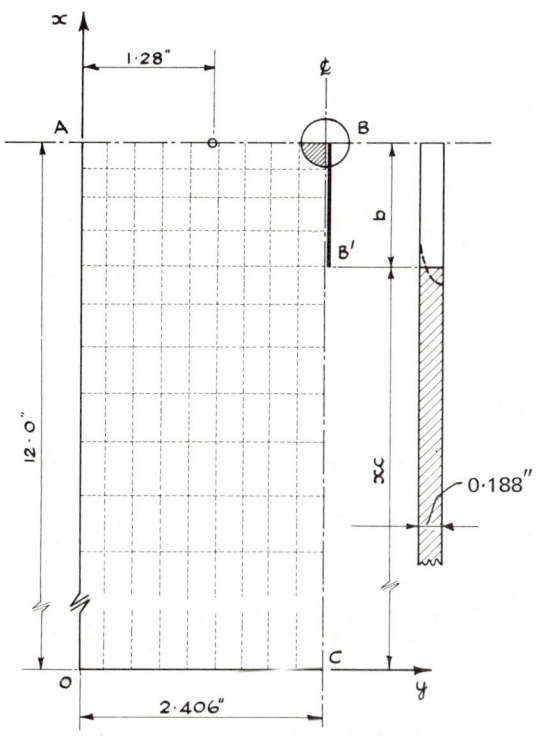

FIG. 1 GEOMETRY OF SPECIMEN

The advantages of this form of displacement function are that not only is it capable of representing exactly any displacement of the type

$$\overline{w}(x, y) = \sum_{r=0}^{3} \sum_{s=0}^{3} A_{rs}. x^r.y^s, \qquad (2)$$

but also that the deformed state on any element boundary is completely defined by nodal values on that boundary. This ensures compatibility of deflections and slopes on all inter-element boundaries within the mesh.

The inclusion of the twist as a degree of freedom, although increasing the size of the computation, permits a more precise specification of boundary conditions. On boundaries such as the axis of symmetry AB (Fig. 1) the condition of zero normal slope may be defined completely by imposing $w_x = 0$ and $w_{xy} = 0$ at each of the nodes on the boundary.

The introduction of a crack in the position BB' as shown in the figure will not affect conditions on the boundary AB except that at the point B the slope w_y must be unrestrained. Similarly for nodes on BC within the region of the crack ($x > xc$) both the normal slope (w_y) and the twist (w_{xy}) must be released. At the crack tip B' as at all nodes for which $x \leqslant xc$, the conditions $w_y = w_{xy} = 0$ are maintained.

These conditions imply a sharp vertical crack front in contrast to the characteristic elliptical shape of the actual crack which is shown in Fig. 2. This discrepancy and the effects of plasticity in a small region around the crack tip B', are considered to have a negligible effect on conditions at the reference point (R) and have been disregarded in the analysis.

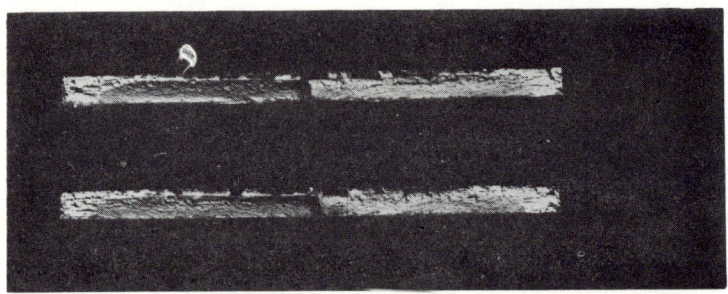

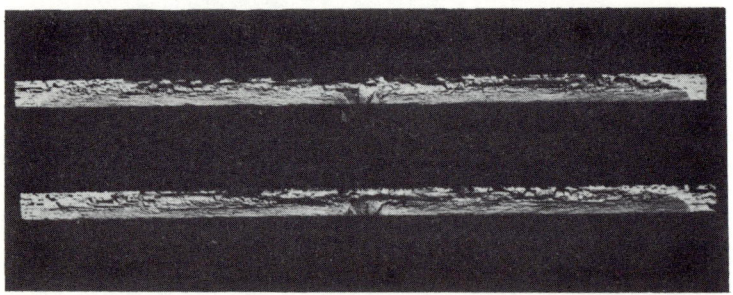

FIG. 2 DETAIL OF FATIGUE CRACKS

The condition of elastic support along the boundary AO can be satisfied, as far as the degrees of freedom w and w_x are concerned, by the introduction of compatible beam elements between nodes on this boundary. The displacement states for such elements are generated in a manner similar to that for plate elements.

$$\overline{w}(x) = \sum_{j=1,2} \left[H_{0j}^{(1)}(x) \cdot w_j + H_{1j}^{(1)}(x) \cdot (w_x)_j \right] \tag{3}$$

As these elements are intended to represent the beam action of the specimen as a whole the stiffness term, used in the calculation of the matrix expression for strain energy, must be a half of the value of the overall flexural stiffness of the complete tube. The elastic restraint imposed on the degrees of freedom w_y and w_{xy} by the bending stiffness of the side wall is allowed for by selection of appropriate terms of the stiffness matrix for this plate.

The stiffness matrices for the various elements are used in the usual manner to form a 'global' stiffness matrix $[K]$ for the whole plate. The matrix equation

$$\{Q\} = [K] \cdot \{q\}, \tag{4}$$

in which $\{Q\}$ represents a vector of applied forces, may be solved by inversion of the stiffness matrix to obtain values for the nodal degrees of freedom $\{q\}$, which are then substituted back into the assumed deflected forms of the elements (Eqn. 1). Strains at the surface of the plate are then found by differentiation.

$$\epsilon_x = -\frac{t}{2} \cdot \frac{\partial^2 \overline{w}}{\partial w^2}(x, y)$$

$$\epsilon_y = -\frac{t}{2} \cdot \frac{\partial^2 \overline{w}}{\partial y^2}(x, y) \tag{5}$$

$$\gamma_{xy} = -t \cdot \frac{\partial^2 \overline{w}}{\partial x \partial y}(x, y)$$

The strains at a particular node in the mesh evaluated from each of the elements connected at that node will not in general be equal, as the assumed deflected form of the elements does not provide for inter-element compatability of curvatures. The strain at the reference point is taken as a mean of values interpolated from the nodal values over a quarter inch gauge length centred on R. Figure 3 compares calculated strain at

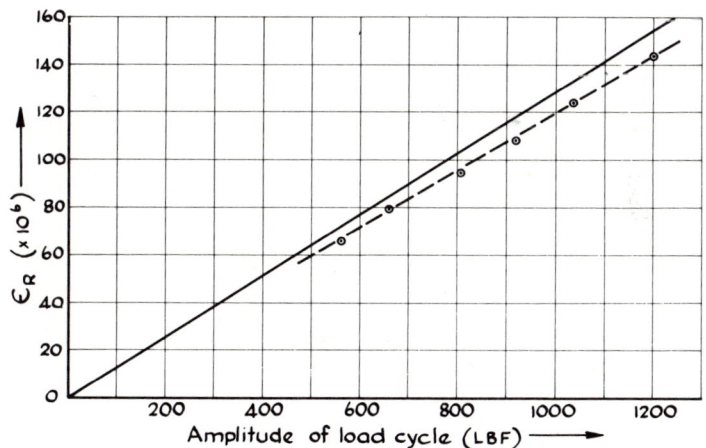

FIG. 3 COMPARISON OF VARIATIONS OF CALCULATED AND OBSERVED VALUES OF ϵ_R WITH LOAD

the reference point with experimental results for various values of applied load. The contours of the surface strain component ϵ_y (i.e. the component measured at the reference point) shown in Fig. 4 are drawn from average computed nodal values and compare the distribution for an uncracked plate with that for one having a crack 2·56 in (65 mm) long.

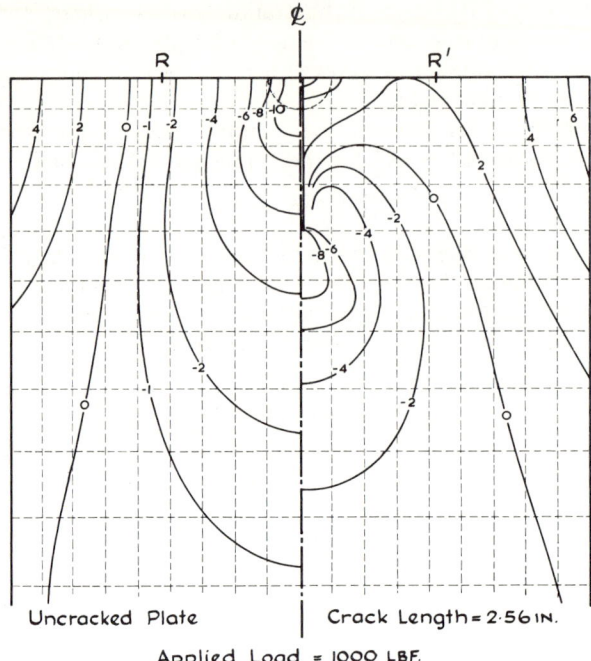

FIG. 4 STRAIN CONTOURS FOR CRACKED AND UNCRACKED PLATES

Selection of the Reference Point

The selection of the reference point is governed by:

(a) The type of sensor to be used. In this case, where it is necessary to record rapidly varying strains over an extended period of time, the most convenient type is the electrical resistance strain gauge. The amplitude of the strain cycle at the reference point must therefore be as low as possible to minimise the possibility of fatigue damage to the gauge before the appearance of a crack in the structure.

(b) The desirability of a rapid rate of change of ϵ_R with increasing crack length to give high sensitivity.

(c) The theoretical consideration that, as overall elastic behaviour of the structure has been assumed in the analysis, the reference point should be sufficiently remote from the crack tip so as not to be affected by the small area of local yielding there.

A prediction of the point of origin of the crack and its direction of propagation is thus required, which in this case was obtained from the theoretical tensile strain distribution. In the absence of such a solution an intuitive prediction must be made. A good compromise of these requirements is obtained if the reference point R is near the point of contraflexure on the transverse axis of symmetry of the panel shown in Fig. 1.

The strain at this reference point was evaluated for a number of different values of the crack length and the results plotted as a continuous curve, as shown in Fig. 5. The amplitude of the strain cycle is expressed as a ratio to the initial value, that is the constant value of the amplitude before the first appearance of a crack. This procedure avoids the necessity of making accurate quantitative measurements of strain.

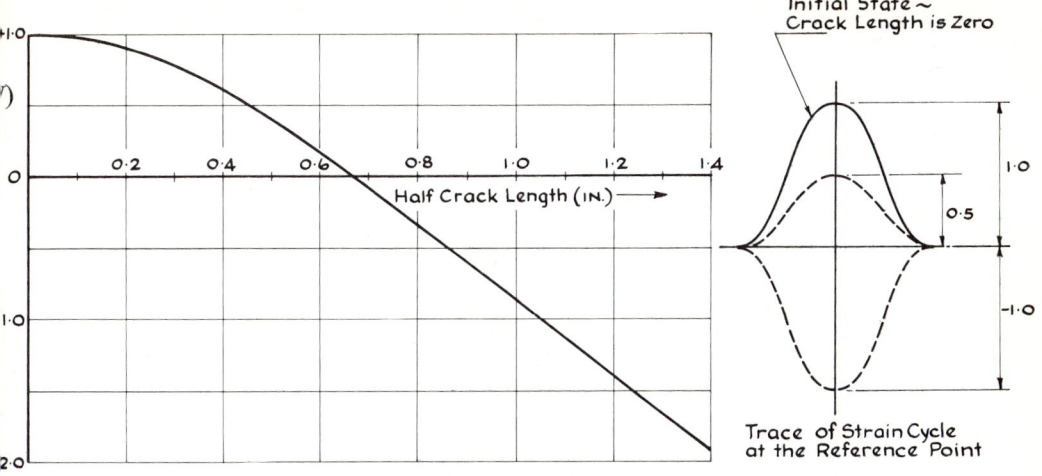

FIG. 5, CALCULATED RELATION BETWEEN ϵ_R AND CRACK LENGTH

However, the ability to make such measurements remains an advantage since it serves as a check on the accuracy of the analysis.

The relationship between crack length and strain amplitude provided by the analysis was then used in conjunction with measured values of strain to plot crack length against the number of load cycles.

Experimental Procedure and Results

The specimens selected for testing consisted of a 24 in (600 mm) length of Rectangular Hollow Section 5 in x 5 in x 3/16 in (127 mm x 127 mm x 5 mm) in mild steel to BS 15. These tubular sections were bolted to rigid end plates supported on ball race bearings mounted on the neutral axis of the section to minimise longitudinal restraint (Fig. 6).

A pulsating load (zero to maximum) was supplied by a hydraulic jack driven by the pulsator of a Losenhausen UHS 60 fatigue testing machine. The jack was mounted vertically above the specimen with its lower load pad resting on a steel plate partially supported by six springs. The purpose of this device was to enable a small preload to be maintained in the hydraulic system when the lowest point of the load cycle was reduced to near zero. The bulk of the load from the jack was transmitted through a ¾ in (19 mm) diameter steel ball to the piston of an oil filled load cell, sliding freely in a vertical sleeve. The load was transmitted directly to the specimen through the base of the load cell where an adjustable steel foot, 0·5 in (13 mm) diameter, rested on the centre of the upper surface of the tube (Fig. 7). A short length of flexible armoured hose connected the load cell to a pressure transducer consisting of a 2 in (50 mm) diameter diaphragm of 16 SWG (1·6 mm) steel plate, above which was mounted the probe of a Wayne-Kerr vibration measurement bridge. This instrument, when calibrated, was used to maintain a continuous check on the load cycle. In fact it was found that the apparatus was able to maintain the load to an accuracy which made manual correction unnecessary. The maximum speed of loading that could be achieved was 500 cpm (8 Hz).

A few preliminary tests showed that, under the stress conditions of the experiment, a crack would originate on the underside of the plate beneath the concentrated load and would extend longitudinally in a plane normal to the surface of the plate. An examination of the strain pattern on the surface of an uncracked tube (Fig. 4) indicated that a strain gauge positioned at the point R would initially record a small compressive strain and that the appearance and growth of a crack would tend to alter the strain pattern so that the recorded strain became increasingly tensile. Accordingly

FIG. 6 GENERAL VIEW OF APPARATUS

strain gauges were fixed both at R and at the symmetrically opposite point R'. In the interests of cost and ease of application it was decided to use paper-backed wire gauges with a quarter inch gauge length. It was found that under the conditions of low compressive strain the performance of these gauges was perfectly satisfactory up to at least 8×10^6 cycles.

One of the gauges fixed to the specimen was connected to a dynamic strain measuring bridge with a visual display which was used as a monitor on the progress

FIG. 7 DETAIL OF LOADING FOOT

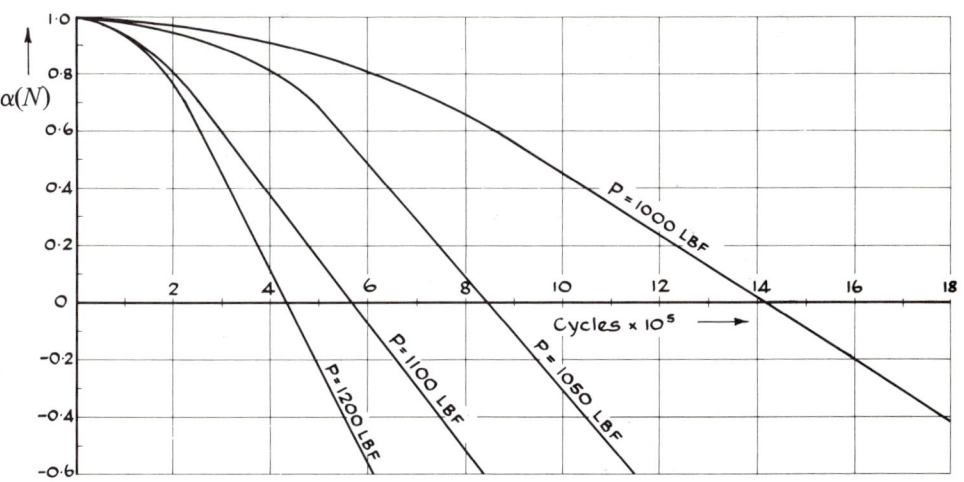

FIG. 8 VARIATION OF MEASURED ϵ_R WITH N

of the experiment and as a means of making quantitative measurements of strain. The signal from the second gauge was fed to an oscilloscope, with an amplifier suited to the low frequencies involved, where it appeared as a sinusoidal trace. This trace was photographed at intervals of 5 minutes (approximately 2500 cycles) by a Bolex H16 cine-camera fitted with a single-frame actuator connected to a time lapse device. A specially constructed hood containing a system of mirrors enabled a digital counter, showing the current number of load cycles completed, to be recorded on the same frame of film. Once developed the cine film was examined and the amplitudes of the traces were measured with a travelling microscope.

The results obtained for four different values of the applied load are shown in Fig. 8. As this part of the investigation is concerned with crack growth the results are plotted in terms of the number of cycles after the start of the crack; the curves for different loads thus have a common origin. From these curves and the computed relation between strain and crack length (Fig. 5) the curves of Fig. 9 showing the increase in crack length with number of cycles of load were constructed for various values of the applied load. Fig. 10 showing rates of crack extension is derived from Fig. 9.

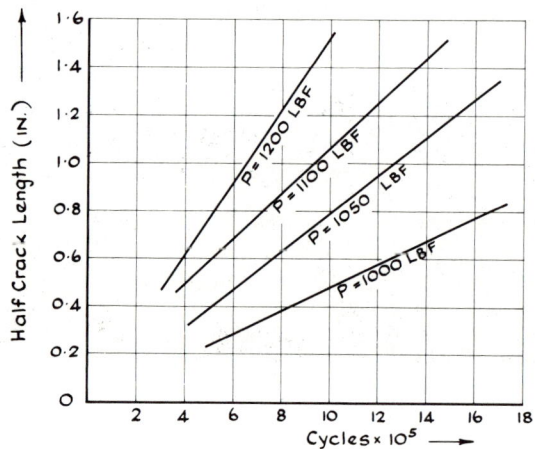

FIG. 9 CORRELATION OF CRACK LENGTH WITH NUMBER OF LOAD CYCLES

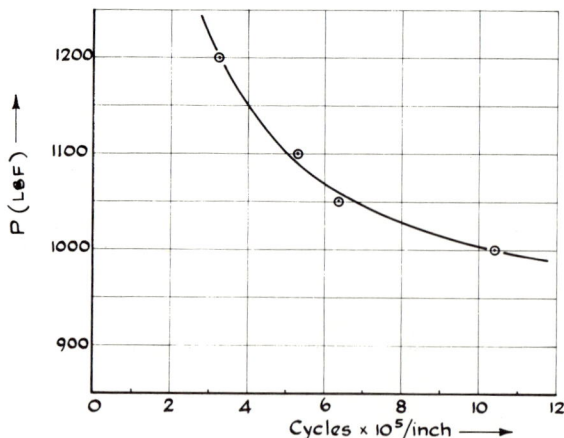

FIG. 10 RATE OF CRACK EXTENSION

Comments

The most obvious feature of these results is the linear relation between crack length and numbers of load cycles. As might be expected the rate of crack growth decreases as the load is decreased and at loads below about 900 lbf (4 kN) no cracking was observed up to 8.0×10^6 cycles of load. However, further tests have shown that loads below this figure are able to propagate macroscopic cracks, albeit at the very slow rates predicted by extrapolation of Fig. 10.

In these model tests the crack tip moved away from the load, the position of which was fixed. A bridge deck, however, is subjected to moving wheel loads so that, in effect, the load is always at the crack tip. It is considered that, for cracks in a limited area around the centre of a deck panel in which variations of the position of the load cause only minor changes in the stress pattern, there will be an approximately linear relation between crack length and number of cycles. Thus, if the spectrum of wheel loadings is known for a particular bridge, the rate growth of a crack might be predicted by a process of summation from the appropriate rate of growth curve. This could seem to indicate that, if the state of failure for a deck panel is defined as a particular crack length at which the panel becomes unserviceable, then the useful life remaining after the first appearance of a crack will depend on a linear rule of cumulative damage.

REFERENCES

1. Head, A. K. The propagation of fatigue cracks. *J. appl. Mech.*, Vol. 23, 1956, pp. 407–410.
2. Frost, N. E. and Dugdale, D. S. The propagation of fatigue cracks in sheet specimens. *J. Mech. Phys. Solids*, Vol. 23, 1956, pp. 407–410.
3. Bogner, F. K., Fox, R. L. and Schmit, L. A. The generation of inter-element-compatible stiffness and mass matrices by use of interpolation formulas. *Proceedings of the Conference on Matrix Methods in Structural Mechanics* (*Wright-Patterson Air Force Base, Ohio, 1965*), 1967.

THE FOURTH DANUBE BRIDGE IN VIENNA—DAMAGE AND REPAIR

R. HECKEL

Waagner-Biro

SYNOPSIS

The paper deals with the buckling of a section of the 4th Bridge across the Danube and the subsequent repair operations.

1. Design

The 4th Bridge across the Danube at Vienna is the largest section of a 3 km bridge system forming part of the North Eastern motorway. The bridge which supports six traffic lanes and two sidewalks has a total width of 31·38 m.

The bridge was submitted for a design and tender quotation and the lowest offer received was for a concrete bridge with two main spans of 165 m each. After the following relevant factors had been duly considered, navigation requirements, the presence of a railway station on the right bank, river regulations, foundation risks, economy and appearance, a steel bridge with a central span of 210 m was ordered since this left the navigable part of the Danube completely free.

Since the navigable pass extends along the right side of the river bed, the only pier in the river had to be suitably located in the shallow dike area near the left bank. This has resulted in a highly asymmetric design of the bridge. The bridge is a continuous steel girder deck bridge and the superstructure extends across the Danube in two spans of 120 m and 210 m; an adjoining 82 m span leaving the quay, railroad track and road areas free from intermediate supports.

With a gradient radius of 40,000 m the crest of the motorway is extremely flat and, due to the navigational profile, the total depths allowed were also very low. The bridge has a web plate depth of 5 m at the left bank abutment, 7·28 m immediately above the river pier, 4·60 m above the right bank pier, and 3·75 m at the right-hand abutment, and the structure therefore represents one of the world's most slender steel girder bridges. Considering the large width and the low total depth available, a cross-section was chosen which consists of two torsion-stiff box girders, each having a width of 7,560 m (see Fig. 4(*a*)).

These boxes are spaced at a distance of 8·10 m apart, and are linked together by the orthotropic deck slab, which guarantees the combined integral action of both boxes even if only one is fully loaded. The omission of diaphragms or transverse bracings between the two boxes is an important feature of this design and it is the first time that such a design has been used on such a large bridge.

The roadway deck comprises an orthotropic steel plate of 10—25 mm thickness with 5 cm of asphalt coating. The plate is reinforced by flat bar stringers 200 by

10 mm to 300 by 20 mm spaced at intervals of 0·36 m, the cross girders being at
2·00 m centres. The bottom plates of the boxes have a thickness of 10—30 mm and
the web plates a thickness of 12—16 mm. The bridge which is constructed with
St 52, St 44 and St 37 steels has a total weight of 5730 tonnes, i.e. 430 kg/m².

2. Erection

Without the use of auxiliary trestles on the 210 m main span, the bridge was erected by
the free cantilever method from both banks; owing to the indicated asymmetry of
the superstructure the closing point is not in the centre.

On the left bank, starting from the abutment, the superstructure was set up along
a length of 64 m on two auxiliary trestles whereupon it was carried forward in free
cantilever erection for another 56 m up to the river pier. From there it was cantilevered without auxiliary support for another 121 m into the central span. On the
right bank erection began on the bank pier, and using only one auxiliary trestle the
bridge was freely constructed over the railway area towards the abutment. Subsequently free cantilever erection was carried forward from the bank pier 87 m into
the central span, see Fig. 1 which shows the bridge in the course of erection. A free

FIG. 1 MIDDLE SPAN DURING ERECTION

cantilever width of 121 m with an average effective depth of only 5 m represents
something of a record. Therefore, only a very lightweight type of erection procedure
could be employed, with the result that it was not possible to use the method of
assembling whole prefabricated box sections — a method frequently used these days.

The cross-section at the cantilever end was assembled out of structural elements
with a maximum weight of 25 tonnes by a special derrick crane weighing 30 tonnes.
Due to the working range of the derrick the two inner web plates were mounted
initially and connected by the central strip of the deck slab. Then the derrick was
moved forward to erect the outer web plates and to complete the two boxes.

Three different joining methods were applied during erection. Riveting still
proved the cheapest type of connection for the horizontal web plate joints which

were prefabricated on the banks, and also for the bottom plate. The field joints of the web plates and cross girders were connected by means of High Strength Friction Grip bolts whilst all deck plate joints and the transverse joints of the bottom plates, including their stiffening ribs, were welded.

The considerable deflection of both cantilever ends required that the bridge be supported on erection stacks having a height of approx. 4 m on the river pier, see Fig. 5, and approx. 3 m on the left bank pier. As a result the camber of the deflection line was such that it was possible to close the central joint without the application of force. In addition in order to save weight the bottom and deck plates were not built into the last 12 m of the left-hand cantilever end. The derrick was withdrawn and the superstructure was finally closed by inserting four fitting pieces each 2 m long, which were brought forward from the right bank by mobile cranes.

It was intended to complete this central section by inserting the bottom and deck plates (still missing over a length of 14 m), and afterwards to lower the superstructure onto the piers in order to obtain the final gradient and the moment curve.

3. The Accident

Early in the afternoon of November 6th the four joints between both halves of the superstructure were finally connected. The same evening at 8.30 p.m. several explosion-like bangs indicated that serious damage had occurred to the bridge. Navigation along the river and rail and road traffic below the bridge was stopped

FIG. 2(a)

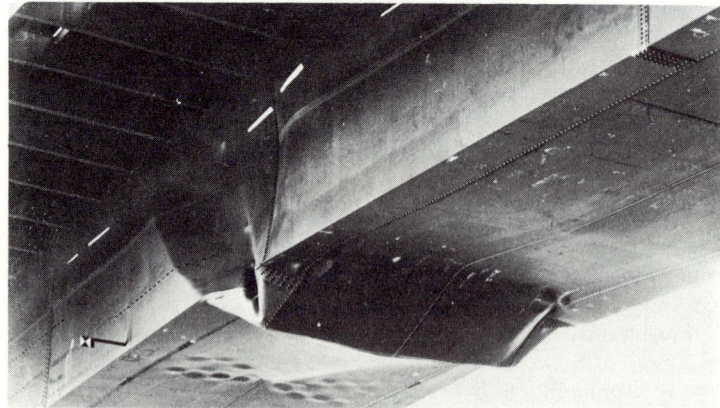

FIG. 2(b)

immediately. A TV-bulletin broadcast the news and as a result nearly all engineers and some of the site working crew were back at the site of the accident by 10.30 p.m.

The analyses made during the next hours established that the following damage had occurred:

At two points — at the centre of the 120 m span and approx. 60 m from the right bank pier in the 210 m span — both box girders had buckled, see Fig. 2. The resultant shortening of the superstructure damaged the abutment bearings and the 4 m high stacks on the river pier. Within seconds the failure of the bending resistance at both of the points at which buckling had occurred thoroughly changed the moment curve obtained in the course of erection (see Fig. 3). The central joint which had not been under stress during closing on that day suddenly had to bear the full dead weight moment of the steel weight of the bridge although the bottom and deck plates had not yet been mounted at this point. Along a length of 14 m there were only four web plates with small flanges (see Fig. 4(b)). These were now being subjected to stresses beyond their elastic limit. Under such enormous stress the upper flanges were twisted and the upper parts of the webs plastically buckled. The energies which were released when the boxes buckled — corresponding approximately to a drop of the bridge by about 3 m — caused some minor seismic waves. Seismographs located about 5 miles from the bridge showed three readings at approximate intervals of five seconds.

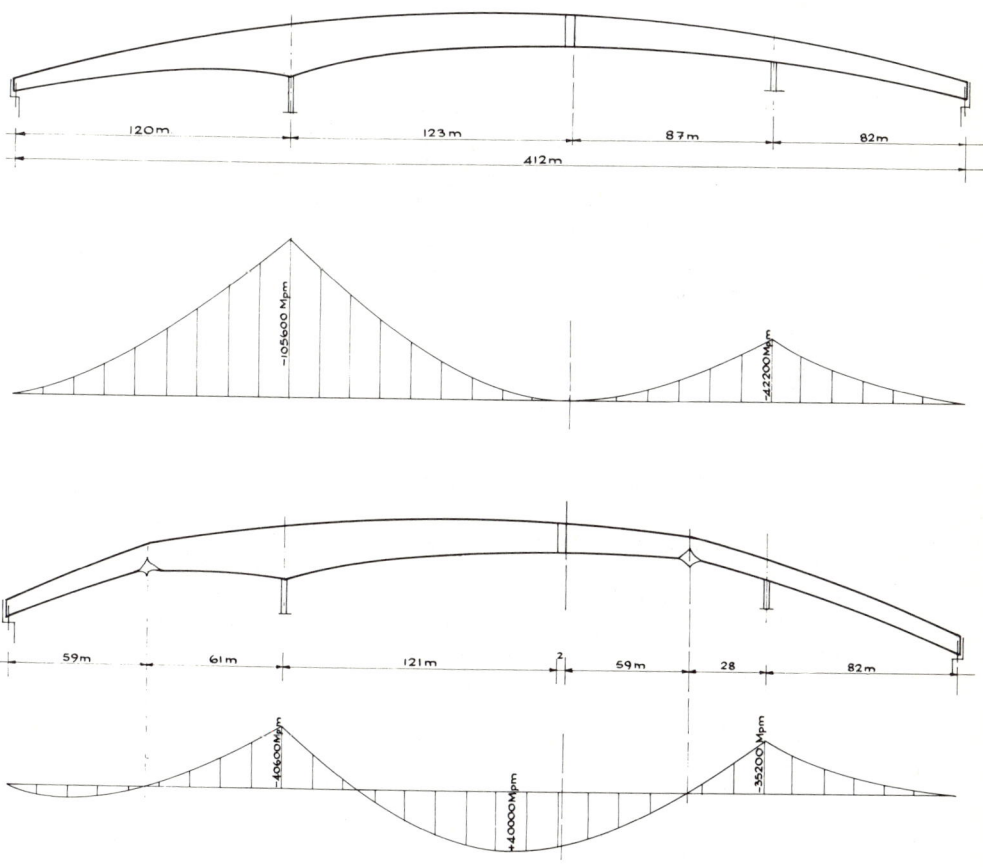

FIG. 3   SYSTEM AND CURVE OF BENDING MOMENTS BEFORE AND AFTER ACCIDENT

The shock waves can be explained as the result of the buckling of the box girders and of the consequent horizontal movement of the bridge due to shortening of the bottom chords.

The cause of the accident could not be traced back to any single mistake. Similar to many previous disasters in engineering, several causes — each of these well below the safety limit — coincided in the same negative sense, and all the provided reserves for safety were used up.

The safety factor required for states of erection is always below the one stipulated for the finished object. In the present case this factor was rated at 1·25, and full use was made of this margin by the achieved cantilever length. In addition, the final

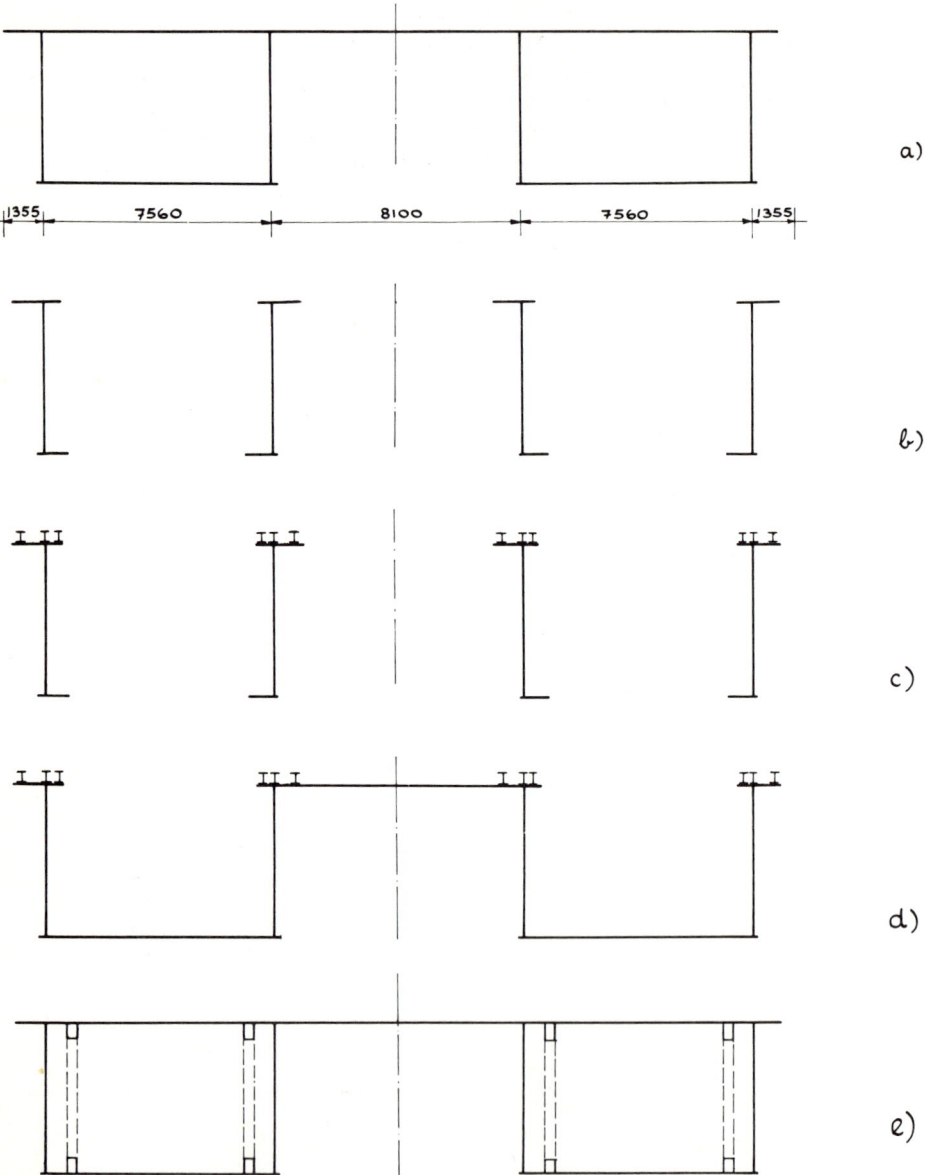

FIG. 4   CROSS SECTIONS OF THE 210m SPAN AT THE CLOSING POINT.

dimensioning of the bridge brought about a different steel weight distribution than that assumed for the calculation. The correct distribution of the steel weights was considered for the calculation of the exact deflection line (which is necessary for the erection procedure) but it was thought that the resulting maximum increase of 4% of the bending moments would be allowable. The following additional causes were discovered subsequently:

1.   The theoretical buckling stress curve serving as the basis for safety calculations is up to 7% higher than the actual buckling stresses in the elastic-to-plastic transitional range. The bottom plates of the torsion boxes at the points where the damage occurred were precisely within this range of maximum deviation.

2.   Deformations due to the welding seams in the bottom plate and its ribs, as well as variations in the thickness of the plates and of rib dimensions at the joints, represent deviations from the ideal straight plane of the panel and discontinuities of the gravity axes. Although workmanship and assembly had been carried out very precisely, and despite the fact that deviations were far below the limits of customary practice, their dominant and actual influence became obvious by the very shape of the buckling figure. At the welding seam the bottom plates had buckled downwards with a rather sharp edge and not — as they should have done theoretically — in the centre of the panel. One of the insights gained as a result of this accident was that even minimum working inaccuracies and cross-sectional discontinuities within the most stringent tolerances may reduce the theoretical buckling safety by 5 − 7%.

3.   The immediate cause which triggered off subsequent events was undoubtedly the temperature drop during the evening. The central joint of the bridge had been closed early in the afternoon. Normally, November temperatures are hardly inclined to rise or fall sharply during the day, and the measurements made a few days previously for the manufacture of the fitting pieces did not show any substantially uneven distribution of the temperature in the superstructure. However, November 6th turned out to be a surprisingly clear and sunny day. Mainly due to radiation, the roadway deck was heated to above air temperature when the bridge was closed. Cooling down to even temperature after the sun had set caused additional compressive stress in the bottom plates of the boxes; this stress finally used up the last reserves of safety.

The experts who analysed the damaged bridge independently of one another were not unanimous in their opinions regarding the proportional rate by which each of the above causes had contributed to the diminishing of the safety factor. However, some marginal values can be stated: for instance the bending moment produced by the erection derrick in the statically determined cantilever girder is known precisely. This derrick was withdrawn prior to closing. Accordingly, the effect of the imperfections discussed in Item 2 must have been definitely smaller than the effect of the erection derrick because otherwise the incident would have occurred at an earlier stage. The temperature effect must have been bigger because it caused the buckling. On the other hand, there is a geometrical limit for the uneven distribution of temperature, since it would not have been possible to insert the fitting pieces if there had been more deformation. After thoroughly considering all these circumstances, the engineers of Waagner-Biró arrived at the conclusion that, most probably causes 1 to 3 contributed to the decrease of the safety factor with an equal share of about 7% each.

4. Repair and Completion

The fact that the state of the heavily damaged superstructure could be reliably analysed the very same night must be regarded as one of the rare and lucky circumstances of this disaster. This analysis provided the opportunity to start with security measures immediately the next day at the proper spots, and by the most suitable methods. There was no need during subsequent months to alter any essential point of the repair schedule which had been programmed during the days immediately following the disaster, nor to deviate from the replanned erection procedure for the completion of the bridge.

At first the actual points of buckling as such required no action. Due to buckling the bending resistance of the box girders had ceased at both these points, apart from some indefinite residual. In order to prevent the bridge from collapsing it was imperative to maintain the transmission of shearing forces. This required watching the buckling points day and night; the cracks continuing to extend from the crumpled plates and ribs were drilled off immediately. After some days all was quiet; no more cracks appeared, even if temperatures rose or dropped considerably.

The critical spot was obviously at the centre of the bridge at the closing point where the as yet incomplete cross section (comprising the four web plates with only small flanges stressed up to the limit of their capacity), already showed signs of bad plastic deformation. It was crucial for the further life of the superstructure to deal with this situation promptly. In order to overcome this dangerous situation I-beams (INP450) were moved onto the damaged upper flanges (see Fig. 4(c)). The beams were used to rectify, by means of hydraulic jacks, the twisted flanges, and while this work proceeded they were immediately welded onto the flanges as soon as these had been straightened out. In other words, the material used for rectification served also to strengthen the flanges thus the optimum safety effect was combined with the lowest additional weight being applied to the bridge.

After three such beams had been welded onto each of the four upper flanges, the immediate danger was over and as soon as this work was concluded, it was possible to mount the bottom plates of both box girders in this central cross section and the deck slab in between the inner web plates (see Fig. 4(d)). This made the cross section sufficiently strong for the operations subsequently required. For the time being, the boxes had to be left open at the top in order to enable the later installation of reinforcements to replace the damaged parts.

In order to be able to proceed with the following operations the supports of the superstructure had to be rectified. It had not been difficult to install temporary supports on the abutments and hydraulic jacks for controlling the longitudinal movement of the bridge. However, as a result of the shortening of the bottom chords due to buckling, the 4 m stacks on the river pier (bearing load of 2600 tonnes) had also been shifted towards the middle of the river by several inches (see Fig. 5). The stacks acting as rocker posts on their edges were thereby overstressed and showed distinct yield phenomena and the cold weather that was due to come could hardly be expected to improve the situation. As a first step, the pressure of the hydraulic jacks was set so as to distribute the weight on all the stacks as evenly as possible. This measure, however, blocked every chance of manoeuvrability for the final lowering of the bridge. Structural modifications of the stacks for the purpose of obtaining reliable rocker posts were neither advisable in this situation nor really possible. It was therefore decided to insert Teflon between stainless steel plates under the stacks, and to brace these once more against the superstructure, so as to obtain an efficient sliding bearing on the river pier. After two nights work involving considerable trouble the Teflon was finally inserted.

As a result, the stage was now set for lowering the superstructure and for starting on the actual repair of the girder elements deformed by buckling. The main girders had buckled at their weakest point, i.e. near the zero point of the deadweight moment curve. In order to mount the replacements without any application of force the bridge had first to be lowered to its final position. Prior to lowering, however, the points of

FIG. 5 ERECTION STACKS (HEIGHT 4 m) ON THE RIVER PIER

FIG. 6 SUPERIMPOSED GIRDERS

damage had to be cut open since it was obviously impossible to bend the buckled steel back into shape. The shearing forces at these points – 400 tonnes at the left and 1180 tonnes on the right point of damage – had to be borne by means of auxiliary devices which had to be constructed in such a manner as to enable a controllable rotation of these 'hinges' during lowering.

For this purpose four large superimposed girders were installed in line with the web plates on the bridge deck over each point of damage. They were rigidly anchored within the boxes on one side and spanned the damaged parts of the boxes which had to be removed by cutting. They were supported by means of hydraulic jacks on the superstructure beyond the point of damage (see Figs 6 and 7). The distribution of forces was thus subjected to dual control, i.e. the deflection of the girder and the oil pressure of the jacks, which was measured. The buckles were then cut open, and as a result the shearing forces were transposed onto the stiffening girders. The setting of the jacks had been calculated correctly, and, hence, the transfer of load was effected without any significant jolting. However, since as a result of cutting

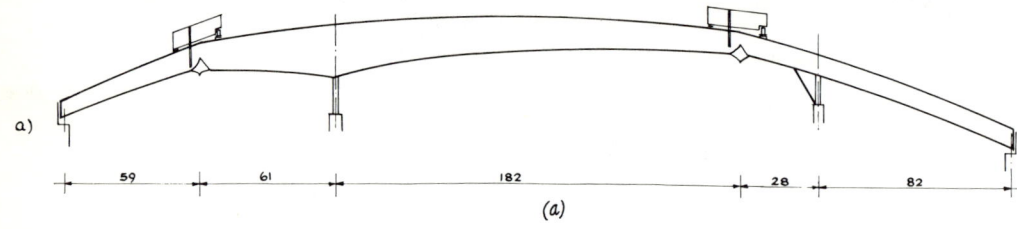

(a)

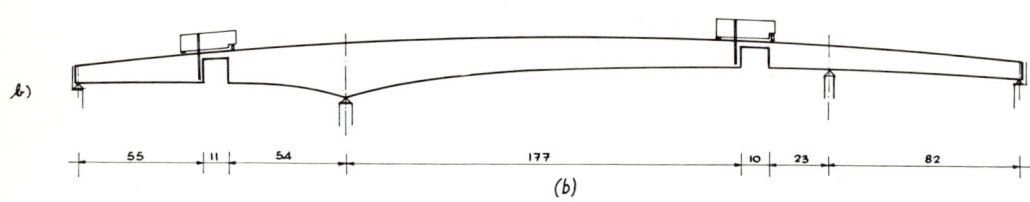

(b)

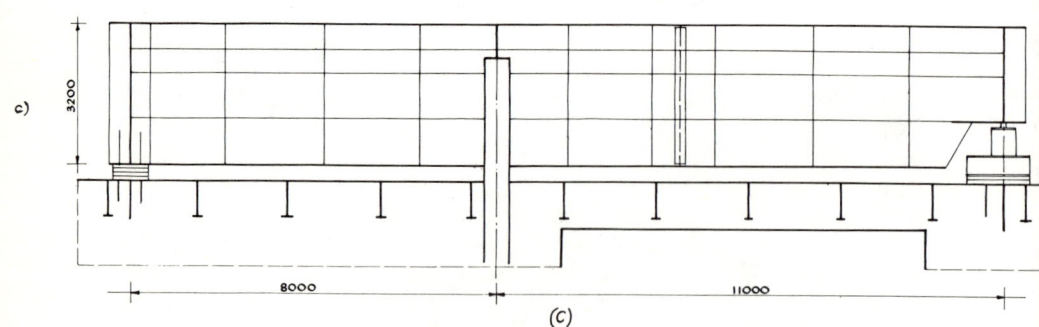

(c)

FIG. 7 BRIDGE SYSTEM WITH SUPERIMPOSED GIRDERS
   (A) BEFORE LOWERING
   (B) AFTER LOWERING AND CUTTING OUT OF DAMAGED PARTS
   (C) DIMENSIONS OF SUPERIMPOSED GIRDERS

the boxes the theoretical hinge was shifted to a point somewhat nearer to the jacks, minor effects were inevitable. For the purpose of lowering, the buckles below the web plate centre were initially cut open in a semicircle (see Fig. 8), in order to prevent additional cracks during lowering and to effect a continuous straightening of the orthotropic slab which was only elastically deformed. It was also hoped that the slightly bent plate portions above the web plate centre would stretch back into shape during lowering, this expectation was realised.

FIG. 8 DAMAGED PARTS OF BOX GIRDERS CUT OUT FOR LOWERING

The lowering procedure as such had to be very carefully controlled. Each downward movement on either stack meant an angular rotation on the damaged points. However, having no real hinges there, the angles of rotation that had to be adjusted by the hydraulic jacks were dependent on the deflection control of the superimposed girders and the straightening deformation of the roadway slabs.

In addition, the structure and the superimposed beams exposed to sun radiation produced considerable thermal effects throughout the whole system. In spite of all these obstacles the lowering operation proceeded without a hitch and was terminated within the scheduled time. In the final position rectangular sections of the lower portions of the torsion boxes were cut out and thus all damaged parts removed.

For the accurate adjustment before the replacement parts were inserted, both bridge ends had to be raised slightly once again; the left hand end because the point of damage in the left span was not precisely in the dead weight zero point of moments; the right hand end because the plastic deformation at the central point required a correction of the deflection line of the superstructure.

The welding of the replacement parts into the boxes was again a problem of temperatures; at this stage, however, the superstructure was not in any extreme stress situation any more. After these operations had been completed the original performance of both boxes at the buckling points had been re-established.

There remained the repair of the central portion of the bridge. The material of the top flanges had been loaded beyond the elasticity point of the material and the repair of the badly deformed web plate elements immediately below also required heating and, possibly, local cutting. It did not appear feasible to relieve these sections of stress as it would have meant lifting the bridge back into the position it had occupied when the damage occurred. In order to get complete freedom for repair operations, steel reinforcements, capable of fully replacing the overstrained flange elements were installed in the orthotropic slab and in the box bottoms. These reinforcements were placed next to each flange and, by means of struts, combined to form trusses (see Fig. 4(*e*)). These trusses were able to replace the action of the web plates in respect of transferring the shearing forces. The repairs of the damaged flange and web plate parts were then carried out and no attention had to be paid to the state of stress in the superstructure.

Thus the bridge superstructure was properly completed and ready to carry the specified load (see Fig. 9). Nevertheless, the plastic compression of the flanges at the centre of the bridge as a result of the disaster led to a minor deviation from the planned gradient. A genuine correction would have required either an auxiliary

FIG. 9 BRIDGE COMPLETED AND REPAIRED

trestle in the middle of the river (impediment to navigation) or another lifting operation by 4 m on the river pier and 3 m on the right bank pier which would have meant a return to the most critical stage of erection, or the installation of hydraulic hoisting cylinders with a capacity of several thousand tons in the box bottoms.

Such an effort for the rectification of just a small geometric deviation seemed hardly defensible. It was therefore decided to improve this optical fault by other means and the sidewalk brackets, edge beams, railings and lighting masts were mounted in such a manner that the deviation from the gradient is now only visible on the bottom chords of the bridge, and is apparent only to those aware of the events and even then from special vantage points only.

The steelwork on the 4th Danube Bridge was thus completed. The badly damaged superstructure was repaired and completed without the aid of auxiliary trestles. Fortunately, there was no delay to the overall motorway scheme since the approach roads were still not ready. The costs of the damage which were covered by insurance amounted to 12% of the costs of the steel superstructure, i.e. only approx. 3·5% of the total cost of the Danube Bridge.

HOW IMPORTANT IS THE CONSTRUCTION METHOD?

F. J. HANSEN

Harris & Sutherland

So many factors influence the design and construction of a bridge that it is very difficult to make fair comparisons. Each bridge site has its own particular problems imposing constraints on the design or the method of construction.

Occasionally a designer may have the opportunity of comparing how different contractors tackle similar bridges, or a contractor can compare how different designers tackle rather similar bridge problems, but it seldom happens that the same designer-contractor team tackles the same bridge problem again and decides to approach it in a new way just for the sake of science and progress.

First of all, there isn't such a thing as a designer-contractor team in our present system and there wouldn't be much incentive to fruitful interchange of ideas between the two sides of such a team, but should the unlikely situation arise that the same team came up against the same problem again, it is most likely that an entirely new labour force and supervisory staff would be involved and a comparison would still be difficult.

However, comparisons have to be made if we want to make progress.

When a designer tries to achieve the cheapest solution to a bridge problem he soon realises that the only thing which is really in his power, is the amount of steel and concrete in the bridge. He appreciates, of course, that the temporary supports to the bridge, until it becomes self-supporting, and the plant and the labour involved is just as important as the permanent materials, but they are outside his control and with some feeling of frustration he resigns himself to minimising the permanent materials content.

When a contractor is tendering for a bridge, he soon finds himself in a situation similar to the designers, but one that is perhaps less frustrating. The permanent materials do not really represent a problem and the plant and temporary works are items he can deal with as competently as the designer deals with the permanent materials — but he fully realises that the most important item, the labour content, does not lend itself to the same exact analysis as the other items, and what is worse, he is also fully aware of the fact that a wrong assessment of the labour content can have the most serious repercussions on temporary works, plant and supervision.

The labour force, in the same way as the contractor and the consulting engineer, wants to make a living out of building bridges and tries therefore to get paid as much as possible for its work. Each time a new operation starts up there is a period of hard bargaining. It naturally takes some time for the employer to get the work organised, and only when every man knows his job can the operation run smoothly and efficiently. It is impossible for the employer to achieve production and for the men to earn money during these starting up periods, so the sooner one can reach agreement on reasonable targets and get down to repetitive production the better for everybody.

It is therefore in everybody's interest that the design of the bridge is such that it permits a continuous flow of a minimum of simple and repetitive operations — preferably independent of weather conditions.

This is, of course, wishful thinking in very many cases, but not if one considers multispan viaducts over difficult or inaccessible ground or elevated motorways in built-up areas with serious traffic conditions during construction.

A number of such structures have already been built in this country, and considerable ingenuity has been exercised in the designs and the adopted construction methods, but they have not reached the degree of mechanisation and assembly belt technique which has been achieved in some countries on the continent.

There are two main reasons for that.

The design itself is no problem; on the contrary. The basic freedom of the designer will probably not be restricted, but having a well defined construction method and sequence at the design stage will solve many problems of position and construction joints and detailing of reinforcement and prestressing. But having a design which is ideal for 'factory-production' of the bridge does not automatically lead to maximum degree of mechanisation.

A conventional approach to the problem will give the estimators an idea of how much money there is in the bridge for temporary works, plant and labour. To go to extreme mechanisation will probably mean two things. Some very special plant will have to be fabricated and more money will have to be spent on plant and temporary works. This extra money must be more than saved by a very substantial reduction in the labour force, and this again puts a very heavy burden on the estimators and managers who have to make their minds up and take some very far-reaching decisions.

It is quite easy to follow the path of cheapest solutions for each individual problem as it crops up, but it is very hard to decide to spend a large sum of money at the start of the contract on a new and unproven piece of equipment, which is only backed by a belief that it will save in labour costs more than its own fabrication, erection and dismantling costs, especially as the labour costs would have to be so much lower than previously experienced, that it appears unrealistic.

It is a fact that several firms on the Continent have gone to the length of fabricating a fully enclosed, fully automatic 'bridge-factory' which travels along on top of the supporting piers and leaves behind itself the completed bridge deck constructed in-situ (see Fig. 1).

It must therefore have happened in this country that progressive contractors also have looked into the question of a 'bridge-factory'. Why has nobody yet taken this step?

The very first reason is probably a simple one of time. It is too late to start planning and building the 'factory' when the bridge contract has been awarded, and past experience with specialised plant has not encouraged anybody to build the thing in advance.

The next reason is the question of probability of ever using the tool again.

The first bridge might not be quite big enough to absorb the total cost of the 'factory' and the contractor does not feel sufficiently confident that he will get another opportunity to use it again.

There are, however, other people than bridge contractors who could have an interest in this approach to bridge construction, e.g. the various clients who can feel

quite confident that they will have more bridges to be built, who would also like to
have them built as cheaply and as fast as possible.

It is always rather difficult to get a clear picture of actual construction costs, even
the contractors find it difficult at times to explain where the money has gone, and if
one tries to study some priced Bills of Quantity the picture becomes even more con-
fused, but it is not impossible and some useful facts are readily available.

FIG. 1

Let us consider a long viaduct with spans of, say, 80 ft. The price for that bridge
appears to be approximately £12 per ft^2 of deck area out of which £5 per ft^2 represents the price for the deck itself.

This price of £5 comprises the actual cost and the profit. If we assume Head
Office charges and profits amount to 10%, the remaining site costs of £4·50 per ft^2
cannot be very far out.

If we thereafter consider the permanent materials in the deck it will appear that
there is approximately 2 ft^3 of concrete and something equivalent to 20—25 lb of
mild steel per ft^2 of deck. The price for these materials brought to site is of the
order of £1, leaving £3·50 per ft^2 of bridge to cover all the remaining items of labour,
plant, temporary works, supervision and so on.

Of all these items, labour is the only one which is not disputed. The labour cost
is without doubt a direct expense whereas the other items could conceivably contain
a hidden profit which confuses the issue. Let us therefore concentrate on the labour
cost.

The labour content, as already mentioned, is the most difficult item to assess at
the tender stage, but undoubtedly the best documented cost item after the construction has been completed. The total might be far in excess of the estimated figure
but there will be plenty of evidence of where the money has gone. Depending upon
the adopted method of construction, the complexity of the structure and efficiency

of the site management, there may be quite a variation in the labour content, but it has to be an exceptionally simple bridge to get down to £1·50 per ft² of deck, leaving £2·00 per ft² to cover all other site costs, excluding the permanent materials.

Introducing a high degree of mechanisation and automation in the method of construction, one would of course expect to reduce the labour and increase the plant, and the question is quite simple: 'Is it possible to save so much labour that it can pay for the necessary increased plant?'.

It is quite obvious that the saving in labour has to be quite dramatic, and it is understandable if no estimator feels tempted to depart from the more conventional methods. It is, however, a fact that some firms on the Continent have invested very large sums of money on bridge construction plant and gone to the extreme in the way of automation and mechanisation; and they have done it more than once, so one can safely assume that they have not lost on the experiment.

A contractor who takes such a bold step and has the satisfaction of succeeding is not very keen on disclosing his secrets to prospective competitors, so it is rather difficult to obtain much inside information, but it is nevertheless possible to form quite an accurate picture of the situation by a few simple observations.

When the labour force is spread over a vast construction site it is very difficult for a visitor, or for that matter for the site agent, to get a clear picture of how many men there are on the site, what they are doing and what progress they are making. If, on the other hand, you have all your men under one roof, and all your progress can be related to the speed with which your 'factory' is advancing, it is quite easy for the site agent, and for that matter also for an inquisitive visitor, to assess the productivity on the site.

By studying published literature, visiting a few sites and asking some pertinent questions, it becomes quite obvious that there are great possibilities for rationalisation of bridge construction.

One approach to this problem would be to apply a high degree of prefabrication, but the use of pre-cast units does not eliminate some in-situ construction and might still lead to extensive temporary supports.

Another approach would be to go to the other extreme of 100% in-situ construction and simply to concentrate on simplifying and mechanising falsework, formwork, concreting, reinforcing and prestressing. If the men can work on top of the bridge in conditions just as good, or better, than those in a precasting yard, then it makes more sense to cast the concrete where you want it than where you don't want it. In a precasting yard there is a limit to the size of unit you can make; restricted by handling, lifting and transport consideration. In-situ there is a natural limit when the construction unit is equivalent to a complete span from pier to pier.

Fig. 2 shows a typical cross-section through a bridge which lends itself to in-situ 'factory' construction, a solid slab of varying thickness on rigid single columns. The column spacings can vary and the bridge can go through curves, and spans up to 150 ft have been dealt with. It appears that the solid slab is competitive up to spans of 120 ft, after which it becomes more economic to introduce a grid of beams and slabs. The only restriction seems to be the single columns, but that is probably not very much of a restriction when one can cope with an overall width of over 100 ft. For anything more than that, it might in any case be just as well to have two independent bridges on single columns.

Figs 3 and 4 show a cross-section and a longitudinal section through the 'factory' and Fig. 5 shows four stages of the operation of this formwork carriage.

There are, of course, many small but important details in the design and operation of such a piece of plant, which have been worked out from years of experience, but the main features will appear from the above figures.

The backbone of the formwork carriage is a heavy longitudinal steel box girder which can span from pier to pier and carry the total weight of one span of wet concrete plus all the necessary formwork, falsework and construction equipment. It can, furthermore, cantilever forward until it picks up the support on the next pier, so the complete assembly of falsework, formwork and plant can move forward to the next span.

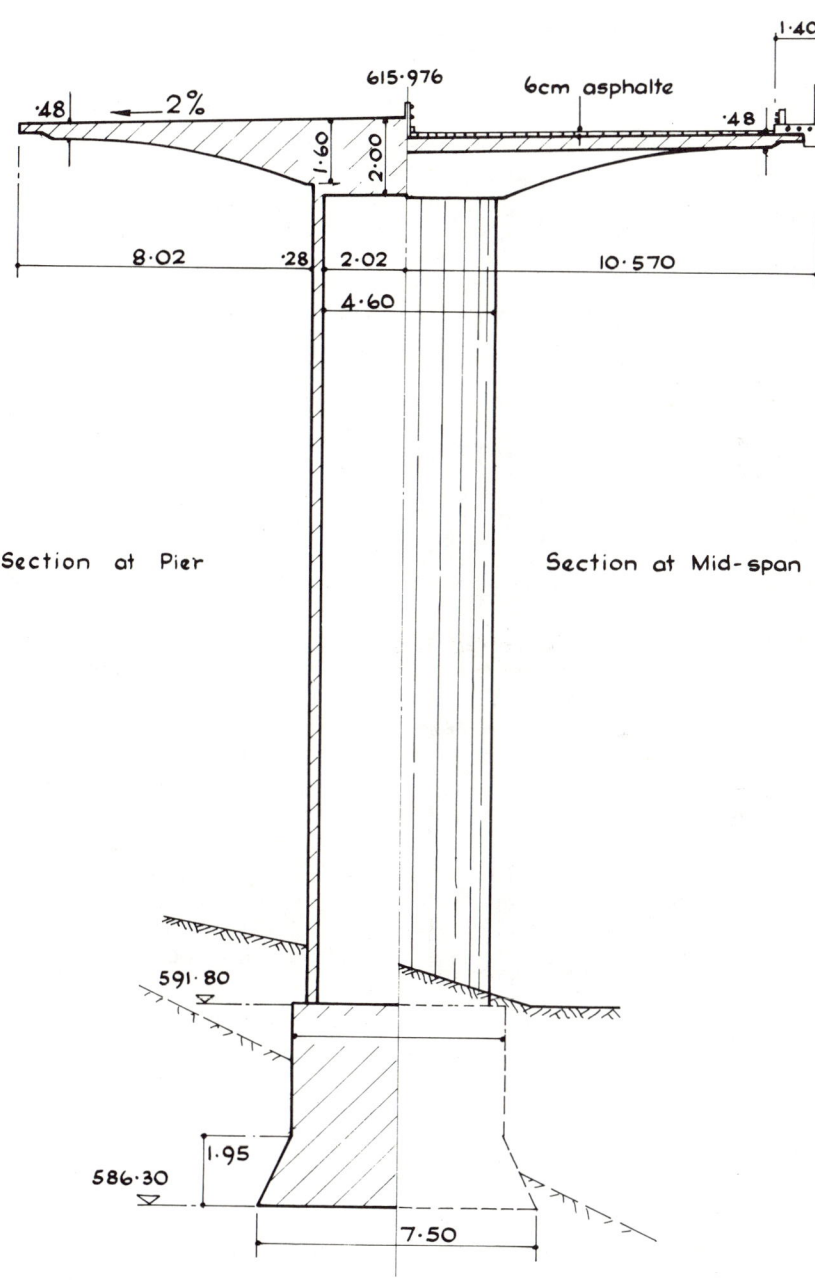

Section at Pier

Section at Mid-span

FIG. 2

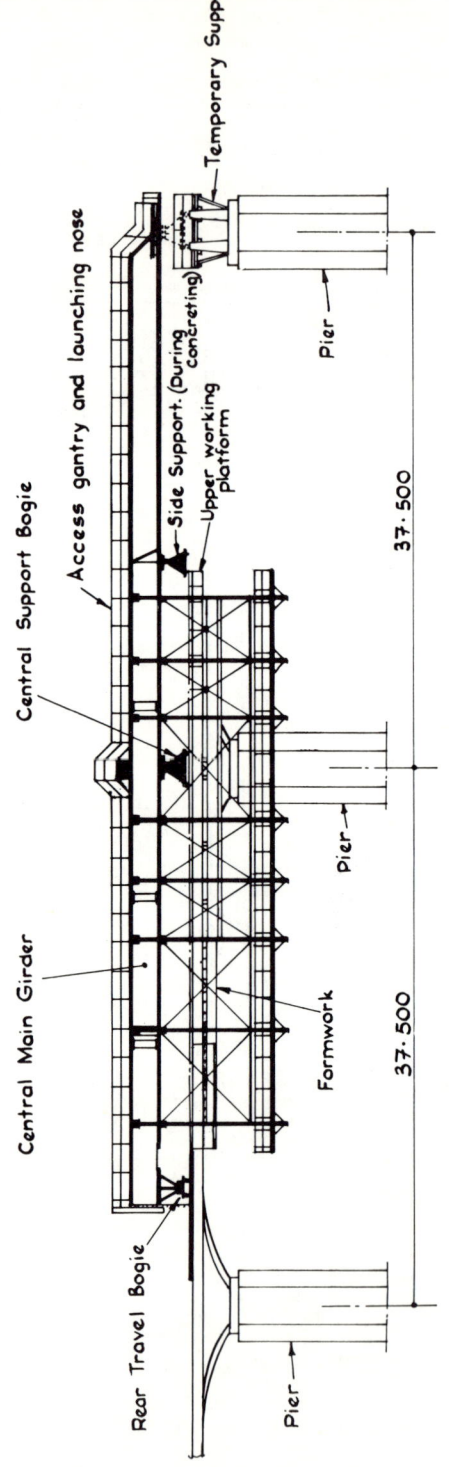

FIG. 3

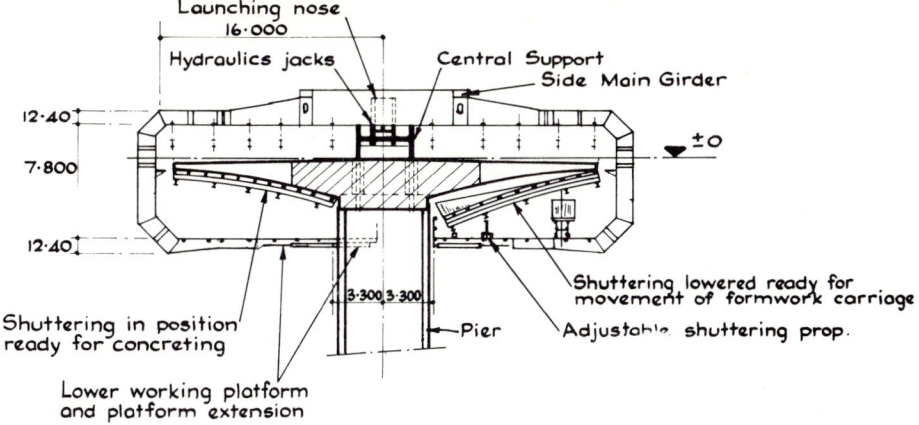

FIG. 4

From this longitudinal spinebeam a number of transverse ribs are cantilevering out, covering the full width of the bridge. They are thereafter turned down vertically and back again horizontally, surrounding the bridge deck completely, with the exception of a clear path for the column. The whole structure is covered in with roof and wall cladding so that people can work inside it protected from the weather.

The formwork is supported from above by means of suspension rods from the top ribs through sleeves in the deck concrete when the formwork carriage is in its concreting position. When it is in position to move forward the formwork is resting on the lower part of the ribs and the section in the path of the piers has been withdrawn.

All transport to the 'factory' takes place along the completed bridge deck. The deck construction therefore is independent of ground conditions, or will cause a minimum of interference with any traffic at ground level. Transport inside the 'factory' is taken care of by means of a number of longitudinal overhead cranes operated from a central control room. Concrete arriving at the door of the 'factory' is tipped into a hopper, from which a conveyor system takes it to the various overhead cranes and fills it into skips which are controlled from the central control room. The concrete therefore gets to its final destination in the bridge deck 'untouched by human hands' and only then is the concreting gang going into action with their vibrators.

The whole operation is obviously extremely smooth and efficient, the only remaining problem is — 'When does it in fact pay to go to such length in mechanisation and automation'.

It is impossible to give a definite and clear cut answer to this question. It will, of course, depend quite a lot upon local conditions, such as relative costs of labour, plant and materials, or the availability of the required skilled labour. It is, however, certain that the size of bridge required to make the 'factory' a paying proposition will shrink rapidly as the cost of labour increases.

Still, even if an absolute answer is impossible, it is quite easy to get a pretty good idea of the answer by means of a very simple analysis.

One particular bridge was 3000 ft long and 100 ft wide, made up of 30 spans 100 ft long. The progress on this bridge was one completed span in 11 working days. The 'factory' worked in 2 shifts, each of 10 hours and the total number of men employed was 45.

Manhours per span — 45 x 10 x 11 = 4950

Formwork Carriage in Operation

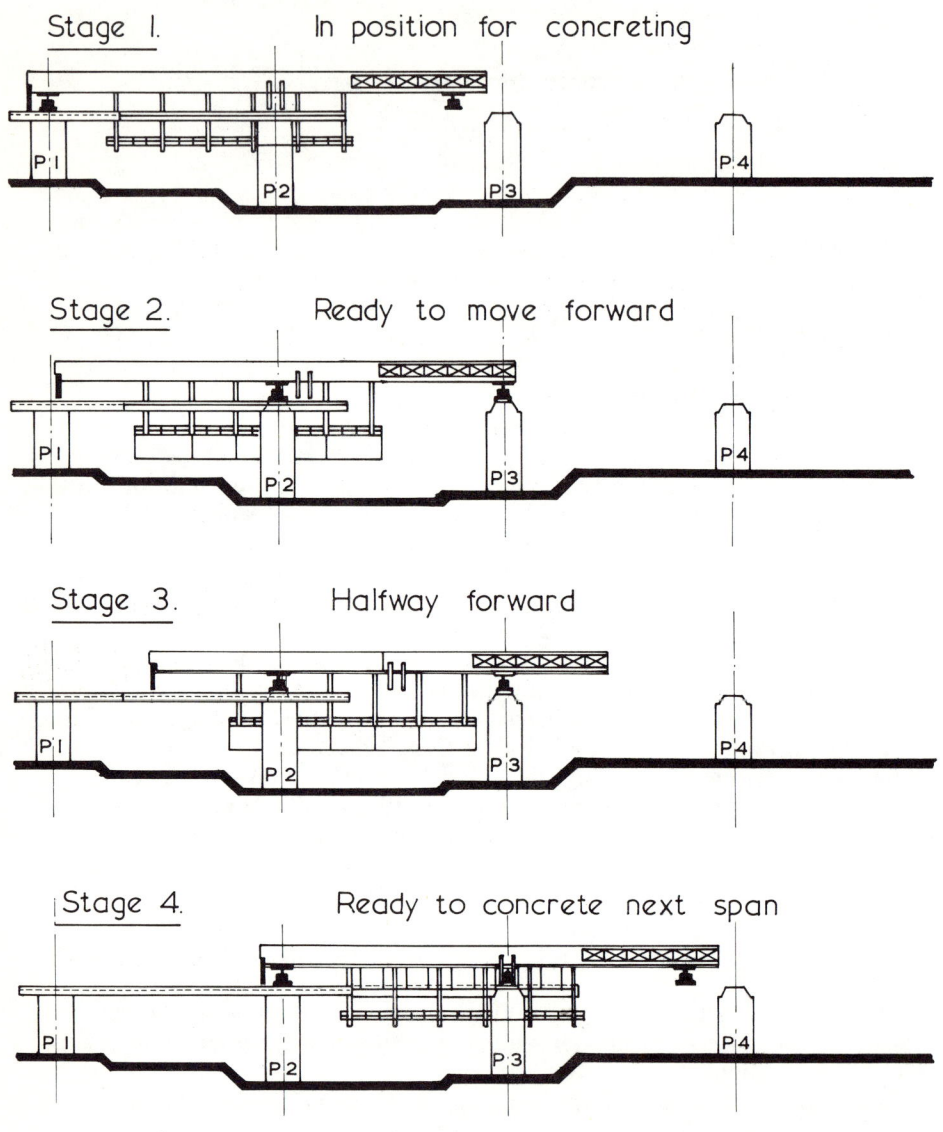

Stage 1.   In position for concreting

Stage 2.   Ready to move forward

Stage 3.   Halfway forward

Stage 4.   Ready to concrete next span

FIG. 5

The area of one span was 100 ft x 100 ft = 10,000 ft^2 so the labour therefore amounted to 0·5 mhr per ft^2. This is a very low figure and if we compare this figure with the figure mentioned earlier, and furthermore assume that the cost of 1 mhr is equal to £1, then it means a saving of £1 per ft^2 of deck.

The saving is a very substantial saving which without doubt can pay the full cost of the 'factory', its erection and dismantling, and still leave something in hand to make the effort worth while. But before we look a little closer at the possible cost of the 'factory' we had better consider the other cost items in the deck.

It is possible that the permanent materials content has gone up in the simplification process, but even a 25% increase would only amount to £0·25 per ft^2. The remaining site costs, which in our example amounted to £2·00 per ft^2 included temporary falsework and formwork. The formwork carriage replaces all that, so it does not seem unduly optimistic to assume a saving in excess of £0·25 per ft^2.

It must therefore be on the safe side to put the saving in labour against the cost of the 'factory'.

If one just considers the structural problem of carrying some 1000 tons of wet concrete, it becomes quite apparent that the formwork carriage will contain 3—400 tons of structural steelwork. The fabrication cost of the carriage, including formwork, overhead crane beams and conveyors, etc., may therefore quite easily amount to, say, £150,000. If we add a further £25,000 for erection and dismantling and running in, it still leaves a saving of £125,000, which, in the contractor's books, might mean more than a doubling of his profits, and the prospects for the next bridge would be very rosy indeed.

The above cost considerations show quite clearly that the construction method is extremely important. There are greater savings to be made by paying attention to the method of construction than by attacking the permanent materials.

It is too late to tackle the construction problem properly when the bridge goes out for tender. If full advantage of mechanisation and automation is to be achieved it is necessary to think big and to do it at the earliest possible date.

The earliest possible date must be at the conception of the designs, putting it firmly on the designer's plate to incorporate the construction methods in his scheme and to include the necessary equipment and temporary supports in the Bill of Quantities.

This will, of course, not apply to every bridge that goes out for tender, but for most major schemes in excess of, say, £2,000,000, containing a substantial amount of repetition, no doubt considerable savings could be achieved by integrating design and construction.

INDEX

The papers are indexed alphabetically (in bold type) and the subject matter then alphabetically indexed within papers.

NOTES